Principles of Enhanced Heat Transfer

SECOND EDITION

Ralph L. Webb
Nae-Hyun Kim

Taylor & Francis
Taylor & Francis Group
Boca Raton   London   New York   Singapore

Published in 2005 by
Taylor & Francis Group
270 Madison Avenue
New York, NY 10016

Published in Great Britain by
Taylor & Francis Group
4 Park Square
Milton Park, Abingdon
Oxon OX14 4RN

© 2005 by Taylor & Francis Group, LLC
10 9 8 7 6 5 4 3 2 1

International Standard Book Number-10: 1-59169-014-5 (Hardcover)
International Standard Book Number-13: 978-1-59169-014-6 (Hardcover)
Library of Congress Card Number 2004021960

Library of Congress Cataloging-in-Publication Data

Webb, Ralph L., 1934-
    Principles of enhanced heat transfer / by Prof. Ralph L. Webb and Prof. Nae-Hyun Kim.-- 2nd ed.
      p. cm.
    Includes bibliographical references and index.
    ISBN 1-59169-014-5 (alk. paper)
    1. Heat--Transmission. 2. Heat exchangers. I. Kim, Nae-Hyun, 1957- II. Title.

TJ260.W36 2005
621.402'2--dc22

2004021960

Taylor & Francis Group
is the Academic Division of T&F Informa plc.

Visit the Taylor & Francis Web site at
http://www.taylorandfrancis.com

Principles of Enhanced Heat Transfer

TABLE OF CONTENTS

PREFACE

Many changes have occurred in the field of enhanced heat transfer since the publication of the first edition in 1993. While the use of relatively complex geometries was initially limited by manufacturing processes, new manufacturing methods now allow the assemblage of many complex surface geometries. Some enhanced surfaces (e.g., boiling and condensing tubes) are in their fourth generation. As noted in Chapter 1, nearly all heat exchangers used in the air-conditioning and automotive industries are enhanced geometries. Further inroads are being seen in the electronic cooling, process, and power industries.

Fundamental and even graduate level courses in heat transfer typically focus on simple geometries, such as flat plates and circular tubes. Enhanced heat transfer surfaces typically involve complex geometries. Study of enhanced heat transfer concepts will improve the student's ability to address such complex geometries. A key element of the material in the book is the inclusion of tools to address or analyze such geometries. This textbook is primarily directed at practitioners of high-performance heat transfer concepts. However, it is also suitable for use as a text in a second level (or graduate level) heat transfer course. I taught a course on Enhanced Heat Transfer for a number of years at Pennsylvania State University. The present book and included problems are products of this course.

The second edition provides a general update to the initially published 17 chapters and adds two new chapters: Micro-Channels (Chapter 18) and Cooling of Electronic Equipment (Chapter 19). The net additions to the book include 126 new figures, 18 tables, and a doubling of the number of problems at the end of the book. These problems can be used by readers to test their understanding or by instructors teaching an academic course. The book also contains an increased number of solved examples.

A major addition to the book is the inclusion of a Reference CD-ROM containing 9,500 references. We have attempted to include all published journal papers since 1960, with a number of earlier references. The bibliography on the CD is in a pdf format, listing references alphabetically by the first author's last name. One may use the CD to search by author, journal, year, or key word in the article title. The format of a typical reference is given below:

Adamek, T. A., and Webb, R. L., 1990. "Prediction of Film Condensation on Horizontal Integral-fin Tubes," *Int. J. Heat Mass Transfer*, Vol. 33, pp. 1721–1735.

The book is now jointly co-authored with Professor Nae-Hyun Kim, who completed his Ph.D. with me in 1987. Prof. Kim has devoted his research to enhanced heat transfer and is now an international authority on the subject. I am extremely grateful to Prof. Kim for his collaboration and hard work on this second edition.

Ralph L. Webb

INTRODUCTION TO ENHANCED
HEAT TRANSFER

1.1 INTRODUCTION

The subject of enhanced heat transfer has developed to the stage that it is of serious interest for heat exchanger application. The refrigeration and automotive industries routinely use enhanced surfaces in their heat exchangers. The process industry is aggressively working to incorporate enhanced heat transfer surfaces in its heat exchangers. Virtually every heat exchanger is a potential candidate for enhanced heat transfer. However, each potential application must be tested to see if enhanced heat transfer "makes sense." The governing criteria are addressed later.

Heat exchangers were initially developed to use plain (or smooth) heat transfer surfaces. An "enhanced heat transfer surface" has a special surface geometry that provides a higher hA value, per unit base surface area than a plain surface. The term "enhancement ratio" (E_h), is the ratio of the hA of an enhanced surface to that of a plain surface. Thus,

$$E_h = \frac{hA}{(hA)_p} \qquad (1.1)$$

Consider a two-fluid counterflow heat exchanger. The heat transfer rate for a two-fluid heat exchanger is given by

$$Q = UA\Delta T_m \tag{1.2}$$

To illustrate the benefits of enhancement, we will multiply and divide Equation 1.2 by the total tube length, L

$$Q = \frac{UA}{L} L\Delta T_m \tag{1.3}$$

The term L/UA is the overall thermal resistance, per unit tube length, and is given by

$$\frac{L}{UA} = \frac{L}{\eta_1 h_1 A_1} + \frac{Lt_w}{k_w A_m} + \frac{L}{\eta_2 h_2 A_2} \tag{1.4}$$

where subscripts 1 and 2 refer to fluids 1 and 2, respectively. The term η is the surface efficiency, should extended surfaces be employed. For simplicity, Equation 1.4 does not include fouling resistances, which may be important. The performance of the heat exchanger will be enhanced if the term UA/L is increased. An enhanced surface geometry may be used to increase either or both of the hA/L terms, relative to that given by plain surfaces. This will reduce the thermal resistance per unit tube length, L/UA. This reduced L/UA may be used for one of three objectives:

1. Size reduction: If the heat exchange rate (Q) is held constant, the heat exchanger length may be reduced. This will provide a smaller heat exchanger.
2. Increased UA: This may be exploited either of two ways:
   a. Reduced ΔT_m: If Q and the total tube length (L) are held constant, the ΔT_m may be reduced. This provides increased thermodynamic process efficiency, and yields a savings of operating costs.
   b. Increased heat exchange: Keeping L constant, the increased UA/L will result in increased heat exchange rate for fixed fluid inlet temperatures.
3. Reduced pumping power for fixed heat duty. Although it may seem surprising that enhanced surfaces can provide reduced pumping power, this is theoretically possible. However, this will typically require that the enhanced heat exchanger operates at a velocity smaller than the competing plain surface. This will require increased frontal area, which is normally not desired.

The important principle to be learned is that *an enhanced surface can be used to provide any of three different performance improvements*. Which improvement is obtained depends on the designer's objectives. Thus, Designer A may seek a smaller heat exchanger, and Designer B may want improved thermodynamic process efficiency.

Although the size reduction of Objective 1 may be valued, the more important objective may be cost reduction. In many cases, the designer requires that the size reduction be accompanied by cost reduction. Another factor to consider for Objective 1 is that the fluid volume in the heat exchanger will also be reduced. This may be an important consideration for a manufacturer of refrigeration equipment, because a smaller volume of expensive refrigerant will be required.

Objectives 2 and 3 are important if "life cycle" costing is of interest. For example, Objective 2 for refrigeration condensers and evaporators will result in reduced compressor power costs. Objective 3 is important for upgrading the capacity of an existing heat exchanger. This may allow plant output to be increased.

Pressure drop (or pumping power) is always of concern to the heat exchanger designer. Hence, a practical enhanced surface must provide the desired heat transfer enhancement and meet the required flow rate and pressure drop constraints. Chapters 3 and 4 discuss how one meets these objectives for single-phase (Chapter 3) and two-phase (Chapter 4) flows. A surface geometry that provides a given heat transfer enhancement level with the lowest pressure drop is definitely preferred.

1.2 THE ENHANCEMENT TECHNIQUES

Bergles et al. [1983] have identified 13 enhancement techniques. These techniques are segregated into two groupings: "passive" and "active" techniques. Passive techniques employ special surface geometries, or fluid additives for enhancement. The active techniques require external power, such as electric or acoustic fields and surface vibration. The techniques are listed in Table 1.1, and are described below.

1.2.1 Passive Techniques

Coated surfaces involve metallic or nonmetallic coating of the surface. Examples include a nonwetting coating, such as Teflon, to promote dropwise condensation, or a hydrophilic coating that promotes condensate drainage on evaporator fins, which reduces the wet air pressure drop. A fine-scale porous coating may be used to enhance nucleate boiling. Figure 1.1a shows the cross section of a sintered porous metal coating for nucleate boiling. The particle size is on the order of 0.005 mm. Figure 1.1b shows larger particles (approximately 0.5 mm) sintered to the surface. These may be used to enhance single-phase convection or condensation.

Rough surfaces may be either integral to the base surface, or made by placing a "roughness" adjacent to the surface. Integral roughness is formed by machining, or "restructuring" the surface. For single-phase flow, the configuration is generally chosen to promote mixing in the boundary layer near the surface, rather than to increase the heat transfer surface area. Figure 1.2a shows two examples of integral roughness. Formation of a rough surface by machining away metal is generally not an economically viable approach. Figure 1.2b shows an enhanced "rough" surface for nucleate boiling. The surface structuring forms artificial nucleation sites, which provide much higher performance than a plain surface. Figure 1.2c shows the wire coil insert, which periodically disturbs the boundary layer. A wire coil insert is an example of a non-integral roughness.

Extended surfaces are routinely employed in many heat exchangers. As shown in Equation 1.4, the thermal resistance may be reduced by increasing the heat transfer coefficient (h), or the surface area (A), or both h and A. Use of a plain fin may provide only area increase. However, formation of a special shape extended surface

Table 1.1 Summary of References on Enhancement Techniques as of November 1983

	Single-Phase Natural Convection	Single-Phase Forced Convection	Pool Boiling	Forced-Convection Boiling	Condensation	Mass Transfer
Passive Techniques (No External Power Required)						
Treated surfaces	—	—	149	17	53	+
Rough surfaces	7	418	62	65	65	29
Extended surface	23	416	75	53	175	33
Displaced enhancement devices	+	59	4	17	6	15
Swirl flow devices	+	140	—	83	17	10
Coiled tubes	+	142	—	50	6	9
Surface tension devices	30	—	12	1	—	+
Additives for liquids	3	22	61	37	—	6
Additives for gases	+	211	—	—	5	0
Active Techniques (External Power Required)						
Mechanical aids	16	60	30	7	23	18
Surface vibration	52	30	11	2	9	11
Fluid vibration	44	127	15	5	2	39
Electric or magnetic fields	50	53	37	10	22	22
Injection or suction	6	25	7	1	6	2
Jet impingement	—	17	2	1	—	2
Compound techniques	2	50	4	4	4	2

Note: — Not applicable; + no citations located.

Source: Bergles and Webb (1985).

may also provide increased h. Current enhancement efforts for gases are directed toward extended surfaces that provide a higher heat transfer coefficient than that of a plain fin design. Figure 1.3 shows a variety of enhanced extended surfaces used for gases. The Figure 1.3a through e surfaces involve repeated formation and destruction of thin thermal boundary layers. Extended surfaces for liquids typically use much smaller fin heights than those used for gases. Shorter fin heights are used for liquids, because liquids typically have higher heat transfer coefficients than gases. Use of high fins with liquids would result in low fin efficiency and result in poor material utilization. Examples of extended surfaces for liquids are shown on Figure 1.4. Figure 1.4a shows an external finned tube, and Figure 1.4b shows an internally finned tube. The internally finned tubes in Figure 1.4c are made by multiple, concentric internally finned tubes. The Figure 1.4d tube contains a five-element

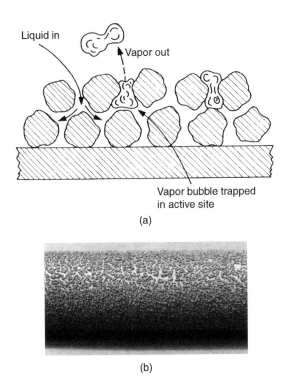

Liquid in

Vapor out

Vapor bubble trapped
in active site

(a)

(b)

Figure 1.1 (a) Illustration of cross section of porous boiling surface, (Courtesy of UOP Corporation, Tonawanda, NY.) (b) Attached particles used for film condensation.

extruded aluminum insert. The surrounding tube is compressed onto the insert to provide good thermal contact. The Figure 1.4 geometries have also been used for forced convection vaporization and condensation. The "microfin" tubes discussed in Chapters 8, 13, and 14 consist of closely spaced triangle-shaped fins approximately 0.25 mm high. These are very important for enhancement of condensation and vaporization. Small hydraulic diameter "micro-channels" ($50 < D_h < 200$ μm) connected in parallel are now used for cooling electronic equipment. These are typically operated in the laminar regime and provide significant enhancement by virtue of the small hydraulic diameter (see Table 2.2). They may be used in either single or two-phase flow.

Displaced insert devices are devices inserted into the flow channel to improve energy transport at the heated surface indirectly. They are used with single- and two-phase flows. The Figure 1.5a and b devices mix the main flow, in addition to that in the wall region. The Figure 1.5c wire coil insert is placed at the edge of the boundary layer, and is intended to promote mixing within the boundary layer, without significantly affecting the main flow.

Swirl flow devices include a number of geometrical arrangements or tube inserts for forced flow that create rotating or secondary flow. Such devices include

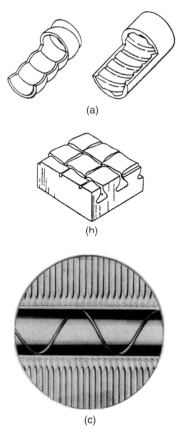

Figure 1.2 (a) Tube-side roughness for single-phase or two-phase flow, (b) "rough" surface for nucleate boiling, (c) wire-coil insert.

full-length twisted-tape inserts (Figure 1.6a), or inlet vortex generators, and axial core inserts with a screw-type winding (Figure 1.6b). Figure 1.6c shows a flow invertor or static mixer intended for laminar flows. This device alternately swirls the flow in clockwise and counterclockwise directions.

Coiled tubes (Figure 1.7) may provide more compact heat exchangers. Secondary flow in the coiled tube produces higher single-phase coefficients and improvement in most boiling regimes. However, a quite small coil diameter is required to obtain moderate enhancement.

Surface tension devices use surface tension forces to drain or transport liquid films. The special "flute" shape of Figure 1.8 promotes condensate drainage from the surface by surface tension forces. The film condensation coefficient is inversely proportional to the condensate film thickness. Heat pipes (not contained in the listing) typically use some form of capillary wicking to transport liquid from the condenser section to the evaporator section.

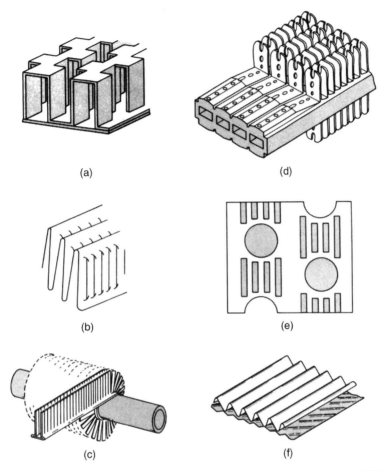

Figure 1.3 Enhanced surfaces for gases. (a) Offset strip fins used in plate-fin heat exchanger, (b) louvered fins used in automotive heat exchangers, (c) segmented fins for circular tubes, (d) integral aluminum strip-finned tube, (e) louvered tube-and-plate fin, (f) corrugated plates used in rotary regenerators. (From Webb and Bergles [1984].)

Additives for liquids include solid particles or gas bubbles in single-phase flows and liquid trace additives for boiling systems.

Additives for gases are liquid droplets or solid particles, either dilute-phase (gas-solid suspensions) or dense-phase (packed tubes and fluidized beds).

1.2.2 Active Techniques

Mechanical aids involve stirring the fluid by mechanical means or rotating the surface. Mechanical surface scrapers, widely used for viscous liquids in the chemical process industry, can be applied to duct flow of gases. Equipment with rotating heat exchanger ducts is found in commercial practice.

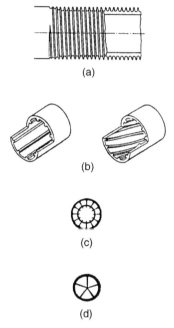

(a)

(b)

(c)

(d)

Figure 1.4 (a) Integral fins on outer tube surface, (b) internally finned tubes (axial and helical fins), (c) cross sections of multiply internally finned tubes, (d) tube with aluminum star insert.

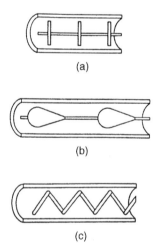

(a)

(b)

(c)

Figure 1.5 (a) Spaced disk devices, (b) spaced streamline-shaped insert devices, (c) displaced wire-coil insert. (From Webb [1987].)

Surface vibration at either low or high frequency has been used primarily to improve single-phase heat transfer. A piezoelectric device may be used to vibrate a surface and impinge small droplets onto a heated surface to promote "spray cooling." This is discussed in Chapter 19.

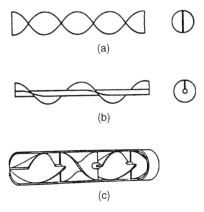

Figure 1.6 Three types of swirl flow inserts: (a) twisted-tape insert, (b) helical vane insert, (c) static mixer.

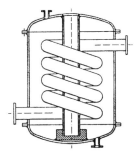

Figure 1.7 Helically coiled tube heat exchanger. (From Bergles and Webb [1985]. With permission.)

Fluid vibration is the more practical type of vibration enhancement because of the mass of most heat exchangers. The vibrations range from pulsations of about 1 Hz to ultrasound. Single-phase fluids are of primary concern.

Electrostatic fields (direct current, d.c., or alternating current, a.c.) are applied in many different ways to dielectric fluids. Generally speaking, electrostatic fields can be directed to cause greater bulk mixing of fluid in the vicinity of the heat transfer surface.

Injection is utilized by supplying gas through a porous heat transfer surface to a flow of liquid or by injecting the same liquid upstream of the heat transfer section. The injected gas augments single-phase flow. Surface degassing of liquids may produce similar effects.

Suction involves vapor removal, in nucleate or film boiling, or fluid withdrawal in single-phase flow through a porous heated surface.

Jet impingement forces a single-phase fluid normally or obliquely toward the surface. Single or multiple jets may be used, and boiling is possible with liquids.

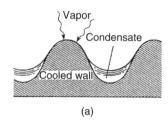

(a)

(b)

Figure 1.8 (a) Illustration of surface tension drainage from the flutes into drainage channels. (b) Fluted tube used for condensation in the vertical orientation.

1.2.3 Technique vs. Mode

Enhancement is applicable to both heat and mass transfer processes, or simultaneous heat and mass transfer. Modes of heat (or mass) transfer of potential interest include

1. Single-phase flow: Natural and forced convection inside or outside tubes
2. Two-phase flow: Boiling and condensation inside tubes, and tube banks
3. Radiation
4. Convective mass transfer

Compound enhancement results from simultaneous use of two or more of the previously discussed techniques. Such an approach may produce an enhancement that is larger than either of the techniques operating separately.

The majority of commercially interesting enhancement techniques are currently limited to passive techniques. However, current work on electro-hydrodynamic (EHD) enhancement of boiling and condensation suggests significant potential. The lack of use of the active techniques is related to the cost, noise, safety, or reliability concerns associated with the enhancement device.

This book primarily focuses on the passive techniques, derived from special configuration of the surface geometry. Chapters 5 through 14 (except Chapter 10) discuss passive techniques for specific heat transfer modes. Chapter 10 discusses fouling of enhanced surfaces. Chapter 15 surveys EHD enhancement, and Chapter 17 considers fluid additives.

1.3 PUBLISHED LITERATURE

1.3.1 General Remarks

Bergles et al. [1983, 1991; Manglik and Bergles, 2004] have reported the journal publications on enhanced heat transfer. Webb et al. [1983] provide a similar report on the U.S. patent literature. Bergles and Webb [1985] summarize the information presented in the literature and patent bibliography reports. Figure 1.9 from Manglik and Bergles [2004] shows the journal and conference publications per year between 1870 and 2001, which total 5676 publications. The figure shows that rapid growth began about 1950. The patent literature is an equally important source for information on enhancement techniques. Figure 1.10, taken from the Webb et al. [1983] survey of the U.S. patent literature, shows 486 patents issued as of June 1982.

The effectiveness of the technique may depend on the heat (or mass) transfer mode. Table 1.1 classifies the enhancement references according to enhancement technique and heat transfer mode. A dash in the table indicates that the technique is not applicable to that particular heat transfer mode.

Some comments are in order regarding the classifying and cataloging procedures for the report on technical literature. Difficulties in classifying some techniques are

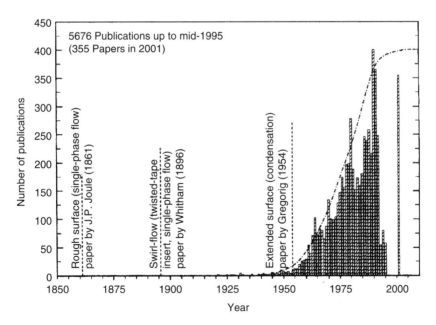

Figure 1.9 Citations on heat transfer augmentation vs. year of publication. Illustrates status as of 2001. From Manglik and Bergles [2004].

acknowledged. For example, it is somewhat arbitrary whether certain of the new structured boiling surfaces are considered *treated surfaces*, *rough surfaces*, or *extended surfaces*. For example, an enhanced boiling surface may appear to have only a slight surface *roughness*, yet it contains a subsurface microstructure that promotes boiling nucleation. Although the surface may provide a moderate surface area increase, the surface area increase is not responsible for the high enhancement level.

Another example is in the important area of fouling of enhanced surfaces. Papers on this subject are included in the *mass transfer* classification. However, the goal is obviously not to increase the fouling rate!

Several topics are not thoroughly referenced for a variety of reasons. Only a few papers are listed on the theory of dropwise condensation, under *treated surfaces*, because the emphasis of the classification is on promoters for dropwise condensation. Papers on use of extended surfaces in single-phase natural convection may not be included in the *extended surfaces* listing. Rather, the primary interest is in enhanced extended surfaces. As noted above, the category of *surface tension devices* does not include the vast literature on heat pipes. *Additives for gases* may include some information on fluidized beds. However, the primary interest of the listing is on enhancement of gas-to-wall heat transfer by the particles, rather than toward improvement of gas-to-particle heat transfer. Literature that is "supporting," e.g., details of the flow over a protuberance or pressure drop in a rough tube without accompanying heat transfer area, is not included in the literature file.

The literature reports of Bergles et al. [1983, 1991] are organized according to the following coding:

00 Surveys
01 Treated Surfaces
02 Rough Surfaces
03 Extended Surfaces
04 Displaced Enhancement Devices
05 Swirl Flow Devices
06 Surface Tension Devices
07 Additives for Liquids
08 Additives for Gases
09 Mechanical Aids
10 Surface Vibration
11 Fluid Vibration
12 Electric or Magnetic Fields
13 Injection or Suction
14 Compound Enhancement
15 Performance Evaluation
16 Coiled Tubes
17 Jet Impingement

Each citation includes the following information:

Author(s)
Title
Publication (or source) and date
Several key words describing the work

The English translations of foreign-language titles are provided except for French and German works. The key words include reference number, year of publication, journal, technique, and mode of heat transfer. The modes of heat transfer are coded as follows (mass transfer is included for completeness):

1. Single-Phase Natural Convection
2. Single-Phase Forced Convection
3. Pool Boiling
4. Forced-Convection Boiling
5. Condensation
6. Mass Transfer

Occasional difficulty arises when multiple techniques and modes are considered in a single paper, as not every mode may be considered for all the techniques. Because there was no convenient way to couple the mode and technique keywords, many of the ambiguities are left unresolved.

Reference numbers are formed according to technique (first two digits), mode (third digit), and number in that category (last three digits). Citations are set up on a minicomputer, recorded on a disc, and transferred to a mainframe computer that exercises the information retrieval program.

The literature is presented according to enhancement technique and mode of heat transfer, and the citations are listed in alphabetical order within each category. Table 1.1 summarizes the number of citations in each category.

1.3.2 U.S. Patent Literature

The second bibliography described covers U.S. patent literature. This report by Webb et al. [1983] contains 554 references to U.S. patents. Patents with issue dates up to June 30, 1983 are included.

The present report classifies the patents in three ways. The first two are by patent number and by name of the first named inventor. The third compilation is similar to that used in Bergles et al. [1983], in that the patents are classified by enhancement technique, mode of heat transfer, and geometry.

There are several changes in the classification utilized for the first bibliography. Specifically, two new classifications that were primarily subsets of other classifications are introduced below.

There are several differences between the enhancement technique classifications used in patent and journal literature reports. The patent literature report includes "formed channels," which are typified by corrugated plates used in plate-fin heat exchangers and deep spirally fluted tubes. The surface distortion is much greater than that typical of roughness. The surface deformation causes mixing by secondary flows. Surface area increase usually is not a significant factor, although the three-dimensional deformed tubes, also included in this classification, can provide a surface area increase of up to 30%. Formed channels provide enhancement to the fluid flowing on each side of the channel.

The patent report separately identifies "attached promoters," which are discrete roughness that are not integral to the heat transfer surface. A good example is the wire coil insert used in circular tubes.

In addition, the categories of Surveys, Coiled Tubes, Jet Impingement, and Performance Evaluation are not included in the patent classification. Coiled Tube concepts are included in the Formed Channels category.

The patents cited in the report are organized according to the following coding system:

Technique
01 Treated Surfaces
02 Rough Surfaces
03 Attached Promoters
04 Extended Surfaces
05 Displaced Enhancement Devices
06 Formed Channels
07 Swirl Flow Devices
08 Surface Tension Devices
09 Additives for Liquids
10 Additives for Gases
11 Mechanical Aids
12 Surface Vibration
13 Fluid Vibration
14 Electrostatic Fields
15 Injection or Suction
16 Compound Enhancement

Mode of Transfer
1 Single-Phase Natural Convection
2 Single-Phase Forced Convection
3 Pool Boiling
4 Flow Boiling
5 Condensation
6 Mass Transfer

Geometry Classification
A Inside Tubes
B Outside Tubes — Gases
C Outside Tubes — Liquids

D   Plate-Type Heat Exchangers
E   Fin-Type Heat Exchangers
F   Other Geometries

For example, a code of 023C indicates a rough surface for pool boiling, with the roughness on the outer surface of a tube. Note that the patent file contains a Geometry Classification, which is not contained in the literature file.

The report contains patents issued through June 1983. An intensive "hands-on" search of patents on file at the U.S. Patent Office was performed for the 1980 edition of the report.

Because there are more than 10,000 patent subclasses, it was necessary to develop a logical approach to establish which subclasses were worthy of search. The patent classifications searched at the U.S. Patent Office included Classes 29, 62, 72, 113, 165, 202, 219, 260, 264, 357, and 427. Only certain subclasses were of interest. An illustration of the U.S. Patent Office classification system is shown in Table 1.2. This table lists the subclasses within Class 165, "Heat Exchange," which were searched. Study of Table 1.2 shows that the patent literature uses a key wording system that significantly differs from key wording familiar to engineers. Hence, special considerations are required for patent searches.

The patent subclasses selected for search were established by several parallel effort. First, the *U.S. Patent Office Class Definitions*, which lists and describes the

Table 1.2 Selected Subclass Titles in Patent Class 165, Heat Exchange

Subclass	Subclass Title
1	Process
2	Heating and cooling
68	With external support lets
76	With repair or assembly means
86	Movable heating or cooling surface
109	With agitating or stirring structure
110	With first fluid holder or collector open to second fluid
111	Separate external discharge port for each fluid
115	Trickler
133	With coated, roughened, or polished surface
134	With projector or protective agent
172	Side-by-side tubular structures or tube sections
174	With internal flow director
177	Tubular structure
179	Projecting internal and external heat transfer means
180	Diverse materials
181	With discrete heat transfer means
184	Helical
185	Heat transmitter
DIG. 3	Electrical devices — cooled
DIG. 18	Condensation

subclasses within each patent class, was reviewed. Study of this document suggested potentially fruitful areas.

The final and most successful method involved an iterative search method. At the beginning of the intensive search, there were approximately 100 patents on file. A "frequency distribution table" that defined the classes/subclasses of these patents was made. The patents themselves also provided some cross-reference information. This worksheet table, along with the previously indicated fruitful areas, identified the initial classes/subclasses to be searched. In addition, current issues of the biweekly *Official Patent Gazette*, which publishes bimonthly abstracts of new patents, were monitored during the search period. This publication was the primary source of recent patents for the second edition.

Copies of approximately 900 patents were ordered from the Commissioner of Patents and Trademarks, Washington, D.C. The patents were then carefully reviewed for relevance and cataloged by the previously described coding system. Many patents were discarded.

The final list in this report contains 486 entries obtained from 454 separate patents. There are 31 patents for which more than 1 technique mode is appropriate. One patent actually has three technique mode entries.

A patent entry, typical of those listed in this bibliography film, is given below.

3,587,730 Milton, R. M.	(patent number and inventor)
1971 June 28	(date of patent issue)
165/110	(patent class/sub-class)
013C	(technique-mode-geometry classification)
Heat Exchange System with Porous Boiling Layer	(title)

Assignees are also identified (opposite the date of issue) for the patents added subsequent to the first edition of the report.

The computer file was used to generate the three output listings contained in Webb et al. (1983):

1. Chronological listing by patent number
2. Alphabetical listing by first listed inventor
3. Technique-mode-geometry classification listing

Table 1.3 shows the distribution of patents within each technique mode classification. The 32 multiple listings are reflected in this table. Enhancement techniques for single-phase forced convection comprise 60% of the patents listed in the file; 88% of the patents involve the "passive techniques." Virtually all of the "passive techniques" involve special surface geometries.

Figure 1.10, which shows the number of patents issued by year since 1928, indicates that U.S. patent activity relevant to heat transfer enhancement has continued

Table 1.3 Application of Enhancement Techniques

		Mode				
	Avail.	CV	B	C	Materials	Performance Potential
Inside Tubes						
Metal coatings	Yes	5	2	5	Al, Cu, St	Hi
Integral fins	Yes	2	1	1	Al, Cu	Hi
Flutes	Yes	4	4	1	Al, Cu	Mod
Integral roughness	Yes	1	2	2	Cu, St	Hi
Wire coil inserts	Yes	1	2	2	Any	Mod
Displaced promoters	Yes	2	2	2	Any	Mod (Lam)
Twisted tape inserts	Yes	1	2	2	Any	Mod
Outside Circular Tubes						
Coatings						
Metal	Yes	5	1	3	Al, Cu, St	Hi (B)
Nonmetal	No	5	4	4	"Teflon"	Mod
Roughness (integral)	Yes	3	2	2	Al, Cu	Hi (B)
Roughness (attached)	Yes	2	3	3	Any	Mod (CV)
Axial fins	Yes	1	2	2	Al, St	Hi (CV)
Transverse fins						
Gases	Yes	1	5	5	Al, Cu, St	Hi
Liquids, two-phase	Yes	1	1	1	Any	Hi
Flutes						
Integral	Yes	5	5	1	Al, Cu	Hi
Nonintegral	Yes	5	5	3	Any	Hi
Plate-Fin Heat Exchanger						
Metal coatings	Yes	5	3	5	Al	Hi (B)
Surface roughness	Yes	4	4	4	Al	Hi (B)
Configured channel	Yes	1	1	1	Al, St	Hi
Interrupted Fins	Yes	1	1	1	Al, St	
Flutes	No	5	5	4	Al	Mod
Plate-Type Heat Exchanger						
Metal coatings	No	5	4	5	St	Hi (B)
Surface roughness	No	4	4	4	St	Mod (CV)
Configured channel	Yes	1	3	3	St	Hi (CV)

Use Code	Heat Transfer Mode	
1 Common use	CV	Convection
2 Limited use	B	Boiling
3 Some special cases	C	Condensation
4 Essentially no use		
5 Not applicable		

at a high level since 1965. During the 5-year period 1978–1982, more than 20 patents per year were granted.

For those who wish to identify new patents, review of the *Official Patent Gazette* is recommended. The *Gazette* is issued biweekly, and it contains patent abstracts, which are segregated by patent class. The most fruitful classes to review are 29 (metal working), 62 (refrigeration), 138 (pipes with flow regulators and/or baffles), and 165 (heat exchange). Approximately 60% of the enhancement patents were found in Class 165.

Several computer patent search programs are available for search purposes. Engineers must learn keywords favored by patent attorneys to use the patent databases effectively.

1.3.3 Manufacturer's Information

Manufacturer's product information is an excellent source for information on enhanced heat transfer. Such sources include manufacturers of enhanced heat transfer tubing and of insert devices. However, much information on enhanced heat transfer products is not reported in manufacturer's product brochures. An example is the enhanced heat transfer products used in automobiles. Although the automotive industry is a

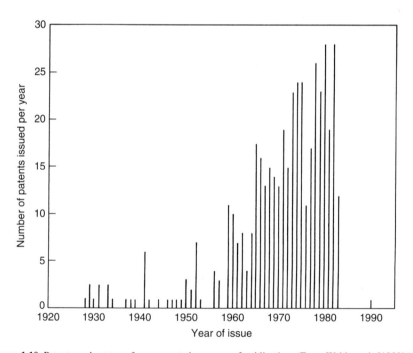

Figure 1.10 Patents on heat transfer augmentation vs. yr of publication. (From Webb et al. [1983].)

prominent user of enhanced heat transfer, heat exchanger technology is not discussed in its consumer-oriented product information. The heat exchangers used in the refrigeration industry typically use enhancement. Some of its product information may discuss their enhanced surfaces. However, if the enhancement is not visible in the product, the manufacturer may choose not to reveal its existence. This is because it is considered as "proprietary information." Chemical process plants also use enhanced surfaces, and these are most certainly proprietary. Hence, an interested person will not gain a very complete perspective of the broad range of enhanced heat transfer technology from product information.

An excellent source for industrial information is the patent literature. Industry uses patents to protect proprietary interests. Unfortunately, one must consciously seek out the patent information, as much of it is not widely published. Some trade magazines publish current patent information in special industrial areas.

1.4 BENEFITS OF ENHANCEMENT

Special surface geometries provide enhancement by establishing a higher hA per unit base surface area. Three basic methods are employed to increase the hA value:

Method 1: Increase h without an appreciable physical area (A) increase. An example is surface roughness inside the tube of Figure 1.2b.
Method 2: Increase of A without appreciably changing h. An example is the Figure 1.4b internally finned tube.
Method 3: Increase of both h and A. The interrupted fin geometries shown in Figures 1.3a through e provide a higher heat transfer coefficient than a plain fin. They also provide increased surface area.

For application to a two-fluid tubular heat exchanger, enhancement may be desired for the inner, the outer, or both sides of the tube. If enhancement is applied to the inner and outer tube surfaces, a doubly enhanced tube results. Applications for such doubly enhanced tubes include condenser and evaporator tubes. Depending on the design application, the two-phase heat transfer process may occur on the tube side or shell side. Consider, for example, a shell-and-tube condenser with cooling water on the tube side. The preferred enhancement geometry for the condensing side may be substantially different from that desired for the water side. Hence, one must be aware of possible manufacturing limitations or possibilities that affect independent formation of the enhancement geometry on the inner and outer surfaces of a tube. Figure 1.11 illustrates five basic approaches that may be employed to provide doubly enhanced tubes. Figures 1.11a through e allow independent selection of the shell-side and tube-side enhancement geometries. However, the forming process used to make the Figure 1.11d tube-side ridges also deforms the outer tube

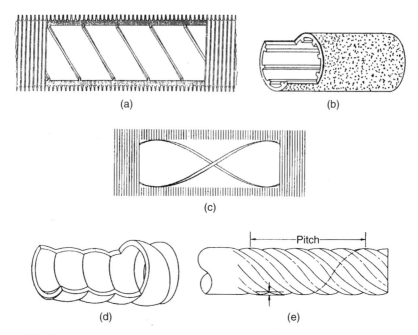

Figure 1.11 Methods used to make doubly enhanced tubes. (a) Helical rib roughness on inner surface and integral fins on outer surface. (b) Internal fins on inner surface and coated (porous boiling surface) on outer surface. (c) Insert device (twisted tape) with integral fins on outer surface. (d) Corrugated inner and outer surfaces. (e) Corrugated strip rolled in tubular form and seam welded. (From Webb [1987].)

surface. This Figure 1.11a tube will give good tube-side enhancement for water flow and good shell-side condensation performance. The Figure 1.11a approach can be made in copper alloys and steel, but is not very practical for titanium, because of the hardness of the material and high fins (e.g., greater than 1.0 mm) would be of little value, because of its low thermal conductivity. Fins up to 1.0 mm high can be made on the outer surface of a titanium tube.

Webb [1981] discusses possibilities for manufacture of doubly enhanced condenser tubes. Users should also be aware that it may not be possible to make a given enhancement geometry in all materials of interest. For example, the Figure 1.11b internally finned tube can easily be made as an aluminum extrusion, but it would be extremely difficult to make from a very hard material, such as titanium. Webb [1980] discusses particular enhancement geometries that have been commercially manufactured and their materials of manufacture. More discussion of current manufacturing technology is presented in later chapters. Table 1.1 and Table 1.2 show that a variety of enhancement techniques are possible. In a practical sense, the various enhancement techniques may be considered to be in competition with one another. The favored method will provide the highest performance at minimum cost for the material of interest.

1.5 COMMERCIAL APPLICATIONS OF ENHANCED SURFACES

1.5.1 Heat (and Mass) Exchanger Types of Interest

Consider the application of enhancement techniques to four basic heat exchanger types: (1) shell-and-tube or tube banks, (2) fin-and-tube, (3) plate-fin, and (4) plate-type. The heat-transfer modes to be considered are single-phase forced convection of liquids and gases, boiling, and condensation. These heat exchanger types require enhancement for four basic flow geometries:

1. Internal flow in tubes, with circular the most important
2. External flow along or across tubes, with circular tubes most common
3. Plate-fin-type heat exchangers made by a stacked construction, which involve flow in noncircular passages
4. Plate-type heat exchangers, which involve flow between formed parallel plates

Table 1.3 summarizes the present and potential use of special surface geometries and other enhancement techniques for the applications of interest. The first part of the table considers application of special surface geometries to the inner and outer surfaces of circular tubes. The latter part of the table considers plate-type and plate-fin heat exchangers. A coding system is employed to define the degree to which an enhancement technique has found application. A dash entry indicates not applicable or not intended to augment that heat transfer mode. The third column lists typical materials used for commercial manufacture. The last column indicates the authors' assessment of the performance potential, independent of whether a commercial manufacturing technique exists. It is evident that a significant number of the available enhancement techniques are commercially utilized.

Enhancement techniques are not limited to heat exchangers. Equipment, such as cooling towers and distillation columns, involve simultaneous heat and mass transfer processes. Enhancement techniques viable for enhancement of heat transfer to gases are equally applicable for convective mass transfer to gases. Convective mass transfer to a gas controls the performance of a cooling tower.

1.5.2 Illustrations of Enhanced Tubular Surfaces

Some of the major commercially used enhanced surfaces are illustrated and discussed. These are broadly representative of commercial technology, but are not meant to be comprehensive.

Corrugated Tubes. Figure 1.2a shows corrugated, or "roped," tubes. This tube is formed by pressing a forming tool on the outer surface of the tube, which produces an internal ridge roughness. The Figure 1.2a tube is the popular Wolverine Korodense™ tube.

Integral-Fin Tubes. Figure 1.4a shows a standard integral-fin tube. Tubes having 19 to 26 fins/in. (748 to 1024 fins/m), and 1.5 mm fin height are routinely used in shell-and-tube exchangers for enhancement of single-phase liquids, for boiling, and for condensation of low-surface-tension fluids, such as refrigerants. A 35 fins/in. (1378 fins/m) integral-fin tube with 0.9 mm fin height is used for condensation of refrigerants. When the outside heat transfer enhancement is sufficiently high, tube-side enhancement is also beneficial. This constitutes a "doubly enhanced" tube. Figure 1.12a shows the Wolverine Turbo-Chil™ tube which has 1024 fins/m. (26 fins/m) on the outside surface and a 10 start internal rib roughness at 49° helix angle. Jaber and Webb (1993) describe tubes developed by Wolverine and Wieland for steam condensation, a high-surface-tension fluid. High-surface-tension fluids require increased fin pitch. A steam condenser tube may have 11 fins/m (433 fins/in.) with 0.80- to 1.0-in. fin height. Figure 1.12b shows the Wieland NW™ tube, with 433 fins/in. and a wavy inside surface enhancement.

Enhanced Condensing Tubes. The tubes shown in Figure 1.13d have a special sawtooth fin geometry, which provides higher condensation coefficients than the standard integral-fin tubes. They were developed for use with refrigerants. Note that the tubes also contain water-side enhancement.

Enhanced Boiling Tubes. At least six enhanced boiling tubes are commercially available, three of which are shown in Figure 1.14. One example is the Figure 1.14a Wolverine Turbo-B™ surface geometry, which is made by a high-speed thread-rolling process. The tubes promote high nucleate boiling at much lower heat fluxes than the integral fin tubes of Figure 1.4a. The enhanced boiling tubes are typically made with water-side enhancements, as illustrated in Figure 1.13 and Figure 1.14. The Figure 1.1a and Figure 1.14 tubes are also used in process applications. The Figure 1.1a porous coating is commercially known as the UOP High-Flux™ tube, and is frequently applied to a corrugated tube.

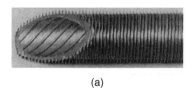

(a)

(b)

Figure 1.12 (a) Wolverine Turbo-Chil tube (integral fin outside surface and 10 start helical rib internal rib roughness). (b) Wieland NW tube (11 fins/m on outer surface and a wavy inside roughness).

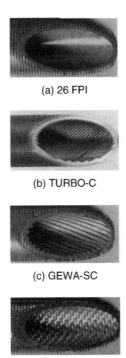

(a) 26 FPI

(b) TURBO-C

(c) GEWA-SC

(d) TRED-D

Figure 1.13 Tubes for refrigerant condensation having enhancement on the condensing refrigerant side (outside) and the water side (inside). (a) Standard integral fin (26 fins/in, or 1024 fins/m), (b) Wolverine Turbo-C. (c) Wieland GEWA-SC. (d) Sumitomo Tred-19D.

Tube-Side Enhancements. A variety of tube side enhancements exist. Figure 1.13 and Figure 1.14 illustrate several of those commercially used. Internally finned tubes having high fins (e.g., 1 mm or more) are quite expensive to make in copper, but are cheaply made in aluminum by extrusion. Figure 1.15 shows the "microfin" tube, made by Wieland, which is widely used in direct expansion refrigerant evaporators and condensers. This tube has very short triangular-cross-section fins (0.20 mm high), which are typically made with a helix angle of approximately 15°. Extruded aluminum microfin tubes are also commercially available.

1.5.3 Enhanced Fin Geometries for Gases

Gas-side fins are typically used for heat transfer between a tube-side liquid (or two-phase fluid) and a gas, e.g., air. Air-side fins are used because the air-side heat transfer coefficient is much smaller than that of liquids or two-phase fluids. Plain fins are "old technology." Enhanced surface geometries, such as wavy fins or interrupted fins (Figure 1.3), give higher performance than plain fins. Louvered fins (Figure 1.3b) are widely used in automotive heat exchangers with rectangular-cross-section tubes. Air conditioning heat exchangers may use the Figure 1.3c or 1.3e

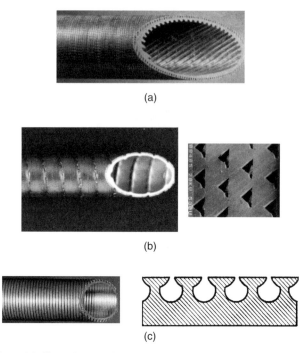

(a)

(b)

(c)

Figure 1.14 Enhanced boiling tubes: (a) the Wolverine Turbo-B, (b) the Hitachi ThermoExcel-E, and (c) the Wieland TW.

Figure 1.15 The Ripple-fin, or "microfin" tube having 0.2-mm-high triangular fins. (Photograph courtesy of Wieland-Werke AG.)

interrupted fins on round tubes. Brazed aluminum, plate-and-fin heat exchangers typically use the Figure 1.3a or 1.3b fin geometries. The Figure 1.3a through 1.3e interrupted fin geometries provide high heat transfer coefficients via the repeated growth and destruction of thin boundary layers.

The Figure 1.3f geometry is typical of enhancements used in rotary regenerators. Use of an interrupted fin geometry would allow leakage between the hot and cold streams, which reduces performance and allows contamination of the "clean" stream.

1.5.4 Plate-Type Heat Exchangers

These heat exchangers are used to exchange heat between liquids, or between a liquid and a two-phase fluid. Plate exchangers do not use extended surface on the

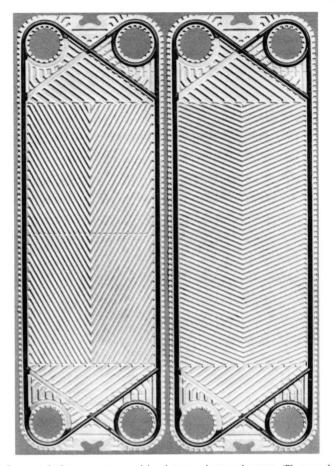

Figure 1.16 Corrugated plate geometry used in plate-type heat exchangers. (Photograph courtesy of Alfa-Laval.)

heat transfer plates. Rather, they use corrugated surfaces, which provides secondary flow and mixing to obtain high heat transfer performance (Figure 1.16).

1.5.5 Cooling Tower Packings

Cooling towers cool a liquid (gravity-drained liquid film) by transferring heat from the water film to air passing through the cooling tower "packing." This involves a simultaneous heat and mass transfer process. The water is cooled principally by evaporation of a small fraction of the water. High heat and mass transfer coefficients are provided for the airflow by special packing geometries, such as illustrated in Figure 1.17. The limiting mass transfer resistance is in the gas phase. The Figure 1.17a packing enhances the mass transfer coefficient by the development and destruction of thin gas boundary layers using the same principle employed in the Figure 1.3a

(a)

(b)

Figure 1.17 Enhanced film-type cooling tower packing geometries. (a) Ceramic block type. (Photograph courtesy of Ceramic Cooling Tower Co.) (b) Plastic corrugated packing. (Photograph courtesy of Munters Corp.)

offset strip-fin heat exchanger geometry. The Figure 1.17b corrugation provides mixing of the airflow, while providing a large surface area of liquid film. The Figure 1.17a geometry provides interrupted gas boundary layers, quite similar to that of the Figure 1.3a offset strip fin heat exchanger geometry.

1.5.6 Distillation and Column Packings

Distillation columns also involve heat and mass transfer between two fluid mixtures. One fluid exists as a liquid film and the other fluid is a gas. Typical column packings are shown in Figure 1.18. These packings provide enhancement for the gas flow in the same manner as for cooling tower packings.

1.5.7 Factors Affecting Commercial Development

It is appropriate to view enhancement techniques as "second generation" heat transfer technology. Such new technology normally requires several phases of development for successful commercialization. These include:

1. Basic performance data for heat transfer and pressure drop, if applicable, must be obtained. General correlations should be developed to predict heat transfer and pressure drop as a function of the geometric characteristics.
2. Design methods must be developed to facilitate selection of the optimum surface geometry for the various techniques and particular applications.
3. Manufacturing technology and cost of manufacture must be available for the desired surface geometry and material.
4. Pilot plant tests of the proposed surface are required to permit a complete economic evaluation and to establish the design, including long-term fouling and corrosion characteristics.

Some enhancement techniques have completed all steps required for commercialization. Examples are extended surface for fin-and-tube heat exchangers, and external low-fin tubing. However, techniques such as internally roughened tubes require further development. Table 1.4 attempts to summarize the status of technology development for several of the more important enhancement techniques applied to circular tubes. Commercialization represents the ultimate stage of development; however, even commercial products require additional development work.

It is recognized that several enhancement techniques are in competition with one another. For application to a given heat-transfer mode, some techniques will appear more interesting than others. Therefore, a continued high level of interest for all competing techniques cannot be expected.

The development of extended surfaces on the outside of tubes has proceeded on a proprietary basis, and has been successfully commercialized. However, additional work directed at developing design methods for optimum surface selection is worthwhile. Extended surface for boiling and condensing applications is similarly well established. Enhanced boiling and condensing surface geometries are now in their third generation. Most of the previous work has focused on the outer surface of the tube. The microfin tube has been very successful for vaporization and condensation of refrigerants inside tubes. Much of the recent work has focused on incremental performance improvements and material reduction.

In regard to internally roughened tubes, a considerable amount of experimental data exists for single-phase forced convection. The major need is development of

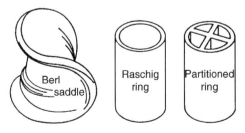

Figure 1.18 Packings for distillation columns. (From Edwards et al. [1973].)

Table 1.4 Status of Technology Development

Geometry	Forced Convection	Boiling	Condensation
Inside Tubes			
Coatings		3,5	1
Roughness	2,3,5	1	1
Internal fins	1,2,3,5	1,5	1
Flutes	5	5	1,5
Insert devices	5	1,5	1,5
Outside Tubes			
Metal coatings		3,5	1
Nonmetal coatings		1	4
Roughness	1,2	1,2,5	1
Extended surface	5	5	5
Flutes			5
Code			
1. Basic performance data			
2. Design methods			
3. Manufacturing technology			
4. Heat exchanger application			
5. Commercialization			

analytical design methods for optimized surface geometries, with additional work on manufacturing technology.

The further development of any technique will depend on two factors: (1) its performance potential, and (2) its manufacturing cost. At present, cost-effective manufacturing technology is probably the most significant barrier to commercial application of high-performance enhancement techniques. Researchers appear intent on identifying the surface geometries that offer the highest performance. It is suggested that more communication between researchers and manufacturing engineers is needed to identify the most cost-effective concepts.

1.6 DEFINITION OF HEAT TRANSFER AREA

Enhanced tubes may involve an area increase, relative to that of a plain tube. In this work, we define the heat transfer coefficient as follows:

1. **Tube Side Enhancement:** The h value is defined on the basis of the *nominal* surface area. The nominal surface area is defined as $A/L = \pi d_i$, where d_i is the tube inside diameter to the base of the enhancement. This definition allows direct comparison of the h value of tube-side enhancements, relative to the h value of a plain tube. Some authors define the h value for an internally finned tube in terms of the total surface area.

2. **Extended Surfaces for Gases:** We define the h value on the basis of the total surface area (A) of the fins and base surface to which the fins are attached. This is consistent with industrial practice.

1.7 POTENTIAL FOR ENHANCEMENT

For two-fluid heat exchangers, the ratio of the thermal resistances between the two fluid streams is of primary importance in determining whether enhancement will be beneficial. Enhancement should be considered for the stream that has the controlling thermal resistance. If the thermal resistances of both streams are approximately equal, one should consider enhancing both sides of the heat exchanger. Practical considerations should also include the effect of design fouling resistances on the potential for enhancement.

In addition to two-fluid heat exchangers, cooling is required for equipment or devices having internal heat generation. Examples are nuclear fuel elements, electrical heaters, or electronic components, such as circuit boards. Use of enhancement here will result in either of two benefits. The device may support a greater heat flux for fixed surface temperature, or the device may operate at a lower temperature, while operating at the same heat flux. The second benefit is of great importance for electronic equipment, because the component service life is inversely related to the operating temperature. This problem involves operation with a specified heat flux boundary condition. Because a second fluid is not involved, as in PEC Example 1.1 in next section, any increase of the heat transfer coefficient will be beneficial for the case of interest.

In Chapter 3, we discuss "Performance Evaluation Criteria" used to evaluate the benefits of enhancement. We provide a number of examples in the book that illustrate applications of enhanced surfaces. The example problems are called "PEC Examples."

1.7.1 PEC Example 1.1

Consider a shell-and-tube heat exchanger that cools engine oil on the shell side with cooling water on the tube side. The exchanger currently uses 19 mm O.D. (outside diameter) plain copper tubes (0.8 mm wall thickness). Would enhancement be beneficial? Which side of the tube should be enhanced? What type of enhancement geometry would offer potential?

We must first determine the heat transfer coefficients for each of the fluid streams. Using available correlations, the designer has determined that the water-side heat transfer coefficient is 10 kW/m²-K, and the heat transfer coefficient for the oil is 1.0 kW/m²-K. For simplicity, the tube wall and fouling thermal resistances are ignored. The overall thermal resistance is

$$L/UA = L/(h_i A_i) + L/(h_o A_o)$$

$$L/UA = 1/(10 \times 0.0174\pi) + 1/(1 \times 0.0190\pi)$$

$$18.58 = 1.83 + 16.75$$

Thus, 0.902 (16.75/18.58) of the total thermal resistance is on the shell (oil) side. There would be no benefit derived from adding enhancement to the tube (water) side. However, substantial benefit would result from shell-side enhancement. A possible candidate geometry is the low, integral-fin tube. It is likely that the shell-side heat transfer coefficient may be enhanced 200% or more. Assuming a 200% increase of the shell-side heat transfer coefficient, the total thermal resistance (L/UA) would then be

$$L/UA = 1/(10 \times 0.0174\pi) + 1/(3 \times 0.019\pi)$$

$$7.41 = 1.83 + 5.58$$

The total tubing length in the enhanced tube heat exchanger would be 0.40 (7.41/18.58) that of the original design. No consideration was given to the oil pressure drop. However, pressure drop considerations are invariably part of the design analysis for enhancement considerations.

1.7.2 PEC Example 1.2

Consider a microchip on an electronic circuit board. The device is rated for 0.1 W at 60°C. The chip currently dissipates heat to ambient air at 30°C. If the surface temperature can be lowered to 50°C, the expected life of the chip may be increased by 25%. Assume that an enhancement concept will increase the surface heat transfer coefficient by 50%. Calculate the operating surface temperature of the chip.

$$Q/A = h_{old}(60 - 30) = h_{new}(T_{s,new} - 30)$$

Using $h_{new}/h_{old} = 1.5$ in the above equation gives $T_{s,new} = 52°C$. The chip surface temperature will be reduced from 60°C to 52°C. By being a little more clever in the enhancement design, we may be able to get the surface temperature down to 50°C!

NOMENCLATURE

A	Heat transfer surface area, m² or ft²
c_p	Specific heat, J/kg-K or Btu/lbm-F
d	Tube diameter, m or ft
E_h	h/h_p at constant Re, dimensionless
e	Roughness height, m or ft
h	Heat transfer coefficient, W/m²-K or Btu/hr-ft²-F
L	Length of flow path in heat exchanger, m or ft
ΔT_m	Mean temperature difference, K or F
Q	Heat transfer rate, W or Btu/hr
T	Temperature, K or °F

t Tube wall thickness, m or ft
U Overall heat transfer coefficient, $W/m^2\text{-}K$ or $Btu/hr\text{-}ft^2\text{-}F$

Greek Symbols

η Surface efficiency, dimensionless

Subscripts

1   Fluid 1
2   Fluid 2
m Tube inner surface
o Tube outer surface
p Plain surface
s Smooth tube
w Tube wall

REFERENCES

Bergles, A.E. and Webb, R.L., 1985. A guide to the literature on convective heat transfer augmentation, in *Advances in Enhanced Heat Transfer—1985*, S.M. Shenkman, J.E. O'Brien, I.S. Habib, and J.A. Kohler, Eds., ASME Symp. Vol. HTD, 43, 81–90.

Bergles, A.E., Nirmalan, V., Junkhan, G.H., and Webb, R.L., 1983. Bibliography on augmentation of convective heat and mass transfer II, *Heat Transfer Laboratory Report HTL-31*, ISU-ERI-Ames-84221, Iowa State University, Ames, December.

Bergles, A.E., Jensen, M.K., Somerscales, E.F.C., and Manglik, R.M., 1991. Literature Review of Heat Transfer Enhancement Technology for Heat Exchangers in Gas Fired Applications, GRI Report GRI 91-0146, Gas Research Institute, Chicago.

Edwards, D.K., Denny, V.E., and Mills, A.F., 1973. *Transfer Processes—An Introduction to Diffusion, Convection, and Radiation*, Hemisphere, New York, 320.

Jaber, M.H. and Webb, R.L., 1993. An experimental investigation of enhanced tubes for steam condensers, *Exp. Heat Transfer*, 6, 35–54.

Manglik, R. and Bergles, A.E., 2004. Enhanced heat and mass transfer in the new millennium: a review of the 2001 literature, *J. Enhanced Heat Transfer*, 11, 87–118.

Webb, R.L., 1980. Special surface geometries for heat transfer augmentation, in *Developments in Heat Exchanger Technology—I*, D. Chisholm, Ed., Applied Science, London, chap. 7.

Webb, R.L., 1981. The use of enhanced surface geometries in condensers, in *Power Condenser Heat Transfer Technology*, P.J. Marto and R.H. Nunn, Eds., Hemisphere, Washington, D.C., 287–324.

Webb, R.L., 1987. Enhancement of single-phase heat transfer, in *Handbook of Single Phase Heat Transfer*, S. Kakac, R.K. Shah, and W. Aung, Eds., John Wiley & Sons, New York, chap. 17, 62 pp.

Webb, R.L. and Bergles, A.E., 1983. Heat transfer enhancement: second generation technology, *Mech. Eng.*, 105, 60–67.

Webb, R.L. Bergles, A.E., and Junkhan, G.H., 1983. Bibliography of U.S. Patents on Augmentation of Convective Heat and Mass Transfer, Heat Transfer Laboratory Report HTL-32, ISU-ERI-Ames-81070, Iowa State University, Ames, December.

HEAT TRANSFER FUNDAMENTALS

2.1 INTRODUCTION

This chapter presents a summary of heat transfer fundamentals, heat exchanger design theory, and introduces the designer to the broad possibilities for the use of enhancement technology. Heat transfer enhancement concepts may be considered for application to virtually any type of heat exchanger or heat exchange device.

The heat exchanger design problem is considerably more complex than the calculation of the heat transfer and pressure drop characteristics of the exchanger. Heat exchanger calculations may be classified by two basic problems:

1. Rating problem: The heat exchanger type, size, and surface geometry are specified. The process conditions are also specified (e.g., flow rate, entering fluid conditions, and fouling factors). This problem requires calculation of the heat transfer rate and pressure drop of each stream.
2. Sizing problem: This problem involves calculation of the heat exchanger size required for specified process requirements (flow rate, entering fluid conditions, and allowable pressure drops).

Both problems require use of fundamental relations to determine the thermal design and pressure drop characteristics of the exchanger. The sizing problem is more complex than the rating problem because the heat exchanger type and surface geometry must be selected before the thermal analysis can be performed.

Table 2.1 Heat Exchanger Design Considerations

Design Specifications	Design Selection
Process Requirements	**Heat exchanger materials**
a. Fluid compositions, and inlet flow conditions (flow rate, temperature, and pressure)	a. Fluid temperatures and pressures
b. Heat duty or required exit temperatures	b. Corrosive characteristics of fluid-material combination
c. Allowable pressure drops	
Operating and Maintenance Considerations	**Heat Exchanger Type**
a. Fouling potential and method of cleaning	a. Design pressure and fluid temperatures
b. Failure due to corrosion, thermal stress, vibration, or freezing	b. Corrosion, stress, vibration, and freezing considerations
c. Repair of leaks	c. Fouling potential and cleaning possibilities
d. Part load operating characteristics	d. First cost, operating, and maintenance costs
Size and Weight Restrictions	**Heat Transfer Surface Geometries**
a. Frontal area, length, or height	a. Thermal resistance ratio of fluids
b. Possible weight restrictions	b. Potential for use of enhanced surfaces
	c. Fouling potential and cleaning possibilities
	d. Unit cost of heat transfer surface

The preliminary decisions required are (1) heat exchanger materials, (2) heat exchanger type and flow arrangement, and (3) heat transfer surface geometries.

Table 2.1 summarizes the considerations necessary to establish the heat exchanger design concept. These selections result from consideration of given or implied "design specifications" and the "design selections" as influenced by the "design specifications."

Having established the exchanger type, the designer then performs the thermal and hydraulic analysis. However, a single calculation will not directly yield the "optimum" design because different fluid velocities or heat transfer surface geometries will yield different results. Typically, the designer must evaluate several possibilities in the search for an "optimum" design. The selection of an optimum design requires that the designer consciously establish which variables (the objective function) are to be optimized (e.g., first cost, operating cost, or alternatively "lifecycle costs," and size dimensions).

The design optimization methods are discussed by Shah et al. [1978]. The optimization process involves a parametric analysis to determine how the specified objective function is influenced by trade-offs among the possible design variables. Shah et al. presents a quite detailed discussion of the general methodology of the heat exchanger sizing problem and the selection of an optimum design for a given application.

2.2 HEAT EXCHANGER DESIGN THEORY

This section presents a summary of heat exchanger thermal analysis. The development of the fundamental relations may be found in any heat transfer text.

2.2.1 Thermal Analysis

Consider the design of a two-fluid heat exchanger as illustrated in Figure 2.1. Although a counterflow exchanger is illustrated, our discussion applies to any flow configuration.

The heat exchanged between the two fluids is dependent on three specifications:

1. Thermodynamic specifications as defined by the hot and cold fluid flow rates and their entering and exit temperatures. Thus,

$$Q = C_h(T_{h1} - T_{h2}) = C_c(T_{c2} - T_{c1})$$ (2.1)

Heat exchanger analysis normally uses the definition $C_h = W_h c_{ph}$ because these two terms occur as a product. The term C_h is defined as the "capacity rate" of the hot fluid.

2. The heat transfer rate equation $dQ = UdA(T_h - T_c)$. The rate of heat exchange is dependent on the overall heat transfer coefficient between the two fluids. The heat exchanger surface area required to satisfy the thermodynamic specifications of Equation 2.1 is obtained by integrating the heat transfer rate equation over the length (or area) of the exchanger. Thus, for constant U,

$$Q = U \int_A (T_h - T_c) dA$$ (2.2)

The "effective mean temperature difference," averaged over the total surface area is

$$\Delta T_M = \frac{1}{A} \int_A (T_h - T_c) dA$$ (2.3)

Substitution of Equation 2.3 in Equation 2.2 gives

$$Q = UA\Delta T_M$$ (2.4)

3. The product of the UA terms is the "overall heat conductance" of the heat exchanger. Stated in thermal resistance terms, $1/UA$ is the overall thermal resistance to heat transfer between the two fluids.

The overall resistance, $1/UA$, is calculated by summing the individual thermal resistances given in Equation 2.5 below.

$$\frac{1}{UA} = \frac{1}{(\eta hA)_h} + \left(\frac{R_f}{A}\right)_h + \frac{t}{k_w A_w} + \frac{1}{(\eta hA)_c} + \left(\frac{R_f}{A}\right)_c$$ (2.5)

Equation 2.5 contains five individual resistances where subscripts h and c refer to the hot and cold streams. The first and fourth terms account for the convective resistance between the flowing fluids and the pipe wall. The third term is the conduction resistance of the solid wall that separates the streams. The second and

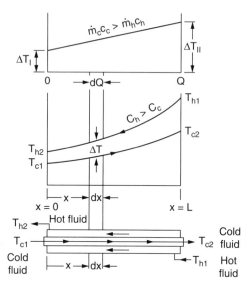

Figure 2.1 Temperature distribution in counterflow heat exchanger.

fifth terms are the fouling resistance on the hot and cold heat transfer surfaces. The overall coefficient, U, may be defined either in terms of the surface area of the hot or cold surface areas; thus, $UA = U_h A_h = U_c A_c$.

The effective temperature difference ΔT_m is a function of the heat exchanger flow geometry (e.g., counterflow or cross flow) and the degree of fluid mixing within each flow stream. For counterflow and parallel flow geometries, substitution of Equation 2.1 in Equation 2.3 and integration yields the following result for ΔT_m, which is commonly known as the *logarithmic mean temperature difference* (LMTD).

$$\text{LMTD} = \frac{\Delta T_I - \Delta T_{II}}{\ln(\Delta T_I / \Delta T_{II})} \tag{2.6}$$

Using the terminology ΔT_I and ΔT_{II} as the difference between the fluid temperatures at each end of the exchanger, Equation 2.6 applies to both parallel and counterflow configurations. For more complex geometries such as crossflow or multipass flow configurations, integration of Equation 2.3 yields more complex expressions for ΔT_m. For such cases it is customary to define a correction factor:

$$F = \frac{\Delta T_m}{\text{LMTD}} \tag{2.7}$$

where the LMTD is calculated for counterflow. The F correction is presented in graphical terms as illustrated in Figure 2.2 for one fluid "mixed" and the other fluid "unmixed." The term unmixed means that a fluid stream (e.g., the cold stream) passes through the heat exchanger in individual flow channels or tubes with no fluid mixing between adjacent flow channels. The F-correction factor for mixed flow

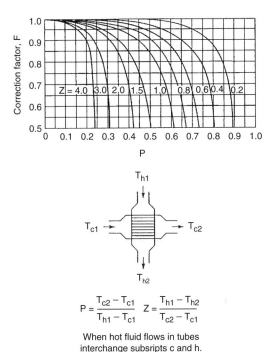

$$P = \frac{T_{c2} - T_{c1}}{T_{h1} - T_{c1}} \quad Z = \frac{T_{h1} - T_{h2}}{T_{c2} - T_{c1}}$$

When hot fluid flows in tubes
interchange subsripts c and h.

Figure 2.2 Correction factor for cross flow heat exchanger — one fluid mixed and one fluid unmixed.

means there are no thermal gradients normal to the flow. The parameters P and Z are defined as

$$Z \equiv \frac{C_c}{C_h} = \frac{T_{h1} - T_{h2}}{T_{c2} - T_{c1}} \tag{2.8}$$

$$P \equiv \frac{T_{c2} - T_{c1}}{T_{h1} - T_{c1}} \tag{2.9}$$

The term Z is the ratio of the capacity rates of the cold and hot streams. The term P is defined as the temperature effectiveness of the cold stream, which is the temperature rise of the cold stream divided by the difference between the inlet temperatures of the two fluids. Multipass exchangers have a combination of parallel and counterflow in alternate passes. Thus, F will depend on the pass arrangement, whether fluid mixing occurs within a pass and the conditions of mixing between passes. Charts to determine F are given in heat transfer textbooks, such as Holman [2001], Incropera and DeWitt [2001], or in Kays and London [1984].

2.2.2 Heat Exchanger Design Methods

There are alternate methods for calculating the heat transfer rate between the two fluids. These are the UA-LMTD and the effectiveness-NTU methods.

The UA-LMTD Method: The UA-LMTD method follows directly from Equations 2.4 and 2.7. Substitution of Equation 2.7 in Equation 2.4 gives

$$Q = F \cdot UA \cdot \text{LMTD} \tag{2.10}$$

For a rating calculation, one calculates UA using Equation 2.5. Or, for a sizing problem, one calculates UA per unit area or length. Second, the LMTD is calculated. Third, the geometry correction, F, is determined as noted below Equation 2.7.

Effectiveness-NTU Method: The heat exchanger effectiveness, $\in$, is defined as

$$\in = \frac{\text{actual heat transfer}}{\text{maximum possible heat transfer}} = \frac{Q_{\text{actual}}}{Q_{\text{max}}} \tag{2.11}$$

where Q_{actual} is given by the heat balance equations (Equation 2.1). The maximum possible heat transfer will occur in a counterflow heat exchanger of infinite area if one fluid undergoes a temperature change equal to the maximum temperature difference available $\Delta T_{\text{max}} = T_{h1} - T_{c1}$.

The calculation for Q_{max} is based on the fluid having the smaller capacity rate, C_{min}, in order to satisfy the first law of thermodynamics. Thus,

$$Q_{\text{max}} = C_{\text{min}} \Delta T_{\text{max}} = C_{\text{min}} (T_{h1} - T_{c1}) \tag{2.12}$$

Substituting Equation 2.12 in Equation 2.11, we obtain

$$Q_{\text{actual}} = \varepsilon \, C_{\text{min}} (T_{h1} - T_{c1}) \tag{2.13}$$

The heat exchanger effectiveness depends on the flow geometry and pass arrangement (e.g., counterflow or cross flow). For a given flow geometry, the effectiveness is a function of two dimensionless quantities, $UA/C_{\text{min}} = \text{NTU}$ and $C_{\text{min}}/C_{\text{max}} = R$. The effectiveness equation is obtained by algebraic manipulation of the equations developed in calculation of the LMTD. The development of the effectiveness equations for parallel and counterflow are given in heat transfer texts (e.g., Holman [2001] or Incropera and DeWitt [2001]). The effectiveness for a counterflow heat exchanger is shown on Figure 2.3 and is given by the algebraic relation

$$\varepsilon = \frac{1 - e^{-\text{NTU}(1-R)}}{1 - R e^{-\text{NTU}(1-R)}} \tag{2.14}$$

The dimensionless ratio UA/C_{min} is called the number of heat transfer units (NTU). It is indicative of the heat exchanger physical size. The capacity rate ratio, R, is equal to zero for an evaporator or condenser if the fluid remains at a constant temperature during the phase change. If $R = 0$, the $\in$–NTU relation is independent of the heat exchanger flow geometry. Thus, all flow geometries have the same $\in$–NTU relation when R = 0. Figure 2.3 shows that the required NTU increases for increasing values of effectiveness. The NTU asymptotically approaches infinity as the effectiveness approaches its thermodynamic limit ($\in = 1$ for counterflow).

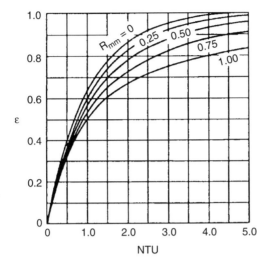

Figure 2.3 Heat exchanger effectiveness for counterflow exchanger.

For a given NTU, counterflow provides the highest effectiveness, and parallel flow the lowest effectiveness, with crossflow providing intermediate values. Crossflow is frequently used for gas–liquid heat exchangers, because it is difficult to design them for counterflow. Shell-and-tube heat exchangers frequently use some form of multipass cross-counterflow. Analytical forms of $\epsilon = fcn$ (NTU) may be found in heat transfer textbooks, such as Incropera and DeWitt [2001] for many different flow configurations.

The F-correction factors and ϵ–NTU equations for "mixed" and "unmixed" flow are simply limiting thermodynamic definitions. Heat transfer text books typically consider banks of tubes as "mixed" flow. It is unlikely that the flow in a tube bank has no temperature gradients normal to the flow. Further, interrupted fin geometries may permit some mixing between lateral flow channels, but it is unlikely that it is fully mixed. Use of the "unmixed" relations will result in overprediction, if some lateral fluid mixing occurs. Webb and DiGiovanni [1989] discuss this concern, and provide a method to predict the performance for partially mixed conditions. However, the designer must estimate the mixing fraction.

2.2.3 Comparison of LMTD and NTU Design Methods

The two design methods are equivalent and differ only in the algebraic form of the resulting equations. Which method is used may depend on the designer's preference. Both design methods offer the same relative ease for the sizing problem. However, the NTU is much easier to use for the rating problem. Because at least one fluid exit temperature is unknown in a rating problem, the LMTD is not directly calculable. In this case a trial-and-error solution is required to obtain the LMTD and the correction factor F.

The NTU method allows physical interpretation of the thermodynamic performance of the heat exchanger not provided by the LMTD method. Also, the ε–NTU relations are readily available in algebraic form, needed for digital computer calculations.

2.3 FIN EFFICIENCY

Extended surfaces or fins are frequently employed in heat exchangers. Fins will be beneficial if they are applied to the fluid stream having the dominant thermal resistance. The fins provide reduced thermal resistance for this flow stream by providing increased surface area. The heat transfer coefficient on the extended surfaces may be either higher or lower than that which would occur on the unfinned surface. For example, the use of low radial fins on horizontal tubes with condensation provides both an area increase and an increased heat transfer coefficient. However, fins used in single-phase forced convection may offer a modest reduction of the coefficient depending on the fin spacing. Because of the temperature gradient in the fin material over its length, a finned surface will not transfer as much heat as a fin of infinite thermal conductivity material. Thus, the heat conductance of a finned surface (hA) must be multiplied by a fin efficiency factor to account for the temperature gradient in the fin. The fin efficiency, η_f, is defined as the ratio of the actual heat transfer from the fin to that which would occur if the entire fin were at its base temperature.

$$\eta_f = \frac{Q}{hA_f(T_w - T_\infty)} \tag{2.15}$$

The efficiency of a fin is a function of its cross-sectional shape, its length, and the geometry of the base surface. Figure 2.4 gives the efficiency of plain and circular fins of uniform cross section. This chart is valid for an adiabatic fin tip boundary condition. The fin efficiency for a plane fin (or for $r_o/r_i = 1.0$ having an insulated fin tip) is given by

$$\eta_f = \frac{\tanh(mL)}{mL} \tag{2.16}$$

Schmidt [1949] has shown that the fin efficiency of circular fins may be calculated by

$$\eta_f = \frac{\tanh(mr_i\phi)}{mr_i\phi} \tag{2.17}$$

where the term ϕ is given by the empirical correlation

$$\phi = \left(\frac{r_o}{r_i} - 1\right)\left(1 + 0.35\frac{r_o}{r_i}\right) \tag{2.18}$$

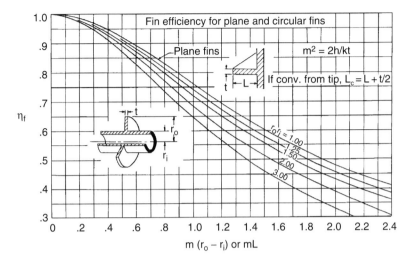

Figure 2.4 Fin efficiency of straight and circular fins, with insulated fin tip.

Harper and Brown [1922] have shown that fins which have convection from their tips may be determined from Figure 2.4 providing the fin length is modified to the value ($L_c = L + t/2$). Fin efficiency equations for a number of specialized fin geometries are discussed in detail by Kern and Kraus [1972]. Equation 2.16 is applicable to spine fins on round tubes (Figure 6.2a) and louver fins on flat tubes (Figure 6.2e).

A finned surface heat exchanger consists of the secondary finned surface and the primary surface to which the fins are attached. A second term called the "total surface efficiency" is defined to account for the efficiency of the composite structure consisting of the fins and the base surface. The total surface efficiency is defined as

$$\eta \equiv 1 - (1 - \eta_f)\frac{A_f}{A} \tag{2.19}$$

This definition assumes the heat transfer coefficient on the finned surface is the same as on the base surface.

2.4 HEAT TRANSFER COEFFICIENTS AND FRICTION FACTORS

The performance of enhanced heat transfer surfaces is frequently compared to that of a plain surface. This section provides recommended design equations for the heat transfer and the friction factor of smooth surfaces in single-phase forced convection, and the heat transfer coefficient for boiling and condensation inside and outside tubes. The equations presented are those commonly used in heat exchanger design applications and may be found in Holman [2001] or Incropera and DeWitt [2001].

2.4.1 Laminar Flow over Flat Plate

The Blasius solution for the average Stanton number with laminar flow over a flat plate of length L, as reported in Chapter 7 of Incropera and DeWitt [2001], is

$$\overline{St} = 0.664\,\text{Re}_L^{-0.5}\,\text{Pr}^{-2/3} \tag{2.20}$$

The average friction factor for the plate is given by

$$\overline{f} = 1.328\,\text{Re}_L^{-0.} \tag{2.21}$$

Writing Equation 2.20 in terms of the j-factor (StPr$^{2/3}$) shows that $j = f/2$, which satisfies the Reynolds analogy. The Reynolds analogy is discussed in Section 2.6.

2.4.2 Laminar Flow in Ducts

Table 2.2 provides friction factors and Nusselt numbers for circular and noncircular ducts. The Nusselt number is shown for two thermal boundary conditions: constant heat flux (subscript H) and constant wall temperature (subscript T). The Nusselt number values (based on hydraulic diameter for the noncircular ducts) are listed in order of decreasing Nusselt number. By comparing the Nusselt number with the duct cross-sectional shape, one may quickly discern how the duct shape affects Nusselt number. The table also shows that the ratio of j/f decreases with decreasing Nusselt number. High-aspect-ratio rectangular channels yield higher heat transfer coefficients and higher heat transfer per unit pressure drop (j/f) than circular tubes and triangular channels.

Figure 7.6 in Chapter 7 shows entrance region solutions (simultaneous velocity and thermal boundary layer development) for laminar flow in circular tubes for q_w = constant and T_w = constant. Figure 5.37 in Chapter 5 shows $\text{Nu}_m/\text{Nu}_{fd}$ vs. the entrance region parameter, $x^*_d [= x/(d_i\text{Re}_d\text{Pr})]$, for circular and noncircular ducts with the T_w = constant boundary condition. Figure 5.38 shows the parameter $K/K(\infty)$, which is the pressure drop increment associated with the entrance region. This term is used with Equation 5.34. Bhatti and Shah [1987] give numerous solutions for developing and developed laminar flow for a variety of geometries, and for several thermal boundary conditions.

2.4.3 Turbulent Flow in Ducts

The Petukhov [1970] equation is recommended as the most precise available equation. It is

$$St = \frac{f/2}{1.07 + 12.7(f/2)^{1/2}(\text{Pr}^{2/3} - 1)} \tag{2.22}$$

Webb [1971] compares the ability of Equation 2.22 to predict heat transfer data in smooth tubes. He shows that Equation 2.22 predicts the smooth-tube,

Table 2.2 Fully Developed Laminar Flow Solutions

Geometry	Nu_H	Nu_T	$f\mathrm{Re}$	K_∞	j_H/f	j_T/f	L_{hy}^+
$\dfrac{2b}{2a} = 0$	8.235	7.541	24.00	0.686	0.386	0.354	0.0056
$\dfrac{2b}{2a} = \dfrac{1}{8}$	6.490	5.597	20.585	0.879	0.355	0.306	0.0094
$\dfrac{2b}{2a} = \dfrac{1}{6}$	6.049	5.137	19.702	0.945	0.346	0.294	0.0010
$\dfrac{2b}{2a} = \dfrac{1}{4}$	5.331	4.439	18.233	1.076	0.329	0.274	0.0147
○	4.364	3.657	16.00	1.24	0.307	0.258	0.038
$\dfrac{2b}{2a} = \dfrac{1}{2}$	4.123	3.391	15.548	1.383	0.299	0.245	0.0255
$\dfrac{2b}{2a} = 1$	3.608	3.091	14.227	1.552	0.286	0.236	0.0324
$\dfrac{2a}{2b} = \dfrac{\sqrt{3}}{2}$	3.111	2.47	13.333	1.818	0.263	0.209	0.0398
$\dfrac{2b}{2a} = \dfrac{\sqrt{3}}{2}$	3.014	2.39	12.630	1.739	0.269	0.214	0.0408
$\dfrac{2b}{2a} = 2$	2.880	2.22	13.026	1.991	0.249	0.192	0.0443
$\dfrac{2b}{2a} = .25$	2.600	1.99	12.622	2.236	0.232	0.178	0.0515

$$j/f \equiv \frac{\mathrm{Nu}\,\mathrm{Pr}^{-1/3}}{f\,\mathrm{Re}} \qquad j_H,\ j_T \text{ for } \mathrm{Pr} = 0.7$$

$H \sim$ constant heat flux boundary condition
$T \sim$ constant temperature boundary condition

constant-property data of four investigators ($0.7 \leq \mathrm{Pr} \leq 75$) within $\pm 8\%$. One may account for the effect of fluid property variations across the boundary layer using the correction factors given in Table 2.3 and Table 2.4 (presented below), and explained in Section 2.5. Petukhov [1970] recommends that the smooth-tube friction factor be predicted by

$$f = (1.58 \ln \mathrm{Re}_d - 3.28)^{-2} \tag{2.23}$$

The Dittus–Boelter equation is frequently used for quick, approximate calculations. However, it is not as accurate as the Petukhov equation. It is

$$\mathrm{Nu}_d = 0.023\,\mathrm{Re}_d^{0.8}\,\mathrm{Pr}^n \tag{2.24}$$

where the Prandtl number exponent (n) is 0.4 for heating and 0.3 for cooling. The friction factor is frequently predicted by the approximate Blasius equations,

$$f = 0.079\,\mathrm{Re}_d^{-0.25} \quad \mathrm{Re}_d < 50,000 \tag{2.25}$$

$$f = 0.046\,\mathrm{Re}_d^{-0.2} \quad \mathrm{Re}_d > 50,000 \tag{2.26}$$

The turbulent duct flow equations are also applicable to noncircular ducts and axial flow in rod bundles by using the hydraulic diameter in place of the tube diameter.

2.4.4 Tube Banks (Single-Phase Flow)

The correlations of Zukauskas [1972] for heat transfer and friction for flow normal to tube banks are recommended. These correlations are given by Incropera and DeWitt [2001], and will not be repeated here.

2.4.5 Film Condensation

The Nusselt analysis as described by Incropera and DeWitt [2001] provides equations for film condensation on vertical plates, inclined plates, and horizontal tubes. The heat transfer coefficient is the average value over the plate length, or tube diameter. The average condensation coefficient on a vertical plate with a gravity drained laminar film is given by

$$\overline{h} = 0.943 \left(\frac{k^3 g(\rho_l - \rho_v)\lambda}{v_l \Delta T_{vs} L} \right)^{1/4} \tag{2.27}$$

For an inclined plate, one simply replaces the gravity force by ($g \sin \theta$), where θ is the plate inclination angle from the vertical direction. Equation 2.27 may be written in terms of the condensate Reynolds number draining from the bottom of the plate, Re_L

$$\overline{h} = 1.47 \left(\frac{k^3 \rho_l (\rho_l - \rho_v) g}{\mu_l^2} \right)^{1/3} \mathrm{Re}_L^{-1/3} \tag{2.28}$$

The condensate Reynolds number is defined as $\mathrm{Re}_L = 4\Gamma/\mu_l$, where Γ is the condensate flow rate per unit plate width.

The condensation coefficient for laminar film condensation on a single, horizontal tube is given by

$$\overline{h} = 0.728 \left(\frac{k_g^3 (\rho_l - \rho_v)\lambda}{v_l \Delta T_{vs} d} \right)^{1/4} \tag{2.29}$$

Equation 2.29 may be written in terms of the condensate Reynolds number draining from the tube.

$$\bar{h} = 1.51 \left(\frac{k^3 \rho_l (\rho_l - \rho_v) g}{\mu_l^2} \right)^{1/3} \mathrm{Re}_L^{-1} \tag{2.30}$$

The condensate Reynolds number is again defined as $\mathrm{Re}_L = 4\Gamma/\mu_l$, where Γ is the condensate flow rate per unit tube length.

The Chun and Seban [1971] correlation is recommended for the local turbulent film condensation coefficient on a vertical plate. It is

$$h = 0.0038 \left(\frac{k^3 \rho_l (\rho_l - \rho_v) g}{\mu_l^2} \right)^{1/3} \mathrm{Re}_L^{0.4} \mathrm{Pr}^{2/3} \tag{2.31}$$

The nomenclature for Equations 2.27 through 2.31 is given in Chapter 12. Equations 2.27 through 2.31 are also applicable to evaporation of laminar or turbulent films. The equations presented here are for the case of zero interfacial shear stress on the condensate film. Interfacial shear stress on the liquid–vapor interface will increase the condensation coefficient, if the vapor and liquid film flow directions are the same. However, vapor shear will decrease the condensation coefficient if the vapor flow direction is opposite to the condensate flow.

A text on two-phase flow and heat transfer should be consulted for equations appropriate to convective condensation inside tubes. The appropriate equations are sensitive to the geometric orientation of the tube, as well as to the entering and leaving vapor quality. A recommended source is Carey [1992].

2.4.6 Nucleate Boiling

The Cooper [1984] correlation, is a well-accepted correlation for prediction of the nucleate boiling coefficient on horizontal plain tubes. This correlation is

$$h = 90 q^{2/3} M^{1/2} p_r^m (-\log_{10} p_r)^{-0.55} \tag{2.32}$$

where $m = 0.12 - 0.2 \log_{10} R_p$.

The term, R_p is the surface roughness expressed in μm. Webb and Pais [1992] found that the Cooper correlation (with $R_p = 0.3$) predicted their data for 5 refrigerants boiling at 4.4 and 26.7°C within ±10%.

Recommended equations for convective vaporization are not given here. As for convective condensation, the equations are sensitive to geometric orientation of the tube, and to the entering and leaving vapor quality. A recommended source is Carey [1992]. Webb and Gupte [1992] give a survey of correlations for convective vaporization in tubes (horizontal and vertical) and for tube banks.

2.5 CORRECTION FOR VARIATION OF FLUID PROPERTIES

Simplified heat exchanger analysis assumes (1) the overall heat transfer coefficient is constant along the exchanger length, and (2) the individual heat transfer coefficients are functions only of the mean fluid temperature. In reality, the heat transfer coefficients depend on the physical properties of the fluid, which change with the fluid temperature — both in the flow direction and across the boundary layer thickness. If the heat exchanger is so short that an entrance region exists, the heat transfer coefficient may also vary with length. In cases involving polymers or viscous oils, substantial property variations may exist, thus introducing errors in the calculation. We consider the effect of changing fluid temperature and local property variation effects on the overall heat transfer coefficient and outline a design method to account for these effects.

2.5.1 Effect of Changing Fluid Temperature

The derivation of the LMTD equation assumes that U is constant. Colburn [1933] has shown that if U varies linearly with temperature difference, $U = a + b(T_h - T_c)$, then

$$\frac{Q}{A} = \frac{U_I \Delta T_{II} - U_{II} \Delta T_I}{\ln \dfrac{U_I \Delta T_{II}}{U_{II} \Delta T_I}} \tag{2.33}$$

where subscripts I and II indicate the inlet and exit ends of the exchanger. Rather than assuming a linear variation of U, one may perform an incremental design, which is well-adapted to computer methods. The heat exchanger is divided into N increments of equal length. An average value of U is calculated for each increment, and the heat duty (q) is determined for each increment. The summation of q values for all increments will provide an accurate design. This procedure is outlined by Holman [2001] and by others.

2.5.2 Effect Local Property Variation

Certain fluid properties vary significantly with temperature, notably fluid viscosity. Hence, the fluid properties at the wall temperature and the local mixed temperature will differ. A method to account for the fluid property variation across the boundary layer is required. Either of two methods is frequently used to correct heat transfer and friction correlations for fluid property variation across the boundary layer. The first is to evaluate the properties at the "film temperature," defined as the average of the local, mixed fluid temperature and the wall temperature. The second is to correct by a fluid property ratio evaluated at the mixed fluid and the wall temperatures. In this method, the fluid properties in the correlation are evaluated at the mixed fluid temperature. We use the second method.

Define $h = h_{cp}F_c$, where h_{cp} is the heat transfer coefficient assuming no temperature difference between the wall and bulk fluid temperature. The F_c is a correction factor to account for a finite temperature difference between the wall and the bulk temperature. For liquids, $F_c = (\mu/\mu_w)^n$, where n depends on the flow conditions (flow regime and geometry) and whether the fluid is being heated or cooled. For gases, F_c is assumed to be dependent on $(T/T_w)^n$, where T is the absolute temperature. Summaries of the property correction term have been provided by Petukhov [1970] and Kays and London [1984]. Our recommendations come from these two sources.

Table 2.3 shows F_c for turbulent flow of gases and liquids for both heating and cooling. The values shown in Table 2.3 are for flow in tubes. If only one entry is provided for F_c, it is from Petukhov [1970]. If a second entry is given (below the first value), it is from Kays and London [1984]. These F_c values are typically used for other flow geometries (e.g., tube banks). Table 2.4 shows F_c for laminar flow of gases and liquids for both heating and cooling. These F_c values are taken from Kays and London [1984] and are for flow in tubes. However, these values are also frequently used for other flow geometries.

Table 2.3 Correction for Fluid Properties (Turbulent Flow)

	Heating	Cooling
Liquids		
$\dfrac{St}{St_{cp}} = \left(\dfrac{\mu}{\mu_w}\right)^n$	$n = 0.11$	$n = 0.25$
$\dfrac{f}{f_{cp}} = \left(\dfrac{\mu}{\mu_w}\right)^m$	$m = \dfrac{1}{6}\left(7 - \dfrac{\mu}{\mu_w}\right)$ $m \simeq -0.25$	$m = -0.24$
Gases		
$\dfrac{St}{St_{cp}} = \left(\dfrac{T}{T_w}\right)^n$	$n = 0.3 \log_{10}\dfrac{T_w}{T} + 0.36$ $n \simeq -0.5$	$n = 0.36$
$\dfrac{f}{f_{cp}} = \left(\dfrac{T}{T_w}\right)^m$	$m = 0.6 - 5.6\left(Re\dfrac{\rho_w}{\rho}\right)^{-0.38}$ $m \simeq -0.1$	$m = 0.6 - 7.9\left(Re\dfrac{\rho_w}{\rho}\right)^{-0.11}$ $m \simeq -0.1$

Sources: Petukhov [1972] and Kays and London [1984].

Table 2.4 Correction for Fluid Properties (Laminar Flow)

	Heating	Cooling
Liquids		
$\dfrac{St}{St_{cp}} = \left(\dfrac{\mu}{\mu_w}\right)^n$	$n = 0.14$	$n = 0.14$
$\dfrac{f}{f_{cp}} = \left(\dfrac{\mu}{\mu_w}\right)^m$	$m = -0.58$	$m = -0.50$
Gases		
$\dfrac{St}{St_{cp}} = \left(\dfrac{T}{T_w}\right)^n$	$n = 0.0$	$n = 0.0$
$\dfrac{f}{f_{cp}} = \left(\dfrac{T}{T_w}\right)^m$	$m = -1.0$	$m = -1.0$

Source: Kays and London [1984].

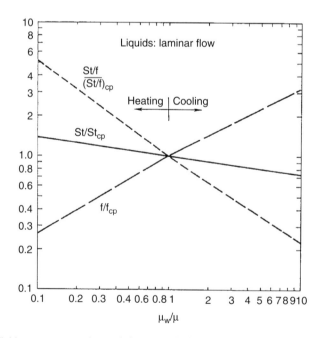

Figure 2.5 Fluid property correction variation vs. μ_w/μ for laminar flow of liquids. (From Kays and London [1964].)

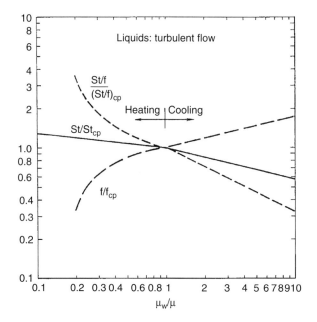

Figure 2.6 Fluid property correction variation vs. μ_w/μ for turbulent flow of gases.

Figure 2.5 and Figure 2.6 show the fluid property corrections for laminar and turbulent flow of liquids, respectively. Figures 2.5 and 2.6 were prepared using the F_c equations of Petukhov [1970], presented in Tables 2.3 and 2.4, respectively. Note that the abscissa is written as μ_w/μ rather than μ/μ_w

To determine the viscosity ratio correction, one must know the wall temperature. This is not explicitly known. It is necessary to iteratively solve for the wall temperature by iteration using

$$Q = h_i A_i (T_i - T_w) = h_o A_o (T_w - T_o) \qquad (2.34)$$

The required calculation procedure is (1) calculate h_i and h_o for the constant properties condition (based on the mean fluid temperature); (2) calculate T_w using Equation 2.34; (3) evaluate F_c from Table 2.3 or 2.4; (4) calculate $h = h_{cp}F_c$. It may be necessary to repeat steps 1 through 4 several times until the assumed and calculated values of T_w agree.

A sophisticated heat exchanger design would account for variations of U due to both changing mean fluid temperature and local property variation effects.

2.6 REYNOLDS ANALOGY

Reynolds analogy provides a useful concept for the evaluating the performance of heat transfer enhancement concepts. By using $Nu \propto Pr^{1/3}$, Colburn's [1933] statement of Reynolds analogy is

$$St\,Pr^{2/3} \equiv j = f/2 \qquad (2.35)$$

Equation 2.35 applies to:

1. Laminar and turbulent flow over flat plates.
2. Turbulent flow in smooth ducts or rod bundles. It may also be applied to noncircular channels using the hydraulic diameter concept.

The equation $2j/f = 1$ establishes a relation between friction due to surface shear and heat transfer. The analogy does not hold when flow separations occur. With flow across a tube bank or over a rough surface, the total drag force will consist of two components — surface shear and form drag. Because form drag is basically a parasitic loss, it contributes little to heat transfer. In this situation $2j/f < 1$, where f includes momentum losses due to flow separations.

Enhanced heat transfer surfaces yield increased heat transfer coefficients, thus $j/j_s > 1$, where the s subscript refers to the smooth surface. For comparison at equal Reynolds numbers, Equation 2.32 implies that the enhanced surface must have a friction increase at least as large as the j/j_s increase. Enhanced surfaces, in general, show $(j/j_s)/(f/f_s) < 1$, with exceptions in special cases. We may define the relative j/f ratio of enhanced-to-smooth surfaces as an "efficiency index"; thus,

$$\eta_e \equiv \frac{j/j_s}{f/f_s} \qquad (2.36)$$

This efficiency index is useful to evaluate the quality of the enhancement concept. The goal is to obtain high values for j/j_s with η_e as close as possible to 1.0. Assuming $j/j_s = 2$, a very good enhanced surface may have an η_e value of 0.8 to 0.9. Smaller values of η_e mean an increased friction penalty to establish a given enhancement level, j/j_s. In some special cases (turbulent flow of high Prandtl number liquids over roughness) it is possible to obtain $\eta_e > 1$. This is because the Prandtl number dependency of the rough surface flow is greater than that of smooth surfaces. For Pr < 1, enhanced surfaces have not shown η_e values that exceed one.

The $2j/f$ analogy does not hold for laminar flow in ducts. However, the concept may be applied in a qualitative sense. In laminar duct flow, both Nu and f depend on the duct shape as shown in Table 2.2. This table lists values of j_H/f, and j_T/f for Pr = 0.7 (gases). The table shows that the parallel plate channel provides the highest j/f, or $2j/f = 0.708$ for constant wall temperature. Compared to $2j/f = 1$ for turbulent flow, the "efficiency index" of this laminar flow geometry is 70.8%. As one moves down the table of laminar flow solutions, the other duct shapes show smaller values of j/f, and thus they have lower values of the efficiency index. For laminar flow in a circular tube, j/f is only 51.6% of that for turbulent flow in the same duct shape. This shows that turbulent flow provides greater heat transfer per unit friction than is possible with laminar duct flow. As the Prandtl number increases, the j/f characteristic of laminar flow decreases, since $j/f \propto Pr^{1/3}$ for laminar duct flows, and j/f is independent of Pr for turbulent flow.

These concepts provide important tools for evaluating the quality of heat transfer enhancement or to select laminar flow duct shapes that yield high heat transfer per unit friction.

2.7 FOULING OF HEAT TRANSFER SURFACES

After a period of operation, the heat exchanger surfaces may become coated with a deposit of solid material. The thermal resistance of such deposits cause a reduced overall heat transfer coefficient. Heat exchangers are normally overdesigned to compensate for the anticipated fouling. The fouling factor may be measured by testing the initially clean and fouled conditions. Then, the overall fouling factor, R_f, is calculated as

$$R_f = \frac{1}{U_{\text{fouled}}} - \frac{1}{U_{\text{clean}}} \tag{2.37}$$

A given fouling resistance will usually cause a greater performance reduction with liquids than with gases. There are two reasons for this. First, heat transfer coefficients for liquids are higher than for gases. Second, gases frequently use extended surfaces, so the fouling deposit is apportioned over a larger surface area. To illustrate these effects, consider a finned tube liquid cooler having $A_o/A_i = 20$, where subscripts o and i refer to the air and liquid sides. Assume the heat transfer coefficient on the liquid and air sides are $h_i = 6000$ W/m²-K and $h_o = 85$ W/m²-K, and that $R_f = 0.0002$ m²-K/W. The liquid side fouling deposit will reduce the effective liquid side heat transfer coefficient $1/(1/h_i+R_c)_f$ by 55%, relative to the unfouled condition. However, the effective air-side heat transfer coefficient $1/(1/h_o+R_{fo})$ is reduced only 2.4%. By using Equation 2.5, U for the fouled condition is only 46% as large as for the unfouled surfaces. Further, the air-side fouling accounts for only 0.23% of the reduced U value. This is due to the much smaller air-side heat transfer coefficient and because the air-side fouling deposit is distributed over 20 times the surface area as compared to the liquid-side fouling. Fouling may cause a different problem for gases. The fouling deposit on the heat exchanger surface increases the gas pressure drop. The increased pressure drop may result in a lower gas flow rate, due to the balance point on the fan curve. Hence, the heat transfer rate is reduced, because of the lower gas flow rate through the heat exchanger.

Fouling raises serious implications for enhanced heat transfer surfaces. Assume that internal roughness is employed to obtain a 100% higher tube-side heat transfer coefficient than that of the smooth tube considered in the previous example for $R_f = 0.0002$ m²-K/W; also assume that the effective liquid side coefficient, $1/(1/h_i+R_f)$, is only 29% as large as that of the clean tube. Thus, a given fouling factor causes a more severe penalty as the heat transfer coefficient is increased.

For the same fouling factor, an internally finned tube may suffer a smaller penalty due to fouling because the fouling effect is distributed over a larger area. Assume that a clean, internally finned tube provides the same heat conductance, $h_i A_i/L$ as the internally roughened tube. If the internal fins provide a 100% surface area increase with $R_f = 0.0002$ m^2-K/W, the effective $h_i A_i / L$ including fouling, will be 45.6% as large as that of the unfouled surface.

Fouling deposits may be removed by mechanical, chemical, or thermal cleaning. Various types of mechanical cleaning systems are used. The outside surfaces of tubes may be cleaned with high-pressure air, water, or steam. Inner tube surfaces are cleaned by forcing special plugs or brushes through the tubes. In most cases the exchanger must be taken out of service for cleaning. However, an innovative tube-side cleaning system is described by Leitner [1980] in which a four-way flow valve is used to force cleaning brushes through the tubes. Because the direction of flow may be reversed, the cleaning system is automatic and performed while the exchanger is in service. Chemical cleaning includes flushing the system with chemical cleaning agents selected to dissolve the particular fouling deposit. Thermal cleaning is accomplished by heating the exchanger to a sufficiently high temperature to vaporize the deposits. This method is limited in application and is mainly used for small gas–gas heat exchangers. Closed water cooling systems (e.g., cooling towers and boilers) use chemical treatment to minimize the fouling characteristics of the water. Fouling of enhanced surfaces is addressed in Chapter 10.

2.8 CONCLUSIONS

This chapter provides basic fundamentals relating to heat transfer from plain surfaces and heat exchanger design fundamentals. The performance of enhanced surfaces is typically compared to that of plain surfaces. The heat exchanger design information is needed to evaluate the performance of enhanced surfaces used in two-fluid heat exchangers. Chapters 3 and 4 apply this knowledge to develop "performance evaluation criteria" for evaluation of the performance of enhanced surfaces used in a heat exchanger, and compared to a heat exchanger using plain surfaces.

The Reynolds analogy is an important tool that shows how flow friction may be related to the heat transfer coefficient. Although enhanced surfaces provide an increased heat transfer coefficient, they may also be expected to demonstrate increased flow friction, especially for single-phase flow. Applied to enhanced surfaces, the Reynolds analogy indicates the minimal friction increase that may be expected from an enhanced surface. The efficiency index provides a measure of how well one has used increased friction to obtain a given heat transfer enhancement ratio.

Fouling can degrade the performance of a heat exchanger, particularly for water flow in tubes. Hence, it is important to understand the fundamentals of fouling, and how it may affect heat exchanger performance. Fouling of enhanced surfaces is treated in Chapter 10.

NOMENCLATURE

A	Heat transfer surface area, m^2 or ft^2
A_f	Area of extended surface, m^2 or ft^2

C Capacity rate (Wc_p), W/K or Btu/hr-°F

c_p Specific heat, kj/kg-K or Btu/lbm-°F

d Tube diameter, m or ft

D_h Hydraulic diameter, m or ft

f Fanning friction factor, $\Delta p = (4fL/d)(\rho u^2/2)$, dimensionless

F Correction factor for log-mean temperature difference, dimensionless

F_c Correction for fluid property variation across boundary layer, dimensionless

h Heat transfer coefficient, W/m²-K or Btu/hr-ft²-°F

k Thermal conductivity, W/m-K or Btu/hr-°F

K Pressure drop increment to account for flow development, $K(\infty)$ (full entrance region), $K(x)$ (over length x), shown on Figure 5.38, dimensionless

L Fin length, m or ft

LMTD   Log-mean temperature difference, K or °F

M Molecular weight, kg/kmol

Nu   Nusselt number (hd/k), dimensionless

p_{cr} Critical pressure, kPa or lbf/ft²

p_r p/p_{cr}, dimensionless

Pr Prandtl number, dimensionless

q Heat flux, W/m² or Btu/hr-ft²

Q Heat transfer rate, W or Btu/hr

r_o Radius over fin tips, m or ft

r_i Radius at base of fins m or ft

Re   Reynolds number, Re_d (based on tube diameter), Re_{Dh} (based on hydraulic diameter), Re_L (based on strip length L) dimensionless

Re_L Condensate film Reynolds number, $4\Gamma\mu_l$, dimensionless

R Capacity rate ratio (C_{min}/C_{max}), dimensionless

R_f Fouling resistance, m²-K/W or hr-ft²-°F/Btu

R_p Surface roughness (Equation 2.32), μm

St   Stanton number, dimensionless

t Tube wall thickness, m or ft

T_c Cold fluid temperature, T_{C1} (entering) and T_{C2} (leaving), K or °F

T_h Hot fluid temperature, T_{h1} (entering) and T_{h2} (leaving), K or °F

T_w Wall temperature, K or °F

T_∞ Fluid temperature, K or °F

U Overall heat transfer coefficient, W/m²-K or Btu/hr-ft²-°F

x^d $x/(d_i Re_d Pr)$, dimensionless

W Mass flow rate of fluid, kg/s or lbm/s

Greek Symbols

ε Heat exchanger effectiveness, dimensionless

Γ Condensate flow rate per unit plate width (or tube length), dimensionless

η_f Fin efficiency, surface efficiency (Equation 2.15), dimensionless

η Surface efficiency (Equation 2.19), dimensionless

μ Dynamic viscosity, kg/s-m or lbm/s-ft

ΔT_m Mean temperature difference, K or °F

Subscripts

c Refers to cold fluid in heat exchanger

fd Fully developed flow

h Refers to hot fluid in heat exchanger

i Refers to conditions inside tube

m Average value over flow length

o Refers to conditions external to tube

H Constant heat flux thermal boundary condition

T Constant wall temperature thermal boundary condition

w At wall

REFERENCES

Bhatti, M.S. and Shah, R.K., 1987. Laminar convective heat transfer in ducts, Chapter 3 in *Handbook of Single-Phase Heat Transfer*, S. Kakaç, R.K. Shah, and W. Aung, Eds., John Wiley & Sons, New York.

Carey, V.P., 1992. *Liquid-Vapor Phase-Change Phenomena*, Hemisphere, Washington, D.C.

Chun, K.R. and Seban, R.A, 1971. Heat transfer to evaporating liquid films. *J. Heat Transfer*, 93, 391–396.

Colburn, A.P., 1933. A method of correlating forced convection heat transfer data and a comparison with fluid friction, *Trans. AIChE*, 29, 174–210.

Cooper, M.G., 1984. Saturation nucleate, pool boiling — a simple correlation. *Int. Chem. Eng. Symp. Series*, 86, 785–792.

Harper, W.B. and Brown, D.R., 1922. Mathematical Equations for Heat Conduction in the Fins of Air-cooled Engines, NACA Report 158.

Holman, J.P., 2001. *Heat Transfer*, 9th ed., McGraw-Hill, New York.

Incropera, F.P. and DeWitt, D.P., 2001. *Fundamentals of Heat and Mass Transfer*, 5th ed., John Wiley & Sons, New York.

Kays, W.M. and London, A.L., 1984. *Compact Heat Exchangers*, 3rd ed., McGraw-Hill, New York.

Kern, D.Q. and Kraus, A.D., 1972. *Extended Surface Heat Transfer*, McGraw-Hill, New York, 168.

Leitner, G.F., 1980. Controlling chiller tube fouling, *ASHRAE J.*, 22(2), 40–43.

Petukhov, B.S., 1970. Heat transfer in turbulent pipe flow with variable physical properties," in *Advances in Heat Transfer*, Vol. 6, T.F. Irvine and J.P. Hartnett, Eds., Academic Press, New York, 504–564.

Schmidt, T.E., 1949. Heat transfer calculations for extended surfaces, *J. ASRE, Refrig. Eng.*, 4, 351–357.

Shah, R.K., Afimiwala, K.A., and Mayne, R.W., 1978. Heat exchanger optimization, *Proc. 6th Int. Heat Trans. Conf.*, 4, 193–199.

Webb, R.L., 1971. A critical evaluation of analytical and Reynolds analogy equations for turbulent heat and mass transfer in smooth tubes, *Wärme Stoffübertrag.*, 4, 197–204.

Webb, R.L. and DiGiovanni, M.A., 1989. Uncertainty in effectiveness-NTU calculations for crossflow heat exchangers, *Heat Transfer Eng.*, 10(3), 61–70.

Webb, R.L. and Gupte, N.S., 1992. A critical review of correlations for convective vaporization in tubes and tube banks, *Heat Transfer Eng.*, 13(3), 58–81.

Webb, R.L. and Pais, C., 1992. Nucleate boiling data for five refrigerants on plain, integral-fin and enhanced tube geometries, *Int. J. Heat Mass Transfer*, 35(8), 1893–1904.

Zukauskas, A., 1972. Heat transfer from tubes in crossflow, *Advances in Heat Transfer*, J.P. Hartnett and T.F. Irvine, Eds., Academic Press, New York, 8, 93–160.

PERFORMANCE EVALUATION CRITERIA FOR SINGLE-PHASE FLOW

3.1 PERFORMANCE EVALUATION CRITERIA (PEC)

A quantitative method is required to evaluate the performance improvement provided by a given enhancement. Normally, one compares the performance of an enhanced surface with that of the corresponding plain (smooth) surface.

There are three considerations in surface performance:

1. The performance objective: The four basic objectives of interest were discussed in Section 1.1.
2. The operating conditions: This includes the fluid flow rate and the entering fluid temperature.
3. The constraints: This includes the allowable pressure drop (or fan or pumping power), and the frontal area (or velocity) constraints.

The "basic performance characteristic" of an enhanced surface for single-phase heat transfer is defined by the j-factor ($= StPr^{2/3}$) and f vs. Reynolds number curves. One possibility to quantify the performance improvement is to calculate the ratios j/j_s and f/f_s where subscript s (or p) is for a smooth (or plain) surface at the same Re. Generally, the friction factor of an enhanced surface in single-phase flow is higher than that of the smooth surface, when operated at the same velocity (or Reynolds number). However, this method is not recommended because it does not define the

55

actual performance improvement, subject to specific operating constraints. If one simply calculated the enhancement ratio, j/j_s at equal velocities, an unfair comparison may result. This is because the enhanced surface would be allowed to operate at a higher pressure drop. The plain surface would give a higher h value if it were allowed to operate at a higher velocity, giving the same pressure drop as the enhanced surface. Thus, the pressure drop constraint is a very important consideration for calculating the performance benefits of an enhanced surface in single-phase flow.

Several measures of "performance" are possible, and have been proposed in the literature. The first may compare the "performance" of the enhanced surface with that of a smooth surface. This may be described as "surface performance comparison," as it does not account for the effects of the second fluid stream in the heat exchanger. The second may determine the effect of the enhanced surface on the overall performance of a two-fluid heat exchanger. We describe and discuss the merits of several possible comparison methods.

PEC analysis for enhancement of a two-phase flow (boiling or condensation) should be handled differently from that for single-phase flow. This is because the pressure drop of a two-phase fluid also reduces the local saturation temperature of the fluid. Thus, the driving potential for heat transfer is also affected. The special PEC for two-phase heat transfer is treated in Chapter 4.

The PEC presented here are based on those previously described by Webb and Eckert [1972], Webb [1981a], Bergles et al. [1974a, b], Bergles [1981], and Webb and Bergles [1983]. The algebraic formulations of the PEC are applicable to single-phase laminar or turbulent flows in tubes or normal to tube banks.

3.2 PEC FOR HEAT EXCHANGERS

The primary interest is to determine how the enhanced surface (or doubly enhanced surfaces) will affect the performance of the heat exchanger. The preferred evaluation method sets a performance objective (e.g., reduced surface area) and calculates the performance improvement relative to a reference design (e.g., smooth tubes) for a given set of operating conditions and design constraints (e.g., constant pumping power). Hence, this method defines the improvement of the objective function (e.g., percent surface area decrease) relative to a smooth surface heat exchanger. Possible performance objectives of interest are summarized in Section 1.1. These objectives assume fixed heat exchanger flow rate and entering fluid temperature. The four objectives stated in Section 1.1 are:

1. Reduced heat transfer surface material for fixed heat duty and pressure drop
2. Reduced LMTD for fixed heat duty and surface area
3. Increased heat duty for fixed surface area
4. Reduced pumping power for fixed heat duty and surface area

Objective 1 allows a smaller heat exchanger size and, one hopes, reduced capital cost. Objective 2 offers reduced operating cost. The reduced LMTD of Objective 2 will affect improved system thermodynamic efficiency, yielding lower system operating cost. Objective 3 is important if the heat exchange capacity of a given heat exchanger is to be increased. Objectives 2 and 3 will typically result in a more expensive heat exchanger, because the enhanced and smooth tube exchangers have the same total tubing length. A more costly enhanced surface will be justified if the operating cost savings are sufficiently high.

The major operational variables include the heat transfer rate, pumping power (or pressure drop), heat exchanger flow rate, and the fluid velocity (or the flow frontal area). A PEC is established by selecting one of the operational variables for the performance objective, subject to design constraints on the remaining variables.

The design constraints placed on the exchanger flow rate and velocity cause key differences among the possible PEC relations. The increased friction factor of enhanced surfaces may require reduced velocity to satisfy a fixed pumping power (or pressure drop) constraint. If the exchanger flow rate is held constant, it may be necessary to increase the flow frontal area to satisfy the pumping power constraint. However, if the mass flow rate is reduced, it is possible to maintain constant flow frontal area at reduced velocity. When the exchanger flow rate is reduced, it must operate at higher thermal effectiveness to provide the required heat duty. This may significantly reduce the performance potential of the enhanced surface if the design thermal effectiveness is sufficiently high. In many cases the heat exchanger flow rate is specified and thus flow rate reduction is not permitted.

3.3 PEC FOR SINGLE-PHASE FLOW

3.3.1 Objective Function and Constraints

Table 3.1 defines PEC for 12 cases of interest with flow of a single-phase fluid inside enhanced and smooth tubes of the same envelope diameter. The PEC are segregated by three different geometry constraints.

1. **FG criteria:** The cross-sectional flow area and tube length are held constant. The FG criteria may be thought of as a retrofit situation in which there is a one-for-one replacement of smooth surfaces with enhanced surfaces of the same basic geometry (e.g., tube envelope diameter, tube length, and number of tubes for in-tube flow). The FG-2 criteria have the same objective functions as FG-1 but require the enhanced surface design to operate at the same pumping power as the reference smooth tube design. In most cases, this will require the enhanced exchanger to operate at reduced flow rate. The FG-3 criterion seeks reduced pumping power for fixed heat duty.

2. **FN criteria:** These criteria maintain fixed cross-sectional flow area and allow the length of the heat exchanger to be a variable. These criteria seek reduced surface area (FN-1) or reduced pumping power (FN-2) for constant heat duty.

3. **VG criteria:** In many cases, a heat exchanger is sized for a required thermal duty with a specified flow rate. In these situations, the FG and FN criteria are not applicable. Because the tube-side velocity must be reduced to accommodate the higher friction characteristics of the enhanced surface, it is necessary to increase the flow area to maintain constant flow rate. This is accomplished by using a greater number of parallel flow circuits. Maintenance of a constant exchanger flow rate avoids the penalty encountered in the previous FG and FN cases of operating at higher thermal effectiveness.

Calculation of the performance improvement for any of the 12 cases in Table 3.1 requires algebraic relations that quantify the objective function and constraints. The concept of the PEC analysis is explained for the case of a prescribed wall temperature or a prescribed heat flux. It is convenient to develop the algebraic relations relative to a smooth surface operating at the same fluid temperature. This allows cancellation of the fluid properties from the equations. The equations applicable to the PEC cases of Table 3.1 are presented below.

3.3.2 Algebraic Formulation of the PEC

The different cases listed in Table 3.1 are derived for flow inside enhanced and smooth tubes of the same inside diameter. Consider a shell and tube heat exchanger of length L, having N tubes in each pass, and N_p passes. The total tube-side surface area in the heat exchanger is $A = \pi d_i L N N_p$, where N is the number of tubes in the

Table 3.1 Performance Evaluation Criteria for Single-Phase Heat Exchange System with d_i = Constant

| Case | Geometry | Fixed | | | | Objective |
		W	P	Q	ΔT_i	
FG-1a	N,L[a]	×			×	$\uparrow Q$
FG-1b	N,L	×		×		$\downarrow \Delta T_i$
FG-2a	N,L		×		×	$\uparrow Q$
FG-2b	N,L		×	×		$\downarrow \Delta T_i$
FG-3	N,L			×	×	$\downarrow P$
FN-1	N	×	×	×	×	$\downarrow L$
FN-2	N	×		×	×	$\downarrow L$
FN-3	N	×		×	×	$\downarrow P$
VG-1		×	×	×	×	$\downarrow NL$
VG-2a	N,L[b]	×	×		×	$\uparrow Q$
VG-2b	N,L[b]	×	×	×		$\downarrow \Delta T_i$
VG-3	N,L[b]	×	×	×	×	$\downarrow P$

[a] N and L are constant in all FG cases.

[b] The product NL is constant for VG-2 and VG-3.

flow passage, and N_p is the number of tube passes. The cross-sectional flow area in the tube is A_c. The basic heat transfer and friction performance characteristics of the enhanced and smooth tubes are normally presented as j ($=$ StPr$^{2/3}$) and f vs. Re $= d_i G/\mu$. Because the tube inside diameter is held constant, one may write

$$h = c_p \text{Pr}^{\ 2/3} jG \tag{3.1}$$

Of interest is the hA value of the enhanced surface, relative to that of the competing smooth surface. Writing Equation 3.1 as the ratio, relative to a smooth surface gives

$$\frac{hA}{h_s A_s} = \frac{j}{j_s} \frac{A}{A_s} \frac{G}{G_s} \tag{3.2}$$

The pumping power is calculated as

$$P = \left(\frac{fA}{A_c} \frac{G^2}{2\rho} \right) \left(\frac{GA_c}{\rho} \right) \tag{3.3}$$

Writing Equation 3.3 as the ratio, relative to the smooth surface, gives

$$\frac{P}{P_s} = \frac{f}{f_s} \frac{A}{A_s} \left(\frac{G}{G_s} \right)^3 \tag{3.4}$$

Elimination of the term G/G_s from Equations 3.2 and 3.4 gives

$$\frac{hA/h_s A_s}{(P/P_s)^{1/3} (A/A_s)^{2/3}} = \frac{j/j_s}{(f/f_s)^{1/3}} \tag{3.5}$$

The variables on the left side of Equation 3.5 are $hA/h_s A_s$, P/P_s, and A/A_s. One of these variables is set as the objective function, and the remaining two are set as operating constraints, with the value 1.0. It is necessary to determine the G/G_s ratio that satisfies Equation 3.5. We assume that the Reynolds (Re$_s$) number and the j_s and f_s of the smooth tube heat exchanger design are known. We will also assume that the equations for enhanced surface j and f as functions of Re are known.

Continuation of the solution for the general case requires specification of (1) the heat exchanger flow rate, (2) the flow frontal area, and (3) the heat transfer coefficient on the outer and inner tube surfaces for the smooth surface heat exchanger. Solution of the general problem involves performing a complete heat exchanger analysis. Before discussing the general problem, we first discuss a simple case, which avoids much of the complexity of the general case.

3.3.3 Simple Surface Performance Comparison

Assume that enhancement is applied to the tube side of a shell-and-tube heat exchanger. The simplest case exists if the total thermal resistance is on the tube side.

Thus, $UA = hA$. Two cases are illustrated. The first is for constant total flow rate, and the second will be for a tube-for-tube replacement (the same number of tubes in the enhanced and smooth tube exchangers). The tube-for-tube replacement corresponds to fixed flow frontal area. The PEC analysis is developed in detail for Case VG-1 in Table 3.1.

3.3.4 Constant Flow Rate

For Case VG-1, Q/Q_s = constant. Because the flow rate is assumed constant, and all of the thermal resistance is on the tube side, the Q/Q_s constraint is satisfied by $hA/h_s A_s = 1$. The pumping power constraint of Case VG-1 is $P/P_s = 1$. Substitution of these ratios in Equation 3.5 gives

$$\frac{A}{A_s} = \left(\frac{f}{f_s} \right)^{1/2} \left(\frac{j_s}{j} \right)^{3/2} \tag{3.6}$$

The G/G_s ratio required to meet both the heat transfer and pressure drop constraints is obtained by substituting A/A_s from Equation 3.6 in Equation 3.4 and solving for G/G_s. The result is

$$\frac{G}{G_s} = \left(\frac{j}{j_s} \frac{f_s}{f} \right)^{1/2} \tag{3.7}$$

It is necessary to iteratively solve Equations 3.6 and 3.7 for the ratios G/G_s and A/A_s. Using the assumed known expressions for $f(\text{Re})$, and $j(\text{Re})$, we may guess Re/Re_s ($= G/G_s$), solve for j/j_s and f/f_s and then calculate G/G_s and A/A_s. This iterative process is repeated until the solutions converge. With G/G_s, j/j_s and f/f_s known, A/A_s is calculated using Equation 3.2 or 3.6. For fixed internal tube diameter, the flow rate constraint $W/W_s = 1$ implies

$$\frac{W}{W_s} = 1 = \frac{G}{G_s} \frac{N}{N_s} \tag{3.8}$$

Solution of Equation 3.8 gives $N/N_s = G_s/G$, which defines the number of tubes required to satisfy the pumping power constraint. Typically, the enhanced tube exchanger must operate at a lower mass velocity, which means it will have more tubes in per pass, and thus a larger shell diameter. The tube pass length, L, is obtained from the equation

$$\frac{A}{A_s} = \frac{N}{N_s} \frac{L}{L_s} \tag{3.9}$$

3.3.5 Fixed Flow Area

If the number of tubes in the exchanger is fixed, the flow frontal area remains constant. The reduced velocity in the enhanced tube exchanger will result in reduced flow rate. The previous solution for G/G_s still applies. Because $N/N_s = 1$, Equation 3.8 gives $W/W_s = G/G_s$, which defines the flow rate in the enhanced tube exchanger. This heat exchanger must operate at higher thermal effectiveness to provide $Q/Q_s = 1$, which will require greater surface area than for the previous case. This shows that, if the heat exchanger flow rate is reduced, the benefits of enhancement will be decreased. Webb [1981a] provides details on the methodology to account for the effect of reduced flow rate on the performance improvement. The designer should avoid flow rate reduction, if possible.

3.4 THERMAL RESISTANCE ON BOTH SIDES

The simple example analyzed above assumed that the total thermal resistance is on the side to which enhancement is applied. This is an unrealistic situation. Webb [1981a] extends the PEC analysis to account for the thermal resistance of both fluid streams, the wall resistance, and fouling resistances. Equations 3.10 and 3.11 below are the generalized relations necessary for the performance improvement provided by an enhanced surface used in an actual heat exchanger. These equations include the possibility that enhancement may be provided to both fluid streams. The dimensionless variables β are β_s defined in Table 3.2.

$$\frac{P}{P_s} = \frac{f}{f_s} \frac{A}{A_s} \left(\frac{G}{G_s} \right)^3 \tag{3.10}$$

$$\frac{UA}{U_s A_s} = \frac{1 + \beta_s}{\dfrac{St_s}{St} \left[\dfrac{f}{f_s} \dfrac{P_s}{P} \left(\dfrac{A_s}{A} \right)^2 \right]^{1/3} + \beta \left(\dfrac{A_s}{A} \right)} \tag{3.11}$$

Equation 3.11 is obtained by normalizing each of the component resistance in $1/UA$ and $1/U_s A_s$ by $R_s = 1/h_s A_s$. Equations 3.10 and 3.11 were developed for tube-side enhancement. However, they are equally applicable to shell-side enhancement or air-side enhancement of a fin-and-tube heat exchanger, if the various terms are appropriately defined. This extension is discussed in Section 3.8.

Table 3.2 Dimensionless Ratios to Be Used in Equations 3.10 through 3.12

Definition	Reference Exchanger	Enhanced Exchanger
Outer surface conductance	$r_{os} = h_s A_s / h_{os} A_{os}$	$r_o = h_s A_s / h_o A_o$
Metal resistance	$r_{ws} = h_s t\, A_s / k A_m$	$r_w = h t A / k A_m$
Fouling resistance	$r_{fs} = h_s R_f$	$r_f = h R_f$
Composite resistance	$\beta_s = r_{os} + r_{ws} + r_{fs}$	$\beta = r_o / E_{ho} + r_w A / A_s + r_f$

Consider the application of Equations 3.10 and 3.11 to tube-side enhancement in a shell-and-tube or fin-and-tube exchanger. The equations give the UA and pumping power (P) ratios of the enhanced exchanger relative to a smooth tube exchanger that operates at specified G. The thermal resistance terms $\beta_{tot,s}$ and β contain known or specified information. Equation 3.11 includes the possibility that the enhanced exchanger may also contain an enhanced outer tube surface, $E_{ho} = h_o/h_{os}$. The parameters of interest in Equation 3.11 are A/A_s, P/P_s, and UA/U_sA_s. For each case of interest in Table 3.1, two of these parameters are set as constraints and the third is the objective function.

The tube diameter of the enhanced exchanger may be different from that of the reference smooth-tube exchanger. The definition of geometrical parameters for air-side extended surfaces is more complicated, as discussed subsequently.

As an example, assume tube-side enhancement in a shell-and-tube exchanger for Case FG-2. Here $P/P_s = A/A_s = 1$ and the objective is to achieve $UA/U_sA_s > 1$. Equation 3.11 becomes

$$\frac{UA}{U_sA_s} = \frac{1+\beta}{\dfrac{St_s}{St}\left(\dfrac{f}{f_s}\right)^{1/3} + \beta} \tag{3.12}$$

and from Equation 3.10, one obtains

$$\frac{G}{G_s} = \left(\frac{f_s}{f}\right)^{1/3} \tag{3.13}$$

These equations may be solved for specified G_s and E_{ho} once appropriate equations are established for the St and f of the enhanced surface.

Figure 3.1 shows the performance benefits of tube-side transverse-rib roughness in a condenser, relative to a condenser having a smooth inside tube surface as reported by Webb [1981b]. The tube-side operating conditions are Pr = 3.0 and Re$_s$ = 40,000. The curves are prepared for case VG-1 (constant total tube-side flow rate, pumping power, heat duty, and entering fluid temperature). To satisfy the constant pumping power constraint, it is necessary to operate the rough tube at reduced velocity, or Re/Re$_s$ < 1. This means the enhanced tube exchanger will have a greater number of tubes of shorter length. The parameter r_o/r_s is the outside-to-inside thermal resistance ratio for the exchanger having a smooth inner tube surface. The line labeled $R_o/R_{ip} = 0$ is for all of the thermal resistance on the tube-side. For $R_o/R_{ip} = 0$, a 45% reduction of total tubing length is possible. However, a practical heat exchanger would not be expected to operate with $R_o/R_{is} \cong 0$. The curves show that the tube-side enhancement is less beneficial as $R_o/R_{is} \geq 0$. For $R_o/R_{is} = 0.5$, the enhanced tube provides 27% reduction of total tubing length. Figure 3.1 suggests that Re$_s$/Re = 1.3 provides a good compromise between heat exchanger flow frontal area and material savings. The table on Figure 3.1 shows the dimensionless roughness height

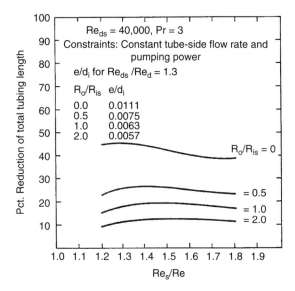

Figure 3.1 Percent reduction of total heat exchanger tubing length obtained by use of transverse ribs ($p/e = 10$). (From Webb [1981b].)

(e/d_i), which satisfies the design constraints at $\mathrm{Re}_s/\mathrm{Re} = 1.3$. The roughness height was selected using methods described in Chapter 9.

3.5 RELATIONS FOR St AND f

Most of the available data for enhanced surfaces are given in graphic form, rather than equation form. However, this is only a minor inconvenience in the implicit solution of the above equations. With these equations, caution should be exercised regarding generality. Because a given enhancement type will usually involve several possible characteristic dimensions, the nominal tube diameter or hydraulic diameter is insufficient for a generalized correlation in the form of St and f vs. Re. Therefore, the resulting "correlation" may be limited to a specific surface geometry and tube diameter (or fin spacing). Further, the data may have been taken with a single fluid, so application to other fluids may be questionable unless the Prandtl number dependency is known. Bergles [1981] also discusses problems associated with applying single-tube data to a "scaled up" situation that involves flow outside a large number of tubes. This situation is probably a greater problem for boiling and condensing than for single-phase heat transfer.

3.6 HEAT EXCHANGER EFFECTIVENESS

The enhanced and smooth exchangers may not operate at the same effectiveness ($\in$). For those cases when the objective is increased heat duty, the $\in$–NTU design method gives

$$\frac{Q}{Q_s} = \frac{W}{W_s} \frac{\in}{\in_s} \frac{\Delta T_i}{\Delta T_{is}} \tag{3.14}$$

where ΔT_i is the temperature difference between the two inlet streams. For fixed inlet temperatures ($\Delta T_i = \Delta T_{is}$), Equation 3.14 yields

$$\frac{Q}{Q_s} = \frac{W}{W_s} \frac{\in}{\in_s} \tag{3.15}$$

Because the operating conditions of the smooth tube exchanger are known, NTU_s ($= U_s A_s / W_s c_{ps}$) is known and $\in$ can be calculated. Once $UA/U_s A_s$ and W/W_s for the enhanced exchanger are known, its NTU is calculated from

$$NTU = NTU_s \frac{UA}{U_s A_s} \frac{W_s}{W} \tag{3.16}$$

Then the $\in$ of the enhanced exchanger may be calculated, and Q/Q_s obtained from Equation 3.15.

When the objective function is increased UA with $Q/Q_s = 1$, Equation 3.14 shows that $\Delta T_i / \Delta T_{is} < 1$.

3.7 EFFECT OF REDUCED EXCHANGER FLOW RATE

The FN and FG cases maintain $N/N_s = 1$. For $P/P_s = 1$, it is necessary to operate at reduced exchanger flow rate ($W/W_s < 1$). The reduced flow rate of these cases may penalize the enhanced exchanger. Such a penalty does not occur for the VG cases, which maintain $W/W_s = 1$. Let us compare this effect for the FN-1 and VG-1 cases, which seek reduced surface area for $Q/Q_s = P/P_s = 1$. Because the FN exchanger operates at higher thermal effectiveness, additional surface area is required to compensate for the reduced LMTD.

Figure 3.2 may be used to illustrate the penalty associated with the $W/W_s < 1$ exchanger. This figure is constructed for shell-side condensation ($C_{min}/C_{max} = 0$), which operates with constant UA and ΔT_i. The $\in$ is given by $\in = 1 - \exp(-NTU)$. To obtain $Q/Q_s = 1$ for $UA =$ constant, the surface area of the $W/W_s < 1$ exchanger must be increased. The area penalty increases with larger NTU of the reference smooth tube exchanger. The $W/W_s < 1$ and $W/W_s = 1$ enhanced exchangers have the same U value if $\Delta p / \Delta p_s = 1$. If $P/P_s = 1$, the $W/W_s = 1$ exchanger will have a slightly higher U value.

To summarize the effect, assume that the smooth tube exchanger is designed for $\in_s = 0.632$ (NTU = 1.0) and that the enhanced exchanger provides a 50% larger U value. For the VG case ($W/W_s = 1$), $U/U_s = 1.5$ thus yields $A/A_s = 1/1.5$.

Figure 3.2 shows $Q/Q_s = 0.903$ for the FN case ($W/W_s = 0.80$). The FN exchanger will require $A/A_s = 0.667/0.903 = 0.74$ to satisfy $Q/Q_s = 1$. Thus, the VG exchanger provides 10% greater surface area reduction than that allowed by the

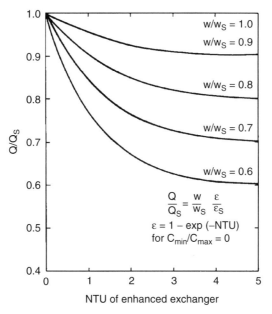

Figure 3.2 Effect of reduced flow rate on Q/Q_s for $UA/U_sA_s =1$. (From Webb [1981a])

FN exchanger. Of course, there may be benefits with reduced flow rate, such as a smaller pump, reduced pressure drop in the supply pipe and header, or smaller supply pipe size.

3.8 FLOW NORMAL TO FINNED TUBE BANKS

The PEC defined in Table 3.1 may also be interpreted for flow normal to tube banks (bare or finned) and for plate fin heat exchangers. The FN and FG cases maintain constant cross-sectional flow area (N/N_s = 1). Cases FG-2 and FN-1 constrain P/P_s = 1 for which a reduced flow rate (W/W_s < 1) may be required to satisfy the P/P_s = 1 constraint. It is possible that the potential benefits of an enhanced surface will be lost if the heat exchanger is required to operate with W/W_s < 1 as discussed by Webb [1981a]. When the flow rate is reduced, the LMTD is reduced (for constant ΔT_i), and additional surface is required to compensate for the reduced LMTD. The VG cases avoid this problem by increasing the flow cross-sectional area sufficiently to maintain W/W_s = 1; this is accomplished by adding tubes in parallel. Hence, when a pumping power constraint is applied, the greatest performance benefit of an enhanced surface will be realized by use of the VG criteria. This implies that enhanced surfaces will yield smaller benefits in retrofit applications with fixed flow area than in new designs, for which the flow frontal area may be increased over that of the corresponding smooth tube design.

The PEC defined in Table 3.1 may also be interpreted for flow normal to tube banks (bare or finned) and for plate-fin heat exchangers, which are discussed in

Table 3.3 Interpretation of Table 3.1 PEC for Finned Tube Banks

Variable	Flow in Tubes	Tube Banks
Flow area	$A_c N$	A_{fr}
Mass flow rate (W)	$A_c NG$	$A_{fr} G_{fr}$
Surface area	$\pi d_i NL$	βV
Flow rate ratio (W/W_s)	$NG/(N_s G_s)$	$A_{fr} G_{fr}/(A_{fr} G_{fr})_s$

Chapter 5 and illustrated by Figure 5.1. Table 3.3 taken from Webb [1982] defines how the variables used in Table 3.2 (flow in tubes) should be interpreted for the geometries shown in Figure 5.1.

3.9 VARIANTS OF THE PEC

The flow resistance constraint used in Table 3.1 and in the previous PEC equations is the flow friction power. One may choose to use different constraints, as is appropriate to the situation. For example, one may choose to fix the pressure drop, rather than the pumping power. An important third possible constraint is the balance point on the fan or pump curve.

Consider Case FG-1 involving a tube-for-tube retrofit of a shell-and-tube condenser. The analysis for this problem is described by Webb et al. [1984]. Cooling water from a lake or cooling tower supplies the condenser cooling water. The enhanced tubes have a higher friction factor than that of the smooth tubes they are replacing. Hence, the circulating water pump must work against a higher system friction resistance characteristic. The flow rate in the condenser will be determined by the balance point between the pump and system characteristics. Figure 3.3 shows this situation, as analyzed by Webb et al. [1984]. The system resistance consists of the sum of the friction loss in the condenser tubes and the piping and fittings external to the condenser. Using the known friction characteristics of the external piping and the condenser tubes, the designer constructs polynomial functions, which represent the system characteristic curves. Assume that $H_o = f_o(W)$ and $H_s = f_s(W_s)$ represent the system characteristics of the enhanced and smooth tubes, respectively. The pump curve is represented by $H_p = f_p(W)$. The smooth tube flow rate at the balance point on the pump curve is given by iterative solution of $f_s(W_s) = f_p(W_s)$. Similarly, the balance point for the enhanced tubes is given by iterative solution of $f_e(W) = f_p(W)$. Because the number of tubes is constant, $G/G_s = W/W_s$. Using the known values of G and G_s, one continues the solution for the h and UA of the enhanced surface using Equation 3.11.

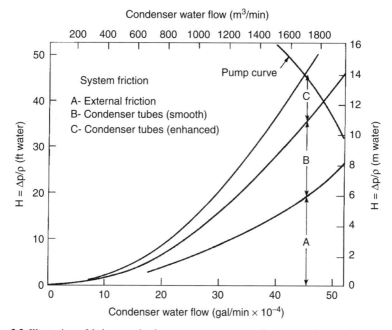

Figure 3.3 Illustration of balance point between pump curve and system resistance for smooth and enhanced tubes in steam condenser of a nuclear plant.

3.10 COMMENTS ON OTHER PERFORMANCE INDICATORS

Many authors have proposed alternate performance comparison methods. Shah [1978] surveys more than 30 comparison methods, and more have been published since 1978. Most of the methods assume that the total thermal resistance is on one side of the heat exchanger. A recent offering by Cowell [1990] describes a methodology to compare compact heat exchanger surfaces, which is basically equivalent to that presented in Section 3.3. However, he allows the hydraulic diameter to be either fixed or variable. Two popular methods are discussed below, and are compared to the methodology of Section 3.3.

3.10.1 Shah

Kays and London [1984] and Shah [1978] describe a PEC method that they term a "goodness factor" comparison. This dimensional method allows direct comparison of the air-side performance ($\beta = \beta_{tot,s} = 0$) of different surface geometries. Rather than writing Equations 3.1 and 3.3 in the form of a ratio, the heat transfer coefficient–surface efficiency product ($h\eta_o$) and friction power (P) are written in dimensional terms as

$$h\eta_o = \frac{c_p\mu}{\mathrm{Pr}^{2/3}}\,\frac{\eta_o j\,\mathrm{Re}}{D_h} \tag{3.17}$$

$$\frac{P}{A} = \frac{\mu^3}{2\rho^2}\,\frac{f\,\mathrm{Re}^3}{D_h^3} \tag{3.18}$$

The fluid properties in Equations 3.17 and 3.18 are evaluated at any preferred "standard condition," and Equation 3.17 is plotted vs. Equation 3.18 giving $(h\eta_o)_{\mathrm{STD}}$ vs. $(P/A)_{\mathrm{STD}}$. For fixed $(P/A)_{\mathrm{STD}}$, the surface having the highest h will require the least heat transfer surface area (A) for equal $h\eta_o A$. Alternatively, one may modify the equations to allow comparison on a volume basis. Multiplying each equation by $\beta(= A/V)$ allows comparison of $(h\eta_o A/V)_{\mathrm{STD}}$ vs. $(P/V)_{\mathrm{STD}}$. Thus, at fixed P/V, one compares the $h\eta_o A$ per unit volume. Geometries having large $h\eta_o A/V$ will require the least core volume for a given air-side performance $(h\eta_o A)$. Equations 3.17 and 3.18 may be written in terms of j, f, and Re to allow direct substitution from a graph or table of the dimensionless surface performance data. The author suggests that the effects of the geometric parameters can be better visualized by writing the P/V and $h\eta_o A/V$ equations in forms containing the mass velocity,

$$\frac{h\eta_o A}{V} = \frac{4\sigma c_p}{\mathrm{Pr}^{2/3}}\,\frac{jG}{D_h} \tag{3.19}$$

$$\frac{P}{V} = \frac{2\sigma}{\rho^2}\,\frac{fG^3}{D_h} \tag{3.20}$$

The above form clearly shows the effect of fin pitch, which is influenced by D_h. As the fin pitch decreases, D_h decreases. Thus, both $h\eta_o A/V$ and P/V will increase in direct proportion to fin pitch. Comparison methods should not confuse the effect of fin pitch and surface geometry. One way to avoid this is to evaluate Equations 3.17 and 3.18 at a fixed D_h for all surfaces. Although the surfaces being compared may have different hydraulic diameters, the j and f vs. Re values of all surface geometries can be evaluated at a common D_h. The PEC equations presented in Section 3.3 use fixed tube diameter to show only the effect of the surface configuration on the performance.

The goodness factor method may be compared to the cases given in Table 3.1:

1. The equations supporting Table 3.1 are written in the form of a ratio, relative to a reference surface. Although this is not necessary, it provides simplicity because the fluid property terms drop out.
2. The Table 3.1 equations allow inclusion of the total resistance between the hot and cold streams.
3. The surfaces are compared at equal D_h in Table 3.1.

Case VG-2a with $\beta = \beta_{tot,s} = 0$ directly corresponds to the surface area goodness factor comparison method. This case yields h/h_s for $P/P_s = A/A_s = 1$ with $W/W_s = 1$. This is equivalent to $P/A = P_s/A_s$.

3.10.2 Soland and Colleagues

Soland et al. [1976, 1978] have compared the air-side performance of 17 parallel plate-fin surface geometries given in Kays and London [1984]. Their analysis is for $\beta = \beta_s = 0$ and corresponds approximately to cases FG-l, VG-la, VG-2a, and VG-3 of Table 3.1. They redefine the j and f factors in terms of the area of the primary surface, and the characteristic dimension in f and Re is given by the spacing between the parting sheets, which contain the extended surface. Figure 3.4, taken from figure 11 of Soland et al. [1976], shows V/V_s for the VG-l case. The reference surface is a plain channel having 6.4 mm between parting sheets, with no fins between the parting sheets. The VG-l comparison used by Soland et al. [1976, 1978] seeks minimum heat exchanger volume (V/V_s), and geometries having small fin pitches and greater distance between parting sheets will be favored in the comparison. Thus, the V/V_s ratio is strongly influenced by the basic geometric parameters, independent of the heat transfer coefficient. Although this method does indicate compactness, it gives no indication of the relative heat transfer surface area or fin material requirements.

3.11 EXERGY-BASED PEC ANALYSIS

A method to evaluate the merit of enhanced surfaces based on thermodynamic second-law analysis was proposed by Bejan [1982, 1996]. It has also been used by others. The method is known as the exergy analysis or entropy generation minimization method. It evaluates the advantage of a given enhanced surface by comparing the rate of entropy generation in an enhanced passage with that in a smooth one. The entropy generation rate per unit length of heat exchanger passage (S_{gen}) is given by Bejan [1982]:

$$S_{gen} = \frac{q'\Delta T}{T^2} + \frac{W}{\rho\Delta T}\left(-\frac{dP}{dx}\right) \qquad (3.21)$$

where q' is the heat transfer per unit length, W is the mass flow rate, and ΔT is the wall-to-fluid temperature difference. The first term on the right-hand side of Equation 3.21 represents the irreversibility due to heat transfer ($S_{gen,T}$), and the second term represents the irreversibility caused by fluid friction ($S_{gen,f}$). Equation 3.21 can be written as, in terms of irreversibility distribution ratio $\phi = S_{gen,f}/S_{gen,T}$,

$$S_{gen} = S_{gen,T}(1+\phi) \qquad (3.22)$$

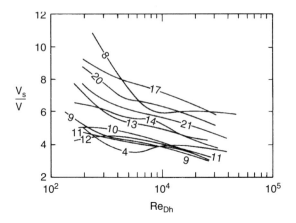

Plate-and-Fin Surfaces Compared

Surface geometry	Plate spacing, b (mm)	Surface code	Surface number
Plain	6.4	11.1	4
	6.4	19.86	8
Louvered	6.4	3/8–6.06	9
	6.4	3/8(a)–6.06	10
	6.4	1/2–6.06	11
	6.4	1/2(a)–6.06	12
	6.4	3/8–8.7	13
	6.4	3/8(a)–8.7	14
	6.4	1/4(b)–11.1	17
	6.4	1/2–11.1	20
	6.4	3/4–11.1	21

Figure 3.4 V/V_s for constant hA and P using Case VG-1. All geometries are the parallel plate-fin heat exchanger geometry having 6.35 mm plate spacing. (From Soland et al. [1976].)

The advantage of enhanced surface is quantified by the entropy generation number, defined by

$$N_s = S_{gen} / S_{gen,s} \qquad (3.23)$$

where S_{gen} is the entropy generation rate in the enhanced passage and $S_{gen,s}$ is the entropy generation rate in a smooth passage. Any enhanced surface yielding value of N_s less than unity reduces the irreversibility of the heat exchanger. Bejan [1982] shows that, for fixed q' and W (FG-1b of Table 3.1), the entropy generation number can be rewritten as

$$N_s = \frac{N_T + \phi_s N_p}{1 + \phi_s} \qquad (3.24)$$

where

$$N_T = \frac{\mathrm{St}_s}{\mathrm{St}} \frac{D_h}{D_{h,s}} \qquad (3.25)$$

$$N_p = \frac{f}{f_s} \frac{D_{h,s}}{D_h} \left(\frac{A_s}{A}\right)^2 \qquad (3.26)$$

$$\mathrm{Re} = \frac{D_h}{D_{h,s}} \frac{A_s}{A} \mathrm{Re}_s \qquad (3.27)$$

$$\phi_s = \left(\frac{T}{\Delta T}\right)_s^2 \left(\frac{G^2}{A_c c_p T}\right)_s \frac{f_s / 2}{\mathrm{St}_s} \qquad (3.28)$$

where ϕ_s is the irreversibility distribution ratio of the smooth surface.

Bejan [1982] provides an example, where Equation 3.24 is used to calculate the entropy generation number for a sand-grain-roughened tube. Figure 3.5 shows the entropy generation number as a function of Re_s and ϕ_s. The figure was constructed for Pr = 0.72 using Nikuradse [1950] and Dipprey and Sabersky [1963] data for f and St, and the Karman–Nikuradse (provided in Kays and Perkins [1973]) and the Petukhov (Equation 2.22) correlations for f_s and St_s. Figure 3.5 shows that the ability of sand-grain roughness to reduce irreversibility depends on ϕ_s and Re_s. Increasing ϕ_s increases the entropy generation number, and the effect is more pronounced at higher Re_s, or larger e/D. The Bejan [1982] study was extended to obtain additional PEC equations by Chen and Huang [1988], Zimparov and Vulchanov [1994], and Zimparov [2000, 2001].

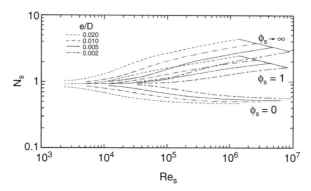

Figure 3.5 Entropy generation number vs. smooth tube Reynolds number for sand-grain roughness. (From Bejan [1982].)

3.12 CONCLUSIONS

The performance evaluation criteria described in this chapter have wide application to single-phase heat exchangers. The PEC equations allow one to make a simple surface performance comparison, or to account for the thermal resistance of both fluid streams. The PEC analysis allows the designer to quickly identify the most effective surfaces or tube inserts of a variety of available choices. Use of the criteria equations with consideration of external thermal resistances and changes in temperature difference along the flow path allows the designer to assess the performance of actual heat exchangers.

The designer should be careful in applying PEC methods that compare two surface geometries having different diameter (or hydraulic diameter). The resulting comparison shows the effect of changing both variables, rather than the effect of surface geometry alone. The effect of surface geometry alone is determined if the diameter is maintained constant.

NOMENCLATURE

A	Heat transfer surface area, m^2 or ft^2
A_c	Flow area at minimum cross section, m^2 or ft^2
A_{fr}	Frontal area, m^2 or ft^2
A_m	Mean conduction surface area, m^2 or ft^2
B	External-to-internal surface area ratio, dimensionless
b	Distance between parting sheets, m or ft
c_p	Specific heat, J/kg-K or Btu/lbm-°F
d_i	Tube inside diameter, m or ft '
D_h	Hydraulic diameter, m or ft
E_h	h/h_s at constant Re, dimensionless
e	Roughness height, m or ft
f	Fanning friction factor, dimensionless
G	Mass velocity, kg/s-m^2 or lbm/s-ft^2
G_{fr}	Mass velocity based on heat exchanger frontal area, kg/s-m^2 or lbm/s-ft^2
g	Gravitational acceleration, m/s^2 or ft/s^2
h	Heat transfer coefficient, W/m^2-K or Btu/hr-ft^2-°F
j	Heat transfer factor, $StPr^{2/3}$, dimensionless
k	Thermal conductivity, W/m-K or Btu/hr-ft^2-°F
L	Length of flow path in heat exchanger, m or ft
LMTD	Log-mean temperature difference, K or °F
N	Number of tubes in each pass of heat exchanger, dimensionless
NTU	Number of heat transfer units, $UA/c_p W_{min}$, dimensionless
p	Roughness axial pitch, m or ft
Δp	Pressure drop, Pa or lbf/ft^2
P	Pumping power, W or HP

Pr   Prandtl number, $c_p\mu/k$, dimensionless
Q Heat transfer rate, W or Btu/hr
q Heat flux, W/m^2 or Btu/hr ft^2
r Resistance ratios defined in Table 3.2, dimensionless
Re   Reynolds number: smooth or rough tube (du/ν); internally finned tube ($D_h u/\nu$), dimensionless
R_f Fouling factor, m^2-K/W or ft^2-hr-°F/Btu
R_i $1/h_i A_i$, m^2-K/W or ft^2-hr-°F/Btu
R_0 $1/h_o A_o$, m^2-K/W or ft^2 hr-°F/Btu
St   Stanton number, $h/c_p G$ dimensionless
T Temperature, K or °F
ΔT Average temperature difference between hot and cold streams, K or °F
ΔT_i Temperature difference between inlet hot and cold streams, K or °F
t Tube wall thickness, m or ft
u Average flow velocity, m/s or ft/s
U Overall heat transfer coefficient, W/m^2-K or °F
V Heat exchanger volume, $A_{fr}L$, m^3 or ft^3
W Flow rate, kg/s or lbm/s

Greek Symbols

β Composite resistance ratio defined in Table 3.2, dimensionless
β Ratio of surface area to exchanger volume, m^{-1} or ft^{-1}
β_T Volume coefficient of thermal expansion, $-(1/p)(\partial p/\partial T)_p$, K^{-1} or R^{-1}
$\in$ Heat exchanger thermal effectiveness, dimensionless
η_o Surface efficiency, dimensionless
μ Dynamic viscosity, kg/s-m or lbm/s-ft
ν Kinematic viscosity, m^2/s or ft^2/s
ρ Density, kg/m^3 or lbm/ft^3
σ Area ratio, A_{fr}/A_c dimensionless

Subscripts

f Evaluated at film temperature
o Tube outer surface
p Plain surface
s Smooth tube
ref   Reference geometry
STD   Reference condition used in Equations 3.17 and 3.18
w Wall condition

Note: Properties are evaluated at the bulk fluid temperature unless otherwise noted.

REFERENCES

Bejan, A., 1982. *Entropy Generation through Heat and Fluid Flow*, Wiley, New York.

Bejan, A., 1996. *Entropy Generation Minimization*, CRC Press, Boca Raton, FL.

Bergles, A.E., 1981. Applications of heat transfer augmentation, in *Heat Exchangers: Thermal Hydraulic Fundamentals and Design*, S. Kakaç, A.E. Bergles, and F. Mayinger, Eds. Hemisphere, Washington, D.C.

Bergles, A.E., Blumenkrantz, A.R., and Taborek, J., 1974a. Performance evaluation criteria for enhanced heat transfer surfaces, in *Heat Transfer 1974, Proc. 5th Int. Heat Transf. Conf.*, Vol. 2, Japan Society of Mechanical Engineers, Tokyo, 234–238.

Bergles, A.E., Bunn, R.L., and Junkhan, G.K., 1974b. Extended performance evaluation criteria for enhanced heat transfer surfaces, *Lett. Heat Mass Transfer*, 1, 113–120.

Chen, B.H. and Huang, W.H., 1988. Performance evaluation criteria for enhanced heat transfer surfaces, *Int. Commun. Heat Mass Transfer*, 15, 55–72.

Cowell, T.A., 1990. A general method for the comparison of compact heat transfer surfaces, *J. Heat Transfer*, 112, 288–294.

Dipprey, D.F. and Sabersky, R.H., 1963. Heat and momentum transfer in smooth and rough tubes at various Prandtl numbers, *Int. J. Heat Mass Transfer*, 6, 329.

Kays, W.M. and London, A.L., 1984. *Compact Heat Exchangers*, 3rd ed., McGraw Hill, New York.

Kays, W.M. and Perkins, H.C., 1973. *Handbook of Heat Transfer*, W.M. Rohsenow and J.P. Hartnett, Eds., McGraw-Hill, New York, chap. 7–4.

Nikuradse, J., 1950, Laws for Flow in Rough Pipes, NACA TM 1292.

Shah, R.K., 1978. Compact heat exchanger surface selection methods, in *Heat Transfer 1974, Proc. 5th Int. Heat Transf. Conf.*, Vol. 4, Hemisphere, New York, 193–199.

Soland, J.G., Mack, W.M., Jr., and Rohsenow, W.B., 1976. Performance Ranking of Plate-Fin Heat Exchanger Surfaces, ASME Paper 76 WA/HT 31.

Soland, J.G., Mack, W.M., Jr., and Rohsenow, W.B., 1978. Performance ranking of plate-fin heat exchanger surfaces, *J. Heat Transfer*, 100, 514–519.

Webb, R.L., 1981a. Performance evaluation criteria for use of enhanced heat transfer surfaces in heat exchanger design, *Int. J. Heat Mass Transfer*, 24, 715–726.

Webb, R.L., 1981b. The use of enhanced heat transfer surface geometries in condensers, in *Power Condenser Heat Transfer Technology: Computer Modeling, Design, Fouling*, P.J. Marto, and R.H. Nunn, Eds., Hemisphere, Washington, D.C., 287–324.

Webb, R.L., 1982. Performance evaluation criteria for air-cooled finned tube heat exchanger surface geometries, *Int. J. Heat Mass Transfer*, 25, 1770.

Webb, R.L. and Bergles, A.E., 1983. Performance evaluation criteria for selection of heat transfer surface geometries used in low Reynolds number heat exchangers, in *Low Reynolds Number Flow Heat Exchangers*, S. Kakaç, R.K. Shah, and A.E. Bergles, Eds., Hemisphere, Washington, D.C., 735–752.

Webb, R.L. and Eckert, E.R.G., 1972. Application of rough surfaces to heat exchanger design, *Int. J. Heat Mass Transfer*, 15, 1647–1658.

Webb, R.L., Haman, L.L., and Hui, T.S., 1984. Enhanced tubes in electric utility steam condensers, in *Heat Transfer in Heat Rejection Systems*, S. Sengupta and Y.G. Mussalli, Eds., ASME Symp. Vol. 37, HTD, 17–26.

Zimparov, V., 2000. Extended performance evaluation criteria for enhanced heat transfer surfaces: heat transfer through ducts with constant wall temperature, *Int. J. Heat Mass Transfer*, 43, 3137–3155.

Zimparov, V., 2001. Extended performance evaluation criteria for enhanced heat transfer surfaces: heat transfer through ducts with constant heat flux, *Int. J. Heat Mass Transfer*, 44, 169–180.

Zimparov, V.D. and Vulchanov, N.L., 1994. Performance evaluation criteria for enhanced heat transfer surfaces, *Int. J. Heat Mass Transfer*, 37, 1807–1816.

PERFORMANCE EVALUATION CRITERIA FOR TWO-PHASE HEAT EXCHANGERS

4.1 INTRODUCTION

This chapter describes quantitative methods to define the performance benefits of enhanced heat transfer surfaces in a heat exchanger, for which at least one of the fluids is evaporating or condensing. Pressure drop of a two-phase fluid may reduce the mean temperature difference for heat exchange. This situation does not exist for heat exchange between single-phase fluids. This chapter shows how the performance evaluation criteria (PEC) previously defined for single-phase flow may be modified to account for the effect of pressure drop. Generalized PEC are defined for three basic applications of two-phase heat exchangers, and two examples of PEC analysis are presented. The PEC analysis for two-phase heat transfer was originally formulated by Webb [1988] and is the only PEC analysis advanced to date for two-phase heat transfer.

4.2 OPERATING CHARACTERISTICS OF TWO-PHASE HEAT EXCHANGERS

A "two-phase heat exchanger" is defined here as one for which at least one of the fluids undergoes a phase change. The heat exchangers of interest here are vaporizers (evaporators, boilers, and reboilers) and condensers. Use of two-phase heat exchangers include three general application areas:

1. Work-producing systems, such as a Rankine power cycle
2. Work-consuming systems, such as a vapor compression refrigeration system or a vacuum distillation process, which use a compressor to support the process
3. Heat-actuated systems, such as an absorption refrigeration cycle, or a distillation process (as defined here, a turbine or compressor would not be in this system)

The quantitative relations given in Chapter 3 assume that the LMTD is not affected by the pressure drop if the total flow rate is held constant. This is not true for two-phase flow. In two-phase flow, pressure drop causes a reduction of the saturation temperature, causing reduced LMTD for fixed entering condenser temperature, or for fixed leaving evaporator temperature. This is illustrated in Figure 4.1. Hence, the previously proposed quantitative relations must be modified to account for the effect of two-phase pressure drop on the LMTD. The effect of a given pressure drop on the decrease of saturation temperature (ΔT_{sat}) depends on the value of dT/dp at the operating pressure. The Clausius–Clapeyron equation may be employed to establish the magnitude of dT/dp

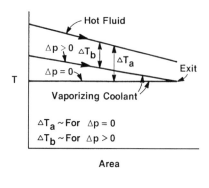

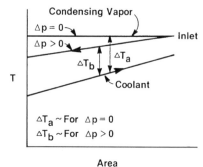

Figure 4.1 Illustration of the effect of two-phase pressure drop on the LMTD. (From Webb [1988]).

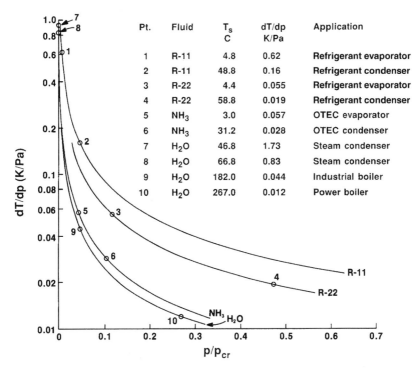

Pt.	Fluid	T_s C	dT/dp K/Pa	Application
1	R-11	4.8	0.62	Refrigerant evaporator
2	R-11	48.8	0.16	Refrigerant condenser
3	R-22	4.4	0.055	Refrigerant evaporator
4	R-22	58.8	0.019	Refrigerant condenser
5	NH_3	3.0	0.057	OTEC evaporator
6	NH_3	31.2	0.028	OTEC condenser
7	H_2O	46.8	1.73	Steam condenser
8	H_2O	66.8	0.83	Steam condenser
9	H_2O	182.0	0.044	Industrial boiler
10	H_2O	267.0	0.012	Power boiler

Figure 4.2 dT/dp vs. p/p_{cr}. (From Webb [1988].)

$$\frac{dT}{dp} = \frac{T(v_v - v_l)}{\lambda} \qquad (4.1)$$

Figure 4.2 shows dT/dp vs. p/p_{cr} for several fluids. This figure shows that dT/dp increases as p/p_{cr} decreases. Typical operating points for condensers and evaporators of refrigeration and power cycles are noted on Figure 4.2. Examination of the figure shows that refrigerant evaporators and power cycle condensers will suffer the greatest LMTD reduction due to two-phase pressure drop.

For single-phase flow in heat exchangers, the pump or fan power requirement (P) is directly proportional to the fluid pressure drop for a fixed flow rate. This is why the pumping power is included as an objective function (or constraint) in Table 3.1. However, the same relationship between pumping power and pressure drop does not exist for two-phase heat exchanger applications.

4.3 ENHANCEMENT IN TWO-PHASE HEAT EXCHANGE SYSTEMS

This section describes how enhanced heat transfer may be employed in the previously defined work-producing, work-consuming, and heat-actuated systems.

4.3.1 Work-Consuming Systems

The purpose of a refrigeration cycle is to cool air, water, or a process fluid. Mechanical power is consumed by the compressor, which "pumps" the evaporated refrigerant from the evaporator pressure to the condenser pressure.

Performance improvements that may be affected by the use of enhanced surfaces are as follows:

1. Reduced heat transfer surface area for fixed compressor power (P_c).
2. Increased evaporator heat duty for fixed compressor lift (pressure difference between condenser and evaporator).
3. Reduced compressor power for fixed evaporator heat duty. This would be affected by reducing the LMTD of the evaporator and/or condenser. This would increase the suction pressure to the compressor and reduce the inlet pressure to the condenser.

For the first case, one would seek reduced surface area in the evaporator and condenser for fixed compressor inlet and outlet conditions. The compressor power is not influenced by pressure drop in either the evaporator or condenser. However, pressure drop in either heat exchanger will reduce the LMTD, and reduce the net

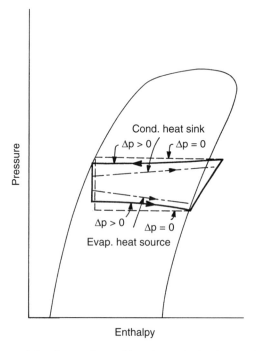

Figure 4.3 Pressure vs. enthalpy diagram for a refrigeration cycle showing the effect of pressure drop on the LMTD in each heat exchanger. (From Webb [1988].)

performance improvement. Figure 4.3 illustrates a refrigeration cycle for this case, and shows the effect of two-phase pressure drop on the LMTD.

The second case would maintain constant heat transfer surface area and take advantage of the increased UA to obtain increased heat duty. Again, pressure drop would act to reduce the increase of evaporator heat duty. A larger compressor capacity would be required, because the refrigerant flow rate would increase in proportion to the heat transfer increase. The surface area is maintained constant for the third case and the UA increase is employed to reduce the LMTD. This allows the compressor lift to be reduced for the fixed evaporator heat duty.

A "vacuum distillation" process may occur at a sufficiently low temperature that a refrigeration system is required to provide condenser coolant (an evaporating liquid) at the required low condensing temperature. Use of an enhanced heat transfer surface in the distillation reflux condenser may allow the compressor of the refrigeration system to operate at a higher suction pressure, and thereby save compressor power.

4.3.2 Work-Producing Systems

Next, consider the case of a Rankine power cycle. The boiler feed pump raises the pressure of the working fluid from the condenser pressure to the boiler pressure. This pumping power is calculated by

$$P = (p_b - p_c)Wv_l$$

(4.2)

The pressure drop in the boiler is negligible compared to the pressure difference between the boiler and the condenser $(p_b - p_c)$. Hence, the mechanical energy consumed by frictional pressure drop in the boiler is of negligible concern. However, pressure drop in the boiler or condenser will decrease the thermodynamic efficiency of the power cycle, because of the effect of dT/dp on the LMTD. Performance improvements that may be affected by the use of enhanced surfaces are twofold:

1. Reduced boiler and/or condenser surface area for constant turbine output (P_t).
2. Increased turbine output for fixed boiler heat input, or fixed condenser heat rejection. This would increase the boiler pressure or reduce the condensing temperature.

The first case maintains fixed boiler exit pressure and/or fixed condenser inlet pressure. Increased pressure drop in either heat exchanger will reduce the LMTD, and hence reduce the performance improvement. Based on the information presented in Figure 4.2, one sees that the pressure drop in the boiler will have a very small effect on the boiler performance improvement. However, the same pressure drop in the condenser may significantly reduce the LMTD.

The second case maintains constant heat transfer surface area, and it employs the increased UA to allow reduction of the LMTD. If the LMTD in the condenser is reduced, the turbine can expand to a lower back pressure and produce more work.

It is unlikely that the LMTD reduction in the boiler would have a significant effect on the turbine power output.

4.3.3 Heat-Actuated Systems

These systems may have a pump to transport liquids, but they do not use compressors or turbines in the process operations. A good example of a heat-actuated system is the absorption refrigeration cycle. This cycle has four heat exchangers that involve condensation or boiling processes (evaporator, concentrator, condenser, and absorber). A pump is used to transport the working fluid between the absorber and the concentrator, but a compressor is not employed. In the LiBr-water absorption cycle, water is the refrigerant. Heat, in the form of steam or combustion products, is added in the concentrator to boil off water and concentrate the LiBr-water solution. If the LMTD in any of the four heat exchangers can be reduced, the thermodynamic irreversibility of the heat exchanger processes will be reduced.

A second example of a heat-actuated system is the oil refining process. A number of cascaded process condensers may be employed in a refinery "crude train." The heat rejected by one condenser may provide heat input to an evaporator. Energy, in the form of combustion products, is employed to provide heat input to the top crude. Energy consumption will be decreased if the LMTD to each cascaded heat exchanger is reduced. This is probably more important than using the higher performance of the enhanced surface to reduce the capital cost of the heat exchanger.

A third example is distillation processes, in which heat rejection occurs to cooling water. Heat, in the form of steam, is added to a reboiler and heat is rejected in a water-cooled condenser. Enhanced surfaces may be employed for the same purposes as in the first two examples. Mechanical energy (e.g., compressors or turbines) is not used in the above examples. Performance improvements may be affected by use of enhanced surfaces for three different purposes:

1. Reduced heat transfer surface area for fixed operating temperatures
2. Increased heat exchange capacity for fixed amount of heat exchange surface area
3. Reduced LMTD for fixed amount of heat exchange surface area, which will increase the thermodynamic efficiency of the process or cycle

Pressure drop of either two-phase stream will act to reduce the LMTD. The smaller LMTD will decrease the surface area reduction possible, or will decrease the efficiency improvement. The above discussion shows that the different operating conditions of two-phase exchangers used in refrigeration and power cycles establish a need for a different set of PEC than those of Table 3.1, which are established for single-phase flow. There are three basic differences:

1. The LMTD is affected by the two-phase pressure drop.
2. The mechanical power of interest in work-producing/consuming cycles is not a pump or fan; rather it is a turbine or compressor. The power consumed by the circulation pump of a power cycle is negligible compared to that produced by

the turbine. Hence, the effect of the heat exchanger performance on the work produced by the turbine or on the work input to the compressor is the key point of concern.

3. In heat-actuated systems, the heat exchanger size or the thermodynamic efficiency of the system is influenced by the LMTD in the heat exchangers. Hence, the effect of the two-phase fluid pressure drop on the LMTD is the point of concern.

4.4 PEC FOR TWO-PHASE HEAT EXCHANGE SYSTEMS

Having defined the performance improvements that may be affected in two-phase heat exchange systems, it is now possible to prepare a table similar to Table 3.1. Table 4.1 shows PEC corresponding to the FG, FN, and VG cases for the previously discussed two-phase heat exchanger applications.

Table 4.1 is laid out using the same cases as listed in Table 3.1. However there are several key differences between Table 3.1 and Table 4.1:

1. The pumping power variable (P) of Table 3.1 is not used in Table 4.1. This is replaced in work-consuming (or work-producing) systems by the variable P_w, which means work performed by (or on) the system. For heat-actuated systems, one may wish to constrain or reduce the pressure drop, because of its effect on the LMTD. The variable Δp applies for heat-actuated systems.

2. The condensers and evaporators considered here are for complete evaporation or condensation. In this case, the heat duty is given by $q = W\lambda\Delta x$. Because Δx will be a constant for the evaporator or condenser, $q \propto W$. Hence, W and q are not independent variables, as they are treated in Table 3.1.

3. Several of the cases in Table 3.1 do not apply for the present considerations. Hence, they do not appear in Table 4.1.

4. Because the variable P_w applies to work-producing or work-consuming systems and Δp applies to heat-actuated systems, separate tables should apply to these two system types. To avoid this complication, Table 4.1 is prepared using an asterisk code (*) to differentiate between the two system types.

Table 4.1 is equally applicable for vaporization or condensation inside tubes, or on the outside of tubes in a bundle.

4.5 PEC CALCULATION METHOD

An algebraic calculation must be performed to determine the value of the objective function for the case of interest in Table 4.1. Before any PEC calculations can be made, it is necessary to know the heat transfer coefficient and friction factor (or Δp characteristics) of the enhanced tube as a function of the flow and geometry

Table 4.1 Performance Evaluation Criteria for Two-Phase Heat Exchange System

Case	Geometry	W	P_w	Q	ΔT_i^*	Objective
FG-1a	N,L			X		$\uparrow Q$
FG-1b	N,L		X	X	X	$\downarrow \Delta T_i^*$
FG-3	N,L	X		X	X	$\downarrow P_w^{\,b}$
FN-1	N	X		X	X	$\downarrow L$
FN-2	N	X		X	X	$\downarrow L$
FN-3	N	X		X		$\downarrow P_w^{\,b}$
VG-1		X	X	X		$\downarrow NL$
VG-2a	NL		X		X	$\uparrow Q$
VG-2b	NL		X		X	$\downarrow \Delta T_i$
VG-3	NL	X		X		$\downarrow P_w^{\,b}$

*ΔT_i is defined as the temperature difference between the leaving boiling fluid and the entering process fluid for vaporizers. For condensers, it is the difference between the entering vapor temperature and the entering coolant temperature.

bReplace by Δp for heat-actuated systems.

conditions. The heat transfer and friction characteristics for vaporization or condensation on enhanced surfaces depend on the following variables:

1. Geometry: The tube diameter and the specific geometric features of the enhanced surface
2. The operating pressure (or saturation temperature)
3. The mass velocity (G), the vapor quality (x), and the heat flux (q) or the wall-to-fluid temperature difference $(T_w - T_s)$.

The data may be measured as a function of the local vapor quality, or as average values over a given inlet and exit vapor quality. Jensen [1988] provides a review of correlations for forced convection vaporization and condensation inside tubes (plain and enhanced). Webb and Gupte [1992] provide a more recent review for forced convection vaporization inside tubes and across tube bundles. Chapters 12 and 13 treat convective vaporization and condensation in enhanced tubes, respectively.

With the basic heat transfer and friction correlations in hand, for the particular enhanced geometry of interest, one is prepared to perform the PEC analysis. Two examples of the PEC calculation are presented.

4.5.1 PEC Example 4.1

Consider Case FN-2 for a refrigeration evaporator with tube-side vaporization. Assume that an evaporator design for vaporization in smooth tubes has been

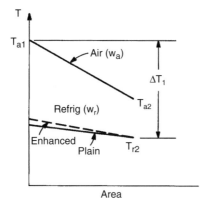

Figure 4.4 Illustration of operating conditions for an air-cooled refrigerant evaporator with tube side vaporization. (From Webb [1988]).

completed. Figure 4.4 illustrates the operating conditions for the smooth-tube design. Air enters the evaporator at flow rate W_{a1} and T_{a1} and is cooled to T_{a2}. The total heat duty of the evaporator is Q_{tot}. Assume that the plain tube design has N refrigerant circuits in parallel. The heat duty per circuit is given by $Q_c = Q_{tot}/N$ and the refrigerant mass flow rate per circuit is given by $W_{rc} = W_r/N_r$. The refrigerant mass velocity is given by $G = W_{rc}/A_c$ where A_c is the flow cross section area for the circuit.

The enhanced tube design operates at the conditions shown on Figure 4.4, except for the entering refrigerant temperature (T_{r1}), which is a dependent variable. The objective is to reduce the evaporator size (circuit length) for fixed compressor suction pressure, which fixes T_{r2}. It is assumed that the frontal area of the finned tube heat exchanger is held constant, and that the evaporator length reduction is accomplished by removing tube rows in the airflow direction. If the pressure drop in the enhanced evaporator tube is greater than that of the plain tube, the LMTD will be reduced, as is qualitatively illustrated by the dashed line of Figure 4.4. The evaporator is designed to operate between entering vapor quality x_1 and 1.0 exit vapor quality. Assume that a correlation exists to calculate the "circuit average" refrigerant heat transfer coefficient (h_i) at the given operating pressure. This is given by

$$h_i = fcn(G,q) \tag{4.3}$$

Similarly, the pressure gradient is assumed to be known in the form

$$\Delta p = fcn(G,L) \tag{4.4}$$

The PEC calculation proceeds as follows:

1. Guess the required circuit length and calculate the circuit averaged heat flux from the required heat load (Q_c) as

$$q = \frac{Q_c}{\pi d_i L} \tag{4.5}$$

2. Calculate the refrigerant-side h and Δp using the known refrigerant mass velocity (G), heat flux (q), and the assumed circuit length (L) using Equations 4.3 and 4.4.

3. Calculate the UA value of the enhanced tube exchanger from

$$\frac{1}{UA} = \frac{1}{h_i A_i} + \frac{t_w}{k_w A_w} + \frac{1}{h_o A_o} \tag{4.6}$$

4. From the calculated refrigerant pressure drop, calculate the entering refrigerant temperature (T_{r1}). Then calculate the LMTD for the heat exchanger.

5. Calculate the "available heat transfer rate" (Q_{av}) by

$$Q_{av} = UA \cdot \text{LMTD} \tag{4.7}$$

6. If $Q_{av} > Q_c$ (the required value), one assumes a shorter circuit length and repeats steps 1 through 5. The procedure is repeated until $Q_{av} = Q_c$. The material savings offered by the enhanced surface are L/L_p, where L_p is the evaporator circuit length of the plain tube design.

It is possible that the enhanced tube design would provide better performance characteristics if it were designed to operate at a different flow rate per circuit than used for the plain tube design. This question must be resolved by performing a parametric study of the enhanced tube design as a function of the number of circuits. This calculation procedure would follow the previously outlined steps.

4.5.2 PEC Example 4.2

Assume the same plain tube heat exchanger design employed in PEC Example 4.1. For the present problem, the enhanced evaporator tube will be used to increase the compressor suction pressure, and hence reduce the compressor power. This calculation involves case FN-3 of Table 4.1. The enhanced tube operating conditions are the same as for PEC Example 4.1, except now the refrigerant temperature leaving the evaporator (T_{r2}) is not fixed. One may be tempted to fix the entering evaporator temperature (T_{r1}) to the value of the plain tube design. However, this has little meaning, because the entering temperature inherently adjusts to meet the compressor suction pressure and the evaporator circuit pressure drop. A more direct approach is to specify the desired compressor suction pressure, which fixes T_{r2} and to calculate the circuit length of the enhanced tube that gives the required heat duty and the desired T_{r2}. If the desired T_{r2} is set too high, one will find that a solution does not exist for the smaller LMTD corresponding to the increased T_{r2}. As T_{r2} increases, a longer evaporator circuit will be required, and the pressure drop will increase with increasing evaporator circuit length. It is possible that the increased UA of the enhanced tube design will not be sufficiently high to overcome the smaller LMTD resulting from the increased T_{r2} and the possible larger pressure drop of the enhanced tube design.

The recommended approach for this problem is then to specify the desired T_{r2} and proceed to calculate the required evaporator circuit length. This calculation

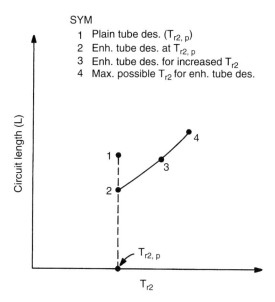

Figure 4.5 Illustration of the effects of enhanced evaporator tubes on the evaporator performance. (From Webb [1988]).

would proceed following the same procedures outlined for PEC Example 4.1. For a specified T_{r2} one calculates the circuit length (L) required to satisfy the specified evaporator load. One may repeat the calculations for several specified values of T_{r2} and plot a graph as indicated by Figure 4.5. This graph provides information to evaluate the trade-offs between evaporator cost and compressor power.

4.6 CONCLUSIONS

PEC are defined for application to heat exchangers, for which at least one of the fluids is a two-phase fluid. Because pressure drop of the two-phase fluid may affect the mean temperature difference in the heat exchanger, the PEC equations must account for the effect of this situation. An assessment of two-phase heat exchanger applications shows that the "reduced pumping power" objective function used for single-phase flow is not generally applicable to the two-phase flow situation. Three generic applications for two-phase heat exchangers are defined and provide a basis for establishing relevant objective functions and operating constraints. The 11 PEC cases defined for two-phase heat exchangers are closely related to the 12 single-phase PEC cases. A new objective function is defined for two-phase exchangers and relates to the work done on or by the system in which the two-phase heat exchanger is installed. This objective function replaces the reduced pumping power objective function for single-phase flow. Two examples using the two-phase PEC are presented.

NOMENCLATURE

A_c	Cross-sectional flow area, m² or ft²
A_i	Inside tube surface area, m² or ft²
A_o	Outside tube surface area, m² or ft²
d_i	Tube inside diameter, m or ft
FG	Criteria used in Table 4.1
FN	Criteria used in Table 4.1
G	Mass velocity, kg/s-m² or lbm/s-ft²
h	Heat transfer coefficient, W/m²-K, or Btu/hr-ft-°F
k_w	Thermal conductivity of tube, W/m-K, or Btu/hr-ft-°F
L	Circuit length, m or ft
LMTD	Log-mean temperature difference, K or °F
N	Number of flow circuits in parallel, dimensionless
P	Power, P_c (compressor), P_t (turbine) P_w (work done on/by system), W or hp
p	Pressure, p_b (boiler), p_c (condenser), p_{cr} (critical), Pa
Δp	Pressure difference, Pa or lbf/in.²
Q	Heat load, W or Btu/hr
Q_{av}	Available heat transfer rate, W or Btu/hr
Q_c	Heat load per circuit, W or Btu/hr
Q_{tot}	Heat load on all circuits, W or Btu/hr
q	Heat flux, W/m², or Btu/hr ft²
R_w	Thermal resistance of tube wall, m²-K/W or hr-ft²-°F/Btu
T	Temperature, T_r (refrigerant), T_w (wall), T_{sat} (saturation), K or °F
ΔT_i	Difference between entering fluid temperatures, K or °F
T_s	Saturation temperature, K or °F
ΔT_{sat}	$T_w - T_s$, K or °F
T_w	Wall temperature, K or °F
U	Overall heat transfer coefficient, W/m²-K or Btu/hr-ft²-°F
VG	Criteria used in Table 4.1
W	Mass flow rate, kg/s or lbm/s
x_i	Inlet vapor quality to heat exchanger, dimensionless
Δx	Vapor quality change, dimensionless

Greek Symbol

v_l	Specific volume of liquid, m³/kg or lbm/ft³
v_g	Saturated vapor specific volume, m³/kg or ft³/lbm
λ	Latent heat of vaporization, J/kg or Btu/lb

Subscripts

a	air
c	Per refrigerant circuit

o	Refers to value on the outside of the tube
p	Plain or smooth surface geometry
r	Refrigerant
t	Refers to tube-side value
1	Inlet to heat exchanger
2	Exit of heat exchanger

REFERENCES

Jensen, M.K., 1988. Enhanced forced convective vaporization and condensation inside tubes, in *Heat Transfer Equipment Design*, R.K. Shah, E.C. Subbarao, and R.A. Mashelkar, Eds., Hemisphere, Washington, D.C., 681–696.

Webb, R.L., 1988. Performance evaluation criteria for enhanced surface geometries used in two-phase heat exchangers, in *Heat Transfer Equipment*, R.K. Shah, E.C. Subbarao, and R.A. Mashelkar, Eds., Hemisphere, Washington, D.C., 697–706.

Webb, R.L. and Gupte, N.S., 1992. A critical review of correlations for convective vaporization in tubes and tube banks, *Heat Transfer Eng.*, 13(3), 58–81.

PLATE-AND-FIN EXTENDED SURFACES

5.1 INTRODUCTION

This chapter discusses enhanced extended surface geometries for the plate-and-fin heat exchanger geometry, which is shown in Figure 5.1. Normally, at least one of the fluids used in the plate-and-fin geometry is a gas. In forced convection heat transfer between a gas and a liquid, the heat transfer coefficient of the gas is typically 5 to 20% that of the liquid. The use of extended surfaces will reduce the gas-side thermal resistance. However, the resulting gas-side resistance may still exceed that of the liquid. In this case, it will be advantageous to use specially configured extended surfaces, which provide increased heat transfer coefficients. Such special surface geometries may provide heat transfer coefficients 50 to 150% higher than those given by plain extended surfaces. For heat transfer between gases, such enhanced surfaces will provide a substantial heat exchanger size reduction. There is a trend toward using enhanced surface geometries with liquids for cooling electronic equipment. Data taken for gases may be applied to liquids if the Prandtl number dependency is known. Brinkman et al. [1988] provide data for water and a dielectric fluid (FC-77), but they do not provide a Prandtl number dependency for their data. In the absence of specific data on Prandtl number dependency, one may assume $St \propto Pt^{-2/3}$.

Figure 5.2 shows six commonly used enhanced surface geometries. Typical fin spacings are 300 to 800 fins/m. Vortex generators illustrated (see Figure 5.28, below) and discussed in Section 5.9 can also provide significant enhancement. Because of the small hydraulic diameter and low density of gases, these surfaces are usually operated with $500 < Re_{Dh} < 1500$ (hydraulic diameter basis). Although increased

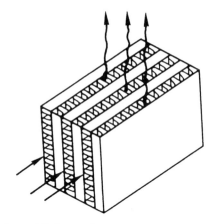

Figure 5.1 A cross-flow plate-and-fin heat exchanger geometry.

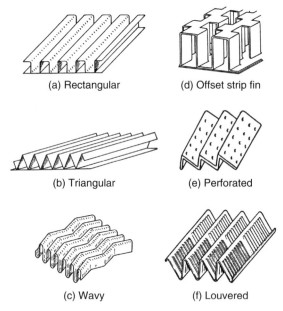

(a) Rectangular   (d) Offset strip fin

(b) Triangular   (e) Perforated

(c) Wavy   (f) Louvered

Figure 5.2 Plate-fin exchanger surface geometries: (a) plain rectangular fins, (b) plain triangular, (c) wavy, (d) offset strip, (e) perforated, (f) louvered. (From Webb [1987].)

performance will exist at higher Reynolds numbers (turbulent regime), fan-power limitations generally limit operation to the above low Reynolds number range. To be effective, the enhancement technique must be applicable to low-Reynolds-number flows. The use of surface roughness will not provide appreciable enhancement for such low-Reynolds-number flows. Two basic concepts have been extensively employed to provide enhancement:

1. Special channel or surface shapes that promote fluid mixing by secondary flows. An example is the wavy channel, which provides mixing due to secondary flows (Taylor–Goertler vortices) and boundary layer separation within the channel. Vortex generators are used to generate longitudinal vortices, which wash the downstream surface.
2. Repeated growth and wake destruction of boundary layers. This concept is employed in the offset-strip fin, the louvered fin, and to some extent, in the perforated fin.

Sources or summaries of data on plate-and-fin surface geometries include Kays and London [1984], Creswick, et al. [1964], and Tishchenko and Bondarenko [1983] (plain fins). Other sources are noted in the following sections.

5.2 OFFSET-STRIP FIN

The offset-strip fin (OSF) is possibly the most important enhancement concept that has been developed to enhance heat transfer with gases. It is effective at both high and low Reynolds numbers. The ehancement mechanism of the louver fin discussed in Section 5.3 is basically the same as that of the OSF. The OSF (and louver fin) geometries are used in both plate-and-fin and finned tube heat exchangers.

5.2.1 Enhancement Principle

Figure 5.3 illustrates the enhancement principle of the Figure 5.2d OSF. A laminar boundary layer is developed on the short strip length, followed by its dissipation in the wake region between strips. Typical strip lengths are 3 to 6 mm, and the Reynolds number (based on strip length) is well within the laminar region. The enhancement provided by an OSF of $L_p/D_h = 1.88$ is shown by Figure 5.4. The OSF and plain fins are taken from figures 10–58 and 10–27, respectively, of Kays and London [1984]. The table in Figure 5.4 compares the dimensions with the OSF scaled to the same hydraulic diameter as that of the plain fin. The scaled dimensions (fin height and fin thickness) of the OSF are approximately equal to those of the plain fin

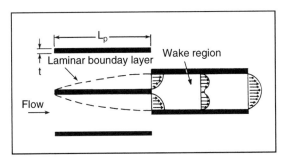

Figure 5.3 Boundary layer and wake region of the OSF. (From Webb [1987].)

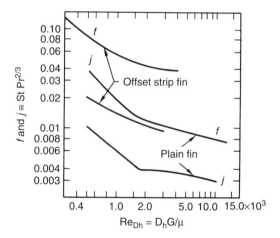

	Surface Geometry	
	Plain	Offset strip
Surface designation	10–27	10–58[a]
Fins/m	437	437
Plate spacing (mm)	12.2	11.9
Hydraulic diam (mm)	3.51	3.51
Fin thickness (mm)	0.20	0.24
Offset strip length in flow dir. (mm)	—	6.6

[a]Dimensions geometrically scaled to give same hydraulic diameter as plain fin.

Figure 5.4 Comparison of j and f for the OSF and the plain fin surface geometries. (From Webb [1987].)

surface. Therefore, the performance difference is due to the enhancement provided by the short strip lengths (6.6 mm) of the OSF. At $Re_{Dh} = 1000$, the j factor of the OSF is 2.5 times higher than that of the plain fin, and the relative friction factor increase is 3.0. Considering the j/f ratio as an "efficiency index" ratio between heat transfer and friction, the OSF yields 150% increased heat transfer coefficient with a heat transfer to friction ratio (j/f) 83% as high as that of the plain fin. Greater enhancement will be obtained by using shorter strip lengths.

The local flow and heat transfer characteristics in an OSF array were investigated by DeJong and Jacobi [1997]. Figure 5.5 shows the flow visualization results. The test sample consists of 8 rows of offset strips with $L_p = 2.54$ cm, $s = 1.28$ cm, and $t = 0.32$ cm. At $Re_{Dh} = 380$, the flow is steady and laminar. At $Re_{Dh} = 550$, a periodic secondary structure is formed in the latter region of the test section, and the wake exhibits a feathery appearance. At $Re_{Dh} = 630$, the wake takes a roughly sinusoidal appearance. At $Re_{Dh} = 720$, large-scale vortex shedding is obvious in the array (from the leading edge of the plate), and the wake closely resembles the classical von Karman vortex street. As the Reynolds number increases ($Re_{Dh} = 850$), the onset of vortex shedding moves upstream. At higher Reynolds numbers, the entire array sheds

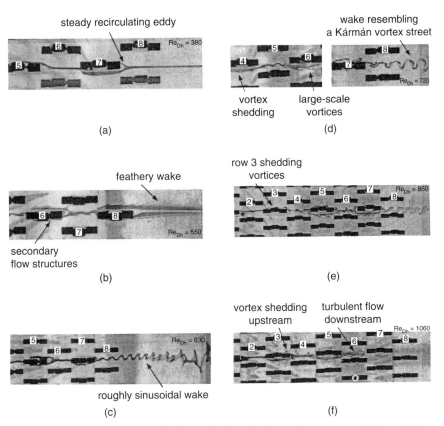

Figure 5.5 Water tunnel flow visualization results for the offset strip geometry with $L_p = 2.54$ cm, $s = 1.28$ cm, $t = 0.32$ cm: (a) $Re_{Dh} = 380$, (b) $Re_{Dh} = 550$, (c) $Re_{Dh} = 630$, (d) $Re_{Dh} = 720$, (e) $Re_{Dh} = 850$, (f) $Re_{Dh} = 1060$. (From DeJong and Jacobi [1997]. Reprinted with permission from Elsevier.)

vortices, and for yet higher Reynolds numbers ($Re_{Dh} = 1060$), a transition to turbulence occurs. They also measured mass transfer coefficients using a naphthalene sublimation test for the same geometry. At low Reynolds numbers, where the flow is laminar and no shedding occurs, the Sherwood number decreases slightly for successive downstream rows. As the Reynolds number increases, a peak in the Sherwood number is observed, which coincides with locations of the onset of vortex shedding illustrated in Figure 5.5.

In Figure 5.6, the heat transfer coefficients (converted from the mass transfer data) are compared with theoretical predictions from Shah and London [1978]. The predictions are for thermal entry length laminar flow between parallel plates at constant temperature. The theoretical values are provided for continuous plates, and interrupted plates. At low Reynolds numbers ($Re_{Dh} < 430$, $v < 0.4$ m/s), where the flow is steady and laminar, the data are in reasonable agreement with theoretical values. At higher Reynolds numbers, the experimental data for the interrupted surface are significantly higher than the theoretical values, because of the vortex shedding.

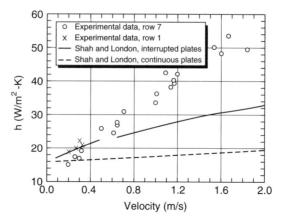

Figure 5.6 Experimental and theoretical average heat transfer coefficients for the offset-strip geometry with $L_p = 2.54$ cm, $s = 1.28$ cm, $t = 0.32$ cm. (From DeJong and Jacobi [1997]. Reprinted with permission from Elsevier.)

The theoretical Shah and London predictions were developed based on simplified approximation to full momentum and energy equations, and do not account for the complex flow characteristics such as vortex shedding.

5.2.2 PEC Example 5.1

A more realistic comparison of the OSF and plain fin performance is obtained using Case VG-1 of Table 3.1. Assume that the plain fin (subscript p) operates at $Re_{Dh,p} = 834$. Equation 3.7, which applies to Case VG-1 of Table 3.1, defines the value of G/G_p at which both surfaces satisfy $1 = hA/h_p A_p = P/P_p = W/W_p$. Hence, me must satisfy

$$\frac{G}{G_p} = \left(\frac{j/j_p}{f/f_p} \right)^{1/2} \tag{5.1}$$

It is necessary to iteratively solve Equation 5.1 for G/G_p, reading the j and f values for the OSF surface from Figure 5.4. The final solution of Equation 5.1 yields $G/G_p = 0.923$. So, $Re_{Dh} = 0.923 \times 834 = 770$. The j and f values of the OSF at $Re_{Dh} = 770$ are now known.

The surface area ratio (A/A_p) is obtained using Equation 3.2, repeated here as Equation 5.2.

$$\frac{hA}{h_p A_p} = \frac{j}{j_p} \frac{A}{A_p} \frac{G}{G_p} \tag{5.2}$$

Substituting the known values of j/j_p and G/G_p in Equation 5.2, one calculates $A/A_p = 0.446$. The OSF geometry requires only 44.6% as much surface area to provide the same hA as the plain fin geometry. Using Equation 3.3, one may show that the flow frontal area must be increased 10% to maintain $P/P_p = 1$.

5.2.3 Analytically Based Models for j and f vs. Re

Kays [1972] proposed a simple, approximate model to predict the j and f vs. Re curves for the OSF. The model assumes (1) laminar boundary layers on the fins, (2) the boundary layers developed on the fin are totally dissipated in the wake region between fins. Using the equations for laminar flow over a flat plate in a free stream, Kays' analysis gives

$$j = 0.664 \, \text{Re}_L^{-0.5} \tag{5.3}$$

$$f = \frac{C_D t}{2L_p} + 1.328 \, \text{Re}_L^{-0.5} \tag{5.4}$$

The first term in Equation 5.4 accounts for the form drag on the plate. The form drag contribution is proportional to the fin thickness (t), and it has a negligible effect on the heat transfer coefficient. Kays suggests use of $C_D = 0.88$ based on potential flow normal to a thin plate. This value is derived by Milne-Thomson [1960]. Note that the Kays model neglects heat transfer and friction on the confining walls of the OSF channel. Although this model is only approximate, it will allow the designer to predict the effect of strip length and thickness.

A rationally based method would numerically solve the momentum and energy equations, or would use a refined analytical model. Patankar and co-workers have published several papers in which the numerical method is used. A summary of this work is given by Patankar [1990]. This approach is generally limited to two-dimensional, laminar flow. Thus, the analysis assumes laminar wakes and $s/h = 0$.

Joshi and Webb [1987] have developed an analytically based model to predict the j and f versus Re_{Dh} characteristics of the OSF array. This model properly accounts for all geometric factors of the array ($\alpha \equiv s/h$, $\delta \equiv t/L_p$, $\gamma \equiv t/s$), heat transfer to the base surface area to which the fins are attached, and is able to account for the nonlaminar region. Writing an energy balance on the cell shown in Figure 5.7, one obtains

$$\eta \text{Nu} = \frac{1-\gamma}{(1+\alpha+\delta)^2} [(1+\delta)\eta_f \text{Nu}_p + \alpha(1+\alpha)\text{Nu}_e] \tag{5.5}$$

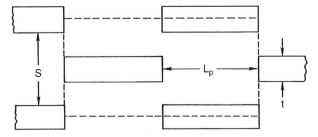

Figure 5.7 Unit cell used to derive the Joshi and Webb [1987] analytical model for the OSF.

The Nu_p $(= h_p 2s/k)$ involves the heat transfer coefficient on the strip fin, and Nu_e $(= h_e D_h/k)$ involves the heat transfer coefficient on the top and bottom channel walls. Similarly, a force–momentum balance on the Figure 5.7 cell gives

$$f = \frac{1-\gamma}{(1+\gamma)(1+\alpha+\delta)}\left(f_p + \alpha f_e + \frac{C_D t}{2L_p} \right)$$ (5.6)

where f_p and f_e are the friction factors on the strip fin and on the top and bottom walls.

Comparison of Equation 5.6 with Equation 5.4 shows that the simple Kays model [1972] neglects the friction on the top and bottom walls, and does not include the expression of geometric factors α, δ, and γ, which multiplies Equation 5.6. The model is segregated into two submodels — one for the laminar region and one for the nonlaminar (or turbulent) region. Equation 5.7 was used as the criterion to select the laminar or nonlaminar region models. The laminar Nu_p and f_p terms use the numerical solution of Sparrow and Liu [1979] for $\alpha = 0$, and zero thickness plates, which is modified to account for the effect of fin thickness. That analysis assumes laminar flow on the fin and wake regions. The terms Nu_e and f_e for the turbulent region are predicted by standard correlations for fully developed turbulent flow in smooth channels. A semiempirical approach was used to develop Nu_p and f_p for the turbulent region. The value 0.88 is used for the drag coefficient (C_D) as was used in the Kays friction model (Equation 5.4).

The Joshi and Webb [1987] model predicted the j factor within 11.5% rms error and the f factor within 16.8% for their 21 core database. Its accuracy is comparable to that of the Manglik and Bergles [1990] and the Muzychka and Yovanovich [2001a] correlations, which are discussed in Section 5.2.4. The Joshi and Webb [1987] model is too complex to use for design purposes. However, its importance is to show that fundamental equations can be used to model heat transfer and friction in the OSF array with good accuracy.

5.2.4 Transition from Laminar to Turbulent Region

Transition from the laminar region occurs at the Reynolds number where the j factor departs from log-linear form, as the Reynolds number is increased. This transition point is defined in Figure 5.8. Joshi and Webb [1987] determined this point from the j vs. Re_{Dh} data for 21 core geometries, and developed a semiempirical correlation to define the transition point. The Reynolds number (GD_h/μ) at which the wake departs from laminar flow is defined as $Re_{Dh,tr}$ is given by

$$Re_{Dh,tr} = 257\left(\frac{L_p}{s}\right)^{1.23}\left(\frac{t}{L_p}\right)^{0.58}\left(\frac{D_h}{\delta_{mom}}\right)$$ (5.7)

where $\delta_{mom} \equiv t + 1.328\, L_p/Re_L^{0.5}$ is fin thickness, plus twice the momentum thickness of the boundary layer at the end of the strip length. The flow visualization measurements of Joshi and Webb show that this transition point occurs before eddies are

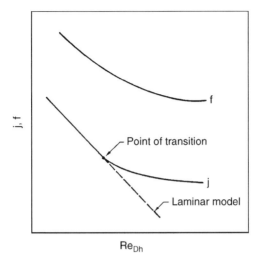

Figure 5.8 Illustration of transition of the j factor from the laminar region. (From Joshi and Webb [1987].)

shed in the wake region. The geometric parameters of the test arrays on which Equation 5.7 was based are $0.13 \leq s/h \leq 1.0$, $0.012 \leq t/L_p \leq 0.048$ and $0.04 \leq t/s \leq 0.20$.

5.2.5 Correlations for j and f vs. Re

Multiple regression, power law correlations have been developed by Wieting [1975] and by Joshi and Webb [1987] to predict the j and f vs. Re_{Dh} characteristics for the OSF array. These correlations were based on data for 21 OSF geometries. Both references present separate correlations for the laminar and nonlaminar wake regions. The Wieting correlation [1975] assumes $Re_{Dh,tr} = 1000$ for all geometries. However, Joshi and Webb [1987] used Equation 5.7 to define the transition region.

Manglik and Bergles [1990] used an asymptotic correlation method developed by Churchill and Usagi [1972] to provide an alternate correlation for the OSF. This method requires that one know the asymptotic values for small and high Reynolds number values. The approach involves developing equations (or correlations) that predict the j and f factors in the laminar and nonlaminar (or the so-called turbulent) regions. Let the laminar and turbulent region friction correlations be described as f_{lam} and f_{turb} (and j_{lam} and j_{turb}), respectively. Then, the equations for the laminar and turbulent regions are patched together to provide a smooth equation for the entire laminar and turbulent regions, which have the form

$$f = (f_{lam}^n + f_{turb}^n)^{1/n} \quad \text{and} \quad j = (j_{lam}^n + j_{turb}^n)^{1/n} \tag{5.8}$$

The n exponent in each equation is empirically chosen to give the best fit of the data set. Notice that as $Re_{Dh} \to 0$, Equations 5.8 approach f_{lam} (or j_{lam}) and for $Re_{Dh} \to \infty$, Equations 5.8 approach f_{turb} (or j_{turb}).

Webb and Joshi [1983] used the Churchill and Usagi [1972] method to correlate their friction data on eight scaled-up OSF model geometries. They correlated 99% of their data within ±15%. Usami [1991] performed a very comprehensive test program using 48 scaled-up OSF geometries. He varied the fin thickness (t), louver pitch (L_p), fin pitch (p_f), and the number of louvers in the flow depth. Some flow visualization was done and friction data were taken for all geometries.

Manglik and Bergles developed a correlation based on Equation 5.8. They first developed linear multiple regression correlations for the laminar and turbulent regimes. They used Equation 5.7 to define the upper limit of the laminar correlation. Their correlation is based on 18 of the 21 surface geometries in the Joshi and Webb [1987] data set. Substituting their correlations for f_{lam} and f_{turb} in Equation 5.8, they obtained the final friction correlation given by

$$f = 9.624 \, \mathrm{Re}_{Dh}^{-0.742} \left(\frac{s}{h}\right)^{-0.186} \left(\frac{t}{L_p}\right)^{0.305} \left(\frac{t}{s}\right)^{-0.266}$$

$$\left[1 + 7.669E - 8 \, \mathrm{Re}_{Dh}^{4.43} \left(\frac{s}{h}\right)^{0.92} \left(\frac{t}{L_p}\right)^{3.77} \left(\frac{t}{s}\right)^{0.236}\right]^{0.1}$$

(5.9)

A similar procedure was used to obtain the final j factor correlation, which is given by

$$j = 0.652 \, \mathrm{Re}_{Dh}^{-0.540} \left(\frac{s}{h}\right)^{-0.154} \left(\frac{t}{L_p}\right)^{0.150} \left(\frac{t}{s}\right)^{-0.068}$$

$$\left[1 + 5.269E - 5 \, \mathrm{Re}_{Dh}^{1.34} \left(\frac{s}{h}\right)^{0.504} \left(\frac{t}{L_p}\right)^{0.456} \left(\frac{t}{s}\right)^{-1.06}\right]^{0.1}$$

(5.10)

The exponent n in Equations 5.8 was chosen as 0.1 for both the f and j correlations, based on best fit of the data. The correlations predict all of the heat transfer data and approximately 90% of the friction data within ±20%. The hydraulic diameter definition applicable to Equations 5.9 and 5.10 is defined as

$$D_h = \frac{4sbL_p}{[2(sL_p + bL_p + tb) + ts]}$$

(5.11)

Muzychka and Yovanovich [2001a] also used Equations 5.8 to develop a correlation for the OSF. They used the full 21 core database of Joshi and Webb [1987] to validate their model. Their equations for f and j in the laminar and turbulent regions use analytically based equations. They compare their correlation with that of Manglik and Bergles [1990] and show that it is in good agreement. However, because the laminar region model is analytically based, they suggest that it may be more accurate for very low Reynolds numbers, below the range of the data in the 21 core data set. Using their equations to correlate the 21 core database resulted in selection of the n exponent in Equation 5.8 to be $n = 3.0$ for the friction factor and

3.5 for the j factor. The j and f factor data in the 21 core database were correlated within an rms error of 16.8% for the f factor and 14.7% for the j factor. Although typical OSF arrays have a rectangular channel shape, the model may be applied to other channel shapes (e.g., sinusoidal or trapezoidal) by using the appropriate f_{fd} and St_{fd} for the shape of interest.

5.2.6 Use of OSF with Liquids

The previously discussed correlations or models were developed based on air data (Pr = 0.7). Hu and Herold [1995a] obtained heat transfer and pressure drop data on seven OSF geometries using water and polyalphaolefin, for which Prandtl number ranged from 3 to 150. Comparison with the Wieting correlation [1975] and the Josh and Webb analytical model [1987] revealed that the Hu and Harold data are highly overpredicted by the correlations. Hu and Harold [1995b] extended the Joshi and Webb model by incorporating the Montgomery and Wilbulswas [1967] thermal developing solution to account for the Prandtl number effect. The modified model then predicted their data within ±20%.

Brazed aluminum automotive oil coolers frequently use oil channels containing the OSF to provide enhancement of the low-Reynolds-number oil. Muzychka and Yovanovich [2001b] developed a variant of their Muzychka and Yovanovich [2001a] correlation applicable to oil coolers. They also took j and f vs. Re data for a 50/50 glycol/water mixture on 10 turbulator insert geometries, on which the correlation is based. Their data spanned $20 < Re_{Dh} < 200$. Data were taken at 2 fluid temperatures, which yielded Pr = 85 and 150. It is interesting to note that their correlation shows Nu $\propto$ Pr$^{1/3}$. This work has singular importance in that it is one of the very few data sets that have been taken on the OSF geometry for liquids, and at high Prandtl number.

5.2.7 Effect of Percent Fin Offset

As conceived, the strips of the OSF fin are offset 1/2 fin spacing, or 50% offset. In this configuration, the wake length is equal to the strip width in the flow direction. However, this may not be the optimum displacement.

It is typically assumed that the thermal and velocity boundary layers are fully dissipated in the wake region. However, there is no rational basis for assuming that full dissipation will occur for a wake length equal to the strip width. Kurosaki et al. [1988] have investigated the benefits of offsetting the strips 30% of the fin pitch, rather than the usual 50%. This provides a wake length of three strip widths. Their work was performed using an array having a total of five rows of strips in the flow direction. They measured the average Nusselt number on each row of strips. Figure 5.9a shows their results for the standard OSF arrangement. This figure shows that the Nu of row 1 equals that of row 2, and the Nu of row 3 equals that of row 4. Row 5 has a smaller Nu than that of rows 3 or 4. Figure 5.9b shows the results for row 3 offset 30% of the fin pitch. In this array, the wake length between rows 1 and 5 is 3 strip widths, rather than 1 strip width. Figure 5.9b shows that the Nu for

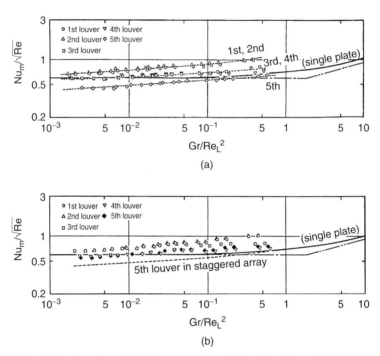

Figure 5.9 Local Nusselt numbers in OSF louver array studied by Kurosaki et al. [1988]. (a) Standard array with 50% offset in rows 2 and 4. (b) Array with 50% offset in rows 2 and 4, and 30% offset in row 3.

row 5 is 10 to 20% higher than for row 5 of the Figure 5.9a array. Hatada and Senshu [1984] have also studied the effects of fin offset for the OSF. Their heat transfer measurements are discussed in Section 5.4.

5.2.8 Effect of Burred Edges

There is a possibility that burrs may be formed on the upstream and downstream fin edges during the shearing operation to make the OSF geometry. This is especially true as the tooling used to shear the strips wears. The existence of such burrs was not established by the original experimenters for the 21 cores in the database. To investigate this possibility, Webb and Joshi [1983] used their asymptotic correlation for the 8 scaled-up, "burr-free" geometries to predict the friction factor of the 21 actual core database. The friction factors of the actual cores were under predicted 10 to 20%. Hence, it appears that the some degree of burrs existed in the 21 actual cores.

5.3 LOUVER FIN

The louvered fin geometry of Figure 5.2f bears a close similarity to the OSF. Rather than offsetting the slit strips, the entire slit fin is rotated 20 to 45°, relative to the airflow direction. The louvered surface is the standard geometry for air-cooled automotive heat exchangers. Currently used louver fin geometries have a louver strip width (in the airflow direction) of 0.9 to 1.5 mm. For the same strip width, the louver fin geometry provides heat transfer coefficients comparable to the OSF. Shah and Webb [1982] provide description of modified louver fin geometries and give references to several data sources. Davenport [1983a, 1983b] provides data on 32 1-row cores. Davenport systematically varied the louver dimensions for two louver heights (12.7 and 7.8 mm). Sunden and Svantesson [1990] investigate louvers that are not oriented 90° to the airflow and find they give lower j factors than the standard 90° orientation. Aoki et al. [1989] and Fujikake et al. [1983] measured the effect of louver angle and louver pitch in specially instrumented laboratory test sections. Kajino and Hiramatsu [1987] and Achaichia and Cowell [1988] provide numerical simulations for laminar flow in a variety of louver arrays. Suga and Aoki [1991] performed an extensive numerical study to determine the optimum value of p_f/L_p for fixed louver angle. Their analysis was performed for louver angles of 20, 26, and 30° and $p_f/L_p \leq 1.125$. They concluded that the optimum design is given by $p_f/L_p = 1.5 \tan \theta$. The optimum designs provide a good balance between high heat transfer performance and pressure drop. Chang and Wang [1996] tested 27 1-row cores with $1.8 \leq p_f \leq 2.2$ mm, $1.32 \leq L_p \leq 1.86$ mm. The louver angle was fixed at 28°.

5.3.1 Heat Transfer and Friction Correlations

Davenport [1983b] tested 32 1-row louver fin geometries (Figure 5.10) and developed multiple regression correlations for the j and f vs. Re_{Dh}. Their correlations are given below.

$$j = 0.249 \, Re_L^{-0.42} \, L_h^{0.33} \, H^{0.26} \left(\frac{L_L}{H} \right)^{1.1} \quad (300 < Re_{Dh} < 4000) \qquad (5.12)$$

$$4f = 5.47 \, Re_L^{-0.72} \, L_h^{0.37} L_p^{0.2} H^{0.23} \left(\frac{L_L}{H} \right)^{0.89} \quad (70 < Re_{Dh} < 1000) \qquad (5.13)$$

$$4f = 0.494 \, Re_L^{-0.39} \, H^{0.46} \left(\frac{L_h}{L_p} \right)^{0.33} \left(\frac{L_p}{H} \right)^{1.1} \quad (1000 < Re_{Dh} < 4000) \qquad (5.14)$$

Equations 5.12 through 5.14 contain dimensional terms; the required dimensions are millimeters. Figure 5.10 illustrates a one-row core and Figure 5.10b defines the terms in the equations. The correlation contains the louver height ($2L_h = L_p \tan \theta$) rather than the louver angle (θ). Notice that the louver length (L_L) is shorter than the fin height (H), which is typical for the louver fin geometry. The characteristic dimension in the Equation 5.13 to Equation 5.14 Reynolds number limits is D_h not L_L as used in the equations. The author has used Equations 5.12 to 5.14 to predict the heat transfer performance of automotive radiators and has found agreement within ±15% of the test data. The Davenport correlation is somewhat flawed, because it does not fully account for all the dimensional variables in the louver fin geometry. Figure 5.10d shows a precise depiction of the louver fin cross section. Dimension S_1 defines the length of the inlet louver, and dimension S_2 the length of the flow redirection louver. The Davenport correlation does not contain dimensions S_1 and S_2, nor does it contain the number of louvers. Dillen and Webb [1993] have developed a rationally based correlation, based on the analytical model of Sahnoun and Webb [1992], which includes the S_1, S_2 dimensions, and the number of louvers.

The state-of the-art correlation is Chang and Wang [1997] for heat transfer and Chang et al. [2000] for friction. They developed heat transfer and friction correlations based on 91 samples of louver fin heat exchangers. The database includes Davenport [1983b], Tanaka et al. [1984], Achaichia and Cowell [1988], Webb [1988], Sunden and Svantesson [1992], Webb and Jung [1992], Rugh et al. [1992], and Chang and

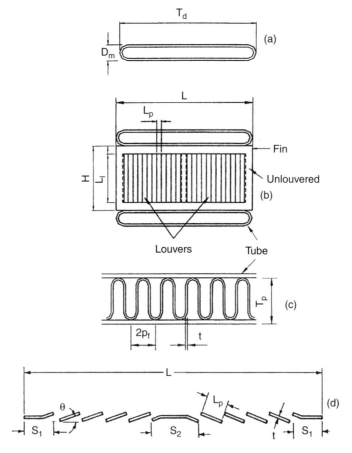

Figure 5.10 Definition of geometric parameters of the louver-fin heat exchanger. (a) Tube cross section, (b) tube-fin array, (c) end view of fins, (d) cross section of louver region. (From Chang and Wang [1997]. Reprinted with permission from Elsevier.)

Wang [1996]. Note that plate fin-and-tube louver fin geometry is included in the database. Their correlations are given below.

$$j = \mathrm{Re}_L^{-0.49}\left(\frac{\theta}{90}\right)^{0.27}\left(\frac{p_f}{L_p}\right)^{-0.14}\left(\frac{H}{L_p}\right)^{-0.29}\left(\frac{T_d}{L_p}\right)^{-0.23}\left(\frac{L_L}{L_p}\right)^{0.68}\left(\frac{T_p}{L_p}\right)^{-0.28}\left(\frac{t}{L_p}\right)^{-0.05} \quad (5.15)$$

$$f = f1*f2*f3 \tag{5.16}$$

$$f1 = 14.39\,\mathrm{Re}_L^{(-0.805\,Pf/H)}[\ln(1.0+(p_f/L_p))]^{3.04} \qquad \mathrm{Re}_L \le 150 \tag{5.17}$$

$$f1 = 4.97\,\mathrm{Re}_L^{(0.605-1.064/\theta^{0.2})}[\ln((t/p_f)^{0.5}+0.9)]^{-0.527} \qquad 150 < \mathrm{Re}_L < 5000 \tag{5.18}$$

$$f2 = [\ln((t/p_f)^{0.48}+0.9)]^{-1.435}(D_h/L_p)^{-3.01}[\ln(0.5\,\mathrm{Re}_L)]^{-3.01} \qquad \mathrm{Re}_L \le 150 \tag{5.19}$$

$$f2 = [(D_h/L_p)\ln(0.3\,\mathrm{Re}_L)]^{-2.966}(p_f/L_L)^{-0.793(T_p/H)} \qquad 150 < \mathrm{Re}_L < 5000 \tag{5.20}$$

$$f3 = (p_f/l_L)^{-0.308}(L/L_L)^{-0.308}(e^{-0.117T_p/D_m})\theta^{0.35} \qquad \mathrm{Re}_L \le 150 \tag{5.21}$$

$$f3 = (T_p/D_m)^{-0.0446}\ln(1.2+(L_p/p_f)^{1.4})^{-3.553}\theta^{-0.447} \qquad 150 < \mathrm{Re}_L < 5000 \tag{5.22}$$

The geometric parameters used in the above correlations are described in Figure 5.10. The heat transfer correlation predicts 89% of the data within ±15%, with 7.5% mean deviation. Inclusion of plate-and-fin louver fin data results in a mean deviation of 8.2%. The friction correlation predicts 83.1% of data within ±15% with a mean deviation of 9.2%. Note that the correlation is applicable only to plate-fin geometries having inline tube arrangements. No correlations are known for plate-fin geometries having staggered tube arrangements. However, this geometry is commercially used.

5.3.2 Flow Structure in the Louver Fin Array

Although louvered surfaces have been in existence since the 1950s, it has only been within the past 20 years that serious attempts have been made to understand the flow phenomena and performance characteristics of the louvered fin. Until the flow visualization studies of Davenport [1980], it was assumed that the flow is parallel to the louvers. At very low values of Re_L, Davenport observed that the main flow stream did not pass through the louvers. But, at high values of Re_L, the flow became nearly parallel to the louvers. Davenport speculated that, at low air velocities, the developing boundary layers on adjacent louvers became thick enough to effectively block the passage, resulting in nearly axial flow through the array. The Davenport results are also discussed by Achaichia and Cowell [1988].

Figure 5.11 shows the Stanton number data of Achaichia and Cowell [1988] for three louver fin geometries used on plate fins (see Figure 6.23) with 11-mm tube pitch. At the highest Re_L, the data are parallel (but lower than) those for laminar, boundary flow over a flat plate (the "flat plate" line). At low Re_L, the data show characteristics similar to laminar, duct flow. The "duct flow" line is for fully developed laminar flow in a rectangular channel with a 0.30 aspect ratio. The aspect ratio for the finned tubes is between 0.25 and 0.37. This behavior is consistent with Davenport's flow observations. Since the heat transfer performance is closely related to the flow structure, we may infer that two types of flow structure exist within the louvered, plate fin array:

1. Duct flow, in which the fluid travels axially through the array, essentially bypassing the louvers
2. Boundary layer flow, in which the fluid travels parallel to the louvers

Kajino and Hiramatsu [1987] conducted a flow visualization study of the louvered fin array using dye injection and hydrogen bubble techniques. Figure 5.12 shows two louver arrays, which have the same louver pitch, but different fin pitches. In the left-hand portion of Figure 5.12a, a significant fraction of the flow bypasses the louvers. This is because the hydraulic resistance of the duct flow region is substantially smaller than that for boundary layer flow across the louvers. When the

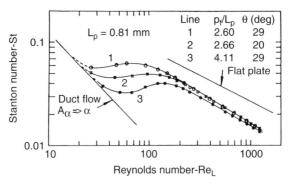

Figure 5.11 Heat transfer characteristics of several louvered plate fin geometries tested by Achiachia and Cowell [1988].

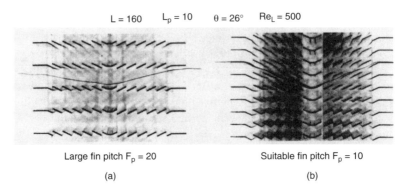

Figure 5.12 Steaklines at $Re_L = 500$ in 160 mm deep flow visualization model of louver fin array ($l = 10$ mm, $\theta = 26°$) tested by Kajino and Hiramatsu [1987]. (a) $p_f = 20$ mm, (b) $p_f = 10$ mm.

fin pitch is reduced, as in the right-hand part of Figure 5.12b, the hydraulic resistance of the duct is increased, so that most of the flow passes through the louvers.

Webb and Trauger [1991] conducted detailed flow visualization experiments, using large-scale models, to accurately simulate the flow in louver fin arrays. The louvered arrays selected permitted study of the effect of louver angle (θ) and the louver to fin pitch ratio (L_p/p_f) on the flow structure within the array. They defined the term flow efficiency (η) to quantify their observations. Figure 5.13 provides a visual definition of the flow efficiency. The flow efficiency (η) is defined using Figure 5.13 as

$$\eta = \frac{N}{D} = \frac{\text{Actual transverse distance}}{\text{Ideal transverse distance}} \tag{5.23}$$

where $D = (L/2 - S_1) \tan \theta$. The flow efficiency is equal to 1 when the flow is parallel to and through the louvers. It is equal to 0 when the flow is axial through the array (100% duct flow). Data were taken for six louver-to-fin pitch ratios ($L_p/p_f = 0.49$, 0.56, 0.66, 0.79, 0.98, and 1.31) and two louver angles ($\theta = 20$ and $30°$). Data were taken over a Reynolds number range of $400 < \text{Re}_L < 4000$. Figure 5.14 shows the flow efficiency for $20°$ louver angle. Figure 5.14 shows that the flow efficiency increases with increasing Re_L. This occurs up to a particular value of Reynolds number, defined as the critical Reynolds number, Re_L^*. Above Re_L^*, the flow efficiency becomes independent of Reynolds number for fixed L_p/p_f. The flow efficiency is given by the following equation developed by Sahnoun and Webb [1992]:

$$\eta = 0.95 \left(\frac{L_p}{p_f} \right)^{0.23} \qquad (\text{Re}_L > \text{Re}_L^*) \tag{5.24}$$

$$\eta = \eta^* - 37.17 \times 10^{-6} (\text{Re}_L^* - \text{Re}_L)^{1.1} \left(\frac{L_p}{p_f} \right)^{-1.35} \left(\frac{\theta}{90} \right)^{-0.61} \qquad (\text{Re}_L < \text{Re}_L^*) \tag{5.25}$$

where $\eta = \eta^*$ at $\text{Re}_L = \text{Re}_L^*$ and $\text{Re}_L^* = 828(\theta/90)^{-0.34}$ with θ in degrees.

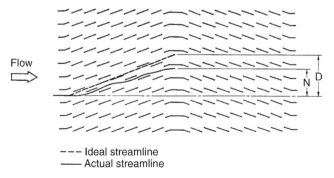

Flow

- - - Ideal streamline
——— Actual streamline

Figure 5.13 Definition of method used to define the flow efficiency used by Webb and Trauger [1991].

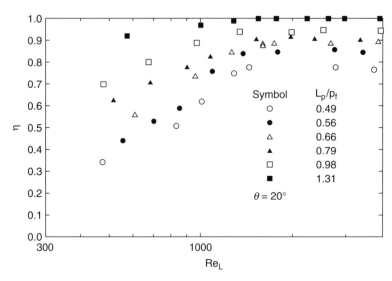

Figure 5.14 Measured flow efficiency vs. Reynolds number for 20° louver angle as reported by Webb and Trauger [1991].

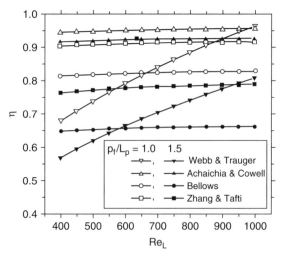

Figure 5.15 The flow efficiency predicted by various correlations. Predictions were made at $p_f = 1.0$ and 1.5 mm, $L_p = 1.0$ mm, $t = 0.1$ mm, and $\theta = 30°$.

The literature reveals three other flow efficiency correlations. The Bellows [1996] correlation is based on flow visualization experiments for 3 samples ($18 \leq \theta \leq 28°$ and $1.09 \leq p_f/L_p \leq 2.2$) having 12 spanwise louver rows. The Webb and Trauger [1991] sample had five spanwise louver rows. Achaichia and Cowell [1988] and Zhang and Tafti [2003] developed correlations based on two-dimensional numerical simulation. The Achaichia and Cowell [1988] simulation assumed zero thickness louvers. The Zhang and Tafti [2003] numerical simulation included fin thickness. The predicted flow efficiencies are shown in Figure 5.15 for 30° louver angle, $p_f = 1.0$ and 1.5 mm and $L_p = 1.0$ mm. The figure shows that correlations based on numerical simulation (Achaichia and Cowell, and Zhang and Tafti) predict higher flow efficiencies compared with those from the experimentally based correlations. The Sahnoun and Webb [1992] and Bellows [1996] correlation predicts approximately equal flow efficiency at $Re_L \approx 600$. The trends are, however, quite different.

Aoki et al. [1989] measured the heat transfer coefficient on individual louver fins using the two-dimensional array illustrated in Figure 5.16a. The louvers were 0.080 mm thick and coated with an electrical insulation layer (0.005 mm Al_2O_3) on which a 0.001-mm-thick layer of nickel was deposited. Electric current in the 0.001-mm nickel coating provided a constant heat flux boundary condition. The heat transfer data were taken with six layers of the Figure 5.16a assembly, which is shown in Figure 5.16b. The data points on Figure 5.16c show the average Nusselt numbers obtained for three louver pitches. The equation for the curve fit of the data is shown on Figure 5.16c. Note that the velocity in the Reynolds number is the local velocity over the louvers ($u/\cos\theta$), where θ is the louver angle. The Pohlhausen solution as

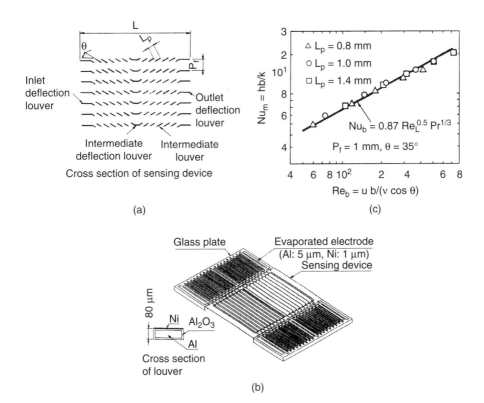

Figure 5.16 Louver fin tests by Aoki et al. [1989]. (a) Louvered heat transfer plate, (b) array of louvered plates tested, (c) average experimental Nusselt numbers.

given by Kays and Crawford [1980] for laminar flow over a flat plate with constant heat flux gives

$$\frac{h_{av}L}{k} = 0.906 \, \mathrm{Re}_L^{1/2} \, \mathrm{Pr}^{1/3} \qquad (5.26)$$

The curve fit of the Figure 5.16c data is 4% lower than the Equation 5.26 theoretical solution. Hence, the average heat transfer coefficient on the louver is well-approximated by the Pohlhausen solution.

The effect of vortex shedding in the louver geometry has been investigated by DeJong and Jacobi [2003]. Similar to the OSF, the shedding starts at the last downstream louver, and moves upstream as the Reynolds number increases. However, the enhancement provided by vortex shedding is not as high for the louver geometry as for the OSF geometry. The critical Reynolds number, where the vortex shedding starts, decreases as the louver angle increases and as p_f/L_p decreases. Their local mass transfer data (22° louver angle and $p_f/L_p = 1.2$) shows that the mass transfer coefficients are approximately constant throughout the array, except near the flow redirection louver. DeJong and Jacobi [2003] attributed this characteristic to the long distance along a streamline from one louver to the next. The small-scale velocity fluctuations generated by the upstream louvers are more likely to die out as the flow passes from louver to louver. For the OSF, the distance between louvers is one louver length (for 50% offset), and the effect of shed vortices is prominent. The effect of the distance between louvers in the flow direction on thermal wakes was also noted by Suga and Aoki [1991] and Lyman et al. [2002]. Greater opportunity for thermal wake dissipation will occur as the wake length increases. However, the heat transfer enhancement by shed vortices will decrease as the wake distance increases. "Good" design should seek the optimum distance between louvers in the flow direction. Suga and Aoki [1991] performed an extensive numerical study to determine the optimum value of p_f/L_p. They concluded that the optimum design is given by $p_f/L_p = 1.5 \tan \theta$. The optimum designs provide a good balance between high heat transfer performance and pressure drop.

5.3.3 Analytical Model for Heat Transfer and Friction

Sahnoun and Webb [1992] developed an analytical model to predict the j and f factor vs. Re_L for the louver fin array. The development is similar to that of the previously discussed Joshi and Webb [1987] model for the OSF array. However, flow in the louver fin is more complex for several reasons. In addition to the internal louvers, Figure 5.10d shows inlet and redirection louvers with louver pitches S_1 and S_2, respectively. The heat transfer coefficient on all louvers is predicted using the Pohlhausen solution for laminar flow over a flat plate (Equation 5.26). As shown on Figure 5.16c, the Pohlhausen solution predicts the heat transfer coefficient on the louvers with good accuracy. The velocity is adjusted to account for the effect of the louver angle ($u/\cos \theta$), and for the flow efficiency factor. Figure 5.10 shows that the louver height (L_h) is less than the fin height (H). The heat transfer coefficient in the unlouvered end regions is predicted using a fully developed laminar flow solution. Similar procedures are used for the friction model. The friction model accounts for pressure drag caused by the finite fin thickness.

The model was validated by predicting the j and f vs. Re data for the 32 one-row cores tested by Davenport [1983b]. The Davenport database are for 2 fin heights — 19 cores with 12.7 mm fin height, and 11 cores with 7.8 mm fin height. At 10.7 m/s air frontal velocity, the model predicted the j factor and f factor within ± 20%.

Dillen and Webb [1994] modified the Sahnoun and Webb [1992] model by empirically accounting for the bypass effect. In the Shanoun and Webb model, the bypass effect was modeled using the flow efficiency equation. Dillen and Webb [1994] showed that their model provided better prediction of the of Davenport [1983b] than was given by the Shanoun and Webb [1992] model. Webb et al. [1995] extended the work of Dillen and Webb [1994] by including the data of Chang and Wang [1996] as well as those of Davernport [1983b].

5.3.4 PEC Example 5.2

Automotive radiators typically use the louver fin geometry, rather than the OSF geometry. Compare the performance of the OSF with the louver fin. Assume the geometric design parameters shown in Table 5.1. The radiator is designed to operate at 27°C with 20 mi/hr (8.94 m/s) air frontal velocity. Compare the j and f factors for the two fin designs at the same air velocity. This corresponds to the simple PEC Case FN-2 of Table 3.1.

The j and f factors are calculated using correlations of Davenport for the louver fin, and Joshi and Webb for the OSF geometries. Use of the correlations gives for the OSF ($j = 0.0124$, $f = 0.0574$) and for the louver fin ($j = 0.0132$, $f = 0.045$).

Using PEC Case FN-2, $A_{osf}/A_{louv} = j_{louv}/j_{osf} = 0.0132/0.0124 = 1.065$. The friction power ratio is proportional to the product, fA for G = constant. Hence, $P_{osf}/P_{louv} = (f_{osf}/f_{louv})(A_{osf}/A_{louv}) = (0.0574/0.045)(1.065) = 1.35$.

Table 5.1 Geometric Design Parameters for PEC Example 5.2

OSF	Louver	Item
472	Same	Fins/m
0.10	Same	Fin thickness (t), mm
7.62	Same	Height of fin array (H), mm
2.03	Same	Width of louver (L_p), mm
NA	26°	Louver angle (for louver fin only)
1.0	0.8	Ratio of louver length to fin height
3.18	Same	Hydraulic diameter (D_h), mm
0.940	Same	Contraction ratio (σ)

NA = not applicable.

This example shows that the louver fin has marginally higher j factor (6.5%, which provides a 6.5% surface area reduction), assuming negligible tube-side thermal resistance. However, the 28% higher friction factor of the OSF, combined with its 6.5% lower j factor, results in 35% higher friction power (and pressure drop).

5.4 CONVEX LOUVER FIN

Hatada and Senshu [1984] have developed a variant of the OSF, which they call the "convex louver fin" geometry. This is a variant of the OSF geometry, which we describe as the offset convex louver fin, and abbreviate as OCLF. Figure 5.17 shows cross-section details of the plate-and-fin geometries they tested, which are numbered

No.	θ (deg)	% Offset
1	0	0
2	12.8	23
3	17.4	33
4	24.6	53
5	9.7	20
6	17.4	33
7	20.7	42
8	24.6	53

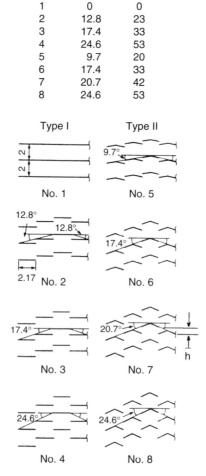

Figure 5.17 Fin geometries tested by Hatada and Senshu [1984].

1 through 8. Hatada and Senshu [1984] describe the OSF and OCLF geometries as Type I and Type II, respectively. All the geometries have 500 fins/m, 2.0 mm fin pitch (p_f), 0.115 mm fin thickness (t), and the louver width (L_p) of arrays 2 through 8 are 2.17 mm. Array 1 has plain fins, and Arrays 2, 3, and 4 are of the OSF type. Array 2 is the "standard" OSF, in which louver is offset 50% of the fin spacing. Arrays 5 through 8 are of convex louver geometry. The several OSF and convex louver geometries differ in the amount of fin offset. Hatada and Senshu use the angle θ shown on Figure 5.17 to define the louver offset. One may translate this to the

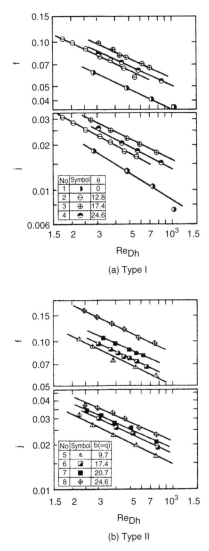

(a) Type I

(b) Type II

Figure 5.18 j and f vs. Re_D for the louver geometries tested by Hatada and Senshu [1984]. (a) OSF (Type I), (b) OCLF (Type II).

physical offset using the table shown on Figure 5.17 using % offset = h/p_f, where h is the offset dimension shown on Figure 5.17 and p_f is the fin pitch.

The heat transfer and friction characteristics of the Figure 5.17 arrays were measured with air flowing in a test section consisting of 38 fin channels, with each plate 43.3 mm wide by 150 mm long. The heat transfer performance was measured using a step-change transient method.

Figure 5.18a shows the effect of the angle θ on the j and f factors for the OSF geometry. This figure shows that the 50% offset (θ = 24.6°) does not provide the highest j factor. At Re_{Dh} = 500, the θ = 17.4° (33% offset) provides a 22% higher j factor than the θ = 24.6° (50% offset) array. This agrees well with the findings of Kurosaki et al. [1988].

For the same angle θ, Figure 5.18b shows that the OCLF provides higher j factor than the OSF geometry. For θ = 24.6°, the j factor of the OCLF array is 48% higher than for the OSF array at Re_{Dh} = 500. However, the friction factor increase is approximately 70%. Figure 5.19 is a plot of j and f vs. θ for the OSF and OCLF geometries for Re_{Dh} = 500. It shows that the maximum values of j and f occur at $\theta \simeq 20°$ for the OSF geometry. The j factor for the OCLF array increases with, at least up to the highest θ tested (24.6°). Figure 5.19 shows that the higher j factor of the OCLF is accompanied by a higher friction factor than in the OSF array. Further, the OCLF friction factor drastically increases above 20° in contrast to the decreasing friction factor of the OSF array. For the same friction power, the j factor of the θ = 20° OCLF geometry is 22% higher than that of the OSF geometry. Figure 5.19 indicates that the j/f ratio of the OCLF array is not favorable for $\theta > 20°$. The most surprising feature of the OCLF array is the small friction increase of the OCLF array, compared to the OSF array, for $\theta \leq 20.7°$. As shown in Figure 5.19, the friction factor for θ = 17.4° in the OCLF geometry is only 7% higher than that of the OSF geometry (θ = 17.4°). For $\theta > 20°$, Figure 5.19 shows that a large friction increase

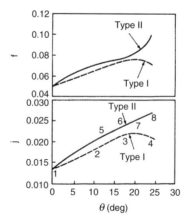

Figure 5.19 Effect of angle θ on j and f of the OSF (Type I) and OCLF (Type II) geometries tested by Hatada and Senshu [1984].

occurs in the OCLF geometry, whereas the friction factor decreases in the OSF geometry.

Hatada and Senshu also performed flow visualization experiments in scaled-up OSF and OCLF arrays, without tubes present. Figure 5.20 shows observed flow patterns for the OSF array ($\theta = 17.4°$) and the OCLF array ($\theta = 17.4°$ and $24.6°$).

No powder region   Powder trace line

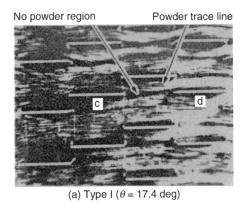

(a) Type I ($\theta = 17.4$ deg)

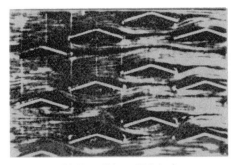

(b) Type II ($\theta = 17.4$ deg)

(c) Type II ($\theta = 24.6$ deg)

Figure 5.20 Streamline flow patterns observed by Hatada and Senshu [1984] in finned arrays without tubes.

Examination of Figure 5.20c suggests that flow separation on the convex louvers causes significant profile drag for the $\theta = 24.6°$ louver angle.

Hitachi [1984] uses the convex louver fin geometry in its commercial plate fin-and-tube heat exchangers. Figure 5.21 is taken from its product brochure.

The results of the Hatada and Senshu study indicate that the convex louver fin may provide higher performance than is yielded by the standard OSF with 50% louver offset. The Hatada and Senshu data suggest that a 23% higher j factor with a 7% friction factor increase is possible. No j and f data have been reported for the OCLF geometry in steady-state tests of plate-and-fin heat exchanger configurations. Hence, confirmation of this speculative possibility must await experimental evaluation. Pauley and Hodgson [1994] conducted flow visualization tests using scaled-up OCLF models ($1.0 \leq p_f/8L_h \leq 3.0$, $\theta = 19°$ and $24°$) in a water tunnel for $400 \leq Re_L \leq 3000$. The mixing efficiency was determined from mixing of the dye, and the largest mixing was obtained $p_f/8L_p = 1.5$.

The local mass transfer characteristics in the OCLF array were investigated by DeJong and Jacobi [1999], who used the naphthalene sublimation technique. The fins had 20° louver angle, 1.6 mm fin thickness, 18.5 mm fin pitch, and a 25.4 mm louver length; thus $p_f/L_p = 0.73$ and $p_f/8L_h = 1.0$. Figure 5.22 shows the average, upstream, and leeward half mass transfer coefficients plotted as a function of Reynolds number. At low Reynolds numbers, the upstream half mass transfer coefficients are higher than those of the leeward half. At high Reynolds numbers, however, the leeward half yields higher mass transfer coefficients, due to the unsteadiness in the shear layer impinging on the leeward surface. Figure 5.22 shows OSF data of the identical geometry for which $\theta = 0$. The mass transfer coefficients of the OSF are lower than those of the convex fin, and the difference increases as the Reynolds number increases. For $1500 \leq Re_{Dh} \leq 3000$, the average Sherwood number of the convex louver fin is 40% higher than that of the OSF. The corresponding increase of the pressure drop is 130%.

Figure 5.21 The OCLF geometry in the Hitachi fin-and-tube heat exchanger. (Courtesy of Hitachi Cable.)

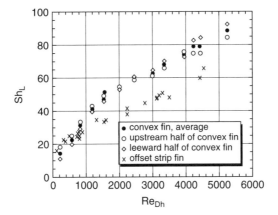

Figure 5.22 Sherwood numbers for OCLF and OSF as reported by DeJong and Jacobi [1999]. (Reprinted with permission of ASME.)

5.5 WAVY FIN

The term wavy or corrugated is used to describe the Figure 5.2c geometry. For a corrugated geometry having constant corrugation angles and sharp wave tips, the key parameters that affect the performance are the wave pitch (p_w), the corrugation angle (θ), and channel spacing (s). Whether the wave geometry has smooth or sharp corners will affect the performance. Goldstein and Sparrow [1977] used a mass transfer technique to measure the local mass transfer coefficient distribution for a herringbone wave configuration. They propose that the enhancement results from Goertler vortices that form as the flow passes over the concave wave surfaces. These are counter-rotating vortices that have a corkscrew-like flow pattern. Flow visualization studies by this author show local zones of flow separation and reattachment on the concave surfaces for a true wavy channel shape. The redeveloping boundary layer from the reattachment point also contributes to heat transfer enhancement. The existence of Goetler vortices in a wavy channel was confirmed from the flow visualization study by Gschwind et al. [1995].

Ali and Ramadhyani [1992] provide a detailed flow visualization study of one corrugated geometry, including local heat transfer and friction factors for $150 \leq \mathrm{Re_{Dh}} \leq 4000$, which is within the typical operating range. Figure 5.23 shows their data with water at Pr = 7. At $\mathrm{Re_{Dh}} = 2000$, the Nusselt number enhancement ratio is 2.3 and 3.2 for the narrow and wide channels, respectively. The corresponding friction increases are 2.3 and 3.8, respectively. The solid line shows the "correlation" of O'Brien and Sparrow [1982], who tested a channel with $\theta = 30°$ and $b/p_w = 0.29$ using water with $4 \leq \mathrm{Pr} \leq 8$. The O'Brien and Sparrow correlation accounts for Pr, but is not expected to account for the effect of θ and b/p_w, as these were not varied in their work. Sparrow and Hossfeld [1984] studied the effect of rounding the corners of the corrugation, as opposed to a true wave shape. Several other studies have been performed, as discussed by Ali and Ramadhyani [1992]; these studies are in the

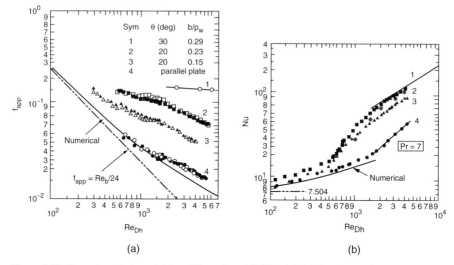

Figure 5.23 Wavy channel data of Ali and Ramadhyani [1992]. (a) Friction factor, (b) Nusselt number for Pr = 7. Curve 1, $\theta = 30°$, $b/p_w = 0.29$, Curve 2, $\theta = 20°$, $b/p_w = 0.23$, Curve 3, $\theta = 20°$, $b/p_w = 0.15$, Curve 4, parallel plate.

$1500 \le Re_{Dh} \le 25,000$ range, and are typically outside the usual heat exchanger operating range. The effect of channel height was investigated by Molki and Yuen [1986] for $4000 \le Re_{Dh} \le 35,000$. Both j and f factors increased as the spacing increased.

Kays and London [1984] provide j and f vs. Re curves for two wavy fin geometries (Figure 5.2c). Their performance is competitive with that of the OSF. No specific correlations exist for the j and f characteristics of wavy or herringbone fins. Two studies, Rosenblad and Kullendorf [1975] and Okada et al. [1972], of wavy channel geometries used in plate-type heat exchangers provide additional data for small-aspect-ratio channels.

5.6 THREE-DIMENSIONAL CORRUGATED FINS

Three-dimensional corrugated channels are used in plate-type heat exchangers and rotary regenerators. Figure 5.39d shows a cutaway view of a corrugated channel. The geometry is specified by p_w (pitch), H (channel height), t (wall thickness), and θ (corrugation angle). Typically, sinusoidal channels are used. The effect of corrugation angle was investigated by a number of investigators (Focke et al. [1985], Stasiek et al. [1996], etc.). Figure 5.24 shows j and f factors (normalized by the $\theta = 30°$ data) obtained by Focke et al. for corrugated channels with $p_w /H = 2$. The electrochemical technique was used to obtain the mass transfer coefficient. The corrugation angle (θ) ranged from 0° (corresponding to parallel straight duct of sinusoidal cross-section) to 90° (corrugations normal to the main flow). Both j and f increase monotonically up to $\theta = 80°$. Beyond 80°, they decrease slightly, and

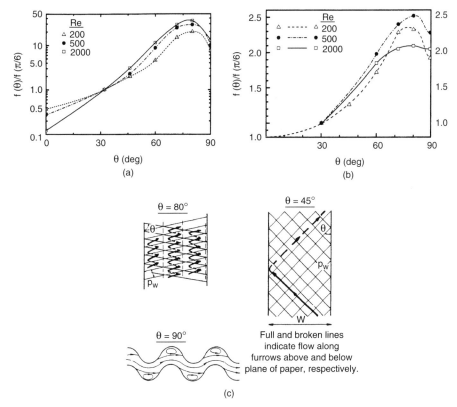

Figure 5.24 The effect of the corrugation angle on the j and f factors (normalized by those of $\theta = 30°$) for corrugated channels with $p_w/H = 2$: (a) friction factor, (b) j factor, (c) sketch of the flow pattern. (From Focke et al. [1985]. Reprinted with permission of Elsevier.)

approach the values of a two-dimensional corrugated channel. The increase is much more pronounced for the friction factors than for the j factors. For θ increasing from 30° to 80°, the friction factors increases 20 to 30 times, while the j factors increase only 2 to 3 times.

The flow patterns observed at the same configuration (Focke and Knibbe [1986]) are shown in Figure 5.24c. At $\theta = 45°$, the fluid flow is predominantly along the furrows. On reaching the plate edge, the fluid streams are reflected and return to the opposite plate edge along the furrows. Focke et al. [1985] suggested similar flow patterns up to $\theta = 60°$. At this angle, they observed two sets of criss-crossing streams, which induce secondary swirling motions. The driving force that produces the swirl in a furrow is the velocity component of the fluid moving along the opposite furrows in a direction perpendicular to the furrow, and is maximum at $\theta = 45°$. For angles below 45°, the interaction between fluid streams is positive, i.e., each of the crossing streams has a velocity component in the same direction as the stream it crosses. For $\theta > 45°$, the interaction is negative, so that cross streams have a retarding effect on each other. Focke et al. speculated that this retarding effect, which increases with increasing θ, eventually leading to a change in flow pattern observed at $\theta = 80°$. At $\theta = 80°$, the reflection occurs between plate contact points, and the flow forms a zigzag pattern. The change of flow pattern causes the decrease of j and f factors above $\theta = 80°$. The complex interactions between fluid streams lead to early transition to turbulence. For $\theta = 80°$, the transition to turbulence was observed at $Re_{Dh} \cong 260$.

Stasiek et al. [1996] investigated the effect of corrugation pitch to height ratio (p_w/H). Local heat transfer coefficients were obtained from liquid crystal thermography. For $2.22 \leq p_w/H \leq 4.0$ the j factor decreased as p_w/H increased. The friction factor increased as p_w/H increased for $2 \leq p_w/H \leq 3$, and remained roughly constant afterward.

Abdel-Kariem and Fletcher [1999] developed j and f correlations, which include the corrugation angle θ. Both laminar ($Re_{Dh} < 500$) and turbulent ($Re_{Dh} \geq 500$) regimes were considered.

For the laminar regime ($Re_{Dh} < 500$):

$$Nu = 0.777 \, Re_{Dh}^{0.444} \, Pr^{0.4} \left(\frac{\theta}{45} \right)^{0.67} \tag{5.27}$$

$$f = 15 \, Re_{Dh}^{-0.3} \left(\frac{\theta}{45} \right)^{2.5} \tag{5.28}$$

For the turbulent regime ($Re_{Dh} \geq 500$):

$$Nu = 0.26 \, Re_{Dh}^{0.67} \, Pr^{0.4} \left(\frac{\theta}{45} \right)^{0.67} \tag{5.29}$$

$$f = 7.3 \, Re_{Dh}^{-0.198} \left(\frac{\theta}{45} \right)^{2.5} \tag{5.30}$$

The correlation predicted the heat transfer data reasonably well except for lower Reynolds numbers, where the deviation was $\pm 30\%$. The deviation was rather large ($\pm 40\%$) for the friction factor.

5.7 PERFORATED FIN

This surface geometry (Figure 5.2e) is made by forming a pattern of spaced holes in the fin material before it is folded to form the U-shaped flow channels. If the porosity of the resulting surface is sufficiently high, enhancement can occur due to boundary layer dissipation in the wake region formed by the holes. Shah [1975] provides a detailed evaluation of the perforated fin based on his study of test data on 68 perforated fin geometries. Shah concludes that little enhancement occurs for $Re_{Dh} < 2000$, if the heat transfer coefficient is based on the plate area before the holes were punched. Moderate enhancement may occur in the transition and turbulent flow regimes, $Re_{Dh} > 2000$, depending on the hole size and the plate porosity.

Shen et al. [1987] tested flat plate-and-fin channels having round holes or rectangular cutouts. They found no benefit in the laminar range, but holes promote earlier transition and they observed moderate j increase in the turbulent range. Substantially higher performance was obtained with the rectangular slots.

The performance of the perforated fin is less than that of a good OSF, and is rarely used today, except as a flow distributor fin. Further, it represents a wasteful way of making an enhanced surface, because the material removed in making the perforated hole is relegated to the scrap barrel. Figure 5.25a shows an interesting variant of the perforated fin, which was tested by Fujii et al. [1988]. This geometry has 2.0-mm-diameter holes in corrugated plates. The plates are aligned, such that the channels have expanding and contracting flow areas. As shown in Figure 5.25b, secondary flow through the perforations is speculated to be an important contributor to the enhancement level. Figure 5.26a shows the geometry of the six surfaces, for which Nu and f vs. Re_{Dh} curves are shown in Figure 5.26b. The fins are 0.5 mm and the fin pitch is 5.5 mm for all geometries. The dimensions of the corrugated plate geometries (NP-1, and P-2, P-2a, P-2b) defined in Figure 5.26a are $l = 15$ mm, $S_1 = 8.6$ m, $S_2 = 2.4$ mm, and 0.145 porosity (fractional hole area of fin plate). The performance of the different geometry variants, relative to the plain fin (NP-1) are (1) perforations alone (P-1) provide no enhancement over NP-1, (2) the NP-2 geometry, without perforations, provides $j/j_p = 1.90$ and $f/f_p = 4.7$, (3) the corrugated/perforated P-2 geometry provides an impressive $j/j_p = 2.7$ and $f/f_p = 6.1$, and (4) the P-2a and P-2b geometries provide performance intermediate between P-2 and NP-2.

Although the P-2 geometry provides high heat transfer enhancement, the resulting efficiency index ($\eta = 0.44$) is small compared to other high-performance enhanced surfaces. The authors provide data on ten other geometry variants. However, these variants provide negligible performance advantage over P-2. They also provide empirical Nu and f correlations for the various geometries.

Fujii et al. [1991] also tested the Figure 5.26 fin geometry as a plate fin-and-tube heat exchanger. These data are discussed in Section 6.5.5.

The Figure 5.26 concept has been applied to a wavy fin by Xuan et al. [2001]. The sample had 2.8-mm fin pitch, 15° corrugation angle, and 10% perforation ratio (perforated area-to-total area). Their data were compared with those of a wavy surface (surface 11.5-3/8W with $p_f = 2.2$ mm, $\theta = 22.6°$) in Kays and London [1984]. The wavy perforated surface provided 20 to 40% higher heat transfer coefficient. The friction factors were approximately the same at low Reynolds numbers. For $Re_{Dh} > 2000$, however, the friction factor of the wavy perforated fin was 15 to 40% higher.

5.8 PIN FINS AND WIRE MESH

A pin fin surface geometry may be made of a special weave of "screen wire." Forming this screening into the shape of U-shaped channels results in the pin fin surface geometry. The wire may have a round, elliptical, or square cross-section

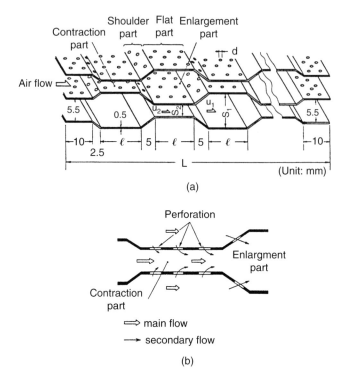

Figure 5.25 (a) Perforated plate fin geometry tested by Fujii et al. [1988], (b) illustration of secondary flows through perforations.

shape. Although such pin fin geometries may have high performance, they are not widely used, because the cost of such surfaces is significantly higher that the cost of the thin sheet used to make geometries such as the OSF and louver fins. Performance data (j and f vs. Re) on five round pin geometries are given by Kays and London [1984] and on nine square pin geometries by Theoclitus [1966]. The Theoclitus data were for inline arrangements ($2 \leq S_t/d \leq 4$) having fin height (e) to pin diameter (d) ratios of $2 \leq e/d \leq 12$. Theoclitus [1966] showed that performance of the square pin fin geometries were approximately predicted using correlations for tube banks with round tubes. The measured heat transfer coefficients were typically 20% below the predicted values. This prediction is reasonably good considering that the pin fins were of square cross section, and that end effects existed on the short e/d pin lengths.

Hamaguchi et al. [1983] tested a regenerator matrix made of multiple layers of woven screen wire made with circular wires. The flow was normal to the wires. They tested a variety of matrix configurations having wire diameters between 0.04 and 0.5 mm, with 20 to 80 screen layers. Torikoshi and Kawabata [1989] developed a "wire mesh" fin geometry. The matrix is better described as "expanded metal mesh." This was made of 0.2-mm-thick sheet, in which short, parallel slits are made. The sheet is then pulled in one direction, and the slits open to form the expanded

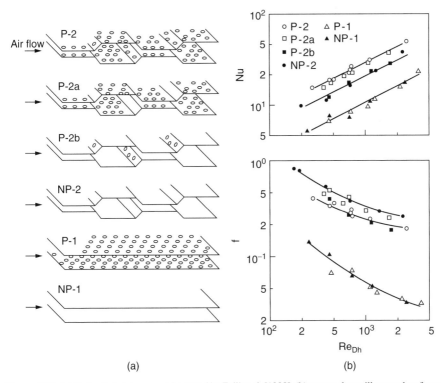

Figure 5.26 (a) Perforated plate geometries tested by Fujii et al. [1988], (b) test results on illustrated surfaces.

metal mesh fin shown in Figure 5.27. Boundary layers form on the square wires of thickness d and dissipation occurs in the void region between the wire attachment points. By increasing the slit length l, the porosity of the matrix is increased (dimension m). It would be possible to corrugate this matrix material to form the packing of a plate-and-fin heat exchanger. However, their tests were performed by compressing layers of the expanded metal to conform to the tube shape, and soldering it to parallel tubes, as shown in Figure 5.27. For 6 layers of mesh having dimensions d = 0.2 mm, m = 2.3 mm, l = 5.0 mm joined to 4-mm-diameter tubes at 24-mm pitch, their test results are represented by

$$\frac{hd}{k} = 0.84 \left(\frac{dG}{\mu} \right)^{0.5} \tag{5.31}$$

Hamaguchi's woven screen wire matrix provided approximately 20% higher heat transfer coefficient than given by Equation 5.31. Note that the Figure 5.27 matrix will have less surface area per unit volume (A/V) than the OSF, or the louver fin array, for the same fin pitch. Extending the work of Torikoshi and Kawabata [1989], Ebisu [1999] investigated the effect of offsetting the fin array and the tube arrangement, which is discussed in Section 6.5.6.

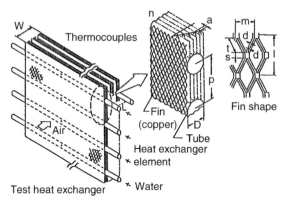

Figure 5.27 Expanded metal (copper) matrix geometry tested by Torikoshi and Kawabata [1989].

5.9 VORTEX GENERATORS

Streamwise vortices will be shed from geometric shapes attached to the wall, which attack the flow at an angle. Figure 5.28 from Fiebig et al. [1993] illustrates several types of vortex generators that have been investigated. Work has been done on vortex generators for plate-and-fin and finned tube geometries. This section discusses fundamental aspects of vortex generators and vortex generators applied to the plate-and-fin geometry.

5.9.1 Types of Vortex Generators

A system of vortices forms on the protrusion, bends around it, and is carried downstream in a longitudinal vortex pattern. Longitudinal vortices were found by Eibeck and Eaton [1987] to persist for more than 100 protrusion heights downstream. To be a practical vortex generator, the protrusion height should be comparable to the local boundary layer thickness. If it is higher, it will cause a significantly higher pressure drop increase than the heat transfer enhancement. As illustrated in Figure 5.28b, vortex generators may be arranged in pairs forming V and Λ in an alternate manner. The interaction of adjacent vortices can be divided into two-types: "common flow down" and "common flow up," as described by Mehta et al. [1983]. When the direction of the secondary flow between two counter-rotating vortices is toward the wall, the vortices are called common flow-down, and when the direction is away from the wall, they are called common flow-up. The Λ pairing of vortex generators produces common flow-down vortices and V pairing produces common flow-up vortices. The wing-type vortex generators in Figure 5.28a will generate common-flow-down vortices.

Figure 5.29 shows a sketch of longitudinal vortices behind a vortex generator placed in a laminar boundary layer on a flat plate (Torii et al. [1994]). The main vortex is formed by the flow separation on the leading edge of the wing, while the corner vortex is formed by deformation of near-wall vortex lines at the

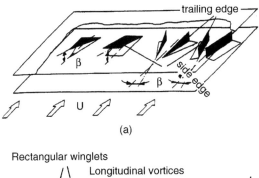

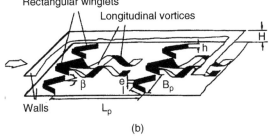

Figure 5.28 Various types of vortex generators. (a) Wing-type vortex generators: from left, delta wing, rectangular wing, delta winglet, and rectangular winglet pair; (b) illustration of vortices and vortex renewal. (From Fiebig et al. [1993].)

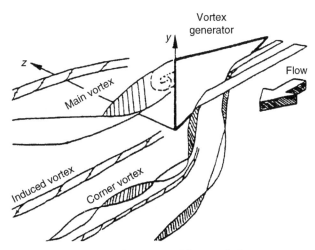

Figure 5.29 A sketch of longitudinal vortices generated from a winglet vortex generator. (From Torii et al. [1994]. Reprinted with permission from Elsevier.)

pressure-side of the wing. Sometimes, an induced vortex rotating opposite to the main and corner vortices are observed.

Figure 5.30a shows a contour plot of the streamwise velocity for the common-flow-down case (Pauley and Eaton [1988]). Two delta winglet vortex generators having 18° angle of attack and 4-cm spacing were used to generate longitudinal

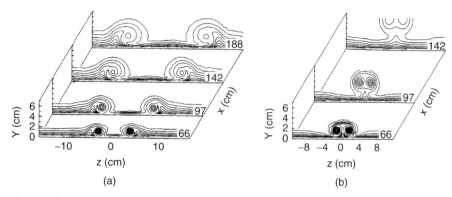

Figure 5.30 Streamwise veloctiy contours downstream of two delta winglet vortex generators with (a) common-flow-down and (b) common-flow-up configuration. Two delta winglet pairs have 18° angle of attack and 40-mm spacing. (From Pauley and Eaton [1988]).

vortices. The vortex pair moves apart as it develops, producing an ever-widening region of boundary layer thinning. The boundary layer at the upwash side is thick. The contour plot of streamwise velocity for the common-flow-up case is shown in Figure 5.30b. The vortices move toward the centerline, and convect each other away from the wall by a strong interaction of vortices. The boundary layer is thin at the downwash side.

5.9.2 Vortex Generators on a Plate-Fin Surface

Vortex generators mounted in a laminar flow channel significantly enhance the heat transfer with managable pressure loss. The level of heat transfer enhancement and pressure loss depends on the vortex generator geometry (size, shape, angle of attack, aspect ratio, etc.) and the Reynolds number. Rigorous research efforts by Fiebig and co-workers yielded the following conclusions (Fiebig [1995]). In their investigations, local heat transfer coefficients were obtained from liquid crystal thermography, and the pressure drops were obtained from drag force measurements. The heights of the vortex generators were comparable with the channel height. Their observations are listed below:

1. Vortex generators generate appreciable heat transfer enhancement (on the average better than 30%) over an area several hundred times the vortex generator area.
2. The two most important parameters, that influence heat transfer and pressure drop characteristics are the angle of attack and vortex generator to primary surface area ratio (A_{VG}/A). Delta forms are more effective than rectangular forms.
3. Winglets provide higher heat transfer and pressure loss than wings (configuration illustrated in Figure 5.28a).
4. Heat transfer increases with angle of attack up to maximum angle, which depends on the form of vortex generator, A_{VG}/A, and Re.

5. Pressure loss increase is mainly due to form drag of the vortex generators. Pressure loss increases monotonically with angle of attack.
6. Both heat transfer and pressure loss increase with Reynolds number and A_{VG}/A.
7. Transition to turbulence occurs at smaller Reynolds number than in plain channel flow.
8. Vortex "breakdown", frequently observed from wings in external flow, does not occur in constricted internal flow.

In the plate-fin geometry, vortex generators are arranged in a multirow configuration. The effect of longitudinal spacing between rows was investigated by Tiggelbeck et al. [1994] for a two-row configuration. Maximum enhancement was found for a row spacing from seven to ten channel heights. In their investigation, the delta winglet pair was used, whose height was the same as the channel spacing, an aspect ratio of 2, the distance between tips of winglet pairs 1/5 of the channel height, and 45° angle of attack, which was an approximate optimum configuration for a single row arrangement.

Brockmeier et al. [1993] performed a numerical simulation on the delta winglet vortex generator applied to a plate-fin geometry. Their analysis was performed for a winglet geometry having an isosceles triangular shape (height/base = 2) at a 30° of attack. The height of the winglet was the same as channel height. The winglet had a nondimensional lateral spacing of 2.0 (based on channel height) and a nondimensional axial pitch of 10. The numerical predictions, however, were not compared with experimental data. They compared the performance of the delta winglet vortex generator with several compact plate-fin geometries, including plain fins, louvered fins, and OSFs. The performance data of these fins were obtained from Kays and London [1984]. One of these was an OSF having 2.1-mm fin depth, 2.38-mm louver pitch, 12.3-mm fin height, and 3.41-mm hydraulic diameter. Figure 5.31 compares the performance of the geometries. The vortex generator geometry yields its best performance at low Reynolds number. At Reynolds number of approximately 700, the vortex generator and the OSF have approximately the same friction factor. However, the vortex generator surface has a j factor approximately 50% higher than the OSF surface. For the same hydraulic diameter and mass flow rates, both geometries will have the same frontal area and area/volume ratio. The vortex generator surface will require only 68% as much surface area for the same heat transfer and pressure drop at $Re_{Dh} = 700$. At the highest Reynolds number ($Re_{Dh} = 3000$), the vortex generator geometry still provides superior heat transfer, but the j/f ratio is a little lower than the OSF geometry. They also did a performance comparison analysis to compare the performance of vortex generators with typically used enhanced plate-and-fin heat exchanger geometries. They used VG-1 of Table 3.1 for the comparison. This VG-1 criterion fixes the flow rate, the heat transfer rate (Q), and the pumping power, and calculates the area ratio needed to transfer the specified Q. The heights of the vortex generators tested by Fiebig and co-workers were comparable with the channel height. Vortex generators have also been applied to externally finned tubes, which are discussed in Sections 6.5.7 and 6.7.4.

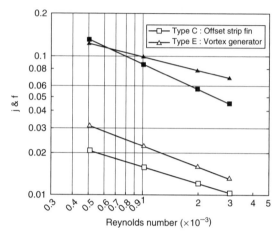

Figure 5.31 The j and f factors of a channel mounted with vortex generators (from numerical study) compared with OSFs. (From Kays and London [1984], as reported by Brockmeier et al. [1993].)

5.10 METAL FOAM FIN

Metallic foam has high surface-area-to-volume ratio, and the flow mixing is enhanced due to the tortuous passage. The pressure drop, however, is larger compared with other enhanced fin geometries, because of the typically small hydraulic diameter. Figure 5.32 shows an enlarged photo of an aluminum foam having 5.0 pores/in. (PPI). Any metal can be foamed. Use of metallic foams in a plate-fin heat exchanger has been studied by Kim et al. [2000]. The effects of the pore density (PPI) and porosity (ϵ) were investigated ($10 \leq$ PPI ≤ 40, $0.89 \leq \epsilon \leq 0.96$). The foam was installed in a wind tunnel, and the heat was supplied by two hot water jackets, which were compression-clamped to the foam. They also tested louver plate fins having $L_p = 1.0$ mm and $p_f = 1.88$ mm. Both the friction and the j factors increased as the PPI decreased or the porosity increased. The primary reason was attributed to the decreasing surface area for decreasing PPI or increasing porosity. The friction factor of the metal foam was larger than that of the louvered fin, and the difference increased as the Reynolds number increased. This implies that the frontal velocity of the aluminum foam is smaller than that of the louvered fin for a fixed pumping power. The reduced velocity will require large frontal area to meet a given airflow rate.

Figure 5.33 shows the results of volume goodness factor comparison for the Kim et al. [2000] samples with $\epsilon = 0.96$. The air-side performances per unit volume ($\eta h \beta$) are plotted as a function of power consumption per unit volume (P/V). The geometry having the larger PPI yields higher air-side performance at fixed pumping power. The air-side performance of aluminum foams is lower than that of the louvered fin, although the performance of 40 PPI approaches that of the louvered fin at large values of pumping power.

Recently, metal foams find application to electronic heat sinks. Bhattacharya and Mahajan [2000] applied metal foams to finned channel heat sinks. The air

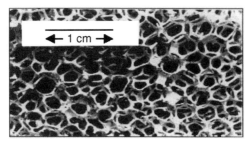

Figure 5.32 Photo of a metal foam (5 PPI) as described by Bhattacharya and Mahajan [2000].

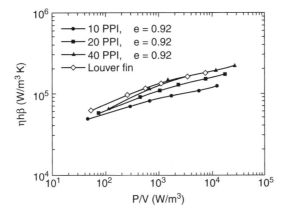

Figure 5.33 The air-side performances per unit volume ($\eta h\beta$) plotted as a function of power consumption per unit volume (P/V) for porous fins ($\epsilon = 0.92$) and a louver fin ($L_p = 1.0$ mm, $p_f = 1.88$ mm) as reported by Kim et al. [2000]. (Reprinted with permission of ASME.)

passages of the finned channels were filled with aluminum metal foam (20 pores/in.). Thermal epoxy was used to attach the foam to the fins. As a result of the large pressure drop of the metal foam, the frontal velocity for the metal foam heat sink was greatly reduced. Figure 5.34 shows the air pressure drop vs. airflow rate for the plain fins and the metal foam–filled fin channels. The figure also shows the balance point for each fin geometry on the fan curve. Note that the air velocity at the balance point for the foam metal is only ⅛ that for the plain fins. Although Bhattacharya and Mahajan [2000] report that the hA value of the foam geometry is higher than that of the plain fin, the foam fins will not transfer as much heat, because of the very large flow rate reduction caused by its high air pressure drop. This conclusion is consistent with other work on enhanced fin geometries having fixed frontal area, which are discussed in Chapter 19. This issue is also addressed in PEC Example 19.1.

Klett et al. [2000] developed a high-thermal-conductivity carbon foam material. Carbon foam has much higher thermal conductivity than aluminum foam. Hence, it will operate at considerably higher fin efficiency. The carbon foam heat exchanger had a heat transfer coefficient of 280 to 500 W/m²-K for 1.5 to 4.5 m/s frontal air velocity. This value is approximately three times that of a louver fin geometry.

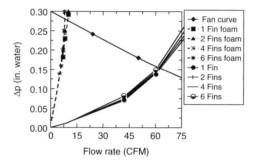

Figure 5.34 Pressure drop vs. flow rate curves for various heat sinks having longitudinal fins, and fins with fin gap filled with 20 PPI metal foam, along with the fan curve as reported by Bhattacharya and Mahajan [2000].

The carbon foam had very large heat transfer surface area per unit volume ($\beta = 1.88 \times 10^6$ m²/m³), which compares to $\beta = 2 \times 10^3$ m²m³ for a louver fin geometry with 25 fins/in. However, this carbon foam had a pore size of 60 to 325 μm, which results in a very small hydraulic diameter. For the same airflow depth, their tests showed that the pressure drop across the foam is approximately 400 times higher than that of the enhanced plate-type heat transfer surface for the same air frontal velocity. Even if the thickness of the graphite foam heat exchanger could be reduced by a factor of 30 (because of its high hA), the pressure drop is still more than 10 times higher than that of the enhanced plate-type heat exchanger. The principal problem with this carbon foam is that the pore size is too small to meet practical design requirements. It may be possible to develop a foam structure with a larger pore size that would be more practical. For an isotropic porous structures (metal foam or mesh), the fin efficiency can be derived similarly to the solid fin geometry. For an isotropic metal foam, the fin efficiency is given by $\eta = [\tanh(mL)]/mL$ where L is the porous fin height and m is defined as

$$m = \left(\frac{h\beta}{k_{eff}} \right)^{0.5}$$

(5.32)

where $\beta = A/V$ (surface-area-to-volume ratio), and k_{eff} is the effective bulk thermal conductivity of the porous foam. Figure 5.35 shows the fin efficiencies of aluminum, copper, and carbon foams having the same porosity. The figure is for 10-mm-high fins, $h = 300$ W/m²-K, and β 2000 m²/m³. The effective thermal conductivities of the materials were calculated based on the thermal conductivities of the parent materials. Comparing the aluminum mesh and the copper mesh, the fin efficiency of the copper mesh is approximately 13% higher than that of the aluminum mesh. The carbon foam mesh shows a higher fin efficiency than copper mesh. However, the carbon foam has relatively low yield strength compared with the other materials. Thus, a problem with carbon foam is that it may be very brittle and subject to breakage. This problem will be worse if the pore size is increased in an attempt to reduce the

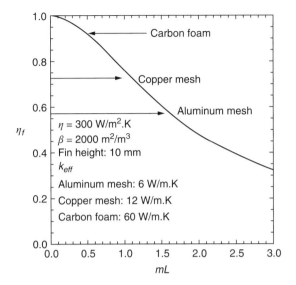

Figure 5.35 Comparison of fin efficiency of three mesh materials.

pressure drop. Copper mesh may be preferred, because of the high strength and thermal conductivity.

5.11 PLAIN FIN

If plain fins are used (e.g., Figure 5.2a and b), the flow channel will have a rectangular or triangular cross section. If the flow is turbulent, standard equations for turbulent flow in circular tubes may be used to calculate j and f, provided Re is based on the hydraulic diameter (D_h). If $\text{Re}_{Dh} < 2000$, one may use theoretical laminar flow solutions for j and f. Values of j and f for developing and for fully developed laminar flow are given in Section 2.4 for a variety of duct shapes. However, many of these solutions assume constant properties. Criteria to establish the value of x/D_h at which the flow becomes fully developed are also given in Table 2.2 of Chapter 2. Shah and London [1978] provide detailed results of many of the published solutions.

Sources of published data on plain fin geometries include Kays and London [1984] and Tishchenko and Bondarenko [1983]. The latter reference provides data on 19 plain fin geometries.

The Reynolds number (Re_{Dh}) is typically below 2200 in the channels of plate-and-fin heat exchangers. Theoretical solutions for laminar have been developed for many channel shapes. The cross-sectional geometry of the plain channel can have a very significant effect of the Nu and f. This is illustrated by PEC Example 5.3 given below.

5.11.1 PEC Example 5.3

This example is presented using the FN-3 criterion. Two plain fin geometries with the same fin pitch are compared: a triangular geometry (T) and a rectangular geometry (R). Figure 5.36 shows the flow cross section of the two geometries. The FN-3 criterion constrains the two geometries to operate at the same mass flow rate with equal frontal velocity and hA values. Because the velocities are known, the j and f values are directly calculable. The calculations are made for airflow at 27°C and 4 m/s frontal velocity. The j and f factors are given by the fully developed laminar solution for constant wall temperature and are taken from Table 2.2. By using the calculated h and f values,

$$\frac{A_T}{A_R} = \frac{h_R}{h_T} = \frac{39.80}{18.53} = 2.148$$

$$\frac{V_T}{V_R} = \frac{A_T}{A_R}\frac{\beta_R}{\beta_T} = 2.148 \times \frac{1225}{1143} = 2.30 \tag{5.33}$$

$$\frac{P_T}{P_R} = \frac{f_T}{f_R}\frac{A_T}{A_R}\left(\frac{D_{hT}Re_T}{D_{h,T}Re_R}\right)^3 = \frac{0.0167}{0.024} \times 2.148 \times \left(\frac{2.94 \times 781}{2.96 \times 776}\right)^3 = 1.49$$

For the same frontal area, flow rate, and heat transfer rate, this Case FN-3 example shows that the triangular geometry requires 115% more surface area and 49% greater pumping power.

5.12 ENTRANCE LENGTH EFFECTS

Extended surface geometries for gases are typically designed for operation at $Re_{Dh} < 2000$. If the flow length is sufficiently short, it is possible that the average j and f over the flow length are higher than the fully developed values, which are given in Table 2.2 of Chapter 2 for a number of channel shapes. Plain fin geometries are more susceptible to entrance length effects than are the enhanced geometries of

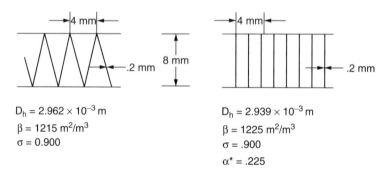

Figure 5.36 Rectangular and triangular fin geometries used for PEC Example 5.3. (From Webb [1983].)

Figure 5.3c and f. Because of the periodic flow interruptions, it is unlikely that entrance region effects would exist for interrupted fin heat exchangers. However, this may not be a good assumption for plain fin geometries. If developing flow exists over more than, say, 20% of the flow length, it is possible that the average Nu and f are moderately higher than the fully developed values. In this case, one should determine the average Nu and f over the airflow length.

The entrance region effect on heat transfer for several plain fin channel geometries is illustrated in Figure 5.37. This figure shows the ratio of the average Nu to the fully developed Nu for a constant wall temperature boundary condition with a developed velocity profile. Consider, for example, airflow (Pr = 0.7) in an equilateral triangle–shaped channel whose dimensionless flow length is $x^* = 0.10$. The mean Nu over the $x^* = 0.10$ flow length is 42% greater than the fully developed value. If $x^* \leq 0.2$, it appears that the entrance region solutions should be used. The figure shows that triangular channels have a longer entrance region than circular or rectangular channels.

The entrance region pressure drop is also higher than the fully developed value. It is calculated by the equation:

$$\Delta p = \left(\frac{4 f_{fd} L}{D_h} + K(\infty) \right) \frac{G^2}{2\rho} \tag{5.34}$$

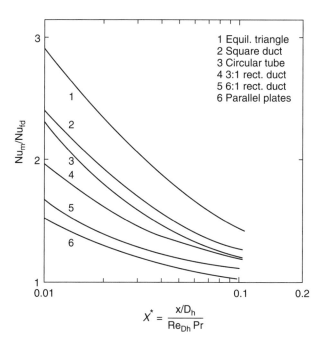

Figure 5.37 Nu_m/Nu_{fd} for laminar entrance region flow with developed velocity profile in different channel shapes for constant wall temperature boundary condition. (From Webb [1987].)

where $K(\infty)$ is the pressure drop increment to be added to account for the increased friction in the development region. Figure 5.38 shows the ratio $K(x)/K(\infty)$ for several duct shapes. The $K(x)$ is the increment to be added for a duct, whose profile is less than fully developed at the duct exit. The hydraulic entrance length (L^+_{hy}) may be determined from Figure 5.38, and is the value of $x/(D_h\mathrm{Re})$ at which $K/K(\infty) \simeq 1$. The $K(\infty)$ and L^+_{hy} values for the channel shapes on Figure 5.38 are listed in Table 2.2 of Chapter 2. To calculate the pressure drop in an entrance region duct, use Equation 5.34 with $K(\infty)$ replaced by K from Figure 5.38.

5.13 PACKINGS FOR GAS–GAS REGENERATORS

Regenerators, either rotary or valved, are commonly used to transfer heat from combustion products to inlet combustion air. The valved type has two identical packings, which alternately serve the hot and cold streams and are switched by quick operating valves. Any of the corrugated plate-fin geometries discussed in this chapter may conceivably be applied to regenerators. Because the hot and cold streams are normally of different pressures, any packing geometry that allows significant transverse flow leakage will reduce a rotary regenerator performance. The louvered and OSF geometries would be susceptible to this problem. However, one may design around this problem by including full-height, continuous fins aligned with the flow at discrete

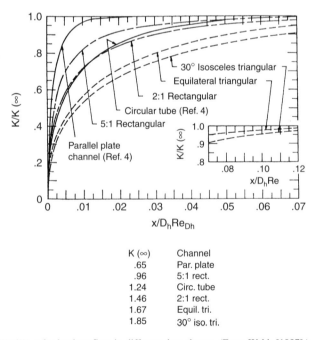

K (∞)	Channel
.65	Par. plate
.96	5:1 rect.
1.24	Circ. tube
1.46	2:1 rect.
1.67	Equil. tri.
1.85	30° iso. tri.

Figure 5.38 $K(x)/K(\infty)$ for laminar flow in different duct shapes. (From Webb [1987].)

spacings. The "brick checkers" geometry commonly used in the valved-type glass furnace regenerator is essentially an OSF (Figure 5.3d) with a large fin thickness.

Packings with small hydraulic diameter (small s) will provide a higher heat transfer coefficient than a packing with a larger hydraulic diameter; for laminar flow, $h \propto 1/D_h$. However, the fouling characteristics of the hot gas may limit the size of the flow passage. Considerably smaller passage size may be used in building ventilation heat-recovery regenerators than those used for heat recovery from coal-fired exhaust gases or glass furnace exhausts. The OSF is not a totally viable candidate. This is because it would permit mixing and contamination between the hot and cold streams.

Electric utility plants frequently use notched plate packings, such as that illustrated in Figure 5.39. Three-dimensional corrugated channels are also used in rotary regenerators. Figure 5.39d shows a cutaway view of a corrugated channel. Considerable work has been done on flow visualization and numerical simulation in three-dimensional corrugated channels. This work is discussed in Section 5.6 and 5.14.4.

5.14 NUMERICAL SIMULATION

Considerable advances have been made in numerical modeling for enhanced surface geometries, and numerous publications have appeared since the first edition of this book. This section reviews some of the important works.

5.14.1 Offset-Strip Fins

Sparrow et al. [1977] were the first to perform simulation of the OSF geometry. Zero fin thickness and an infinite fin height were assumed. Patankar and Prakash [1981] extended the analysis to finite fin thickness. They compared the numerical results with one set of experimetal data by Kays and London [1984]. The prediction of the friction factor was in reasonable agreement with experimental data, but the predicted j factor was approximately 100% high. No satisfactory explanation was provided on this overprediction. Kelkar and Patankar [1989] performed a numerical analysis on a three-dimensional OSF geometry assuming a zero fin thickness. The three-dimensional aspect of the flow was negligible for aspect ratios less than 0.2. The numerical solution underpredicted the friction factors by approximately 15%

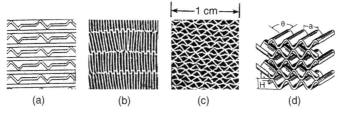

(a) (b) (c) (d)

Figure 5.39 Matrix geometries for rotary regenerators: (a) notched plate matrix, (b) deep fold rectangular matrix, (c) triangular matrix. (Courtesy of Combustion Engineering Air Preheater Division, Wellsville, NY). (d) Corrugated channels (From Ciofalo et al. [1996]. Reprinted with permission from Elsevier.)

and overpredicted the Nusselt numbers by 15% compared to those predicted by the Wieting [1975] correlation.

Suzuki et al. [1994] studied unsteady flow field in an inline array of three fins. The predicted flow field was compared with experimental data obtained by Xi et al. [1995]. A good agreement was found for both the mean and fluctuating velocities. Analysis of the flow pattern and heat transfer near the fin surface has led to the conclusion that heat transfer is enhanced by local instabilities near the surface, which are created by upstream vortices that impinge on the fin.

Zhang et al. [1997] compared two- and three-dimensional, steady and unsteady solutions for the OSF geometry. They showed that, in the unsteady laminar flow regime, steady flow calculations do not adequately represent enhanced large-scale mixing. They also showed that as the Reynolds number increases, intrinsic three-dimensional effects become important.

Mercier and Tochon [1997] performed a two-dimensional time-dependent analysis of turbulent flow using pseudo-direct numerical simulation (DNS). For DNS, the mesh size must be smaller than the smallest turbulence scale, and thus requires a great amount of computer memory. They considered an OSF array under developing and fully developed flow conditions. The time-dependent solution provided fundamental information on the flow structure, such as the flow separation, the creation of recirculation zones, etc. The simulations were in good agreement with the flow visualization data obtained from similar geometries. The predicted friction and j factors were compared with several correlations from the open literature. The numerical prediction underestimated the friction factor by 5 to 30%, and underestimated the j factor by 30 to 50%. Mercier and Tochon [1997] attributed the discrepancy to uncertainties inherent in applying a periodic boundary condition for a fully developed solution.

Xi and Shah [1999] conducted unsteady two- and three-dimensional numerical analysis for the OSF geometry. The fin thickness was included, and the computational domain included the complete OSF. A comparison with the experimental data of Mochizuki et al. [1988], obtained using a scaled up model, showed good agreement. A comparison with an actual OSF array geometry of London and Shah [1968] showed lower predictions for f factors and higher predictions for j factors. They attributed the underprediction of the f factor to burrs on the real OSF. The overprediction of the j factor, however, was not adequately explained. A comparison of two- and three-dimensional results indicated that the effect of three-dimensional analysis is small in the laminar flow region ($Re_{Dh} < 1600$). The two-dimensional computations yielded acceptable results at low Reynolds numbers.

5.14.2 Louver Fins

Kajino and Hiramatsu [1987] solved the problem of steady laminar two-dimensional flow over louver fins. They obtained stream lines, velocity profiles, and Nusselt number distributions for one louver geometry. Achaichia and Cowell [1988] modeled one louver in the fully developed region assuming periodic boundary conditions. Their results confirmed the existence of duct-directed flow at low Reynolds numbers and louver-directed flow (see Section 5.3.2) at high Reynolds numbers. This flow behavior was related to the flow efficiency by Webb and Trauger [1991]. Suga and Aoki [1991], using a two-dimensional laminar flow model over the complete louver fin array, investigated the effect of fin geometry on heat transfer performance. Based on numerical results, they proposed an equation for the optimum L_p/p_f. The optimum value occurs when the thermal wake behind the louver flows along a line halfway between two louvers farther downstream.

Atkinson et al. [1998] conducted a time-dependent two- and three-dimensional numerical study using a commercial computational fluid dynamics (CFD) code (STAR-CD). The results showed vortex shedding from the trailing edges of the first two louvers. Comparison was made with the experimental data of Achaichia and Cowell [1988]. The friction factors agreed well with experimental values. The j factors were highly (90%) overpredicted by the two-dimensional model. However, the overprediction reduced to 25% with the three-dimensional model.

Tafti et al. [2000] investigated the flow transition in a louver fin geometry by conducting time-dependent two-dimensional analysis. The initial instability was observed at $Re_L = 400$ at the exit louver. As the Reynolds number increases, the instabilities moved upstream, and, at $Re_L = 1300$, most of the louver exhibited unsteadiness. These findings are in close agreement with the experimental results by DeJong and Jacobi [2003].

A review article by Heikal et al. [1999] provides a detailed overview of work on numerical simulation of the louver fin geometry. Their discussion includes mesh development strategies, treatment of boundary conditions, and solution interpretation procedures. Results are also provided for two- and three-dimensional models for a number of louver fin geometries. The numerical results are compared with experimental results.

5.14.3 Wavy Channels

Asako and Faghri [1987] solved the problem of laminar flow and heat transfer in a corrugated duct. A nonorthogonal coordinate transformation mapped the corrugated channel into a rectangular domain. The governing equations were solved assuming constant thermophysical properties. Yang et al. [1995] extended the work of Asako and Faghri [1987] to a transition region by using a low-Reynolds-number turbulence model. They also conducted flow visualization experiments. The numerical model predicted the size of the recirculation area reasonably well. The size of the recirculation region reached a maximum at $R_{Dh} = 500$, and then decreased, because of the high diffusion in turbulent flow for higher Reynolds numbers.

Ergin et al. [1997] investigated two turbulence models; the standard k–$\in$ model and the low-Re turbulence model. For $500 \leq Re_{Dh} \leq 3000$, the low-Re turbulence model yielded better results, whereas the k-$\in$ model was more accurate for higher Reynolds numbers. Kouidry [1997] performed direct numerical simulation of turbulent flow in a corrugated channel. The advantage of DNS methods compared to time-average methods such as k-$\in$ models is that they are independent of flow configurations and that local instantaneous information is available. The numerical results were compared with experimental data. The mean velocity profile, the flow pattern, and the vortices were well predicted.

McNab et al. [1998] computed the flow over herringbone corrugated channels using a commercial STAR-CD CFD code. A three-dimensional approach was adopted. A laminar model was used for $Re_{Dh} < 1500$, and the k-$\in$ model was used for higher Reynolds numbers. The computed j and f factors were compared with the measurements of Abou-Madi [1998]. They agreed relatively well in the turbulent region (17 to 27%), but in the laminar region, the friction factors were 33% under-predicted and the j factors were 54% overpredicted.

Xu and Min [2004] investigated four flow models — steady laminar, unsteady laminar, steady turbulent (RNG k-$\in$), and unsteady turbulent — to predict the experimental data by Ali and Ramadhyani [1992]. Based on the comparison of the predicted values with the experimental data, Xu and Min [2004] suggested a laminar flow model for $Re \leq 600$, and a turbulent flow model for $Re_{Dh} > 600$. The unsteady and steady models yielded little difference both for the laminar and for the turbulent models. The computed Nusselt numbers and friction factors agreed with the Ali and Ramadhyani [1992] data within 5.0 and 10.8% with most of the data underpredicted.

5.14.4 Chevron Plates

As observed from experimental work, the flow remains mainly in the furrow regions for corrugation angles below 30°. For higher corrugation angles, however, the flow is mostly three-dimensional and highly mixed, even for very low Reynolds numbers (e.g., $Re_{Dh} = 200$).

Ciofalo et al. [1996] conducted numerical investigations on the flow between chevron plates. Several numerical models (laminar, k-$\in$ turbulence, low-Re k-$\in$ turbulence, and large eddy simulation) were compared. Results show that laminar and k-$\in$ models were unsatisfactory. They underpredicted and overpredicted both the friction factors and j factors, respectively. The best agreement with the experimental data was obtained with the low-Re turbulence model or LES. The difference was within ± 50% for the complete range of Reynolds numbers and for various corrugation angles. They also compared the LES and the DNS methods; the DNS did not improve the results significantly.

Sunden [1999] investigated the thermal and hydraulic performances of plate heat exchangers for two different corrugation angles (22.5° and 45°). A steady laminar model was applied for Reynolds numbers up to 2000 and a low-Reynolds k-$\in$ model was used for turbulent flows. A three-dimensional body-fitted mesh was employed. The numerical results were compared with experiments and showed that the Nusselt number was underpredicted by 25% and the friction factor was overpre-dicted by 17 to 40%.

5.14.5 Summary

Although numerical analysis has progressed considerably in the past 20 years, most of the numerical analysis has been limited to 2-dimensional modeling. Most of the flows encountered in a plate-fin geometry are unsteady, often with flow separation and reattachment zones. These complex characteristics cannot be accurately predicted with conventional time-average modeling such as k-$\in$ models. The computationally demanding LES or DNS models provide local instantaneous information, and predict the average heat transfer and pressure loss better than time-average models.

5.15 CONCLUSIONS

Plate-and-fin heat exchanger geometry has become an increasingly important design. The high-performance offset strip and louver fins provide quite high heat transfer coefficients for gases and two-phase applications. It offers significant advantages over the traditional fin-and-round-tube geometry. Key advantages are lower gas pressure drop than circular tube designs, and the ability to have the fins normal to the gas flow over the full gas flow depth. Vortex generators or metallic foams are emerging as new enhancement geometries. Significant advances have been made on numerical analysis of the complex plate-fin geometries.

Early variants were applied to gas to gas applications. It is now used for gases, liquids, or two-phase fluids on either side. Designs using extruded aluminum tubes with internal membranes allow quite high tube-side design pressure, e.g., 150 atm. Further innovative designs, applications, and advanced fin geometries are expected. Currently, aluminum, steel, and even ceramics are used.

NOMENCLATURE

A Total heat transfer surface area (both primary and secondary, if any) on one side of a direct transfer type exchanger; total heat transfer surface area of a regenerator, m^2 or ft^2

A_c Flow cross-sectional area in minimum flow area, m^2 or ft^2

b Distance between plates in a plate-fin exchanger, or channel height, m^2 or ft^2

C_D Drag coefficient, dimensionless

c_p Specific heat of fluid at constant pressure, J/kg-K or Btu/lbm-°F

d Diameter of pin fins, m or ft

D Mass diffusion coefficient, kg/m-s or lbm/ft-s

D_h Hydraulic diameter, m or ft

D_m Minor diameter of a flat tube, defined in Figure 5.10, m or ft

D_{sh} Hydraulic diameter based on $A_c = sh$, m or ft

f Fanning friction factor, $\Delta p_f D_{ph}/2LG^2$, dimensionless

G Mass velocity based on the minimum flow area, kg/m^2-s or lbm/ft^2-s

H Louver fin height, defined on Figure 5.10, or corrugation height, m or ft

h Heat transfer coefficient based on A, W/m^2-K or Btu/hr-ft^2-°F

h Fin height ($h = b - t$), m or ft

h_m Mass transfer coefficient, kg/m^2-s or lbm/ft^2-s

j $StPr^{2/3}$, dimensionless

k Thermal conductivity of fluid, W/m-K or Btu/hr-ft-°F

K Pressure drop increment to account for flow development, $K(\infty)$ (full entrance region), $K(x)$ (over length x), shown on Figure 5.38

L   Fluid flow (core) length on one side of the exchanger, m or ft

L_p Strip flow length of OSF or louver pitch of louver fin, m or ft

L_h Louver height shown on Figure 5.10, m or ft

L_L Louver length shown on Figure 5.10, m or ft

L_s^+ $(L_p/2s)/(2su/v)$, dimensionless

Nu   Nusselt number = hD_h/k, dimensionless

P Fluid pumping power, W or hp

Pr   Prandtl number = $c_p\mu/k$, dimensionless

p_f Fin pitch, m or ft

p_w Axial wave pitch of wavy fin, m or ft

Re_{Dh} Reynolds number based on hydraulic diameter, $Re_{Dh} = D_h G/\mu$, dimensionless

$Re_{Dh,tr}$ Transition from laminar to turbulent flow, dimensionless

Re_s Reynolds number based on S_l, dimensionless

Re_L Reynolds number based on the interruption length = GL_p/μ dimensionless

Sh_L Sherwood number based on interruption length, $h_m L_p/D$, dimensionless

S_t Transverse tube pitch, m or ft

St   Stanton number = h/Gc_p, dimensionless

s Spacing between two fins (= $p_f - t$), m or ft

t Fin thickness, m

T_d Major diameter of a flat tube, defined on Figure 5.10, m or ft

T_p Transverse tube pitch of a flat-tube heat exchanger, defined on Figure 5.10, m or ft

u Velocity based on $A_c = s(b - t)$, m² or ft²

v Velocity based on $A_c = (s - t)(b - t)$, m² or ft²

x Cartesian coordinate along the flow direction, m or ft

x^* $x/(D_h RePr)$, dimensionless

Greek Letters

α Aspect ratio of rectangular duct (s/b), dimensionless

β Surface area to volume ratio (A/V), m⁻¹ or ft⁻¹

δ t/l, dimensionless

γ t/s, dimensionless

δ_{mom} $t + 1.328 L_p/(Re_L)^{0.5}$

θ Corrugation angle in corrugated fin geometry; also included half angle of internal fin cross section normal to flow, rad or deg

ν Kinematic viscosity, m²/s or ft²/s

η_f Fin efficiency or temperature effectiveness of the fin, dimensionless

η Surface efficiency of finned surface = $1 - (1 - \eta_f)A_f/A$, dimensionless

η Flow efficiency (Figure 5.13), dimensionless

μ Fluid dynamic viscosity coefficient, Pa-s or lbm/s-ft
ρ Fluid density kg/m^3 or lbm/ft^3

Subscripts

fd   Fully developed flow
L Laminar region
m Average value over flow length
p Plain tube or surface
T Turbulent region
w Evaluated at wall temperature
x Local value

REFERENCES

Abdel-Kariem, A.H. and Fletcher, L.S., 1999. Comparative analysis of heat transfer and pressure drop in plate heat exchangers, in *Proceedings of the 5th ASME/JSME Thermal Engineering Conference*, San Diego, CA.

Abou-Madi, M., 1998. A Computer model for mobile air-conditioning system, Ph.D thesis, University of Brighton, Brighton, U.K.

Achaichia, A. and Cowell, T.A., 1988. Heat transfer and pressure drop characteristics of flat tube and louvered plate fin surfaces, *Exp. Thermal Fluid Sci.*, 1, 147–157.

Ali, M.M. and Ramadhyani, S., 1992. Experiments on convective heat transfer in corrugated channels, *Exp. Heat Transfer*, 5, 175–193.

Aoki, H., Shinagawa, T., and Suga, K., 1989. An experimental study of the local heat transfer characteristics in automotive louvered fins, *Exp. Thermal Fluid Sci.*, 2, 293–300.

Asako, Y. and Faghri, M., 1987. Finite volume solutions for laminar flow and heat transfer in a corrugated duct, *J. Heat Transfer*, 109, 627–634.

Atkinson, K.N., Drakulic, R., Heikal, M.R., and Cowell, T.A., 1998. Two and three-dimensional numerical models of flow and heat transfer over louvered fin arrays in compact heat exchangers, *Int. J. Heat Mass Transfer*, 41, 4063–4080.

Bellows, K.D., 1996. Flow Visualization of Louvered-Fin Heat Exchangers, Masters thesis, University of Illinois, Urbana Champaign.

Bhattacharya, A. and Mahajan, R., 2000. Finned metal foam heat sinks for electronics cooling in forced convection, in *Proc. Thirty-Fourth National Heat Transfer Conference*, Pittsburgh, PA.

Brinkman, R., Ramadhyani, S., and Incropera, F.P., 1988. Enhancement of convective heat transfer from small heat sources to liquid coolants using strip fins, *Exp. Heat Transfer*, 1, 315–330.

Brockmeier, U., Guntermann, Th., and Fiebig, M., 1993. Performance evaluation of a vortex generator heat transfer surface and comparison with different high performance surfaces, *Int. J. Heat Mass Transfer*, 36, 2575–2587.

Chang, Y.-J. and Wang, C.-C., 1996. Air side performance of brazed aluminum heat exchangers, *J. Enhanced Heat Transfer*, 3, 15–28.

Chang, Y.-J., and Wang, C.-C., 1997. A generalized heat transfer correlation for louver fin geometry, *Int. J. Heat Mass Transfer*, 40, 533–544.

Chang, Y.-J., Hsu, K.-C., Lin, Y.-T., and Wang, C.-C., 2000. A generalized friction correlation for louver fin geometry, *Int. J. Heat Mass Transfer*, 43, 2237–2243.

Churchill, S.W. and Usagi, R., 1972. A general expression for the correlation of rates of transfer and other phenomena, *AIChE J.*, 18(6), 1121–1128.

Ciofalo, M., Stasiek, J., and Collins, M.W., 1996. Investigation of flow and heat transfer in corrugated passages II numerical results, *Int. J. Heat Mass Transfer*, 39, 165–192.

Creswick, F.A., Talbert, S.G., and Bloemer, J.W., 1964. Compact Heat Exchanger Study, Battelle Memorial Institute Report, Columbus, OH, April 15.

Davenport, C.J., 1980. Heat Transfer and Fluid Flow in the Louvered-Fin Heat Exchanger, Ph.D. thesis, Lanchester Polytechnic, Lanchester, U.K.

Davenport, C.J., 1983a. Heat transfer and flow friction characteristics of louvered heat exchanger surfaces, in *Heat Exchangers: Theory and Practice*, J. Taborek, G.F. Hewitt, and N. Afgan, Eds., Hemisphere, Washington, D.C., 387–412.

Davenport, C.J., 1983b. Correlations for heat transfer and flow friction characteristics of louvered fin, in *Heat Transfer—Seattle 1983, AIChE Symposium*, Series, No. 225, 79, 19–27.

DeJong, N.C. and Jacobi, A.M., 1997. An experimental study of flow and heat transfer in parallel-plate arrays: local, row-by-row and surface average behavior, *Int. J. Heat Mass Transfer*, 40, 1365–1378.

DeJong, N.C. and Jacobi, A.M., 1999. Local flow and heat transfer behavior in convex-louver fin arrays, *J. Heat Transfer*, 121, 136–141.

DeJong, N.C. and Jacobi, A.M., 2003. Localized flow and heat transfer interactions in louvered fin arrays, *Int. J. Heat Mass Transfer*, 46, 443–445.

Dillen, E.R. and Webb, R.L., 1994. Rationally Based Heat Transfer and Friction Correlations for the Louver Fin Geometry, SAE Paper 940504, Warrendale, PA.

Ebisu, T., 1999. Development of new concept air-cooled heat exchanger for energy conservation of air-conditioning machine, in *Heat Transfer Enhancement of Heat Exchangers*, S. Kakaç et al., Eds., Kluwer Academic, Dordrenchts, 601–620.

Eibeck, P.A. and Eaton, J.K., 1987. Heat transfer effects of a longitudinal vortex embedded in a turbulent shear flow, *J. Heat Transfer*, 109, 16–24.

Ergin, S., Ota, M., Yamaguchi, H., and Sakamoto, M., 1997. Analysis of periodically fully developed turbulent flow in a corrugated duct using various turbulent models and comparison with experiments, *JSME Centennial Grand Congress, Int. Conf. on Fluid Eng*, Tokyo, Japan, 1527–1532.

Fiebig, M., 1995. Vortex generators for compact heat exchangers, *J. Enhanced Heat Transfer*, 2, 43–62.

Fiebig, M., Valencia, A., and Mitra, N.K., 1993. Wing-type vortex generators for fin-and- tube heat exchangers, *Exp. Thermal Fluid Sci.*, 7, 287–295.

Focke, W.W. and Knibbe, P.G., 1986. Flow visulaization in parallel plate ducts with corrugated walls, *J. Fluid Mechanics*, 165, 73–77.

Focke, W.W., Zachariades, J., and Oliver, I., 1985. The effect of the corrugation inclination angle on the thermohydraulic performance of plate heat exchanger, *Int. J. Heat Mass Transfer*, 28, 1469–1479.

Fujii, M., Seshimo, Y., and Yamananaka, G., 1988. Heat transfer and pressure drop of the perforated surface heat exchanger with passage enlargement and contraction, *Int. J. Heat Mass Transfer*, 31, 135–142.

Fujii, M., Seshimo, Y., and Yoshida, T., 1991. Heat transfer and pressure drop of tube-fin heat exchanger with trapezoidal perforated fins, in *Proc. 1991 ASME-JSME Joint Thermal Engineering Conf.*, J.R. Lloyd and Y. Kurosake, Eds., Vol 4, ASME, New York, 355–360.

Fujikake, K., Aoki, H., and Mitui, H., 1983. An apparatus for measuring the heat transfer coefficients of finned heat exchangers by use of a transient method, *Proc. Japan 20th Symposium on Heat Transfer*, 466–468.

Goldstein, L.J. and Sparrow, E.M., 1977. Heat/mass transfer characteristics for flow in a corrugated wall channel, *J. Heat Transfer*, 99, 187–195.

Gschwind, P., Regele, A., and Kottke, V., 1995. Sinusoidal wavy channels with Taylor-Goetler vortices, *Exp. Thermal Fluid Sci.*, 11, 270–275.

Hamaguchi, K., Takahashi, S., and Miyabe, H., 1983. Heat transfer characteristics of a regenerator matrix (case of packed wire gauzes), *Trans. JSME*, 49B(445), 2001–2009.

Hatada, T. and Senshu, T., 1984. Experimental Study on Heat Transfer Characteristics of Convex Louver Fins for Air Conditioning Heat Exchangers, ASME paper 84-HT-74, New York.

Heikal, M.R., Drakulic, R., and Cowell, T.A., 1999. Multi-louvered fin surfaces, in *Recent Advances in Analysis of Heat Transfer for Fin Type Surfaces*, B. Sunden and P.J. Heggs, Eds., Computational Mcchanics, Billerica, MA, 277–293.

Hitachi, 1984. Hitachi High-Performance Heat Transfer Tubes. Cat. No. EA-500, Hitachi Cable Co., Tokyo, Japan.

Hu, S. and Herold, K.E., 1995a. Prandtl number effect on offset fin heat exchanger performance: experimental results, *Int. J. Heat Mass Transfer*, 38, 1053–1062.

Hu, S., and Herold, K.E., 1995b. Prandtl number effect on offset fin heat exchanger performance: predictive model for heat transfer and pressure drop, *Int. J. Heat Mass Transfer*, 38, 1043–1051.

Joshi, H.M. and Webb, R.L., 1987. Prediction of heat transfer and friction in the offset-strip fin array, *Int. J. Heat Mass Transfer*, 30, 69–84.

Kajino, M. and Hiramatsu, M. 1987. Research and development of automotive heat exchangers, in *Heat Transfer in High Technology and Power Engineering*, W.J. Yang and Y. Mori, Eds., Hemisphere, Washington, D.C., 420–432.

Kays, W.M., 1972. Compact Heat Exchangers, AGARD Lecture Series on Heat Exchangers, 57, J.J. Ginoux, Ed., AGARD-LS-57-72, January.

Kays, W.M. and Crawford, M.E., 1980. *Convective Heat and Mass Transfer*, McGraw-Hill, New York, 151.

Kays, W.M. and London, A.L., 1984. *Compact Heat Exchangers*, 3rd ed., McGraw-Hill, New Year.

Kelkar, K.M. and Patankar, S.V., 1989. Numerical prediction of heat transfer and fluid flow in rectangular offset fin arrays, *Numerical Heat Transfer A*, 15, 149–164.

Kim, S.Y., Paek, J.W., and Kang, B.H., 2000. Flow and heat transfer correlations for porous fin in a plate-fin heat exchanger, *J. Heat Transfer*, 122, 572–578.

Klett, J., Ott, R., and McMillan, A., 2000. Heat Exchangers for Heavy Vehicles Utilizing High Thermal Conductivity Graphite Foams, SAE Paper 2000-01-2207, Warrendale, PA.

Kouidry, F., 1997. Etude des Ecoulements Turbulents Charges de Particules: Application a Lencrassment Particulaire des Echangeurs a Plaques Corruguees, Ph.D thesis, University Joseph Fouirier, Grenoble, France.

Kurosaki, Y., Kashiwagi, T., Kobayashi, H., Uzuhashi, H., and Tang, S.-C., 1988. Experimental study on heat transfer from parallel louvered fins by laser holographic interferometry, *Exp. Thermal Fluid Science*, 1, 59–67.

London, A.L. and Shah, R.K., 1968. Offset rectangular plate-fin surfaces heat transfer and flow friction characteristics, *ASME J. Eng. Power*, 90, 218–228.

Lyman, A.C., Stephan, R.A., Thole, K.A., Zhang, L.W., and Memory, S.B., 2002. Scaling of heat transfer coefficients along louvered fins, *Exp. Thermal Fluid Sci.*, 26, 547–563.

Manglik, R.M. and Bergles, A.E., 1990. The thermal-hydraulic design of the rectangular offset strip-fin compact heat exchanger, in *Compact Heat Exchangers*, R.K. Shah, A.D. Kraus, and D. Metzger, Eds., Hemisphere, Washington, D.C., 123–150.

McNab, C.A., Atkinson, K.N., Heikal, M.R., and Taylor, N., 1998. Numerical modeling of heat transfer and fluid flow over herringbone corrugated fins, *Heat Transfer 1998, Proceedings of 11th Int. Heat Transfer Conference*, 6, 119–124.

Mehta, R.D., Shabaka, I.M.M.A., Shibl, A., and Bradshaw, P., 1983. Longitudinal Vortices Imbedded in Turbulent Boundary Layers, AIAA Paper 83-0378, Albuquerque, NM.

Mercier, P. and Tochon, P., 1997. Analysis of turbulent flow and heat transfer in compact heat exchangers by pseudo direct numerical simulation, in *Compact Heat Exchangers for Process Industries*, R.K. Shah, Ed., Begell House, New York, 223–230.

Milne-Thomson, L.M., 1960. *Theoretical Hydrodynamics*, 4th ed., Macmillan, New York, 319.

Mochizuki, S., Yagi, Y., and Yang, W.J., 1988. Flow pattern and turbulent intensity in stacks of interrupted parallel plate surfaces, *Exp. Thermal Fluid Science*, 1, 51–57.

Molki, M. and Yuen, C.M., 1986. Effect of interwall spacing on heat transfer and pressure drop in a corrugated wall channel, *Int. J. Heat Mass Transfer*, 29, 987–997.

Montgomery, S.R. and Wilbulswas, P., 1967. Laminar flow heat transfer for simultaneously developing velocity and temperature profiles in ducts of rectangular cross-section, *Appl. Sci. Res.*, 18, 247–259.

Muzychka, Y.S. and Yovanovich, M.M., 1998. Modeling Nusselt numbers for thermally developing laminar flow in non-circular ducts, 98, 2586, *Proc. 7th AIAA/ASME Joint Thermophysics Conf.* AIAA paper, Albuquerque, NM.

Muzychka, Y.S. and Yovanovich, M.M., 2001a. Modeling the f and j characteristics of the offset-strip fin array, *J. Enhanced Heat Transfer*, 8, 261–278.

Muzychka, Y.S. and Yovanovich, M.M., 2001b. Modeling the f and j characteristics for transverse flow through an offset-strip fin at low Reynolds number, *J. Enhanced Heat Transfer*, 8, 243–260.

O'Brien, J.E. and Sparrow, E.M., 1982. Corrugated-duct heat transfer, pressure drop and flow visualization, *J. Heat Transfer*, 104, 410–416.

Okada, K., Ono, M., Tomimura, T., Okuma, T., Konno, H., and Ohtani, S., 1972. Design and heat transfer characteristics of new plate type heat exchanger, *Heat Transfer-Japanese Research*, 1(1), 90–95.

Patankar, S.V., 1990, Numerical prediction of flow and heat transfer in compact heat exchanger passages, in *Compact Heat Exchangers*, R.K. Shah, A.D. Kraus, and D. Metzger, Eds., Hemisphere, Washington, D.C., 191–204.

Patankar, S.V. and Prakash, C., 1981. An analysis of the effect of plate thickness on laminar flow and heat transfer in interrupted plate passages, *Int. J. Heat Mass Transfer*, 24, 1801–1810.

Pauley, W.R. and Eaton, J.K., 1988. Experimental study of the development of longitudinal vortex pairs embedded in a turbulent boundary layer, *AIAA J.*, 26, 816–823.

Pauley, L.L. and Hodgson, J.E., 1994. Flow visualization of convex louver fin arrays to determine maximum heat transfer conditions, *Exp. Thermal Fluid Sci.*, 9(1), 53–60.

Rosenblad, G. and Kullendorf, A., 1975. Estimating heat transfer rates from mass transfer studies on plate heat exchanger surfaces, *Wärme Stoffubertrag.*, 8, 187–191.

Rugh, J.P., Pearson, J.T., and Ramadhyani, S., 1992. A study of a very compact heat exchanger used for passenger compartment heating in automobiles, in *Compact Heat Exchangers for Power and Process Industries*, ASME Symp. Ser., HTD. 201, 15–24.

Sahnoun, A. and Webb, R.L., 1992. Prediction of heat transfer and friction for the louver fin geometry, *J. Heat Transfer*, 114, 893–900.

Shah, R.K., 1975. Perforated Heat Exchanger Surfaces: Part 2—Heat Transfer and Flow Friction Characteristics, *ASME Paper*, 75-WA/HT-9, New York.

Shah, R.K. and London, A.L., 1978. *Laminar Flow Forced Convection in Ducts*, Supplement 1 to *Advances in Heat Transfer*, Academic Press, New York.

Shah, R.K. and Webb, R.L., 1982. Compact and enhanced heat exchangers, in *Heat Exchangers: Theory and Practice*, J. Taborek, G.F. Hewitt, and N.H. Afgan, Eds., Hemisphere, Washington, DC., 425–468.

Shen, J., Gu, W., and Zhang, Y., 1987. An investigation on the heat transfer augmentation and friction loss performances of plate-perforated fin surfaces, in *Heat Transfer Science and Technology*, B.-X. Wang, Ed., Hemisphere, Washington, D.C., 798–804.

Sparrow, E.M., and Hossfeld, M., 1984. Effect of rounding protruding edges on heat transfer and pressure drop in a duct, *Int. J. Heat Mass Transfer*, 27, 1715–1723.

Sparrow, E.M., and Liu, C.H., 1979. Heat transfer, pressure drop and performance relationships for in-line, staggered, and continuous plate heat exchangers, *Int. J. Heat Mass Transfer*, 22, 1613–1625.

Sparrow, E.M., Baliga, B.R., and Patankar, S.V., 1977. Heat transfer and fluid flow analysis of interrupted-wall channels with application to heat exchangers, *J. Heat Transfer*, 99, 4–11.

Stasiek, J., Collins, M.W., Ciofalo, M., and Chew, P.E., 1996. Investigation of flow and heat transfer in corrugated passages — I. experimental results, *Int. J. Heat Mass Transfer*, 39, 149–164.

Suga, T., and Aoki, H., 1991. Numerical study on heat transfer and pressure drop in multilouvered fins, in *Proc. 1991 ASME/JSME Joint Thermal Engineering Conference*, Vol. 4, J.R. Lloyd and Y. Kurosake, Eds., ASME, New York, 361–368.

Sunden, B., 1999. Flow and heat transfer mechanisms in plate-frame heat exchangers, in *Heat Transfer Enhancement of Heat Exchangers*, S. Kakaç, Ed., Kluwer Academic, Dordrecht, 185–206.

Sunden, B. and Svantesson, J. 1990. Thermal hydraulic performance of new multilouvered fins, *Proc. 9th Int. Heat Trans. Conf.*, 5, 91–96.

Sunden, B. and Svantesson, J., 1992, Correlation of j and f factors for multilouvered heat transfer surfaces, *Proceedings of the 3rd UK National Heat Transfer Conference*, 805–811.

Suzuki, K., Xi, G.N., Inaoka, K., and Hagiwara, Y., 1994. Mechanism of heat transfer enhancement due to self sustained oscillation for an in-line fin array, *Int. J. Heat Mass Transfer*, 37(Suppl. 1), 83–96.

Tafti, D.K., Wang, G., and Lin, W., 2000. Flow transition in a multilouvered fin array, *Int. J. Heat Mass Transfer*, 43, 901–919.

Tanaka, T., Itoh, M., Kudoh, M., and Tomita, A., 1984. Improvement of compact heat exchangers with inclined louvered fins, *Bull. JSME*, 27, 219–226.

Theoclitus, G., 1966. Heat transfer and flow-friction characteristics of nine pin-fin surfaces, *J. Heat Transfer*, 88, 383–390.

Tiggelbeck, St., Mitra, N.K., and Fiebig, M., 1994. Comparison of wing-type vortex generators for heat transfer enhancement in channel flows, *J. Heat Transfer*, 116, 880–885.

Tishchenko, Z.V. and Bondarenko, V.N., 1983. Comparison of the efficiency of smooth-finned plate heat exchangers, *Int. Chem. Eng.*, 23(3), 550–557.

Torii, K., Nishino, K., and Nakayama, K., 1994. Mechanism of heat transfer augmentation by longitudinal vortices in a flat plate boundary layer, in *Heat Transfer 1994, Proc. 10th Int. Heat Trans. Conf.*, 5, 123–128.

Torikoshi, K. and Kawabata, K., 1989. Heat transfer and flow friction characteristics of a mesh finned air-cooled heat exchanger, in *Convection Heat Transfer and Transport Processes*, R.S. Figliola, M. Kaviany, and M.A. Ebadian, Eds., ASME Symp. HTD, 116, 71–77.

Usami, H., 1991. Pressure drop characteristics of OSF surfaces, in *Proc. 1991 ASME/JSME Joint Thermal Engineering Conference*, 4, J.R. Lloyd and Y. Kurosake, Eds., ASME, New York, 425–432.

Webb, R.L., 1983. Enhancement for extended surface geometries used in air-cooled heat exchangers, in *Low Reynolds Number Flow Heat Exchangers*, Hemisphere, Washington, D.C., 721–734.

Webb, R.L., 1987. Chapter 17, in *Handbook of Single-Phase Heat Transfer*, S. Kakaç, R.K. Shah, and W. Aung, Eds., John Wiley & Sons, New York, 17.1–17.62.

Webb, R.L., 1988, PSU unpublished data for five radiators.

Webb, R.L. and Joshi, H.M., 1983. Prediction of the friction factor for the offset strip-fin matrix, in *ASME-JSME Thermal Eng. Joint Conf.*, 1, ASME, New York, 461–470.

Webb, R.L. and Jung, S.H., 1992. Air-side performance of enhanced brazed aluminum heat exchangers, *ASHRAE Trans.*, 98(2), 391–401.

Webb, R.L. and Trauger, P., 1991. The flow structure in the louver fin heat exchanger geometry, *Exp. Thermal Fluid Sci.*, 4, 205–217.

Webb, R.L., Chang, Y.-J., and Wang, C.-C., 1995. Heat Transfer and Friction Correlations for the Louver Fin Geometry, *IMechE Symp.* C496/081/95.

Wieting, A.R., 1975. Empirical correlations for heat transfer and flow friction characteristics of rectangular offset fin heat exchangers, *J. Heat Transfer*, 97, 488–490.

Xi, G., Hagiwara, Y., and Suzuki, K., 1995. Flow instability and augmented heat transfer of fin arrays, *J. Enhanced Heat Transfer*, 2, 23–32.

Xi, G.N. and Shah, R.K., 1999. Numerical analysis of OSF heat transfer and flow friction characteristics, in *Proc. Int. Conf. Computational Heat and Mass Transfer*, A.A. Mohamad and I. Sezai, Eds., Eastern Mediterranean University Printinghouse, Eazimaguse, Cyprus, 75–87.

Xu, W. and Min, J., 2004. Numerical predictions of fluid flow and heat transfer in corrugated channels, *Proc. Intl. Symp. on Heat Transfer Enhancement and Energy Conservation*, Guangzhou, China, Jan. 12–15, 1, 714–721.

Xuan, Y.M., Zhang, H.L., and Ding, R., 2001. Heat transfer enhancement and flow visualization of wavy-perforated plate-and-fin surface, in *Proc. Third International Conference on Compact Heat Exchangers and Enhancement Technology for the Process Industries*, R.K. Shah, A.W. Deakin, H. Honda, and T.M. Rudy, Eds., Begell House, New York, 215–222.

Yang, L.C., Asako, Y., Yamaguchi, Y., and Faghri, M., 1995. Numerical prediction of transitional characteristics of flow and heat transfer in a corrugated duct, in *Heat Transfer in Turbulent Flows*, ASME Symp. Ser. HTD, 318, 145–152.

Zhang, L.W., Tafti, D.K., Najjar, F.M., and Balachandar, S., 1997. Computations of flow and heat transfer in parallel plate fin heat exchangers on the CM-5: effects of flow unsteadiness and three-dimensionality, *Int. J. Heat Mass Transfer*, 40, 1325–1341.

Zhang, X. and Tafti, D.K., 2003. Flow efficiency in multi-louvered fins, *Int. J. Heat Mass Transfer*, 46, 1737–1750.

EXTERNALLY FINNED TUBES

6.1 INTRODUCTION

Finned-tube heat exchangers have been used for heat exchange between gases and liquids (single or two phase) for many years. Figure 6.1a and b shows two important finned-tube heat exchanger construction types. Figure 6.1a is the plate fin-and-tube geometry, and Figure 6.1b shows individually finned tubes. Although round tubes are shown in Figure 6.1, oval or flat tubes are also used, for example, in automotive radiators. A plain air-side geometry is shown in the Figure 6.1 geometries. Figure 6.1a and b show a staggered tube arrangement, which provides higher performance than an inline tube arrangement. Externally finned tubes are also frequently used for liquids. Figure 6.1c shows an integral-fin tube used for liquids. However, extended surfaces for liquids typically use lower fin height than for gases. Because liquids have higher heat transfer coefficients than gases, fin efficiency considerations require shorter fins with liquids than with gases. When used with liquids, the fin height is typically in the 1.5 to 3 mm range. The dominant amount of material in this chapter is applicable to high fins, which are used for gases.

Because the gas-side heat transfer coefficient is typically much smaller than the tube-side value, it is important to increase the air-side hA-value. A plain surface geometry will increase the air-side hA value by increasing the area (A). Use of enhanced fin surface geometries will provide higher heat transfer coefficients than a plain surface. To maintain reasonable friction power with low-density gases, the gas velocity is usually less than 5 m/s.

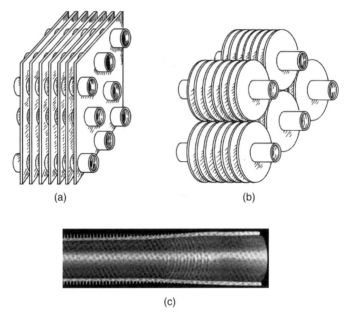

(a)   (b)

(c)

Figure 6.1 Finned tube geometries used with circular tubes: (a) plate fin-and-tube used for gases, (b) individually finned tube having high fins, used for gases. (From Webb [1987]) (c) Low, integral-fin tube.

Important basic enhancement geometries include wavy and interrupted fins. Variants of the interrupted strip fin are also used with finned-tube heat exchangers for heat exchange to a tube-side fluid. Figure 6.2 shows four variants of the interrupted strip fin applied to gas-fluid heat exchangers. Figure 6.2a through d are applied to circular tubes, and Figure 6.2e and f are used with flat, extruded aluminum tubes. These extruded tubes have internal membranes for pressure containment. Figure 6.2a through 6.2d are commonly used in commercial air-conditioning equipment. In Figure 6.2a, the segmented aluminum fin (or "spine fin") is spiral-wound on the tube, and is affixed using an adhesive as described by Abbott et al. [1980] and by Webb [1983a, 1987]. In Figure 6.2b through d, a copper or aluminum tube is mechanically expanded on aluminum plate fins. The Figure 6.2e version is typically made with aluminum fins brazed on flat extruded aluminum tubes. Figure 6.2f shows the "skive fin" design, as described by O'Connor and Pasternak [1976], for which the fins are slit from the thick wall of an aluminum extrusion and bent upward. These constructions are described by Webb [1983a, 1987], and Shah and Webb [1982]. The Figure 6.2e and f geometries on flat aluminum tubes have not found wide commercial acceptance as that of the Figure 6.2a through 6.2d geometries for residential air-conditioning applications. This may be due in part to the cost of brazing or of the extruded tubes. However, the Figure 6.2e and f designs have been recently introduced for use in automotive air-conditioning evaporators and condensers. Recent developments in automotive brazed aluminum manufacturing technology have made the costs of the Figure 6.2e heat exchanger construction more favorable.

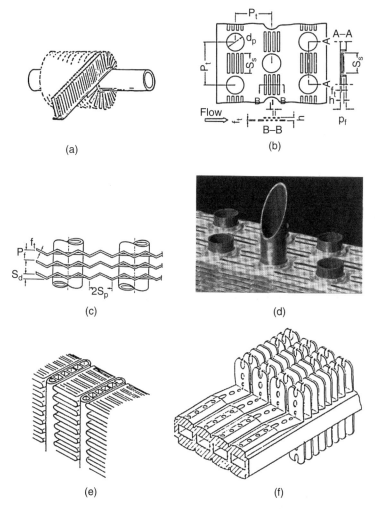

Figure 6.2 Air-side geometries used in finned tube heat exchangers: (a) spine-fin, (b) slit type OSF, (c) wavy fins, (d) convex louver fin, (e) Louver fins brazed to extruded aluminum tube, (f) interrupted skive fin integral to extruded aluminum tube.

Figure 6.14c shows a recent fin design, where winglet-type vortex generators are formed radially (see Section 6.5.7).

Because the gas-side heat transfer coefficient may be 5 to 20% that of the tube-side fluid, the use of closely spaced, high fins is desirable. High fin efficiency can be obtained, if the fin material has high thermal conductivity, e.g., aluminum or copper. If steel fins are required, fin efficiency considerations will dictate shorter or thicker fins. Operational constraints, such as gas-side fouling may limit the fin density. Air-conditioning applications use 500 to 800 fins/m while process air coolers are usually limited to 400 fins/m. Dirty, soot-laden gases may limit the fin density to 200 fins/m. Different correlations are required for the Figure 6.1a and b geometries.

The fin material used also depends on the operating temperature and the corrosion potential. Listed below are the fin and tube materials used in a variety of applications:

1. Residential air conditioning: Aluminum fins and copper or aluminum tubes
2. Automotive air conditioning: Aluminum fins and aluminum tubes
3. Automotive radiators: Aluminum fins brazed to aluminum tubes, or copper fins soldered to brass tubes
4. Process industry heat exchangers: Air-cooled condensers that may use aluminum fins on copper or steel tubes
5. Boiler economizers and heat-recovery exchangers: Steel fins on steel tubes required by the higher operating temperature

In addition to describing the various fin geometries and their performance characteristics, this chapter compares the performance of alternative heat exchanger and fin configurations. Then heat exchanger and enhanced fin geometries that will yield the highest performance per unit heat exchanger core weight are identified. Finally, possible improvements in the air-side surface geometry are considered.

6.2 THE GEOMETRIC PARAMETERS AND THE REYNOLDS NUMBER

6.2.1 Dimensionless Variables

The flow pattern in finned-tube heat exchangers is very complex, due to its three-dimensional nature and flow separations. The use of enhanced fin geometries introduces further complications. Good progress has been made in attempts to analytically or numerically predict the heat transfer coefficient and friction factor. The numerical achievements on finned tubes are discussed in Section 6.11.

Equations to predict the heat transfer coefficient and friction factor are usually based on power-law correlations using multiple regression techniques. Use of this method requires that one know the geometric and flow variables involved. The geometric and flow variables that affect the heat transfer coefficient and friction factor are the following.

1. Flow variables: Air velocity (u), viscosity (μ), density (ρ), thermal conductivity (k), and specific heat (c_p).
2. Tube bank variables: Tube root diameter (d_o), transverse tube pitch (S_t), row pitch (S_l), tube layout (staggered or inline), and the number of rows (N).
3. Fin geometry variables: For a plain fin, these are the fin pitch (p_f), fin height (e), fin thickness (t). If, for example, an enhanced wavy fin geometry is used the added variables are the wave height (e_w), the wave pitch (p_w), and the wave shape.

Thus, there are seven geometry variables for a plain fin (excluding the tube layout) and five flow variables. Two additional variables are introduced to account for the wavy fin geometry. Dimensional analysis specifies that the number of possibly important dimensionless groups is the number of variables minus the number of dimensions. Because four dimensions are involved for heat transfer (mass, length, time, and heat transfer), there are eight dimensionless variables for the plain fin and ten dimensionless variables for the wavy fin. The dimensionless flow variables typically used in correlations are the Reynolds number and the Prandtl number. For heat transfer, one has the Nusselt number or the Stanton number. For pressure drop, one uses the friction factor.

There are no "rules" for selecting the appropriate dimensionless geometric variables. This is simply "cut-and-try" to select the ones that give the best correlation of the data set. Further, power-law correlations have no rational basis and provide only an empirical correlation of the data set. It is dangerous to extrapolate such correlations beyond the range of the variables used to develop the correlation.

6.2.2 Definition of Reynolds Number

The basic definition of the Reynolds number is $L_c G/\mu$, where L_c is a characteristic dimension and G is usually defined as the mass velocity in the minimum flow area. For fully developed flow inside a plain tube, there is only one possible characteristic dimension — the tube diameter. As previously stated, there are seven dimensions associated with a plain fin geometry, and nine for the wavy fin geometry. Hence, there is no unique characteristic dimension. Hence, there are eight possible values of L_c for definition of the Reynolds number of the plain fin geometry. One approach to defining Reynolds number is to identify a characteristic dimension that appears to dominate over the other possible choices. For a bare tube bank, the possible choices are S_t, S_l, and d_o. However, there is no uniform agreement on the characteristic dimension used to define the Reynolds number. Two different characteristic dimensions have been used to define L_c in the Reynolds number. They are the tube diameter (d_o) or the hydraulic diameter (D_h). Kays and London [1984] choose to use the hydraulic diameter for the characteristic dimension for all situations, including bare and finned-tube banks. There is no evidence to suggest that hydraulic diameter is a better choice. In fact, there is evidence that the tube diameter may be a better choice for finned-tube banks. This is shown later. The conclusion is that the choice of characteristic dimension is arbitrary.

For fully developed flow in tubes, one defines laminar and turbulent regimes. Do such regimes also exist for finned-tube banks? To evaluate this, consider the case of the Figure 6.1a geometry. Assume that the geometry uses $d_o = 19$ mm with an equilateral triangular pitch of $S_t = 44.45$ mm, and 472 fins/m with 0.2 mm thickness. Assume air enters the exchanger at 3 m/s and 20°C. The mass velocity in the minimum flow area (G) is 7.34 kg/m²s and the hydraulic diameter is 3.68 mm. The Reynolds numbers based on d_o and D_h are 8710 and 1640, respectively. Is the flow laminar or turbulent? Based on Re_{Dh}, one would say it is laminar. But, based on tube diameter (Re_d), one would say it is turbulent. In reality, it exhibits some of both characteristics.

If the tubes were not present, the flow geometry would be a parallel plate channel, for which D_h = 3.82 mm. The Reynolds number is 1588, which is clearly laminar. However, the tubes shed eddies, which wash over the fin surface and provide mixing of the flow.

If the Reynolds number based on hydraulic diameter were dominant over the Reynolds number based on tube diameter, one would expect that the Nu and f data for different fin pitches would tend to fall on one line. It is shown below that this is not the case.

6.2.3 Definition of the Friction Factor

This book strives to use only the Fanning friction factor (f), defined in the Nomenclature at the end of the chapter. However, other friction factor definitions are frequently used for tube banks (bare and finned). A common definition for tube banks is given the symbol f_{tb}. It is related to the Fanning friction factor by the equation

$$f_{tb}N = \frac{fL}{D_h} \tag{6.1}$$

where N is the number of tube rows in the flow direction and L is the flow depth. For bare or finned tubes, $L = S_l (N - 1) + d_e$, where d_e is diameter over the fins. For a bare tube bank, $d_e = d_o$.

6.2.4 Sources of Data

Much of the data on finned tube heat exchangers were developed by industrial organizations, and is thus proprietary. However, some have been published in the open literature. There are two sources for compilations of published data. One is the book by Kays and London [1984] and a report by Rozenman [1976a]. Both are relatively old, and do not contain state-of-the-art data on enhanced surfaces. Kays and London present data for 22 geometries and Rozenman provides data for 161 geometries. The data of both authors are presented in the format of j and f vs. Re_{Dh}, and complete geometry details are provided. Rozenman [1976b] provides empirical correlations for j and f vs. Re_{Dh}. Since publication of these two references, various journal and conference publications have provided additional data on numerous geometries, including enhanced fin designs. These data are discussed in the following sections.

6.3 PLAIN PLATE-FINS ON ROUND TUBES

Figure 6.1a shows the finned-tube geometry with continuous, plain plate fins in a staggered tube layout. An inline tube geometry is seldom used because it provides substantially lower performance than the staggered tube geometry. The performance difference between inline and staggered tube layouts is discussed in Section 6.8.

6.3.1 Effect of Fin Spacing

Rich [1973] measured heat transfer and friction data for the Figure 6.1a geometry having plain fins, four rows deep, on 12.7-mm-diameter tubes equilaterally spaced on 32-mm centers. The tubes and fins were made of copper, and the fins were solder-bonded to minimize contact resistance. The geometry of all heat exchangers was identical, except the fin density $(1/p_f)$, which was varied from 114 to 811 fins/m. All fins were 0.25 mm thick.

Figure 6.3 shows the friction factor and the Colburn j factor $(StPr^{2/3})$ data (smoothed curve fit) as a function of Reynolds number (based on D_h) for the eight fin spacings tested. Entrance and exit losses were subtracted from the pressure drop, and are not included in the friction factor. Figure 6.3 clearly shows that the hydraulic diameter based Re (Re_{Dh}) does not correlate either the j or f data.

Rich proposed that the friction drag force is the sum of the drag force on a bare tube bank (Δp_t) and the drag caused by the fins (Δp_f). The difference between the total drag force and the drag force associated with the corresponding bare tube bank is the drag force on the fins. Thus, the friction component resulting from the fins is given by

$$f_f = (\Delta p - \Delta p_t) \frac{2A_c\rho}{G^2 A_f} \tag{6.2}$$

The term Δp_t is that measured for a bare tube bank of the same geometry, without fins. Both Δp drop contributions are evaluated at the same minimum area mass

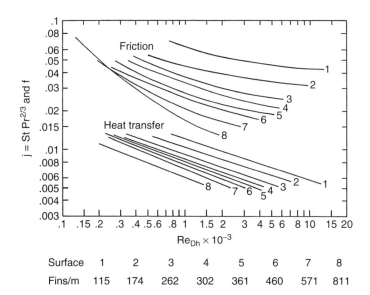

Figure 6.3 Heat transfer and friction characteristics of a four-row plain plate fin heat exchanger for different fin spacings.

velocity. Figure 6.4 shows the same j factor data and the fin friction factor calculated by Equation 6.2 plotted vs. the Reynolds number based on the longitudinal row pitch (S_l). The row pitch (S_l) is constant for all of the test geometries. Figure 6.4 shows that the j factor is a function of velocity in the minimum flow area (G_c), and is essentially independent of fin spacing. At the same mass velocity (G_c), the bare tube bank heat transfer coefficient is 40% larger than that of the finned-tube bank. Figure 6.4 shows that the resulting friction correlation is reasonably good, except for the closest fin spacings. The friction factor data of surfaces 7 and 8 may be questionable, since these surfaces show smaller j/f values than for the other fin spacings. This behavior is unexpected. Normally, the j/f ratio will increase as the fin spacing is reduced, because the fractional parasitic drag associated with the tube is reduced. Use of the Reynolds number based on S_l has no real significance, as all geometries tested had the same S_l. The same degree of correlation would result from use of a Reynolds number based on the tube diameter (d_o), which was also constant. Figure 6.4 may be regarded as evidence that the Reynolds number based on hydraulic diameter will not correlate the effect of fin pitch.

In a later study, Rich [1975] used the same heat exchanger geometry with 551 fins/m to determine the effect of the number of tube rows on the j factor. Figure 6.5 shows the average j factor (smoothed data fit) for each exchanger as a function of Re_{SI}. The numbers on the figure indicate the number of rows in each coil. The row effect is greatest at low Reynolds numbers and becomes negligible at $Re_{SI} > 15,000$.

Following Rich [1973, 1975], a number of studies have been performed on plain finned-tube heat exchangers (McQuiston [1978], Seshimo and Fujii [1991], Kayansayan [1993], Wang et al. [1996b], Abu Madi et al. [1998], Yan and Sheen [2000],

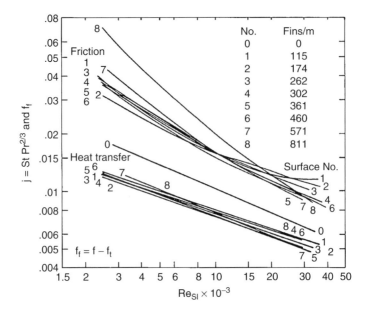

Figure 6.4 Plot of the j factor and the fin friction vs. Re_{SI}.

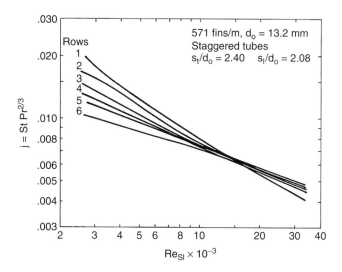

Figure 6.5 Average heat transfer coefficients for plain plate-finned tubes (571 fins/m) having one to six rows. Same geometry dimensions as Figure 6.4.

Wang and Chi [2000]). These studies generally confirmed Rich's observation that the j factor shows negligible effect of fin pitch, but does show significant row effect at low Reynolds numbers. Wang et al. [1996b] and Wang and Chi [2000] reported that the friction factor is independent of the number of tube rows.

6.3.2 Correlations for Staggered Tube Geometries

Correlations to predict the j and f factors vs. Reynolds number for plain fins on staggered tube arrangements were developed by McQuiston [1978], Gray and Webb [1986], Kim et al. [1999], and Wang et al. [2000a]. The McQuiston correlation is based on the data of Rich [1973, 1975] shown on Figure 6.3 and Figure 6.5, and three other investigators. Gray and Webb used the same data set as McQuiston, plus that of two additional investigators. The McQuiston and the Gray and Webb heat transfer correlations are comparable in accuracy. However, the Gray and Webb friction factor correlation is much more accurate than that of McQuiston.

The Gray and Webb [1986] heat transfer correlation (for four or more tube rows of a staggered tube geometry) is

$$j_4 = 0.14 \, \mathrm{Re}_d^{-0.328} \left(\frac{S_t}{S_l} \right)^{-0.502} \left(\frac{s}{d_o} \right)^{0.031} \tag{6.3}$$

Equation 6.3 assumes that the heat transfer coefficient is stabilized by the fourth tube row, hence the j factor for more than four tube rows is the same as that for a four-row exchanger. The correction for rows less than four is based on correlation of the Figure 6.5 data, and is given by

$$\frac{j_N}{j_4} = 0.991 \left[2.24 \, \text{Re}_d^{-0.092} \left(\frac{N}{4} \right)^{-0.031} \right]^{0.607(4-N)} \tag{6.4}$$

Equations 6.3 and 6.4 correlated 89% of the data for 16 heat exchangers within ±10%. The McQuiston [1978] correlation gives comparable results.

The Gray and Webb [1986] friction correlation assumes that the pressure drop is composed of two terms. The first term accounts for the drag force on the fins, and the second term accounts for the drag force on the tubes. The validity of this model was previously established in the discussion of Figure 6.4. The friction factor of the heat exchanger is given by

$$f = f_f \frac{A_f}{A} + f_t \left(1 - \frac{A_f}{A} \right) \left(1 - \frac{t}{p_f} \right) \tag{6.5}$$

The friction factor associated with the fins (f_f) is given by Equation 6.6:

$$f_f = 0.508 \, \text{Re}_d^{-0.521} \left(\frac{S_t}{d_o} \right)^{1.318} \tag{6.6}$$

The friction factor associated with the tubes (f_t) is obtained from a correlation for flow normal to a staggered bank of plain tubes. Gray and Webb used the Zukauskas [1972] tube bank correlation, also given in Incropera and DeWitt [2001], to calculate the tube bank contribution, Δp_t. The f_t is calculated at the same mass velocity (G) that exists in the finned-tube exchanger. Equation 6.5 correlated 95% of the data for 19 heat exchangers within ±13%. The equation is valid for any number of tube rows. McQuiston [1978] also developed a friction correlation using the same data set; however, his friction correlation has quite high error limits, +167/–21% for the same data.

The range of dimensionless variables used in the development of the Gray and Webb correlations are $500 \leq \text{Re}_d \leq 24{,}700$, $1.97 \leq S_t/d_o \leq 2.55$, $1.7 \leq S_l/d_o \leq 2.58$ and $0.08 \leq s/d_o \leq 0.64$.

Recent work by Seshimo and Fujii [1991] provides more generalized correlations for staggered banks of plain fins having one to five tube rows. They tested 35 heat exchangers, having systematically changed geometric parameters. They used three tube diameters (6.35, 7.94, and 9.52 mm) with the multirow designs using an equilateral triangular pitch. Data were obtained for four fin densities, from 454 to 1000 fins/m. They tested one-row designs with different transverse tube pitch and fin depth and show that the one- and two-row data may be separately correlated using an entrance length parameter. Their data were correlated using a Reynolds number (Re_{Dv}) defined in terms of the volumetric hydraulic diameter (D_v). The D_v is given by

$$D_v = \frac{4 A_m L}{A} \tag{6.7}$$

where A_mL is defined as the total volume of the exchanger, less the volume of the tube bank. The one- and two-row data were correlated in terms of the entrance length parameter, $X_{Dv}^+ \equiv Re_{Dv} PrD_v / L$. The correlations are

$$Nu = 2.1(X_{Dv}^+)^n \tag{6.8}$$

$$fLD_v = c_1 + c_2(X_{Dv}^+)^{-m} \tag{6.9}$$

where the constants and exponents differ for one- or two-row exchangers:

1. One-Row: $n = 0.38$, $m = 1.07$, $c_1 = 0.43$, and $c_2 = 35.1$.
2. Two-Row: $n = 0.47$, $m = 0.89$, $c_1 = 0.83$, and $c_2 = 24.7$.

For three or more rows, these entrance length-based correlations did not work very well over the entire Reynolds number range ($200 < Re_{Dh} < 800$), because vortex shedding from the tubes seems to be an important factor. For $Re_{Dh} > 400$, the data were correlated using conventional Nusselt number and Reynolds number (Re_{Dh}) and flow based on the minimum flow area. For $Re_{Dh} < 400$, the one-row variant of Equations 6.8 and 6.9 correlated the data for one to five rows.

Historically, one finds that the tube diameter used in finned-tube heat exchangers is decreasing. Some window air conditioners use tube diameters as small as 5.0 mm. Kim et al. [1999] improved the Gray and Webb [1986] correlation by including the data of Wang and Chi [2000] and Youn [1997] for heat exchangers having 7-mm-diameter tubes. The Kim et al. correlation predicted the data for tube diameters larger than 7 mm with approximately the same accuracy as Gray and Webb [1986]. However, the improvement was significant for the 7-mm tube data. Another general correlation has been developed by Wang et al. [2000a], which includes tube diameters as small as 6.7 mm. The Kim et al. and Wang et al. correlations were compared at $Re_d = 2500$ for $1 \le N \le 3$, $1.3 \le p_f \le 3.0$ mm. For heat exchangers having 9.5-mm-OD tubes, the predicted j factors by Kim et al. correlation agree with those by Wang et al. correlation within 10%. For the 7.0-mm tube configuration, the two correlations yielded approximately the same j factors for $N = 3$. However, the difference increased as the row number decreased. Kim et al. correlation generally predicts larger friction factors than Wang et al. correlation. The Kim et al. [1999] correlation (for three or more tube rows) is

$$j_3 = 0.163 \, Re_d^{-0.369} \left(\frac{S_t}{S_l}\right)^{0.106} \left(\frac{s}{d_o}\right)^{0.0138} \left(\frac{S_t}{d_o}\right)^{0.13} \quad (N \ge 3) \tag{6.10}$$

$$\frac{j_N}{j_3} = 1.043 \left[Re_d^{-0.14} \left(\frac{S_t}{S_l}\right)^{-0.564} \left(\frac{s}{d_o}\right)^{-0.123} \left(\frac{S_t}{d_o}\right)^{1.17} \right]^{(3-N)} \quad (N = 1,2) \tag{6.11}$$

$$f_f = 1.455 \, \text{Re}_d^{-0.656} \left(\frac{S_t}{S_l} \right)^{-0.347} \left(\frac{s}{d_o} \right)^{-0.134} \left(\frac{S_t}{d_o} \right)^{1.23}$$ (6.12)

Kim et al. used the Jakob [1938] correlation for the friction factor due to tubes (f_t), which is

$$f_t = \frac{\pi}{4} \left(0.25 + \frac{0.118}{[S_{\pm}/d_o - 1]^{1.08}} \, \text{Re}_d^{-0.16} \right) \left[\left(\frac{S_t}{d_o} \right) - 1 \right]$$ (6.13)

Equation 6.5 is used to calculate the friction factor of the heat exchanger.

6.3.3 Correlations for Inline Tube Geometries

Schmidt [1963] reports data and a correlation for the inline geometry. However, little use exists for an inline tube arrangement. This is because tube bypass effects substantially degrade the performance of an inline tube arrangement. The degree of performance degradation for inline circular fins is discussed in Section 6.4.1.

6.4 PLAIN INDIVIDUALLY FINNED TUBES

6.4.1 Circular Fins with Staggered Tubes

Extruded fins or helically wrapped fins on circular tubes, as shown by Figure 6.1b, are frequently used in the process industries and in combustion heat-recovery equipment. Both plain and enhanced fin geometries are used. A staggered tube layout is used, especially for high fins ($e/d_o > 0.2$). A substantial amount of performance data has been published, and several heat transfer and pressure drop correlations have been proposed. The dominant amount of data was taken with a staggered tube arrangement, six or more tube rows deep. The correlations must account for the three tube bank variables (d_o, S_t, and S_l), the fin geometry variable (t, e, and s), and the number of tube rows. Webb [1987] provides a survey of the published data and correlations.

The recommended correlations for a staggered tube layout are made by Briggs and Young [1963] for heat transfer and Robinson and Briggs [1966] for pressure drop. Both correlations are empirically based and are valid for four or more tube rows. The heat transfer correlation is

$$j = 0.134 \, \text{Re}_d^{-0.319} \left(\frac{s}{e} \right)^{0.2} \left(\frac{s}{t} \right)^{0.11}$$ (6.14)

Equation 6.10 is based on airflow over 14 equilateral triangular tube banks and covers the following ranges: $1100 \leq \text{Re}_d \leq 18,000$, $0.13 \leq s/e \leq 0.63$, $1.0 \leq s/t \leq 6.6$, $0.09 \leq e/d_o \leq 0.69$, $0.01 \leq t/d_o \leq 0.15$, $1.5 \leq S_t/d_o \leq 8.2$. The standard deviation was 5.1%.

The isothermal friction correlation of Robinson and Briggs [1966], which we have rewritten in terms of the tube bank friction factor, is

$$f_{tb} = 9.47 \, \mathrm{Re}_d^{-0.316} \left(\frac{S_t}{d_o} \right)^{-0.927} \left(\frac{S_t}{S_d} \right)^{0.515} \tag{6.15}$$

Equation 6.15 is based on isothermal airflow data over 17 triangular pitch tube banks (15 equilateral and 2 isosceles). The data span the ranges $2000 \le \mathrm{Re}_d \le 50{,}000$, $0.15 \le s/e \le 0.19$, $3.8 \le s/t \le 6.0$, $0.35 \le e/d_o \le 0.56$, $0.01 \le t/d_o \le 0.03$, $1.9 \le S_t/d_o \le 4.6$. The standard deviation of the correlated data was 7.8%. Equation 6.11 is recommended with strong reservations, because it does not contain any of the fin geometry variables (e, s, or t). Because only a small range of s/e was covered in the tests, it is probable that the correlation will fail outside the s/e range used for developing the correlation. Gianolio and Cuti [1981] compared their data for 17 tube bank geometries containing 1 to 6 rows with the Briggs and Young [1963] and the Robinson and Briggs [1966] correlations. For induced draft, their six-row data are 0 to 10% above that of Briggs and Young. Their data for $N < 6$ were increasingly underpredicted as the number of rows decreases. For $N < 6$, they recommended that the Briggs and Young value be multiplied by the factor $(1 + G/\rho N^2)^{-0.14}$, where G is in kg/m²s units. Gianolio and Cuti [1981] also state that their induced draft h values are 10 to 40% higher than those for forced draft. No explanation is provided for this unexpected result. The Robinson and Briggs [1966] correlation did not predict their data very well.

Although the data on which Equation 6.14 is based included low fin data, e.g., $e/d_o < 0.1$, Rabas et al. [1981] developed more accurate j and f correlations for low fin heights and small fin spacings. The correlations are given below with the exponents rounded off to two significant digits.

$$j = 0.292 \left(\frac{d_o G_c}{\mu} \right)^n \left(\frac{s}{d_o} \right)^{1.12} \left(\frac{s}{e} \right)^{0.26} \left(\frac{t}{s} \right)^{0.67} \left(\frac{d_e}{d_o} \right)^{0.47} \left(\frac{d_e}{t} \right)^{0.77} \tag{6.16}$$

where $n = -0.415 + 0.0346(d_e/s)$. The friction correlation is

$$f = 3.805 \left(\frac{d_o G_c}{\mu} \right)^{-0.234} \left(\frac{s}{d_e} \right)^{0.25} \left(\frac{e}{s} \right)^{0.76} \left(\frac{d_o}{d_e} \right)^{0.73} \left(\frac{d_o}{S_t} \right)^{0.71} \left(\frac{S_t}{S_l} \right)^{0.38} \tag{6.17}$$

The equations are valid for staggered tubes with $N \ge 6$, $5000 \le \mathrm{Re}_d \le 25{,}000$, $1.3 \le s/e \le 1.5$, $0.01 \le s/t \le 0.06$, $e/d_o \le 0.10$, $0.01 \le t/d_o \le 0.02$, and $1.3 \le S_t/d_o \le 1.5$. The equations predicted 94% of the j data and 90% of the f data within ±15%. Rabas and Taborek [1987] present a survey of correlations, row correction factors, and other issues concerning low, integral-fin tube banks. They compare the ability of Equations 6.11 and 6.17 to predict other data sets, e.g., Groehn [1977]. Other correlations have been developed by Groehn [1977] and ESDU [1985], which were tested to higher Re_d. Apparently, no correlations exist for $\mathrm{Re}_d \le 1000$.

A staggered tube layout gives higher values for j at the same Re_d, especially for high fins ($e/d_o > 0.3$). Hence, the inline tube layout is not recommended for $e/d_o > 0.3$. Rabas and Huber [1989] discuss the reduction of the j factor with increased number of tube rows. Designers interested in inline finned-tube banks should refer to Schmidt [1963], who developed a heat transfer correlation based on data from 11 sources.

6.4.2 Low Integral-Fin Tubes

The Rabas et al. [1981] correlation given by Equations 6.16 and 6.17 is recommended for a staggered layout of low integral fins. Corresponding equations have not been developed for inline tube layouts. However, Brauer's j data [1964] for the $e/d_o = 0.07$ inline tube bank shown in Figure 6.28 is within 20% of that of a staggered bank having the same e/d_o, S_t/d_o, and S_l/d_o. The inline friction factor was approximately 35% smaller than that of the staggered bank. It appears that the performance decrement of inline banks having low fins, e.g., $e/d_o = 0.1$, is not nearly as severe as for high fins, e.g., $e/d_o = 0.4$.

6.5 ENHANCED PLATE FIN GEOMETRIES WITH ROUND TUBES

The wavy (or herringbone) fin and the offset strip fin (also referred to as parallel louver) geometries are the major enhanced surface geometries used on circular tubes. Figure 6.2c shows the wavy fin geometry applied to circular tubes. This figure shows the geometrical dimensions that influence the heat transfer and friction characteristics. The combination of tubes plus a special surface geometry establishes a very complex flow geometry. The heat transfer coefficient of the wavy fin is typically 50 to 70% greater than that of a plain (flat) fin.

6.5.1 Wavy Fin

There are two basic variants of the wavy fin geometry as illustrated in Figure 6.6. They are the smooth wave and the herringbone configurations. Much work has been done on the herringbone wave geometry. However, very limited work has been done on the smooth wave geometry. Goldstein and Sparrow [1977] used a mass transfer technique to measure local and average mass transfer coefficients on a model having herringbone configuration. At Re = 1000 (based on fin spacing), the wave configuration yielded a 45% higher mass transfer coefficient compared with the plain fin counterpart. They proposed that the enhancement results from Goetler vortices that form on concave wave surfaces.

Beecher and Fagan [1987] published heat transfer data for 20, 3-row plate fin-and-tube geometries having the Figure 6.6a wavy fin geometry. Figure 6.6a defines the geometric parameters of the wavy fin. All cores had staggered tubes with three rows with $S_t/S_l = 1.15$. Two tube diameters were tested, $d_o = 9.53$ and 12.7 mm. The fin pitch was varied from 244 fins/m (6.2 fins/in.) to 510 fins/m (13 fins/in.) with

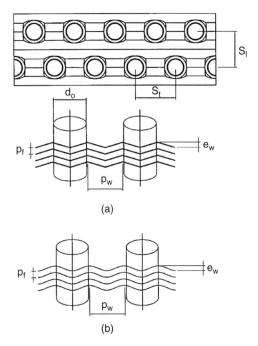

Figure 6.6 Two basic geometries of the wavy fin; (a) herringbone wave, (b) smooth wave.

dimensionless wave heights of $0.076 \le e_w/d_o \le 0.25$ and wave pitches of $0.058 \le e_w/p_w \le 0.346$. The wavy fins have a 3.18-mm-wide (0.125 in.) flat region around the fin collar. The Nusselt number data were presented as Nu_a ($= h_a D_h/k$) vs. the Graetz number, $\mathrm{Gz} = \mathrm{Re_{Dh}Pr}D_h/L$. The Nu_a is based on the arithmetic mean temperature difference (AMTD), rather than the LMTD.

Webb [1990] developed a multiple regression correlation of the Beecher and Fagan [1987] wavy fin data. Because the curve of Nu_a vs. Gz was not a straight line on log-log coordinates, a two-region correlation was used. The correlations are

$$\mathrm{Nu}_a = 0.5\,\mathrm{Gz}^{0.86}\left(\frac{S_t}{d_o}\right)^{0.11}\left(\frac{s}{d_o}\right)^{-0.09}\left(\frac{e_w}{S_l}\right)^{0.12}\left(\frac{p_w}{S_l}\right)^{-0.34} \qquad \mathrm{Gz} \le 25 \qquad (6.18)$$

$$\mathrm{Nu}_a = 0.83\,\mathrm{Gz}^{0.76}\left(\frac{S_t}{d_o}\right)^{0.13}\left(\frac{s}{d_o}\right)^{-0.16}\left(\frac{e_w}{S_l}\right)^{0.25}\left(\frac{p_w}{S_l}\right)^{-0.43} \qquad \mathrm{Gz} > 25 \qquad (6.19)$$

For $5 \le \mathrm{Gz} \le 180$, 96% of the data were correlated within ±10%. The Nusselt number is traditionally based on the LMTD rather than the AMTD; however, Beecher and Fagan discovered that, at low air velocities, small errors in air temperature measurement led to large errors in calculation of the LMTD and they chose to base the Nusselt number on the AMTD. The Nu based on the AMTD may be converted to the LMTD-based Nusselt number (Nu_l), by the following equation:

$$\text{Nu}_l = \frac{\text{Gz}}{4} \ln\left(\frac{1 + 2\text{Nu}_a / \text{Gz}}{1 - 2\text{Nu}_a / \text{Gz}}\right) \tag{6.20}$$

Much data on the herringbone wave fin geometry were published by Wang and co-workers (Wang et al. [1997b, 1998a, 1999a, 1999d], Abu Madi et al. [1998]), mostly on staggered layout. The effect of fin pitch and the effect of tube rows were generally similar to those of the plain finned tubes. The j factors were approximately independent of the fin pitch. The effect of tube rows, however, was not as pronounced as that of the plain finned tubes. The reason was attributed to the turbulence generated by the wave configuration. The effect of corrugation depth was investigated by Wang et al. [1999d]. The j and f factors increased as the corrugation depth increased. Wang et al. [1997b] provide j and f data on 6 inline geometries with row numbers varying from 2 to 4.

General j and f correlations for the herringbone wave configuration were developed by Kim et al. [1997]. The database included Beecher and Fagan [1987] and Wang et al. [1997b]. A procedure similar to that used by Gray and Webb [1986] was taken for the development of the correlation. Correlations include staggered, as well as inline geometries. For the staggered layout, 92% of the heat transfer data were correlated within ±10%, and 91% of the friction data were correlated within ±15%. Wang et al. [1999c] also provide j and f correlations for the herringbone wave geometry. The correlation has been developed based on their own data. Listed below are the Kim et al. [1997] correlations.

$$j_3 = 0.394\,\text{Re}_d^{-0.357}\left(\frac{S_t}{S_l}\right)^{-0.272}\left(\frac{s}{d_o}\right)^{-0.205}\left(\frac{p_w}{2e_w}\right)^{-0.558}\left(\frac{e_w}{s}\right)^{-0.133} \tag{6.21}$$

$$\frac{j_N}{j_3} = 0.978 - 0.01N \quad \text{Re}_d > 1000 \tag{6.22}$$

$$\frac{j_N}{j_3} = 1.35 - 0.162N \quad \text{Re}_d < 1000 \tag{6.23}$$

$$f_f = 4.467\,\text{Re}_d^{-0.423}\left(\frac{S_t}{S_l}\right)^{-1.08}\left(\frac{s}{d_o}\right)^{-0.034}\left(\frac{p_w}{2e_w}\right)^{-0.672} \tag{6.24}$$

Kim et al. used the Zukauskas [1972] correlation for the friction factor due to tubes (f_t). Equation 6.5 is used to calculate the friction factor of the heat exchanger.

Limited data on the smooth wave configuration are provided Mirth and Ramadhyani [1994] for the staggered tube layout. They tested five heat exchangers with two different wave fin patterns. Three samples had smooth continuous wave fins, while the other two had a relatively flat area between each pair of waves. The samples had four and eight rows. A correlation was developed based on their own data. Youn et al. [1998] provided additional data on two-row heat exchangers.

Kang and Webb [1998] compared j and f factors of the herringbone and the smooth wave geometries. The two configurations had the same wave depth and wave pitch. Data were taken from scaled-up (1.37 times) models. The smooth wave geometry yielded 4% higher j factors and 10% higher f factors. The smooth wavy geometry had an approximate sinusoidal fin pattern.

6.5.2 Offset Strip Fins

The Figure 6.2b OSF concept (also known as "slit fins") has been applied to finned-tube heat exchangers with plain fins for dry cooling towers and for refrigerant condensers. Figure 6.2b shows one such geometry, which was studied by Nakayama and Xu [1983]. Figure 6.7 shows the heat transfer coefficients of the OSF and a plain fin used in a two-row staggered tube heat exchanger having 525 fins/m on 10-mm-diameter tubes. At 3 m/s air velocity, the OSF provides a 78% higher heat transfer coefficient than the plain fin. For the same louver geometry, the OSF will provide a higher heat transfer coefficient, when used in the plate-and-fin-type heat exchanger. The OSF shown in Figure 5.4 provides 150% higher heat transfer coefficient than the plain fin at the same velocity. Comparison of plain fin geometries in Figures 5.4 and 6.7 shows that the heat transfer coefficient of the Figure 6.7 plain fin is 90% greater than that of the Figure 5.4 plain fin geometry. Thus, the flow acceleration and fluid mixing in the wake of the tube provide a substantial enhancement for plain fins on tubes.

Generalized empirical correlations for j and f vs. Re have not been developed for OSF geometry on round tubes. However, Nakayama and Xu [1983] propose an

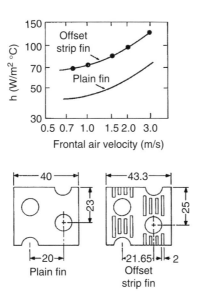

Figure 6.7 Comparison of the heat transfer coefficient for the OSF and plain fin geometries for 9.5-mm-diameter tubes, 525 fins/m, and 0.2-mm fin thickness as reported by Nakayama and Xu [1983].

empirical correlation to define the enhancement level (h/h_p) of an OSF geometry having 2.0-mm strip width (in the flow direction) and 0.2-mm fin thickness.

Recent data on OSF geometry have been provided by Wang and Chang [1998], Wang et al. [1999b], Kang and Webb [1998], Yun and Lee [2000], and Du and Wang [2000]. One important issue of the strip design is the direction of the strip, relative to the airflow direction. Radial strips will provide better heat conduction path compared with the normal strips, and will improve the heat transfer. Radial strips, however, will face the airflow at an oblique angle. This will lengthen the effective strip width, and may slightly reduce the heat transfer. Youn et al. [2003] investigated the performance of the radial strip geometry having 1.5-mm strip width. The data were compared with those of Du and Wang [2000], which had 1.0-mm width strips formed normal to the flow direction. The slit area fractions of both geometries were approximately the same (0.47 for the radial strips and 0.45 for the normal strips). The results showed that the j and f factors were approximately equal. It is likely that the pros and cons of the two geometries cancel each other, yielding approximately the same j and f factors.

Several correlations are available to predict the j and f factors of OSF heat exchangers (Nakayama and Xu [1983], Kang and Webb [1998], Wang et al. [1999b], Du and Wang [2000], Youn et al. [2003]). The geometry range of the correlations are, however, very limited, because most of them were developed using a small database of the investigators data. An interesting correlation concept has been proposed by Kang and Webb [1998]. They correlated their data using the strip area fraction. They show that the j factor increases as the strip area increases. The concept has also been used by Youn et al. [2003]. Currently, no general correlation is available.

6.5.3 Convex Louver Fins

Hitachi [1984] uses the convex louver fin geometry in its commercial plate fin-and-tube heat exchangers. Figure 6.2d is taken from the Hitachi [1984] product brochure. The performance of the convex louver fin plate-and-fin surface geometry was compared to the OSF geometry in Section 5.4. Hatada et al. [1989] report performance data of the Figure 6.2d type fin geometry for a one-row heat exchanger. Figure 6.8 illustrates the finned tube geometry, and Figure 6.9 shows the air-side h and Δp values vs. air velocity for three geometries tested. The geometry details of the three Figure 6.9 geometries are defined on Figure 6.8. Figure 6.9 shows data for two variants of the convex louver fin geometry. Fin number 2 has a uniform convex louver shape (louver angle $\theta_1 = 12.5°$). Fin number 1 has $\theta_1 = 17.5°$ in the regions between the tubes (between sections B-B and C-C on Figure 6.8), and $\theta_2 = 40°$ adjacent to the tubes. The reduced louver angle near the tubes allows more airflow in the vicinity of the tubes. The $17.5°$ louver angle in the fin region between the tubes was found to give high j and j/f by Hatada and Senshu's [1984] studies of the plate-and-fin geometry, as discussed in Section 5.4. Figure 6.9 shows that the number 1 fin geometry gives approximately 10% higher h value than the number 2 fin geometry. The h value of the number 1 louver fin geometry is 2.85 times that

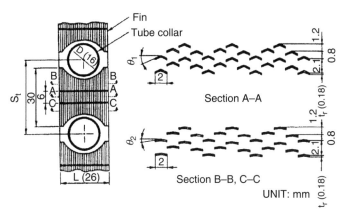

Figure 6.8 Convex louver plate fin-and-tube geometry tested by Hatada et al. [1989].

of the plain fin (number 3) at the same air velocity. When compared at the same air friction power, the *h* value of the number 1 fin is 2.3 times that of the number 3 plain fin.

The effect of fin pitch and tube rows on the *j* and *f* factors of the convex louver geometry have been investigated by Wang et al. [1996a, 1998a]. The samples had 15.5° convex angle. The trends were similar to those of the wavy fin geometry, which showed that the *j* factors were independent of fin pitch. The row effect on the *j* factors was relatively weak compared with that of the plain fin geometry. The friction factors were independent of the number of tube rows. The *j* and *f* factors of the convex louver fin heat exchangers showed a 21 to 41% and 60 to 72% increase as compared to the corresponding wavy fin geometry. The performance of the convex louver fin geometry was compared with the louver and wavy fin geometries. Based on the volume goodness comparison method (see Section 3.10.1), the convex louver geometry yielded the best performance, followed by the louver and the wavy fin geometries.

6.5.4 Louvered Fin

The louver geometry discussed in Section 5.3 has been applied to finned-tube heat exchangers. Louver patterns are formed on the fin area between tubes. Care must be exercised in louvering the fin surface, because the louvers can cut the conduction path from the tube. The air-side performance of louvered fin heat exchangers has been investigated by Wang and co-workers (Chang et al. [1995], Wang et al. [1998b, 1999e]). Their study included six different louver geometries, 1.21 mm $\leq p_f \leq$ 2.49 mm, one to six rows. Similar to the other geometries such as plain or wavy fin, the *j* factors were independent of fin pitch. The effect of the number of tube rows was negligible for $Re_d > 2000$. However, significant reduction of the *j* factor with increasing number of tube row was found for the lower Reynolds numbers. The row effect is discussed in detail in Rich [1975] for plain fins, and also shown in Figure 6.5.

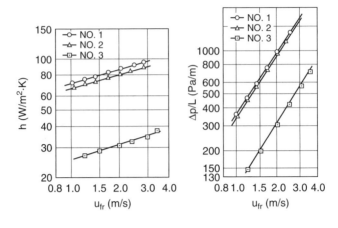

Heat Exchanger Dimensions

	Convex strip fin		Plain plate fin
Feature	Fin no. 1	Fin no. 2	Fin no. 3
Transverse tube pitch S_t (mm)	38	38	36
Fin depth L (mm)	26	26	42
Number of rows	1	1	1
Tube diameter d (mm)	16	16	16
Fin pitch p_f (mm)	2.2	2.2	2.1
Fin thickness t (mm)	0.18	0.18	0.18
Ramp angle θ_1 (Degrees)	17.5	12.5	—
Ramp angle θ_2 (Degrees)	40	12.5	—

Figure 6.9 Performance data for the Figure 6.7 convex louver surface geometries of Hatada et al. [1989].

The friction factors were independent of the tube rows. Wang et al. [1999e] developed j and f correlations based on their data.

The two most widely used enhanced geometries in the air-conditioning and refrigeration heat exchangers are the slit and louvered fins. Wang et al. [2001] compared the performances of the state-of-the-art configuration of the two geometries. Various comparison methods were tried; the volume goodness comparison, VG-1 cirterion in Table 3.1, etc. The two geometries yielded comparable results. Hence, one cannot conclude that one basic geometry is better than the other. Their performance will be dependent on the louver (or slit) pitch, and the fraction of the fin area on which louvers (or slits) exist.

6.5.5 Perforated Fins

Fujii et al. [1991] tested a plate-fin geometry made of corrugated, perforated plates whose fin geometry is illustrated in Figure 5.25. Section 5.7 discusses the performance of this geometry as a plate-and-fin heat exchanger configuration. Fujii et al. [1991] applied the Figure 5.25 surface to a one-row plate-fin heat exchanger having 0.5-mm-thick copper fins, and obtained experimental results. Their heat exchanger was made of 28-mm-diameter tubes at 76-mm tube pitch and 66-mm fin depth. Figure 6.10 defines the geometry details, and presents the test results for the 2 fin geometry variants, each for 6.0- and 8.0-mm fin pitch. Fin geometries number 1 and number 3 are the same, as are number 2 and number 4. Both fin geometry variants provide approximately equal Nu. However, geometry number 3 provides a lower friction factor than geometry number 4 (6-mm fin pitch). Although the Nu is

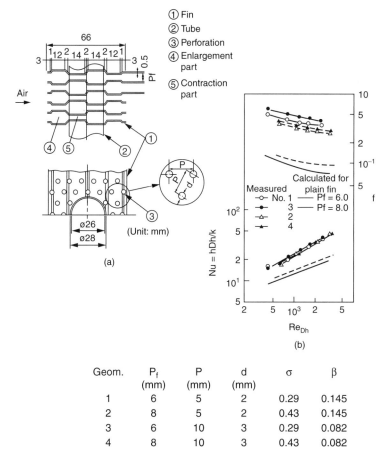

Geom.	P_f (mm)	P (mm)	d (mm)	σ	β
1	6	5	2	0.29	0.145
2	8	5	2	0.43	0.145
3	6	10	3	0.29	0.082
4	8	10	3	0.43	0.082

Figure 6.10 (a) Illustration of one-row finned-tube heat exchanger tested by Fujii et al. [1991], (b) air-side test results.

increased approximately 100%, the friction factor of geometry number 3 is increased a factor of 2.3, relative to a plain fin. This friction performance is not competitive with other high-performance fin geometries discussed in this chapter. Note that the Figure 6.10 data may be scaled to other tube diameters by scaling all the dimensions in the ratio of the new and original tube diameters.

6.5.6 Mesh Fins

The mesh fin geometry described in Chapter 5.8 has been applied to circular finned-tube heat exchangers. A good design copper mesh fin heat exchanger using 4-mm-diameter tubes yields approximately 100% higher heat transfer at the same pumping power than conventional louver fin heat exchangers (Ebisu [1999]). The high heat transfer coefficient of the mesh fin, small tube diameter (4 to 5 mm), and large surface-to-volume ratio of the mesh fin heat exchanger lead to the high performance. Extending the work of Torikoshi and Kawabata [1989] for a mesh fin heat exchanger with inline fin coniguration, Ebisu [1999] investigated the effect of offsetting the fin array ($0.0 \leq a/m \leq 0.5$) for 6 layers of mesh having dimensions $d = 0.2$ mm, $m = 2.3$ mm, $l = 5.0$ mm, where m is the distance between slits and a is the offset length of the succeeding fin (shown in Figure 5.27). The mesh fins were soldered to 4.0-mm-diameter tubes having 20-mm pitch. The heat transfer coefficient increased as the degree of offset increased. The $a/m = 0.0$ corresponds to inline fin configuration. The heat transfer enhancement obtained by 50% offsetting the fin array ($a/m = 0.5$) was 134%, when compared with results for the inline arrangement ($a/m = 0$). The corresponding increase of the pressure drop was 40%. Ebisu [1999] also investigated the effect of tube arrangement ($0.0 \leq y/P_t \leq 0.5$), where y is the offset distance of the downstream tube from inline position. The $y/P_t = 0.0$ and 0.5 corresponds to inline and staggered tube layouts, respectively. The largest heat transfer coefficient was obtained for $0.25 \leq y/P_t \leq 0.35$. The pressure drop was also the highest in that range. Figure 6.11 shows the flow visualization results for three-row tube bundles having different offsets. Figure 6.11 shows that, for $y/P_t = 0.0$ and 0.1, the tubes at the second and subsequent rows are surrounded by the separated flow from upstream tubes. For $y/P_t = 0.5$, the tubes in the third row are under the influence of the wakes from the first row. For $y/P_t = 0.25$, the downstream tubes are not influenced by the wakes.

In Figure 6.12, the performance of copper mesh finned heat exchangers are compared with copper or aluminum louver fin heat exchangers ($d_o = 9.5$ mm, $P_t = 25.4$ mm). Note that the mesh fins are attached to a 4.0-mm tube. In Figure 6.12, hA/V is the heat transfer per unit volume, and P/V is the pumping power per unit volume. Figure 6.12 shows that the hA/V values of good-design mesh finned heat exchangers having offset fin array and staggered tube layout (shown as large symbols in the figure) are approximately twice as high as that for the aluminum louver fin heat exchanger at the same pumping power.

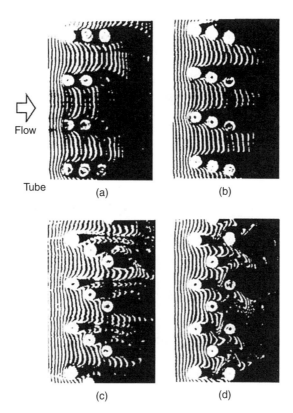

Figure 6.11 Flow visualization results of three-row tube bundles having different tube offsets: (a) $y/P_t = 0$, (b) $y/P_t = 0.1$, (c) $y/P_t = 0.25$, (d) $y/P_t = 0.5$. (From Ebisu [1999]. Reprinted with kind permission of Springer Science and Business Media.

6.5.7 Vortex Generators

For circular finned tubes, a low heat transfer coefficient exists in the wake region behind the tubes, especially at low Reynolds numbers. Vortex generators on the fin surface help to reduce the width of the wake zone, and thus improve heat transfer in the wake region. Studies of vortex generators on circular finned tubes have shown that the performance improvement is not as great (relative to a plain fin surface) as for flow in channels without any tubes. An explanation for this is that in the circular finned tube geometry, horseshoe vortices generated in front of the tube cause longitudinal horseshoe vortices, which increase the heat transfer over the fins along the path. These horseshoe vortices provide significant enhancement on the fins, relative to that provided by vortex generators.

Optimum location of vortex generators on the circular finned tube geometry was studied by Fiebig et al. [1990]. A single tube configuration with $d = 50$ mm, $H = 20$ mm in a laminar flow channel was experimentally investigated. The vortex generators used were a winglet pair having $\wedge$ configuration (common flow down),

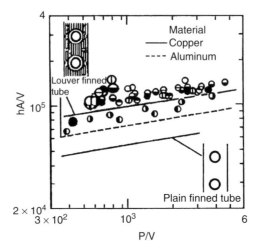

Figure 6.12 Heat transfer per unit volume (*E*) vs. pumping power per unit volume (*P*) for mesh fin heat exchangers. Louver fin and plain fin data shown as lines. (From Ebisu [1999]. Reprinted with kind permission of Springer Scientific and Business Media.)

aspect ratio of 2, 10 mm height arranged with 45° angle of attack. An approximate optimum location was found right behind the tube with winglets one tube diameter apart in the spanwise direction. In the Reynolds number range (based on channel height) 2000 to 5000, the configuration yielded a heat transfer enhancement up to 20%, relative to a tube with plain fins. Interestingly, the pressure drop decreased up to 10%. They attributed this to the delayed boundary layer separation on the tube by longitudinal vortices generated by the vortex generators, which introduce high-momentum fluid into the region behind the cylinder.

Fiebig et al. [1993] extended the study to a three-row heat exchanger geometry. Figure 6.13 shows the geometry they tested. Note that the height of the winglet is the same as the height of the channel. The average heat transfer enhancement was 55 to 65% for the inline tube arrangement, and 9% for the staggered arrangement. The corresponding pressure drop increase was 20 to 45%, and 3%, respectively. In the Fiebig et al. study, the vortex generators were in common-flow-down configuration.

Torii et al. [2002] tested a three-row geometry with vortex generators mounted in common flow-up configuration. As illustrated in Figure 6.14, vortex generators with 165° angle of attack were installed behind the first row only for the three-row configuration. Significant decrease of the pressure loss was achieved. For the staggered geometry, the heat transfer enhancement was 10 to 30%. However, the pressure loss was reduced 34 to 55%. For the inline vortex generator geometry, the heat transfer enhancement was 10 to 20%, and the pressure loss reduction was 8 to 15%. Torii et al. [2002] attributed the significant performance improvement to the separation delay, reduction of form drag, and removal of poor heat transfer zone behind the tube.

Lozza and Merlo [2001] tested two-row finned-tube heat exchangers having various enhanced geometries and vortex generators. Figure 6.15 shows the louver

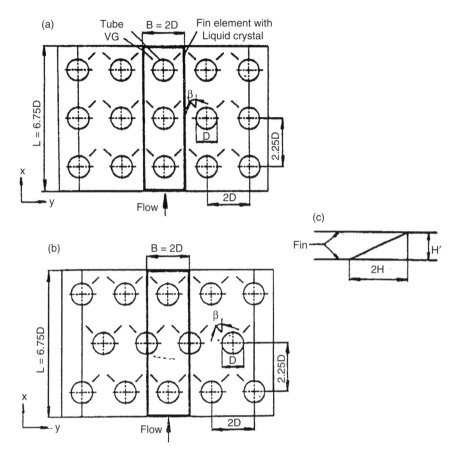

Figure 6.13 The three-row fin-and-tube geometry tested by Fiebig et al. [1993]: (a) inline, (b) staggered arrangement, (c) shape of the vortex generator, d = 32 mm, H = 7 mm, 45° angle of attack.

fin geometries tested. All geometries had 2.0-mm fin pitch. Figure 6.15b may be considered a high-performance conventional louver fin geometry, which has a significantly greater fraction of the surface covered by louvers. The Figure 6.15b surface provides 14% higher j factor than does the Figure 6.15a geometry. Figure 6.15c shows a louver fin geometry with winglet vortex generators 1.6 mm high. The Figure 6.15c geometry provides approximately the same j factor as the Figure 6.15a geometry and 12% smaller j factor than the Figure 6.15b geometry. Thus, it appears that the addition of winglet vortex generators to a louver fin geometry is not as good as allocating the same area to louvers. Further, the j/f ratio of the Figure 6.15c geometry is not as good as the Figure 6.15b geometry. Lozza and Merlo [2001] also tested two-row plain fin geometries, with and without winglet vortex generators. The plain fin/vortex generator surface (like Figure 6.15c without louvers) provided 50% j factor increase and 60% friction increase compared to the plain fin geometry. This enhancement is greater than that found by Fiebig et al. [1990], who used vortex

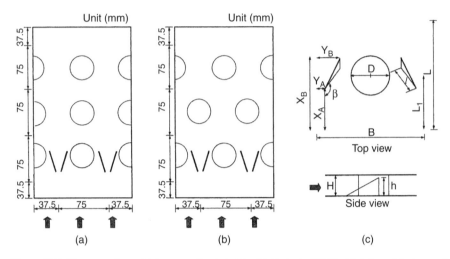

Figure 6.14 Three-row fin-and-tube geometry tested by Torii et al. [2002]: (a) inline, (b) staggered arrangement, (c) shape of the vortex generator. Vortex generators are arranged in common flow-up configuration, $d = 30$ mm, $H = 5.6$ mm, $h = 5.0$ mm, 165° angle of attack. (From Torii et al. [2002] Reprinted with permission from Elsevier.)

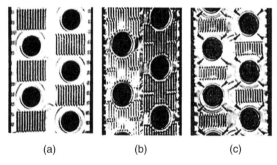

Figure 6.15 Fin configurations tested by Lozza and Melo [2001]: (a) louver fin A, (b) louver fin B, (c) louver fin with vortex generator. (From Lozzar and Melo [2001]. Reprinted with permission from Elsevier.)

generators only in the tube wake region. Based on the above comparison, it does not appear that vortex generators provide greater enhancement than can be obtained from conventional slit or louver fin geometries, when applied to round tubes. All the advanced fin geometries will reduce the fin efficiency by cutting the fins to form louvers, slits, vortex generators, etc.

O'Brien et al. [2003] tested four-row individually finned bundles having annular fins of 25.4 mm inside diameter and 57.15 mm outside diameter for $100 \leq Re_{Dh} \leq 3000$. Two winglet pairs, shown in Figure 6.16, were tested; (a) a winglet pair having 2:1 aspect ratio located on the downstream side of the tube in a 45° common-flow-down position, (b) a winglet pair having 4:1 aspect ratio located on the downstream side of the tube in a 45° common-flow-up position. The height of the winglet was the same as the fin pitch (2.54 mm). The tube bundle had a staggered equilateral

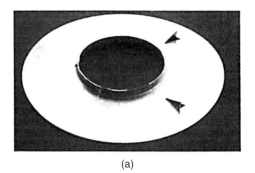

(a)

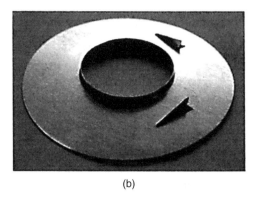

(b)

Figure 6.16 Individual fins having a pair of winglet vortex generators: (a) common-flow-down, (b) common-flow-up configuration. (From O'Brien et al. [2003].)

geoemtry, with tube pitch of 63.5 mm. The single-blow technique was used to obtain the heat transfer coefficient. The j factors of the enhanced geometries were higher than those of the plain geometry — 28% higher for the common-flow-down geometry and 40% higher for the common-flow-up geometry. The corresponding friction factor increase was 8.9 and 24%, respectively. The pressure drop results do not agree with the findings by Fiebig et al. [1990], who reported a decrease of pressure drop with vortex generators. The vortex generator height nondimensionalized by the channel height was 1.0 for O'Brien et al. [2003] and 0.5 for Fiebig et al. [1990]. The high vortex generator of O'Brien et al. [2003] is believed to have caused the higher pressure drop.

6.6 ENHANCED CIRCULAR FIN GEOMETRIES

6.6.1 Illustrations of Enhanced Fin Geometries

Figure 6.17 shows some of the enhanced fin geometries that have been used on circular tubes. In table 2 of Webb [1980], references are provided for information on performance of the Figure 6.17 fin geometries. All the geometries provide

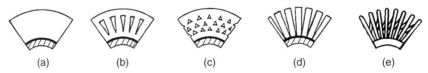

Figure 6.17 Enhanced circular fin geometries: (a) plain circular fin; (b) slotted fin; (c) punched and bent triangular projections; (d) segmented fin; (e) wire loop extended surface. (From Webb [1987].)

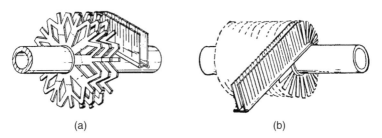

Figure 6.18 Segmented or spine fin geometries used in air conditioning applications. (a) (From LaPorte et al. [1979].) (b) Described by Abbott et al. [1980] and tested by Eckels and Rabas [1985].

enhancement by the periodic development of thin boundary layers on small-diameter wires or flat strips, followed by their dissipation in the wake region between elements. Perhaps the most popular enhancement geometry is the "segmented fin" or "spine fin" (Figure 6.17d), which is similar in concept to the offset strip fin shown in Figure 6.7. The segmented fin is used in a wide range of applications, from air conditioning to boiler economizers. Figure 6.18 shows two versions of the segmented fin used in air-conditioning applications. In Figure 6.18b, a 0.15-mm-thick aluminum fin strip is tension-wound on the tube and bonded to the tube with an epoxy resin. The fin segment width is 0.75 mm. Figure 6.19 shows the j and f performance of the Figure 6.18a and b fin geometries and compares their performance with that of plain fins.

6.6.2 Spine or Segmented Fins

The data of Figure 6.19 are for eight rows on a triangular pitch with $d_o = 12.7$ mm and $t = 0.51$ mm. Note that the Figure 6.19 configurations have a negative fin tip clearance. Figure 6.19 shows that the Figure 6.18a and b geometries have approximately the same j and f for the same fin spacing. At $Re_d = 4000$, $j/j_p \simeq 2.1$ and $f/f_p \simeq 4.2$, where subscript p refers to the plain fin. The substantial heat transfer enhancement is accompanied by a relatively high friction factor increase. One is interested in increasing the value of hA/L by the use of an enhanced fin geometry. The Figure 6.18a fin has only 43% as much surface area as a plain fin of the same d_e. With $j/j_p = 2.1$, one would obtain $hA/h_pA_p = 2.1 \times 0.43 = 0.90$. For the same hA, the Figure 6.18a geometry would yield a 53% saving of fin material $(1 - 0.43/0.90)$.

A steel segmented fin geometry has been used for boiler economizers and waste heat recovery boilers. Figure 6.20 shows the j and f vs. Re_d curves for a four-row

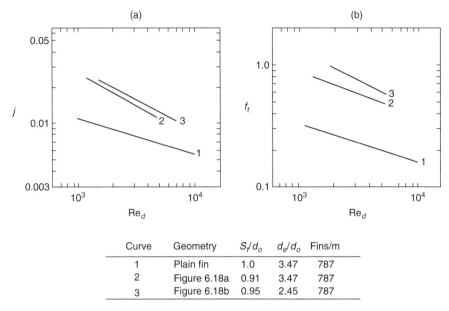

Curve	Geometry	S_f/d_o	d_e/d_o	Fins/m
1	Plain fin	1.0	3.47	787
2	Figure 6.18a	0.91	3.47	787
3	Figure 6.18b	0.95	2.45	787

Figure 6.19 j and f vs. Re_d characteristics of the Figure 6.18 enhanced fin geometries compared with a plain, circular fin geometry as reported by Eckels and Rabas [1985].

staggered and a seven-row inline tube segmented fin geometry as reported by Weierman et al. [1978]. Also shown are the j and f curves for a staggered plain fin geometry having the same geometrical parameters as the staggered segmented geometry. The plain fin j and f values were calculated using Equations 6.14 and 6.15, respectively. For a staggered tube geometry, Figure 6.20 shows that the j factor of the segmented fin is 40% greater than that of the plain fin geometry. Note that the heat transfer performance of the inline segmented fin geometry is much lower than that of the staggered, segmented fin geometry. The poor performance of the inline geometry was previously noted, and is further discussed in Section 6.8.

Steel fin geometries are used for boiler economizers and heat-recovery boilers. Steel is preferred because of the high gas temperature and the corrosive potential of the combustion products. Data on several types of fin geometries are reported. Weierman [1976] gives empirical design correlations for steel segmented and plain fin geometries for staggered and inline tube layouts. Rabas et al. [1986] present additional data on steel segmented fin tubes. The Rabas et al. [1986] provides data for staggered and inline arrangements. Breber [1991] provides a review of data and new data on "stud fin" circular tubes. This geometry consists of steel studs welded to a base tube. The stud cross-sectional shape may be round, rectangular, or elliptical. Breber recommends appropriate correlations to predict the heat transfer coefficient and friction factor.

Holtzapple and Carranza [1991] and Holtzapple et al. [1990] provide friction and heat transfer data, respectively, on a spine fin tube made of copper tubes and

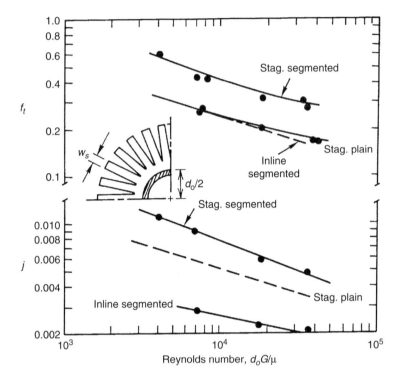

Figure 6.20 Comparison of segmented fins (staggered and inline tube layouts) with plain, staggered fin tube geometry as reported by Weierman et al. [1978]. $S_t/d_o = 2.25$, $e/d_o = 0.51$, $s/e = 0.12$, $w_s/e = 0.17$.

fins. The spine fins are cut from the thick wall tube, and bent so they project from the tube wall. Their tests were performed on 9.4- and 12.6-mm-diameter tubes with 6.8 and 9.3 mm spine lengths, respectively. There were 26 spines around the circumference, with an axial spacing of 4.76 mm. This geometry provides considerably less total finned surface area than typical of the Figure 6.18 geometries. An advantage of the Holtzapple et al. spine fin is that the fins are integral to the tube wall. However, the geometry is relatively expensive to manufacture. Data are provided on several tube pitch layouts. Carranza [1990] provides an empirical pressure drop correlation.

6.6.3 Wire Loop Fins

Benforado and Palmer [1964] provide data on the Figure 6.17e wire loop fin geometry. The data were taken for equilateral triangular and inline arrangements on 25.4-mm-diameter tubes on approximately 66-mm transverse tube pitch. The wire loops were made of 0.71-mm-diameter wire, with 12.9-mm fin height. They also tested a plain circular fin geometry having the same fin pitch and height. The wire loop geometry gave an approximately 50% higher heat transfer coefficient (based on actual surface area) and the same pressure drop as the plain fin.

6.7 OVAL AND FLAT TUBE GEOMETRIES

6.7.1 Oval vs. Circular Individually Finned Tubes

Oval and flat cross-sectional tube shapes are also applied to individually finned tubes. Figure 6.21 compares the performance of staggered banks of oval and circular finned tubes tested by Brauer [1964]. Both banks have 312 fin/m, 10-mm-high fins on approximately the same transverse and longitudinal pitches. The oval tubes gave 15% higher heat transfer coefficient and 25% less pressure drop than the circular tubes. The performance advantage of the oval tubes results from lower form drag on the tubes and the smaller wake region on the fin behind the tube. The use of oval tubes may not be practical unless the tube-side design pressure is sufficiently low.

Min and Webb [2004] numerically investigated the effect of tube aspect ratio of an oval tube on the air-side heat transfer and pressure drop characteristics of an infinite row heat exchanger having herringbone wavy fins. The numerical calculations were performed in three dimensions for 2 frontal air velocities 2.54 and 3.39 m/s yielding hydraulic diameter Reynolds numbers of 770 and 1150. Investigated were five tube geometries including a round tube, three elliptical oval tubes, and a flat tube. The tube configurations are shown in Figure 6.22, and the geometric details are provided in Table 6.1. The 15.88-mm round tube served as the baseline tube. The longitudinal and transverse tube pitches were 31.75 and 30.48 mm. The corrugation height (e_w) and the projected corrugation pitch (p_w) of the fin was 1.67 and 10.58 mm. The fin pitch was 2.117 mm. The four oval tubes were made by reforming the baseline round tube and had the same perimeter as that of the round tube. The aspect ratio of the three elliptical tubes were 2.0, 3.0, and 4.29, respectively, and the aspect ratio of the flat tube was 3.0. As the tube aspect increased, the air-side heat transfer coefficient and pressure drop decreased. At 2.54-m/s frontal velocity, the 3.0-aspect-ratio elliptical tube yielded a 9.0% lower air-side heat transfer coefficient and a 48.7% lower air-side pressure drop than the round tube. As compared to the 3.0-aspect-ratio elliptical tube, the same aspect ratio flat tube had a 2.3% higher air-side heat transfer coefficient and a 6.0% higher pressure drop.

Webb and Iyengar [2000] compared the air-side performance of the 1024 fins/m Figure 6.23 oval tube geometry with that of a two-row finned-tube heat exchanger having 8.0-mm-diameter round tubes and convex louver fins (CLF) at 807 fins/m, tested by Wang et al. [1996a]. Operating at the same frontal velocity, the comparison shows that oval tube h value is 10% higher and the Δp is 17% lower than that of the 8.58-mm-diameter round tube design. The oval tube design has approximately 20% greater surface area ratio (A/A_{fq}) than the 807 fins/m CLF design. Hence, the oval tube design offers 32% greater hA/A_{fr} (1.10×1.20) than the round tube design. The 32% greater hA/A_{fr} and 17% lower Δp offers a significant performance improvement.

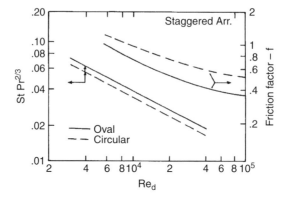

Tube Bank Dimensions (mm)

	Circular	Oval
Tube dia. (d)	29	19.9/35.2
Fin. ht. (e)	9.8	10/9.3
Fin. thk. (t)	0.4	0.4
Face pitch (S_t/d)	1.03	1.05
Row pitch (S_l/d)	1.15	1.04
Fins/m	312	312

Figure 6.21 Heat transfer and friction characteristics of circular and oval finned tubes in a staggered tube layout as reported by Brauer [1964].

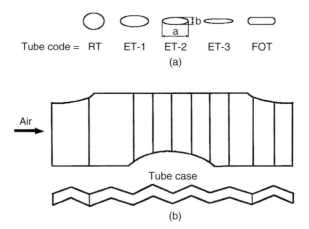

Figure 6.22 Tube geometries considered by Min and Webb for numerical calculation: (a) cross-sectional shape, (b) computational domain for the ET-2 oval tube case. (From Min and Webb, [2004].)

Webb and Iyengar [2000] also performed three-dimensional stress analysis of the Figure 6.23 oval tube geometry. The purpose was to determine if the oval tube geometry would be suitable for use as a refrigerant condenser. Hence, the analysis sought to determine the maximum allowable internal tube pressure. This work defined two failure modes:

1. Permanent deformation defined by the elastic limit of the material. This is the practical design limit.
2. Limiting load, at which the structure will not carry increased load (internal pressure). This failure occurs in the plastic range, and large material deformation occurs.

Table 6.1 Tube Dimensions Considered by Min and Webb [2004] for Numerical Calculation

Tube Code	a (mm)	b (mm)	a/b	D_h (mm)	$D_h/(D_h)_{RT}$
RT	15.88	15.88	1	14.86	1
ET-1	20.59	10.3	2	12.44	0.82
ET-2	22.37	7.45	3	9.26	0.62
ET-3	23.34	5.44	4.29	6.55	0.45
FOT	20.95	6.98	3	9.54	0.64

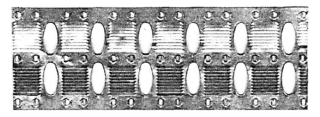

Figure 6.23 Photo of oval tube fins used in analysis of Webb and Iyengar. (From Webb and Iyengar [2000].)

Their analysis was performed for different fin thickness and fin pitch values. A key finding was that use of 707 fins/m and 140 μm fin thickness will satisfy the elastic limit at a tube internal pressure of 5.8 MPa. This is within the design requirements of R-22 and R-410A. Use of more than 707 fins/m will satisfy higher internal pressures.

6.7.2 Flat Extruded Aluminum Tubes with Internal Membranes

Higher design pressures are possible using flattened aluminum tubes made by an extrusion process. Figure 6.2f shows a patented finned tube concept made from an aluminum extrusion, as described by O'Connor and Pasternak [1976]. Such tubes can be made with internal membranes, which strengthen the tube and allow for a high tube-side design pressure. A variety of fin-and-tube shapes may be made in aluminum extrusions of different shapes. These "skive fins" are formed from the thick wall using a modified high-speed punch press without creation of scrap material. The punch press slits the thick aluminum wall and simultaneously bends the chip outward to form the fin. The process is applicable to virtually any cross-section geometry of the extrusion. Designs have been made with circular and flattened tube geometries. Haberski and Raco [1976] show photographs of a number of geometries that have been fabricated. Cox [1973] provides test data on a circular tube geometry, and Cox and Jallouk [1973] give test results on the Figure 6.2f geometry.

The Figure 6.2e flat tube geometry offers significant advantages over the Figure 6.2a or Figure 6.2b strip fin geometries on round tubes.

1. The airflow is normal to all of the narrow strips on the Figure 6.2e geometry, which is not the case for the Figure 6.2a spine fin design. Further, the wake dissipation length decreases in the direction of the fin base in the Figure 6.2a geometry.
2. A low-velocity wake region does not occur behind the tubes of the Figure 6.2e flat tube geometry. The low wake velocity of the Figure 6.2a and b geometries causes a substantial reduction of the heat transfer coefficient, as documented by Webb [1980].
3. The fraction of the Figure 6.2e surface that is louvered is substantially greater than in the Figure 6.2b geometry. If a greater area distribution of louvers were provided in the Figure 6.2b geometry, the fin efficiency would substantially decrease. This is because the slits would cut the heat conduction path from the base tube.
4. Further, the low projected area of the Figure 6.2e flat tube will result in lower profile drag.

Some automotive air-conditioning condensers and evaporators use brazed aluminum heat exchangers having flat extruded aluminum tubes with internal membranes. Figure 14.13 shows such heat exchangers. The internal membranes are required to meet the 11,000 kPa (1600 psi) burst pressure required for R-12 or R-134a used in automotive air conditioners. These heat exchangers typically use the louver fin geometry shown in Figure 5.2f. Webb and Gupte [1990] compare the performance of this heat exchanger construction with the wavy plate fin-and-tube, and the spine fin geometries. These comparisons are presented in Section 6.10.1.

6.7.3 Plate-and-Fin Automotive Radiators

Figure 6.24 shows an automotive radiator geometry having louvered plate fins on flat tubes. The automotive radiator is designed to operate at low pressure, so internal membranes are not required in the tubes. The outer tube cross-section dimensions of brass tubes are typically 1.0 to 1.5 mm × 12.0 to 19.0 mm. Achaichia and Cowell [1988] developed a dimensional correlation for the Figure 6.24 inline louver fin geometry, based on their tests of 16 core geometries. Their test data span $1.7 \leq p_f \leq 3.44$ mm, $0.81 \leq L_p \leq 1.4$ mm, $8 \leq S_t \leq 14$ mm, and $22 \leq \theta \leq 30°$. Their j and f factor correlations are

$$\gamma = \frac{1}{\theta}\left(0.936 - \frac{243}{Re_L} - 1.76\frac{P_f}{L_p} + 0.995\,\theta\right) \qquad (6.25)$$

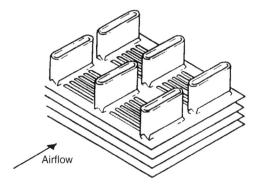

Airflow

Figure 6.24 Illustration of the louvered plate fin automotive radiator with inline tubes. (From Achaichia and Cowell [1988].)

$$ j = 1.234\gamma\, \mathrm{Re}_L^{-0.59} \left(\frac{S_t}{L_p} \right)^{-0.09} \left(\frac{p_f}{L_p} \right)^{-0.04} \tag{6.26} $$

$$ f = 533 p_f^{-0.22} L_p^{0.25} S_t^{0.26} H^{0.33} \left[\mathrm{Re}_L^{(0.318\log_{10}\mathrm{Re}_L - 2.25)} \right]^{1.07} \tag{6.27} $$

Where all dimensions are in mm.

6.7.4 Vortex Generators on Flat or Oval Fin-Tube Geometry

The strength of horseshoe vortex generated by a tube in a finned-tube heat exchanger increases with the diameter of the tube. However, the pressure drop and the size of the wake region, where the heat transfer is poor, also increase. One may reduce the wake size by using, instead of round tubes, flat or oval tubes with their major axis parallel to the main flow direction. The heat transfer deterioration in the wake region will be smaller for a flat tube of the same cross-sectional area as for a round tube. Further, the drag force on a flat tube will be smaller than that on a round tube of the same perimeter. The heat transfer coefficient of the flat tube geometry may be lower than that of the round tube geometry as previously noted and discussed by O'Brien et al. [2002]. Additional enhancement may be obtained by vortex generators.

Valencia et al. [1996] investigated the effect of vortex generator location for the three configurations shown in Figure 6.25. The minor diameter of the flat tube was 12 mm, and the major diameter was 70 mm. The best performance (50% heat transfer enhancement and 36% pressure drop increase) was obtained for their Configuration 1, where vortex generators are located upstream of the tube with the closest spacing between vortex generators equal to twice the tube width. Fiebig et al. [1994] extended the study to three-row staggered tube configuration shown in Figure 6.26, and compared the results with those from round tube geometry ($d_o = 32$ mm). The vortex generators increased the heat transfer dramatically (100%) for the flat tube, but only marginally (10%) for the round tube. At $\mathrm{Re}_H = 3000$ (based on channel height H),

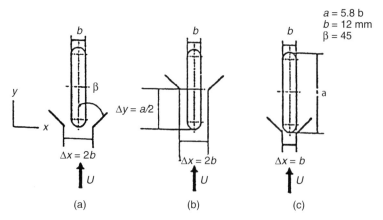

Figure 6.25 The locations of delta winglet pairs: (a) upstream of the tube with the closet winglet spacing equal to twice the tube width, (b) at the middle of the tube, (c) upstream of the tube with the closet winglet spacing equal to the tube width, as reported by Valencia et al. [1996]. (Reprinted with permission of ASME.)

the flat tube with vortex generators yielded 60% higher heat transfer coefficient and only half the pressure loss as the round tube with vortex generators.

O'Brien et al. [2002] investigated the effect of vortex generators on an oval tube geometry. Tests were conducted in a narrow rectangular duct fitted with an oval tube. The oval tube geometry had 3:1 aspect ratio and major diameter to channel height ratio $a/H = 8.66$. Pairs of winglets were used as vortex generators. Heat transfer coefficients were determined from transient thermal images obtained using infrared camera. The measured Nu_H (based on channel height) vs. Re_H results are shown in Figure 6.27. Addition of the single winglet pair to the oval tube geometry yielded average 38% heat transfer enhancement with a 5 to 10% pressure drop increase. The highest heat transfer (highest pressure drop as well) was observed for the case of the circular tube with winglets. The circular tube geometry had $d_o/H = 5.0$. Figure 6.27 also shows that the Nusselt numbers for the round tube geometry without winglets are larger than oval tube geometry without winglets.

6.8 ROW EFFECTS — STAGGERED AND INLINE LAYOUTS

The published tube bank correlations are generally for deep tube banks and do not account for row effects. The heat transfer coefficient will decrease with rows in an inline bank due to the bypass effects. However, the coefficient increases with number of tube rows in a staggered bank. This is because the turbulent eddies shed from the tubes cause good mixing in the downstream fin region. As an approximate rule, one may assume that the heat transfer coefficient for a staggered tube bank of the Figure 6.2a and b finned tubes has attained its asymptotic value at the fourth tube row.

Inline tube banks generally have a smaller heat transfer coefficient than staggered tube banks. At low Re_d ($Re_d < 1000$) with deep tube banks ($N \geq 8$), Rabas and Huber [1989] show that the heat transfer coefficient may be as small as 60% of the staggered tube value. The heat transfer coefficient of the inline bank increases as the Re_d is increased; at $Re_d = 50,000$ with $N \geq 8$, the inline to staggered ratio may approach 0.80.

There is a basic difference in the flow phenomena in staggered and inline finned-tube banks. Figure 6.28 compares the performance of inline and staggered banks of plain, circular finned tubes (Figure 6.1b) as reported by Brauer [1964]. If the fins are short ($e/d_o = 0.07$) the staggered bank gives 30% higher heat transfer coefficient. However, when the fin height is increased to $e/d_o = 0.53$, the heat transfer coefficient for the staggered tube arrangement is as much as 100% higher than for the inline arrangement. Brauer [1964] argues that bypass effects in the inline arrangement are

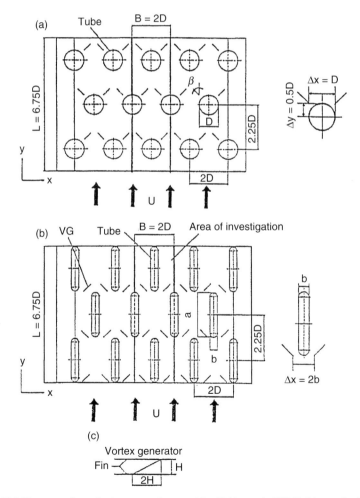

Figure 6.26 Three-row fin-and-tube geometries tested by Fiebig et al. [1994]: (a) round tube, (b) flat tube, (c) shape of the vortex generator, $d_o = 32$ mm, $H = 7$ mm, $a = 70$ mm, $b = 12$ mm, 45° angle of attack.

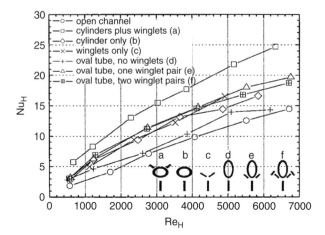

Figure 6.27 Nu_H vs. Re_H (H: channel height) of various geometries tested by O'Brien et al. [2002], $a/H = 8.66$, $d/H = 5.0$, where a is the major diameter of the oval tube and d is the diameter of the circular tube.

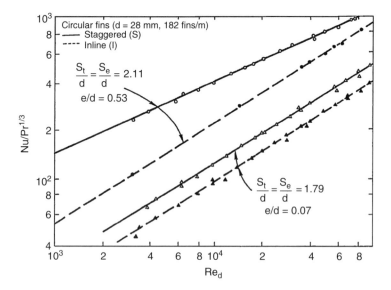

Figure 6.28 $NuPr^{-1/3}$ vs. Re_d for inline and staggered banks of circular finned tubes with plain fins, as reported by Brauer [1964].

responsible for the poor performance. Figure 6.29, taken from Brauer [1964], shows the flow patterns in staggered and inline tube arrangements. The streamlines are shown by the dashed lines. The low-velocity wake regions, or "dead spaces," are shown by the shaded area with fine dots. A much greater fraction of the fin surface area is contained in the low-velocity wake region for the inline arrangement

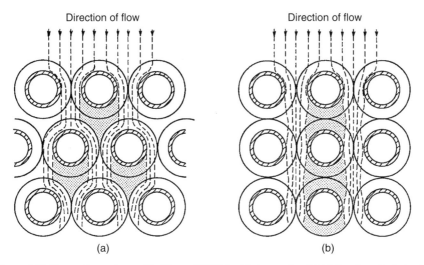

Figure 6.29 Flow patterns observed by Brauer [1964] for (a) staggered and (b) inline finned tube banks.

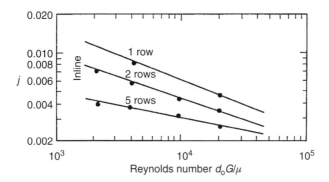

Figure 6.30 Row effect of inline tube banks for the Figure 6.20 segmented fin tubes ($S_t/d_o = S_l/d_o = 2.25$). (From Webb [1987].)

compared to the staggered arrangement. Consequently, the inline arrangement will have a lower surface average heat transfer coefficient. Outside the shaded zone, particularly between the fin tips, a strong bypass stream exists. Because of poor mixing between the wake stream and bypass stream, the weaker wake stream is quickly heated; its mixed temperature is greater than that of the bypass stream. Thus, the actual temperature difference between the surface and the wake stream is much less than indicated by an overall LMTD based on the mixed outlet temperature. The staggered arrangement provides a good mixing of the wake and bypass streams after each tube row.

The superiority of the staggered fin geometry is shown in Figure 6.20 for $e/d_o = 0.51$ segmented fins tested by Weierman et al. [1978]. The staggered bank (st) has four rows on an equilateral triangular pitch ($S_t/d_o = 2.25$) and the seven row inline (il) bank has $S_t/d_o = S_l/d_o = 2.25$. At $Re_d = 10,000$, $j_{st}/j_{il} = 2.17$ and $f_{st}/f_{il} = 1.73$.

Figure 6.30 shows the row effect of the inline segmented fin geometry. It shows that the bypass effect reduces the performance the inline geometry as the number of rows increases. Rabas and Huber [1989] have performed tests of the performance differences of staggered and inline banks of tubes having plain fins. Their work shows that the inline bank attains asymptotic values of j and f. In general, their work shows that plain, inline finned-tube banks yield the lowest performance for (1) low Re_d (2) large S_t/d_o, and (3) small s/d_o. Rabas et al. [1986] provide additional data on inline vs. staggered layout for plain and the Figure 6.18a segmented fin geometer.

6.9 HEAT TRANSFER COEFFICIENT DISTRIBUTION (PLAIN FINS)

6.9.1 Experimental Methods

Although the standard fin efficiency calculation assumes the heat transfer coefficient is constant over the fin surface area, local measurements have shown this is not the case. The flow accelerates around the tube and forms a wake region behind the tube. This causes local variations of the heat transfer coefficient. Several researchers have measured the distribution of the local heat transfer coefficient on plain fins. Such measurements under steady-state heat transfer conditions are very difficult to perform, because the local fin temperature and local heat flux are required.

Neal and Hitchcock [1966] used the steady-state method to determine the local values of the heat transfer coefficient. Jones and Russell [1980] used a transient heating method for the circular fin geometry. The local heat flux is calculated from the slope of the temperature vs. time curve at each local fin temperature. Jones and Russell also used a different transient method, which avoided measurement of the local fin temperature distribution. They measured the time required to melt a thermal paint of known melting temperature. Others have used mass transfer techniques to infer the local heat transfer coefficients from the heat-mass transfer analogy. Saboya and Sparrow [1974] cast solid naphthalene plates in the form of a plate fin-and-tube flow passage. The local mass transfer coefficients were defined by measuring the thickness of naphthalene lost by sublimation, during a timed test run. Krückels and Kottke [1970] used another mass transfer technique. Their method involves a chemical reaction between a surface coating and ammonia added to the air stream. The reaction caused permanent discoloration of the coated surface. A photometric method is used to establish the local mass transfer coefficient, which is related to the degree of discoloration.

6.9.2 Plate Fin-and-Tube Measurements

Saboya and Sparrow [1974, 1976a, 1976b] used the naphthalene mass transfer method to measure the local coefficients for one-, two-, and three-row plate fin-and-tube geometries. Figure 6.31 shows the geometry of the two-row layout. All geometries had the same tube size, pitches, and fin spacing (1.65 mm). Figure 6.32 shows

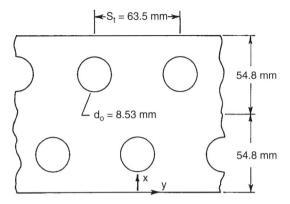

Figure 6.31 Two-row finned tube geometry simulated by Saboya and Sparrow [1974].

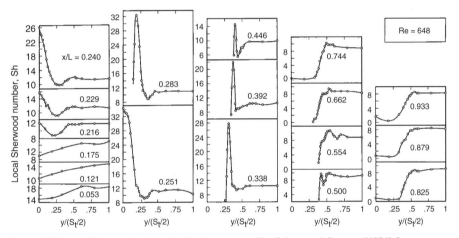

Figure 6.32 Local Sherwood number distribution measured by Saboya and Sparrow [1974] for a one-row plate fin geometry (see Figure 6.31).

the local Sherwood numbers for the one-row layout at Re = 648. The Sherwood number corresponds to the Nusselt number for heat transfer. The tube is located at $0.27 \leq x/L \leq 0.73$. Near the leading edge of the fin ($x/L \leq 0.2$) the developing boundary layer produces a relatively high Sh (12 to 18). Just upstream of the tube at $x/L = 0.221$, Sh, shows a peak beginning to form. This is caused by the vortex that develops on the front of the tube and is then swept around the tube. This vortex produces a natural heat transfer augmentation. Following the vortex around the tube, Sh attains a peak (34) at $x/L = 0.257$ and decreases to 14 at the tube centerline, where the boundary layer separates from the tube. The curves do not extend to $y = 0$, because of the presence of the tube. Away from the tube-induced vortex, $8 < Sh < 10$ for $0.283 \leq x/L \leq 1.0$, except for the wake region behind the tube. Aside from the peak Sh caused by the tube vortex, the maximum Sh does not occur at the tube

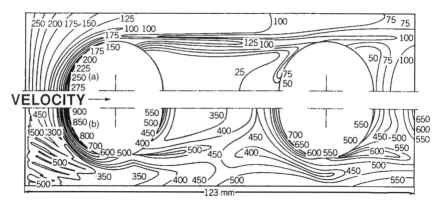

Figure 6.33 Distribution of mass transfer coefficients (m³/m²-hr) on a two-row plate finned tube measured by Krückels and Kottke [1970]. (a) $Re_d = 1160$, (b) $Re_d = 5800$.

centerline ($x/L = 0.5$), where the minimum flow area exists. A very low performance wake region exists behind the tube ($x/L > 0.73$). For $x/L > 0.744$, $1 \leq Sh \leq 4$ in the wake region, compared with $Sh = 8$ outside of the wake. Data taken at 100% higher velocities showed no appreciable increase of Sh in the tube wake region. However, the peak values of Sh caused by the tube vortex were 50% larger and persisted until the end of the fin, where $Sh = 16$. The $Re = 648$ data show that the local coefficient varies by a 3:1 factor when the wake region is excluded, and by 28:1 when the wake regions are included. The variations are even larger at high Re.

Figure 6.33 shows the local mass transfer coefficient measure by Krückels and Kottke [1970] for a two-row inline plate fin geometry. The upstream air Reynolds numbers (Re_d) are 1160 and 5800 for the upper and lower portions of the figure, respectively. The results are in qualitative agreement with the observations of Saboya and Sparrow.

6.9.3 Circular Fin-and-Tube Measurements

The local heat (or mass) transfer coefficients on this geometry have been measured by Neal and Hitchcock [1966] and Krückels and Kottke [1970] for single circular finned tubes and for banks of staggered arrangement. Figure 6.34 shows the mass transfer results of Krückels and Kottke for a single circular finned tube. Neal and Hitchcock's results for row 2 and row 6 of a staggered bank show qualitatively similar coefficient distributions as indicated by Figure 6.34. Salient among the findings of these researchers follow:

1. Markedly higher heat transfer occurs on the upstream area of the fin than on the downstream area. Maximum coefficients on the fin occur 70° to 90° from the forward stagnation point.
2. Higher heat transfer coefficients occur near the fin tip than near the base of the fin.

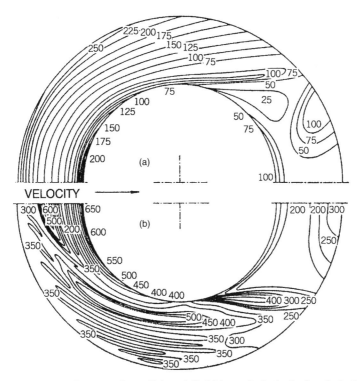

Figure 6.34 Distribution of mass transfer coefficients (m³/m²-hr) on a single circular finned tube (212 fins/m) measured by Krückels and Kottke [1970]. (a) Re$_d$ = 1940, (b) Re$_d$ = 9700.

3. Stagnation flow on the front of the tube produces high heat transfer at the fin root. Slightly smaller coefficients near the fin tip, at the front, are caused by flow separation.
4. Radial heat flow occurs only near the front of the fin. A general migration of heat toward the front of the tube, especially as the fin tip is approached.
5. The measured fin efficiency is less than that calculated assuming uniform heat transfer coefficients and radial heat flow (0.6 vs. 0.7 for the case studied).

Hu and Jacobi [1993] obtained local mass transfer coefficients for single row of circular finned tubes using the naphthalene sublimation technique. The finned-tube sample was 76.2-mm fin diameter, 38.1-mm tube diameter, and 141 fins/m. The mass transfer data revealed similar local characteristics as those observed by Neal and Hitchcock [1966]. Kearney and Jacobi [1996] extended the work of Hu and Jacobi to a two-row geometry. Both inline and staggered configurations were investigated. For the Reynolds number range studied (5000 ≤ Re$_{Dh}$ ≤ 28,000), the mass transfer coefficients were not sensitive to the bundle arrangement, due to the small number of tube rows. If the number of tube rows were large, the inline geometry would have yielded a smaller mass transfer coefficient due to the bypass effect as described by Brauer [1964] and shown in Figure 6.29. The fin efficiencies calculated from the mass transfer data were in good agreement with those obtained assuming uniform heat transfer coefficients.

6.10 PERFORMANCE COMPARISON OF DIFFERENT GEOMETRIES

6.10.1 Geometries Compared

Webb and Gupte [1990] compared the performance of the six enhanced surface geometries described in Table 6.2. The first four geometries have fins on 9.52-mm-diameter round tubes. The last two geometries (OSF and louver) are brazed aluminum heat exchangers and are illustrated in Figure 6.35. The extruded aluminum tube used in the Figure 6.35a brazed aluminum exchanger has 4.0-mm minor tube diameter with 0.78-mm wall thickness and internal membranes at 5.0-mm pitch.

The analysis was performed to compare candidate geometries for residential air-conditioning condensers. A key objective was to compare the performance of enhanced fin geometries using round tubes with brazed aluminum exchangers that use flat aluminum tubes. Aluminum fin-and-tube material was used in all the heat exchangers.

The first four geometries in Table 6.2 are commercially used, and employ round tubes. The louver fin geometry is used in automotive radiators and air-conditioning heat exchangers. A fin diameter (diameter over the fins) of 25.4 mm was used for the Figure 6.18b spine fins. The extruded aluminum tube used with the OSF and louver geometries (Figure 6.35) had 3.46-mm minor diameter with 0.75-mm wall thickness, and can be made with any tube depth. Ideally, one would calculate the heat exchanger depth (in the airflow direction) needed to meet the required heat exchange duty, and then form the aluminum extrusion having this depth. One-row of the plate-and-fin geometry is defined as the tube depth required to provide the same value of $A/A_{fr}N$ as that of the spine fin, with both having the same fin pitch.

Table 6.2 Heat Exchanger Geometries Compared by Webb and Gupte [1990]

	Spine	Wavy	Slit	CLF	OSF	Louver
Figure	6.2a	6.2c	6.2b	6.2d	6.35b	6.35c
d_o (mm)	9.52	9.52	9.52	9.52		
b (mm)					3.46	3.46
a (mm)					7.87	8.38
t (mm)	0.76	0.76	0.76	0.76	0.76	0.76
S_t (mm)	25.4	23.62	25.4	25.4	21.84	21.84
S_l (mm)	25.4	20.60	21.60	21.60		
L_p (mm)			1.98		1.59	1.59
n_L		4			5	5
θ (degrees)					20	20

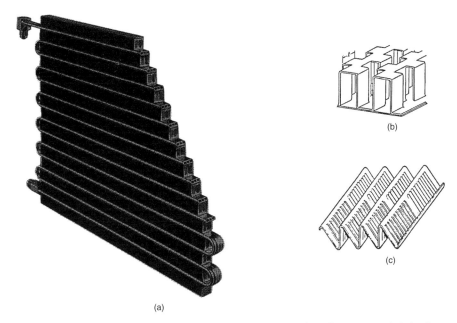

Figure 6.35 Brazed aluminum heat exchanger having (a) 4.0-mm-minor-diameter extruded aluminum tube (courtesy Modine Manufacturing Co.) used with either the (b) OSF or (c) louver fins.

The potential advantages of the Figure 6.35 flat tube OSF geometry, relative to the Figure 6.2a and b strip fin geometries are discussed in Section 6.6.2. The 21.84-mm transverse tube used for the Figure 6.35 geometries resulted from an optimization study.

6.10.2 Analysis Method

Following are the sources of the various correlations, or data, used to predict the air-side heat transfer and friction characteristics:

1. Figure 6.2a spine fin: Scaled data of Eckels and Rabas [1985]
2. Figure 6.2b slit fin: Mori and Nakayama [1980] data
3. Figure 6.2c wavy fin: Webb [1990] wavy fin correlation
4. Figure 6.2d CLF: Hatada and Senshu [1984] data
5. Figure 6.35b OSF: Wieting [1975] correlation
6. Figure 6.35c louver fin: Davenport [1984] correlation

The Kandlikar [1987] correlation was used to predict the tube-side heat transfer coefficient for vaporization of R-22 in plain tubes. The same correlation is used to calculate the tube-side heat transfer coefficient for the flat tubes using the hydraulic diameter in the Reynolds number definition.

Figure 6.36a and b show h_o and Δp vs. air frontal velocity (u_{fr}) for the six geometries. The row depth of the OSF and louver plate-and-fin geometry is

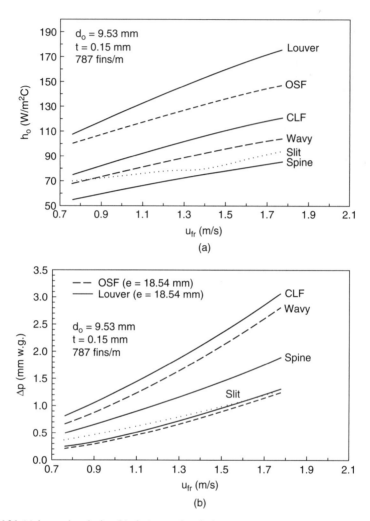

Figure 6.36 (a) h_o vs. air velocity, (b) air Δp vs. air velocity.

arbitrarily defined as 15.24 mm, which provides the same $A_o/A_{fr}N$ as the spine fin, finned-tube geometry. Figure 6.36a shows that the OSF and louver fins have much higher heat transfer coefficients than any other fin geometries, and that the heat transfer coefficient of the louver fins is slightly above that of the OSF at 1.3 m/s frontal velocity. Figure 6.36b also shows that the Δp for the CLFs is higher than for the other geometries.

The calculation methodology used the VG-1 criterion described in Chapter 3. The objective is to reduce the heat exchanger size and weight. Reduced airflow frontal area is also desirable. The performance evaluation criterion (PEC) determined the heat exchanger frontal area and fin pitch needed to meet the required heat transfer rate for fixed fan power and constant air mass flow rate, with a given number of tube rows in the airflow direction. For the case of the Figure 6.36 flat tubes, the tube depth (in the airflow direction) may be made any desired value. Hence, the tube depth was taken as an independent variable. The frontal area and fin pitch was adjusted to meet the required heat duty and friction power constraints.

The case of an R-22 evaporator is used to compare the performance of the various geometries. The evaporator operates at the following conditions: heat duty (2408 W), airflow rate (0.53 m³/s), refrigeration saturation temperature (2.77°C), inlet air temperature (10°C), fins/m for the reference spine fin (787), air Δp (0.74 mm w.g.). The air frontal velocity was allowed to vary from 0.76 to 1.78 m/s, and the fin/m is allowed to vary from 314 to 867 for each air-side surface geometry. The calculation procedure used to size the heat exchanger for a specified frontal velocity is as follows:

1. The required *UA* is calculated for the specified airflow rate, heat duty, inlet air temperature, and refrigerant saturation temperature.
2. Set the fins per meter at the lowest value (314 fins/m) and the number of tube rows in the airflow direction (*N*) to one.
3. Calculate the heat exchanger frontal area, and the air-side and tube-side heat transfer coefficients for the specified air frontal velocity.
4. Calculate the available *UA* from

$$\frac{A_{fr}N}{UA} = \frac{A_{fr}N}{h_iA_i} + \frac{A_{fr}N}{\eta h_{oA}o} \tag{6.28}$$

5. Calculate the air pressure drop.
6. If the available *UA* is less than the required value, increase the fins/m value by 20 fins/m, and repeat steps 2 through 4. If the available *UA* ≥ required *UA*, then the pressure drop is checked. If this is within the specified 0.74 mm w.g. limit, an acceptable solution exists. If no design is obtained, the number of rows is increased by one and the calculations are repeated from step 2. This methodology results in selecting the heat exchanger having the minimum number of tube rows, which should yield the minimum material cost.

The heat exchanger calculation for each geometry was repeated for 0.025 m/s increments of frontal velocities between 0.76 and 1.78 m/s. For all the designs, the number of refrigerant circuits was set to one. The refrigerant mass velocity in the round tubes was 336 kg/s-m². This refrigerant mass velocity is typical of that used in commercial heat pump evaporators.

6.10.3 Calculated Results

Table 6.3 compares the various performance parameters of the six heat exchanger configurations. The Figure 6.18b spine fin is taken as the reference for comparison. Each design listed in Table 6.3 provides the same heat duty and operates at the same air-side pressure drop and airflow rate. The higher pressure drops of the finned-tube heat exchangers make them operate at much lower air frontal velocities, and thus the frontal area requirements are increased. For example, the spine fin design operates at 0.96 m/s frontal velocity, as compared to 1.42 m/s for the OSF plate-and-fin design. Since air-side frontal area is inversely proportional to frontal velocity, the OSF exchanger requires 31% less frontal area than the spine fin design.

The convex louver fin (Figure 6.2d) provides the highest performance (lowest weight) of the round tube exchangers — a 22% weight reduction relative to the spine fin. The weight reductions of the Figure 6.35 OSF and louver plate-and-fin designs (without splitter plates) are 53 and 48%, respectively.

The OSF and louver plate-and-fin exchangers are not constrained to an integer number of rows, as is required for the round tube design. The flat aluminum tubes may be extruded to any desired width using internal membranes to contain the internal pressure.

6.11 PROGRESS ON NUMERICAL SIMULATION

With ever-increasing computing power, significant progress has been made on the numerical investigation of finned-tube heat exchangers. Torikoshi et al. [1994] performed three-dimensional unsteady numerical calculations for a model of one-row plate fin-and-tube heat exchanger. The numerical results were in good agreement with the experimental data by Saboya et al. [1974, 1984]. It was found that the flow field downstream of the fin trailing edge became more destabilized as the fin pitch increased. However, the heat transfer on the fin surface was not affected by such change of flow pattern, because the flow instability was restricted in the region downstream of the fin. Torikoshi and Xi [1995] extended the analysis to investigate the effect of tube diameter. Increasing the tube diameter did not improve the heat transfer performance, but it increased the pressure drop. The analysis was extended to conjugate heat transfer by Onishi et al. [1999] and Tsai et al. [1999].

Table 6.3 Comparison of All-Aluminum Heat Exchangers (Plain Tubes with Two Refrigerant Circuits)

	Spine	Wavy	Slit	CLF	OSF	Louver
Rows	1	1	1	1	0.59	0.59
fins/m	728	433	590	433	866	866
u_{fr} (m/s)	0.96	1.07	1.32	1.19	1.39	1.44
h_o (W/m^2-C)	64.8	81.7	77.7	93.1	130.0	133.1
η	0.93	0.91	0.90	0.89	0.83	0.83
G_{ref} (kg/m^2-s)	336	336	336	336	1127	1107
h_i (W/m^2-°C)	3531	3554	3690	3667	10288	9732
w_{fin} (kg)	1.96	1.58	1.85	1.49	0.97	0.93
w_{tub} (kg)	1.23	1.19	0.96	0.99	0.84	0.84
w_{tot} (kg)	3.19	2.77	2.81	2.48	1.81	1.77

Jang and Chen [1997] performed three-dimensional laminar, steady calculations for wavy finned tube heat exchangers. The effect of tube rows, corrugation angles, and corrugation depths were investigated. The results generally confirmed experimental findings; the j and f factors increased as the corrugation angle increased, and, for a fixed corrugation angle, they decreased as the wavy height increased. Min and Webb [2001b] solved the three-dimensional steady conjugate heat transfer problem for a model of the wavy heat exchanger. The fin efficiency was determined from the temperature distribution of the fin, which agreed with the Schmidt equation within ±2.3%.

Attempts have been made to simulate more complex fin geometries such as slit fins (Bae et al. [1999], Sheui et al. [1999]), louver fin (Leu et al. [2001]), and convex louver fin (Jang et al. [2001]). The simulations are, however, limited to steady flow assumptions. Experimental findings clearly reveal the importance of flow unsteadiness in such enhanced geometries. Future research efforts should be directed to the incorporation of flow unsteadiness by complex fin geometries.

6.12 RECENT PATENTS ON ADVANCED FIN GEOMETRIES

In addition to the prior conventional enhanced fins, there are many enhanced designs that are conceptually feasible. Wang [2000] examined 51 patents issued during 1981 to 1999. Several of them are described in this section. Figure 6.37a illustrates a refined version of the CLF (Bemisderfer and Wanner [1991]), where the windward and the leeward length of the louver is different. Shorter windward louver will increase heat transfer at the leading edge. Figure 6.37b shows another improvement of convex louver fin by Ueda et al. [1994]. The convex louver has slits on the windward and leeward sides. The flow at the convex side will communicate with the recirculating flow at the concave side through the slits, and will reduce the recirculation zone.

Figure 6.37c illustrates the Beamer and Cowell patent [1998], where the louver angles were varied along the flow direction. The airflow is turned in a successive and incremental manner. As a result, the deflected airflow is steeper and higher. Beamer and Cowell claimed that the benefits of increased heat transfer would compensate for the increase in the pressure drop.

Figure 6.37d shows the radial slit design by Youn and Kim [1998]. As discussed in Section 6.5.4, radial slits will not cut the heat conduction path from the tube. The effective slit width, however, will increase. For typical residential air-conditioning units, heat exchangers have a small number of tube rows (e.g., four or fewer). In such cases, a variation of enhancement level along the flow direction is helpful because the heat transfer performance is quite high in the entrance region. Figure 6.37e shows the inventions by Yun and Kim [1997] and Jung and Jung [1999] where the concept has been adopted.

Esformes [1989] obtained the first patent on use of vortex generators with finned surfaces (Figure 6.37f). It consisted of an array of winglets that are bent out of the smooth wavy fin surface. The vortex generators are expected to generate additional counter-rotating vortices on top of those generated by the wavy surface, which will enhance the fluid mixing. Itoh et al. [1986] proposed an "accordion fin," which was later updated by Tanaka et al. [1994]. Itoh et al. [1995] tested the fin pattern shown in Figure 6.37g, and claimed a 30% increase of heat transfer coefficient compared with those of the conventional louver fin geometry. They also argued that the accordion fin can provide better condensate drainage for evaporators. The condensate trapped in the special V/W slits would drain better, due to the unbalance of the surface tension force caused by the unequal radius.

Ikejima et al. [1998] developed a special wire-tube geometry. As shown in Figure 6.37h, fine wires are wound helically around two adjacent tubes. When fabricating the heat exchanger, either the tube outside, or the wires, are coated with a metal film formed by plating, and they are interlaced to form a heat exchanger.

6.13 HYDROPHILIC COATINGS

Condensate is formed on evaporator fins when the surface temperature drops below the dew point temperature of the ambient air. Bare aluminum fins are typically nonwetting, causing the condensate to form in droplets. If the condensate stands up as droplets and forms bridges between fins, a substantially higher pressure drop will be observed in dehumidification than in the dry case because of the blockage of the airflow passage by the condensate drops. However, if the condensate wets the fins and forms a thin film on the fin surface, there will be very little increase in pressure drop over the dry fin value. The appearance of the condensate formed on a surface is controlled by the wetting properties of the surface. Dropwise condensation will occur on a surface having poor wettability, while the filmwise condensation takes place on a surface having good wettability. The receding contact angle is commonly used for indicating surface wettability as shown by Min and Webb [2001a]. The advancing and receding contact angles define the maximum and minimum contact angles of a surface, respectively. The static contact angle is typically a value between the advancing and receding contact angles. Min et al. [2000] investigated the long-term performance of cooling coils having different surface treatments and found that the wet coil pressure drop was an approximate function of the receding contact angle.

To improve the wettability, hydrophilic coatings are applied to the surface of dehumifying heat exchangers. Hong and Webb [2000] describe various types of coatings (or surface treatments). They conducted dry/wet cycling tests on coupon samples to check the ability of the coatings to maintain small contact angles. Prior to initiating the cycling test, static contact angles as small as 10° to 15° were measured. The static contact angle increased with increasing number of dry/wet cycles, with a maximum in the 55° to 65° range after 1000 cycles. The uncoated aluminum surface showed improved wettability with increased number of dry/wet cycles due to surface oxidation and contamination. Min and Webb [2001a] investigated the effect of surface treatment on four typical fin surface materials, including aluminum, copper, and two commercial coatings on aluminum. The treatments

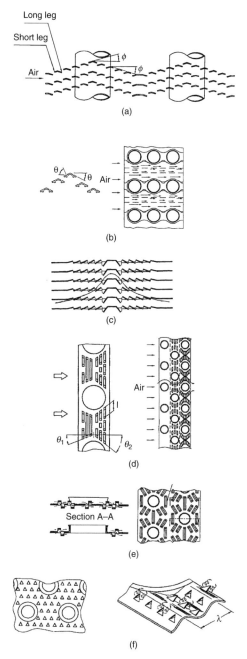

Figure 6.37 Patented enhanced fin geometries: (a) convex louver fin by Bemisderfer and Wanner [1991], (b) convex louver fin by Ueda et al. [1994], (c) louver fin by Beamer and Cowell [1998], radial slit fin patented by Youn and Kim [1998], (d) slit fin patented by Yun and Kim [1997] and Jung and Jung [1999], (f) vortex generators patented by Esformes [1989], (g) accordion fin patented by Tanaka et al. [1994], (h) woven wire fin patented by Ikejima et al. [1998]. (From Wang [2000].)

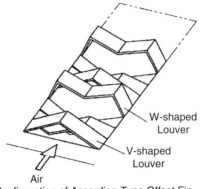

W-shaped Louver

V-shaped Louver

Air

Configuration of Accordion-Type Offset Fin

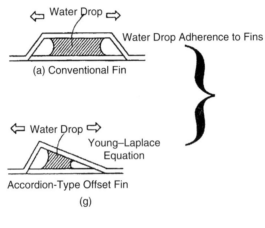

Water Drop

Water Drop Adherence to Fins

(a) Conventional Fin

Water Drop

Young–Laplace Equation

Accordion-Type Offset Fin

(g)

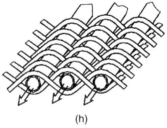

(h)

Figure 6.38 (cont.) Patented enhanced fin geometries: (a) convex louver fin by Bemisderfer and Wanner [1991], (b) convex louver fin by Ueda et al. [1994], (c) louver fin by Beamer and Cowell [1998], radial slit fin patented by Youn and Kim [1998], (d) slit fin patented by Yun and Kim [1997] and Jung and Jung [1999], (f) vortex generators patented by Esformes [1989], (g) accordion fin patented by Tanaka et al. [1994], (h) woven wire fin patented by Ikejima et al. [1998]. (From Wang [2000].)

included acetone cleaning, surface grinding, and oil contamination. Surface grinding yielded small receding contact angles, but acetone cleaning was not effective for bare aluminum and copper fin material. Oil contaminations increased the receding contact angles of all four surfaces. However, it is noted that some fin press oils include a surfactant, which will promote good short-term wettability, until the oil is depleted or evaporated. Kim et al. [2002] measured the contact angles of a hydrophilic surface made by plasma coating on aluminum fin stock. The plasma-coated surface maintained nearly the same value of contact angle (22°) throughout the 1000 wet/dry cycles. The uncoated aluminum fin stock had 90° contact angle at the start and decreased to 53° after 1000 cycles. The heat transfer and pressure drop data were obtained for the heat exchangers with and without the plasma treatment. The plasma-treated heat exchanger yielded 25% lower air pressure drop than the untreated heat exchanger. The heat transfer coefficients were approximately the same for both heat exchangers.

The j and f factors of finned-tube heat exchangers under dehumidifying conditions have been studied by many investigators. Different from the dry surface heat transfer, where standard data reduction procedures are established, there exists some ambiguity on the reduction of wet surface heat transfer. This issue is well addressed by Mirth and Ramadhyani [1993]. They compared the log mean enthalpy method (LMED) and the McQuiston [1978] method. The two methods differ in determining the heat transfer rate to the surface and in determining the fin efficiency. The McQuiston method is believed to be incorrect. Thus, great care is required when j factors are compared. McQuiston [1978] presented data for five plate fin-and-tube heat exchangers. The j factor under wet conditions was lower than that of the dry condition. The McQuiston method was used for the data reduction. Wang et al. [1997a] obtained j and f factors under wet conditions for nine plain fin-and-tube heat exchangers. The effects of fin spacing, the number of tube rows, and the inlet conditions were investigated. The j factors were strongly influenced by the inlet relative humidity when McQuiston's method was used, whereas they were independent of the relative humidity when LMED method was used. Based on this observation, Wang et al. [1997a] recommended LMED method. The j factors under wet conditions were comparable with those under the dry condition. Similar to the dry case, the effect of fin spacing on the j factor was negligible. The row effect, however, was less pronounced compared with the dry case. The friction factors of wet coils were much greater than those of the dry coils. Similar conclusions were drawn for louvered fin-and-tube heat exchangers (Wang et al. [2000b]).

Hong and Webb [1999] investigated the effect of hydrophilic coating for three enhanced fin geometries (wavy, lanced, louver). For all three fin geometries, the wet-to-dry pressure drop ratio was 1.2 at 2.5 m/s frontal air velocity. The coatings had no influence on the wet or dry heat transfer coefficient.

6.14 CONCLUSIONS

Enhanced surfaces are routinely used for application to gases. The dominant enhancement types are wavy or some form of interrupted strip fin. When used on circular tubes, a heat transfer enhancement of 80 to 100% is practically achieved.

Some advances were made on analytical or numerical models to predict the heat transfer performance of banks of high finned tubes. Power-law empirical correlations have been developed for plain fins. Although some work has been done to develop correlations for enhanced fins on circular tubes, they are not sufficiently general to account for the many geometric variables involved. The problem is complicated by the row effect of inline and staggered tube arrangements. Inline layouts provide significantly lower heat transfer performance than the staggered tube layout. Few correlations are available for the inline layout. Low, integral fins used with liquids are not as susceptible to tube bank layout effects. The row effect difference diminishes as e/d_o decreases.

Both the individually finned-tube and plate fin-and-tube geometries are used. Enhanced surface geometries are available for both geometries. Preference of the geometry type is dependent on application and manufacturing cost considerations, rather than performance. High performance, per unit weight, can be obtained from both geometry types.

Higher performance can be obtained from oval or flat tubes. Oval tubes may not be practical for high tube-side design pressure. If aluminum is an acceptable material, extruded aluminum tubes having internal membranes offer potential for significant performance improvement, relative to round tube designs. High tube-side design pressures can be met. The resulting brazed aluminum heat exchangers offer many advanced technology possibilities. High-performance fin geometries are quite adaptable to this heat exchanger concept. Considerable work has been done on vortex generators for fin-and-tube heat exchangers. However, it does not appear that vortex generators provide greater enhancement than can be obtained from conventional slit or louver fin geometries, when applied to round tubes. The advantages offered by the extruded aluminum tube having internal membranes can be achieved using different methods of manufacture. For example, a flat tube can be made by roll forming. By brazing a corrugated strip inside the tube, high design pressure can be realized. Virtually no work has been done on this. Dehumidifying heat exchangers operate under wet surface conditions, for which the pressure drop can be significantly higher than under dry conditions. Special "hydrophilic" coatings have been developed that significantly reduce the wet pressure drop.

Gas-side fouling may limit the permissible enhanced fin geometry, and the fin spacing. Gas-side fouling is discussed in Chapter 10.

NOMENCLATURE

A	Total heat transfer surface area (both primary and secondary, if any) on one side of a direct transfer type exchanger; total heat transfer surface area of a regenerator, m² or ft²
a	Major diameter for rectangular tube cross section, m or ft
A_c	Flow cross-sectional area in minimum flow area, m² or ft²
AMTD	Arithmetic mean temperature difference, K
A_f	Fin or extended surface area on one side of the exchanger, m² or ft²
A_{fr}	Heat exchanger frontal area, m² or ft²
$A_\pm$	Tube outside surface area considering it as if there were no fins, m² or ft²

b	Minor diameter for rectangular tube cross section, m or ft
c_p	Specific heat of fluid at constant pressure, J/kg-K or Btu/lbm-°F
D_{ab}	Diffusion coefficient for component a through component b, m²/s
D_h	Hydraulic diameter of flow passages, $4LA_c/A$, m or ft
D_v	Volumetric hydraulic diameter, m or ft
d_e	Fin tip diameter for a finned tube, m or ft
d_i	Tube inside diameter, or diameter to the base of internal fins or roughness, m or ft
d_o	Tube outside diameter, fin root diameter for a finned tube, minor diameter of rectangular cross section tube, m or ft
e	Fin height, m or ft
e_w	Wave height of wavy fin, m or ft
f	Fanning friction factor, $\Delta p \rho D_h/2LG^2$, dimensionless
f_f	Friction factor of fins in Equations 6.5 and 6.6 (= $2\Delta p_\pm \rho A_c/A_f G^2$), dimensionless
f_t	Friction factor of tubes in Equations 6.5 (= $2\Delta p_\pm \rho A_c/A_t G^2$), where $A_t = A - A_f$, dimensionless
f_{tb}	Tube bank friction factor = $\Delta P \rho/2NG^2$, dimensionless
G	Mass velocity based on the minimum flow area, kg/m²-s or lbm/ft²-s
G_{ref}	Refrigerant side mass velocity, kg/m²-s or lbm/ft²-s
Gz	Graetz number = $RePrD_h/L$, dimensionless
H	Louver fin height (= $S_t - d_{o,min}$) or channel height, m or ft
h_a	h based on the arithmetic mean temperature difference, W/m²-K or Btu/hr-ft²–°F
j	Colburn factor = $StPr^{2/3}$, dimensionless
K	Thermal conductance = hA, W/K or Btu/hr°F
K_m	Mass transfer coefficient, kg/m²-s or lbm/ft²-hr
k	Thermal conductivity of fluid, W/m-K or Btu/hr-ft-°F
L	Fluid flow (core) length on one side of the exchanger, m or ft
L_p	Strip flow length of OSF or louver pitch of louver fin, m or ft
LMTD	Log-mean temperature difference, K or °F
L_p	Strip flow length of OSF or louver pitch of louver fin, m or ft
N	Number of tube rows in the flow direction, dimensionless
Nu_a	Nusselt number based on $h_a D_h/k$, dimensionless
Nu_{Dh}	Nusselt number = hD_h/k, dimensionless
Nu_d	Nusselt number = hd_o/k, dimensionless
Nu_H	Nusselt number = hH/k, dimensionless
NTU	Number of heat transfer units = $UA/(Wc_p)_{min}$, dimensionless
n_L	Number of louvers in airflow depth, dimensionless
Pr	Prandtl number = $c_p\mu/k$, dimensionless
p_f	Fin pitch, center to center spacing, m or ft
p_w	Wave pitch of wavy fin, m or ft
Δp	Air-side pressure drop, Pa or lbf/ft²
Δp_f	Pressure drop assignable to fin area in finned tube exchanger, Pa or lbf/ft²

Δp_t	Pressure drop assignable to tubes in finned tube exchanger, Pa or lbf/ft^2
Q	Heat transfer rate in the exchanger W or Btu/hr
Re_{Dh}	Reynolds number based on the hydraulic diameter = GD_h/μ, dimensionless
Re_{Dv}	Reynolds number based on D_v, dimensionless
Re_d	Reynolds number based on the tube diameter = Gd/μ, $d = d_i$ for flow inside tube and $d = d_o$ for flow outside tube, dimensionless
Re_H	Reynolds number based on channel height = GH/μ, dimensionless
Re_L	Reynolds number based on the interruption length = GL_p/μ, dimensionless
Re_l	Reynolds number based on S_l, dimensionless
S_f	Flow frontal area of heat exchanger, m^2 or ft^2
Sh	Sherwood number for mass transfer ($= K_m D_h/D_{\text{ab}}$), dimensionless
S_l	Longitudinal tube pitch, m or ft
$S_\pm$	Transverse tube pitch, m or ft
St	Stanton number = h/Gc_p, dimensionless
s	Spacing between two fins = $p_f - t$, m or ft
ΔT	Temperature difference between hot and cold fluids, K or °F
t	Thickness of fin, m or ft
u	Fluid velocity, u_{fr} frontal velocity, u_c in minimum flow area, m/s or ft/s
U	Overall heat transfer coefficient, W/m^2-K or Btu/hr-ft^2-°F
w_s	Width of segmented fin (Figure 6.20), m or ft
w_{tub}	Weight of tubes in heat exchanger, kg or lbm
w_{fin}	Weight of fins in heat exchanger, kg or lbm
w_{tot}	Weight of tubes and fins in heat exchanger, kg or lbm
W	Mass flow rate, kg/s or lbm/s
x	Coordiante direction from leading edge of fin, m or ft
X_{DV}^+	$\text{RePr}D_v/L$, dimensionless

Greek Letters

ν	Kinematic viscosity, m/s^2 or ft/s^2
η_f	Fin efficiency or temperature effectiveness of the fin, dimensionless
η_o	Surface efficiency of finned surface = $1 - (1 - \eta_f)/A_f/A$, dimensionless
μ	Fluid dynamic viscosity coefficient, Pa-s or lbm/s-ft
ρ	Fluid density kg/m^3 or lbm/ft^3
θ	Louver angle for louver fin, radians

Subscripts

i	Tube inside surface
il	Inline tube arrangement
o	Outside (air-side) surface
p	Plain tube or surface

st   Staggered tube arrangement
w   Evaluated at wall temperature

REFERENCES

Abbott, R.W., Norris, R.H., and Spofford, W.A., 1980. Compact heat exchangers in General Electric products — sixty years of advances in design and in manufacturing technologies, in *Compact Heat Exchangers — History, Technology, Manufacturing Technologies*, R.K. Shah, C.F. McDonald, and C.P. Howard, Eds., ASME Symp. Vol. HTD, 10, 37–56.

Abu Madi, M., Johns, R.A., and Heikal, M.R., 1998. Performance characteristics correlation for round tube and plate finned heat exchangers, *Int. J. Refrig.,* 21, 507–517.

Achaichia, A. and Cowell, T.A., 1988. Heat transfer and pressure drop characteristics of flat tube and louvered plate fin surfaces, *Exp. Thermal Fluid Sci.*, 1, 147–157.

Bae, J.H., Park, M.-H., and Lee, J.-H., 1999. Local flow and heat transfer of a 2-row offset strip fin-tube heat exchanger, *J. Enhanced Heat Transfer*, 6, 13–29.

Beamer, H.E. and Cowell, T.A., 1998. Heat Exchanger Cooling Fin with Varying Louver Angle, U.S. patent 5,730,214.

Beecher, D.T. and Fagan, T.J. 1987. Effects of fin pattern on the air-side heat transfer coefficient in plate finned-tube heat exchangers, *ASHRAE Trans.*, 93(2), 1961–1984.

Bemisderfer, C. and Wanner, J., 1991, Chevron Lanced Fin Design with Unequal Leg Lengths for a Heat Exchanger, U.S. patent 5,062,475.

Benforado, D.M. and Palmer, J., 1964. Wire loop finned surface — a new application (heat sink for silicon rectifiers), *Chem. Eng. Prog. Symp. Ser.*, No. 57, 61, 315–321.

Brauer, H., 1964. Compact heat exchangers, *Chem. Prog. Eng.* (London), 45(8), 451–460.

Breber, G., 1991. Heat transfer and pressure drop of stud finned tubes, *Chem. Eng. Prog. Symp. Ser.*, No. 283, 87, 383–390.

Briggs, D.E. and Young, E.H., 1963. Convection heat transfer and pressure drop of air flowing across triangular pitch banks of finned tubes, *Chem. Eng. Prog. Symp. Ser.*, No. 41, 59, 1–10.

Carranza, R.G. and Holtzapple, M.T., 1991. A generalized correlation for pressure drop across spined pipe in cross-flow. Part I, *ASHRAE Trans.*, 97(2), 122–129.

Chang, W.R., Wang, C.C., Tsi, W.S., and Shyu, R.J., 1995. Air-side performance of louver fin heat exchanger, *Proceedings of the ASME/JSME Thermal Engineering Joint Conference*, Vol. 4, L.S. Fletcher and T. Aihara, Eds., ASME, New York, 367–372.

Cox, B., 1973. Heat Transfer and Pumping Power Performance in Tube Banks — Finned and Bare, ASME Paper 73-HT-27, New York.

Cox, B. and Jallouk, P.A., 1973. Methods for evaluating the performance of compact heat exchanger surfaces, *J. Heat Transfer*, 95, 464–469.

Davenport, C.J., 1984. Correlations for heat transfer and flow friction characteristics of louvered fin, *Heat Transfer—Seattle 1983*, AIChE Symposium Series No. 225, 79, 19–27.

Du, Y.-J. and Wang, C.-C., 2000. An experimental study of the airside performance of the superslit fin-and-tube heat exchangers, *Int. J. Heat Mass Transfer,* 43, 4475–4482.

Ebisu, T., 1999. Development of new concept air-cooled heat exchanger for energy conservation of air-conditioning machine, in *Heat Transfer Enhancement of Heat Exchangers*, S. Kakaç et al., Eds., Kluwer Academic, Dordrecht, 601–620.

Eckels, P.W. and Rabas, T.J., 1985. Heat transfer and pressure drop performance of finned tube bundles, *J. Heat Transfer*, 107, 205–213.

ESDU, 1985. Low-Fin Staggered Tube Banks: Heat Transfer and Pressure Loss for Turbulent Single-Phase Crossflow, Engineering Sciences Data Unit, ESDU Item 84016.

Esformes, J.L., 1989. Ramp Wing Enhanced Plate Fin, U.S. patent 4,817,709.

Fiebig, M., Mitra, N.K., and Dong, Y., 1990. Simultaneous heat transfer enhancement and flow loss reduction on fin-tubes, *Heat Transfer 1990, Proc. 9th Int. Heat Transfer Conference*, Jerusalem, 4, 51–56.

Fiebig, M., Valencia, A., and Mitra, N.K., 1993. Wing-type vortex generators for fin-and-tube heat exchangers, *Exp. Thermal Fluid Sci.*, 7(4), 287–295.

Fiebig, M., Valencia, A., and Mitra, N.K., 1994. Local heat transfer and flow losses in fin-and-tube heat exchangers with vortex generators: a comparison of round and flat tubes, *Exp. Thermal Fluid Sci.*, 8(1), 35–45.

Fujii, M., Seshimo, Y., and Yoshida, T., 1991. Heat transfer and pressure drop of tube-fin heat exchanger with trapezoidal perforated fins, in *Proc. 1991 ASME-JSME Joint Thermal Engineering Conf.*, vol. 4, J.R. Lloyd and Y. Kurosake, Eds., ASME, New York, 355–360.

Gianolio, E. and Cuti, F., 1981. Heat transfer coefficients and pressure drops for air coolers under induced and forced draft, *Heat Transfer Eng.*, 3(1), 38–48.

Goldstein, L.J. and Sparrow, E.M., 1977. Heat/mass transfer characteristics for flow in a corrugated wall channel, *J. Heat Transfer*, 99, 187–195.

Gray, D.L. and Webb, R.L., 1986. Heat transfer and friction correlations for plate fin-and-tube heat exchangers having plain fins, *Heat Transfer 1986, Proc. Eighth Int. Heat Transf. Conf.*, 2745–2750.

Groehn, H.G., 1977. Flow and Heat Transfer Studies of a Staggered Tube Bank Heat Exchanger with Low Fins at High Reynolds Numbers, Central Library of the Julich Nuclear Research Center, Julich, Germany.

Haberski, R.J. and Raco, R.J., 1976. Engineering Analysis and Development of an Advanced Technology, Low Cost, Dry Cooling Tower Heat Transfer Surface, Curtiss-Wright Corporation Report C00-2774-1.

Hatada, D., Ueda, U., Oouchi, T., and Shimizu, T., 1989. Improved heat transfer performance of air coolers by strip fins controlling air flow distribution, *ASHRAE Trans.*, 95(1), 166–170.

Hatada, T. and Senshu, T., 1984. Experimental Study on Heat Transfer Characteristics of Convex Louver Fins for Air Conditioning Heat Exchangers, ASME paper, 84-H-74, New York.

Hitachi Cable Ltd., 1984. Hitachi High-Performance Heat Transfer Tubes, Cat. No. EA-500, Hitachi Cable, Ltd., Tokyo, Japan.

Holtzapple, M.T. and Carranza, R.G., 1990. Heat transfer and pressure drop of spined pipe in cross flow. Part 1: Pressure drop studies, *ASHRAE Trans.*, 96(2), 122–129.

Holtzapple, M.T., Allen, A.L., and Lin, K., 1990. Heat transfer and pressure drop of spined pipe in cross flow. Part III: Heat transfer studies, *ASHRAE Trans.*, 96(2), 130–135.

Hong, K. and Webb, R.L., 1999. Performance of dehumidifying heat exchangers with and without wetting coatings, *J. Heat Transfer*, 121, 1018–1026.

Hong, K. and Webb, R.L., 2000. Wetting coatings for dehumidifying heat exchangers, *Int. J. Heating Ventilation Air Conditioning Refrig. Res.* 6(3), 229.

Hu, X. and Jacobi, A.M., 1993. Local heat transfer behavior and its impact on a single-row, annularly finned tube heat exchanger, *J. Heat Transfer*, 115, 66–74.

Ikejima, K., Gotoh, T., Yumikura, T. Takeshita, M., and Yoshita, T., 1998. Heat Exchanger and Method of Fabrication the Heat Exchanger, U.S. patent 5,769,157.

Incropera, F.P. and DeWitt, D.P., 2001. *Fundamentals of Heat Mass Transfer,* 5th ed., John Wiley & Sons, New York.

Itoh, M., Kogure, H., Iino, K., Ochiai, I., Kitayama, Y., and Miyagi, M., 1986. Fin-and-Tube Type Heat Exchanger, U.S. patent 4593756.

Itoh, M., Kogure, H., Miyagi, M., Mochizuki, S. Yagi, Y., and Kunugi, Y., 1995. Development of an accordion-type offset fin heat exchanger, *Trans. Jap. Assoc. Refrig.*, 12(2), 219–224 [in Japanese].

Jakob, M., 1938. Heat transfer and flow resistance in cross flow of gases over tube banks, *Trans. ASME*, 60, 384.

Jang, J.-Y. and Chen, L.-K., 1997. Numerical analysis of heat transfer and fluid flow in a three-dimensional wavy-fin and tube heat exchanger, *Int. J. Heat Mass Transfer*, 40, 3981–3990.

Jang, J.-Y., Shieh, K.-P., and Ay, H., 2001. Three-dimensional thermal-hydraulic analysis in convex louver finned-tube heat exchangers, *ASHRAE Trans.*, 107(2), 503–509.

Jones, T.V. and Russell, C.M.B., 1980. Heat Transfer Distribution on Annular Fins, ASME Paper 78-H-30, New York.

Jung, G.H. and Jung, S.H., 1999. Heat Exchanger Fin Having an Increasing Concentration of Slits from an Upstream to a Downstream Side of the Fin, U.S. patent 5,934,363.

Kandlikar, S.G., 1987. A general correlation for saturated two-phase flow boiling heat transfer inside horizontal and vertical tubes, *Boiling and Condensation in Heat Transfer Equipment*, ASME Symp. Vol. HTD, 85, 9–19.

Kang, H.C. and Webb, R.L., 1998. Evaluation of the wavy fin geometry used in air-cooled finned tube heat exchangers, *Heat Transfer 1998, Proceedings of the 11th Int. Heat Transfer Conf.*, Kyongju, Korea, 6, 95–100.

Kayansayan, N., 1993. Heat transfer characterization of flat plain fins and round tube heat exchangers, *Exp. Thermal Fluid Sci.* 6(3), 263–272.

Kays, W.M. and London, A.L., 1984. *Compact Heat Exchangers*, McGraw-Hill, New York.

Kearney, S.P. and Jacobi, A.M., 1996. Local convective behavior and fin efficiency in shallow banks of in-line and staggered annularly finned tubes, *J. Heat Transfer*, 118, 317–324.

Kim, G.-R., Lee, H., and Webb, R.L., 2002. Plasma hydrophilic surface treatment for dehumidifying heat exchangers, *Exp. Thermal Fluid Sci*, 27, 1–10.

Kim, N.H., Youn, B., and Webb, R.L., 1999. Air-side heat transfer and friction correlations for plain fin-and-tube heat exchangers with staggered tube arrangements, *J. Heat Transfer*, 121(3), 662–667.

Kim, N.-H., Yun, J.-H., and Webb, R.L., 1997. Heat transfer and friction correlations for wavy plate fin-and-tube heat exchangers, *J. Heat Transfer*, 119(3), 560–567.

Krückels, S.W. and Kottke, V., 1970. Investigation of the distribution of heat transfer on fins and finned tube models, *Chem. Eng. Tech.*, 42, 355–362.

LaPorte, G.E., Osterkorn, C.L., and Marino, S.M., 1979. Heat Transfer Fin Structure, U.S. patent 4,143,710.

Leu, J.S., Liu, M.-S., Liaw, J.-S., and Wang, C.-C., 2001. A numerical investigation of louvered fin-and-tube heat exchangers having circular and oval tube configurations, *Int. J. Heat Mass Transfer*, 44, 4235–4243.

Lozza, G. and Merlo, U., 2001. An experimental investigation of heat transfer and friction losses of interrupted and wavy fins for fin-and-tube heat exchangers, *Int. J. Refrig.*, 24, 409–416.

McQuiston, F.C., 1978. Correlation for heat, mass and momentum transport coefficients for plate-fin-tube heat transfer surfaces with staggered tube, *ASHRAE Trans.*, 84(1), 294–309.

Min, J. and Webb, R.L., 2001a. Condensate formation and drainage on typical fin materials, *Exp. Thermal Fluid Sci.*, 25, 101–111.

Min, J. and Webb, R.L., 2001b. Numerical predictions of wavy fin coil performance, *J. Enhanced Heat Transfer*, 8(3), 159–174.

Min, J., Webb, R.L., and Bemisderfer, C.H., 2000. Long-term hydraulic performance of dehumidifying heat-exchangers with and without hydrophilic coatings, *Int. J. Heating Ventilation Air Conditioning Refrig. Res.*, 6(3), 257.

Min, J.C. and Webb, R.L., 2004. Numerical analyses of effects of tube shape on performance of a finned-tube heat exchanger, *J. Enhanced Heat Transfer*, 11, 63–76.

Mirth, D.R. and Ramadhyani, S., 1993. Comparison of methods of modeling the air-side heat and mass transfer in chilled water cooling coils, *ASHRAE Trans.*, 99(2), 285–299.

Mirth, D.R. and Ramadhyani, S., 1994. Correlations for predicting the air-side Nusselt numbers and friction factors in chilled-water cooling coils, *Exp. Heat Transfer*, 7, 143–162.

Mori, Y. and Nakayama, W., 1980. Recent advances in compact heat exchangers in Japan, in *Compact Heat Exchangers — History, Technology, Manufacturing Technologies*, R.K. Shah, C.F. McDonald, and C.P. Howard, Eds., ASME Symp. Vol. HTD, 10, 5–16.

Nakayama W. and Xu, L.P., 1983. Enhanced fins for air-cooled heat exchangers—heat transfer and friction factor correlations, *Proc. 1983 ASME-JSME Thermal Eng. Conf.*, 1, 495–502.

Neal, S.B.H.C. and Hitchcock, J.A. 1966. A study of the heat transfer processes in banks of finned tubes in cross flow, using a large scale model technique, *Proc. Third Int. Heat Transfer Conf.*, Chicago, 3, 290–298.

O'Brien, J., Sohal, M., Foust, T., and Wallstedt, P.C., 2002. Heat transfer enhancement for finned-tube heat exchangers with vortex generators: experimental and numerical results, *Heat Transfer 2002, Proc. 12th Int. Heat Transfer Conf.*, 4, 207–212.

O'Brien, J.E., Sohal, M.S., and Wallstedt, P.C., 2003. Heat transfer testing of enhanced finned tube bundles using the single blow technique, in *Proceedings of ASME Summer Heat Transfer Conference*, Las Vegas, NV, HT2003-47426.

O'Connor, J.M. and Pasternak, S.F., 1976. Method of Making a Heat Exchanger, U.S. patent 3,947,941.

Onishi, H., Inaoka, K., Matsubara, K., and Suzuki, K., 1999. Numerical analysis of flow and conjugate heat transfer of two-row plate-finned tube heat exchanger, in *Proceedings of the International Conference on Compact Heat Exchangers and Enhancement Technology for the Process Industries*, R.K. Shah, K.J. Bell, H. Honda, and B. Thonon, Eds., Begell House, New York, 175–183.

Rabas, T.J. and Huber, F.V., 1989. Row number effects on the heat transfer performance of inline finned tube banks, *Heat Transfer Eng.*, 10(4), 19–29.

Rabas, T.J. and Taborek, J., 1987. Survey of turbulent forced-convection heat transfer and pressure drop characteristics of low-finned tube banks in cross flow, *Heat Transfer Eng.*, 8(2), 49–62.

Rabas, T.J., Eckels, P.W., and Sabatino, R.A., 1981. The effect of fin density on the heat transfer and pressure drop performance of low finned tube banks, *Chem. Eng. Commun.*, 10(1), 127–147.

Rabas, T.J., Myers, G.A., and Eckels, P.W., 1986. Comparison of the thermal performance of serrated high-finned tubes used in heat-recovery systems, in *Heat Transfer in Waste Heat Recovery and Heat Rejection Systems*, J.P. Chiou and S. Sengupta, Eds., ASME Symp. Vol. HTD, 59, 33–40.

Rich, D.G., 1973. The effect of fin spacing on the heat transfer and friction performance of multi-row, plate fin-and-tube heat exchangers, *ASHRAE Trans.*, 79(2), 137–145.

Rich, D.G., 1975. The effect of the number of tube rows on heat transfer performance of smooth plate fin-and-tube heat exchangers, *ASHRAE Trans.*, 81(1), 307–319.

Robinson, K.K. and Briggs, D.E., 1966. Pressure drop of air flowing across triangular pitch banks of finned tubes, *Chem. Eng. Prog. Symp. Ser.*, No. 64, 62, 177–184.

Rozenman, T., 1976a. Heat Transfer and Pressure Drop Characteristics of Dry Cooling Tower Extended Surfaces, Part I: Heat Transfer and Pressure Drop Data, Report BNWL-PFR 7-100, Battelle Pacific Northwest Laboratories, Richland, WA, March 1.

Rozenman, T., 1976b. Heat Transfer and Pressure Drop Characteristics of Dry Cooling Tower Extended Surfaces, Part II: Data Analysis and Correlation, Report BNWL-PFR 7-102, Battelle Pacific Northwest Laboratories, Richland, WA.

Saboya, F.E.M. and Sparrow, E.M. 1974. Local and average heat transfer coefficients for one-row plate fin and tube heat exchanger configurations, *J. Heat Transfer*, 96, 265–272.

Saboya, F.E.M. and Sparrow, E.M. 1976a. Transfer characteristics of two-row plate fin and tube heat exchanger configurations, *Int. J. Heat Mass Transfer*, 19, 41–49.

Saboya, F.E.M. and Sparrow, E.M. 1976b. Experiments on a three-row fin and tube heat exchanger, *J. Heat Transfer*, 98, 26–34.

Saboya, F.E.M., Rosman, E.C., and Carajilescov, P., 1984. Performance of one- and two-row tube and plate fin heat exchangers, *J. Heat Transfer*, 106, 627–632.

Schmidt, E., 1963. Heat transfer at finned tubes and computations of tube bank heat exchangers, *Kaltetechnik*, No. 4, 15, 98; No. 12, 15, 370.

Seshimo, Y. and Fujii, M., 1991. An experimental study on the performance of plate and tube heat exchangers at low Reynolds numbers, *Proc. 1981 ASME-JSME Thermal Eng. Conf.*, 4, 449–454.

Shah, R.K. and Webb, R.L., 1982. Compact and enhanced heat exchangers, in *Heat Exchangers: Theory and Practice*, J. Taborek, G.F. Hewitt, and N. Afgan, Eds., Hemisphere Washington, D.C., 425–468.

Sheui, T.W.H., Tsai, S.F., and Chiang, T.P., 1999. Numerical study of heat transfer in two-row heat exchangers having extended fin surfaces, *Numerical Heat Transfer A*, 35(7), 797.

Tanaka, T., Hatada, T., Itoh, M., Senshu, T., Katsumata, N., Michizuki, Y., Terada, H., Izushi, M., Sato, M., Tsuji, H., and Nagai, M., 1994, Fin-Tube Heat Exchanger, U.S. patent 5,360,060.

Torii, K., Kwak, K.M., and Nishino, K., 2002. Heat transfer enhancement accompanying pressure-loss reduction with winglet-type vortex generators for fin-tube heat exchangers, *Int. J. Heat Mass Transfer*, 45, 3795–3801.

Torikoshi, K. and Kawabata, K., 1989. Heat transfer and flow friction characteristics of mesh finned air-cooled heat exchangers, in *Convection Heat Transfer and Transport Processes*, HTD vol. 116, R.S. Figliola, M. Kaviany, and M.A. Ebadian, Eds., American Society of Mechanical Engineers, New York, pp. 71–77.

Torikoshi, K. and Xi, G., 1995. A numcrical study of flow and thermal fields in finned tube heat exchangers, in *Proceedings of the IMECE*, HTD, 317-1, 453–458.

Torikoshi, K., Xi, G.N., Nakazawa, Y., and Asano, H., 1994. Flow and heat transfer performance of a plate fin-and-tube heat exchanger (First report: Effect of fin pitch), *Heat Transfer 1994, Proceedings of the 10th Int. Heat Transfer Conf.*, 4, 411–416.

Tsai, S. F., Sheu, T.W.H., and Lee, S.M., 1999. Heat transfer in a conjugate heat exchanger with a wavy fin surface, *Int. J. Heat Mass Transfer*, 42, 1735–1745.

Ueda, H., Hatada, T., Kunugi, N., Ooucgi, T., Sugimoto, S., Shimizu, T., and Kohno, K., 1994. Heat Transfer Fins and Heat Exchanger, U.S. patent 5,353,886.

Valencia, A., Fiebig, M., and Mitra, N.K., 1996, Heat transfer enhancement by longitudinal vortices in a fin-tube heat exchanger element with flat tubes, *J. Heat Transfer,* 118(1), 209–211.

Wang, C.-C., 2000. Technology review — a survey of recent patents of fin-and-tube heat exchangers, *J. Enhanced Heat Transfer*, 7(5), 333.

Wang, C.-C. and Chang, C.-T., 1998. Heat and mass transfer for plate fin-and-tube heat exchangers with and without hydrophilic coating, *Int. J. Heat Mass Transfer*, 41, 3109–3120.

Wang, C.-C. and Chi, K.-Y., 2000. Heat transfer and friction characteristics of plain fin-and-tube heat exchangers. Part I: New experimental data, *Int. J. Heat Mass Transfer*, 43, 2681–2691.

Wang, C.-C., Chen, P.-Y., and Jang, J.-Y., 1996a. Heat transfer and friction characteristics of convex-louver fin-and-tube heat exchangers, *Exp. Heat Transfer*, 9(1), 61.

Wang, C.-C., Chang, Y.-J., Hsieh, Y.-C. and Lin, Y.-T., 1996b. Sensible heat and friction characteristics of plate fin-and-tube heat exchangers having plane fins, *Int. J. Refrig.*, 19(4), 223–230.

Wang, C.-C., Hsieh, Y.-C., and Lin, Y.-T., 1997a. Performance of plate finned tube heat exchangers under dehumidifying conditions, *J. Heat Transfer*, 119(1), 109–117.

Wang, C.-C., Fu, W.-L., and Chang, C.-T., 1997b. Heat transfer and friction characteristics of typical wavy fin-and-tube heat exchangers, *Exp. Thermal Fluid Sci.*, 14, 174–186.

Wang, C.-C., Tssai, Y.-M., and Lu, D.-C., 1998a. Comprehensive study of convex-louver and wavy fin-and-tube heat exchangers, *J. Thermophys. Heat Transfer*, 12(3), 423–430.

Wang, C.-C., Chi, K.-Y., Chang, Y.-J., and Chang, Y.-P., 1998b. An experimental study of heat transfer and friction characteristics of typical louver fin-and-tube heat exchangers, *Int. J. Heat Mass Transfer*, 41, 817–822.

Wang, C.-C., Lin, Y.-T., and Chang, C.-J., 1999a. Investigation of wavy fin-and-tube heat exchangers: a contribution to databank, *Exp. Heat Transfer,* 12(1), 73.

Wang, C.-C., Tao, W.-H., and Chang, C.-J., 1999b. An investigation of the airside performance of the slit fin-and-tube heat exchangers, *Int. J. Refrig.*, 22, 595–603.

Wang, C.-C., Jang, J.-Y., and Chiou, N.-F., 1999c. Heat transfer and friction correlation for wavy fin-and-tube heat exchangers, *Int. J. Heat Mass Transfer*, 42, 1919–1924.

Wang, C.-C., Chang, J.-Y., and Chiou, N.-F., 1999d. Effects of waffle height on the air-side performance of wavy fin-and-tube heat exchangers, *Heat Transfer Eng.*, 20(3), 45–56.

Wang, C.-C., Lee, C.-J., Chang, C.-T., and Lin, S.-P., 1999e. Heat transfer and friction correlation for compact louvered fin-and-tube heat exchangers, *Int. J. Heat Mass Transfer*, 42, 1945–1956.

Wang, C.-C., Chi, K.-Y., and Chang, C.-J., 2000a. Heat transfer and friction characteristics of plain fin-and-tube heat exchangers. Part II: Correlation, *Int. J. Heat Mass Transfer*, 43, 2693–2700.

Wang, C.-C., Lin, Y.-T., and Lee, C.-J., 2000b. Heat and momentum transfer for compact louvered fin-and-tube heat exchangers in wet conditions, *Int. J. Heat Mass Transfer*, 43, 3443–3452.

Wang, C.-C., Lee, W.-S., and Sheu, W.-J., 2001. A comparative study of compact enhanced fin-and-tube heat exchangers, *Int. J. Heat Mass Transfer*, 44, 3565–3573.

Webb, R.L., 1980. Air-side heat transfer in finned tube heat exchangers, *Heat Transfer Eng.*, 1(3), 33–49.

Webb, R.L., 1983a. Enhancement for extended surface geometries used in air-cooled heat exchangers, in *Low Reynolds Number Flow Heat Exchangers*, Hemisphere, Washington, D.C., 721–734.

Webb, R.L., 1983b. Heat transfer and friction characteristics for finned tubes having plain fins, in *Low Reynolds Number Flow Heat Exchangers*, Hemisphere, Washington, D.C., 431–450.

Webb, R.L., 1987. Enhancement of single-phase heat transfer, in *Handbook of Single-Phase Heat Transfer*, S. Kakaç, R.K. Shah, and W. Aung, Eds., John Wiley & Sons, New York, 17.1–17.62.

Webb, R.L., 1990. Air-side heat transfer correlations for flat and wavy plate fin-and-tube geometries, *ASHRAE Trans.*, 96(2), 445–449.

Webb, R.L. and N. Gupte, 1990. Design of light weight heat exchangers for air-to-two phase service, in *Compact Heat Exchangers: A Festschrift for A.L. London*, R.K. Shah, A. Kraus, and D.E. Metzger, Eds., Hemisphere, Washington, D.C., 311–334.

Webb, R.L. and Iyengar, A., 2000. Oval finned tube heat exchangers — limiting internal operating pressure, *J. Enhanced Heat Transfer*, 8, 147–158.

Weierman, C., 1976. Correlations ease the selection of finned tubes, *Oil Gas J.* 74, 94–100.

Weierman, C., Taborek, J., and Marner, W.J., 1978. Comparison of inline and staggered banks of tubes with segmented fins, *AIChE Symp. Ser.*, 74(174), 39–46.

Wieting, A.R. 1975. Empirical correlations for heat transfer and flow friction characteristics of rectangular offset fin heat exchangers, *J. Heat Transfer*, 97, 488–490.

Yan, W.-M. and Sheen, P.-J., 2000. Heat transfer and friction characteristics of fin-and-tube heat exchangers, *Int. J. Heat Mass Transfer*, 43, 1651-1659.

Youn, B., 1997. Internal Report, Samsung Electronic Co.

Youn, B. and Kim, Y.S., 1998. Heat Exchanger Fins of an Air Conditioner, U.S. patent 5725625.

Youn, B., Kil, Y.-H., Park, H.-Y., Yoo, K.-C., and Kim, Y.-S., 1998. Experimental study of pressure drop and heat transfer characteristics of 10.07 mm wave and wave-slit fin-tube heat exchangers with wave depth of 2 mm, in *Heat Transfer 1998, Proceedings of 11th Int. Heat Transfer Conf.*, Vol. 6, Kyongju, Korea.

Youn, B., Kim, Y.-S., Park, H.-Y., and Kim, N.-H., 2003. An experimental investigation on the airside performance of fin-and-tube heat exchangers having radial slit fins, *J. Enhanced Heat Transfer*, 10(1), 61–80.

Yun, J.Y. and Kim, H.-Y., 1997. Structure of Heat Exchanger, U.S. patent 5697432.

Yun, J.Y. and Lee, K.-S., 2000. Influence of design parameters on the heat transfer and flow friction characteristics of the heat exchanger with slit fins, *Int. J. Heat Mass Transfer*, 43, 2529–2539.

Zukauskas, A., 1972. Heat transfer from tubes in cross flow, in *Advances in Heat Transfer*, J.P. Hartnett and T. Irvine, Jr., Eds., Vol. 8, Academic Press, San Diego, CA, 93–160.

INSERT DEVICES FOR SINGLE-PHASE FLOW

7.1 INTRODUCTION

Chapter 7 through Chapter 9 address techniques to enhance the heat transfer coefficient inside a tube for single-phase flow. Hence, these three chapters discuss three different approaches to enhancement of tube-side convective heat transfer. Insert devices involve various geometric forms that are inserted in a smooth, circular tube. These devices are in competition with internal fins (Chapter 8) and integral roughness (Chapter 9). Which of the three methods is preferred depends on two factors—performance and initial cost. Integral internal fins and roughness require deformation of the material on the inside surface of a long tube. Cost-effective manufacturing technology to deform the inner surface of a tube has only recently been developed. Insert devices represent an early approach to tube side enhancement, which allowed use of a plain tube. Anticipating the material to be discussed in Chapters 7 through 9, this chapter shows that insert devices are generally not competitive with the current performance and cost of internal roughness for turbulent flow. Only in laminar flow are insert devices an effective solution. However, insert devices may be used to upgrade the performance of an existing heat exchanger.

Work to conceive insert devices has been largely empirical. Figure 7.1 through Figure 7.5 show the various insert devices. The dominant literature on insert devices involves the following five concepts:

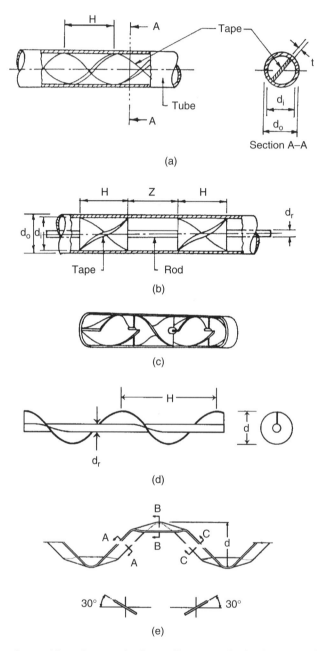

Figure 7.1 Tape inserts: (a) continuous twisted tape, (b) segmented twisted tape, (c) Kinex mixer, (d) helical tape, (e) bent strips.

1. Twisted tape inserts (Figure 7.1a), which cause the flow to spiral along the tube length. The tape inserts generally do not have good thermal contact with the tube wall, so the tape does not act as a fin.
2. Extended surface inserts (Figure 7.2a), which are an extruded shape inserted in the tube. The tube is then drawn to provide good thermal contact between the wall and the insert. The insert reduces the hydraulic diameter and acts as an extended surface.
3. Wire coil inserts (Figure 7.3a), which consist of a helical coiled spring that functions as a non-integral roughness.
4. Mesh or brush inserts, as shown in Figure 7.4.
5. An insert device that is displaced from the tube wall and causes periodic mixing of the gross flow. Figure 7.5a and b are representative devices.

For turbulent flow, it is more effective to mix the flow in the viscous boundary layer at the wall (e.g., Figure 7.3a) than it is to mix the gross flow using the Figure 7.4 or 7.5 devices. This is because the dominant thermal resistance is very close to the wall. Integral internal roughness (Chapter 9) can generally provide a given

(a)

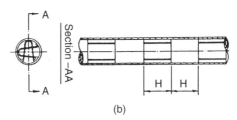

(b)

Figure 7.2 Extended surface inserts: (a) extruded insert. (Courtesy Wieland-Werke AG). (b) Interrupted sheet metal. (From Jayaraj et al. [1989].)

$E_i = h/h_n$ with a higher efficiency index (η) than is provided by the Figure 7.3a wire coil insert.

For laminar flow, the dominant thermal resistance is not limited to a thin boundary layer adjacent to the flow. So, devices that mix the gross flow are more effective in laminar flow than in turbulent flow.

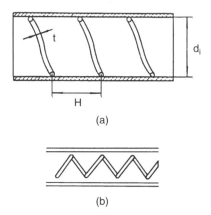

(a)

(b)

Figure 7.3 Wire coil inserts: (a) wires touching tube wall, (b) wires displaced from tube wall.

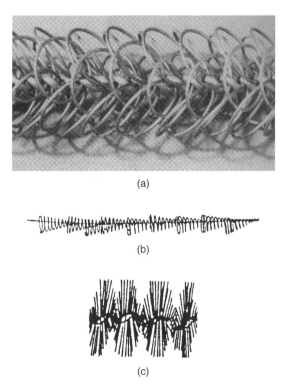

(a)

(b)

(c)

Figure 7.4 Mesh or brush inserts: (a) mesh insert, (b) helical coil insert, (c) brush insert.

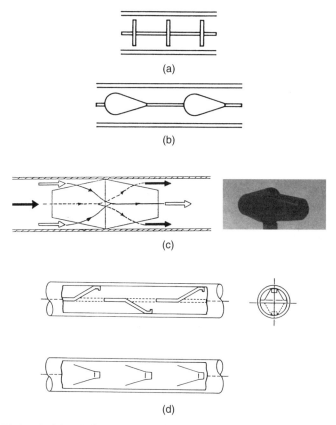

Figure 7.5 Displaced mixing devices: (a) spaced disks, (b) spaced streamline shapes, (c) flow eversion device (from Maezawa and Lock [1978]), (d) louvered strip.

Although the extended surface insert of Figure 7.2a is an effective enhancement concept for both laminar and turbulent flow, it suffers from cost concerns. The twisted tape (Figure 7.1a) and mesh inserts (Figure 7.4a) are of primary interest for laminar flow, but their potential performance is diminished, because the tape is not in good thermal contact with the wall.

We discuss the relative merits of the various devices, and provide information that may be used to predict their heat transfer and friction characteristics.

7.2 TWISTED TAPE INSERT

The continuous twisted tape insert shown in Figure 7.1a has been extensively investigated for both laminar and turbulent flow. Variants of the twisted tape that have been evaluated include short sections of twisted tape at the tube inlet, or periodically spaced along the tube length (Figure 7.1b and c). Bergles and Joshi [1983] present a survey of the performance of the different types of swirl flow devices for laminar flow.

The insert shown in Figure 7.1a consists of a thin twisted strip that is slid into the tube. The axial distance for 180° twist is dimension H. Authors describe their tapes by the "twist ratio", $y = H/d_i$. Alternate to the y parameter, one may use the helix angle of the tape, given by

$$\tan \alpha = \frac{\pi d_i}{2H} = \frac{\pi}{2y} \tag{7.1}$$

To allow easy insertion of the tape, there is usually a small clearance between the tape width and the tube inside diameter. This clearance results in poor thermal contact between the tape and the tube wall, so the heat transfer from the tape may be quite small. The blockage caused by the finite tape thickness increases the average velocity. Heat transfer enhancement may occur for three reasons:

1. The tape reduces the hydraulic diameter (D_h), which causes an increased heat transfer coefficient, even for zero tape twist.
2. The twist of the tape causes a tangential velocity component. Hence, the speed of the flow is increased, particularly near the wall. The heat transfer enhancement is a result of the increased shear stress at the wall and mixing by secondary flow.
3. Heat is transferred from the tape, if good thermal contact with the wall exists. However, little heat transfer is expected from a loosely fitting tape.

Thorsen and Landis [1968] show that centrifugal forces, caused by the tangential velocity component, provide enhancement by mixing fluid from the core region with fluid near the wall. However, this will occur only when the flowing fluid is being heated. The colder, high-density core region fluid is forced outward to mix with the warm, low-density fluid near the wall. If the fluid is being cooled, the centrifugal force acts to maintain thermal stratification of the fluid.

Although much work has been done on twisted tapes, recent papers by Manglik and Bergles [1992a, 1992b] present an updated understanding of the heat transfer and friction characteristics for both laminar and turbulent flow. The papers provide correlations valid for deep laminar flow through the turbulent regime, including the transition regime for constant wall temperature boundary condition. In the laminar flow regime, Manglik and Bergles [1992a] show that the laminar flow Nusselt number is influenced by entrance length and buoyancy forces (natural convection); these effects are controlled by the magnitude of the Graetz number ($Gz_d = \pi d_i Re_d Pr/4L$) and the Rayleigh number ($Ra = GrPr$), respectively. They show that Nu is a function of Sw, Gz, Ra, Pr, μ/μ_w. The friction factor for fully developed flow depends on Sw and μ/μ_w. The "swirl number," is defined as $Sw = Re_{Sw} y^{-1/2}$, where $y = H/d_i$. The Manglik and Bergles laminar and turbulent flow correlations are presented in Sections 7.2.2 and 7.2.3, respectively.

Note that there is no common agreement on the characteristic dimension used in the Reynolds number definition for twisted tapes. Manglik and Bergles [1992a, 1992b] use Re_d (based on d_i), while some others use Re_{Dh} (based on D_h).

7.2.1 Laminar Flow Data

Laminar flow is a relatively complex subject, because the flow is influenced by the following conditions: (1) the thermal boundary condition, (2) entrance region effects, (3) natural convection at low Reynolds number, (4) fluid property variation across the boundary layer, and (5) the duct cross-sectional shape. Further, the local Nusselt will be different for simultaneously developing velocity and temperature profiles, as compared to a fully developed velocity profile. Figure 7.6, which shows theoretical solutions for laminar, entrance region flow in a circular tube, is provided to aid understanding of the twisted tape data. These solutions are taken from Shah and Bhatti [1987], and assume fully developed velocity profile entering the heating region, constant properties in the tube, and no natural convection effects. Figure 7.6 shows different results for the q = constant and T_w = constant boundary conditions. Because Figure 7.6 shows the entrance region, the Nusselt number is plotted vs. the entrance region parameter, Nu_x vs. $x_d^*[= x/(d_i \mathrm{Re}_d \mathrm{Pr})]$. The figure shows the local Nusselt number (Nu_x) and the Nusselt number averaged over the length x/d_i (Nu_m) for constant heat flux (subscript H) and constant wall temperature (subscript T). Figure 7.6 shows that the fully developed condition is not obtained until $x_d^* \cong 0.01$.

Laminar flow data are taken with either a constant heat flux (q = constant) or an approximate constant wall temperature (T_w = constant) thermal boundary condition, usually with a developed velocity profile entering the heated test section. For the q = constant boundary condition, the local heat transfer coefficient (h_x) is obtained using the known heat flux, and the wall temperature is measured by thermocouples

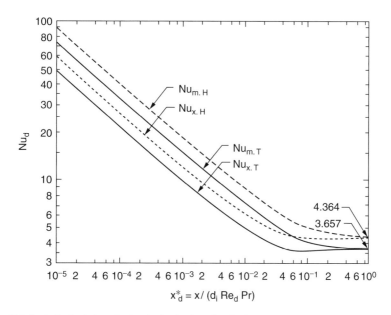

Figure 7.6 Analytical solutions for developing laminar flow in circular tube with fully developed velocity profile. (From Shah and Bhatti [1987].)

along the tube length. It is usually possible to obtain the local Nusselt number for fully developed flow ($Nu_{fd} = h_{fd}d_i/k$) from wall thermocouples sufficiently far from the heated inlet. With the T_w = constant boundary condition, it is nearly impossible to obtain data for the Nu_{fd} using a moderate test section length. In the T_w = constant test, the local heat flux is not known. So, experimenters measure the UA value over the test section length, and subtract the shell-side resistance to obtain the average tube-side heat transfer coefficient over the test section length. This measurement corresponds to the $Nu_{m,T}$ solid line on Figure 7.6. Even if Nu_{fd} could be obtained in the T_w = constant tests, the Nu_{fd} would be different for q = constant and T_w = constant data, as shown by Figure 7.6.

There are two additional complicating factors in obtaining laminar flow data. First is the effect of fluid property variation across the boundary layer. Laminar flow is much more susceptible to the effect of fluid property variation than is turbulent flow. As shown in Tables 2.3 and 2.4, the effect of property variation across the boundary layer is accounted for by the term $(\mu/\mu_w)^{0.14}$. For heating with $\mu/\mu_w = 1.5$, the h_x would be increased 5.5%. For cooling with $\mu/\mu_w = 1/1.5$, h_x would be reduced 5.5%. A second complication is the effect of buoyancy force (natural convection), which is determined by the Rayleigh number, Ra = GrPr. Whether natural convection effects are important for a twisted tape insert depends on the ratio Gr/Sw^2, which is the ratio of the buoyancy and centrifugal forces. Hence, one must be careful in comparing laminar flow data, since the data are affected by entrance effects, thermal boundary condition, fluid property variation across the boundary layer, and buoyancy force.

Figure 7.7 from Bergles and Joshi [1983] shows Nusselt number data for the two boundary conditions. Figure 7.7a shows the average $Nu_{d,m}$ for T_w = constant,

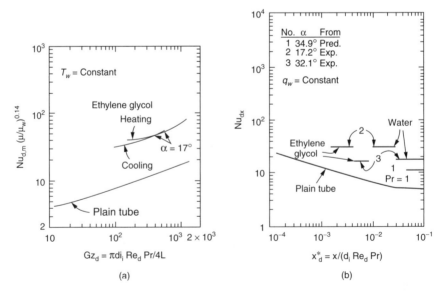

Figure 7.7 Nusselt number for laminar flow in tubes with twisted tape insert: (a) constant wall temperature, (b) constant heat flux. (From Bergles and Joshi [1983].)

and Figure 7.7b shows the local Nu_x for $q =$ constant. The data are presented as a function of entrance region parameters, because not all of the data points are for the fully developed condition. The Bergles and Joshi [1983] survey uses the independent variable x^{*d} for the $q =$ constant data and $Gz = \pi/(4x^{*d})$ for $T_w =$ constant data.

Figure 7.7a shows the $T_w =$ constant ethylene glycol data ($20 < Pr < 100$) for $\alpha = 17°$ ($y = 5.1$) of Marner and Bergles [1985], which provides 300% enhancement over the smooth tube value. Note that the data are not fully developed, and the cooling data fall below the heating data at the lower Gz values. Figure 7.8a shows the friction factor data for the Figure 7.7a data. The friction data of Marner and Bergles [1978] are shown by curve 3. Figure 7.8b shows the numerically predicted fully developed friction factor of Date [1974]. As indicated on Figure 7.8b, the $\alpha = 0$

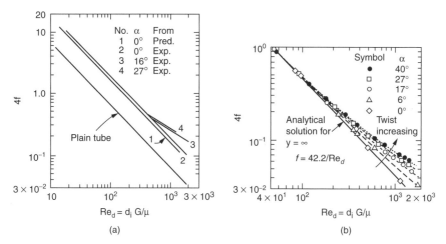

Figure 7.8 Friction factor in tubes with twisted tape insert for laminar flow: (a) experimental and predicted, (b) predicted by Date [1974] for zero tape thickness. (From Bergles and Joshi [1983].)

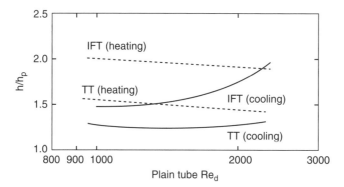

Figure 7.9 Performance comparison of internally finned tube for laminar flow of ethylene glycol (IFT, 16 fins, $e/D_i = 0.084$, $\alpha = 27°$, and twisted tape insert (TT, $\alpha = 30°$, $y = 5.4$, $t/D_i = 0.053$) for laminar flow using case FG-2a.

curve is for a tape of zero thickness and no twist. Increasing the twist causes $f\,\mathrm{Re}_d$ to depart from the 42.2 value.

Marner and Bergles [1978] also tested an internally finned tube in laminar flow. Figure 7.9 compares h/h_p for the twisted tape and the internally finned tube for PEC Case FG-2a of Chapter 3. Figure 7.9 shows that the performance of the twisted tape is better in heating than in cooling. It also shows that the performance of the internally finned tube is better than that of the twisted tape.

7.2.2 Predictive Methods for Laminar Flow

Date [1974] has performed numerical predictions for the twisted tape in fully developed laminar flow for q = constant with constant properties. His analysis (corrected) for a loosely fitting tape with $\alpha = 34.9°$ and $Pr = 1$ is shown by curve 1 on Figure 7.7b. Date's analysis also shows that the Nusselt number increases with increasing Prandtl and Reynolds numbers, contrary to laminar flow in plain tubes, which is independent of Re and Pr. Note that Date's [1974] reported Nu values are 50% low, due to a computational error, which is discussed by Hong and Bergles [1976]. Date and Singham [1972] developed empirical power-law correlations to fit their numerical predictions of the friction factor.

The recommended friction factor correlation is the recent work of Manglik and Bergles [1992a]. Figure 7.10 shows the basis of the correlation and shows the ratio of $(f\,\mathrm{Re}_d)_{sw}/(f\,\mathrm{Re}_d)_{sw,\,y\,=\,\infty}$, where the $(f\,\mathrm{Re}_d)_{sw,\,y\,=\,\infty}$ is for a tape having no twist.

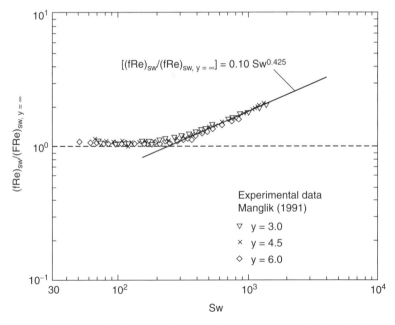

Figure 7.10 Asymptotic behavior of isothermal friction factor for fully developed laminar flow with twisted tape insert. (From Manglik and Bergles [1992a].)

The figure shows that the ratio approaches 1.0 as $Sw \rightarrow 0$ and the $f\,Re_d$ ratio approaches an asymptotic value $(0.10\,Sw^{0.425})$ for $Sw \equiv Re_{sw}/y^{1/2} \gg 1$. An asymptotic method is used to define a correlation, which approaches each of the two limiting asymptotes. The final correlation is given by

$$\frac{(f\,Re_d)_{sw}}{(f\,Re_d)_{sw,y=\infty}} = (1+10^{-6}Sw^{2.55})^{-1/6} \tag{7.2}$$

where

$$(f\,Re_d)_{sw,y=\infty} = 15.767\left(\frac{\pi+2-2t/d_i}{\pi-4t/d_i}\right)^2 \tag{7.3}$$

If $Sw = Re_{sw}/y^{1/2} \rightarrow 0$, the friction factor approaches the value for fully developed flow in a tape having zero twist (Equation 7.3). The term in brackets of Equation 7.3 accounts for the effect of the finite tape thickness on the velocity. The friction factor, f_{sw} is defined as

$$f_{sw} = \frac{\Delta p d_i}{2\rho u_{sw}^2 L_s} = f\frac{L}{L_s}\left(\frac{u_c}{u_{sw}}\right)^2 \tag{7.4}$$

Thus, f_{sw} is defined in terms of the swirl velocity (u_{sw}) and the length of the helically twisting streamline at the tube wall $(L_s = L/\cos\alpha)$. The second form of Equation 7.4 shows how f_{sw} is related to the conventional friction definition, based on the axial velocity (u_c) and pipe length (L).

Constant Heat Flux. Hong and Bergles [1976] have developed an empirical correlation based on their q = constant, twisted tape data for water and ethylene glycol. Two twisted tapes were used that had helix angles of $\alpha = 17°$ and $32°$. The measured local heat transfer coefficient corresponds to fully developed condition, Nu_{fd}. Unfortunately, they did not use the measured wall temperature to determine the $(\mu/\mu_w)^{0.14}$ correction. Their heat transfer correlation is given by Equation 7.5, in which $y = H/d_i = \pi/(2\tan\alpha)$.

$$Nu_d = 5.172\left[1+0.005484\left(\frac{Re_d}{y}\right)^{1.25}Pr^{0.7}\right]^{0.5} \tag{7.5}$$

Xie et al. [1992] provide data for an oil $(40 \le Pr \le 80)$ in an electrically heated tube for continuous tape with seven different twist ratios (y) between 1.67 and 5.0. The data were taken for $4000 \le Re_d \le 30,000$. They provided the following empirical correlations for their data.

$$Nu_d = 0.0149\,Re_{d,f}^{0.833}\,Pr^{2/3}\,y^{-0.493}(\mu/\mu_w)^{0.14} \tag{7.6}$$

$$f = (3.61 + 8.92y - 1.52y^2)\mathrm{Re}_{d,f}^{-(0.28+0.14y-0.021y^2)}(\mu/\mu_w)^{1/3} \tag{7.7}$$

where Re_{df} is the Reynolds number evaluated at the film temperature. Using PEC FG-2a of Table 3.1, they show that the highest performance (h/h_p) is provided by $y = 2.5$ at $\mathrm{Re}_d = 6000$ and by $y = 3$ for $\mathrm{Re}_d = 20,000$.

Saha and Dutta [2001] investigated the effect of shortening the twist tape length for oil in laminar flow $(205 < \mathrm{Pr} < 518)$ by installing a short length tape (at the inlet end) in their electrically heated test section. The twist ratio was varied from $y = 2.5$ to 10. Typical graphs for the friction factor and the Nusselt number are shown in Figure 7.11a and b, respectively, for $y = 5$. The figures show that both friction factors and Nusselt numbers (based on tube inner diameter) decrease as the tape length decreases. The PEC analysis using FG-2a (constant pumping power) in Table 3.1 revealed some increase of performance (10%, 21% and 22% heat transfer increase for half length, one-third length, and one-fourth length twist tapes) compared with the full-length tape. Saha and Dutta [2001] also tested twist tapes having gradually decreasing pitches (increased y values). Approximately 15% reduction in both the friction factors and the Nusselt numbers were observed compared with those of the uniform pitch.

Constant Wall Temperature. Manglik and Bergles [1992a] developed a correlation for the $T_w =$ constant case using the data of Marner and Bergles [1978, 1985] for ethylene glycol and polybutene $(1000 < \mathrm{Pr} < 7000)$, and the water and ethylene glycol data of Manglik [1991]. They show that $\mathrm{Nu}_{d,m} = \mathrm{Nu}_{d,m}(\mathrm{Sw}, \mathrm{Gz}, \mathrm{Ra}, \mathrm{Pr}, \mu/\mu_w)$. The Sw parameter accounts for swirl, Gz accounts for the entrance length, and Ra accounts for natural convection. This correlation is also based on limiting asymptotes. This correlation is built as follows. For fully developed flow in a tube containing a tape with $y = \infty$, $\mathrm{Nu}_{d,m} = 4.612$. Accounting for entrance effects (Gz), and fluid property variation, the theoretical equation for flow in a tube containing a $y = \infty$ tape is

$$\mathrm{Nu}_{d,m} = 4.612(1 + 0.0951\mathrm{Gz}^{0.894})^{0.5}(\mu/\mu_w)^{0.14} \tag{7.8}$$

For fully developed flow $(\mathrm{Gz} \ll \mathrm{Sw})$, the data are correlated by

$$\mathrm{Nu}_{d,m} = 0.106\mathrm{Sw}^{0.767}\,\mathrm{Pr}^{0.3}(\mu/\mu_w)^{0.14} \tag{7.9}$$

An asymptotic correlation is obtained for the limit $\mathrm{Sw} \to \infty$, $\mathrm{Gz} \to 0$ by combining the fully developed asymptote $(\mathrm{Nu}_{d,m} = 4.162)$ with Equation 7.9.

$$\mathrm{Nu}_{d,m} = 4.612[1 + 6.413 \times 10^{-9}(\mathrm{Sw}\,\mathrm{Pr}^{0.391})^{3.835}]^{0.2}(\mu/\mu_w)^{0.14} \tag{7.10}$$

Using the asymptotic correlation method, they included thermal entrance effects (Gz) by combining Equations 7.8 and 7.10 to obtain

$$\mathrm{Nu}_{d,m} = 4.612[(1 + 0.0951\mathrm{Gz}^{0.894})^{2.5}$$

$$+6.413 \times 10^{-9}(\mathrm{Sw}\,\mathrm{Pr}^{0.391})^{3.835}]^{0.2}(\mu/\mu_w)^{0.14} \tag{7.11}$$

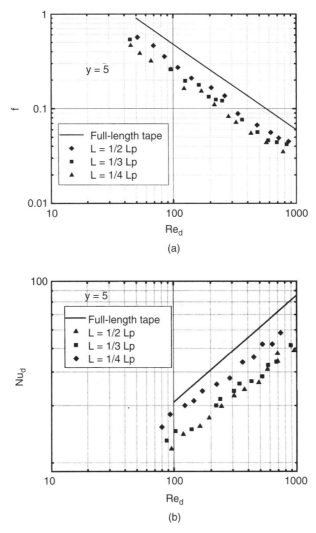

Figure 7.11 Effect of shortening the twist tape length for oil in laminar flow ($205 \leq Pr \leq 518$) for $y =$ 5. (a) Friction factor vs. Reynolds number, (b) Nusselt number vs. Reynolds number. (From Saha and Dutta [2001].) (Reprinted with permission of ASME.)

Natural convection effects will dominate, when $Gr > Sw^2$. The natural convection influence were determined by correlating the data vs. $Re_d Ra$ to obtain

$$\frac{Nu_{d,m}}{4.612} = 4.294 \times 10^{-2} (Re_d\, Ra)^{0.223} \tag{7.12}$$

The final correlation is obtained by using the following two asymptotes: (1) the swirl flow asymptote for $Ra \rightarrow 0$ given by Equation 7.11, and (2) the buoyancy effects asymptote given by Equation 7.12. Using the asymptotic correlation method, the final form of the correlation is obtained.

$$Nu_{d,m} = 4.612\langle[(1+0.0951Gz^{0.894})^{2.5} + 6.413\times10^{-9}(Sw\,Pr^{0.391})^{3.835}]^2$$
$$+2.132\times10^{-14}(Re_d\,Ra)^{2.23}\rangle^{0.1}(\mu/\mu_w)^{0.14}$$

(7.13)

Figure 7.12 shows predicted values using Equation 7.13, and includes some of the Manglik and Bergles [1992a] data. Figure 7.12a shows the effect of natural convection and swirl flow. Figure 7.12b shows the effect of entrance effects and swirl flow. Note that Equation 7.13 is limited to constant wall temperature, and is not applicable to a constant heat flux boundary condition.

Patil [2000] investigated the effect of reducing the tape width in laminar flow of a pseudo- plastic power-law fluid under constant wall temperature condition. Twisted tapes of four different widths (w = 0.95, 0.79, 0.66, and 0.44, where $w = W/d_i$) were tested in a 25-mm-ID tube. The twist ratio was varied from 2.56 to 5.40. Compared with full width tapes, reduced width tapes yielded 18 to 56% lower friction factors and 5 to 25% lower Nusselt numbers. The friction factors and the Nusselt numbers decreased as the tape width decreased. The performance analysis using FG-2a in Table 3.1 revealed that the reduced tapes yielded lower heat transfer than full-width tapes.

7.2.3 Turbulent Flow

Smithberg and Landis [1964] developed a semiempirical model for the friction factor, which is based on adding the momentum losses for helical flow and tape induced fluid mixing. They showed that the model reasonably predicted their data for tapes having $3.62 \le y \le \infty$. Zhuo et al. [1992] found that the Smithberg and Landis friction model underpredicted their friction data (y = 3, 4.25, 5.25) approximately 15%. Based on his evaluation of published data, Date [1973] proposed an empirical correlation for the friction factor for the range $5000 < Re_{Dh} < 70,000$ and $1.5 < y < \infty$, or $0 < \alpha < 46°$. The correlation is

$$\frac{f}{f_p} = \left(\frac{y}{y-1}\right)^m$$

(7.14)

where

$$m = 1.15 + \frac{1.25(70,000 - Re_{Dh})}{65,000}$$

(7.15)

and f_p is the friction factor in a plain tube, given by

$$f_p = 0.046\,Re_d^{-0.2}$$

(7.16)

The recommended friction correlation is from Manglik and Bergles [1992b].

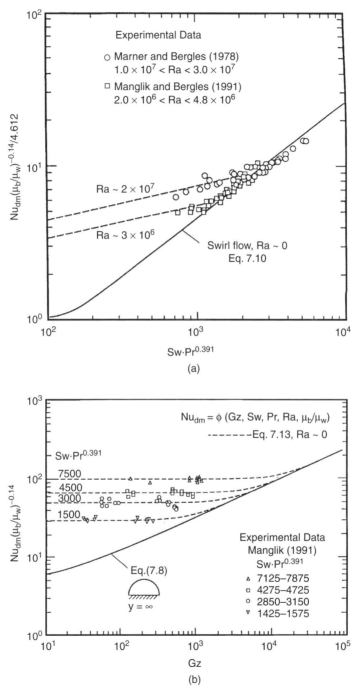

Figure 7.12 Comparison of predictions using Equation 7.13 with data illustrating the effect of (a) natural convection (Ra) and (b) entrance length (Gz). (From Manglik and Bergles [1992a].)

$$f = \frac{0.079}{\text{Re}_d^{0.25}}\left(\frac{\pi}{\pi - 4t/d_i}\right)^{1.75}\left(\frac{\pi + 2 - 2t/d_i}{\pi - 4t/d_i}\right)^{1.25}\left(1 + \frac{2.752}{y^{1.29}}\right) \tag{7.17}$$

Equation 7.17 accounts for tape thickness. The authors show that this is in good agreement with data and correlations published by others.

Smithberg and Landis [1964] developed the first semianalytical model for turbulent heat transfer with a twisted tape. The heat transfer model was outdated by Thorsen and Landis [1968], who suggested that buoyancy effects arising from density variations in the centrifugal field have an effect on heat transfer. They showed that the swirl flow-induced buoyancy effect should depend on the dimensionless group $\text{Gr}/\text{Re}_{Dh}^2$, which may be written as

$$\frac{\text{Gr}}{\text{Re}_{Dh}^2} = \frac{2D_h\beta_T\Delta T\tan\alpha}{d_i} \tag{7.18}$$

Thorsen and Landis measured the heat transfer coefficient for heating and cooling of water in tubes having tapes with three different helix angles, $\alpha = 11.1°$, $16.9°$, and $26.5°$. The heating data were correlated by

$$\text{Nu} = 0.021F\left(1 + 0.25\sqrt{\frac{\text{Gr}}{\text{Re}_{Dh}}}\right)\text{Re}_{Dh}^{0.8}\,\text{Pr}^{0.4}\left(\frac{T_w}{T_h}\right)^{-0.32} \tag{7.19}$$

and the cooling data by

$$\text{Nu} = 0.023F\left(1 - 0.25\sqrt{\frac{\text{Gr}}{\text{Re}_{Dh}}}\right)\text{Re}_{Dh}^{0.8}\,\text{Pr}^{0.3}\left(\frac{T_w}{T_h}\right)^{-0.1} \tag{7.20}$$

where

$$F = 1 + 0.004872\frac{\tan^2\alpha}{d_i(1 + \tan^2\alpha)} \tag{7.21}$$

and d_i is in meters.

Lopina and Bergles [1969] attempted to account for the increased speed of the flow (caused by the spiral flow) and the centrifugal buoyancy effect using a superposition model. The model also accounts for the possibility that the tape may act as an extended surface. The model may be expressed as

$$q = q_{sc} + q_{cc} + q_f \tag{7.22}$$

where q_{sc} is swirl convection, q_{cc} is centrifugal convection, and q_f is from the fin.

The swirl convection (q_{sc}) is predicted using an appropriate equation for turbulent flow in plain tubes with the Reynolds number calculated in terms of D_h and a modified velocity (u_{sw}) account for the speed of the swirl flow at the wall. The heat transfer coefficient for the swirl convection term (q_{sc}) is given by

$$\text{Nu}_{\text{sc}} = \frac{h_{\text{sc}}D_h}{k} = 0.023\,\text{Re}_{\text{Dh}}^{0.8}\,\text{Pr}^{0.4} \tag{7.23}$$

where

$$\text{Re}_{\text{Dh}} = \frac{u_{\text{sw}}D_h}{\nu} \quad \text{where} \quad u_{\text{sw}} = u_c\sqrt{1+\tan^2\alpha} \tag{7.24}$$

The q_{cc} term in Equation 7.22 represents the centrifugal convection effect identified by Thorsen and Landis. Lopina and Bergles [1969] use an equation for turbulent natural convection from a horizontal plate and replace the gravity force (g) by a radial acceleration (g_r), which is

$$g_r = \frac{2u_\theta^2}{d_i} = \frac{2}{d_i}\left(\frac{u_c\pi}{2y}\right)^2 = \frac{4.94}{d_i}\left(\frac{u_c}{y}\right)^2 \tag{7.25}$$

The heat transfer coefficient for the centrifugal convection term is calculated by

$$\text{Nu}_{\text{cc}} = \frac{h_{\text{cc}}D_h}{k} = 0.12(\text{Gr}_{\text{Dh}}\,\text{Pr})^{1/3} \tag{7.26}$$

where

$$\text{Gr}_{\text{Dh}} = \frac{4.94\beta_T\Delta T_{fs}D_h\,\text{Re}_{\text{Dh}}^2}{d_i y^2} \tag{7.27}$$

The term q_f in Equation 7.22 accounts for heat transfer from the tape as an extended surface. Evaluation of this term requires knowledge of the contact resistance between the tube wall and the tape, and is usually unknown. For the poor thermal contact situation, $q_f = 0$. Refer to Lopina and Bergles [1969] for details on calculation of q_f, should this be of interest.

The most recent heat transfer correlation is by Manglik and Bergles [1992b]. Their correlation is based on the asymptotic method, and is valid for $T_w = $ constant and $q = $ constant with $\text{Re}_d > 10{,}000$. The correlation is

$$\frac{\text{Nu}_d}{\text{Nu}_{d,y=\infty}} = 1 + \frac{0.769}{y} \tag{7.28}$$

where $\text{Nu}_{d,y=\infty}$ is for a tape having no twist, and is given by

$$\text{Nu}_{d,y=\infty} = 0.023\,\text{Re}_d^{0.8}\,\text{Pr}^{0.4}\left(\frac{\pi}{\pi-4t/d_i}\right)^{0.8}\left(\frac{\pi+2-2t/d_i}{\pi-4t/d_i}\right)^{0.2}\phi \tag{7.29}$$

The term ϕ accounts for fluid property variation. For liquids, $\phi = (\mu/\mu_w)^n$ with $n = 0.18$ for heating and $n = 0.30$ for cooling. For gases, $\phi = (T/T_w)^m$, with $m = 0.45$ for heating and $m = 0.15$ for cooling with the temperature T in Kelvin (or Rankine).

Manglik and Bergles [1992b] also propose a friction factor correlation for the laminar–turbulent transition region. They show that the transition is smooth, and their laminar (Equation 7.2) and turbulent correlations (Equation 7.17) may be combined to obtain a correlation for the transition region. Their recommended correlation is

$$f = (f_l^{10} + f_t^{10})^{0.1} \tag{7.30}$$

where f_l and f_t are given by Equations 7.2 and 7.17, respectively.

7.2.4 PEC Example 7.1

A heat exchanger to preheat 0.542 kg/s of combustion air from 26.7°C to 165.6°C using combustion products at 271°C is required. The capacity rate ratio is $C_a/C_g = 0.9$, where subscript a is air and subscript g is combustion gas. The heat exchanger has 100 steel tubes ($k = 41.5$ W/m-K) 25.4 mm inside diameter and 1.5 mm wall thickness. The hot gas flows normal to the tubes and air flows inside 100 tubes in parallel.

You have completed a plain tube design which has a pass length of $L_p = 2.51$ m and requires $UA = 863$ W/K with $h_a = 55.9$ and $h_g = 170$ W/m²-K. Because $h_a < h_g$, you plan to use a twisted-tape ($H/d_i = 3.14$) inside the tube. Use PEC FN-2 of Table 3.1 ($Q/Q_p = W/W_p = 1$) to determine the tube length saving. For simplicity, assume zero tape thickness and neglect the tube wall resistance and any fouling resistances. Use the analytical model of Lopina and Bergles [1969] for the heat transfer coefficient.

Solution. This problem is straightforward, because the Reynolds number in the tube is directly calculable. The Nu for the tube is obtained using Equation 7.22, which requires calculating the Nu_{sc} (swirl convection) and Nu_{cc} (centrifugal convection) terms. Assume zero heat transfer from the tape ($q_f = 0$).

The hydraulic diameter is $\pi d_i /(\pi + 2) = 15.5$ mm, and the Reynolds number ($Re_{Dh} = GD_h/\mu$) is 8594. The Nu_{sc} is directly calculated using Equation 7.23, giving $Nu_{sc} = 27.9$. The Nu_{cc} given by Equation 7.26, and requires an iterative calculation, since ΔT_{fs} (wall-to-fluid temperature difference) is not known. Substituting the known terms in Equation 7.26, $Nu_{cc} = 0.12(33,468 \ \Delta T_{fs})^{1/3}$. Guess $\Delta T_{fs} = 0.6$ $\Delta T_{lm} = 0.6 \times 112.2$ K $= 67.3$ K, and thus obtain $Nu_{cc} = 15.73$. So, $Nu = Nu_{sc} + Nu_{cc} = 27.9 + 15.73 = 43.63$, and $h_i = Nuk/D_h = 88.46$ W/m²-K. Solving for L/UA gives

$$\frac{L}{UA} = \frac{L}{h_i A_i} + \frac{L}{h_o A_o}$$

$$= \frac{1}{88.46 \times 0.0254\pi} + \frac{1}{170 \times 0.0284\pi} = 0.208 \tag{7.31}$$

Using $UA/h_i A_i = \Delta T_{fs}/\Delta T_{lm}$, yields $\Delta T_{fs} = 59.0$ K as compared to the guessed value of 67.3 K. So, Nu_{cc} is recalculated using $\Delta T_{fs} = 59.0$ K, which yields $Nu_{cc} = 44.33$, and results in $h_i = 89.9$ W/m²-K. Solving Equation 7.26 again gives $L/UA = 0.205$ and $\Delta T_{fs} = 58.6$ K, which is close enough to the second assumed $\Delta T_{fs} = 59.0$ K.

The required pass length is $L = (UA/100) \times (L/UA) = 8.63 \times 0.205 = 1.77$ m, which compares to 2.51 m for the plain tube design.

Because the tube-side flow is turbulent, use Equation 7.14 and obtain $f/f_p = 1.635$. Then, the pressure drop ratio is

$$\frac{\Delta p}{\Delta p_p} = \frac{f}{f_p} \frac{L}{L_p} = 1.635 \times \frac{1.77}{2.51} = 1.152 \tag{7.32}$$

7.2.5 Twisted Tapes in Annuli

Gupte and Date [1989] report data for three tape geometries in an annulus. Their data were taken for air with heat applied to the outer wall of the annulus for two radius ratios ($r_i/r_o = 0.41$ and 0.61). The three tapes tested had $y = 2.66$, 5.30 ($r_i/r_o = 0.41$), and 5.04 ($r_i / r_o = 0.61$). The Reynolds number for the heat transfer data spanned $10,000 < \mathrm{Re_{Dh}} < 50,000$.

They also developed a semianalytical model to predict their data. The model reasonably predicted the friction data for the three tapes. The heat transfer model worked well for the $y \cong 5$, but underpredicted the $y = 2.66$ data.

Because only the outer wall was heated, centrifugal forces will act to stratify the flow. In practical applications, one would expect heat exchange to occur at only the inner wall. If the inner wall is heated, centrifugal forces will act to mix the flow. Hence, it is likely that the performance for a heated inner wall will be higher than the data of Gupte and Date [1989].

Coetzee et al. [2001] provide additional data for $y = 0.73$, 1.78, and 2.88, all with $r_i / r_o = 0.69$. The Reynolds number range was $3000 \leq \mathrm{Re_{Dh}} \leq 25,000$. Compared with the smooth annulus counterpart, approximately threefold increase in the Nusselt number (with approximately equal pressure drop increase) was obtained for $y = 0.73$ geometry. The enhancement decreased as the twist ratio (y) increased.

7.2.6 Twisted Tapes in Rough Tubes

Zimparov [2001] report data for five tapes with $4.7 \leq y \leq 15.3$ in two three-start corrugated tubes ($e/d_i = 0.0407$, 0.0569 with $p/e = 7.45$ and 10.73 respectively). The Reynolds number spanned $3000 \leq \mathrm{Re}_d \leq 60,000$. The heat transfer coefficients and friction factors increased as the e/d_i increased and y decreased. Although the heat transfer enhancement was significant for these tubes, the pressure drop penalty was enormous. For example, for the tube with $e/d_i = 0.0569$ and $y = 4.7$ at $\mathrm{Re}_d = 30,000$, the heat transfer enhancement was 8.4 as compared with the smooth tube. The pressure drop increase, however, was 168 times. The PEC analysis yielded conflicting results. Application of FG-1a of Table 3.1 resulted in heat transfer improvement for the corrugated tube with tape inserts compared with the corrugated tube alone. Application of FG-2a, however, resulted that the corrugated tube (no tape) yielded superior heat transfer performance. The study was extended by Zimparov [2002] to two single-start corrugated tubes ($e/d_i = 0.0371$ and 0.0441 with $p/e = 11.6$ and 13.5) for five tapes ($2.4 \leq y \leq 7.7$). Similar results to those of the previous study (Zimparov

[2001]) were obtained. Shivkumar and Rao [1998] provide additional data for three tapes ($2.92 \leq y \leq 9.73$) in a corrugated tube ($e/d_i = 0.0245$ and $p/e = 47.6$).

An analytical model was developed by Zimparov [2004a, 2004b] to predict the heat transfer coefficients and friction factors for turbulent flow in a corrugated tube with twisted tape insert. The flow field is divided into two regions — a core region and a wall region. The helicoidal core flow was modeled following the suggestion of Smithberg and Landis [1964], and the mixing-length model was used to account for the wall roughness effect. The model reasonably predicted his own data (Zimparov [2001, 2002]).

Zhang et al. [1997] report data for three tapes ($y = 6, 7.5, 10$) in circular tubes having axially interrupted ribs. Three internal rib configurations were investigated: $e/d_i = 0.0626$ inline rib, $e/d_i = 0.0625$ staggered rib and $e/d_i = 0.125$ staggered rib. The Reynolds number spanned $16,000 \leq \text{Re}_d \leq 82,000$. The staggered rib configuration yielded higher heat transfer and pressure drop than the inline configuration. The maximum heat transfer enhancement of 2.2 to 3.2 (compared to a smooth tube) was obtained for the tube with $y = 6$ and $e/d_i = 0.125$ staggered rib. The pressure drop increase, however, was 13 to 14 times. When compared with the tube having twisted tape only, these tubes yielded 25 to 40% higher heat transfer at a constant pumping power (FG-2a in Table 3.1).

7.3 SEGMENTED TWISTED TAPE INSERT

Saha et al. [1989] have conducted an extensive experimental program of the Figure 7.1b segmented twisted tape in laminar flow of water with constant heat flux. Their tape geometries had twist ratios (H/d_i) of 3.18, 5, 7.5, 10, and ∞, or helix angles (α) of 26.3°, 17.4°, 8.82°, and 0°. Elements of twisted tape were spaced a distance $z = Z/d_i$. For each twist ratio y, they tested dimensionless spacings of $z = 2.5, 5, 7.5$, and 10. Figure 7.13 shows their results for $y = 5$. The Nusselt number is the average value over the 1.8-mm-long, 11-mm-ID test section. The line labeled $z = 0$ on Figure 7.13a is the Hong and Bergles [1976] correlation, Equation 7.5. The figure shows that the smallest z value gives the highest Nu, and Nu decreases for increasing z. At $z = 2.5$, and $\text{Re}_d = 1000$, the Nu is increased 15%, relative to the continuous twisted tape. The friction factor data are shown on Figure 7.13b. The solid line is for a tape with $y = \infty$ and has the value $f = 46.45/\text{Re}_d$. The dashed line is from a correlation for continuous tapes with $y = 5$. The highest friction factor also occurs for the smallest z. The segmented tapes yield friction factors higher than the continuous tape ($z = 0$) for $y \leq 7.5$. But, for $z = 10$, the friction factor is smaller than the continuous tape. Properly spaced tape segments provide higher enhancement than a continuous tape, because the wake mixing region between the tape segments dissipates the thermal and velocity boundary layers. This mixing process soon dies out, and the reinitiation of the swirl flow is required. For $y = 3.46$, Saha et al. shows that z values of 7.5 and 10 give smaller Nu values than a continuous tape.

Table 7.1 compares the enhancement provided by the segmented tapes, relative to a continuous tape using PEC FG-2b in Chapter 3. The Nu_{st} is for the segmented

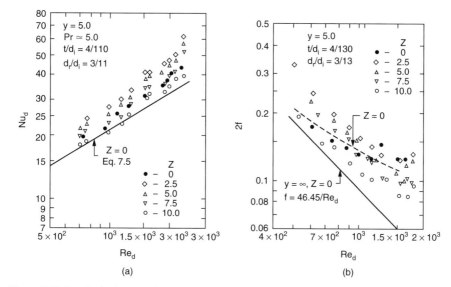

Figure 7.13 Developing laminar flow of water with q = constant in a tube having a segmental twisted tape of $y = 5$ and $0 < z < 10$: (a) Nusselt number, (b) friction factor. (From Saha et al. [1989].)

Table 7.1 Case FG-1 PEC for Segmented Twisted Tapes

	Nu_{st} / Nu_{ct} for y values of:			
z	3.18	5.0	7.5	10.0
2.5	1.44	1.47	1.28	1.08
5.0	1.34	1.31	1.13	1.05
7.5	1.17	1.16	1.12	1.04
10	1.09	1.11	1.09	1.00

tape, and Nu_{ct} for a continuous tape. The enhancement ratios are constant for $679 \leq Re_{d,c} \leq 1918$, where $Re_{d,c}$ is the Reynolds number for the continuous tape.

The Reynolds number of the segmented tape is smaller than that of the continuous tape, in order to meet the constant pumping power criterion. The Reynolds numbers are related by the equation

$$\frac{Re_{st}}{Re_{ct}} = \left(\frac{f_{st}}{f_{ct}} \frac{A_{c,ct}}{A_{c,st}} \right)^{1/3} \tag{7.33}$$

The cross-sectional flow area and hydraulic diameters used in Equation 7.33 are based on the volumetric average values.

Saha et al. [1989] give a series of rather complex empirical correlations to predict the Nu and f as a function of Re, y, and z. The Nu correlation is an extension of Equation 7.5, and converges to the Equation 7.5 value if $z = 0$.

Saha and Dutta [2001] investigated the effect of number of tape module twist sections for laminar oil flow ($205 \leq Pr \leq 518$). Typical graphs for the friction factors and the Nusselt numbers for $y = 5$ and $z = 2.5$ are shown in Figure 7.14a and b. In the graphs, m denotes the number of segmental twist sections. The graphs show a significant reduction both in the friction factor and the Nusselt number as the number of twists increases from one to two. This can be appreciated from the fact that the number of tape modules in the test section for $m = 2$ is less than that for $m = 1$. The figures also show that the friction factor and the Nusselt number do not decrease much from $m = 2$ to $m = 3$. They reported 14 to 47% reduction in friction factors with comparable reduction in Nusselt numbers for the test range ($2.5 \leq y \leq 5.0$ and $2.5 \leq z \leq 5.0$) when the number of twists increased from one to two.

Xie et al. [1992] provide data on the Figure 7.1c static mixer for an oil at $Pr = 41$ in an electrically heated tube with $2000 \leq Re_d \leq 20,000$. Data are provided for $y = 3, 3.5$, and 4.0. No correlation was given.

Date and Saha [1990] numerically predicted the friction and heat transfer characteristics of laminar flow in a circular tube fitted with regularly spaced twisted tape elements. The predictions reasonably agreed with the experimental data of water (Saha et al. [1989]). They also reported that increasing the number of turns on the tape element or reducing the connecting rod diameter would increase the thermal performance.

7.4 DISPLACED ENHANCEMENT DEVICES

7.4.1 Turbulent Flow

These are some of the very earliest devices investigated, and are probably the least effective for turbulent flow. Figure 7.5a and b illustrates two types of displaced insert devices tested by Koch [1958] and by Evans and Churchill [1963] in laminar and turbulent flow. Colburn and King [1931] tested the Figure 7.4b device and a variant of the Figure 7.5a device. These devices periodically mix the gross flow structure and accelerate the local velocity near the wall. Koch found that the Figure 7.5a and b devices have substantially higher pressure drop than the twisted tape (Figure 7.1a) or the Figure 7.3a wire coil insert. Theoretical reasoning argues that, for turbulent flow, the fluid should be mixed in the viscous dominated region near the wall, where the thermal resistance is large. The Figure 7.4 and 7.5 devices mix the flow in the core region and experience quite high profile drag forces, which substantially increase the pressure drop. Other displaced insert devices that have been tested in turbulent flow include mesh or brush inserts of Figure 7.4, and bristle brushes, static mixer devices (Figure 7.1c), and flow-driven propellers, all of which promote mixing across the total flow cross section. Mergerlin et al. [1974] found that bristle brushes (Figure 7.4c) provide h/h_p as high as 8.5, but the pressure drop was increased by a factor of 2800. Bergles [1985] provides a PEC evaluation of Koch's [1958] data for the Figure 7.5a and b displaced insert devices using PEC FG-2a of Table 3.1. PEC FG-2a calculates h/h_p for constant pumping power, tube diameter ($d_o = 50$ mm), and tube length. The devices provide very poor performance, relative to other competing

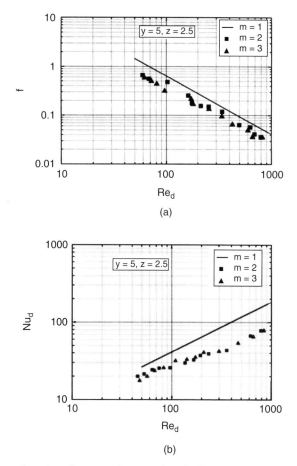

Figure 7.14 Effect of number of segmented tape sections for laminar oil flow ($205 \leq \mathrm{Pr} \leq 518$) with $y = 5$ and $z = 2.5$ for short length twisted tapes. (a) Friction factor vs. Reynolds number, (b) Nusselt number vs. Reynolds number. (From Saha and Dutta [2001].)

devices and are not recommended for operation in the turbulent regime. The velocity in the enhanced tube must be reduced so low that little benefit, if any, occurs.

The Figure 7.3b displaced wire coil insert device is somewhat different from the other displaced insert devices. It is similar to the Figure 7.3a wall-attached wire coil insert, except it causes mixing in a narrow region close to the tube wall. Thomas [1967] tested displaced wires for turbulent flow of water in an annulus. He concluded the most favorable St/f performance was obtained when the wires were axially spaced in pairs approximately 9 wire diameters, followed by a second pair separated approximately 75 wire diameters from the first pair. The enhancement was provided by an increased velocity gradient at the surface and by interaction of the cylinder wake with the fluid in the boundary layer. Tests of this displaced wire coil in a circular tube have not been reported. It would be relatively difficult to install such a displaced wire coil in a tube.

Xie et al. [1992] provide data on the Figure 7.5d louvered insert for an oil ($Pr = 41$) in an electrically heated tube with $2000 \le Re_d \le 30,000$. This consists of a flat strip, which is slit and bent at approximately a 45° angle. The strips, spaced at distance p, are bent in alternate directions, so that when it is slipped in the tube, the flat strip is centered in the tube. Their data were taken for an interrupted element pitch of $y = p/d_i = 1$, 2, 4, and 6.

$$Nu_d = 0.0202 \, Re_{d,f}^{0.767} \, Pr^{2/3} \, y^{-0.282} (\mu/\mu_w)^{0.14} \qquad (7.34)$$

$$f = (5.68 - 2.07y + 0.43y^2) \, Re_{d,f}^{-(0.34-1.41y-0.021y^2)} (\mu/\mu_w)^{1/3} \qquad (7.35)$$

where Re_{df} is evaluated at the film temperature.

Park et al. [2000] tested 8 Figure 7.5d type inserts using air in an electrically heated tube for $20,000 \le Re_d \le 80,000$. The eight inserts included two single straight tapes with different tape width, two cross tapes with different tape width, a wire mesh tape, a Figure 7.5d tape with delta wings, a Figure 7.5d tape with rectangular winglets (refer to Figure 5.28), and a helical coil insert of Figure 7.4b. The highest heat transfer coefficient (also the highest friction factor) was obtained for the delta wing tape, followed by the helical coil insert and the wire mesh tape. The straight and the cross tapes yielded low heat transfer coefficients and friction factors. The performance analysis was conducted using FG-2a in Table 3.1, and the results are shown in Figure 7.15. Figure 7.15 shows h/h_p vs. the equivalent smooth tube Reynolds number $Re_s \, [=(f/f_p) \, Re_d]$ for equal pumping power. The figure shows that the enhancement ratio decreases as the Reynolds number increases except for the straight tape, where the performance is relatively flat. High performance is obtained for the helical coil insert (or conical coil), rectangular winglet (or ribbon type), and delta wing tapes. Additional data on straight or cross tapes for turbulent airflow are reported by Hsieh and Kuo [1994].

7.4.2 Laminar Flow

Because the thermal resistance is not confined to a thin boundary layer region near the wall in laminar flow, the Figure 7.4 and 7.5 devices offer greater potential than for turbulent flow. Such devices may be helpful for cooling of viscous fluids, such as an oil. When oil is cooled, the higher viscosity at the wall causes reduced velocity and temperature gradients at the tube wall. If the insert promotes bulk fluid mixing or increases the temperature and velocity gradient at the wall, enhancement will result. There are two principal questions; (1) Which devices give the highest heat transfer enhancement? (2) What is the pressure drop penalty for the enhancement?

Bergles and Joshi [1983] and Bergles [1985] compare the performance of such displaced insert devices and twisted tapes for laminar flow. The comparisons are somewhat inconclusive, as they are not made for the same fluids, e.g., air vs. ethylene glycol.

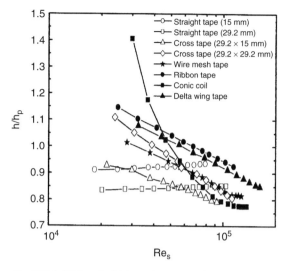

Figure 7.15 h/h_p vs. Re_s (FG-2a PEC analysis) for various types of twisted tapes for air in an electrically heated tube with $20,000 \leq Re \leq 80,000$. (From Park et al. [2000].)

The commercially available HEATEX mesh insert device shown in Figure 7.4a provides significantly higher enhancement than a twisted tape for cooling oil in laminar flow. Compare the HEATEX insert (subscript H) with a twisted tape insert (subscript T). The geometries compared are HEATEX Geometry D in Oliver and Aldington [1986] with a twisted tape ($y = 4.8$) that operates at $Re_d = 500$ using PEC VG-1 of Table 3.1. PEC VG-1 seeks A_H/A_T at $P_H/P_T = Q_H/Q_T = 1$, assuming all of the thermal resistance is on the tube side. The analysis shows $G_H/G_T = 0.89$, $A_H/A_T = 0.38$, and $L_H/L_T = 0.34$. Thus, to obtain equal hA and friction power, the HEATEX device requires only 38% as much surface area. The tube length is reduced 66%, and the number of tubes is increased 12%.

7.4.3 PEC Example 7.2

Compare the performance of the three insert devices tested by Xie et al. [1992] for oil flow (Pr = 41) at $Re_{d,p} = 5000$. The three insert devices are twisted tapes (Figure 7.1a), the static mixer insert (Figure 7.1c), and the louvered strip insert device (Figure 7.5d). The performance is compared for PEC FG-2a of Table 3.1. This PEC compares h/h_p for fixed flow rate, tube length, and pumping power. The flow rate in the enhanced tube may be reduced to compensate for the increased pressure drop in the enhanced tube. The specific geometries compared are listed in Table 7.2. Figure 7.16 shows that the Figure 7.5d louver strip insert with $y = 1.0$ (Code C-1) provides the best performance, followed by the Figure 7.1c static mixer with $y = 3.0$ (Code B-1). The lowest performance is provided by the twisted tape. At $Re_{d,p} = 5000$, the Code C-1 louvered strip provides $h/h_p = 1.65$, as compared to only 20% for the $y = 1.67$ twisted tape. At $Re_{d,p} = 40,000$, only the C-1 device is better than a plain tube.

Table 7.2 Case FG-2a PEC Comparison for Insert Devices Tested by Xie et al. [1992]

Geometry	Figure	Code	y
Twisted tape	7.1a	A-1	1.67
		A-2	2.5
		A-3	3.5
Static mixer	7.1c	B-1	3.0
		B-2	3.5
		B-3	4.0
Louver strip	7.5d	C-1	3.0
		C-2	2.0
		C-3	4.0

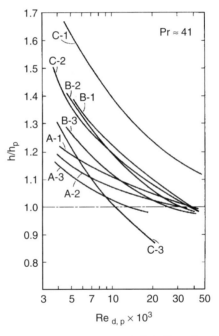

Figure 7.16 FG-2a comparison of twisted tapes (A-1, A-2, A-3), static mixers (B-1, B-2, B-3), and louvered strip inserts (C-1, C-2, C-3). The insert geometry dimensions are defined in Table 7.2. (From Xie et al. [1992].)

7.5 WIRE COIL INSERTS

The Figure 7.3a wire coil insert is made by tightly wrapping a coil of spring wire (wire diameter e) on a circular rod. The coil outside diameter, d_c, is made slightly larger than the tube inside diameter, d_i. When the coil spring is pulled through the tube, the wires form a helical roughness of height e at helix angle $\sin^{-1}(d_i/d_c)$ and spacing $p = \pi d_c \cos\alpha$. It is necessary that the coil spring forces the wire tightly against the tube wall to hold the wire in place and prevent tube wall erosion. This requires a helix angle of 25° or more. To obtain $\alpha = 25°$ in a 17.6-mm-diameter tube using wire of diameter $e = 1.0$ mm would require a wire coil diameter of 19.36 mm. The resulting coil insert would have $e/d_i = 0.057$, $p/d_i = 3.47$, and $p/e = 61$. The dimensionless geometric parameters that influence the heat transfer and friction characteristics are α, e/d_i, and p/e (or p/d_i). This example is given to show that practical considerations limit the dimensions that influence the performance of the wire coil insert.

The wire coil insert provides enhancement by flow separation at the wire, causing fluid mixing in the downstream boundary layer. Because the boundary layer mixing will dissipate downstream from the wire, the local enhancement quickly dissipates. Hence, rational selection of the wire diameter and spacing requires knowledge of how the local heat transfer coefficient varies with dimensionless distance (p/e) downstream from the trip wire. Two measurements of the local heat transfer coefficient downstream from a wall attached wire have been reported. Edwards and Sheriff [1961] used a boundary layer on a flat plate, and Emerson [1961] worked with pipe flow. In both airflow studies, the thermal and velocity boundary layers were large compared to the rib height. Figure 7.17 shows h_{max}/h_p vs. x/e at $Re_x = u_\infty x/\nu = 530,000$, where h_p is for the undisturbed boundary layer flow. The maximum local enhancement ratio of 1.45 is insensitive to Re_x, and is attained 10 rib heights downstream from the rib, which corresponds to the boundary layer reattachment point. Figure 7.18 shows the local enhancement ratio for $0.7 \leq e \leq 6.5$ mm plotted vs. e/δ_L, where δ_L is the laminar sublayer thickness of the undisturbed boundary layer at the rib, and is given by $\delta_L u^*/\nu = 5$. Figure 7.18 shows a linear increase of h_{max}/h_p with increasing e/δ_L for $2 \leq e/\delta_L \leq 8$, and only a small increase of h_{max}/h_p for $e/\delta_L > 8$. The dimensionless thickness of the viscous dominated boundary layer is typically assumed to be defined by $yu^*/\nu \simeq 30$, or $6\delta_L$. Hence, the maximum of h_{max}/h_p occurs when the wire diameter is 1.3 times that of the viscous-dominated

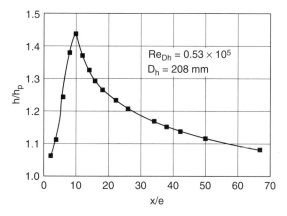

Figure 7.17 Enhancement of the local heat transfer coefficient provided by a 1.58-mm-diameter transverse wire for boundary layer flow of air over a flat plate at $Re_L = 530,000$. (From Edwards and Sheriff [1961].)

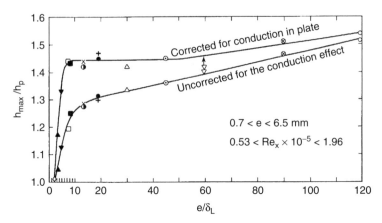

Figure 7.18 Effect of Re_L and wire diameter on h_{max}/h_p for boundary layer flow of air over a flat plate for wire sizes between 0.40 and 6.35 mm. (From Edwards and Sheriff [1961].)

boundary layer thickness ($6\delta_L$). Larger wire diameters protrude into the turbulence-dominated boundary layer and produce very little additional benefit. However, the larger wires will significantly increase the pressure drop, because of profile drag. These studies suggest an approximate rule for picking the maximum effective wire size. The wire coil insert amounts to "wall-attached roughness" of spaced helical ribs. Roughness is discussed in detail in Chapter 9. A wire coil insert with a helix angle large enough to fix the wire coil in the tube will have a p/e ratio of 40 or higher. Figure 7.17 shows that h/h_p drops to 1.18 at $p/e = 30$. So, little enhancement will exist in the range of $p/e > 30$, unless a large-diameter wire is used. Generally, integral roughness geometries discussed in Chapter 9 provide better heat transfer and pressure drop performance than can be obtained from wire coil inserts, which typically have a p/e ratio of 50 or higher.

7.5.1 Laminar Flow

Uttawar and Raja Rao [1985] tested seven different wire coil insert geometries in laminar flow ($30 < Re_d < 675$) for heating of an oil ($300 < Pr < 675$). The range of insert geometries tested were $0.08 < e/d_i < 0.13$ and $32 < \alpha < 61°$. Because the heated tube length was only 60 diameters, the flow was not fully developed. The measured enhancement levels were $1.5 \leq Nu_d/Nu_{d,p} \leq 4.0$. The friction increases were considerably less than the Nusselt number increase. The heat transfer data were correlated by

$$Nu_{Dv} = 1.65 \tan \alpha \, Re_{Dv}^m \, Pr^{0.35} \left(\frac{\mu}{\mu_w} \right)^{0.14} \tag{7.36}$$

where $m = 0.25(\tan \alpha)^{-0.38}$. The Nu, f and Re are based on the volumetric hydraulic diameter, D_v. A friction factor correlation was not developed. However, when defined in terms of D_v friction factor was only 5 to 8% higher than the smooth tube value for $Re_{D_v} < 180$. Chen and Zhang [1993] tested seven wire coil insert geometries for laminar flow ($273 \le Re_d \le 245$) of oil ($194 \le Pr \le 464$) under heating condition. The range of insert geometries were $0.056 \le e/d_i \le 0.133$ and $3.75 \le p/e \le 24$. The heated tube length was 100 diameters. The following correlations were proposed based on their own data.

$$Nu_d = 1.258 \ Re_d^{0.566} \ Pr^{0.169} (p/d_i)^{0.186} (p/e)^{-0.408} \tag{7.37}$$

$$f = 95.049 \ Re_d^{-0.129} \ Pr^{-0.230} (p/d_i)^{0.848} (p/e)^{-1.428} \tag{7.38}$$

7.5.2 Turbulent Flow

The correlations for turbulent flow are discussed in Section 9.4.3. This is because the wire coil insert essentially acts as a "wall-attached roughness." Its enhancement mechanism is the same as that of integral roughness, which is covered in detail in Chapter 9. The correlation of Sethumadhavan and Raja Rao [1983] given in Section 9.4.3 is recommended for turbulent flow of liquids and gases. The lowest Re_d tested by Sethumadhavan and Raja Rao [1983] for their turbulent flow correlation is 4000. and the largest Re_d tested by Uttawar and Raja Rao [1985] for their laminar flow correlation is 675. What correlation should be used in the range $675 \le Re_d \le 4000$? It is not possible to make a firm recommendation, but it is probable that the turbulent flow correlation should be applicable to Re values lower than 4000 — possibly as low as 1500.

7.6 EXTENDED SURFACE INSERT

Figure 7.2a shows this device. The insert device is formed as an aluminum extrusion. After inserting the extrusion in the tube, a tube-drawing process is employed to obtain a tight mechanical joint between the tube and the insert. The aluminum extrusion is normally formed with five legs, although the number of legs is a design choice, and may be between four and eight. By twisting the extrusion before its insertion in the tube, one may also promote a swirling flow. However, the helix angles would be $15°$ or less.

This insert device is not often used for liquids and gases. Apparently, this is because the pressure drop and cost are unfavorable compared to other enhancement devices (e.g., roughness). Hilding and Coogan [1964] provide j and f vs. Re data for a six-legged straight extrusion in turbulent flow. For turbulent flow, one may predict the turbulent flow j and f characteristics using an appropriate turbulent flow equation for smooth tubes with the tube diameter replaced by the hydraulic diameter (D_h)

Trupp and Lau [1984] have predicted the Nu and f for laminar flow in a tube having full-height fins of infinite thermal conductivity. The included angle between the fin legs was varied from $8°$ to $180°$.

Analytical predictions for the Nusselt number of the device must account for the fin efficiency and consider the possibility that a thermal contact resistance may exist between the aluminum and tube contact surfaces. The contact resistance may be negligible for flow of gases but may be appreciable for liquids.

7.7 TANGENTIAL INJECTION DEVICES

This concept involves tangentially injecting part of the flow at locations around the tube circumference, at the tube inlet end. This provides a swirl flow along the tube length. One method of fluid injection is shown by Figure 7.19. Hay and West [1975] found that the enhancement depends on the ratio of the injected momentum (M_t) to the total momentum of the axial flow (M_T). Although high enhancement levels are obtained at the tube inlet, they decay along the tube length. Razgatis and Holman [1976] provide an early survey of heat transfer to swirling flows. Dhir et al. [1989] and Dhir and Chang [1992] provide recent information on such swirl flows. Figure 7.20

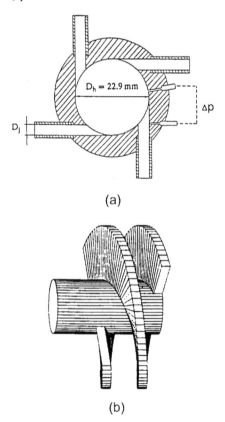

(a)

(b)

Figure 7.19 Swirl flow injector designs. (a) Used by Dhir et al. [1989]. (b) An alternate injector design from Dhir and Chang [1992].

shows the enhancement ratio obtained by Dhir et al. [1989] for airflow in a 22.9 mm-ID tube at $Re_d = 25{,}000$, for different values of M_t/M_T. They used the injector design shown in Figure 7.19a. The tangential-to-total-momentum ratio is given by

$$\frac{M_t}{M_T} = \frac{W_t^2}{W^2}\frac{A_c}{A_{c,j}}$$ (7.39)

Data by Dhir et al. [1989] show that the enhancement ratio is not significantly affected by either Reynolds number or Prandtl number. For the total flow rate entering four injectors ($W_t/W = 1$) and $M_t/M_T = 5.08$, the enhancement ratio drops from 5.8 at the tube inlet to approach 1.0 near 80 pipe diameters. The enhancement is caused by the high swirl velocity in the boundary layer at the wall. Because this velocity decreases along the pipe length, the enhancement decreases. Figure 7.20 shows that the enhancement ratio drops to 1.5 at $x/d_i = 50$. Consider a 3-m-long, 18-mm-ID heat exchanger tube operated at $Re_d = 16{,}400$. Hence, the 1.5 enhancement factor would be obtained at $x = 0.018 \times 50 = 0.9$ m from the inlet. Significant disadvantages of this enhancement concept are that it does not provide a uniform enhancement level along the tube length and that it dies out after a relatively short length. Further, a very high enhancement level at the tube inlet may not be of much value, if the controlling resistance is on the shell side at this location. Although it is conceivably possible to inject fluid at points along the tube length, this may not be practical.

Dhir et al. [1989] and Dhir and Chang [1992] correlate the enhancement level by

$$\frac{Nu}{Nu_{fd}} = 1 + 1.93 \left(\frac{M_t}{M_T}\right)^{0.6} Pr^{-1/7} \exp[-m(x/d_i)^{0.6}]\, e$$ (7.40)

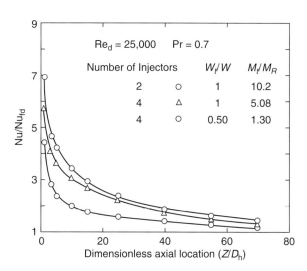

Figure 7.20 Enhancement ratio for tangential injection of air in a 22.9-mm-inside-diameter tube at Re = 25,000. (From Dhir et al. [1989].)

where

$$m = 0.89 \left(\frac{M_t}{M_T} \right)^{0.2} \mathrm{Re}_d^{-0.18} \, \mathrm{Pr}^{-0.083} \qquad (7.41)$$

The correlation is based on data for water and air and correlated 400 data points within ±15%

7.8 CONCLUSIONS

This chapter discusses insert devices for single-phase flow in tubes. In the turbulent regime, insert devices are not competitive with internally finned tubes or roughness. However, they may offer advantages for laminar flow of very viscous liquids. In laminar flow, the dominant thermal resistance is not limited to a very thin thermal boundary layer at the wall. The commercially used integral roughness is generally not a good candidate for laminar flow, because the roughness is too small to enhance a laminar flow.

Insert devices may be used to upgrade the performance of an existing heat exchanger having a plain inner tube surface. In this case, the performance improvement would be calculated using the FG-1 or FG-2 PEC of Table 3.1. A consequence of using insert devices to upgrade an existing heat exchanger is that the pressure drop will be significantly increased (for fixed flow rate), or the flow rate must be decreased to compensate for the increased pressure drop.

Cost-effective manufacturing technology to form tube-side roughness or internal fins is a fairly recent development. Hence, insert devices represent an early approach to tube-side enhancement, which allowed use of a plain tube. A variety of insert devices are available; the major types discussed in this chapter are twisted tapes, wire coil inserts, extended surface inserts, mesh or brush inserts, and insert devices displaced from the tube wall.

Design correlations exist for use of twisted tapes inserts in laminar, transition, and turbulent flow regimes. Although design equations are well established for the twisted tape insert, it may not necessarily be the best insert device. The potential performance of the twisted tape insert is diminished, because the tape is not in good thermal contact with the wall.

Tangential swirl injection is probably not competitive with integral roughness for turbulent flow. A significant disadvantage of swirl injection is that the heat transfer coefficient varies with axial location. This enhancement has not been sufficiently investigated for laminar flow. Practical solutions of how to inject a tangential flow are yet to be identified.

Application of insert devices to laminar flow is a complex issue, because this flow is sensitive to entrance length, thermal boundary conditions, and natural convection effects. Significant performance differences may exist for a given insert device in heating and cooling. The performance characteristics of the twisted tape insert has been well defined for these various characteristics. However, this is not the case for the other insert devices. Potentially interesting insert devices that require additional research to define their laminar flow characteristics are extended surface devices (Figure 7.2a). The wire mesh insert (Figure 7.4a) appears to offer significant enhancement for laminar flow of viscous liquids. Comparisons of several insert devices are presented. However, a complete comparison of all insert devices applied to laminar flow of very viscous liquids for both heating and cooling is yet to be completed.

Virtually no work has been done to establish fouling characteristics of insert devices. It is probable that insert devices, which introduce internal obstructions to the flow (e.g., a mesh insert), would cause fouling problems. The twisted tape insert or extended surface insert probably will not experience fouling problems.

NOMENCLATURE

A	Heat transfer surface area, m^2 or ft^2
A_c	Flow cross-sectional area in minimum flow area, m^2 or ft^2
$A_{c,j}$	Cross-sectional area for tangential injection, m^2 or ft^2
C	Capacity rate ($= Wc_p$), kJ/kg or Btu/hr-°F
c_p	Specific heat of fluid at constant pressure, J/kg-K or Btu/lbm-°F
d_c	Outside diameter of wire coil used to make Figure 7.3a insert, m or ft
D_h	Hydraulic diameter of flow passages, $4LA_c/A$, m or ft
D_v	Volumetric hydraulic diameter, 4 × void volume/total surface area, m or ft
d_i	Tube inside diameter, or diameter to the base of internal fins or roughness, m or ft
d_o	Tube outside diameter, fin root diameter for a finned tube, m or ft
e	Wire diameter for wire coil insert, m or ft
f	Fanning friction factor, $\Delta p_f d_i / 2LG^2$, dimensionless
f_{Dh}	Fanning friction factor based on D_h, $\Delta p_f D_h / 2LG^2$, dimensionless
f_{sw}	Fanning friction factor for swirl flow, $\Delta p_f d_i / 2L(G_{sw})^2$, dimensionless
G	Mass velocity based on the minimum flow area, kg/m^2-s or lbm/ft^2-s
G_{sw}	Mass velocity based on u_{sw} and the minimum flow area, kg/m^2-s or lbm/ft^2-s
Gr_d	Grashof number = $g_r \beta \Delta T d^3 / v^2$, dimensionless
Gr_{Dh}	Grashof number = $g_r \beta \Delta T D_h^3 / v^2$, dimensionless
Gz	Graetz number = $\pi d_i RePr/4L = \pi/4L_d^*$, dimensionless
g	Acceleration due to gravity, 9.806 m/s^2 or 32.17 ft/s^2
g_r	Radial acceleration defined by Equation 17.25, m/s^2 or ft/s^2
H	Length for 180° revolution of twisted tape, m or ft
h	Heat transfer coefficient based on A, h_x (local value), W/m^2-K or Btu/hr-ft^2-°F
j	Colburn factor = $StPr^{2/3}$, dimensionless
k	Thermal conductivity of fluid, W/m-K or Btu/hr-ft-°F
L	Fluid flow length, m or ft

L_s Swirl length of twisted tape ($= L/\cos\alpha$), m or ft

L^* $L/d_i\mathrm{Re}_d\mathrm{Pr}$, dimensionless

LMTD   Log-mean temperature difference, K or °F

m Number of twisted tape modules, dimensionless

M_T Axial momentum ($= W_t^2/\rho A_c$), kg-m/s^2 or lbm-ft/s^2

M_t Tangential injected momentum ($= W_t^2/\rho A_{c,j}$), kg-m/s^2 or lbm-ft/s^2

$\mathrm{Nu}_{\mathrm{Dh}}$ Nusselt number based on D_h ($= hD_h/k$), dimensionless

$\mathrm{Nu}_{\mathrm{Dv}}$ Nusselt number based on D_v ($= hD_v/k$), dimensionless

Nu_d Nusselt number based on d_i ($= hd_i/k$) dimensionless

NTU   Number of heat transfer units $= UA/(Wc_p)_{\min}$, dimensionless

P Fluid pumping power, W or hp

Pr   Prandtl number $= c_p\mu/k$, dimensionless

p Axial pitch of wire or roughness elements, m or ft

p_t $2H$, m or ft

Δp Fluid static pressure drop on one side of a heat exchanger core, Pa or lbf/ft^2

Q Heat transfer rate in the exchanger, W or Btu/hr

Ra   Rayleigh number $= \mathrm{Gr}\,\mathrm{Pr}$, dimensionless

r Tube or annulus radius, m or ft

R_{fi} Tube-side fouling resistance, m^2-K/W or ft^2-hr-°F/Btu

$\mathrm{Re}_{\mathrm{Dh}}$ Reynolds number based on the hydraulic diameter $= GD_h/\mu$, dimensionless

$\mathrm{Re}_{\mathrm{Dv}}$ Reynolds number based on D_v, dimensionless

Re_d Reynolds number based on the tube diameter $= Gd_i/\mu$, dimensionless

Re_s Equivalent smooth tube Reynolds number $= (f/f_p)\,\mathrm{Re}_d$, dimensionless

$\mathrm{Re}_{\mathrm{sw}}$ Swirl Reynolds number $= d_i u_{\mathrm{sw}}\rho/\mu$, dimensionless

Re_x Reynolds number based on axial distance ($= u_\infty x/\nu$), dimensionless

S_l Longitudinal tube or element pitch, m or ft

t Thickness of fin or twisted tape, m or ft

St   Stanton number $= h/Gc_p$, dimensionless

Sw   Swirl number, $\mathrm{Re}_{\mathrm{sw}} y^{-1/2}$, dimensionless

ΔT_{fs} Temperature difference between fluid and surface, K or °F

ΔT_i Temperature difference between hot and cold inlet fluids, K or °F

ΔT_{lm} Log-mean temperature difference between fluid and surface, K or °F

U Overall heat transfer coefficient, W/m^2-K or Btu/hr-ft^2-°F

u_c Fluid mean axial velocity at the minimum free flow area, m/s or ft/s

u_∞ Free stream velocity over flat plate, m/s or ft/s

u_{sw} $u_c(1 + \tan^2\alpha)^{1/2} = u_c[1 + (\pi/2y)^2]^{1/2}$, m/s or ft/s

u_θ Tangential velocity, m/s or ft/s

u^* Friction velocity $= (\tau_w/\rho)^{1/2}$, m/s or ft/s, dimensionless

W Fluid mass flow rate, kg/s or lbm/s, or width of tape, m or ft

W_t Tangentially injected fluid mass flow rate, kg/s or lbm/s

x Cartesian coordinate along the flow direction, m or ft

$\mathrm{x}_{\mathrm{d}}^*$ $x/(d_i\mathrm{Re}_d\mathrm{Pr})$, dimensionless

y Twist ratio $= H/d_i = \pi/(2\tan\alpha)$, dimensionless

z Dimensionless pitch of segmented inserts (Z/d_i) on Figure 7.1b

Z Spacing between tape segments (Figure 7.1b)

Greek Letters

α Helix angle relative to tube axis = $\pi d_i/p_i$, radians (or degrees)

β Volume coefficient of thermal expansion, 1/K or 1/R

δ_L Thickness of laminar sublayer on flat plate, m or ft

ν Kinematic viscosity, m²/s or ft²/s

η_f Fin efficiency or temperature effectiveness of the fin, dimensionless

μ Fluid dynamic viscosity coefficient, Pa-s or lbm/s-ft

ρ Fluid density kg/m³ or lbm/ft³

τ_w Wall shear stress, Pa or lbf/ft²

Subscripts

ct   Continuous tape

fd   Fully developed flow

m Average value over tube cross section or flow length

p Plain tube or surface

H Constant heat flux thermal boundary condition

st   Segmented tape

T Constant wall temperature thermal boundary condition

w Evaluated at wall temperature

x Local value

REFERENCES

Bergles, A.E., 1985. Techniques to augment heat transfer, in *Handbook of Heat Transfer Applications*, 2nd ed. W.M. Rohsenow, J.P. Hartnett, and E.N. Ganic, Eds., McGraw-Hill, New York, Chap. 3.

Bergles, A.E. and Joshi, S.D., 1983. Augmentation techniques for low Reynolds number in-tube flow, in *Low Reynolds Number Flow Heat Exchangers*, Hemisphere, Washington, D.C., 694–720.

Chen, L. and Zhang, H.J., 1993. Convection heat transfer enhancement of oil in a circular tube with spiral spring inserts, in *Heat Transfer Measurements and Analysis*, L.C. Chow and A.F. Emery, Eds., HTD-ASME Symp. Vol. 249, 45–50.

Coetzee, H., Liebenberg, L., and Meyer, J.P., 2001. Heat transfer and pressure drop characteristics of angled spiralling tape inserts in a heat exchanger annulus, in *Proceedings of the ASME Process Industries Division*, R.A. Paper, N.K. Aminemi, O. Toma, R. Rudland, and E. Crain, Eds., ASME New York, 79–86.

Colburn, A.P. and King, W.J., 1931. Relationship between heat transfer and pressure drop, *Ind. Eng. Chem.*, 23(8), 918–923.

Date, A.W., 1973. Flow in tubes containing twisted tapes, *Heating Ventilating Eng.*, 47, 240–249.

Date, A.W., 1974. Prediction of fully-developed flow in a tube containing a twisted tape, *Int. J. Heat Mass Transfer*, 17, 845–859.

Date, A.W. and Saha, S.K., 1990. Numerical prediction of laminar flow in a tube fitted with regularly spaced twisted tape elements, *Int. J. Heat Fluid Flow*, 11(4), 346–354.

Date, A.W. and Singham, J.R., 1972. Numerical Prediction of Friction and Heat Transfer Characteristics of Fully Developed Laminar Flow in Tubes Containing Twisted Tapes, ASME Paper 72-HT-17, ASME, New York.

Dhir, V.K. and Chang, F., 1992. Heat transfer enhancement using tangential injection, *ASHRAE Trans.*, 98(2), 383–390.

Dhir, V.K., Tune, V.X., Chang, F., and Yu, J., 1989. Enhancement of forced convection heat transfer using single and multi-stage tangential injection, in *Heat Transfer in High Energy Heat Flux Applications*, R.J. Goldstein, L.C. Chow, and E.E. Anderson, Eds., ASME Symp. Vol. HTD, 119, 61–68.

Edwards, F.J. and Sheriff, N., 1961. The heat transfer and friction characteristics for forced convection air flow over a particular type of rough surface, in *International Developments in Heat Transfer*, ASME, New York, 415–425.

Emerson, W.H., 1961. Heat transfer in a duct in regions of separated flow, in *Proc. Third Intl. Heat Transfer Conf.*, 1, 267–275.

Evans, L.B. and Churchill, S.W., 1963. The effect of axial promoters on heat transfer and pressure drop inside a tube, *Chem. Eng. Prog. Symp. Ser. 59*, 41, 36–46.

Gupte, N. and Date, A.W., 1989. Friction and heat transfer characteristics of helical turbulent air flow in annuli, *J. Heat Transfer*, 111, 337–344.

Hay, N. and West, P.D., 1975. Heat transfer in free swirl in flow in a pipe, *J. Heat Transfer* 97, 411–416.

Hilding, W.E. and Coogan, C.H., Jr., 1964. Heat transfer and pressure drop in internally finned tubes, in *ASME Symp. on Air Cooled Heat Exchangers*, ASME, New York, 57–84.

Hong, S.W. and Bergles, A.E., 1976. Augmentation of laminar flow heat transfer in tubes by means of twisted tape inserts, *J. Heat Transfer*, 98, 251–256.

Hsieh, S-S. and Kuo, M-T., 1994. An experimental investigation of the augmentation of tube-side heat transfer in a crossflow heat exchanger by means of strip-type inserts, *J. Heat Transfer*, 116, 381–390.

Jayaraj, D., Masilamani, J.G., and Seetharamu, K.N., 1989. Heat Transfer Augmentation by Tube Inserts in Heat Exchangers, SAE Technical paper 891983, Warrendale, PA.

Koch, R., 1958. Druckverlust und Wäermeuebergang bei verwirbeiter Stroemung, *Ver. Dtsch. Ingen. Forschungsheft*, Ser. B, 24(469) 1–44.

Lopina, R.F. and Bergles, A.E., 1969. Heat transfer and pressure drop in tape-generated swirl flow of single-phase water, *J. Heat Transfer*, 91, 434–442.

Maezawa, S. and Lock, G.S.H., 1978. Heat transfer inside a tube with a novel promoter, *Heat Transfer 1978, Proc. 6th Int. Heat Transfer Conf.*, 2, Hemisphere, Washington, D.C., 596–600.

Manglik, R.M., 1991. Heat Transfer Enhancement of In-Tube Flows in Process Heat Exchangers by Means of Twisted-Tape Inserts, Ph.D. thesis, Department of Mechanical Engineering, Rensselaer Polytechnic Institute, Troy, NY.

Manglik, R.M. and Bergles, A.E., 1992a. Heat transfer and pressure drop correlations for twisted-tape inserts in isothermal tubes: Part I, Laminar flows, in *Enhanced Heat Transfer* M.B. Pate and M.K. Jensen, Eds., ASME Symp. Vol. HTD, 202, 89–98.

Manglik, R.M. and Bergles, A.E., 1992b. Heat transfer and pressure drop correlations for twisted-tape inserts in isothermal tubes: Part II, Transition and turbulent flows, in *Enhanced Heat Transfer*, M.B. Pate and M.K. Jensen, Eds., ASME Symp. HTD, 202, 99–106.

Marner, W.J. and Bergles, A.E., 1978. Augmentation of tube-side laminar flow heat transfer by means of twisted tape inserts, static mixer inserts and internally finned tubes, in *Heat Transfer 1978, Proc. 6th Int. Heat Transfer Conf.*, Vol. 2, Hemisphere, Washington, D.C., 583–588.

Marner, W.J. and Bergles, A.E., 1985. Augmentation of highly viscous laminar tubeside heat transfer by means of a twisted-tape insert and an internally finned tube, in *Advances in Enhanced Heat Transfer — 1985*, S.M. Shenkman, J.E. O'Brien, I.S. Habib, and J.A. Kohler, Eds., ASME Symp Vol. HTD, 43, 19–28.

Mergerlin, F.E., Murphy, R.W., and Bergles, A.E., 1974. Augmentation of heat transfer in tubes by means of mesh and brush inserts, *J. Heat Transfer*, 96, 145–151.

Oliver, D.R. and Aldington, R.W. J., 1986. Enhancement of laminar flow heat transfer using wire matrix turbulators, *Heat Transfer — 1986, Proc. Eighth Int. Heat Transfer Conf.* 6, 2897–2902.

Park, Y., Cha, J., and Kim, M., 2000. Heat transfer augmentation characteristics of various inserts in a heat exchanger tube, *J. Enhanced Heat Transfer*, 7, 23–34.

Patil, A.G., 2000. Laminar flow heat transfer and pressure drop characteristics of power-law fluids inside tubes with varying width twisted tape inserts, *J. Heat Transfer*, 122, 143–149.

Razgatis, R. and Holman. J.P., 1976. A survey of heat transfer in confined swirl flows, *Heat Mass Transfer Processes*, 2, 831–866.

Saha, S.K. and Dutta, A., 2001. Thermohydraulic study of laminar swirl flow through a circular tube fitted with twisted tapes, *J. Heat Transfer*, 123, 417–427.

Saha, S.K., Gaitonde, U.N., and Date, A.W., 1989. Heat transfer and pressure drop characteristics of laminar flow in a circular tube fitted with regularly spaced twisted-tape elements, *Exp. Thermal Fluid Sci.*, 2, 310–322.

Sethumadhavan, R. and Raja Rao, M., 1983. Turbulent flow heat transfer and fluid friction in helical wire coil inserted tubes, *Int. J. Heat Mass Transfer*, 26, 1833–1845.

Shah, R.K. and Bhatti, M.S., 1987. Laminar convective heat transfer in ducts, in *Handbook of Single-Phase Heat Transfer*, S. Kakaç, R.K. Shah, and W. Aung, Eds., John Wiley & Sons, New York, 3.20.

Shivkumar, C. and Rao, M.R., 1998. Studies on compound augmentation of laminar flow heat transfer to generalized power law fluids in spirally corrugated tubes by means of twisted tape inserts, *ASME Proc. 96, 1988 National Heat Transfer Conference*, Vol. 1, HTD, H.R. Jacobs, Ed., 96, 685–692.

Smithberg, E. and Landis, F., 1964. Friction and forced convection heat transfer characteristics in tubes with twisted tape swirl generators, *J. Heat Transfer*, 87, 39–49.

Thomas, D.G., 1967. Enhancement of forced convection mass transfer coefficient using detached turbulence promoters, *Ind. Eng. Chem. Process Design Dev.*, 6, 385–390.

Thorsen, R. and Landis, F., 1968. Friction and heat transfer characteristics in turbulent swirl flow subjected to large transverse temperature gradients, *J. Heat Transfer*, 90, 87–89.

Trupp, A.C. and Lau, A.C.Y., 1984. Fully developed laminar heat transfer in circular sector ducts with isothermal walls, *J. Heat Transfer*, 106, 467–469.

Uttawar, S.B. and Raja Rao, M., 1985. Augmentation of laminar flow heat transfer in tubes by means of wire coil inserts, *J. Heat Transfer*, 105, 930–935.

Xie, L., Gu., R., and Zhang, X., 1992. A study of the optimum inserts for enhancing convective heat transfer of high viscosity fluid in a tube, in *Multiphase Flow and Heat Transfer, Second International Symposium*, Vol. 1, X-J. Chen, T.N. Veziroglu, and C.L. Tien, Eds., Hemisphere, New York, 649–656.

Zhang, Y.M., Han, J.C., and Lee, C.P., 1997. Heat transfer and friction characteristics of turbulent flow in circular tubes with twisted-tape inserts and axial interrupted ribs, *J. Enhanced Heat Transfer*, 4 (4), 297.

Zhuo, N., Ma, Q.L., Zhang, Z.Y., Sun, J.Q., and He, J., 1992. Friction and heat transfer characteristics in a tube with a loose fitting twisted-tape insert, in *Multiphase Flow and Heat Transfer, Second International Symposium*, Vol. 1, X-J. Chen, T.N. Veziroglu, and C.L. Tien, Eds., Hemisphere, New York, 657–661.

Zimparov, V., 2001. Enhancement of heat transfer by a combination of three-start spirally corrugated tubes with a twisted tape, *Int. J. Heat Mass Transfer*, 44, 551–574.

Zimparov, V., 2002. Enhancement of heat transfer by a combination of a single-start spirally corrugated tubes with a twisted tape, *Exp. Thermal Fluid Sci.*, 25, 535–546.

Zimparov, V., 2004a. Prediction of friction factors and heat transfer coefficients for turbulent flow in corrugated tubes combined with twisted tape inserts. Part I: Friction factors, *Int. J. Heat Mass Transfer*, 47, 589–599.

Zimparov, V., 2004b. Prediction of friction factors and heat transfer coefficients for turbulent flow in corrugated tubes combined with twisted tape inserts. Part 2: Heat transfer coefficients, *Int. J. Heat Mass Transfer*, 47, 385–393.

INTERNALLY FINNED TUBES AND ANNULI

8.1 INTRODUCTION

Internally finned tubes are primarily used for liquids, although they may also be used for pressurized gases. An example for in-tube gas flow is an air compressor intercooler. Internally finned tubes used for condensation and vaporization inside tubes are discussed in Chapters 13 and 14, respectively. Because liquids have higher heat transfer coefficients than gases, fin efficiency considerations require shorter fins with liquids than with gases. Figure 8.1a and b show internally finned tubes having integral internal fins. These tubes are made with either axial or helical fins. The thermal conductance per unit tube length is $\eta_o h A/L$. Typical finned surfaces provide an A/L in the range of 1.5 to 3 times that of a bare tube. Fin efficiency concerns will limit the practical fin height for use with liquids. For plane fins, the fin efficiency is given by

$$\eta_f = \frac{\tanh(me)}{me} \tag{8.1}$$

where

$$m = \left(\frac{2h}{k_f t}\right)^{1/2} \tag{8.2}$$

Low material thermal conductivity (k_f) and/or high fin height (e) will result in small fin efficiency. Use of water, which has a high heat transfer coefficient, will require low fin height, e.g., less than 2.0 mm. Tube materials other than copper and aluminum may have relatively low thermal conductivity, which will limit the practical fin height. High fins would be applicable only for liquids having low heat transfer coefficients. The Figure 8.1c insert device could be used for gases or liquids having a low heat transfer coefficient. The insert device must have good thermal contact with the tube wall.

8.2 INTERNALLY FINNED TUBES

The flow in internally finned tubes (Figure 8.1) may be either laminar or turbulent. When used to cool viscous fluids, such as oil, it is possible that the flow will be laminar. Figure 8.2 defines the key dimensions of the internally finned tube. In addition to the dimensions defined in Figure 8.2, the fins may be at a helix angle. The helix angle of the fins, relative to the tube axis, is given the symbol α.

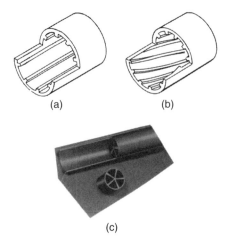

(a)   (b)

(c)

Figure 8.1 Illustration of integral-fin tubes for liquids: (a) axial internal fins, (b) helical internal fins, (c) extruded aluminum insert device.

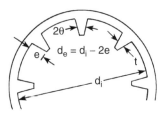

Figure 8.2 Definition of dimensions of the internally finned tube.

After reviewing the enhancement geometries described in Chapters 8 and 9, the reader will observe that some of the roughness geometries discussed in Chapter 9 appear quite similar to an internally finned tube. This is the case for the internally roughened Turbo-C™ tube shown in Figure 9.19a. To fall within the internal fin classification requires that no flow separations exist on the internal fin. As the helix angle is increased, one expects flow separation will occur at some critical helix angle. It is doubtful that this condition is attained for the geometries discussed in this chapter. This may not be the case for the Turbo-C tube, which has a 33° helix angle. Flow separation is a key feature of the enhancement mechanism of rough tubes. However, the theoretical analysis and correlations for internally finned tubes presume separation does not exist. An internally finned tube should provide a significant surface area increase, e.g., 50% or more.

8.2.1 Laminar Flow

Considerable data and numerical predictions have been reported for laminar flow.

Fully Developed Laminar Flow. Watkinson et al. [1975a] report Nu and f data (50 < Re_d < 3000) for steam heating of oil (180 < Pr < 350) in 18 different internally finned tubes. Marner and Bergles [1985] report Nu and f for a tube with 16 axial fins (e/d_i = 0.026) for both heating and cooling conditions (24 < Pr < 85, and 380 < Re_d < 3470) taken with T_w ≃constant. The data are not fully developed, and the reported heat transfer coefficients and friction factors are the average value for over the L/d_i = 96 test section length. Figure 7.9 shows h/h_p of the Marner and Bergles [1985] data for PEC FG-2a (see Table 3.1). This figure also contains results for a twisted tape insert (α = 30°) illustrated by Figure 7.1. Figure 7.9 shows that (1) the internal fin performance is better in heating than in cooling, and (2) the internal fin geometry is superior to the twisted tape. Marner and Bergles [1989] provide further interpretation of the Marner and Bergles [1985] data for polybutene, which has 1260 ≤ Pr ≤ 8130. They provide power-law factors to correct the heating and cooling data for fluid property variation across the boundary layer.

Rustum and Soliman [1988b] report water data on four internally finned tubes and a smooth tube. They report isothermal friction data, and local and fully developed heat transfer coefficients for water flow in four internally finned tubes using electrically heated tubes.

The problem of laminar flow in internally finned channels with axial fins is ideally suited to numerical solution. A number of studies are reported in the literature. Early studies by Hu and Chang [1973], Nandakumar and Masliyah [1975], and Soliman and Feingold [1977] addressed fully developed flow in tubes with zero thickness fins. Patankar and Chai [1991] analyze laminar flow in finned horizontal annuli. Soliman and Feingold [1977] and Soliman et al. [1980] account for finite fin thickness and thermal conductivity for fully developed flow. Prakash and Patankar [1981] and Rustum and Soliman [1988a] account for the effects of natural convection in vertical and horizontal tubes, respectively. Entrance region solutions are provided by Prakash and Liu [1985], Choudhury and Patankar [1985], and Rustum and Soliman [1988a]. Shome and Jensen [1996b] investigated the effect of variable properties in helically finned tubes.

Soliman and co-workers [1977, 1979, 1980] numerically solved the energy equation for fully developed laminar flow in internally finned tubes and predicted the Nusselt number for constant wall temperature (Soliman et al. [1980]) and constant heat flux (Soliman [1979]) boundary conditions. Their heat flux boundary condition has uniform heat input axially, with uniform circumferential wall (and fin) temperature. They assume constant fluid properties and no free convection effects. The range of fin geometries analyzed are $0.1 \leq e/d_i \leq 0.4$, $4 \leq n_f \leq 32$, and $1.5 \leq \theta \leq 3°$, where n_f is the number of fins and 2θ is the fin included angle shown in Figure 8.2. The analysis includes the effect of fin thermal conductivity as defined by the parameter $\theta k_f / k$, where k and k_f are the thermal conductivities of the fluid and fin material, respectively. Soliman and Feingold [1977] have analytically solved the momentum equation for the friction factor. Table 8.1 shows the calculated Nu_d and

Table 8.1 Enhancement Ratios Provided by Internally Finned Tubes for Fully Developed Laminar Flow[a]

n_f	e/d_i	A/A_p	f/f_p	$(Nu_d/Nu_p)_T$ 5	$(Nu_d/Nu_p)_T$ ∞	$(Nu_d/Nu_p)_{H1}$ 5	$(Nu_d/Nu_p)_{H1}$ ∞
4	0.1	1.26	1.24	1.04	1.04	1.05	1.07
	0.2	1.51	1.91	1.28	1.30	1.38	1.45
	0.3	1.76	3.28	2.25	2.44	2.47	2.84
	0.4	2.02	4.80	3.73	4.40	3.61	4.52
8	0.1	1.51	1.57	1.06	1.06	1.08	1.10
	0.2	2.02	3.53	1.29	1.31	1.50	1.56
	0.3	2.58	8.67	2.41	2.50	3.79	4.84
	0.4	3.07	14.5	8.07	9.64	7.81	10.45
16	0.1	2.02	2.02	1.03	1.03	1.06	1.06
	0.2	3.04	5.93	1.09	1.10	1.18	1.21
	0.3	4.06	22.2	1.45	1.47	2.07	2.18
	0.4	5.07	60.8	8.59	8.66	17.4	24.4
24	0.1	2.53	2.26	1.01	1.01	1.02	1.02
	0.2	4.06	6.99	1.02	1.03	1.05	1.06
	0.3	5.58	31.7	1.13	1.13	1.29	1.32
	0.4	7.11	172.0	3.29	3.30	9.31	10.5
32	0.1	3.04	2.36	1.00	1.00	1.00	1.01
	0.2	5.08	6.99	1.00	1.00	1.01	1.02
	0.3	7.11	36.3	1.03	1.03	1.08	1.09
	0.4	9.15	355.0	1.66	1.66	2.81	2.91

[a] Ratios as predicted by Soliman et al. [1979, 1980]. Subscript p refers to plain tube.

f_d, relative to the plain tube value (subscript p). The friction factor and Nu_d refer to the diameter and surface area of a plain tube. Hence, the ratio Nu_d/Nu_p is the ratio of the conductance (hA/L) of the finned tube to that of the plain tube. Similarly, the pressure drop increase at fixed velocity is given by f/f_p. Table 8.1 also shows the area ratio (A/A_p). The table includes Nu_d/Nu_p for $\theta k_f/k$ of 5 and ∞. Several important conclusions may be drawn from Table 8.1:

1. For $e/d_i < 0.3$, the maximum Nu_d/Nu_p occurs with $n_f = 8$. However, f_d/f_p continues to increase for $n_f > 8$. Hence the use of $n_f > 8$ is of no value in providing increased hA/L.
2. The $Nu_d/Nu_p < A/A_p$, except for the highest fins at $4 < n_f < 16$. However, the friction factor increase is greater than the area increase for all geometries.
3. The Nu_d/Nu_p for $\theta k_f/k = 5$ is within 10% of that for $\theta k_f/k = \infty$. The for $\theta k_f/k = \infty$ case corresponds to 100% fin efficiency.

Table 8.2 shows values of $\theta k_f/k$ of interest. Examination of this table shows that the values listed in Table 8.1 for $\theta k_f/k = \infty$ are of primary interest. For $\theta k_f/k = \infty$, one may apply the conventionally used formulas for fin efficiency (η_f) to the Nu for $\theta k_f/k = \infty$ and obtain acceptable accuracy. Soliman et al. [1977, 1979, 1980] have not attempted to compare their predicted Nu and f values for constant-properties, fully developed flow with available data of Watkinson et al. [1975a] and Marner and Bergles [1978, 1985, 1989]. However, Rustum and Soliman [1988b] compare their experimental friction results with the numerical results of Nandakumar and Masliyah [1975] and Soliman and Feingold [1977]. The numerical results generally agree within +10/–25% of the experimental friction factor. It is likely that experimental values for heating may exceed the Nusselt numbers given in Table 8.1.

Bergles and Joshi [1983] provide additional comparisons of theoretical and experimental performance of internally finned tubes in laminar flow. Watkinson et al. [1975a] also present data on helical internally finned tubes. Increase of the fin helix angle adds to the enhancement.

Rabas and Mitchell [2000] report laminar and turbulent region data for 2 22.23-mm-ID internal helical finned tubes, whose dimensions were unspecified for proprietary reasons. The data were taken with propylene glycol–water mixtures for which $107 \le Pr \le 530$. The fins were formed by an embossing method and the tubes were seam-welded. Because of the high Pr, an entrance region flow existed over the 2.13-m tube length. For $Re_d > 10,000$, $2.5 < h/h_p < 3$, and the friction increase was

Table 8.2 Values of $\theta k_f/k$ for Different Fluid–Material Combinations at 25°C

Fluid	Pr	Copper	Aluminum	Steel
Air	0.7	16,000	8000	2000
Water	6.0	670	330	80
Oil	1200	2700	1300	400

about the same. For $2000 < \text{Re}_d < 10{,}000$, the h/h_p attained a value as high as 5.9. Little enhancement was observed in the laminar region.

Entrance Length and Free Convection Effects. Numerical solutions for the entrance region have been performed by Prakash and Liu [1985], Choudhury and Patankar [1985], and Rustum and Soliman [1988a]. Choudhury and Patankar [1985] numerically solve the momentum and energy equations for zero fin thickness in the entrance region flow in internally finned tubes at Pr = 0.7 and 5.0. Their analysis shows that the entrance length (based on tube diameter) is not much different for internally finned and plain tubes. The analysis was performed for 6, 12, and 24 fins with $0.05 < e/d_i \leq 0.3$. Similar analysis is reported by Prakash and Liu [1985] for zero fin thickness.

Rustum and Soliman [1988a] extend the analysis of Soliman et al. [1979, 1980] to predict the local entrance region Nusselt number for hydrodynamically fully developed flow. Although they studied the same n_f and e/d_i range of Soliman et al. [1979, 1980], they used zero fin thickness. The analysis was performed for constant-heat-flux and constant-wall-temperature boundary conditions. Table 8.3 lists the thermal entrance lengths for the two thermal boundary conditions. This is the value of X^+ [$= x/(d_i \text{Re}_d \text{Pr})$] at which the fully developed Nusselt number is attained. For $e/d_i \leq 0.2$, the entrance length L_t^+ for the heat flux boundary condition is not very sensitive to the number of fins (n_f), and L_t^+ is close to that for plain tubes (0.0442). Smaller L_t^+ values exist for the wall temperature boundary condition, and they are also smaller than for plain tubes (0.0357). Figure 8.3 shows their results for 24 internal fins plotted in the form of Nu_d vs. X^+. Note the rapid drop of Nu_d for the constant-temperature boundary condition. The entrance length analysis of Prakash and Liu [1985] is also for zero thickness fins, but is for simultaneously developing

Table 8.3 Thermal Entrance Lengths for Internally Finned Tubes

		n_f			
e/d_i	0	4	8	16	24
Heat Flux Boundary Condition					
0.0	0.0442				
0.1		0.0451	0.0462	0.0475	0.0466
0.2		0.0458	0.0524	0.0544	0.0518
0.3		0.0287	0.0412	0.0596	0.0630
0.4		0.0109	0.0589	0.0028	0.0102
Constant Wall Temperature Boundary Condition					
0.0	0.0357				
0.1		0.0259	0.0209	0.0157	0.0131
0.2		0.0255	0.0197	0.0128	0.0107
0.3		0.0274	0.0234	0.0093	0.0064
0.4		0.0088	0.0055	0.0296	0.0093

temperature and velocity profiles. The thermal entrance length for $e/d_i = 0.15$ found by Prakash and Liu was approximately 50% smaller for the predictions of Rustum and Soliman [1988a]. Prakash and Liu [1985] also solved for the entrance region friction factor. Table 8.4 gives the hydrodynamic entrance lengths (L_h^+) and the incremental pressure drop, $K(\infty)$, in the entrance region calculated by Prakash and Liu. The hydrodynamic entrance length is defined as the length for the local friction factor to fall to a value 1.05 times the fully developed value. The $K(\infty)$ is the ratio of the entrance region drop to that for fully developed value in the same flow length.

The experimental data are typically for a certain tube length, which may include an entrance region and free convection effects. If the fluid is heated, such experimental Nu values may be substantially greater than the values predicted by Rustum

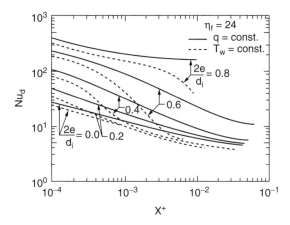

Figure 8.3 Nu_d vs. X^+ for constant wall temperature (T) and constant-heat-flux boundary conditions. (From Rustum and Soliman [1988a].)

Table 8.4 Hydrodynamic Entrance Lengths Parameters for Internally Finned Tubes

	n_f			
e/d_i	0	8	16	24
Hydrodynamic Entrance Length				
0.00	0.0415			
0.15		0.0433	0.0438	0.0417
0.30		0.0320	0.0540	0.0622
0.50		0.0052	0.0024	0.0014
Incremental Pressure Drop, $K(\infty)$				
0.00	1.25			
0.15		2.44	4.11	5.40
0.30		2.85	10.70	23.50
0.50		1.58	1.79	1.93

and Soliman [1998a]. However, cooling tends to reduce the Nusselt number, relative to the enhancement provided by free convection and entrance region effects; this is seen on Figure 7.9. The effects of natural convection are influenced by the magnitude of the Raleigh number (Ra), defined by

$$Ra = Gr\,Pr = \left(\frac{G\beta q d_i^4}{v^2 k}\right)\left(\frac{c_p \mu}{k}\right) \tag{8.3}$$

Figure 8.4 shows the influence of Ra on water flow in the entrance region of a 13.9-mm-ID tube having ten fins 1.5 mm high. The solid line shows the predictions of Rustum and Soliman [1988a] for Ra = 0 (pure forced convection). Figure 8.4 shows that increasing Ra significantly increases Nu_d over the Ra = 0 value. As Ra approaches zero, the data approach the predicted value for Ra = 0. Also note that the entrance length decreases with natural convection effects present. For the four internal fin geometries tested, Rustum and Soliman [1988a] show that the effect of Rayleigh number may be correlated by the empirical expression:

$$\frac{Nu_{fd}}{Nu_p} = 1 + \left(\frac{Ra_m}{C_1}\right)^{C_2} \tag{8.4}$$

where Nu_p is for fully developed forced convection (Ra = 0), Nu_{fd} is the fully developed Nusselt number (Ra > 0), and Ra_m is evaluated at the mean bulk temperature. The constants are geometry dependent. For the Figure 8.4 geometry, $C_1 = 1.98E{-}6$, and $C_2 = 0.44$. Their data also show that Nu_d is insensitive to Re_d for constant Ra.

Zhang and Ebadian [1992] numerically investigated the influence of free convection (mixed flow) on laminar flow in horizontal internally finned tubes with constant heat flux and constant wall temperature. They analyzed tubes having 0, 3, 7, and 11 fins for $0.15 \le e/d_i \le 0.5$. Their analysis shows that buoyancy force enhances the Nusselt number for plain and finned tubes, although the enhancement effects

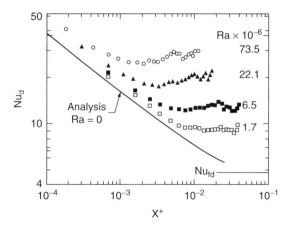

Figure 8.4 Influence of Rayleigh number on Nu for laminar entrance region flow with electric heat input (d_i = 13.9 mm, n_f = 10, e/d_i = 0.11). (From Rustum and Soliman [1988a].)

are smaller for a finned tube. Figure 8.5 shows the effect of Rayleigh and fin height on a tube having 11 fins. The figure plots Nu_{Dh}/Nu_p vs. Ra for $e/d_i = 0.15, 0.3$, and 0.5. The figure shows that Nu_{Dh}/Nu_p increases with increasing Ra, and that the enhancement is greatest for the smaller fin heights. Also note that the friction factor ratio decreases with increasing Ra.

Microfin Tubes. Shome and Jensen [1996a] tested 15 internally finned tubes in laminar flow using ethylene glycol. Of the 15 tubes, 6 tubes were categorized as high-finned tubes, and 6 tubes were categorized as microfin tubes. The distinction between a high-finned tube and a microfin tube is the fin height; a high-finned tube has $e/d_i \geq 0.05$ and a microfin tube has $e/d_i \leq 0.03$. The high-finned tubes had fins ranging $8 \leq n_f \leq 30$, $0.05 \leq e/d_i \leq 0.085$, and $0° \leq \alpha \leq 30°$. The microfin tubes had $22 \leq n_f \leq 54$, $0.0075 \leq e/d_i \leq 0.03$, and $15° \leq \alpha \leq 45°$. The tube outer diameters were approximately 25 mm. The fluid heating or cooling closely simulated the constant-wall-temperature boundary condition. The test range covered $150 \leq Re_d \leq 2000$, $50 \leq Pr \leq 185$, $0.3 \leq \mu_b/\mu_w \leq 3.6$, and two Rayleigh number ranges: $3 \times 10^5 \leq Ra \leq 2 \times 10^6$ (low heat flux) and $3 \times 10^6 \leq Ra \leq 8 \times 10^6$ (high heat flux). The length to tube ID ratio of the test section was approximately 120. The microfin tube is shown in Figure 13.7d. The turbulent flow and heat transfer data in microfin tubes are discussed in Section 8.2.2 Figure 8.6a shows the effect of number of fins for high-finned tubes with $e/d_i = 0.05$ and $\alpha = 30°$. The smooth tube data are also shown as lines. The variable viscosity effect was implemented using the Sieder–Tate [1936] correlation, and the corrected data are plotted vs. $Gz[= \pi/(4X^+)]$. The cooling data are significantly lower than the heating data, which suggests that the Sieder and Tate property correction is not appropriate. Figure 8.6a shows that the Nusselt number increases as the number of fins increases from $n_f = 8$ to 14. When further increasing the number of fins to 30, however, the Nusselt number decreases. This optimum number of fins is in close agreement with the numerical prediction by Soliman et al. [1980] shown in Table 8.1. Figure 8.6a shows that, at a low heat flux, the Nusselt

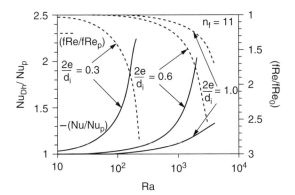

Figure 8.5 Effect of natural convection on $f_{Dh}Re_{Dh}$ and Nu_{Dh} for an internally finned tube having 11 fins, for constant heat flux. (From Zhang and Ebadian [1992].)

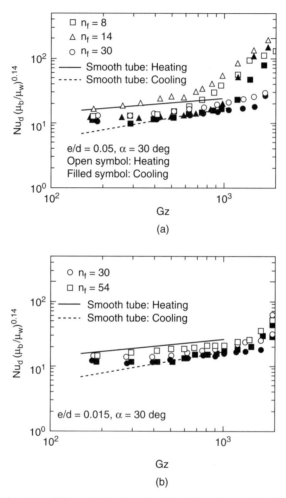

Figure 8.6 Effect of number of fins on the laminar flow Nusselt number at a low heat flux ($3 \times 10^5 \leq$ Ra $\leq 2 \times 10^6$): (a) internal finned tubes with $e/d_i = 0.05$ and $\alpha = 30°$, (b) microfin tubes with $e/d_i = 0.015$ and $\alpha = 30°$. (From Shome and Jensen [1996a].)

numbers of high-finned tubes are slightly lower (for heating) or approximately the same (for cooling) as that of smooth tubes. Shome and Jensen [1996a] also provided the high heat-flux data for the same high-finned tubes. The Nusselt numbers of high-finned tubes were higher than those of the smooth tubes. For $n_f = 14$ tube, the enhancement was approximately 50%. The difference in Nusselt numbers between heating and cooling was also significantly reduced, due to the dominant free convection contrbution. Data for two microfin tubes ($n_f = 30, 54$) for $e/d_i = 0.015$ and $\alpha = 30°$ are shown in Figure 8.6b for a low heat flux. Different from the high-finned tubes, the Nusselt numbers are almost independent of the number of fins. At a low heat flux, the Nusselt numbers of the microfin tubes are lower than the smooth tube values heating, and approximately the same for cooling. At a high heat flux (not shown here), the Nusselt numbers of the microfin tubes are slightly higher than the smooth tube values.

Shome and Jensen [1996a] also investigated the effects of fin height and the helix angle. For a high-finned tube, increasing the fin height or the helix angle increased the Nusselt number, especially at a high heat flux. For a microfin tube, however, the effects were minimal. To summarize, the fin geometry had little effect on heat transfer, particularly for microfin tubes, and for high-finned tubes at a low heat flux. For a microfin tube in laminar flow, the microfins embedded in the viscous sublayer, and no effect of the fins is manifested. At a high heat flux, high-finned tubes yielded 50 to 75% enhancement over the smooth tube. Note that the enhancement level obtained by Shome and Jensen [1996a] using ethylene glycol is significantly lower than the 200 to 300% obtained by Marner and Bergles [1989] using Polybutene-20. The larger entrance effect along with the higher bulk-wall viscosity ratio of the Polybutene-20 may be responsible. Shome and Jensen [1996a] also report the pressure drop data. For high-finned tubes, the increase in pressure drop over the smooth tube is as much as 50%. The pressure drop increases as the number of fins or the fin height increases. The increase is much smaller (10 to 15%) for the microfin tubes. Discussion of laminar flow heat transfer in a rough tube is provided in Section 9.2.

Al-Fahed et al. [1999] investigated the heat transfer with laminar flow of oil in a 14.0-mm-ID microfin tube having 60 fins with 0.2-mm fin height and 15° helix angle. The heat transfer coefficients and the friction factors were virtually the same as those of the smooth tube.

Shome and Jensen [1996b] provide a numerical counterpart of the Shome and Jensen [1996a] experimental investigation. The developing mixed convection laminar flow in helically finned tubes was modeled for a constant-temperature boundary condition. A rotating polar coordinate was used to model the helical fins. The numerical investigation successfully predicted the experimental data. They also investigated the variable viscosity effect, and the results are shown in Figure 8.7 for a tube with $n_f = 30$, $e/d_i = 0.05$, and $\alpha = 30°$. The wall-to-inlet temperature difference (ΔT) was 15 K for heating and –15 K for cooling. The Nusselt number (Figure 8.7a) increases with increasing ΔT for heating and decreases with increasing ΔT for cooling. This is because of the increased free convection in the near-wall region when fluid is heated. The reverse is true for cooling. Figure 8.7b shows that the variable viscosity effect is more pronounced for the friction factor than for the Nusselt number. Shome [1998] extended the study to a constant-heat-flux boundary condition.

Segmented Internally Finned Tubes. Kelkar and Patankar [1990] numerically analyze internally finned tubes whose fins are segmented along their length. The fin segment is of length H in the flow direction, separated by an equal distance H before the next fin. Figure 8.8 illustrates the staggered and inline segmented fins analyzed. Table 8.5 shows their calculated results for $e/d_i = 0.15$, $n_f = 12$ fins with

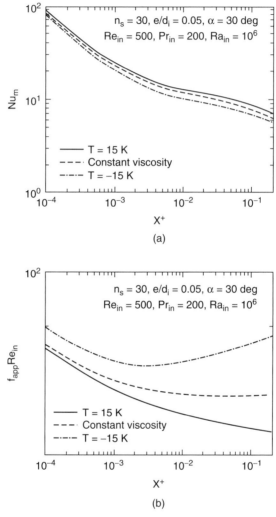

Figure 8.7 Numerical results on the effect of variable viscosity for an an internal-fin tube with $n_f = 30$, $e/d_i = 0.05$, and $\alpha = 30°$. Calculations were done at $Re_{in} = 500$, $Pr_{in} = 200$, and $Ra_{in} = 10^6$. (a) Average Nusselt number, (b) friction factor. (From Shome and Jensen [1996b].)

$L/(d_i/Re_d) = 0.001$ with $Pr = 0.7$. The tabled values are for the Nu averaged over the finned and unfinned axial increments.

Table 8.5 shows that the inline segmented fins give only 6% higher Nusselt than continuous fins, and the staggered arrangement is 6% below that of continuous fins. However, the inline arrangement gives 22% lower friction than continuous fins. Note that staggered arrangement uses the same fin surface area as continuous fins. But, the inline arrangement has half the fin surface area, because of the axial spaces between fins.

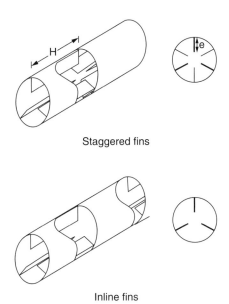

Staggered fins

Inline fins

Figure 8.8 Illustration of inline and staggered segmented internal fins analyzed by Kelkar and Patankar [1990]. (From Kelkar and Patankar [1990].)

Table 8.5 Results for Segmented and Continuous Internal Fins for $e/d_i = 0.15$

Geometry	$f\mathrm{Re}/f\mathrm{Re}_p$	$\mathrm{Nu}_d/\mathrm{Nu}_p$
Continuous fins	2.17	1.18
Inline segmented fins	1.71	1.25
Staggered segmented fins	2.19	1.12

8.2.2 Turbulent Flow

Conventional Internally Finned Tubes. Among the earliest data reported on internally finned tubes are those of Hilding and Coogan [1964], who worked with airflow in 13.6-mm-ID tubes. Their tubes contained four or eight fractional- or full-height fins. They also investigated use of axial interruptions of the internal fins. This improved performance in laminar flow, but not in the turbulent regime.

Carnavos [1980] developed empirical correlations of Nu and f vs. Re_{Dh} for correlations for turbulent flow in internally finned tubes. His correlation was based on Carnavos [1979, 1980] data on 21 surface geometries, including helix angles as great as 30°. The data are for heating of fluids ($6 < \mathrm{Pr} < 30$) and span $10{,}000 < \mathrm{Re}_{Dh} < 60{,}000$. Carnavos attempted to correlate the data using standard heat transfer and friction factor correlations (based on the hydraulic diameter) for turbulent flow in plain tubes. He found that this method overpredicted the Nu and f for axial fins. He then developed geometry-dependent correction factors to correlate the data. Webb and Scott [1980] restated the correlations in terms of the fundamental dimensional geometry parameters. The Carnavos correlations for straight and helical fins as stated by Webb and Scott [1980] are

$$\frac{\text{Nu}_{Dh}}{\text{Nu}_p} = \frac{hD_h/k}{h_p d_i/k} = \left[\frac{d_i}{d_{im}}\left(1 - \frac{2e}{d_i}\right)\right]^{-0.2}\left(\frac{d_i D_h}{d_{im}^2}\right)^{0.5}\sec^3\alpha \qquad (8.5)$$

$$\frac{f_{Dh}}{f_p} = \frac{d_{im}}{d_i}\sec^{0.75}\alpha \qquad (8.6)$$

The term d_{im} is the tube inside diameter that would exist if the fins were melted and the material returned to the tube wall. The friction factor and Nusselt number in the above equations are defined in terms of the total surface area and hydraulic diameter, and $\text{Re}_{Dh} = D_h G/\mu$. Carnavos used the Dittus–Boelter and the Blasius equations to calculate the plain tube Nusselt number (Nu_p) and friction factor (f_p), respectively. These equations are

$$\text{Nu}_p = \frac{h_p d_i}{k} = 0.023\,\text{Re}_p^{0.8}\,\text{Pr}^{0.4} \qquad (8.7)$$

$$f_p = 0.046\,\text{Re}_p^{-0.2} \qquad (8.8)$$

Equations 8.5 and 8.6 correlated the Carnavos data on 21 tube geometries within ±8%. The range of dimensionless geometric parameters covered by the correlated data are $0.03 \le e/d_i \le 0.24$, $1.4 \le \pi d_i/n_f e \le 7.3$, $0.1 \le t/e \le 0.3$, and $0 \le \alpha \le 30°$. If one desires to calculate the enhancement ratio in terms of h based on $A_i/L = \pi d_i$, one writes Nu_d/Nu_p as

$$\frac{\text{Nu}_d}{\text{Nu}_p} = \frac{\text{Nu}_{Dh}}{\text{Nu}_p}\cdot\frac{d_i}{D_h}\left(1 + \frac{2n_f e}{\pi d_i}\right) \qquad (8.9)$$

Similarly, the friction factor ratio f_d/f_p, is given by

$$\frac{f_d}{f_p} = \frac{f_{Dh}}{f_p}\cdot\frac{d_i}{D_h} \qquad (8.10)$$

Webb and Scott [1980] have applied cases VG-1, VG-2, and VG-3 of the Table 3.1 PEC to turbulent flow in internally finned tubes and determined their performance relative to plain tubes. A key purpose of this analysis was to determine preferred internal fin geometries. This analysis assumes all of the thermal resistance is on the tube side. The results for the VG-1 analysis are presented in PEC Example 8.1.

Trupp and Haine [1989] measured turbulent flow heat transfer and friction data in five internally finned tubes having $10 \le n_f \le 16$. These are tube numbers 9, 10, 13, 14, and 20 also tested by Carnavos [1979, 1980]. Electric heat input was provided by heating wire wrapped on the tube circumference, and local heat transfer coefficients were determined using wall thermocouples. The data spanned $500 \le \text{Re}_{Dh} \le$

10,000, and they allow comparison with both the laminar friction results of Watkinson et al. [1975a] and the turbulent flow data and correlation of Carnavos [1979, 1980]. The turbulent friction data may also be compared with that of Watkinson et al. [1973, 1975b]. The geometrical dimensions of the tubes are defined in Table 8.6, and Figure 8.9 shows the friction data. The data show reasonably good agreement with the extrapolated Carnavos [1980] correlation, Equation 8.6. Note that the lowest Reynolds number (Re_{Dh}) on which the Carnavos friction correlation was developed is 10,000. The Watkinson et al. [1973] data are higher than the Trupp and Haine [1989] data and the Carnavos [1980] correlation. Figure 8.10 shows the Trupp and Haine fully developed heat transfer data compared with the Carnavos [1980] correlation, Equation 8.5. The figure shows excellent agreement with Equation 8.5. Figure 8.10 also shows the transition Reynolds number for the internally finned tubes. Transition occurs at Reynolds numbers (based on hydraulic diameter) between 500 and 1000. Trupp and Haine also investigated the effect of natural convection by

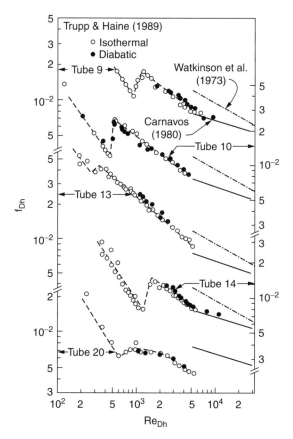

Figure 8.9 Friction data of Trupp and Haine [1989] for the Table 8.6 internally finned tubes. (From Trupp and Haine [1989].)

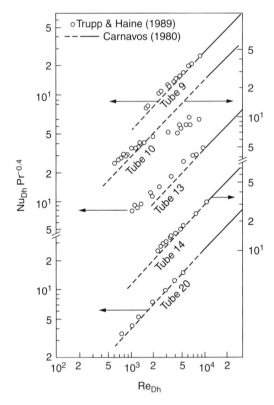

Figure 8.10 Fully developed Nusselt number data of Trupp and Haine [1989] for the Table 8.6 internally finned tubes. (From Trupp and Haine [1989].)

varying the Rayleigh number. Their data showed a very weak influence of Rayleigh number for $0.9 \leq Ra \times 10^{-6} \leq 1.8$.

Kim and Webb [1993] develop an analytical model to predict the turbulent heat transfer coefficient and friction factor in internally finned tubes with axial fins. This model applies the "law of the wall" to the flow in the interfin region. The law of the wall states that the velocity and temperature profile is independent of the channel shape, and is described by Hinze [1975]. The "universal velocity profile" for $y^+ > 26$ is given by

$$\frac{u}{u^*} = 2.5 \ln \frac{yu^*}{v} + 5.5 \tag{8.11}$$

The friction factor is predicted by integrating Equation 8.11 over the interfin region and the core region between the fin tips. Similarly, the dimensionless temperature profile is obtained by integrating the corresponding universal temperature profile over the same flow regions. Kim and Webb used the universal temperature profile ($y^+ > 0$) proposed by Gowen and Smith [1967], which is given by

Table 8.6 Tube Geometries Tested by Trupp and Haine [1989]

No.	9	10	13	14	20
d_o (mm)	12.7	9.53	9.53	15.9	12.7
d_i (mm)	10.3	8.00	7.04	13.9	10.4
e (mm)	1.28	1.27	2.29	1.50	1.47
n_f	10	16	10	10	16
α (deg)	0	0	0	0	2.5

$$\frac{T}{T^*} = 2.5 \ln \frac{yu^*}{v} + 5 \ln \left(\frac{5\,\mathrm{Pr}+1}{30} \right) + 5\,\mathrm{Pr} + 8.55 \qquad (8.12)$$

The symbols in Equation 8.12 are defined in Chapter 9.

Carnavos [1979] friction data for 11 internally finned tubes were predicted within ± 10%. The Carnavos [1979, 1980] airflow and water flow Nusselt numbers were predicted within ± 15%.

Local Flow Structure Measurements. Trupp et al. [1981] performed measurements of the flow structure in a scaled-up internally finned tube with 114.3 mm ID for $50,000 < \mathrm{Re_{Dh}} < 71,000$. The tube had six axial internal fins with $e/d_i = 0.33$ and $t/e = 0.13$. The fin height is considerably higher than would be expected for practical applications. They measured the velocity distribution in the interfin region, secondary velocities, Reynolds stresses, and the wall shear stress. Figure 8.11a shows the measured velocity profile, and Figure 8.11b shows the surface shear stress distribution. Trupp et al. [1981] also observed two counter-rotating corkscrew vortices in the interfin region.

Microfin Tubes. Jensen and Vlakancic [1999] obtained turbulent heat transfer and friction data on the same helically finned tubes used for the laminar flow tests described in Shome and Jensen [1996a]. The tubes consisted of seven high-finned tubes with fin geometry ranging $8 \leq n_f \leq 30$, $0.05 \leq e/d_i \leq 0.085$, and $0° \leq \alpha \leq 30°$ and eight microfin tubes with $22 \leq n_f \leq 54$, $0.0075 \leq e/d_i \leq 0.03$, and $15° \leq \alpha \leq 45°$. Tubes with $e/d_i \leq 0.03$ and 30 or more fins were classified as microfin tubes. Thus, the tube with $e/d_i = 0.03$ and $n_f = 22$, which had been classified as a microfin tube by Shome and Jensen [1996a] for the laminar flow test, was classified as a high-finned tube for the turbulent flow test. This is a wider range of geometries than tested by Carnavos [1980]. The test range spanned $7000 \leq \mathrm{Re_d} \leq 70,000$ and $3.5 \leq \mathrm{Pr} \leq 5$. The increase in Nusselt number over that of the smooth tube was 50 to 150% for high-finned tubes. For microfin tubes, even greater Nusselt number enhancement of 20 to 220% was obtained. The increase in the friction factor was 40 to 170% and 40 to 140% for the high-finned tubes and the microfin tubes, respectively. Both the Nusselt number and the friction factor increased as n_f, e/d_i and α increased. They also developed empirical correlations of their data. Separate correlations were developed for high-finned tubes and for microfin tubes. The correlations are given below.

For high-finned tubes,

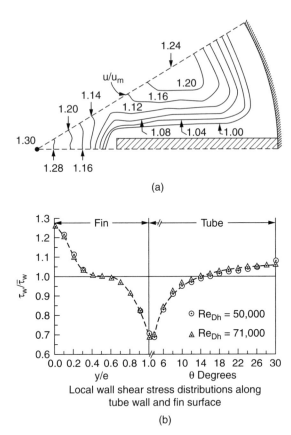

(a)

(b)

Figure 8.11 Experimental measurements of Trupp et al. [1981] $e/d_i = 0.33$, $t/e = 0.13$, $Re_d = 71,000$. (a) Measured velocity profile, (b) surface shear stress distribution. (From Trupp et al. [1981].)

$$\frac{f}{f_p} = \left(\frac{l_c}{d_i}\right)^{-1.25} \left(\frac{A_n}{A_{fa}}\right)^{1.75} \tag{8.13}$$

$$\frac{Nu_d}{Nu_p} = \left(\frac{l_c}{d_i}\right)^{-0.5} \left(\frac{A_n}{A_{fa}}\right)^{0.8} GF \tag{8.14}$$

where

$$\frac{l_c}{d_i} = \left[\frac{A_{core}}{A_{fa}}\left(1-\frac{2e}{d_i}\right) + \frac{A_{fin}}{A_{fa}}\left(\frac{\pi}{n_f}\left(1-\frac{e}{d_i}\right)-\frac{t}{d_i}\right)\right]$$

$$\left[1-0.203\left(\frac{n_f \sin\alpha}{\pi}\right)^{0.65}\left(\frac{2e}{d_i}\right)^{0.2}\right] \tag{8.15}$$

$$\text{GF} = \left(\frac{A}{A_{\text{nom}}}\right)^{0.29}\left[1-1.792\left(\frac{n_f\sin\alpha}{\pi}\right)^{0.64}\left(\frac{2e}{d_i}\right)^{2.76}(\text{Re}_d)^{0.27}\right] \qquad (8.16)$$

For the microfin tube, Equations 8.13 and 8.14 also apply with different characteristic length (l_c/d_i) and geometry factor (GF) expressions.

$$\frac{l_c}{d_i} = \left[1-K\left(\frac{n_f\sin\alpha}{\pi}\right)^b\left(\frac{2e}{d_i}\right)^c\left(\left(\frac{\pi}{n_f}-\frac{t}{d_i}\right)\cos\alpha\right)^d\right] \qquad (8.17)$$

where $K = 1.577$, $b = 0.64$, $c = 0.53$, $d = 0.28$ for $e/d_i \le 0.02$ and $K = 0.994$, $b = 0.89$, $c = 0.44$, $d = 0.41$ for $0.02 < e/d_i \le 0.03$. Equation 8.17 is applicable for $\text{Re}_d > 20{,}000$. Jensen and Vlakancic [1999] provide separate friction factor correlation for $2500 < \text{Re}_d \le 20{,}000$. The GF equation for a microfin is

$$\text{GF} = \left(\frac{A}{A_{\text{nom}}}\right)^{1.0}\left[1-0.059\left(\frac{n\sin\alpha}{\pi}\right)^{-0.31}\left(\left(\frac{\pi}{n_f}-\frac{t}{d_i}\right)\cos\alpha\right)^{-0.66}\right] \qquad (8.18)$$

Jensen and Vlakancic [1999] stated that these correlations predict most of the friction and heat transfer data of Carnavos [1980] within $\pm 15\%$.

Segmented Internally Finned Tubes. El-Sayed et al. [1997] obtained pressure drop data in segmented internally finned tubes of the Figure 8.8 geometry. The finned tubes were made from plain brass tubes, where axial slots were machined to insert brass strips. Both the inline and staggered fin geometry, as well as continuous fins, were tested using air for $5000 \le \text{Re}_{\text{Dh}} \le 50{,}000$. The pressure drop of the tube with continuous fins was higher than that with inline fins, and lower than that with staggered fins. Braga and Saboya [1986] provide Nusselt number and friction data for turbulent airflow in an annulus having fins arranged in a staggered manner.

Numerical Approach. Said and Trupp [1984] developed a two-equation turbulence model to predict the heat transfer coefficient and friction factor for airflow in axial internally finned tubes. Their analysis was performed for $0.1 \le e/d_i \le 0.4$, $6 \le n_f \le 14$, $\theta = 3°$, and $25{,}000 \le \text{Re}_{\text{Dn}} \le 150{,}000$. They developed the following empirical correlations to predict their numerical results:

$$f = 0.21\,\text{Re}_{\text{Dh}}^{-0.216}\left(\frac{p}{D_h}\right)^{0.351}\left(\frac{d_iD_h}{d_{\text{im}}^2}\right)^{0.148} \qquad (8.19)$$

$$\text{Nu} = 0.027\,\text{Re}_{\text{Dh}}^{0.774}\left(\frac{p}{D_h}\right)^{0.397}\left(\frac{d_iD_h}{d_{\text{im}}^2}\right)^{-0.168} \qquad (8.20)$$

Patankar et al. [1979] have also numerically solved the momentum and energy equations for turbulent flow in internally finned tubes having zero fin thickness. They used one adjustable constant from the data of Carnavos [1979]. Although they do not provide a correlation of their results, they show good ability to predict the Carnavos [1979] data. Ivanović et al. [1990] analyzed the same geometries of Patankar et al. [1979] and extended the work using numerical methods, which can reveal secondary flow features. Their analysis used the low Reynolds number version of the k-ϵ turbulence model.

Jensen and co-workers have done considerable recent work on numerically modeling turbulent flow in internally finned tubes having axial or helical internal fins. The previously discussed numerical work considered a rectangular fin cross section and axial fins. Typically, the modeled fins had a relatively high e/d_i, as compared to the low helical internal finned tubes used in tubes such as the current-technology Wieland and Wolverine chiller tubes. The work of Jensen and co-workers has addressed trapezoidal fin cross section and helix angle. Liu and Jensen [1999] used a two-layer turbulence model to capture the near-wall turbulence in two helically finned tubes having a rectangular cross section. The analysis was performed for fully developed turbulent flow. Figure 8.12a shows the basic geometry modeled by Liu and Jensen [1999], and their solid model is shown in Figure 8.12b, which uses a rotated periodic boundary condition. Liu and Jensen [2001] performed a parametric numerical study, in which they predicted the effect of fin pitch, fin height, fin thickness, helix angle, and fin cross-sectional shape (rectangular, rectangular with a round tip, and triangular). Calculations were done for Reynolds numbers between 10,000 and 70,000. Predictions were made for n_f = 10 to 40, e/d_i = 0.015 to 0.05, t = 0.05 to 0.22 mm, and α = 10° to 40°. They showed that their predictions agreed well with the experimental work of Jensen and Vlakancic [1999]. Kim et al. [2002] extended the work to further examine the effect of fin cross-sectional shape, including fins of an asymmetric shape. At present, commercial internally finned tubes do have a trapezoidal fin of an asymmetric shape, which results in performance that is somewhat sensitive to flow direction. This series of work has significantly advanced computational abilities to model a relatively complex three-dimension geometry. Readers interested in such geometries would be well advised to read these papers.

Bhatia and Webb [2001a] conducted a numerical study of turbulent flow and heat transfer in microfin tubes using the commercial CFD program, Fluent. The core flow was modeled using RNG k-ϵ model, and the near-wall flow was modeled using the one-equation model of Wolfstein [1988]. The turbulent heat transfer region was modeled using the Reynolds analogy (Pr_t = 0.85). Friction losses were predicted for five tube geometries, for which experimental data were available. The predicted friction factor was within +13% to -9% of the experimental values. The heat transfer simulation was done for one tube geometry with the helix angle of 17.5°. The simulation was carried out $1 \leq Pr \leq 4$. The Nusselt number was overpredicted, ranging from 29% for Pr = 1 to 55% for Pr = 4. They recommended refinement of the near-wall grid (smaller mesh size, and an initial y^+ value of 0.5) for better prediction for higher-Prandtl-number cases.

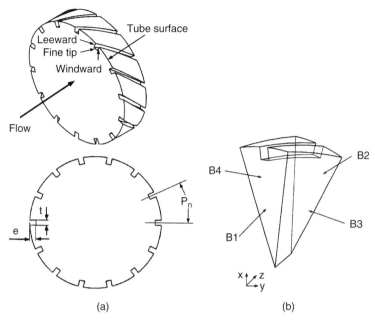

Figure 8.12 Solid model of the internal fin tube by Liu and Jensen [1999]. (a) Basic geometry, (b) computational geometry model. (From Liu and Jensen [1999].)

8.2.3 PEC Example 8.1

Assume that a plain tube heat exchanger has been designed to provide heat duty q with specified flow rates and inlet temperatures. The tubes are 19.2-mm-OD (outside diameter) and 0.70-mm wall thickness. The plain tube design operates at Re_p with N tubes per pass and tube length L_p per pass. One desires to use internally finned tubes to reduce the total length of tubing (NL). The design constraints are those of case VG-1 of Table 3.1. Assuming that the total thermal resistance is on the tube side, the equations in Section 3.3.4 apply. Using Equations 8.7 and 8.8, one solves Equation 3.7 for G/G_p. For constant inside diameter, $G/G_p = (Re_{Dh}/Re_p)(d_i/D_h)$.

The geometric parameters of the finned tube are e, t, n_f, and α. For a given tube geometry, one iteratively solves Equation 3.7 for G/G_p and calculates Re_{Dh}/Re_p. Constraints on Equations 3.1 and 3.2 are $hA/h_pA_p = P/P_p = 1$. With the known Re_{Dh}, one calculates St and f of the finned tube and then solves for A/A_p using Equation 3.2 or 3.4. The $A/A_p = NL/N_pL_p$. The ratio of tube material in the finned and plain tube exchangers is given by

$$\frac{V_m}{V_{m \cdot p}} = \frac{N}{N_p} \frac{L}{L_p} \frac{M}{M_p} \tag{8.21}$$

where M is the tube weight per unit length. Figure 8.13a shows the effect of varying the geometry parameters e/d_i and n_f with $e/t = 3.5$ and $\alpha = 0$. This figure

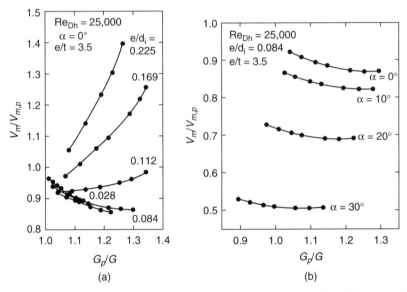

Figure 8.13 Performance comparison of internally finned tubes and plain tube ($d_i = 17.78$ mm) for Case VG-2 of Table 3.1: (a) effect of e and n_f for $e/t = 3.5$ and $\alpha = 0$, (b) effect of n_f and α for $e/d_i = 0.084$ and $e/t = 3.5$ mm. The points on each curve define n_f (from the left, n_f = 5, 8, 12, 16, 25, 32, and 40 fins). (From Webb and Scott [1980].)

shows that the smallest $V_m/V_{m,p}$ occurs with $e = 1$ to 1.5 mm and decreases with increasing n_f when $e = 1$ to 1.5. Figure 8.13b shows the effect of helix angle for $e/d_i = 0.084$ and $e/t = 3.5$. The $V_m/V_{m,p}$ may be reduced nearly 50% with $\alpha = 30°$. The curve for $\alpha = 30°$ shows that n_f has little effect on $V_m/V_{m,p}$. However, as n_f increases, G_p/G must increase to meet the $P/P_p = 1$ constraint. Because heat exchanger cost is quite sensitive to shell diameter, selection of $12 \leq n_f \leq 16$ appears to be a good choice. Because $A_c/A_{c,p} = G_p/G$, selection of the smaller G_p/G will reduce the shell diameter, and hence the shell cost. The preferred tube geometry ($e/d_i = 0.084$, $n_f = 12$, and $\alpha = 30°$) provides a 58% reduction of tubing length and a 48% reduction of tubing material (weight) relative to a plain tube design.

   This example assumed no thermal resistance on the shell side. Webb [1981] describes how the analysis may be extended to account for shell side, fouling, and tube-wall resistances. The value $V_m/V_{m,p}$ will decrease when these additional resistances are present.

8.3 SPIRALLY FLUTED TUBES

Figure 8.14 shows two variants of spirally fluted tubes. These tubes provide an extended surface by deforming the tube wall to form spiral flutes. Note that the tubes have spiral flutes on both the inner and outer surfaces. The tubes shown in Figure 8.14a and b were developed for General Atomics Corp. by Yampolsky [1983,

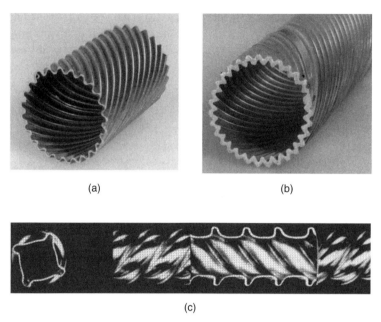

(a)   (b)

(c)

Figure 8.14 Spirally fluted tubes: (a) stainless steel tube developed by Yampolsky [1983], (b) Yampolsky tube made of aluminum, (c) the spirally indented tube.

1984]. This tube is formed by corrugating strip material (Figure 8.14a) and then rolling it in a circular form, such that the corrugations are at a helix angle. It may also be made by extruding aluminum sheet with the fluted pattern, and then twisting the extrusion to form the helix angle as shown in Figure 8.14b. Versions of the tube have been made with helix angles (α) of 30° and 40°.

The Figure 8.14c tube, which can be described as "spirally indented," is manufactured by several companies and is made in a variety of diameters. The tube is pulled through a die to deform the wall into the spirally indented shape. The tubes are in several diameters with different flute heights, helix angles, and number of starts.

Because these tubes provide enhancement on both sides, it is necessary to measure the overall heat transfer coefficient and independently determine the thermal resistance on one side. This is typically done using the modified Wilson plot method, as described by Shah [1990]. Frequently, investigators have measured the overall heat transfer coefficient for condensation on the outer tube surface. Accurate use of the Wilson plot method to derive the tube-side heat transfer coefficient should involve two key elements in the test procedure:

1. Constant condensing saturation temperature and heat flux must be maintained.
2. The tube-side Reynolds number exponent should be a constant over the range of water velocities used for determination of the shell-side heat transfer coefficient.

Careful evaluation method of the test methods employed sometimes reveals errors or inaccuracies in using the Wilson plot method. Hence, readers should be careful in accepting the tube-side coefficients derived for the spirally fluted tubes. For example, the data by Marto et al. [1979] for the tubes shown in Figure 8.14a and c appear to show higher tube-side coefficients than have been measured by others.

8.3.1 Spirally Fluted Tube

Yampolsky [1983, 1984] has developed and tested an aluminum, spirally fluted tube shown in Figure 8.14b. Test results are also reported by Marto et al. [1979], Panchal and France [1986], Ravigururajan and Bergles [1986], and Obot et al. [1991]. Figure 8.15 shows the friction and heat transfer characteristics reported by Obot et al. [1991], which include test results of other investigators. Tubes 1 and 2 are made of stainless steel with 0.5-mm wall thickness (Figure 8.14a) and Tube 3 is aluminum with 1.5-mm wall thickness (Figure 8.14b). The dimensions are given in Table 8.7. The flutes of Tube 3 (Table 8.7) provide a surface area enhancement of 1.50, relative to a plain tube of the same d_i. The Nu_d is defined as $h'd_i/k$, where h' is based on $A/L = \pi d_i$ and the f and Re_d are based on the maximum internal diameter (d_i). The heat transfer data are plotted as $Nu_d/Pr^{0.4}$ vs. Re_d using an assumed Prandtl number dependency. Figure 8.15b shows this reasonably correlates the heat transfer data of Obot et al. [1991] and Panchal and France [1986] for air (Pr = 0.71) and the data of Yampolsky [1984] and Ravigururajan and Bergles [1986] for water (Pr $\simeq$ 5.50). Figure 8.15b shows that the correlated data of all investigators for Tube 3 are in close agreement. The tube provides $Nu_d/Nu_p = 1.9$ and $f_d/f_p = 3.1$ at $Re_d = 30,000$. Compared to a plain tube of $d_o = d_e$, operated at the same mass velocity, the fluted tube has a thermal conductance (hA/L) 1.5 times higher than the plain tube and 2.0 times the pressure drop. However, the fluted tube operates at 10% lower flow rate, because of its smaller flow area (A_c).

The heat transfer data of Yampolsky [1984], taken with condensing steam on the outer surface, show the same Nusselt number for both heating and cooling. Earlier data reported by Yampolsky [1983] show higher heat transfer coefficients for heating than for cooling. A possible explanation of this is the very high condensing heat fluxes used in the heating tests. A potential application problem with this tube is how to make a plain tube end for insertion in a tube sheet.

The data of Obot et al. [1991] show significant enhancement in the laminar flow regime. Barba et al. [1983] numerically study laminar flow in the spirally fluted tube. Their analysis is done for Pr = 0.71, 5.0, and 93 with $Re_d < 2500$. The analysis shows increasing values of h/h_p with increasing Pr.

Baughn et al. [1993] obtained local heat transfer coefficients using liquid crystal thermography in a spirally fluted tube having 30° helix angle. The flutes were machined in a Plexiglas tube. The local heat transfer coefficients were measured from crest to crest on the windward and leeward sides of the flute at two axial locations ($x/d_i = 3.4$ and 4.4). The results show that the heat transfer coefficients are essentially the same at two axial locations, which suggests that thermal development is very rapid for spirally fluted tubes. The heat transfer coefficients on the windward side were higher than those at the leeward side. Perera and Baughn [1994] extended

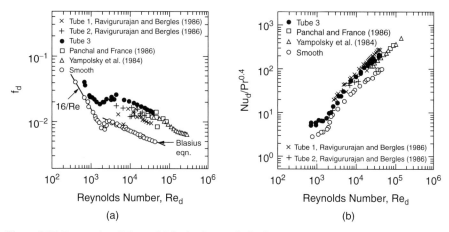

Figure 8.15 Test results of Figure 8.14b aluminum spirally fluted tube. Tube 3 has $d_i = 28.5$ mm, with $n_f = 31$, $\alpha = 30°$, $e/d_i = 0.056$, $p/e = 3.57$, $t = 1.5$ mm, $D_h = 16.7$ mm. (a) Friction factor, (b) $Nu/Pr^{0.4}$. (From Obot et al. [1991].)

Table 8.7 Geometry of Figure 8.15 Tubes Reported by Obot et al. [1991][a]

Tube	d_i	e/d_i	p/e	n_f	α	Material
1	21.54	0.044	3.64	20	40.6	Stainless steel
2	23.96	0.056	4.11	25	30.2	Stainless steel
3	28.49	0.056	3.57	30	30.0	Aluminum

[a]Dimensions are in millimeters.

the study to investigate the effect of helix angle. Slightly higher heat transfer coefficients were obtained for helix angles 15° and 45° than for 30°

8.3.2 Spirally Indented Tube

In-Tube Flow. Ravigururajan and Bergles [1986] provide a literature survey on various types of corrugated tubes, including the spirally indented tube. Blumenkrantz and Taborek [1971] show that the spirally indented tube provides up to 200% enhancement for heating a viscous fluid in laminar flow. However, no enhancement was provided for cooling the same fluid. This is apparently because centrifugal force tends to prevent mixing of the boundary layer, as discussed for twisted tapes in Chapter 7.

Richards et al. [1987] tested 12 different spirally indented tubes in turbulent water flow using condensing steam on the outer surface. They assumed a 0.8 exponent on the Reynolds number for the tube-side flow. The validity of the 0.8 exponent has not been confirmed. Table 8.8 lists the geometric parameters of the 12 tubes tested. Table 8.9 gives the curve-fit coefficients and (C_i and B) and exponent n for the heat transfer and friction equations given below.

$$\frac{hd_e}{k} = C_i \left(\frac{d_e G}{\mu} \right)^{0.8} \mathrm{Pr}^{1/3} \left(\frac{\mu}{\mu_w} \right)^{0.14} \tag{8.22}$$

$$f = B \left(\frac{d_e G}{\mu} \right)^n \tag{8.23}$$

The mass velocity, and the characteristic dimension in the Nusselt number and Reynolds numbers are based on the envelope tube diameter (d_e). Table 8.9 also lists the tube-side enhancement level (j/j_p) and the efficiency index [$\eta = (j/j_p)/(f/f_p)$] for $\mathrm{Re}_d = 10{,}000$. Examination of Table 8.9 shows $1.84 \leq j/j_p \leq 3.14$ and $0.13 \leq \eta \leq 0.5$. The efficiency index (η) values are quite low compared to the performance of the fluted tubes shown in Figure 8.14a and b. In fact, the η values of the Table 8.9 tubes are typically lower than the other tube-side enhancements discussed in Chapters 7 through 9.

Srinivasan and Christensen [1992] extended the study of Richards et al. [1987] to 15 internally indented tubes with $0.1 \leq e/D_{vi} \leq 0.4$, $0.4 \leq p/D_{vi} \leq 7.3$, $30° \leq \alpha \leq 58°$, where D_{vi} is the volume-based tube inner diameter. Tests were conducted for a wide range of Reynolds numbers ($200 \leq \mathrm{Re}_{Dvi} \leq 80{,}000$) using water. The transition to turbulent flow was determined from the friction factor curve. The transition was observed to occur at $300 \leq \mathrm{Re}_{Dvi} \leq 1000$. The Nusselt number curves, however, did not show a change of slope over the Reynolds number range ($500 \leq \mathrm{Re}_{Dvi} \leq 80{,}000$). In the turbulent regime, the increase in Nusselt number over the smooth tube was 1.1 to 2.6. The friction factor increase was much larger (up to ninefold). The heat transfer and friction correlations were developed based on their study, which later modified by Arnold et al. [1993] as follows

For the friction factor, $\mathrm{Re}_{Dvi} \leq 1500$:

Table 8.8 Dimensionless Geometric Parameters for Tubes Tested by Richards et al. [1989]

Tube	d_e/d_c	e/d_e	e/p	e/d_c	p/d_c	A_c/A_e
1	1.56	0.179	0.238	0.278	1.168	0.89
2	1.59	0.186	0.479	0.296	0.618	1.41
3	1.77	0.217	0.416	0.385	0.925	1.48
4	1.23	0.093	0.179	0.114	0.637	0.92
5	1.38	0.139	0.272	0.192	0.706	1.02
6	1.49	0.165	0.515	0.247	0.408	1.11
7	1.93	0.241	0.349	0.465	1.332	1.60
8	1.90	0.237	0.704	0.449	0.638	2.04
9	1.71	0.208	0.275	0.356	1.295	1.22
10	1.83	0.226	0.221	0.414	1.873	1.28
11	2.07	0.258	0.225	0.534	2.373	1.39
12	1.68	0.202	0.500	0.388	0.776	1.36

Table 8.9 Curve Fit and Performance Parameters for Doubly Fluted Tubes as described Table 8.8 j/j_p and η at Re = 10,000)

Tube	C_i	B	n	j/j_p	η
1	0.0442	4.07	0.297	1.84	0.21
2	0.0681	3.79	0.253	2.96	0.24
3	0.0440	6.33	0.276	2.15	0.13
4	0.0455	0.45	0.125	2.45	0.52
5	0.0496	1.37	0.208	2.30	0.34
6	0.0632	0.76	0.117	3.14	0.37
7	0.0596	8.73	0.305	3.02	0.20
8	0.0501	6.81	0.235	2.56	0.12
9	0.0487	4.24	0.244	2.61	0.17
10	0.0480	5.55	0.260	2.25	0.14
11	0.0526	14.73	0.365	2.63	0.30
12	0.0495	3.51	0.220	2.56	0.25

$$f = 0.554 \left(\frac{64.0}{Re_{Dvi} - 45.0} \right) \left(\frac{e}{D_{vi}} \right)^{0.384} \left(\frac{p}{D_{vi}} \right)^{(-1.454 + 2.083 e/D_{vi})} \left(\frac{\alpha}{90} \right)^{-2.42} \tag{8.24}$$

$Re_{Dvi} > 1500$

$$f = 1.209 \left(Re_{Dvi} \right)^{-0.261} \left(\frac{e}{D_{vi}} \right)^{(1.26 - 0.05 p/D_{vi})}$$

$$\left(\frac{p}{D_{vi}} \right)^{(-1.66 + 2.033 e/D_{vi})} \left(\frac{\alpha}{90} \right)^{(-2.669 + 3.67 e/D_{vi})} \tag{8.25}$$

For the Nusselt number, $Re_{Dvi} \leq 5000$:

$$Nu_{Dvi} = 0.014 \, Re_{Dvi}^{0.842} \left(\frac{e}{D_{vi}} \right)^{0.067} \left(\frac{p}{D_{vi}} \right)^{0.293} \left(\frac{\alpha}{90} \right)^{-0.705} Pr^{0.4} \tag{8.26}$$

$Re_{Dvi} > 5000$

$$Nu_{Dvi} = 0.064 \, Re_{Dvi}^{0.773} \left(\frac{e}{D_{vi}} \right)^{-0.242} \left(\frac{p}{D_{vi}} \right)^{-0.108} \left(\frac{\alpha}{90} \right)^{0.599} Pr^{0.4} \tag{8.27}$$

Annular Flow. Garimella and Christenson [1995a] investigated the pressure drop characteristics in annuli with spirally indented inner tubes. Nine spirally indented tubes were tested for laminar, transition, and turbulent flow regimes. Transition to turbulent flow was observed in the $310 \leq Re_{Dvo} \leq 1000$ range. The friction factor increase over that of the smooth tube was between 1.1 and 2.0 in the laminar regime, while the value increased to 10 in the turbulent regime. Empirical friction correlations were developed based on their data. The heat transfer coefficients of the same geometry were obtained by Garimella and Christenson [1995b] for 14 spirally indented tubes for $700 \leq Re_{Dvo} \leq 40,000$. Similar to the in-tube flow, Nusselt number curves did not show a change of slope over the entire Reynolds number range. In the turbulent region, the Nusselt number increase over the smooth tube was from 1.1 to 4.0. The Nusselt number increased with the flute depth, and decreased with the flute pitch and annulus radius ratio. An empirical heat transfer correlation based on their data was developed.

Srinivasan et al. [1994] conducted a numerical analysis of heat transfer and fluid flow in a spirally indented tube using a porous substrate model. The model divided the flow into two regions — the fluted region and the core region. The fluted region was modeled as a porous substrate with directionally dependent permeabilities. Experimental data were used to relate the substrate permeability and the tube geometric parameters. The numerical results reasonably predicted the experimental data.

8.4 ADVANCED INTERNAL FIN GEOMETRIES

One class of tubes, which is difficult to classify as an internal fin or as a roughness, has been developed. Examples are the internal enhancements shown in Figure 1.13b and c. They provide a relatively large internal area increase (e.g., 50%), but have a high helix angle (25° to 45°). These tubes have some of the characteristics of roughness and internal fins. The tube is similar to an internally finned tube, because it provides significant internal surface area increase. Yet it has a higher helix angle than a helical internal finned tube. It is likely that the flow separates at the fin tip, which probably does not occur in an internally finned tube having a helix angle less than 25°. The Figure 1.13 geometries are discussed in Chapter 9. The low fins (e.g., 0.5 mm) are typically much more closely spaced than an internally finned tube.

A quite different internally finned tube, discussed in Chapters 13 and 14, is the "microfin tube," which is shown in Figure 13.7d. This tube has 0.2 to 0.25-mm-high triangular shaped fins spaced at approximately 2.0 mm. The fins are formed with a helix angle of 15° to 30°. These microfins may be formed in a copper tube at high speed by drawing the tube over a grooved slug. This tube has found major application for convective vaporization and condensation of refrigerants. Khanpara et al. [1987] measured the enhancement for subcooled R-22 and R-113 liquid flow in the microfin tube shown in Figure 14.5. The particular tube tested had 8.83-mm-ID 60 fins 0.22 mm high, and 17° helix angle. The heat transfer coefficient was based on $A_i/L = \pi d_i$. For $Re_d > 10,000$, $h/h_p \simeq 2.0$ with the smooth tube data predicted by the Dittus–Boelter equation (Equation 2.24) for cooling, which uses $Pr^{0.3}$. The enhancement was somewhat higher than the 54% internal surface area increase. The enhancement level significantly decreases for $Re_d < 10,000$. Koyama et al. [1996] measured the heat transfer coefficient for heating of water and subcooled liquid refrigerants R-22, R-123, and R-134a in a 8.37-mm-ID microfin tube having 60 fins, 0.168-mm fin height, and 18° helix angle. The heat transfer coefficient of his data, plus that of Khanpara et al. [1987], was reasonably well correlated for $Re_d > 10,000$ by use of the Dittus–Boelter equation adjusted to use a coefficient of 0.028. The heat transfer coefficient (h) is based on the total internal surface area, while the Nu and Re are based on the nominal inside diameter (d_i). Because the tube tested by Koyama et al. had a 52% area increase, relative to a plain tube of the same nominal diameter, the 0.028 coefficient means that the heat transfer coefficient based on the nominal tube diameter is 1.85 times that of a plain tube of the same nominal diameter. Figure 8.16 shows the correlated data. For $Re_d < 10,000$ (down to 3000), their correlation overpredicts the Nusselt number.

Brognaux et al. [1997] measured the effect of Prandtl number ($0.70 \leq Pr \leq 7.85$) for turbulent flow of water in microfin tubes. He tested three single grooved tubes (17.5, 20, and 27° helix angle) and three cross-grooved tubes with different second grooved depths (40, 60, and 80% of the first groove) with first groove angle 17.5° and second groove angle –17.5°. All the tubes were 14.9 mm ID with $e/d_i = 0.0236$, p_n/e =1.66, and 78 microfins. Additional data taken at Penn State University broadened the range of helix angles to values between 0° and 34°. The best performance was obtained for the cross-grooved tube with 80% cross-cut depth. The enhancement ratio was as high as 95% and the efficiency index [$\eta = (j/j_p)/(f/f_p)$] was 4 to 6% better than that of the plain tube. They correlated the data using the roughness similarity functions (Section 9.3). They also obtained a Prandtl number exponent of 0.56 to 0.57, which is in good agreement with most of the research done on rough tubes, as shown in Table 9.11. Of the three single grooved tubes tested, the one with 20° helix angle provided significantly higher efficiency index than the one with 27° helix angle. When the Petukhov equation was used with hydraulic diameter to predict the data, the heat transfer coefficient was overpredicted nearly 30%. This is because the high shear stress in the interfin region slows the velocity causing reduced heat transfer coefficient. Brognaux et al. developed an empirical correlation to account for Prandtl number and helix angle. These data were correlated by the empirical equation:

$$\mathrm{Nu}_{di,m} = C \, \mathrm{Re}_{di,m}^{0.81} \, \mathrm{Pr}_l^{0.55} \tag{8.28}$$

where $C = 0.02271 + 3.72\mathrm{E}\text{-}05 \; \alpha - 9.337\mathrm{E}\text{-}7 \; \alpha^3$, and α is the helix angle. The characteristic diameter is the melt-down diameter, d_{im} (the inside diameter if the fins were melted and returned to the tube wall).

Narayanamurthy [1999] extended the work of Brognaux et al. [1997] and reports the heat transfer coefficient for 12 microfin tubes (14.9 mm ID) using water and the friction factor of 15 tubes with Reynolds numbers between 5000 and 70,000. The heat transfer coefficient is based on the nominal inside surface area defined by A_i/L = πd_i, where d_i is the tube diameter to the rib root. He developed multiple regression correlations for the f and j factors, which are given by

$$j = 0.00563 \, \mathrm{Re}_d^{-0.12} \, n_f^{-0.246} \, (e/d_i)^{0.427} \, (19.63 + \alpha) \tag{8.29}$$

Equations 8.28 and 8.29 may be applied to any tube diameter by using the relationship $n_f = \pi d_i/p_n$.

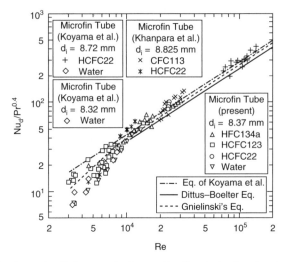

Figure 8.16 Single-phase microfin heat transfer data predicted by single-phase heat transfer correlations. (From Koyama et al. [1996].)

Wang et al. [1996] measured heat transfer and friction in seven microfin tubes (7.0 and 9.52 mm OD) using water. He developed a heat transfer correlation based on the heat transfer similarity model for roughness described in Section 9.3.3. The correlations predicted 85% of the heat transfer data and 96% of the friction data within ± 10%. Narayanamurthy [1999] also reports heat transfer correlations based on use of the heat transfer similarity model for roughness.

Note that the Koyama et al. [1996] and Brognaux et al. [1997] equations do not account for the actual internal dimensions of the microfin tube. These are the microfin height (e), the fin pitch (p), the fin base thickness (t_b), the apex angle (2θ), the fin tip radius (r_t) of the microfin, and the helix angle (α) for a microfin tube. A general predictive model must account for these individual dimensions. The above equations for single-phase heat transfer in microfin tubes are useful in identifying optimum geometries for condensation in microfin tubes using the equivalent Reynolds number model of Moser et al. [1998].

Bhatia and Webb [2001b] numerically investigated the effect of the microfin geometric parameters on the Nusselt number and the friction factor for a 14.9-mm-ID tube using Pr = 1. Their objective was to identify an optimum microfin geometry. The friction factor and the Nusselt number increased as the helix angle and the fin height increased. With increasing number of fins, the Nu/f ratio was approximately constant for a range of 45 to 75 fins. Little dependence on the fin shape (width of the fin tip, apex angle) was noticed as long as the fin cross-sectional area remained constant. However, alteration of any geometric parameter that results in reduction of the interfin channel cross-sectional area increases the amount of flow bypassing the interfin channel. An increase of the bypass flow reduces the interfin shear stress and Nusselt number. Because the heat transfer coefficient was based on the nominal area, there is little benefit in using a high number of fins. The friction factor and

Nusselt number decreases above 75 fins. Hence, based on material utilization, an optimum number of fins exists. Alteration of any geometric parameter that resulted in the reduction of the interfin channel cross-sectional area reduced the friction factor and the Nusselt number. Their results reported in this paper are very useful to determine the optimum internal geometry parameters. Based on the numerical results, a 14.9-mm tube geometry having 45 to 65 fins with fin height above 0.3 mm was recommended. Numerical work on internal fin tubes including microfin geometry is discussed in Section 8.2.2.

8.5 FINNED ANNULI

Figure 8.17a shows axial fins, which are used for a double-pipe heat exchanger, or for axial flow on the outer surface of a tube bundle. When used in a double-pipe heat exchanger, the fins may span the full width of the annulus, as shown by Figure 8.17b. Multitube designs as illustrated in Figure 8.17c are also used. A common use of double-pipe exchangers is for gases or high-viscosity liquids, e.g., oil, for which the flow may be in the laminar or transition regime. Then some form of enhancement is of interest, and various enhanced fin geometries may be employed. Figure 8.17d shows the "cut-and-twist" geometry. This provides intermittent mixing along the length. The length between the cuts is in the range of 0.3 to 1.0 m. Figure 8.17e shows the offset strip fin (OSF) applied to double-pipe geometry. The fin material is wrapped around the inner tube and soldered. Guy [1983] discusses the design and application of this geometry.

The geometry of the annular flow passage varies from trapezoidal (for small d_i/d_o and a small number of fins) to rectangular (for $d_i/d_o \approx 1.0$ with a large number of fins). In the limiting condition of rectangular channels with small clearance between the fin tip and the tube wall, one may reasonably predict the performance using appropriate equations for flow in rectangular channels. Note that the outer annulus tube is not a heat transfer surface.

DeLorenzo and Anderson [1945] presents finned annulus data for 24, 28, and 36 fins on a 48.3-mm-diameter tube for heating and cooling of oils ($11 < Pr < 1600$). Gunter and Shaw [1942] present data for the cut-and-twist geometry shown in Figure 8.14c. The correlations presented in Chapter 5 are applicable to the Figure 8.17d OSF geometry.

Taborek [1997] provides friction and Nusselt number correlations for laminar, transition, and turbulent flow in finned annuli, including the cut-and-twist geometry. The correlations are based on the data of Gunter and Shaw [1942], DeLorenzo and Anderson [1945], and unpublished work of Gardner. For the turbulent region ($Re_{Dh} \geq 15{,}000$), the same correlation as that developed for plain annular geometry is applicable for finned annuli.

$$\mathrm{Nu}_{Dh} = \mathrm{Nu}_p \left[0.86 \left(\frac{D_o}{D_i} \right) \right] \tag{8.30}$$

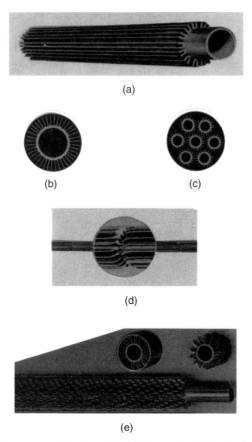

(a)

(b)   (c)

(d)

(e)

Figure 8.17 (a) axial fins on external surface, (b) axial fins used in double-pipe heat exchanger, (c) axial fins with multitubes, (d) cut-and-twist axial fins, (e) offset strip fins. (Parts a to d are from Brown Fin-Tube Brochure. Part e is reproduced courtesy of Wieland-Werke AG.)

where Nu_p is the Nusselt number for the smooth tube, and Petukhov [1970] correlation is recommended. For laminar and transition region ($Re_{Dh} < 15,000$), the Nusselt number is calculated from

$$Nu_{Dh} = \left[Nu_L^z + Nu_x^z \right]^{1/z} \left(\frac{\mu_b}{\mu_w} \right)^n \tag{8.31}$$

$$Nu_L = \left[Nu_\infty^3 + N_{L,a}^{\ 3} \right]^{1/3} \tag{8.32}$$

$$Nu_{L,a} = 2.1 \left(Re_{Dh} \, Pr \, \frac{D_h}{L} \right)^{1/3} \qquad Nu_\infty = 4.12 \tag{8.33}$$

$$\mathrm{Nu}_x = \mathrm{Nu}_{tr}\left(\frac{\mathrm{Re}_{Dh}}{15,000}\right)^{1.25} \qquad (8.34)$$

where Nu_{tr} is the Nusselt number for turbulent flow at $\mathrm{Re}_{Dh} = 15,000$. The exponent z is given by a function of Reynolds number as $z = 0.1\,\mathrm{Re}_{Dh}^{0.4}$ with the limit of $z > 1.2$. The friction factor is given by the following equation, which is applicable for the entire Reynolds number range:

$$f = \left[f_{tub}^3 + f_{lam}^3 \right]^{1/3} \qquad (8.35)$$

$$f_{tub} = \left[1.58\ln\left(\mathrm{Re}_{Dh}\right) - 3.28 \right]^{-2}\left(\frac{\mu_b}{\mu_w}\right)^n \qquad (8.36)$$

$$f_{lam} = \left(\frac{16}{\mathrm{Re}_{Dh}}\right)\left(\frac{\mu_b}{\mu_w}\right)^n \qquad (8.37)$$

Taborek [1997] also provides a friction correlation for the cut-and-twist geometry (Figure 8.17d). Braga and Saboya [1999] provide additional data on fully developed Nusselt numbers and friction factors for turbulent airflow in a finned annulus.

Edwards et al. [1988] provide heat transfer data in the entrance region of inner finned annuli. The geometry included two diameter ratios (0.66 and 0.79) each having 0, 12, 24, and 36 fins. Tests were performed using air with $4000 \leq \mathrm{Re}_{Dh} \leq 15,000$. Local heat transfer coefficients were obtained from liquid crystal images. Nusselt number correlations applicable to the entrance region were developed from their data.

8.6 CONCLUSIONS

Internally finned tubes and annuli are among the earliest tube-side enhancement geometries. The early versions used extruded aluminum insert devices, which provide full height fins. The first integral, internal fin tubes were made of copper using a cold swaging process. Performance can be improved by forming the fins at a helix angle, e.g., up to 25°. Although the copper tubes provide high performance, they are quite expensive. External, axially finned tubes for annuli are made by welding steel fins on a steel tube. The same axial fin geometries may be made at lower cost in aluminum, by a hot extrusion process. Helical fins may be provided by twisting the aluminum tube. The Figure 8.14 spirally indented tubes are variants of internally finned tubes, which provide enhancement on both the inside and outside surfaces. Making a corrugated strip, followed by rolling into a circular shape and seam welding offers a low-cost possibility. However, it is difficult to make a plain tube end.

Advanced internally roughened tubes, as illustrated in Figure 1.13b and c, provide a relatively large internal area increase, approaching that of internally finned tubes. The Figure 1.13 geometries are discussed in Chapter 9. They are similar to an internally finned tube, because they provide significant internal surface area increase. Yet they have a higher helix angle than a helical internal finned tube. It is likely that the flow separates at the fin tip, which does not occur in an internally finned tube. The low fins (e.g., 0.5 mm) are typically much more closely spaced than an internally finned tube.

The microfin tube shown in Figure 13.7d has closely spaced fins 0.20 to 0.25 mm high, and is used for convective vaporization and condensation of refrigerants. It is expected that variants of the Figure 1.13b and d tubes will be developed and investigated for application to single-phase flow of liquids.

Many numerical solutions have been provided for laminar flow in internally finned tubes. These solutions cover both fully developed and entrance region conditions, and constant heat flux and constant wall temperature boundary conditions. Recent efforts accounted for fluid property variation across the boundary layer.

The Carnavos [1980] empirical correlation is applicable for prediction of turbulent flow heat transfer and friction. Jensen and Vlakancic [1999] correlations are applicable to a wider range. Taborek [1997] provides correlations for laminar and turbulent flow in finned annuli. Kim and Webb [1993] have shown that the law of the wall may be used to model turbulent flow in internally finned tubes.

As discussed in Chapter 10, the conventional internally finned tube provides good fouling characteristics.

NOMENCLATURE

A	Total heat transfer surface area (both primary and secondary), m^2 or ft^2
A_c	Cross-sectional flow area, m^2 or ft^2
A_{core}	Core flow area of an internally finned tube, $A_n(1 - 2e/d_i)^2$, m^2 or ft^2
A_f	Fin or extended surface area, m^2 or ft^2
A_{fa}	Actual flow area of an internally finned tube, $A_n - n_f et$, m^2 or ft^2
A_{fin}	Inter-fin flow area of an internally finned tube, $A_{fa} - A_{core}$, m^2 or ft^2
A_n	Nominal flow area of an internally finned tube, $\pi d_i^2/4$, m^2 or ft^2
A_{nom}	Heat transfer surface area based on $A/L = \pi d$, inside tube $(d = d_i)$, outside tube $(d = d_o)$, m^2 or ft^2
c_p	Specific heat of fluid at constant pressure, J/kg-K or Btu/lbm-°F
D_h	Hydraulic diameter of flow passages, $4LA_c/A$, m or ft
D_{vi}	Volume-based tube inner diameter, $(4V/\pi L)^{1/2}$, m or ft
D_{vo}	Volume-based annulus diameter, $D_{vi} + 2t$, m or ft
d_c	Internal core flow diameter between fin tips for internally finned tube (Figure 8.2), m or ft
d_e	External diameter at tip of fins (or flutes) for a finned (or fluted) tube, m or ft

d_i	Plain tube inside diameter, or diameter to the base of internal fins or roughness, m or ft
d_{im}	Internal diameter if enhancement material is uniformly returned to tube wall thickness, m or ft
d_o	Tube outside diameter, fin root diameter for a finned tube, m or ft
e	Fin height or roughness height, m or ft
f_{Dh}	Fanning friction factor based on D_h $(= \Delta p_f d_h/2LG^2)$, dimensionless
f_d	Fanning friction factor based on d_i $((= \Delta p_f d_i/2LG^2)$, dimensionless
G	Mass velocity based on the minimum flow area, kg/m²-s or lbm/ft²-s
Gr	Grashof number $= g\beta\Delta T D_h^3/\nu^2$, dimensionless
Gz	Graetz number $= \pi d_i RePr/4L$, dimensionless
h	Heat transfer coefficient based on A, W/m²-K or Btu/hr-ft²-°F
h'	Heat transfer coefficient based $A/L = \pi d_i$, W/m²/K or Btu/hr-ft²-°F
j	$StPr^{2/3}$, dimensionless
k	Thermal conductivity of fluid, W/m-K or Btu/hr-ft-°F
k_f	Thermal conductivity of fin material, W/m-K or Btu/hr-ft-°F
L	Fluid flow (core) length on one side of the exchanger, m or ft
L_h	Hydraulic entrance length, m or ft
L_t	Thermal entrance length, m or ft
L_h^+	Hydraulic entrance length, L_h/D_hRe, dimensionless
L_t^+	Thermal entrance length, $L_t/D_h Re_{Pr}$ dimensionless
l	Strip flow length of OSF, m or ft
M	Mass of tube material, kg or 1bm
N	Number of tube rows in the flow direction, or number of tubes in heat exchanger, dimensionless
Nu_{Dh}	Nusselt number ($= hD_h/k$), dimensionless
Nu_d	Nusselt number ($= hd/k$), inside tube ($d = d_i$), outside tube ($d = d_o$), dimensionless
n_f	Number of fins in internally finned tube, dimensionless
P	Fluid pumping power, W or hp
Pr	Prandtl number $= c_p\mu/k$, dimensionless
Pr_t	Turbulent Prandtl number, dimensionless
p	Axial pitch of fins or roughness elements, m or ft
p_n	Fin pitch normal to tube axis ($= \pi d_i/n_f$), m or ft
p_f	Fin pitch, center to center spacing, m or ft
Δp	Fluid static pressure drop, Pa or lbf/ft²
Q	Heat transfer rate in the exchanger, W or Btu/hr
Ra	Rayleigh number, GrPr, dimensionless
Re_{Dh}	Reynolds number based on the hydraulic diameter $= GD_h/\mu$, dimensionless
Re_d	Reynolds number based on the tube diameter $= Gd/\mu$, $d = d_i$ for flow inside tube and $d = d_o$ for flow outside tube, dimensionless
Re_{Dvi}	Reynolds number based on D_{vi}, GD_{vi}/μ, dimensionless
Re_{Dvo}	Reynolds number based on D_{vo}, GD_{vo}/μ, dimensionless
R_{fi}	Tube-side fouling resistance, m²-K/W or ft²-°F-hr/Btu
r_t	Fin tip radius, m or ft

St   Stanton number = h/Gc_p, dimensionless
s   Spacing between two fins = $P_f - t$, m or ft
t   Thickness of tube wall, fin or twisted tape, m or ft
t_b Fin base thickness, m or ft
ΔT_i Temperature difference between hot and cold inlet fluids, K or °F
ΔT_{lm} Log-mean temperature difference, K or °F
U   Overall heat transfer coefficient, W/m²-K or Btu/hr-ft²-°F
u^* Friction velocity = $(\tau_w/\rho)^{1/2}$, m/s or ft/s
u_m Fluid mean axial velocity at the minimum free flow area, m/s or ft/s
V   Heat exchanger total volume, m³ or ft³
V_m Heat exchanger tube material volume, m³ or ft³
W   Fluid mass flow rate, kg/s or lbm/s
x   Cartesian coordinate along the flow direction, m or ft
X^+ $x/D_h RePr$, dimensionless

Greek Letters

α Helix angle relative to tube axis radians or degrees
γ Reciprocal of fin pitch, m or ft
η_f Fin efficiency or temperature effectiveness of the fin, dimensionless
η_o Surface efficiency of finned surface = $1 - (1 - \eta_f)A_f/A$, dimensionless
2θ Included angle of fin cross section normal to flow, radians or degrees
θ^* $\theta k_f/k$, dimensionless
μ Fluid dynamic viscosity coefficient, Pa-s or lbm/hr-ft
ν Kinematic viscosity, m²/s or ft²/hr
ρ Fluid density, kg/m³ or lbm/ft³
τ_w Wall shear stress, Pa or lbf/ft²

Subscripts

fd   Fully developed flow
H1   Heat flux boundary condition
m   Average value over flow length
p   Plain tube or surface
T   Wall temperature boundary condition
w   Evaluated at wall temperature
x   Local value

REFERENCES

Al-Fahed, S., Chamra, L.M., and Chakroun, W., 1999. Pressure drop and heat transfer comparison for both microfin tube and twisted-tape inserts in laminar flow, *Exp. Thermal Fluid Sci.*, 18, 323–333.

Arnold, J.A., Garimella, S., and Christensen, R.N., 1993. Fluted Tube Heat Exchanger Design Manual, GRI Report 5092-243-2357.

Barba, A., Bergles, G., Gosman, A.D., and Launder, B.E., 1983. The Prediction of Convective Heat Transfer in Viscous Flow Through Spirally Fluted Tubes, ASME Paper 83-WA/HT-37, New York.

Baughn, J.W., Kapila, K., Perera, C.K., and Yan, X., 1993. An experimental study of local heat transfer in a spirally fluted tube, in *Turbulent Heat Transfer*, ASME HTD, 239, 49–56.

Bergles, A.E. and Joshi, S.D., 1983. Augmentation techniques for low Reynolds number in-tube flow, in *Low Reynolds Number Flow in Heat Exchangers*, S. Kakaç, R.K. Shah, and A.E. Bergles, Eds., Hemisphere, Washington, D.C., 694–720.

Bhatia, R.S. and Webb, R.L., 2001a. Numerical study of turbulent flow and heat transfer in micro-fin tubes. Part 1, Model validation, *J. Enhanced Heat Transfer*, 8, 291–304.

Bhatia, R.S. and Webb, R.L., 2001b. Numerical study of turbulent flow and heat transfer in micro-fin tubes. Part 2, Parametric study, *J. Enhanced Heat Transfer*, 8, 305–314.

Blumenkrantz, A. and Taborek, J., 1971. Heat Transfer and Pressure Drop Characteristics of Turbotec Spirally Deep Grooved Tubes in the Laminar and Transition Regime, Report 2439–300-8, April, Heat Transfer Research, Inc.

Braga, C.V.M. and Saboya, F.E.M., 1986. Turbulent heat transfer and pressure drop in smooth and finned annular ducts, *Eighth International Heat Transfer Conference,* San Francisco, CA, 6, 2831–2836.

Braga, C.V.M. and Saboya, F.E.M., 1999. Turbulent heat transfer, pressure drop and fin efficiency in annular regions with continuous longitudinal rectangular fins, *Exp. Thermal Fluid Science*, 20, 55–65.

Brognaux, L., Webb, R.L., Chamra, L.M., and Chung, B.Y., 1997. Single-phase heat transfer in micro-fin tubes, *Int. J. Heat Mass Transfer*, 40, 4345–4358.

Carnavos, T.C., 1979. Cooling air in turbulent flow with internally finned tubes, *Heat Trans. Eng.,* 1(2), 41–46.

Carnavos, T.C., 1980. Heat transfer performance of internally finned tubes in turbulent flow, *Heat Trans. Eng.,* 4(1), 32–37.

Choudhury, D. and Patankar, S.V., 1985. Analysis of developing laminar flow and heat transfer in tubes with radial fins, in *Advances in Enhanced Heat Transfer — 1985*, ASME Symp. Vol. HTD, 43, S.M. Shenkman, J.E. O'Brien, I.S. Habib, and J.A. Kohler, Eds., 57–63.

DeLorenzo, B. and Anderson, E.D., 1945. Heat transfer and pressure drop of liquids in double pipe fintube exchangers, *ASME Trans.,* 67, 697–702.

Edwards, D.P., Hirsa, A., and Jensen, M.K., 1993. Turbulent air flow in longitudinally finned tubes, in *Heat Transfer in Turbulent Flows*, R.S. Amano, R.H. Pletcher, S.A. Sheriff, R.G. Watts, and A.N. Anand, Eds., ASME, Symp. Vol. HTD, 246, 9–16.

Edwards, D.P., Hirsa, A., and Jensen, M.K., 1992. An experimental investigation of flow in longitudinally finned tubes, in *Fundamentals of Forced Convection Heat Transfer*, M.A. Ebadian and P.H. Oosthuizen, Eds., ASME Symp. Vol. HTD, 210, 141–148.

Edwards, D.P. and Jensen, M.K., 1994. Pressure drop and heat transfer predictions of turbulent flow in longitudinally finned tubes, in *Advances in Enhanced Heat Transfer*, T.J. Rabas and J.E. Bogart, Eds., Asme Symp. Vol. HTD, Vol. 287, 17–24.

Edwards, R.J., Jambunathan, K., and Button, B.L., 1988, Experimental investigation of turbulent heat transfer in simultaneously developing flow in finned annuli, in *Proc. First World Conference on Experimental Heat Transfer, Fluid Mechanics and Thermodynamics*, R.K. Shah, E.N. Ganic, and K.T. Yang, Eds., Elsevier, New York, 543–550.

El-Sayed, S.A., Abdel-Hamid, M.E., and Sadoun, M.M., 1997. Experimental study of turbulent flow inside a circular tube with longitudinal interrrupted fins in the streamwise direction, *Exp. Thermal and Fluid Sci.*, 15, 1–15.

Fabbri, G., 1998. Heat transfer optimization in internally finned tubes under laminar flow conditions, *Int. J. Heat Mass Transfer*, 41, 1243–1253.

Garimella, S. and Christensen, R.N., 1995a. Heat transfer and pressure drop charactersitics of spirally fluted annuli. Part I, Hydrodynamics, *J. Heat Transfer*, 117, 54–60.

Garimella, S. and Christensen, R.N., 1995b. Heat transfer and pressure drop charactersitics of spirally fluted annuli. Part II, Heat transfer, *J. Heat Transfer*, 117, 61–68.

Gowen, R.A. and Smith, J.W., 1967. Heat transfer performance of internally finned tubes in turbulent flow, *Chem. Eng. Sci.*, 22, 1701–1711.

Gunter, A.Y. and Shaw, W.A., 1942. Heat transfer, pressure drop, and fouling rates of liquids for continuous and noncontinuous longitudinal fins, *ASME Trans.*, 64, 795–802.

Guy, A.R., 1983. Double-pipe heat exchangers, in *Heat Exchanger Design Handbook*, Vol. 3, E.U. Schlunder, Ed., Hemisphere, Washington, D.C., Section 3.2.

Hilding, W.E. and Coogan, C.H., 1964. Heat transfer and pressure drop measurements of internally finned tubes, in *Symposium on Air-Cooled Heat Exchangers*, ASME, New York, 57–85.

Hinze, J.O., 1975. *Turbulence*, 2nd ed., McGraw-Hill, New York.

Hu, M.H. and Chang, Y.P., 1973. Optimization of finned tubes for heat transfer in laminar flow, *J. Heat Transfer*, 95, 332–338.

Ivanovi, M., Selimović, R., and Bajramović, R., 1990. Mathematical modeling of heat transfer in internally finned tubes, in *Mathematical Modeling and Computer Simulation of Processes in Energy Systems*, H. Hanjalić, Ed., Hemisphere Washington, D.C., 147–153.

Jensen, M.K. and Vlakancic, A., 1999. Experimental investigation of turbulent heat transfer and fluid flow in internally finned tubes, *Int. J. Heat Mass Transfer*, 42, 1343–1351.

Kelkar, K.M. and Patankar, S.V., 1990. Numerical prediction of fluid flow and heat transfer in a circular tube with longitudinal fins interrupted in the streamwise direction, *J. Heat transfer*, 112, 342–348.

Khanpara, J.C., Pate, M.B., and Bergles, A.E., 1987. Local evaporation heat transfer in a smooth tube and a micro-fin tube using refrigerants 22 and 113, in *Boiling and Condensation in Heat Transfer Equipment*, E.G. Ragi, Ed., ASME Symp. Vol. HTD, 85, 31–39.

Kim, J.H., Jensen, M., and Jansen, K., 2002. Fin shape effects in turbulent heat transfer in tubes with helical fins, *Heat Transfer 2002, Proc. 12th Int. Heat Transfer Conf.*, 4, 183–188.

Kim, N H. and Webb, R.L., 1993. Analytic prediction of the friction and heat transfer for turbulent flow in axial internal fin tubes, *J. Heat Transfer*, 115, 553–559.

Koyama, S., Yu, J., Momoki, S., Fujii, T., and Honda, H., 1996. Forced convective flow boiling heat transfer of pure refrigerants inside a horizontal microfin tube, in *Proceedings of Convective Flow Boiling, An International Conference*, J.C. Chen, Y. Fujita, F. Mayinger, and R. Nelson, Eds., Taylor & Francis, London, 137–142.

Liu, X. and Jensen, M.K., 1999. Numerical investigation of turbulent flow and heat transfer in internally finned tubes, *J. Enhanced Heat Transfer*, 6, 105–119.

Liu, X. and Jensen, M.K., 2001. Geometry effects on turbulent flow and heat transfer in internally finned tubes, *J. Heat Transfer*, 123, 1035–1044.

Marner, W.J. and Bergles, A.E., 1978. Augmentation of tube-side laminar flow heat transfer by means of twisted tape inserts, static mixer inserts and internally finned tubes, *Heat Transfer 1978, Proc. 6th International Heat Transfer Conference*, Vol. 2, Hemisphere, Washington, D.C., 583–588.

Marner, W.J. and Bergles, A.E., 1985. Augmentation of highly viscous laminar tubeside heat transfer by means of a twisted-tape insert and an internally finned tube, in *Advances in Enhanced Heat Transfer — 1985*, S.M. Shenkman, J.E. O'Brien, I.S. Habib, and J.A. Kohler, Eds., ASME Symp. HTD, 43, 19–28.

Marner, W.J. and Bergles, A.E., 1989. Augmentation of highly viscous laminar heat transfer inside tubes with constant wall temperature, *Exp. Thermal Fluid Sci.*, 2, 252–267.

Marto, P.J., Reilly, D.J., and Fenner, J.H., 1979. An experimental comparison of enhanced heat transfer condenser tubing, in *Advances in Enhanced Heat Transfer*, ASME Symp. Vol., J.M. Chenoweth, J. Kaellis, J. Michel, and S.M. Shenkman, Eds., ASME, New York, 1–10.

Moser, K.W., Webb, R.L., and Na, B., 1998. A new equivalent Reynolds number model for condensation in smooth tubes, *J. Heat Transfer*, 120, 410–417.

Nandakumar, K. and Masliyah, H.H., 1975. Fully developed viscous flow in internally finned tubes, *Chem. Eng. J.*, 10, 113–120.

Narayanamurthy, R., 1999. Single Phase Turbulent Flow in Microfin and Helically Ribbed Tubes, M.S thesis, Pennsylvania State University, University Park.

Obot, N.T., Esen, E.B., Snell, K.H., and Rabas, T.J., 1991. Pressure drop and heat transfer for spirally fluted tubes including validation of the role of transition, in *Fouling and Enhancement Interactions*, T.J. Rabas and J.M. Chenoweth, Eds., ASME Symp. Vol. HTD, 164, 85–92.

Panchal, C.B. and France, D.M., 1986. Performance Tests of the Spirally Fluted Tube Heat Exchanger for Industrial Cogeneration Applications, Argonne National Laboratory Report ANL/CNSV-59.

Patankar, S.V., and Chai, J.C., 1991. Laminar natural convection in internally finned horizontal annuli, ASME Paper 91-HT-12, ASME, New York.

Patankar, S.V., Inanović, M., and Sparrow, E.M., 1979. Analysis of turbulent flow and heat transfer in internally finned tubes and annuli, *J. Heat Transfer*, 101, 29–37.

Perera, K.K. and Baughn, J.W., 1994. The effect of pitch angle and Reynolds number on local heat transfer in spirally fluted tubes, in *Optimal Design of Thermal Systems and Components*, ASME HTD, 279, 99–112

Petukhov, B.S., 1970. Heat transfer in turbulent pipe flow with variable physical properties, in *Advances in Heat Transfer*, Vol. 6, T.F. Irvine and J.P. Hartnett, Eds., Academic Press, New York, 504–564.

Prakash, C. and Liu, Y-D., 1985. Analysis of laminar flow and heat transfer in the entrance region of an internally finned circular duct, *J. Heat Transfer*, 107, 84–91.

Prakash, C. and Patankar, S.V., 1981. Combined free and forced convection in internally finned tubes with radial fins, *J. Heat Transfer*, 103, 566–572.

Rabas, T.J. and Mitchell, H., 2000. Internally enhanced carbon steel tubes for ammonia chillers, *Heat Transfer Eng.*, 21, 3–16.

Rabas, T.J., Webb, R.L., Thors, P., and Kim, N.H., 1993. Influence of roughness shape and spacing on the performance of three-dimensional helically dimpled tubes, *J. Enhanced Heat Transfer*, 1, 53–64.

Ravigururajan, T.S. and Bergles, A.E., 1986. Study of Water-Side Enhancement for Ocean Thermal Conversion Heat Exchangers, HTL-44/ERI Project 1718, Iowa State University, Ames, IA.

Richards, D.E., Grant, M.M., and Christensen, R.N., 1987. Turbulent flow and heat transfer inside doubly-fluted tubes, *ASHRAE Trans.*, 93(2), 2011–2026.

Rustum, I.M. and Soliman, H.M., 1988a. Numerical analysis of laminar forced convection in the entrance region of tubes with longitudinal internal fins, *J. Heat Transfer*, 110, 310–313.

Rustum, I.M. and Soliman, H.M., 1988b. Experimental investigation of laminar mixed convection in tubes with longitudinal internal fins, *J. Heat Transfer*, 110, 366–372.

Said, N.M.A. and Trupp, A.C., 1984, Predictions of turbulent flow and heat transfer in internally finned tubes, *Chem. Eng. Commun.*, 31, 65–99.

Shah, R.K., 1990. Assessment of modified Wilson plot techniques used for obtaining heat exchanger design data, *Proc. 1990 Int'l. Heat Transfer Conf.*, Vol.5, Hemisphere, Washington, D.C., 51–56.

Shome, B., 1998, Mixed convection laminar flow and heat transfer of liquids in horizontal internally finned tubes, *Numerical Heat Transfer A*, 33(1) 65–84.

Shome, B. and Jensen, M.K., 1996a. Experimental investigation of laminar flow and heat transfer in internally finned tubes, *J. Enhanced Heat Transfer*, 4, 53–70.

Shome, B. and Jensen, M.K., 1996b. Numerical investigation of laminar flow and heat transfer in internally finned tubes, *J. Enhanced Heat Transfer*, 4, 35–52.

Sieder, E.N. and Tate, G.E., 1936. Heat transfer and pressure drop of liquids in tubes, *Eng. Chem.*, 28, 1429–1435.

Soliman, H.M., 1979. The effect of fin material on laminar heat transfer characteristics of internally finned tubes, in *Advances in Enhanced Heat Transfer*, J.M. Chenoweth, J. Kaellis, J.W. Michel, and S. Shenkman, Eds., ASME, New York, 95–102.

Soliman, H.M. and Feingold, A., 1977. Analysis of fully developed laminar flow in longitudinally internally finned tubes, *Chem. Eng. J.*, 14, 119–128.

Soliman, H.M., Chau, T.S., and Trupp, A.C., 1980. Analysis of laminar heat transfer in internally finned tubes with uniform outside wall temperature, *J. Heat Transfer*, 102, 598–604.

Srinivasan, V. and Christensen, R.N., 1992. Experimental investigation of heat transfer and pressure drop characteristics of flow through spirally fluted tubes, *Exp. Thermal Fluid Sci.*, 5, 820–827

Srinivasan, V., Vafai, K., and Christensen, R.N., 1994. Analysis of heat transfer and fluid flow through a spirally fluted tube using a porous substrate approach, *J. Heat Transfer*, 116, 543–551.

Taborek, J., 1997. Double pipe and multitube heat exchangers with plain and longitudinal finned tubes, *Heat Transfer Eng.*, 18(2), 34–45.

Trupp, A.C. and Haine, H., 1989. Experimental investigation of turbulent mixed convection in horizontal tubes with longitudinal internal fins, in *Heat Transfer in Convective Flows*, R.K. Shah, Ed., ASME HTD, 107, 17–25.

Trupp, A.C., Lau, A.C.Y., Said, N.N.A., and Soliman, H.M., 1981, Turbulent flow characteristics in an internally finned tube, in *Advances in Enhanced Heat Transfer* — 1981, R.L. Webb, T.C. Carnavos, E.L. Park, Jr., and K.M. Hostetler, Eds., ASME Symp. Vol. HTD, 18, 11–20.

Wang, C.C., Chiou, C.B., and Lu, D.C., 1996. Single-phase heat transfer and flow friction correlations for microfin tubes, *Int. J. Heat Fluid Flow*, 17, 500–508.

Watkinson, A.P., Miletti, P.L., and Tarassoff, P., 1973. Turbulent heat transfer and pressure drop in internally finned tubes, *AIChE Symp. Ser.*, 69(131), 94–103.

Watkinson, A.P., Miletti, P.L., and Kubanek, G.R., 1975a. Heat transfer and pressure drop of internally finned tubes in laminar oil flow, *ASME Paper* 75-HT-41, New York.

Watkinson, A.P., Miletti, P.L., and Kubanek, G.R., 1975b. Heat transfer and pressure drop of internally finned tubes in turbulent air flow, *ASHRAE Trans.*, 81(1), 330–349.

Webb, R.L., 1981. Performance evaluation criteria for use of enhanced heat transfer surfaces in heat exchanger design, *Int. J. Heat Mass Transfer*, 24, 715–726.

Webb, R.L. and Scott, M.J., 1980. A parametric analysis of the performance of internally finned tubes for heat exchanger application, *J. Heat Transfer*, 102, 38–43.

Webb, R.L., Narayanamurthy, R., and Thors, P., 2000. Heat transfer and friction characteristics of internal helical-rib roughness, *J. Heat Transfer*, 122, 134–142.

Wolfstein, M., 1988. The velocity and temperature distribution of one dimensional flow with turbulence augmentation and pressure gradient, *Int. J. Heat Mass Transfer*, 12, 301–318.

Yampolsky, J.S., 1983. Sprially fluted tubing for enhanced heat transfer, in *Exchangers—Theory and Practice*, J. Taborek, G.F. Hewitt, and N. Afgan, Eds., Hemisphere, Washington, D.C., 945–952.

Yampolsky, J.S., Libby, P.A., Launder, B.E., and LaRue, J.C., 1984. Fluid Mechanics and Heat Transfer Spirally Fluted Tubing, GA Technologies Report GA-A17833.

Zhang, H.Y. and Ebadian, M.A., 1992. The influence of internal fins on mixed convection inside a semicircular duct, in *Enhanced Heat Transfer*, M.B. Pate and M.K. Jensen, Eds., ASME Symp. Vol. HTD, 202, 17–24.

INTEGRAL ROUGHNESS

9.1 INTRODUCTION

Considerable data exist for single-phase forced convection flow over rough surfaces. Data exist for six different flow geometries:

1. Flat plates
2. Circular tubes
3. Noncircular channels in gas turbine blades
4. Longitudinal flow in rod bundles
5. Annuli having roughness on the outer surface of the inner tube
6. Flow normal to circular tubes

Internally roughened tubes are becoming quite important in commercial applications. The refrigeration industry routinely uses roughness on the water side of evaporators and condensers in large refrigeration equipment, as described by Webb and Robertson [1988] and Webb [1991]. Water-side roughness also offers economic benefits for use in electric utility steam condensers, as described by Webb et al. [1984] and Jaber et al. [1991].

Figure 9.1 shows a modern gas turbine blade, containing roughened channels. Air from the compressor flows in the rough channels. The work of Han [1984, 1988], Han et al. [1991] and Metzger et al. [1983, 1987] has significantly advanced this technology.

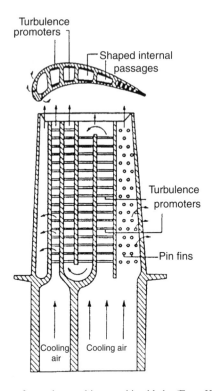

Figure 9.1 Cooling concepts of a modern multipass turbine blade. (From Han et al. [1988].)

Many papers have been published on work related to the use of roughened fuel rods in gas-cooled nuclear reactors, e.g., Dalle-Donne and Meyer [1977].

There are many possible roughness geometries. Figure 9.2 is the author's attempt to catalog the possible geometries. This figure shows three basic roughness families. For any basic type, the key dimensionless variables are the dimensionless roughness height (e/d), the dimensionless roughness spacing (p/e), the dimensionless rib width (w/e), and the shape of the roughness element. The ridge-and-groove-type roughness may also be applied at a helix angle, as shown in Figure 9.2. For a specific roughness type, a family of geometrically similar roughnesses is possible simply by changing e/d, while maintaining constant p/e and w/e. Thus, the designer is faced with choosing among virtually thousands of possible specific roughness geometries and sizes. Figure 9.3 shows two of the early commercially used roughness geometries. Their geometric parameters are e (rib height), p (rib spacing), α (helix angle), and the rib shape. The tube shown in Figure 9.3a has internal helical ribs and is normally made with low integral fins on the external surface. The internal ribs are made by cold deformation of the metal into a grooved internal mandrel to form the helical ridges. The helically corrugated tube shown in Figure 9.3b is made by rolling a sharp-edged wheel on the outer surface of the tube.

Type of roughness or internal surface:	Three-dimensional roughness ("uniform roughness")	Ridge-type two-dimensional roughness ("repeated ribs")	Groove-type two-dimensional roughness
Basic geometry:			
Variations of basic geometry by change of p/e (shows probable range of p/e of interest):			
Possible "element shapes" for basic geometry:			

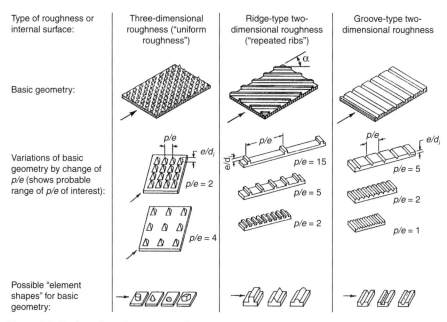

Figure 9.2 Catalog of roughness geometries.

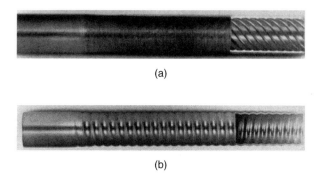

(a)

(b)

Figure 9.3 Illustration of commercially used enhanced tubes: (a) helical rib Turbo-Chil™ tube, (b) corrugated Korodense™ tube. (Courtesy of Wolverine Tube, Decatur, AL.)

9.2 ROUGHNESS WITH LAMINAR FLOW

9.2.1 Laminar Flow in Roughened Circular Tubes

Laminar flow in a tube is influenced by several factors: thermal boundary condition, entrance region effects, and natural convection. The fundamentals are discussed in Section 7.2 and the effect of such variables is addressed in Chapter 8 for laminar flow in internally finned tubes. Vicente et al. [2002a] obtained laminar flow data for ten 16-mm ID helically dimpled tubes using water and glycol for $10^{-4} \le x^+$ $[= x/(d_i\mathrm{RePr})] \le 10^{-2}$, $10^6 \le \mathrm{Ra} \le 10^8$. Figure 9.21 (later in the chapter) shows a drawing of their dimpled tube geometry. The geometric details of the tubes are provided in Table 9.1. Heat was added to the test section by supplying an electric current to the tube wall. A hydrodynamic entrance section preceded the heat transfer section to establish a fully developed velocity profile at the inlet of the heat transfer section. The measured local heat transfer coefficients are thus for fully developed flow with a constant heat flux boundary condition.

Figure 9.4 shows the friction data of tubes 01 to 05; a similar trend was obtained for the other 5 tubes. Figure 9.4 shows that the laminar–turbulent transition Reynolds number is between 1400 and 2000 and it decreases as e/d_i increases. Both the laminar and turbulent friction factors increase as e/d_i increases. The following equations are proposed for the transition Reynolds number and for the friction factor in the laminar region.

Table 9.1 Roughness Dimension of Dimpled Tubes Tested by Vicente et al. [2002a]

Tube No.	d_i (mm)	e (mm)	p (mm)	z (mm)	e/d_i	d_i^2/pz
01	16.0	1.33	13.0	8.85	0.0831	2.225
02	16.0	1.58	13.1	8.99	0.0988	2.175
03	16.0	1.91	13.8	8.89	0.1194	2.085
04	16.0	1.28	14.6	8.91	0.0800	1.975
05	16.0	1.83	14.5	9.02	0.1144	1.964
06	16.0	1.59	17.2	9.02	0.0944	1.652
07	16.0	1.84	16.6	9.06	0.1150	1.700
08	16.0	1.87	16.8	8.90	0.1169	1.709
09	16.0	1.55	10.9	8.90	0.0969	2.646
10	16.0	1.64	11.4	8.76	0.1025	2.575

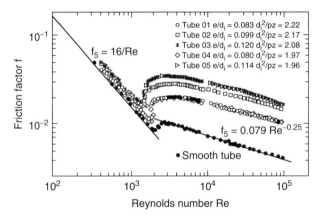

Figure 9.4 Friction factor vs. Reynolds number for five dimpled tubes. (From Vicente et al. [2002a]. Reprinted with permission from Elsevier.)

$$\text{Re}_{\text{crit}} = 2100\left[1 + 7.9 \times 10^7 \, (e/d_i)^{-6.54}\right]^{-0.1} \tag{9.1}$$

$$f = f_s \left[1 + 123.2 \, (e/d_i)^{2.2} \, \text{Re}_d^{0.2}\right] \tag{9.2}$$

In Equation 9.2, f_s is the smooth tube friction factor [= 16/Re]. In the laminar region, Equation 9.2 yields f/f_s = 1.1 to 1.3. Figure 9.4 also shows the friction factor in the turbulent region, where it is two to four times higher than for a smooth tube. The turbulent characteristics of these tubes are discussed in Section 9.6.1.

Vicente et al. [2002a] obtained heat transfer data for three dimpled tubes having fixed dimple density $d_i^2/pz \cong 2.1$ with three different values of e/d_i (= 0.083 to 0.119). These are tubes 01, 02, and 03 in Table 9.1. The heat transfer data for tube 02 are shown in Figure 9.5 in terms of Nu vs. x^+ for different Rayleigh numbers. The dimpled tube behaved similar to that of a smooth tube (compare Figure 7.6). The Nusselt numbers are highly dependent on the Rayleigh number. The Nusselt number of the dimpled tubes were up to 30% higher than that of the smooth tube, and the difference increases as the Rayleigh number increases. They report a negligible difference of Nusselt numbers between the three dimpled tubes. Shome and Jensen [1996] also report heat transfer enhancement by natural convection in internally finned tubes. The fully developed Nusselt numbers of the dimpled tubes were correlated as

$$\text{Nu}_d = 4.36[1 + \text{Ra}/67,000]^{0.24} \tag{9.3}$$

Esen et al. [1994a] measured heat transfer and friction for airflow (Pr = 0.7) in 23 commercial enhanced tubes with helical ridges for $500 \leq \text{Re}_d \leq 60,000$. These tubes were supplied by General Atomics, Hitachi Cable, Wieland-Werke, and Yorkshire Alloys. The same tubes were also tested by Obot et al. [2001a] at Pr = 6.8

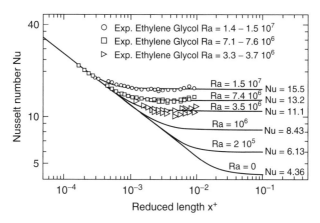

Figure 9.5 Nusselt number vs. x^+ for Tube 02 in Table 9.1. (From Vicente et al. [2002a]. Reprinted with permission from Elsevier.)

and 24.8. A key importance of these data sets is that they also provide data within the laminar Reynolds number range, as well as a wide Prandtl number range. The test section was preceded by a hydraulic entrance section, and the test section was electrically heated. The heat transfer coefficient reported is that averaged over the heated tube length. The L/d_i varied from 9.5 to 39, with the smaller L/d_i for the largest-diameter tubes. Hence, for the laminar flow data, the reported data are within the thermal entrance region.

Such laminar flow data are likely influenced by natural convection effects, but this effect was not quantified. Esen et al. [1994b] and Obot et al. [2001b] sought to correlate the Esen et al. [1994a] and Obot et al. [2001a] laminar flow data using an empirical normalization technique. The enhanced tube data are normalized using the transition Re and f at the transition point (f_m). They plot the normalized friction factor (f_m) vs. the normalized Reynolds number (Re$_m$), which are defined as follows:

$$\mathrm{Re}_m = (\mathrm{Re}_{c,s} / \mathrm{Re}_c)\mathrm{Re} \qquad (9.4)$$

$$f_m = (f_{c,s}/f_c)f \qquad (9.5)$$

$$\mathrm{Nu}_m = (\mathrm{Nu}_{c,s} / \mathrm{Nu}_c)\mathrm{Nu} \qquad (9.6)$$

where subscript c,s refers to the critical transition value for the smooth tube and subscript c is the transition value for the enhanced tube. The smooth tube transition value was obtained from separate experiments ($f_{c,s} = 0.0093$, Nu$_{c,s} = 6.1$, and Re$_{c,s} = 2093$).

The reduced f_m and Nu$_m$ are shown in Figure 9.6. The heat transfer and pressure drop data in the laminar regime are well correlated by the similarity parameters Re$_m$, f_m, and Nu$_m$. Figure 9.6a shows that, when reduced using Equations 9.4 through 9.6, f_m in the laminar region is almost independent of the geometric details of the roughness elements. All the laminar flow friction data are collapsed into a single line ($f = 19.5/\mathrm{Re}_m$) with ±15% scatter. This correlation compares to the fully developed plain tube value of $f = 16/\mathrm{Re}_s$. Their laminar flow heat transfer data were correlated within ±10% by the empirical equation.

$$\mathrm{Nu}_d = 0.008f\,\mathrm{Re}_d^{1.5}\,\mathrm{Pr}^{0.4} \qquad (9.7)$$

They also attempted to provide a similar correlation for the turbulent region, which is discussed in Section 9.7.2.

9.2.2 Laminar Flow in Roughened Flat Tubes

Integral roughness has been used to enhance the tube-side heat transfer coefficient in "low-flow" automotive radiators, for which the flow Reynolds number is small and may be less than 2000. Farrell et al. [1991] tested seven flat radiator tubes (one smooth, one full ribbed, two broken ribbed, and three dimpled). The friction factors

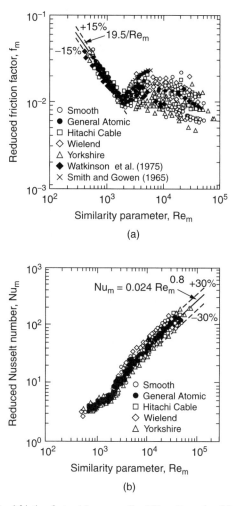

Figure 9.6 The normalized friction factor (a) or normalized Nusselt number (b) vs. normalized Reynolds number for commercial enhanced tubes as reported by Esen et al. [1994b] and Obot et al. [2001b]. (From Esen et al. [1994b].)

were obtained for $200 \leq Re_{Dh} \leq 11{,}000$. The heat transfer coefficients, however, were obtained for the Reynolds numbers larger than 2000. The broken-ribbed tube, which had the highest e/d_i, yielded the largest heat transfer coefficient as well as the highest friction factor.

Olsson and Sunden [1996] tested ten radiator tubes with airflow. The geometries include one smooth, two ribbed, five dimpled, and two tubes having an offset-strip internal fin insert that is soldered to the tube walls. Figure 9.7 shows photos of the outside surfaces and the geometric details are listed in Table 9.2. The air heat transfer data were taken with constant wall temperature and the data provide the average heat transfer coefficient over the tube length. Figure 9.8a shows the friction factor

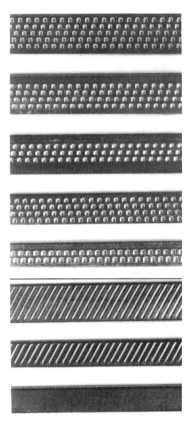

Figure 9.7 Outside surface geometry of the radiator tubes tested by Olsson and Sunden [1996]. From the top d5, d4, d3, d2, r2, r1, and s. (From Olsson and Sunden [1996]. Reprinted with permission from Elsevier.)

Table 9.2 Geometric Dimensions of the Radiator Tubes Tested by Olsson and Sunden [1996]

Tube	D_h (mm)	p (mm)	e (mm)	H (mm)	θ (degree)	W (mm)
Smooth (s)	3.11	—	—	1.74	—	14.5
Rib-roughened (r1)	3.11	4.0	0.18	1.74	30	14.6
Rib-roughened (r2)	3.05	4.0	0.20	1.64	30	21.6
Dimpled (d1)	3.53	4.5	0.45	2.00	30	15.2
Dimpled (d2)	2.81	4.8	0.45	1.52	18	18.3
Dimpled (d3)	3.08	4.8	0.45	1.65	18	23.2
Dimpled (d4)	3.08	4.8	0.45	1.65	18	23.2
Dimpled (d5)	3.08	4.8	0.45	1.65	18	23.2
OSF	D_h (mm)	s (mm)	b (mm)	L_p (mm)	H (mm)	W (mm)
(osf1)	2.82	3.5	2.6	5.0	3.0	26.2
(osf2)	2.82	3.5	2.6	5.0	3.1	43.4

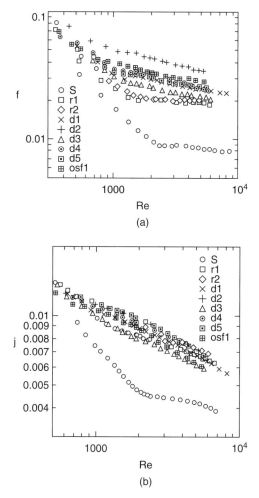

Figure 9.8 Friction factor (a) and j factor (b) vs. Reynolds number for the radiator tubes tested by Olsson and Sunden [1996]. (From Olsson and Sunden [1996]. Reprinted with permission from Elsevier.)

of the tubes. The enhanced tubes show higher friction factors than the smooth tube in both the laminar and turbulent regions. However, as the Reynolds number decreases deep into the laminar region, the friction factors tend to converge and approach the smooth tube value. Also note that the laminar–turbulent transition Reynolds number decreases as the friction factor increases. The d2 tube, which has the largest e/H (= 0.3), yields the highest friction factor. The ribbed tubes (r1 and r2), which have the smallest values of e/H, yield the lowest friction factor. The corresponding j factors are shown in Figure 9.8b. Similar to the friction behavior, the j factors tend to converge at low Reynolds numbers, and approach the smooth tube value. The j factors, however, do not show a clear laminar–turbulent transition as do the friction factors. High values of the j factors are obtained for dimpled tubes

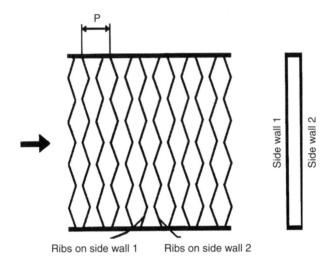

Figure 9.9 Multiple V-ribbed channel tested by Olsson and Sunden [1998b]. (From Olsson and Sunden [1998b]. Reprinted with permission of ASME.)

d2 and d5. The volume goodness comparison (see Equations 3.17) revealed that the highest goodness factors were obtained for ribbed tubes r1 and r2, followed by dimpled and offset strip finned (OSF) tubes.

Olsson and Sunden [1998a] measured the pressure drop and heat transfer in rib-roughened rectangular channels ($\frac{1}{8}$ aspect ratio) having cross ribs, parallel ribs, cross V ribs, parallel V ribs, and multiple V ribs. The ribs were made by plastic tubes of 1.5 mm diameter. The rib pitches were fixed at 32 mm, and the channel height was 14.5 mm. The test was performed in the low-Reynolds-number range (from 500 to 15,000) with possible application to heat exchangers, such as automotive radiators. The f and j factor behavior was similar to that shown in Figure 9.8. Both the f and j factors tend to converge and approach the smooth tube value at low Reynolds numbers. Significant increase of the j factor (also the f factor) was obtained for the multiple-V-rib geometry.

Olsson and Sunden [1998b] investigated the effect of rib configuration for the multiple-V-ribbed channel. Figure 9.9 shows a sketch of the channel with multiple V ribs. Ten different configurations with $e/H = 0.1$ to 0.2, $\alpha = 45°$ to 75°, p/H = 3 to 7 were tested. In the laminar region, the slope of the j factor curves is close to -0.5. If the flow is fully developed and natural convection effects are negligible, the slope of the j factor should be -1.0 for laminar flow. The L/D_h of Olsson and Sunden's test section was 56.2. The -0.5 slope of j vs. Re curve was also reported by Esen et al. [1994a] for commercial rough tubes. The slope of the f factor is close to -0.5. The friction data of the radiator tubes shown in Figure 9.8a also show approximately similar values. However, Esen et al. [1994a] (Figure 9.6a) and Vicente et al. [2001] (Figure 9.4) showed that the friction factor slope of the enhanced tubes is approximately -1.0, the same value as that of the smooth tube.

9.3 HEAT-MOMENTUM TRANSFER ANALOGY CORRELATION

Correlations are needed to predict the turbulent friction and heat transfer characteristics of rough surfaces. Rationally based correlations for friction and heat transfer have been developed for geometrically similar roughness. Consider a two-dimensional roughness, such as illustrated by Figure 9.3a. A geometrically similar family of roughness geometries will exist if p/e, w/e, α, and the rib shape are held constant. Such a family of roughened tubes would differ in their e/d_i values. The friction and heat transfer correlations described in this section are believed to apply to any arbitrary family of geometrically similar roughness.

9.3.1 Friction Similarity Law

A friction similarity law was developed by Nikuradse [1933] and described by Schlicting [1979]. Nikuradse showed the similarity law to be valid for closely packed sand-grain roughness. The velocity profile, valid for $y/e \geq 1$, is given by

$$\frac{u}{u*} = 2.5\ln\frac{y}{e} + B(e^+) \qquad (9.8)$$

The velocity profile is based on the "law of the wall" velocity distribution for rough surfaces, and is discussed by Hinze [1975] and Schlicting [1979]. The law of the wall for rough surfaces states that the velocity profile for $y/e \geq 1$ is independent of the duct shape, and depends on the "roughness Reynolds number," e^+. The e^+ may be written in the following two equivalent forms:

$$e^+ = \frac{eu*}{v} = \frac{e}{d_i}\mathrm{Re}_d\sqrt{\frac{f}{2}} \qquad (9.9)$$

where $u*$ is the friction velocity $(\tau_o\rho)^{1/2}$. The term $B(e^+)$ is the dimensionless velocity at the tip of the roughness elements $(y/e = 1)$, and τ_o is the average apparent wall shear stress. For flow in a tube, τ_o is given by the force balance

$$\tau_o = -\frac{d_i}{4\rho}\frac{dp}{dx} \qquad (9.10)$$

The friction factor is obtained by integrating the velocity profile over the flow area. For a circular pipe, the integration gives

$$\frac{\bar{u}}{u*} = \left(\frac{2}{f}\right)^{1/2} = -2.5\ln\left(\frac{2e}{d_i}\right) - 3.75 + B(e^+) \qquad (9.11)$$

Nikuradse's [1933] data for six different e/d_i values (Figure 9.10a) were correlated using Equation 9.11 as shown in Figure 9.10b. Figure 9.10a shows that the friction factor approaches (and equals) the smooth tube value in the laminar regime.

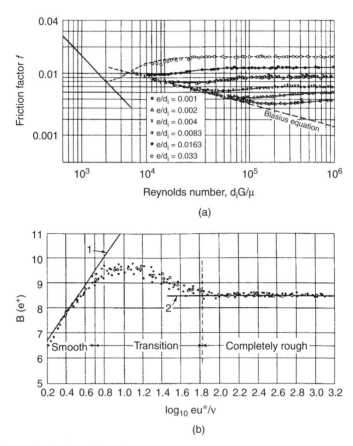

Figure 9.10 (a) Friction factors for artificially roughened tubes, as measured by Nikuradse. (b) Roughness parameter B for Nikuradse's sand roughness. Curve 1, hydraulically smooth; curve 2, completely rough. (From Schlicting [1979].)

In the turbulent regime, the smaller e/d_i data remain on the smooth tube curve to a higher Reynolds number and then drift above the smooth tube line. The friction factors all attain an asymptotic, constant value, which is termed the "fully rough" condition. As shown by Figure 9.10b, the $B(e^+) = 8.48$ for the fully rough condition, which is attained at $\log_{10} e^+ = 1.82$, or $e^+ = 70$.

Equations 9.8 and 9.11 should hold for any family of geometrically similar roughness. But, the function $B(e^+)$ will be different for different basic roughness types.

9.3.2 PEC Example 9.1

Water flows at 2.4 m/s in a 15.9-mm-diameter tube at 27°C having closely packed sand grain roughness. Determine the roughness size (e) required to attain the fully rough condition. Then, calculate f/f_s.

Figure 9.10b shows that the fully rough condition is attained at $\log_{10} e^+ = 1.8$, or $e^+ = 70$. At the specified flow condition ($v = 0.857E\text{-}6$ m^2/s), the $\text{Re}_d = 44,530$. Use Equation 9.9 with $(f/2)^{\frac{1}{2}}$ from Equation 9.11 to solve for e/d_i. Thus, one iteratively solves for $e/d_i = 0.020$ using

$$70 = \frac{e}{d_i}\text{Re}_d\sqrt{\frac{f}{2}} = \frac{(e/d_i)44,530}{8.48 - 2.5\ln\left(\dfrac{2e}{d_i}\right) - 3.75} \tag{9.12}$$

With e/d_i known, $e = 0.020 \times 15.9 = 0.31$ mm. Use Equation 9.9 to calculate the friction factor, $f = 0.01227$. Using the Blasius friction factor for a smooth tube, obtain $f_s = 0.00544$. Hence, $f/f_s = 2.25$.

9.3.3 Heat Transfer Similarity Law

Dipprey and Sabersky [1963] developed a rational correlation based on the heat-momentum transfer analogy for rough surfaces. The model is applicable to any type of geometrically similar surface roughness. The correlating function is obtained by combining the momentum and energy equations for turbulent flow and integrating across the boundary layer thickness. This section briefly describes the basis of the model applied to boundary layer flow over a plate for both smooth and rough surfaces. Including the analogy for smooth surfaces allows a better understanding of the concepts applied for rough surfaces.

Smooth Surfaces. The analogy equation for smooth surfaces is developed in detail by Webb [1971]. Written in dimensionless form, these equations for flow over smooth or rough surfaces are

$$\frac{\tau}{\tau_o} = \frac{v_e}{v}\frac{du^+}{dy^+} \tag{9.13}$$

$$\frac{q}{q_o} = \frac{v_e}{v}\frac{1}{\text{Pr}_e}\frac{dT^+}{dy^+} \tag{9.14}$$

Eliminating the effective viscosity (v_e) from Equations 9.13 and 9.14 to combine the equations, and integrating over the boundary layer thickness (δ^+), gives

$$T_w^+ - T_\infty^+ = \int_0^{\delta^+} \text{Pr}_e \frac{du^+}{dy^+} dy^+ \tag{9.15}$$

The assumptions listed in Table 9.3 are made for the integration of Equation 9.15. For convenience, the following two identities are used:

$$T_w^+ - T_\infty^+ = \frac{(f/2)^{1/2}}{\text{St}}, \quad u_\delta^+ = \frac{u_\delta}{u^*} = \left(\frac{2}{f}\right)^{1/2} \tag{9.16}$$

Table 9.3 Assumptions for Integration of Equation 9.7

Assumption	Smooth	Rough
$\epsilon_m/\nu \gg 1, \epsilon_h/\alpha \gg 1$	$y > y_b$	Same
$Pr_e = 1$	$y > y_b$	Same
$(q/q_w)(\tau_e/\tau) = 1$	$y > 0$	$y > e$

Separate integrals are written for the viscous- and turbulence-dominated regions, and the term du^+/dy^+ integrated over $0 \leq y^+ \leq y_b^+$ is added and subtracted to make use of the second identity of Equation 9.16. The integration for a smooth surface gives

$$\frac{(f_s/2)^{1/2}}{St_s} = \int_0^{y_b^+} (Pr_e - 1)\frac{du^+}{dy^+}dy^+ + \int_0^{\delta^+} \frac{du^+}{dy^+}dy^+ \tag{9.17}$$

Using the second identity of Equation 9.9 for the second integral allows writing Equation 9.17 as

$$\frac{f_s/(2St_s) - 1}{(f_s/2)^{1/2}} = \int_0^{y_b^+} (Pr_e - 1)\frac{du^+}{dy^+}dy^+ \tag{9.18}$$

The integral of Equation 9.18 is a function only of the molecular Prandtl number, which may be written as $F(Pr)$. The $F(Pr)$ is obtained by plotting data in the form of the left-hand side of Equation 9.18 vs. Prandtl number. Petukhov [1970] has shown that the Prandtl number function is given by

$$F(Pr) = 12.7(Pr^{2/3} - 1) \tag{9.19}$$

Combining Equations 9.18 and 9.19 gives the heat-momentum analogy relation for smooth surfaces:

$$\frac{f_s/(2St_s) - 1}{(f_s/2)^{1/2}} = 12.7(Pr^{2/3} - 1) \tag{9.20}$$

Solving Equation 9.20 for the smooth surface Stanton number gives

$$St_s = \frac{f_s/2}{1.0 + 12.7(f_s/2)^{1/2}(Pr^{2/3} - 1)} \tag{9.21}$$

For flow in tubes, the first term in the denominator (1.0) should be replaced by 1.07. Webb [1971] compares the ability of Equation 9.21 to predict heat transfer data in smooth tubes. He shows that Equation 9.21 (using the 1.07 term) predicts the smooth-tube, constant-property data of investigators with $0.7 \leq Pr \leq 75$ within $\pm 8\%$. Petukhov [1970] recommends that the smooth tube friction factor be predicted by

$$f_s = (1.58 \ln Re_d - 3.28)^{-2} \tag{9.22}$$

Rough Surfaces. Development of the Dipprey and Sabersky [1963] analogy model for rough surfaces parallels that of the smooth surface model, and uses the approximations listed in Table 9.3. For $y_b^+ \le e^+$, the integration of Equation 9.17 results in

$$\frac{f/(2\text{St})-1}{(f/2)^{1/2}} + B(e^+) = \int_0^{e^+} (\text{Pr}_e - 1)\frac{du^+}{dy^+}dy^+ \qquad (9.23)$$

As compared with the smooth surface analogy (Equation 9.18), the integral is a function of both e^+ and Pr. The functional dependence is written as $g(e^+, \text{Pr})$, and it is assumed this function can be written as the product of $\bar{g}(e^+)\,\text{Pr}^n$. Then, the heat transfer correlation is

$$\bar{g}(e^+)\text{Pr}^n = \frac{f/(2\text{St})-1}{\sqrt{f/2}} + B(e^+) \qquad (9.24)$$

Figure 9.11 shows Dipprey and Sabersky's [1963] correlation of sand-grain data ($0.0024 \le e/d_i \le 0.049$ and $1.2 \le \text{Pr} \le 5.94$). The B(e$^+$) term is obtained from Figure 9.10b. The Prandtl number dependence, $\text{Pr}^{-0.44}$, is obtained by cross plotting the $g(e^+, \text{Pr})$ vs. e^+ data for each of the three Prandtl numbers. Solving Equation 9.24 for St gives

$$\text{St} = \frac{f/2}{1 + \sqrt{f/2}\left[\bar{g}(e^+)\text{Pr}^n - B(e^+)\right]} \qquad (9.25)$$

where $\text{Pr}^n = \text{Pr}^{0.44}$. One calculates the friction factor and Stanton number as follows:

1. Select e^+ and obtain $B(e^+)$ from Figure 9.10b and $g(e^+)$ from Figure 9.11.
2. Calculate the friction factor using Equation 9.11.
3. Calculate the Stanton number using Equation 9.25.

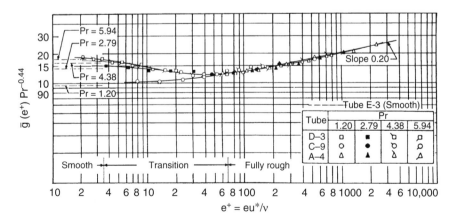

Figure 9.11 Heat transfer correlation of Dipprey and Sabersky [1963] sand-grain roughness data. (From Dipprey and Sabersky [1963].)

9.4 TWO-DIMENSIONAL ROUGHNESS

Substantial work has been done on two-dimensional rib roughness. Figure 9.12 shows five variants of two-dimensional roughness. Figure 9.12a and b shows transverse ribs, which are normal to the flow. Figure 9.12c through e shows helical rib roughness, with helix angles less than 90°. The roughness elements may be integral to the base surface (Figure 9.12c and d), or they may be in the form of wire coil inserts (Figure 9.12e). A wire coil insert is an example of attached helical rib roughness. The integral helical rib roughness may be made as single- or multistart elements, whereas a wire coil insert is a single-start roughness. The element axial pitch (p) may be made with a smaller, dimensionless axial spacing (p/e) than is possible with wire coil inserts. The relation between the axial pitch of the roughness elements (p), the number of starts (n_s), and the helix angle, measured from the tube axis α, is

$$p = \frac{\pi d_i}{n_s \tan \alpha} \tag{9.26}$$

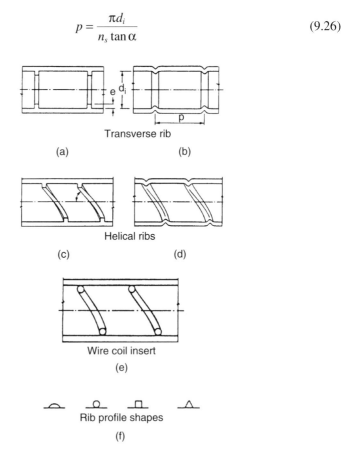

Transverse rib

(a)   (b)

Helical ribs

(c)   (d)

Wire coil insert

(e)

Rib profile shapes

(f)

Figure 9.12 Illustrations of different methods of making two-dimensional roughness in a tube: (a) integral transverse rib, (b) corrugated transverse rib, (c) integral helical rib, (d) helically corrugated, (e) wire coil insert, (f) different possible profile shapes of roughness. (From Ravigururajan and Bergles [1985].)

The shape of the roughness element in a corrugated tube is also different from that of a helical integral roughness or a wire coil insert. The geometric factors that influence the j and f factors are e/d_i, p/e, α or $(e/d_i, n_s, \alpha)$ and the shape of the roughness element. Each of the variants of repeated rib roughness is discussed in separate sections.

9.4.1 Transverse Rib Roughness

Figure 9.13 shows the flow patterns cataloged by Webb et al. [1971] for transverse-rib roughness ($\alpha = 90°$) as a function of the dimensionless rib spacing (p/e). The flow separates at the rib and reattaches six to eight rib heights downstream from the rib. Measurements of Edwards and Sheriff [1961] for a two-dimensional rib on a flat plate show that the heat transfer coefficient attains its maximum value near the reattachment point. The Edwards and Sheriff [1961] data are shown in Figure 7.17.

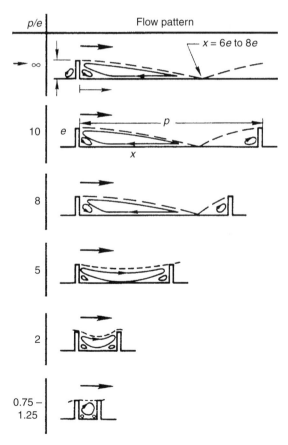

Figure 9.13 Catalog of flow patterns over transverse rib roughness as a function of rib spacing. (From Webb et al. [1971].)

Figure 9.13 shows that reattachment does not occur for $p/e < 8$. The highest average coefficient occurs for $10 \le p/e \le 15$.

The heat transfer correlation of Section 9.3.3 should apply to any family of geometrically similar roughness. Equations 9.11 and 9.24 may be used to correlate the data for any family of geometrically similar roughnesses. However, the $B(e^+)$ and $g(e^+)$ functions may be different for different roughness families. For any geometrically similar roughness type, the data for different e/d_i will fall on the same curve when plotted in the form $\bar{g}(e^+)\,Pr^n$ vs. e^+.

Webb et al. [1971] applied the model to geometrically similar, $p/e = 10$, transverse-rib roughness ($0.01 \le e/d_i \le 0.04$ and $0.7 \le Pr \le 38$). Figure 9.14 shows the $\bar{g}(e^+)$ function determined by Webb et al. [1971], and Figure 9.15 shows the $B(e^+)$ correlation. The figure also contains data for $e/d_i = 0.02$ with $p/e = 20$ and 40, which are not geometrically similar to the $p/e = 10$ data. Figure 9.14 shows that Equation 9.24 does an excellent job of correlating the three e/d_i values (0.01, 0.02, and 0.04) for $p/e = 10$. The figure also shows that the nonsimilar $p/e = 20$ and 40 data also fall on the same correlating line as the $p/e = 10$ data. There is no rational reason to expect this, so it is regarded as a fortuitous occurrence. The data are correlated by a Prandtl number exponent of $n = 0.57$ (compare Equation 9.25) as compared to $n = 0.44$ found by Dipprey and Sabersky [1963] for sand-grain roughness. Further commentary on the Prandtl number dependency is given later. The $B(e^+)$ function for the transverse rib roughness of Webb et al. [1971] is shown in Figure 9.15.

Webb et al. [1972] investigated the effect of the p/e and rib cross-sectional shape on the correlation of transverse-rib data taken by other investigators. They show that the $g(e^+)$ function for different p/e and rib shape reasonably agrees with that of Figure 9.14. However, the $B(e^+)$ function is sensitive to p/e and rib shape.

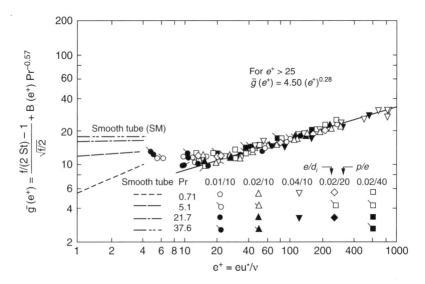

Figure 9.14 Heat transfer correlation for transverse-rib roughness ($\alpha = 90°$). (From Webb et al. [1971].)

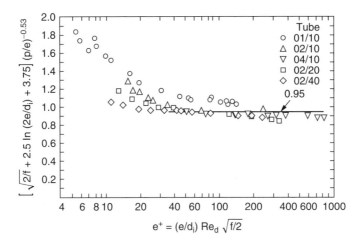

Figure 9.15 Final friction correlation for repeated-rib tubes. (From Webb et al. [1971].)

Almeida and Souza-Mendes [1992] used the naphthalene sublimation technique to measure the length of the thermal entrance region, and the local distribution of the mass transfer coefficient between ribs. They showed that the mass transfer coefficients are fully developed in three tube diameters for $e/d_i = 0.01$.

Hijikata et al. [1987] investigated the effect of the rib shape on performance. Their work was done for transverse corrugations in the wall of a parallel plate channel having $e/D_h = 0.01$ and $p/e = 15$. They investigated arc, sine, and square rib shapes and concluded that the arc shape provides the same heat transfer coefficient as the other shapes, but has a lower pressure drop. The base width of the arc-shaped element was four times its height. Ribs are made by re-forming the mother material, and the resultant shape has round edges. Taslim and Spring [1994] tested ribbed channels with round edges and filleted base ribs, and compared the results with those from sharp-edged ribs. Two rib aspect ratios (1 and 2) were tested. The rounding decreased both the Nusselt number and the friction factor, with the rounding effect more prominent for the high-aspect-ratio ribs. Liou and Hwang [1993] and Chandra et al. [1997] provide additional information on the effect of rib shape.

Tanasawa et al. [1983] tested a perforated form of transverse-rib roughness illustrated in Figure 9.16b and c. The tests were performed with airflow in a rectangular cross section channel. The best performance was provided by the Figure 9.16c geometry, with 3.0-mm-diameter perforations in the lower half of the 9-mm-high rib ($e/D_h = 0.045$), which had 10% open area in the rib. This geometry provided only 3% higher St than the Figure 9.16a rib, but the friction factor was only 80% that of the unperforated rib of the same height. Nearly the same performance was obtained with 1.0-mm-diameter perforations and 10% open area. The friction is smaller because of reduced profile drag. Following Tanasawa et al.'s study, Hwang and Liou [1994] investigated the effect of the rib porosity ($0 \le \epsilon \le 0.5$) of the hole-type perforated rib geometry. The test section was rectangular (40 by 160 mm), and the two long walls were roughened. Laser holographic interferometry was employed to measure the local heat transfer coefficients. For $10{,}000 \leq Re_{Dh} \leq 50{,}000$, the friction factor decreased with increasing rib porosity. However, the maximum Nusselt number was obtained for $\in\ = 0.44$. Hwang and Liou [1995] investigated the effect of rib pitch-to-height ratio ($5 \leq p/e \leq 20$). The test section had the same dimensions as Hwang and Liou [1994]. The hole-type perforated ribs had 50% open-area ratio. For $10{,}000 \leq Re_{Dh} \leq 50{,}000$, the perforated ribbed geometry yielded higher Nusselt number than the solid ribbed geometry; 10 to 40% higher for $p/e = 10$ and 60 to 80% higher for $p/e = 5$. The large increase at $p/e = 5$ was caused by modification of the recirculating flow (shown in Figure 9.13) resulting from the flow through the perforations. The friction factor of the perforated rib geometry was approximately 50% lower than that of the solid rib configuration. Hwang [1998] investigated the effect of rib porosity ($\in\ = 19$, 32, and 48%) of the slit-type rib geometry. The slits were formed horizontally. The test section was rectangular (60×120 mm), and two long walls were roughened with ribs of $e/D_h = 0.167$ and $p/e = 10$. The heat transfer coefficient increased and the friction factor decreased as the rib porosity increased, yielding the best performance for $\in\ = 48\%$. A numerical study by Liou et al. [2002] on the slit-rib geometry generally confirmed the experimental findings. Liou et al. [2002] compared the experimental results on perforated ribs by several investigators (Hwang and Liou [1995, 1994], Tanasawa et al. [1983], Hwang [1998]). They concluded that hole-type perforated ribs provide better thermal performance than the slit-type ribs for the same pumping power (VG-2a of Table 3.1).

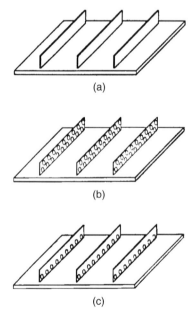

(a)

(b)

(c)

Figure 9.16 Transverse rib promoters investigated by Tanasawa et al. [1983].

9.4.2 Integral Helical-Rib Roughness

How does the helix angle (α) affect the St and f characteristics of repeated rib roughness? Figure 9.17 shows the St and f data of Gee and Webb [1980] for $p/e = 15$ and $e/d_i = 0.01$ as a function of helix angle. As α increases, the friction factor drops faster than does the Stanton number. Gee and Webb found that the maximum St/f occurs at $\alpha = 45°$. They give correlations for $\bar{g}(e^+)$ and $B(e^+)$ vs. e^+. Nakayama et al. [1983] provide additional data for helical ribs, including $0° \leq \alpha \leq 80°$. Han et al. [1978] provide transverse-rib data, and $\bar{g}(e^+)$ and $B(e^+)$ correlations for 17 transverse-rib geometries ($5 \leq p/e \leq 20$) and two helical rib ($\alpha = 40°, 55°$) geometries. An error in their 1978 data reduction is corrected by Han et al. [1979]. The Figure 9.3a helically ribbed tubes is commercially available, and is known as the Turbo-Chil™ tube. The Turbo-Chil tube has ten rib starts, $\alpha = 47°$, $p/e = 11.1$, and $e/d_i = 0.0264$. The $\bar{g}(e^+)$ and $B(e^+)$ functions for this tube are given by Equations 9.27 and 9.28, respectively, which were developed by the author, and are valid for $e^+ > 25$.

$$\bar{g}(e^+) = 7.68(e^+)^{0.136} \tag{9.27}$$

$$B(e^+) = 1.7 + 2.06 \ln e^+ \tag{9.28}$$

Equations 9.27 and 9.28 are used in Equation 9.25 (with $n = 0.57$) to predict the Stanton number of the Turbo-Chil tube. The friction factor is calculated using Equation 9.28 in Equation 9.11. This correlation may be used to calculate St and f for any value of e/d_i, providing geometric similarity is maintained ($p/e = 11.1$, $\alpha = 47°$, and the same rib shape). Withers [1980b] provides additional information on helical-rib roughness for different values of p/e and α.

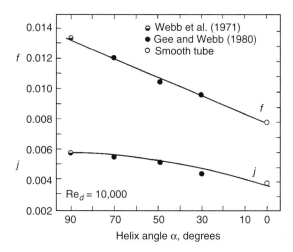

Figure 9.17 Effect of helix angle on f and St for $e/d_i = 0.01$, $p/e = 15$. (From Gee and Webb [1980].)

The Turbo-Chil tube with 10 starts was replaced in 1988 by the Turbo-C™ tube with 30 starts, and followed by the Turbo-CII™ tube in 1995 with 38 starts. These tubes are shown in Figure 9. 19 (later in the chapter). The basic geometric variables of the helically ribbed tube are the rib helix angle (α), the number of starts (n_s), the rib height (e), and the rib pitch (p), which are related as $p = \pi D/(n_s \tan \alpha)$. As the number of starts increases, the axial rib pitch decreases and the ribs provide considerable surface area increase. The 10-start Turbo-Chil has $p/e = 11$, and these ribs provide a surface area increase of only 17%. However, the surface area increase of the 38-start Turbo-CII tube is approximately 60%. For the 10-start tube having p/e =11, the heat transfer is enhanced by the boundary layer separation and reattachment, as illustrated in Figure 9.13. For the 38-start tube having p/e =3.73, however, one would not expect boundary layer reattachment to occur. Because of the large number of starts, the 38-start tube provides significant surface area increase, as does the internally finned tube whose end view is illustrated in Figure 8.2. Traditional internally finned tubes have larger e/d_i than the Figure 9.19a tube, and local flow separation is assumed not to occur if the helix angle is small (e.g., less than 15°). Hence, the Figure 9.19a tube having a large number of starts and a high helix angle cannot be described as either a classic rough tube or an internally finned tube. It probably provides some of the attributes of a rough tube (local flow separations) and some of the internally finned tube (large surface area increase). Webb et al. [2000] tested seven tubes of that category having 15.54-mm ID, 18 to 45 starts, 25° to 45° helix angles and 0.33 to 0.55-mm rib heights. The highest heat transfer enhancement ($h/h_p = 2.32$) and efficiency [$(h/h_p)/(f/f_p) = 1.18$] were obtained for the tube having $e = 0.33$ mm, 45 starts, and $\alpha = 45°$. The enhancement and the efficiency index of the 10-start tube ($e = 0.43$ mm, $\alpha = 45°$) were 1.74 and 0.95, respectively. The Ravigururajan and Bergles [1996] correlation (shown in Section 9.4.4) overpredicted the heat transfer coefficient by 10 to 45% and predicted the friction factors within ±30%. Based on the data, the following roughness functions (see Equations 9.11 and 9.24) were proposed by Webb et al. [2000], which predicted most of the heat transfer and friction data within ±10%.

$$B(e^+) = 4.762(e^+)^{0.214} n_s^{-0.110} \alpha^{-0.297} \tag{9.29}$$

$$\overline{g}(e^+) = 1.714(e^+)^{0.06} n_s^{-0.23} \alpha^{-0.16} \tag{9.30}$$

Heat transfer surface fouling is a concern in the advanced tube geometries. This issue and supporting data are discussed in Chapter 10.

9.4.3 Wire Coil Inserts

Although wire coil inserts were introduced in Section 7.5, their turbulent flow performance is discussed here because they may also be classified as a "wall-attached roughness." Their enhancement mechanism is the same as that of helical-rib roughness. A key difference between integral helical-rib roughness and wire

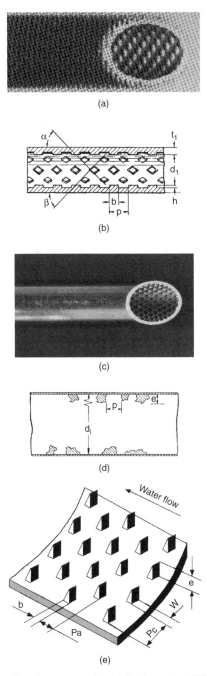

(a)

(b)

(c)

(d)

(e)

Figure 9.18 Three-dimensional roughness geometries: (a) Sumitomo Tred-19D™ tube of Sumitomo [1983], (b) "cross rifled" steel boiler tube of Nakamura and Tanaka [1973], (c) three-dimensional ribs of Cuangya et al. [1991], (d) attached particle roughness of Fenner and Ragi [1979], and (e) three-dimensional ribs by Liao et al. [2000]. (From Liao et al. [2000].)

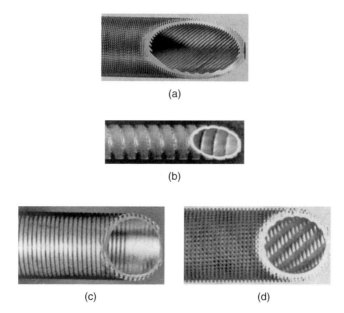

Figure 9.19 Photos of commercially used enhanced tubes. (a) Wolverine Turbo-C™. (Courtesy of Wolverine Tube Division.) (b) Wieland GEWA-TW™. (Courtesy of Wieland-Werke AG.) (c) Hitachi Thermoexcel-CC™. (Courtesy of Hitachi Cable, Ltd.) (d) Sumitomo Tred-26D™. (Courtesy of Sumitomo Light Metal Industries.)

coil inserts is that the wire coil insert is a single-start roughness, and typically has a larger wire pitch (and p/e) than the multistart helical-rib roughness.

The correlation of Sethumadhavan and Raja Rao [1983] is recommended for turbulent flow of liquids, because the data include a wide Prandtl number range ($5.2 \leq \mathrm{Pr} \leq 32$), and because they provide both heat transfer and friction correlations. The heat transfer correlation is probably valid for air ($\mathrm{Pr} = 0.7$). The researchers also measured St and f for a smooth tube, and found good agreement with the accepted correlations. They tested inserts having helix angles of 30°, 45°, 60°, and 75° for $e/d_i = 0.08$ and 0.12 for $4000 \leq \mathrm{Re}_d \leq 100{,}000$. For the same Pr, they found that the St for $e/d_i = 0.08$ was the same as for the $e/d_i = 0.12$ inserts. However, the larger wire size gives significantly higher friction factor for the same Re_d. The friction correlation, based on the friction similarity law (Equation 9.11), is

$$B(e^+) = 7.0(\tan\alpha)^{-0.18}(e^+)^{0.13} \qquad (9.31)$$

The heat transfer data were correlated using the Equation 9.24 heat-momentum transfer analogy model. The correlated results are given by

$$\bar{g}(e^+) = 8.6(\tan\alpha)^{-0.18}\mathrm{Pr}^{-0.55}(e^+)^{0.13} \qquad (9.32)$$

Equations 9.31 and 9.32 correlated data very well. They showed that Equation 9.32 closely predicts the water data of Kumar and Judd [1970]. Although the correlation does not explicitly include e/d_i, it should be applicable to smaller e/d_i values. In this author's opinion, the heat transfer correlation should also be valid for air (Pr $\simeq$ 0.7). Note that the Prandtl number exponent (0.55) agrees very closely with the 0.57 value determined by Webb et al. [1971] for transverse-rib roughness.

Sethumadhavan and Raja Rao [1983] provide a performance evaluation of the wire coil inserts using PEC FN-1 of Chapter 3. They evaluated the effect of helix angle for α = 30°, 45°, 60°, and 75° with e/d_i = 0.08. The highest (and equal) performance is provided by the α = 60° and 75° helix angles.

Two other correlations have also been developed for turbulent flow of water. The two correlations span significantly different ranges of p/d_i. Kumar and Judd [1970] generated electric heat in the wall of a stainless steel tube. The wire used for the inserts was coated with a plastic electrical insulator. The empirical heat transfer correlation was based on their test data for $0.108 \leq e/d_i \leq 0.15$, $1.12 \leq p/d_i \leq 5.5$, and $6000 \leq \mathrm{Re}_d \leq 100{,}000$. The relationship

$$\mathrm{Nu}_d = 0.175 \left(\frac{p}{d_i} \right)^{-0.35} \mathrm{Re}_d^{0.7} \, \mathrm{Pr}^{1/3} \tag{9.33}$$

correlated their data with 7.5% root mean square (rms) deviation. The correlation does not contain the parameter e/d_i. Apparently, they operated in the "fully rough" regime where the heat transfer coefficient is not a function of e/d_i. A dependence on e/d_i should become evident at some unknown lower e/d_i. Hence, care should be used in applying the correlation at e/d_i values less than the smallest they tested (0.108). A friction correlation was not developed.

Prasad and Brown [1988] subtracted the annulus-side resistance to obtain a correlation for their water data $0.058 \leq e/d_i \leq 0.11$ and $0.2 \leq p/d_i \leq 0.6$, and 40,000 $\leq \mathrm{Re}_d \leq 97{,}000$. The p/d_i the wire coils used was so small that a thin, axial wire was soldered to the coil to maintain the wire pitch. Refer to their paper for the heat transfer and friction correlations.

Zhang et al. [1991] measured the wall temperature in an air-to-air heat exchanger for their tests of coil inserts using round and rectangular wire shapes. Their circular wire data are for $0.037 \leq e/d_i \leq 0.09$ and $0.35 \leq p/d_i \leq 2.48$, and $6000 \leq \mathrm{Re}_d \leq 80{,}000$. The empirical correlations for the circular wire data are

$$\mathrm{Nu}_d = 0.253 \, \mathrm{Re}_d^{0.716} \left(\frac{e}{d_i} \right)^{0.372} \left(\frac{p}{d_i} \right)^{0.171} \tag{9.34}$$

$$f = 62.36 \ln \mathrm{Re}_d^{-2.78} \left(\frac{e}{d_i} \right)^{0.816} \left(\frac{p}{d_i} \right)^{-0.689} \quad 6 \leq \mathrm{Re}_a \times 10^{-3} < 15 \tag{9.35}$$

$$f = 5.153 (\ln \mathrm{Re}_d)^{-1.08} \left(\frac{e}{d_i} \right)^{0.796} \left(\frac{p}{d_i} \right)^{-0.707} \quad 15 \leq \mathrm{Re}_d \times 10^{-3} < 100 \tag{9.36}$$

They also give correlations for the rectangular wire coil inserts, whose performance is very close to the circular wire inserts.

9.4.4 Corrugated Tube Roughness

Substantial data have also been reported on single- and multistart corrugated tubes of the type illustrated in Figure 9.3b. The reported data are summarized in Table 9.4, and additional data by Newson and Hodgson [1973] are discussed in Section 12.3.5. The Newson and Hodgson work involves steam condensation on vertical, corrugated tubes with tube-side cooling water flow.

All authors except Sethumadhavan and Raja Rao [1986] and Li et al. [1982] used only single-start roughness. The Sethumadhavan and Raja Rao [1986] tubes were either 1, 2, 3, or 4 starts with $\alpha = 65°$. Four of the Li et al. geometries have 2, 3, or 4 starts. Li et al. provide data for helix angles as small as 45° using multistart roughness. Sethumadhavan and Raja Rao [1986] and Dong et al. [2001] investigated the Prandtl number effect using Newtonian fluids [water (Sethumadhavan and Raja Rao), water and oil (Dong et al.)]. Raja Rao [1988] used water and power law non-Newtonian fluids to obtain $5.1 \leq Pr \leq 82$.

The dimensionless geometric factors that affect the heat transfer coefficient and friction factor are p/e, e/d_i, and α. For a given pitch, $\tan \alpha \propto 1/n_s$, where n_s is the number of starts. All authors used the friction similarity law (Equation 9.11) and the analogy model (Equation 9.24) to correlate their data. Table 9.5 gives the correlations for the $\bar{g}(e^+)$ and $B(e^+)$ of several investigators. The variable e/d_i does not appear in the correlations, because the similarity correlations inherently include the effect of e/d_i.

The term n' in the Table 9.5 Raja Rao [1988] correlation is a "flow behavior power law index," which is 1.0 for water. The Prandtl number dependence of Raja Rao et al. ($n = 0.55$) agrees closely with that found by Webb et al. [1971] for transverse-rib roughness. Comparison of the Withers [1980a] and Mehta and Raja Rao [1988] correlations shows that the two $\bar{g}$ correlations have significantly different exponents for the effect of the helix angle (α). The heat transfer coefficient increases with increasing helix angle. Note that the Sethumadhavan and Raja Rao [1986]

Table 9.4 Data Range for Corrugated Tubes

Ref.	Fluid	No.	e/d_i	p/e	α
Withers [1980a]	Water	14	0.016–0.043	11–24	79–85
Mehta and Raja Rao [1988]	Water	11	0.008–0.097	4.5–49	69–84
Sethumadavan and Raja Rao [1986]	Water	5	0.012–0.030	13–45	65
	Glycerine				
Raja Rao [1988]	Water	12	0.020–0.060	4.6–30	65–82
	Power law				
Li et al. [1982]	Water	20	0.008–0.069	7.7–29	41–85
Dong et al. [2001]	Water and oil	4	0.020–0.040	14.5–31.2	79–82

Table 9.5 Correlations for g and $B(e^+)$

Ref.	$\bar{g}(e^+)\mathrm{Pr}^n$	$B(e^+)$
Withers [1988]	$4.95(\tan \alpha)^{0.33}(e^+)^{0.127}\mathrm{Pr}^{0.5}$	Does not use $B(e^+)$
Mehta and Raja Rao [1988]	$7.92(\tan \alpha)^{0.15}(e^+)^{0.11}\mathrm{Pr}^{0.55}$	$0.465(p/e)^{0.53}(\ln e^+ + 0.25)$
Raja Rao [1988]	$6.06(\tan \alpha)^{0.15}(e^+)^{0.13}\mathrm{Pr}^{0.55}$	$0.465(p/e)^{0.53}(\ln e^+ + 0.25)(n')^{2.5}$
Sethumadavan and Raja Rao [1986]	$8.6(e^+)^{0.13}\mathrm{Pr}^{0.55}$	$0.40(e^2/pd_i)^{-1/3}(e^+)^{0.164}$
Dong et al. [2001]	$7.33(e^+)^{0.17}(\alpha/50)^{-0.16}\mathrm{Pr}^{0.548}$	$0.466(e^2/pd_i)^{-0.317}(e^+)^{0.169}(\alpha/50)^{-0.16}$

Table 9.6 Predicted g Function for $\alpha = 70°$ and $90°$

Correlation	$[\bar{g}(e^+) (\tan \alpha)^n]$ for $\alpha = 70°$	$\alpha = 85°$	Comments
Withers [1980a]	10.38	16.64	Water
Mehta and Raja Rao [1988]	13.11	16.25	Water
Raja Rao [1988]	10.72	13.28	Water and power law
Sethumadavan and Raja Rao [1986]	13.07	NA	Water and water/glycol

NA = not applicable.

correlation does not contain the $(\tan \alpha)$ term, because the data were for $(\alpha = 65°)$. Table 9.6 compares the $\bar{g}$ function predicted by the four correlations.

The Withers [1980a] and Mehta and Raja Rao [1988] correlations give approximately equal values for $\alpha = 85°$. For $\alpha = 70°$, the larger exponent on the $(\tan \alpha)$ term results in a smaller $\bar{g}$ value than given by Mehta and Raja Rao [1988]. The Raja Rao [1988]) $\bar{g}$ correlation is approximately 16% below that of Mehta and Raja Rao [1988]. All the Table 9.5 correlations contain statistical correlating errors. Withers did not choose to use the $B(e^+)$ correlation for the friction factor, because he felt the error was too large. He based his friction correlation on an equation previously developed by Churchill [1973] to correlate the friction factor in commercially rough tubes. The Churchill equation for commercial pipe roughness is

$$\sqrt{\frac{2}{f}} = -2.46 \ln \left[\frac{e}{d_i} + \left(\frac{7}{\mathrm{Re}_d} \right)^{0.9} \right] \tag{9.37}$$

Withers replaced the e/d_i and 0.9 terms with empirical constants. Thus, he fitted his friction data for each tube to determine the curve fit constants r and m. The resulting equation predicted the data for each tube with a standard deviation of 3%. The resulting equation is

$$\sqrt{\frac{2}{f}} = -2.46 \ln \left[r + \left(\frac{7}{\mathrm{Re}_d} \right)^m \right] \tag{9.38}$$

Table 9.7 lists the geometries tested by Withers [1980a], in order of increasing e/d_i, and gives the parameters m and r needed to calculate the friction factor using Equation 9.38. Wolverine provides commercial versions of helically corrugated tubes. These are called the Wolverine Korodense™ LPD and MHT tubes described in the *Wolverine Data Book*, Wolverine [1984], which are also listed in Table 9.7. The heat transfer coefficient of the LPD tube is approximately 25% less than that of the MHT geometry. The friction factor of the LPD version is about half that of the MHT tube.

Ravigururajan and Bergles [1985] empirically correlated the r and m terms as functions of p/e, p/d_i (= $\pi/\tan \alpha$, and α (degrees). It was not necessary to include both p/d_i, and α, since $p/d_i = \pi/\tan \alpha$. The correlations are

$$r = 0.17\left(\frac{e}{d_i}\right)^{-1/3}\left(\frac{p}{d_i}\right)^{0.03}\left(\frac{\alpha}{90}\right)^{-0.29}$$

$$m = 0.0086 + 0.033\left(\frac{e}{d_i}\right) + 0.005\left(\frac{p}{d_i}\right) + 0.0085\left(\frac{\alpha}{90}\right)$$

(9.39)

The previously discussed correlations were developed by each investigator using only their own test data. Rabas et al. [1988] sought to develop an empirical correlation for single-start corrugated roughness using data from eight sources. The data bank included 60 roughness geometries, which spanned $17.3 \le d_i \le 28.4$ mm, $0.13 \le e \le 1.42$ mm, $5 \le p \le 30$ mm. The reader is referred to the published paper for this

Table 9.7 Tubes Tested by Withers [1980a]

Tube No.	d_i (in.)	e/d_i	p/e	α (deg)	m	r
2300	0.892	0.016	20.10	83.03	0.71	0.0001
2100	0.869	0.016	18.76	84.58	0.68	0.00094
2200	0.862	0.020	14.45	84.73	0.564	0
2	0.805	0.024	19.46	81.70	0.649	0
1	0.805	0.024	19.46	81.67	0.622	0
LPD	All	0.025	20.00	81.00	0.61	0.00088
9	1.175	0.030	11.31	83.88	0.50	0.0035
1100	0.547	0.031	14.70	81.72	0.65	0.0023
7	0.922	0.033	12.14	82.73	0.47	0.0015
20	0.855	0.036	12.42	81.85	0.445	0
6	0.921	0.038	11.00	81.53	0.45	0.0073
15	1.152	0.040	10.98	82.06	0.48	0.00995
MHT	All	0.040	12	81	0.44	0.00595
14	0.866	0.043	14.20	78.98	0.52	0.0006
33	0.863	0.047	12.28	79.52	0.47	0.0081
5	0.682	0.052	17.45	73.83	0.51	0.0027

rather complex empirical correlation. The rms error was 10.9 and 9.5% for the Nusselt number and friction factor, respectively. They also compare the ability of several other correlations discussed here to predict their data set. It should be noted that they adjusted the original heat transfer data of Mehta and Raja Rao [1979]; which is discussed in their paper.

Tan and Xaio [1989] measured the Prandtl number dependence of a single-start corrugated roughness ($e/d_i = 0.053$, $p/e = 10.8$) for five Prandtl numbers between 0.70 and 20.2. They found $g(e^+, \text{Pr}) \propto \text{Pr}^{0.55}$, which agrees well with Webb et al. [1971]. Dong et al. [2001] investigated the heat transfer and pressure drop performance of four single-start corrugated tubes ($0.020 \leq p/e \leq 0.040$, $14.5 \leq p/e \leq 31.2$, $78.8 \leq \alpha \leq 82.1$) using water and oil ($2 \leq \text{Pr} \leq 140$). The data were correlated using roughness functions (see Equations 9.11 and 9.24). The results are provided in Table 9.5.

9.4.5 PEC Example 9.2

Water flows at 2.4 m/s in a 15.9-mm-diameter tube having $p/e = 10$ transverse-rib roughness at 27°C($v = 0.857\text{E-6 } \text{m}^2/\text{s}$). Figure 9.15 shows that the fully rough condition is attained at $e^+ = 30$, for which $B(e^+) = 3.5$.

a. Calculate the roughness size and spacing (e and p) for the incipient fully rough condition. Substituting Equation 9.11 in Equation 9.9 and solving for e/d_i gives

$$\frac{e}{d_i} = \frac{e^+}{\text{Re}_d}\left(\frac{2}{f}\right)^{1/2} = \frac{e^+}{\text{Re}_d}\left[-2.5\ln(2e/d_i) - 3.75 + B(e^+)\right] \qquad (9.40)$$

Substituting $e^+ = 30$ and $B(e^+) = 3.5$ iteratively solve for $e/d_i = 0.007$. Then, $e = 0.007 \times 15.9 = 0.111$ mm and $p = 10$, $e = 1.11$ mm.

b. Calculate f/f_s and compare the e and f/f_s with that calculated for PEC Example 9.1. Equation 9.9 gives $f = 0.0183$. Using $f_s = 0.00542$, obtain $f/f_s = 3.37$.

c. What is the "equivalent sand-grain" roughness size? The equivalent sand-grain roughness may be obtained from

$$\left(\frac{2}{f}\right)^{1/2} = B_{tr}(e^+) - 2.5\ln\left(\frac{2e}{d_i}\right) - 3.75 = B_{sg}(e^+) - 2.5\ln\left(\frac{2e}{d_i}\right) - 3.75 \qquad (9.41)$$

where $B_{tr}(e^+) = 3.5$ and $B_{sg}(e^+) = 8.48$ (Figure 9.10b). Solve Equation 9.41 to obtain $e_{sg}/e = 7.38$. This shows that sand-grain roughness size must be 7.38 times the 2-dimensional rib height to obtain the same friction factor!

9.5 THREE-DIMENSIONAL ROUGHNESS

Cope [1945] was the first to test three-dimensional roughness. He used a knurling process to roughen the inner tube surface and provided water flow data on three roughness geometries. Internal knurling is a relatively expensive process. The sand-grain roughness studied by Nikuradse [1933] and Dipprey and Sabersky [1963] did not involve a practical method of making a commercial three-dimensional roughness. Gowen and Smith [1968] embossed a three-dimensional roughness pattern on a thin sheet, which was rolled into tubular form and then the axial seam was soldered. They tested two roughness geometries and provide data for $0.7 \leq Pr \leq 15.5$. Again, the Gowen and Smith fabrication method is not commercially viable.

Significant progress has been made in manufacturing practical three-dimensional roughness configurations, which are shown in Figure 9.18. The Figure 9.18a Tred-19D™ tube has 19 fins/in. on the outer surface and dimples on the inner surface.

The Figure 9.18b steel tube and the Figure 9.18c copper tubes are "cross rifled" using internal mandrels. Takahashi et al. [1988] provide a comprehensive experimental study using 11 variants of the Figure 9.18c geometry with water flow. The best-performing geometries (A6 and A8) are listed in Table 9.8, which compares the three-dimensional roughness with a corresponding two-dimensional helical rib roughness. The table gives the roughness dimensions and compares the tube performance using PEC VG-1 ($Q/Q_s = P/P_s = W/W_s = 1$). The PEC analysis is for a refrigerant evaporator having shell-side boiling ($h_o = 12,800$ W/m²-K) and 10°C water flowing in the tubes. The smooth inside diameter tube operates at $Re_s = 30,000$. A tube-side fouling resistance of $R_{fi} = 8.6E{-}5$ m²-K/W is assumed. The thermal resistance of the smooth inside diameter tube is 1.9 times the shell-side resistance. Tube B3 is a high-performance two-dimensional helical-rib roughness. The three-dimensional roughness Tube A6 has the same e/d_i and p/e as Tube B3. Tubes B3 and A6 have approximately equal heat transfer performance, because they provide the same A/A_s. But, Tube A6 has smaller friction, which allows operation at higher G/G_s. The larger e/d_i of tube A8 provides higher heat transfer performance than tube A6 (lower A/A_s) and operates at approximately equal G/G_s as Tube B3. Thus, at the same Reynolds number, the three-dimensional Tube A8 has slightly lower friction than the two-dimensional B3 tube, but Tube A8 provides significantly higher heat transfer coefficient. The reduction of the recirculation region behind the protrusion and increased turbulent intensity for the three-dimensional geometry may

Table 9.8 Case VG-1 PEC for Takahashi et al. [1988] Tests (Copper Tubes, $d_i = 13.9$ mm)

Geo	Type	e/d_i	α (deg)	p/e	A/A_s	G/G_s
B3	2-D	0.021	80	13.1	0.79	0.71
A6	3-D	0.022	30	12.6	0.78	0.80
A8	3-D	0.036	30	7.60	0.73	0.72

be responsible for the increased heat transfer of the three-dimensional geometry as compared with the two-dimensional geometry.

Cuangya et al. [1991] provide data for low Reynolds number airflow (4000 ≤ Re_d ≤ 9000 for three of the Figure 9.18c geometries. At Re_d = 5000, they obtained a 2.8 enhancement factor over a plain tube. Liao et al. [2000] investigated the effect of roughness patterns on thermal performance for the Figure 9.18e geometry using water and glycol (2.9 ≤ Pr ≤ 13.8). For Pr = 5.5, the thermal performance ratio [$= j/j_s/(f/f_s)^{1/3}$] increased as e/d_i increased, had a maximum at p_d/e = 5, and decreased as p_c/w increased. They correlated the data using roughness similarity functions (Section 9.3). They also obtained the Prandtl number exponent of 0.51, which is in close agreement with the value of repeated rib geometry (0.57) by Webb et al. [1971].

The Figure 9.18d tube is formed by sintering a single layer of metal particles to the interior tube surface. The preferred geometry has 50% area coverage. The Fenner and Ragi [1979] patent presents heat transfer and friction data for water (Pr = 10) in a 17.2-mm-ID tube for different particle diameters (*e*). The best performance was provided by e/d_i = 0.0092, which provides 105% heat transfer enhancement and 96% friction factor increase. McLain [1975] describes a method to emboss a three-dimensional roughness pattern on a flat strip, which is then rolled into a circular strip and axially welded. Modern high-speed tube welding techniques make this a very practical concept for commercial production. Some of the microfin tubes discussed in Chapter 13 are made by this method.

9.6 PRACTICAL ROUGHNESS APPLICATIONS

9.6.1 Tubes with Inside Roughness

Considerable success has been achieved in the commercial development of tubes having inside roughness. This section discusses the commercially used tubes that have tube-side roughness. Figure 9.3 shows the 19-mm-OD Wolverine, Turbo-Chil, and Korodense tubes. The Wolverine [1984] *Engineering Data Book* provides design correlations. Yorkshire Imperial Metals [1982] also provides corrugated tubes, similar to the Korodense tube. The referenced Yorkshire publication also provides design correlations. The Turbo-Chil tube has ten internal helical ribs, and is made with 26, 35, or 40 fins/in. integral-fins on the outside diameter.

Figure 9.19 shows tube-side enhancements currently used in large water-cooled evaporators and condensers. Commercial tube manufacturers have developed a new class of helical rib roughened tubes, which are all similar to Figure 9.19a. They differ only in the number of starts, fin height, and helix angle. The geometries are neither a classic "roughness" (little surface area increase), nor an internally finned tube (significant surface area increase, but no flow separation). They have some of the characteristics of both enhancement types. They are similar to an internally finned tube, because they provide significant internal surface area increase (compared to Figure 9.3b). Yet they have a higher helix angle than the helical internal-finned tubes of Chapter 8. It is likely that the flow separates at the fin tip, which does not occur in an internally finned tube.

The Figure 9.19a Wolverine Turbo-C™ condenser tube has 0.5-mm fin height and 30 internal helical ribs, as compared to the newer Wolverine Turbo-C III™, which has 34 internal ribs. A similar boiling tube is the Turbo-B III™. The tube-side roughness in the Figures 9.19b, 9.19c and 9.19d tubes is formed by applying pressure to the external tube surface. Figure 9.19b shows the Hitachi Thermoexcel-CC™ condenser tube (Hitachi [1984]), which has $e/d_i = 0.025$ and $p/e = 47$ and an enhanced outer surface. Ito et al. [1983] provide Nu and f data for five variants of the Hitachi Thermoexcel-CC™ tube, which they call the Thermoexcel-CPC™ tube. Figure 9.19c shows the Wieland 19 fins/in GEWA-TW™ (used for boiling and condensing). The GEWA-TW™ tube has a wavy inside surface, whose wave pitch is equal to the outside fin pitch. Figures 9.18a and 9.19c show the Sumitomo Tred-19D™ (19 external fins/in) and Tred-26D™ (26 external fins/in), condenser tubes, respectively, having a 3-dimensional roughness.

Equations have previously been given for prediction of the performance of the Turbo-Chil and Korodense tubes. The Wolverine Turbo-B performance (cooling) is given by the following equations, which are taken from Wolverine [1984]:

$$\frac{hd_i}{k} = 0.06 \, \mathrm{Re}_d^{0.8} \, \mathrm{Pr}^{0.4} \tag{9.42}$$

$$f = 0.198 \, \mathrm{Re}_d^{-0.267} \tag{9.43}$$

where the Reynolds number is $\mathrm{Re}_d = d_i u_m / v$, with d_i to the base of the enhancement. The heat transfer coefficient is based on the nominal surface area ($A_i/L = \pi d_i$).

Figure 9.20 was obtained from Thors [2004] as supplementary to the Thors et al. [1997] patent on the Turbo-BIII™ tube. The figure shows the progression of tube-side performance of old-to-new versions of the generic Turbo-B™ tube. The geometric details of the tubes are provided in Table 9.9. The figure shows that Turbo-BIII gives 31% higher friction factor and 32% higher Stanton number than the original Turbo-Chil tube at Re = 25,000. The highest efficiency [$\eta = (\mathrm{St}/\mathrm{St}_p)/(f/f_p)$] is observed for the Turbo-BIII LPD tube.

The authors have calculated the enhancement ratios, h/h_p and f/f_p, for most of the commercially available enhanced tubes. Table 9.10 shows the enhancement ratios for water (Pr = 10.4) at $\mathrm{Re}_d = 25,000$ and lists the internal enhancement dimensions. The tubes are listed in order of increasing h/h_p. The greatest heat transfer enhancement is provided by the Turbo-BIII and the three-dimensional roughness Tred-19D™ tube. The pressure drop of the Tred-19D tube appears to be lower than is expected. Although the three-dimensional roughness Tube A8 of Table 9.8 is not a commercial geometry, it is included in Table 9.10 for comparison. It provides even higher performance than Turbo-BIII or the Tred-19D tube. The A8 tube provides 3.7 times higher heat transfer coefficient and 3.3 times higher friction than a plain tube.

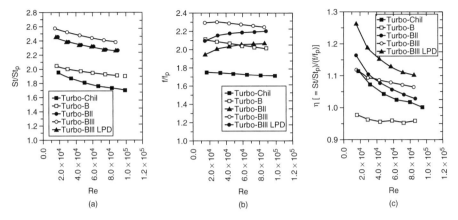

Figure 9.20 (a) f/f_s, (b) St/St$_s$, (c) (St/St$_s$)/(f/f_s) for Turbo-B tubes showing the progression of tube-side performance improvement. (From Thors [2004] and Thors et al. [1997].)

Table 9.9 Geometric Details of the Turbo-B Series Tubes (Dimensions in mm)

Tube No.	I	II	III	IV	V
Product name	Turbo-Chil	Turbo-B	Turbo-BII	Turbo-BIII	Turbo-BIII LPD
Fins/m	1575	1575	1969	2362	2362
Fin height	1.32	0.61	0.69	0.55	0.55
ID	14.55	16.05	16.05	16.38	16.38
Rib height	0.381	0.559	0.381	0.406	0.368
Rib pitch	4.27	2.36	1.07	1.31	1.31
No. of starts	10	30	38	34	34
Helix angle (deg)	46.5	33.5	49	49	49

The method used to make the Tred-19D and Tred-26D™ integral-fin tubes represents an innovative and important manufacturing method. The Figure 9.18a Tred-19D tube has a symmetrical roughness pattern. However, the Figure 9.19g Tred-26D tube has an intermittent helical-rib roughness. Rabas et al. [1993] discuss the method and show how it can be used to make a variety of internal three-dimensional roughness patterns. The dimples on the inner surface are formed by spaced protrusions on the finning disks, which are rolled on the outer tube surface. These protrusions apply pressure to the outer tube surface in the root region of the external fins. The Figure 9.3b Korodense and the Figure 9.3a helical ribs are formed by applying constant external pressure on the outer tube surface. By using the dimpled protrusion method, it is possible to make intermittent helical ribs. The technique is described in the U.S. patent of Kuwahara et al. [1989] and illustrated in Figure 9.21a. The roughness dimensions of the Figure 9.19e Thermoexcel-CC helical-rib roughness are the axial rib pitch (p), the roughness height (e), and the helix angle. The Figure 9.21a tube has the same dimensional variables, plus the

Table 9.10 Performance Comparison of Commercially Enhanced Tubes (d_o = 19 mm, Re = 25,000, Pr = 10.4)

Tube	d_i	e/d_i	n_s	p/e	α	h/h_p	f/f_p	η
GEWA-TW™	15.3	0.016	1	5.3	89	1.40	1.40	1.00
Thermoexcel-CC™	14.97	0.025	1	46.7	73	1.59	1.90	0.84
GEWA-SC™	15.02	0.035	25	2.67	30	1.87	1.65	1.13
Korodense (LPD)™	17.63	0.025	1	20.3	81	1.89	2.26	0.84
Turbo-Chil™	14.60	0.026	10	11.1	47	1.98	1.83	1.08
Korodense (MHT)™	17.63	0.04	1	12.0	81	2.50	4.63	0.54
Tred-26D™	14.45	0.024	10	7.63	45	2.24	1.88	1.19
Turbo-B™	16.05	0.028	30	1.94	35	2.34	2.14	1.09
Turbo BIII LPD™	16.38	0.022	34	3.56	49	2.40	1.98	1.21
Turbo BIII™	16.38	0.025	34	3.22	49	2.54	2.30	1.10
Tred-19D™	14.45	0.024	10	7.63	57	2.55	1.76	1.45
A8 (Table 9.9)	13.5	0.036	2	7.6	30	3.75	3.35	1.11

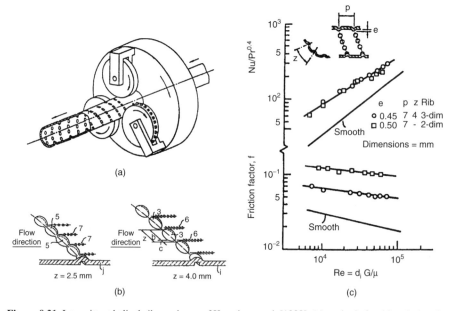

Figure 9.21 Intermittent helical-rib roughness of Kuwahara et al. [1989]: (a) method of making the interior dimples, (b) effect of transverse dimple pitch on local vortices, and (c) performance of intermittent helical ribs. Tubes with similar dimple configurations were tested by Vicente et al. [2002a].

dimple pitch (z) and the dimple width (w) along the rib. Figure 9.21b shows different dimple pitches (z), which result in vortices generated at the dimples. Kuwahara et al. [1989] present heat transfer and friction data that show the effect of varying e, p, and w on performance. Figure 9.21c compares the performance of a 15.8-mm-ID tube having continuous and intermittent helical roughness spaced at p = 7 mm. The

$e = 0.45$-mm intermittent rib provides the same Nu as the $e = 0.5$-mm continuous helical rib. However, the friction factor of the intermittent helical rib is 48% less than that of the continuous helical rib. The solid lines show the smooth tube performance. Vicente et al. [2002b] tested ten helically dimpled tubes ($0.08 \leq e/d_i \leq 0.12$, $0.65 \leq p/d_i \leq 1.1$) illustrated in Figure 9.21 using water and ethylene glycol ($2.5 \leq \mathrm{Pr} \leq 100$) in the turbulent flow. Based on empirical curve fits of the test data, the following roughness functions were derived:

$$B(e^+) = 0.839(e/d_i)^{-0.85}(d_i^2/pz)^{-0.18}(e^+)^{0.11} \qquad (9.44)$$

$$\bar{g}(e^+)\mathrm{Pr}^n = 4.871(e^+)^{0.24}\,\mathrm{Pr}^{0.60} \qquad (9.45)$$

Note that the 0.60 Prandtl number exponent is in close agreement with the 0.57 value of Webb et al. [1971] for the repeated rib geometry (see Table 9.11, later in chapter).

9.6.2 Rod Bundles and Annuli

Considerable data have been reported for transverse- and helical-rib roughness for flow in annuli and rod bundles. The annulus data are for roughness on the outer surface of the inner tube. Flow is axial to the tubes in a rod bundle. Much of this work was done between 1965 and 1978 in support of gas-cooled nuclear reactors. The ribs may be integral with the tube wall (Sheriff and Gumley [1966], Wilkie [1966], Wilkie et al. [1967]) or may be in the form of a helical wire wrap (White and Wilkie [1970], Williams et al. [1970]). White and Wilkie [1970] show that a 45° helix angle is preferred. Webb et al. [1972] show that their correlation for transverse-rib roughness inside tubes predicts the transverse-rib roughness data for axial flow in rod bundles.

   Much of the data supporting the rod bundle geometry has been taken in annuli, in which the inner wall is rough and the outer surface is smooth. Other investigators have used plane channels, having one rough and one smooth wall. It is necessary to "transform" the data, so that they are applicable to the rod bundle geometry, for which all bounding surfaces are rough. Hall [1962] proposes a special definition of equivalent diameter that should be used to convert the measured annulus data to apply to the rod bundle problem. Lewis [1974] provides an excellent assessment of the fundamental principles involved in the "Hall transformation." See the references by Sheriff and Gumley [1966], Wilkie [1966], Wilkie et al. [1967], Maubach [1972], Dalle-Donne and Meyer [1977], Meyer [1980], and Hudina [1979] for data and specific details of the method they used to transform their data. Several of these papers give velocity profile data for rough surfaces.

9.6.3 Rectangular Channels

A gas turbine blade with internal roughness is shown in Figure 9.1, which shows that the hollow interior of the turbine blade is separated into channels by material membranes that join the top and bottom walls of the blade. These membranes are required to provide the structural integrity. The top and bottom walls are rough and the side walls are smooth. Four references on use of roughness to cool the blades are given in Section 9.1. The channel aspect ratio A_r is defined as $A_r = L_s/L_r$, where L_s is the height of the smooth wall and L_r is the width of the rough surface. The composite friction factor (f) and heat transfer coefficient (h) are made up of the rough and smooth wall contributions. Han [1988] shows that the composite friction factor of a channel having rough L_r walls and smooth L_s walls may be calculated by

$$\overline{f} = \frac{L_s f_s + L_r f}{L_s + L_r} \tag{9.46}$$

where f applies to a channel, whose bounding surfaces are rough, and f_s is for a smooth wall channel.

Han [1984] showed that his data for transverse-rib roughness in rectangular channel of $A_r = 1.0$ was well predicted by the correlation of Webb et al. [1971] given in Figure 9.14 with $B(e^+)$ from Figure 9.15. Hence, the friction factor data for a channel bounded by rough walls may be used with Equation 9.46 to predict the composite friction factor for a rough/smooth wall channel of any aspect ratio. Calculation of the Stanton number for a rectangular channel having a combination of rough and smooth surfaces may be similarly predicted, as described by Han [1984]. Han [1988] measured friction and St data for transverse-rib roughened channels having $A_r = \frac{1}{4}, \frac{1}{2}, 1, 2$, and 4. The friction factor for the rough surface wall was extracted using Equation 9.46 and correlated using Equation 9.11. A similar treatment was applied to the heat transfer data.

Much of the recent work on roughness for gas turbine blades has focused on different roughness geometries, or roughness patterns. These data were taken in rectangular channels having roughness on two opposite walls and a smooth surface on the other two opposite walls. Taslim and Spring [1988] tested channels having ribs at $\alpha = 45°$, and Han et al. [1989] investigated angles of 30°, 45°, 60°, and 90°. Han et al. [1991] tested nine rib configurations in a square channel with $p/e = 10$. The nine rib configurations tested are shown in Figure 9.22a. Figure 9.22b shows the enhancement ratio, Nu/Nu$_s$, for the ribbed and smooth sides plotted vs. f/f_s. For fixed f/f_s, this figure shows that the highest enhancement is provided by the V rib configuration. Han and Zhang [1992] extended the study of Han et al. [1991] by investigating the effect of discontinuous ribs, called broken ribs by the authors. The ribs were divided into three to five elements, which were then offset in the flow direction. The parallel broken rib, or V-shaped broken rib geometry, showed 20 to 30% higher heat transfer coefficients, compared with parallel continuous rib or V-shaped continuous rib configuration for the same pressure drop penalty.

Han et al. [1993] investigated the heat transfer and pressure drop characteristics of a square channel having three-dimensional ribs on two opposite walls. Three-dimensional ribs may be envisioned as vortex generators. The vortex generators for laminar channel flow are discussed in Section 5.9. For laminar flow, the vortex

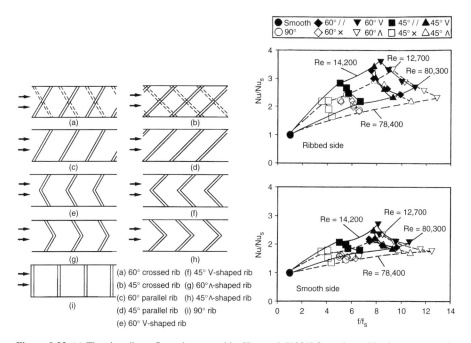

Figure 9.22 (a) The nine rib configurations tested by Han et al. [1991] for a channel having two opposite rough and two opposite smooth walls in a square channel; (b) enhancement ratio vs. f/f_p. (From Han et al. [1991].)

generator height is comparable with the channel height. To use vortex generators for turbulent flow, however, the vortex generator height should be much smaller than the channel height to avoid excessive pressure drop. The roughness configurations tested by Han et al. [1993] are shown in Figure 9.23. The rib height-to-hydraulic diameter ratio was 0.125, and the rib pitch-to-height ratio was 5 or 10. The delta-shaped ribs produced higher heat transfer augmentation and lower pressure drop penalty than the wedge-shaped ribs. The best performance was obtained for the delta-shaped, backward-aligned ribs, which performed as well as the V-shaped broken ribs by Han and Zhang [1992].

Hishida and Takase [1987] measured the local Nusselt number for airflow in a 0.106-aspect-ratio channel with rectangular ribs (one heated rib) on 1 wall. Their data spanned $2.5 \leq p/e \leq 60$ and provide empirical correlations for Nu_x vs. Re_x downstream from the ribs. Kukreja et al. [1993] measured the local mass transfer distribution in square channels with rectangular, oblique (60°), and V-shaped (45° and 135°) ribs for $5 \leq p/e \leq 20$. Two opposite walls were roughened. For oblique and V-shaped rib cases, there were significant spanwise variation of the local mass transfer coefficients, which showed three-dimensional, oblique secondary flow near the ribbed walls. For $p/e = 10$, 60° V-shaped ribs yielded the highest mass transfer. No pressure drop data were reported. Ekkad and Han [1997] and Ekkad et al. [1998] provide a detailed local heat transfer coefficient distribution in a two-pass ribbed duct of various rib configurations. The effect of tube rotation on ribbed channel heat

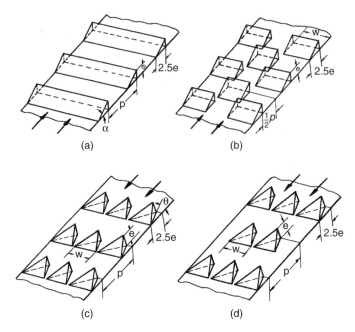

Figure 9.23 Three-dimensional ribs tested by Han et al. [1993]: (a) Wedge-shaped continuous rib, (b) wedge-shaped discontinuous rib, (c) delta-shaped backward aligned rib, (d) delta-shaped backward offset rib. (From Han et al. [1993].)

transfer has been studied by many investigators both experimentally and numerically. Interested readers are advised to consult the monograph by Han et al. [2000].

Recently, Ligrani and co-workers (Mahmood et al. [2001a, 2001b], Ligrani et al. [2001], Mahmood and Ligrani [2002], Burgess et al. [2003]) investigated the heat transfer and pressure drop characteristics of channels having a concave dimpled geometry. The dimpled surfaces provide moderately high heat transfer with relatively low pressure drop increase because the concave dimples do not protrude into the flow, and produce almost no form drag. Figure 9.24 illustrates the concave dimple arrangement by Mahmood et al. [2001a]. The channel aspect ratio was 16, and the dimples were formed on the bottom wall only. The dimples had a depth-to-projected-diameter ratio of 0.2, and the ratio of the channel height to dimple print diameter was 0.5. Note that the dimples are quite large in diameter, relative to the channel height. Their flow visualization study using smoke (shown in Figure 9.25) revealed several longitudinal vortex pairs shed from the edge of the dimples. These vortex structures augment the heat transfer near the downstream rims of the dimples. Figure 9.26 shows the heat transfer enhancement and the pressure drop increase as compared with smooth channel values (Burgess et al. [2003]). The heat transfer coefficient and the friction factor increase as the ratio of dimple depth to dimple diameter increases. A separate study by Mahmood and Ligrani [2002] showed that the heat transfer coefficient and the friction factor increased as the ratio of channel height (H) to dimple diameter (d) decreased. Addition of protrusions to the top wall

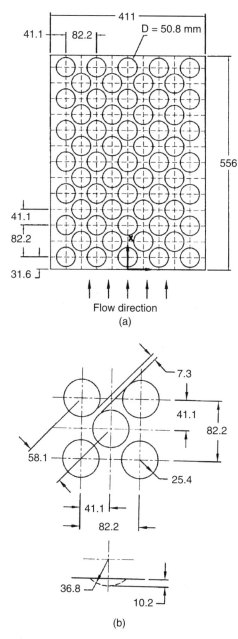

Figure 9.24 Concave dimple geometry tested by Mahmood et al. [2001a]. (a) Entire dimpled test surface; (b) individual dimple geometry. All dimensions are given in mm. (From Mahmood et al. [2001a].)

improved the bottom dimpled surface heat transfer significantly (Ligrani et al. [2001], Mahmood et al. [2001b]). Figure 9.26 shows that the heat transfer enhancement of $e/d = 0.3$ and $H/d = 1.0$ geometry is 2.5 with pressure drop increase of 2.0.

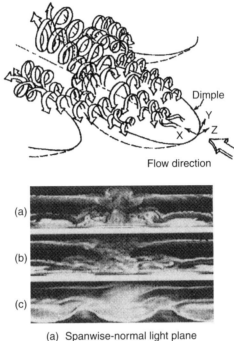

Dimple

Flow direction

(a)

(b)

(c)

(a) Spanwise-normal light plane
(b) Right-diagonal light plane
(c) Left-diagonal light plane

Light plane orientation

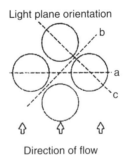

Direction of flow

Figure 9.25 Sketch of three-dimensional flow structure and flow visualization photographs taken at different planes for Re = 1250 and H/d = 0.5 as reported by Mahmood et al. [2001a]. (Reprinted with permision of ASME.)

Note that Chyu et al. [1997] had dimples on two opposite channel surfaces, whereas the others had dimples on only one channel surface. In all cases, the heat transfer coefficients are based on the projected areas. Ligrani et al. [2003] compared the performance of the dimpled surface with that of a ribbed surface. The dimpled surface yielded approximately the same efficiency index [$\eta = (Nu/Nu_p)/(f/f_p)^{1/3}$] as the ribbed geometry.

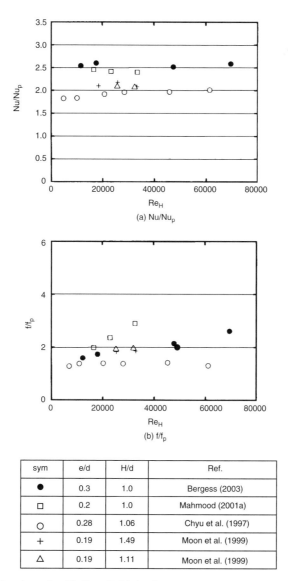

sym	e/d	H/d	Ref.
●	0.3	1.0	Bergess (2003)
□	0.2	1.0	Mahmood (2001a)
○	0.28	1.06	Chyu et al. (1997)
+	0.19	1.49	Moon et al. (1999)
△	0.19	1.11	Moon et al. (1999)

Figure 9.26 (a) Nusselt number, Nu/Nu_p, (b) friction factor; f/f_p, enhancement vs. Reynolds number for various dimpled geometries as reported by Burgess et al. [2003].

9.6.4 Outside Roughness for Cross Flow

Enhancement may be desired on the outside tube surface for shell-and-tube heat exchangers or tube banks. Several studies of knurled roughness on the outside of tubes have been performed using either air or transformer oil as flowing fluid. Groehn and Scholz [1976] tested knurled roughness ($e/d_o = 0.017$ and 0.03) with cross flow of air at 40 bar pressure in an inline tube bank. For $Re_d \leq 25,000$ the roughness

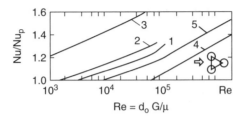

$$\text{Re} = d_o\, G/\mu$$

Figure 9.27 Nu/Nu$_p$ vs. Re for knurled roughness on tube bank in cross flow: Curves 1, 2, and 3 are for Pr$_f$ = 84 with e/d_o = 6.67E-3 (1), 15.0E-3 (2), 15.0E-3 (3), and curves 4 and 5 are Pr$_f$ = 0.7 with e/d_o = 1.0E-3 (4), 8.0E-3 (5).

provided no heat transfer enhancement. At higher Re$_d$, the coefficient was increased above that for the smooth tube bank. For Re$_d$ ≥ 100,000, both tubes provided approximately 50% enhancement. The friction factor was slightly below that of the smooth tubes. Achenbach [1977] performed local measurement on a single knurled cylinder in cross flow using air at 40 bar. He found that roughness decreases the critical Reynolds number, at which the boundary layer transitions to a turbulent boundary layer. Žukauskas and Ulinskas [1983] tested knurled roughness on staggered tube banks with air and transformer oil. They tested four roughness sizes, $0.008 \le e/d_o \le 0.04$. Heat transfer enhancement was obtained for the oil at Re$_d$ ≥ 2500. But enhancement did not occur for airflow until Re$_d$ ≥ 10,000 to 70,000. Figure 9.27 shows Nu/Nu$_s$ vs. Re$_d$ for their tests with air and oil. The authors provide curves of Nu$_d$ and f vs. Re$_d$ for the banks tested, and also give design correlations. A later publication by Žukauskas and Ulinskas [1988] provides curves of Nu$_d$ and f vs. Re$_d$ for the banks tested, design correlations, and recommendations for selection of the roughness size. For $1000 \le$ Re$_d \le$ 100,000 the Nusselt number for knurled roughness is given by

$$\text{Nu}_d = 0.5\,\text{Re}_d^{0.65}\,\text{Pr}^{0.36}\left(\frac{S_t}{S_l}\right)^{0.2}\left(\frac{e}{d_o}\right)^{0.1}\left(\frac{\text{Pr}}{\text{Pr}_w}\right)^{0.25} \tag{9.47}$$

Knurled roughness is not commonly used as shell-side enhancement in tube banks. The favored shell-side enhancement is low, integral-fin tubes, which are discussed in Chapter 6.

9.7 GENERAL PERFORMANCE CHARACTERISTICS

9.7.1 St and f vs. Reynolds Number

The distinct advantage of the rough surface heat-momentum analogy correlation (Equation 9.24) is that data for one e/d_i are applicable to other e/d_i, providing that geometric similarity of the roughness geometry is maintained. However, the correlations do not allow easy interpretation, because the correlated data are plotted as $g(e^+)$ and $B(e^+)$ vs. the roughness Reynolds number e^+. Engineers are accustomed to viewing St and f data plotted vs. the pipe Reynolds number. However, one may use the correlations to generate St, f, vs. Re data for constant e/d_i, which may then be plotted in the familiar format and compared against that of a smooth surface. This will provide a better physical understanding of the Reynolds and Prandtl number characteristics of rough surfaces.

Figure 9.28 shows curves of St, f, and $2St/f$ vs. Re_d for the sand-grain roughness tested by Dipprey and Sabersky [1963] for 4 values of e/d_i at Pr = 5.1. These curves were generated from curve fits of the correlated data shown in Figures 9.10b and 9.11. Figure 9.28 also shows the Stanton number and friction curves for a smooth tube at Pr = 5.1. The smooth tube values are calculated from the Petukhov equation (Equation 9.21). The salient features shown in Figure 9.28 are as follows

1. The St and f values for each roughness size asymptotically approach the smooth surface St and f at sufficiently low Re_d. Above this Re_d, the roughness elements begin to project into the turbulent region, causing a deviation from the smooth tube turbulent St and f curves. When the Re_d is sufficiently high, the friction factor attains a constant value, and the St attains a maximum value. The dashed line corresponding to $e^+ = 42$ is the condition at which the friction factor attains 95% of its "fully rough" value. Here the fully rough condition (FR) is defined as the e^+ or Re_d at which the friction factor attains $0.95f_{FR}$. The St_{max} occurs at approximately e^+_{FR}.
2. The term $2St/f$ describes heat transfer per unit friction. $(St/f)/(St_s/f_s)$ has previously been defined as the efficiency index (η). The $2St/f$ attains a maximum value if Pr > 1. The maximum $2St/f$ may exceed that of a smooth surface. This maximum occurs at a value of Re_d less than the fully rough condition.
3. The Re_d at which St and $2St/f$ attain maxima, and f attains its fully rough value, decreases with increasing e/d_i. However, the e^+ values at which the maxima occur are independent of e/d_i.
4. In the fully rough region, $2St/f$ monotonically decreases, yielding very small values for the efficiency index (η). As e/d_i is increased, f/f_s increases much faster than St/St_s. Thus, at a fixed Re_d in the fully rough region, larger e/d_i will yield smaller values of $2St/f$.

Webb [1979] proposes that all geometrically similar roughnesses of interest will exhibit the above four qualitative characteristics. The minimum Pr at which a maximum of $2St/f$ occurs apparently depends on the roughness type. To illustrate the generality of these basic roughness characteristics, Figure 9.29 is presented for $p/e = 10$ transverse-rib roughness, based on the study of Webb et al. [1971]. The parameters of Figure 9.28 and Figure 9.29 are the same. Both roughness types exhibit the same qualitative behavior, but major differences exist. The transverse-rib roughness gives a higher f and St than the sand-grain type for the same e/d_i (see PEC Example 9.2). For example, with transverse-rib roughness the maximum St for e/d_i = 0.005 is about the same as for e/d_i = 0.02 sand-grain roughness. The "equivalent sand-grain roughness" of the transverse-rib is $e_s/e = 4$, based on equivalent heat transfer performance. The transverse rib does not attain a maximum of $2St/f$ at

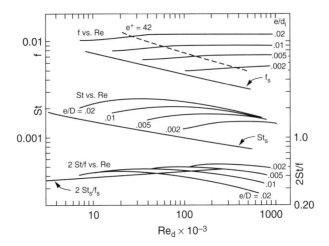

Figure 9.28 Heat transfer and friction characteristics of sand-grain roughness calculated from Dipprey and Sabersky [1963] correlation for Pr = 5.1. (From Webb [1979].)

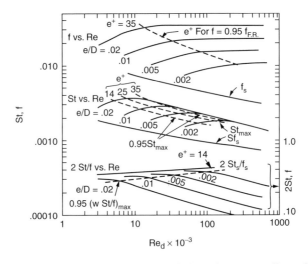

Figure 9.29 Heat transfer and friction characteristics of $p/e = 10$ transverse-rib roughness calculated from Webb et al. [1971] correlation for Pr = 5.1. (From Webb [1979].)

Pr = 5.1; a maximum does not occur until Pr ≥ 10. Thus, the sand-grain roughness yields a higher efficiency index. Dashed lines at $e^+ = 14$ and 35 are drawn in Figure 9.29. Inspection of the figure shows that St and 2St/f are within 5% of their maximum values within the range $14 \le e^+ \le 35$. Hence, this e^+ range is suggested for this roughness geometry.

9.7.2 Other Correlating Methods

Extending our understanding to other roughness types may not be as complex as suggested previously. In their work on transverse-rib roughness, Webb et al. [1971]

also tested $p/e = 20$ and 40 and found that the heat transfer data were correlated with the $p/e = 10$ data, as shown in Figure 9.14. Webb and Eckert [1972] also evaluated the effect of rib geometry (e.g., circular and hemispherical shaped ribs) by testing the correlation with different rib shapes. They found that these data were correlated on the same $\bar{g}(e^+)$ vs. e^+ line, thus accounting for the different value of the friction similarity parameter. That there is no rational reason to expect that nongeometrically similar roughness will fall on the same correlating line as geometrically similar roughness should be emphasezid. Clearly, the $\bar{g}(e^+)$ functions of sand-grain roughness (Figure 9.11) and transverse-rib roughness (Figure 9.14) are not equal. At $e^+ = 50$, the $\bar{g}(e^+) = 11.34$ for sand-grain roughnes and 6.85 for transverse-rib roughness.

Webb et al. [1971] found that the friction similarity function $B(e^+)$ was different for each p/e ratio. They were able to *empirically* correlate the p/e effect. Is there any possibility for correlating heat transfer data for all roughness types on the same line? Burck [1970] attempted to do this. He determined the "equivalent sand-grain roughness" for each roughness geometry (e_{sg}), and calculated the roughness Reynolds number as $(e_{sg})^+ = e_{sg}u^*/v$. The equivalent sand-grain roughness size is determined at the fully rough condition by solving the following equation for e_{sg}:

$$2.5\ln(e_{sg}/e) = 8.48 - B(e^+) \qquad (9.48)$$

Edwards [1966] did this for his repeated-rib airflow data. Webb and Eckert [1972] show the transverse-rib data for $10 \le p/e \le 40$ of Webb et al. [1971] plotted in this format. The correlation was not as good as the original correlation in terms of the actual e^+.

Burck chose to plot the efficiency index (η) vs. $(e_{sg})^+$ for the data of 14 investigators. Although the correlation was not totally satisfactory, it reasonably grouped the data. It would seem that a better way to use the equivalent sand-grain roughness is to maintain the $\bar{g}$ vs. e^+ format, and plot $\bar{g}(e_{sg})$ vs. $(e_{sg})^+$. Although different roughness types may be "normalized" to equivalent sand-grain roughness, the correlation will not be as good as that obtained by maintaining the actual e^+ format.

There are assumptions in the rough surface heat-momentum analogy that can affect the accuracy of the correlation. Use of the law-of-the-wall velocity distribution (Equation 9.8) assumes that the fully developed velocity profile u/u^* is independent of pipe Reynolds number. This is a good approximation, but is not precise for either rough or smooth surfaces. Ye et al. [1987] measured the velocity profile in a tube having corrugated roughness. Their velocity profiles were measured for $12,000 \le \mathrm{Re}_d \le 20,000$, which is not a large enough range to determine the Reynolds number effect.

Some investigators (e.g., Sheriff and Gumley [1966]) have attempted to correlate $\mathrm{St}/\mathrm{St}_s$ vs. e^+, which also provides an approximate correlation. Finally, Ravigururajan and Bergles [1985] attempted to correlate transverse-rib roughness data from many investigators using a linear multiple regression method. A poor correlation was obtained, because the data were correlated within ±50%.

Ravigururajan and Bergles [1996] extended their 1985 study in an attempt to develop a general empirical power-law correlation applicable to internally enhanced tubes of any geometry (e.g., ribbed tube, coiled tube, corrugated tube). The applicable range of the correlation is $0.01 \leq e/d_i \leq 0.02$, $0.1 \leq p/d_i \leq 7.0$, $0.3 \leq \alpha/90 \leq 1.0$, $5000 \leq \mathrm{Re}_d \leq 250{,}000$, and $0.66 \leq \mathrm{Pr} \leq 37.6$. The friction correlation predicted 96% of the data within ±50%, and the heat transfer correlation predicted the 99% of the data within ±50%. Because of the high deviation, this correlation is not recommended for general use. The correlations are as follows.

$$f/f_s = \left[1 + \left[29.1\,\mathrm{Re}_d^{(0.67-0.06p/d_i-0.49\alpha/90)} (e/d_i)^{(1.37-0.157p/d_i)} (p/d_i)^{(-1.66E-6\mathrm{Re}_d-0.33\alpha/90)} \right. \right.$$

$$\left. \left. (\alpha/90)^{(4.59+4.11E-6\mathrm{Re}_d-0.15p/d_i)} (1+2.94\sin(90-\theta)/n) \right]^{15/16} \right]^{16/15} \qquad (9.49)$$

$$\mathrm{Nu}/\mathrm{Nu}_s = \left[1 + \left[2.64\,\mathrm{Re}_d^{0.036} (e/d_i)^{0.212} (p/d_i)^{-0.21} (\alpha/90)^{0.29}\,\mathrm{Pr}^{0.024} \right]^7 \right]^{1/7} \qquad (9.50)$$

In Equation 9.49, θ is the rib profile angle defined in Figure 8.2, and n is the number of sharp corners of the rib facing the flow. These parameters identify the rib profile. The heat-momentum analogy method presented in this chapter has been shown to be far superior to the empirical multiple regression method. Use of the equivalent sand-grain roughness method with the heat-momentum analogy correlation should yield a significantly better correlation of the data set used by Ravigururajan and Bergles [1985, 1996].

Esen et al. [1994b] sought to correlate enhanced tube data using the empirical normalization technique described by Equations 9.4 through 9.6. The enhanced tube data were normalized using the transition Re and f. Comparison with the data obtained from commercial enhanced tubes revealed that the laminar flow friction data are collapsed into a single line ($f = 19.5/\mathrm{Re}_m$) with ±15% scatter. The laminar flow heat transfer data were correlated within ±10% by the empirical equation (Equation 9.3). A similar correlation has been provided for the turbulent Nusselt number, which ±30% scatter, or more. No correlation was provided for the friction factor in the turbulent regime.

9.7.3 Prandtl Number Dependence

Rough surfaces show a significantly different Prandtl number dependency than smooth surfaces. The Pr dependency of rough tubes, relative to smooth tubes, may be observed by examining the behavior of $\mathrm{St}/\mathrm{St}_s$ vs. e^+ for constant e/d_i. Curves of St vs. e^+ have been generated for $p/e = 10$ transverse-rib roughness for different Prandtl numbers using curve-fits of the $\bar{g}(e^+)$ and $B(e^+)$ functions shown in Figures 9.14 and 9.15, respectively. Figure 9.30 shows the results plotted as $\mathrm{St}/\mathrm{St}_s$ vs. e^+ for $0.71 \leq \mathrm{Pr} \leq 200$ with $e/d_i = 0.01$. In this comparison, St and St_s are calculated at the same Re_d. If the rough and smooth surfaces have the same Pr dependence, $\mathrm{St}/\mathrm{St}_s$ will be the same for all Pr. These curves show that $\mathrm{St}/\mathrm{St}_p$ increases with Pr.

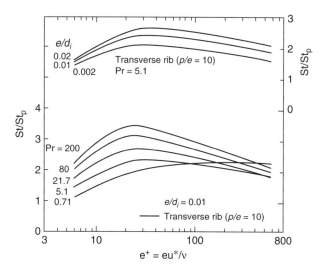

Figure 9.30 St/St$_p$ vs. e^+ for $p/e = 10$ transverse-rib roughness calculated from Webb et al. [1971] correlation for different Prandtl numbers. (From Webb [1979].)

Figure 9.30 also shows that a maximum value of St/St$_p$ is attained at $e^+ = 20$ for Pr > 5.1. However, at large e^+, the curves tend to approach a common value asymptotically. The interpretation of these curves is that (1) at high e^+ (high Re$_d$), the rough and smooth tubes have the same Prandtl number dependency; (2) at lower e^+, the rough surface has a stronger Prandtl number effect than a plain surface. Figure 9.30 shows that the rough surface will be of greater benefit for high-Prandtl-number fluids than for low-Prandtl-number fluids. This means that the heat transfer coefficient increase is greater than the friction increase for higher-Prandtl-number fluids. The curves in the upper part of Figure 9.30 are for different e/d_i at Pr = 5.1.

The basic form of the rough surface heat-momentum analogy correlation (Equation 9.25) is

$$St = \frac{f/2}{1 + \sqrt{f/2}\left[\bar{g}(e^+)Pr^n - B(e^+)\right]} \tag{9.51}$$

At high e^+ and/or Pr, $\bar{g}\,Pr^n \gg B$, so the asymptotic value of Equation 9.51 may be written

$$\frac{f/2St - 1}{\sqrt{f/2}} \simeq \bar{g}(e^+)Pr^n \tag{9.52}$$

This algebraic form for the Prandtl number dependence is unfamiliar to those who use power-law type equations for smooth surfaces, such as the Sieder–Tate equation, which shows St $\propto$ Pr$^{-2/3}$. However, the Petukhov [1970] equation for smooth tubes (Equation 9.20) has an algebraic form similar to that of Equation 9.52. For Pr >> 1, the asymptotic value of Equation 9.20 is

$$\frac{f_s / 2 St_s - 1}{\sqrt{f_s / 2}} \cong 12.7 \, Pr^{2/3} \qquad (9.53)$$

If the asymptotic Pr dependence of rough tubes is the same as that of smooth tubes, the exponent n in Equation 9.52 would have the value 2/3. However, several studies have shown $n \neq 2/3$ for rough tubes. It is not yet fully established how the roughness type may affect the exponent of Pr.

Table 9.11 summarizes the results of work to determine the Prandtl number dependence (Pr^n) for rough surfaces. The Dawson and Trass data are for mass transfer tests, so Sc^n corresponds to Pr^n. Equations 9.51 and 9.52 show that higher St values are obtained if the exponent n of Pr is small.

Table 9.11 shows that rough tubes do not have the same $Pr^{2/3}$ exponent as smooth tubes. The Sethumadhavan and Raja Rao [1983, 1986], Dawson and Trass [1972], Brognaux et al. [1997], Dong et al. [2001], and Vicente et al. [2002a] data give 0.55 to 0.60 exponents, which agree closely with the 0.57 value of Webb et al. [1971] as shown in Table 9.11. Thus, the value $n \simeq 0.57$ appears to be applicable to a wide range of repeated-rib or discrete roughness geometries. Although Withers [1980a, 1980b] shows $n = 0.5$, the vast majority of his data points are for $3 \leq Pr \leq 7$. Hence, his Prandtl number test range may be too small to determine the Prandtl number exponent with good precision. The Webb [1971], Sethumadhavan and Raja Rao [1983, 1986], Dawson and Trass [1972], and Vicente et al. [2002a] data cover a much wider Prandtl number range.

The Pr exponent for sand-grain roughness (sg) is different from that for repeated-rib roughness (tr), 0.44 vs. 0.57. Again, the Prandtl number range tested by Dipprey and Sabersky [1963] was quite small, $1.2 \leq Pr \leq 5.9$. Because both roughness types have $n < 2/3$, the ratio St/St$_s$ will increase with increasing Pr. The smaller exponent for sand-grain roughness (sg) means that the ratio St$_{sg}$/St$_{tr}$ will increase with increas-

Table 9.11 Prandtl Number Exponents for Rough Surfaces

Ref.	n	Pr (or Sc)	Roughness Type
Dipprey and Sabersky [1963]	0.44	1.2–5.94	Sand-grain
Webb et al. [1971]	0.57	0.7–37.6	Ribs ($10 < p/e < 40$)
Sethumadhavan and Raja Rao [1986]	0.55	5.2–32	Corrugated
Sethumadhavan and Raja Rao [1983]	0.55	5.2–32	Wire coil inserts
Dawson and Trass [1972]	0.58	320–4500	Ribs ($p/e \simeq 3.7$)
Withers [1980a]	0.50	2.5–10	Corrugated
Withers [1980b]	0.50	4.7–9.3	Helical rib
Vicente et al. [2002a]	0.60	2.5–100	Three-dimensional helical dimple
Brognaux et al. [1997]	0.56–0.57	0.7–7.85	Microfin
Liao et al. [2000]	0.51	2.9–13.8	Three-dimensional protrusion
Dong et al. [2001]	0.55	2–140	Corrugated

ing Pr. It would be desirable to test the sand-grain roughness over a larger Prandtl number range to verify the 0.44 exponent.

9.8 HEAT TRANSFER DESIGN METHODS

Because there are a large number of different basic roughness types and a range of possible e/d_i, p/e, and α for each type, the designer is faced with a confusing array of choices. To approach the question, one must first define the design objective and establish design constraints. Possible design objectives are listed in Table 3.1. Assume that tube-side roughness will be used to reduce the size of a heat exchanger for case VG-1 of Table 3.1. Thus, one is interested in calculating A/A_p for constant heat duty, flow rate, and pumping power. One may place some limits on the roughness choices by defining a hierarchy of questions:

1. What is the A/A_p for a given roughness type and e/d_i?
2. Will another e/d_i (for the same basic roughness type) give a smaller A/A_p?
3. Will another basic roughness type (and e/d_i) provide further reduction of A/A_p?

First, Question 1 is addressed for the VG-1 criterion, which is described in Section 3.3.4. Assume, for example, that a heat exchanger has been designed that has a plain surface on the tube-side. The tube-side mass flow rate, pressure drop, and the UA value of the plain tube heat exchanger are known. Will a tube-side roughness be of benefit? If the dominant thermal resistance is on the tube side, a tube-side roughness is of potential benefit. Hence, the designer wants to calculate A/A_p for $P/P_p = UA/U_pA_p = 1$.

For simplicity, assume that the total thermal resistance is on the tube side. Then the constraints are satisfied by Equation 3.7. The required velocity in the rough tube that will satisfy the pumping power constraint (Equation 3.6) is unknown. Hence, one must use an iterative calculation procedure to solve Equations 3.6 and 3.7 for A/A_p and G/G_p (or $Re_d/Re_{d,p}$). The solution procedure is complicated by the fact that the equations for the rough tube St and f are written in terms of the roughness Reynolds number (e^+) rather than the pipe Reynolds number (Re_d). Two solution procedures are possible and are described.

9.8.1 Design Method 1

Assume that empirical curve fits of the St and f vs. Re_d (or G) characteristics for the selected e/d_i have been developed. Using these equations, one solves Equation 3.7 for G/G_p. With G/G_p known, the Re_d of the rough tube is calculated and St and f are readily calculated. Next, one calculates the U value for the heat exchanger, and the area ratio (A/A_p) is given by U_p/U. Note that the pumping power constraint is inherently satisfied by Equation 3.7.

9.8.2 Design Method 2

The second method avoids the need for the St and f vs. Re_d curve fits and uses Equation 9.25 Using Equation 9.9 one may write

$$\frac{Re_d}{Re_{d,p}} = \frac{e^+}{Re_{d,p}} \frac{d_i}{e} \sqrt{\frac{2}{f}} \qquad (9.54)$$

If the rough and smooth tubes are of the same inside diameter, $Re_d/Re_{d,p} = G/G_p$. Using Equations 9.25 with $n = 0.57$ and Equation 9.54 in Equation 3.7, $Re_d/Re_{d,p} = [(f/f_p)/(St/St_p)]^{-1/2}$ gives

$$\frac{e^+}{Re_p} \frac{d_i}{e} \sqrt{\frac{2}{f}} = \left(\frac{f_p/(2St_p)}{1 + \sqrt{f/2}\left[\bar{g}(e^+)Pr^{0.57} - B(e^+) \right]} \right)^{1/2} \qquad (9.55)$$

The values of $Re_{d,p}$, St_p, and f_p for the smooth tube and the e/d_i are known. Equation 9.55 is iteratively solved for e^+ using Equation 9.11 to eliminate $2/f$. One uses the $\bar{g}(e^+)$ and $B(e^+)$ equations for the roughness family of interest. The iterative solution gives the value of e^+ that satisfies Equation 9.55. With e^+ known, one easily computes St, f, and Re_d of the rough tube and proceeds to calculate A/A_p as for Method 1.

Using either Method 1 or Method 2, one obtains the predicted A/A_p and G/G_p. One desires $G/G_p \simeq 1$. If the calculated $G/G_p \le 1$, it will be necessary to increase the number of tubes in the heat exchanger. For $Pr \ge 20$, one may find $G/G_p > 1$. Is it possible that a different e/d_i will provide a lower A/A_p or a more desirable G/G_p? This is Question 2. One may repeat the analysis for different e/d_i to answer this question. Webb and Eckert [1972] address this question for $p/e = 10$, $\alpha = 90°$ rib roughness, using PEC VG-1, VG-2, and VG-3 of Table 3.1. For this roughness type, they conclude that the criterion $14 \le e^+ \le 20$ will provide the best value for the objective function. If one can specify the desired e^+, Equation 9.55 may be iteratively solved for the required e/d_i. Such evaluations have not been done for different roughness types, e.g., $p/e = 15$, $\alpha = 45°$ helical-rib roughness. Hence, one cannot state the preferred e^+ for different helix angles (α).

9.9 PREFERRED ROUGHNESS TYPE AND SIZE

9.9.1 Roughness Type

What is the best roughness type (Question 3)? There is no direct way to answer this question without performing a parametric study of candidate roughness types. Webb [1982] uses a case study method to compare the performance improvements of different roughness types. This analysis was performed for case VG-1 of Table 3.1. The design conditions of the reference plain tube heat exchanger are $Re_{d,p} = 46,200$,

$h_o = 27,260$ W/m²-K, $h_i = 6814$ W/m²-K, and $R_{fi} = 0.000044$ m²-K/W (fouling resistance). Five basic types of tube-side enhancements were evaluated, including internal fins. The enhanced tubes are listed in order of decreasing material savings ($V_m/V_{m,p}$), where V_m is the tube material volume in the heat exchanger. The ratio of the tube lengths in the enhanced and plain tube designs is L/L_p, and the relative number of tubes in the exchangers is N/N_p, which is equal to G_p/G for the roughness geometries. A summary of the key results of the study is shown in Table 9.12. Table 9.12 provides several conclusions:

1. The greatest material savings are provided by the helical internal fins. The savings for the helical fins is 12% greater than that for the axial internal fins. The helical fins can be operated at approximately the same velocity as that of the plain tube design.
2. The smallest material savings (42%) is provided by the helical ribs. However, the helical ribs can be operated at a velocity only 4% lower than the plain tube design. The material savings for the corrugated tube is approximately the same as that for the helical ribs. However, the corrugated tube must be operated at a lower velocity.
3. The transverse ribs provide 5% greater material savings than do the helical ribs; however, the velocity must be reduced to 84% of the plain tube value, which requires a larger heat exchanger shell diameter.
4. The heat exchanger length using the helical internal fins is only 37% as long as that of the plain tube design.

Although the helical internal fin geometry appears to be superior to the roughness geometries, one must consider if it can be made in the material of interest, and the relative cost of the enhanced and plain tubes. At present, it appears that the helical rib roughness is of significantly lower cost than an internally finned tube. Second, one must consider whether the design fouling factor (R_{fi}) can be maintained. This is discussed in Chapter 10.

9.9.2 PEC Example 9.3

Your company plans to manufacture tubes having $p/e = 10$, transverse rib roughness [$\alpha = 90°$] applicable to $5000 \leq Re_d \leq 80,000$ and $1 \leq Pr \leq 100$. The tubes will be

Table 9.12 Performance Comparison of Tube-Side Enhancement Geometries (Case VG-1 of Table 3.1)

Geometry	α	e/d_i	N/N_p	L/L_p	$1 - V_m/V_{m,p}$
Helical fins ($n_f = 32$)	30	0.0420	1.01	0.37	0.42
Transverse ribs ($p/e = 10$)	90	0.0060	1.16	0.46	0.53
Axial fins ($n_f = 32$)	0	0.0420	1.04	0.49	0.54
Corrugated ($p/e = 19.5$)	82	0.0090	1.02	0.56	0.57
Helical ribs ($p/e = 15$)	45	0.0069	0.96	0.60	0.58

produced in 19.05-mm-OD with 0.70-mm wall thickness. Select the number of different roughness sizes needed to support the desired operating range, and select the specific roughness dimensions, p and e. Use the information given in Figures 9.14, 9.15, 9.29, and 9.30 to support your recommendations.

Based on the discussion of Figure 9.29 in Section 9.7.1, set $14 \leq e^+ \leq 35$ as the design range for each tube. This will provide St and 2St/f within 0.95 of the maximum values. Using Figure 9.15 with $p/e = 10$, find the values of $B(e^+)$ at $e^+ = 14$ and 35. The $B(e^+)$ values are 4.5 and 3.2, respectively. Use Equation 9.9 with $(f/2)^{1/2}$ defined by Equation 9.11 to solve for the required e/d_i, giving

$$\frac{e}{d_i} = \frac{e^+}{Re_d}\left[-2.5\ln\left(\frac{2e}{d_i}\right) - 3.75 + B(e^+) \right] \tag{9.56}$$

Using $B(35) = 3.2$, iteratively solve for the $e/d_i = 0.0048$ at $Re_d = 80,000$. Then repeat the procedure using $B(14) = 4.5$ to solve for the minimum Re_d at $e^+ = 14$ with $e/d_i = 0.0048$. This gives $Re_d = 36,060$. Thus, a tube having $e/d_i = 0.0048$ is applicable to $36,060 \leq Re_d \leq 80,000$. Then choose the next required e/d_i following the same procedure. For the second tube, solve for the required e/d_i using $e^+ = 35$ at $Re_d = 36,060$ and obtain $e/d_i = 0.00916$. Following the same procedure, the minimum Reynolds number for tube 2 is 16,430. Continuing the procedure for tube 3, obtain $e/d_i = 0.0169$ at $e^+ = 35$ and $Re_d = 16,430$. Tube 3 will have $Re_d = 7635$ at $e^+ = 14$, which approximately satisfies the minimum e^+. Hence, three e/di values (0.0169, 0.00916, and 0.0048) are required to satisfy the operating Re_d. A similar procedure can be applied for any type of enhancement geometry. One may now use Equation 9.25 to generate curves of St and f vs. Re and Pr for each of the selected rough tubes. This example shows that (1) if an enhanced tube is used below its design e^+ range, small St/St$_s$ will result; and (2) if the tube is used above its e^+ range, a small efficiency index will result.

9.10 NUMERICAL SIMULATION

Significant advances have been made in numerically solving the momentum and energy equations for complex flow geometries. As discussed in Chapter 8, the heat transfer and friction characteristics of internally finned tubes have been numerically modeled. Another example of numerical simulation is the OSF geometry discussed in Chapter 6. However, roughness is a more difficult problem, because local flow separations are involved for two-dimensional roughness. Three-dimensional roughness geometries involve further complications. The simplest problem to attack is laminar flow over spaced two-dimensional ribs. This is similar to the OSF problem analyzed by Patankar [1990]. The analysis assumes a periodically, fully developed flow. Hence, the numerical simulation zone is limited to the geometrically repeating pattern. The problem becomes more difficult when the flow is turbulent.

Numerical predictions have been made for laminar and turbulent two-dimensional, transverse ribs. Numerical predictions using two-dimensional models on transverse rib geometry include Benodekar et al. [1985], Fujita et al. [1986], Hijikata et al. [1987], Hung et al. [1987], Durst et al. [1988], Fodemski and Collins [1988], and Becker and Rivir [1989]. The recent three-dimensional models include Prakash and Zerkle [1995], Rigby et al. [1997], Ekkad and Han [1997], and Iacovides and Raisee [1999].

9.10.1 Predictions for Transverse-Rib Roughness

Recent work by Arman and Rabas [1991, 1992] and Rabas and Arman [1992] is described to show the potential for prediction of practical two-dimensional roughness geometries. Figure 9.31 shows the near-wall flow field for turbulent flow over a two-dimensional rib. The near-wall flow consists of stable recirculation and boundary layer development zones. Above this zone is a free shear region and large eddies. The authors used the two-layer turbulence model of Chen and Patel [1988]. This model divides the flow into (1) a near-wall region and (2) the fully developed core region. The standard k-$\in$ model was used in the core region, whereas a one-equation model was used in the near-wall region. The standard k-$\in$ two equation model failed to adequately predict the reattachment length. The problem occurs because the model requires the friction velocity (u^*), for which the local wall shear stress varies with axial position, and changes sign at the reattachment point.

Figure 9.32a of Arman and Rabas [1991] shows the ability to predict the experimental friction factors for five repeated-rib geometries of Webb et al. [1971]. The geometry code used in Figure 9.32 is as follows: the first two digits are the e/d_i, and the second two digits are the p/e of the roughness. Excellent agreement between predictions and experiment was obtained for the smaller e/d_i and the larger p/e. Figure 9.32b shows the predicted axial wall shear stress distribution for $e/d_i = 0.02$ and $p/e = 20$. The figure shows that the shear stress changes sign at $x/e \simeq 7$, which reasonably corresponds to the reattachment point shown in Figure 9.31. The small recirculation zones just upstream and downstream from the rib are responsible for the local variation of the shear stress at these locations.

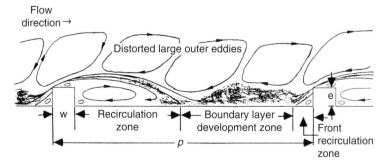

Figure 9.31 Structure of turbulent flow over a two-dimensional rib. (From Arman and Rabas [1991].)

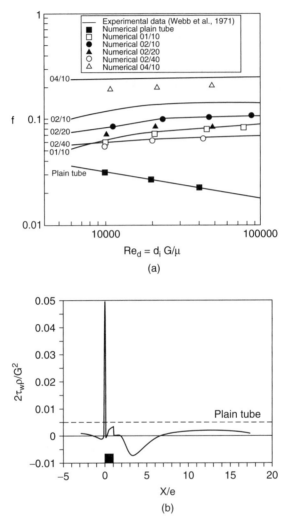

Figure 9.32 (a) Predicted friction factors compared with data of Webb et al. [1971]; (b) predicted axial variation of wall shear stress for $e/d_i = 0.01$ and $p/e = 20$ at Re = 47,000. (From Arman and Rabas [1991].)

Arman and Rabas [1992] showed good ability to predict the Nusselt number and friction factor of the two-dimensional roughness data of Webb et al. [1971]. Figure 9.33 shows a comparison between the predicted and experimental Nusselt number data of Webb et al. [1971] for $e/d_i = 0.02$ and $p/e = 20$ for Pr = 0.71, 5.1, and 21.7. Their paper also presents predictions for other transverse-rib geometries tested by Webb et al. [1971]. The average error of all of the predictions was 15%.

The discussion in Section 9.7.4 showed that the Prandtl number dependency of transverse-rib roughness is significantly different from that for a smooth tube. Rabas and Arman [1992] sought to explain the physical mechanism for this difference.

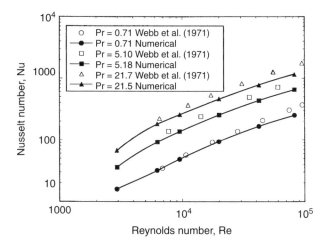

Figure 9.33 Comparison of predicted and Webb et al. [1971] experimental data for transverse-rib roughness [e/d_i = 0.02, p/e = 20] for 0.71 ≤ Pr ≤ 21.7. (From Rabas and Arman [1992].)

They predicted the axial distribution of the local Nusselt number, and their interpretation of these curves for different Pr provides valuable insights on the mechanism of the Prandtl number influence.

Fodemski and Collins [1988] predict heat transfer and friction in a rectangular channel containing square ribs spaced at p/e = 6. They solve the conjugate problem, which includes conduction in the tube wall. The scope of this work parallels that of Arman and Rabas [1991, 1992].

Most of earlier numerical studies on ribbed geometry have used two-dimensional models. In recent years, some researchers have reported three-dimensional studies. Prakash and Zerkle [1995] predicted the flow and heat transfer in a ribbed rectangular duct. Turbulence was modeled with the k-$\in$ model in conjunction with a wall function. They concluded that the Reynolds stress model was necessary to capture the anisotropic nature of flow around the ribs. Rigby et al. [1997] used the k-$\in$ turbulence model to predict the flow and heat transfer in a rectangular duct with rib roughness on the walls. They compared their results with the experimental data by Ekkad and Han [1997]. The heat transfer coefficients between the ribs showed a good agreement with the experimental data, but the heat transfer coefficients on the top of the ribs were underestimated. Iacovides and Raisee [1999] explored turbulence modeling issues related with the flow and heat transfer in a square channel with rib-roughened walls. They compared four turbulence models: a zonal k-$\in$, a low-Re k-$\in$, a zonal differential stress model (DSM), and a low-Re DSM. The zonal models underpredicted the heat transfer data. The low-Re DSM model simulated the data more closely than the low-Re k-$\in$ model.

9.10.2 Effect of Rib Shape

Arman and Rabas [1992] predicted the effect of the rib shape on the Nusselt number and friction factor. The numerical model was validated by predicting the Nu and f

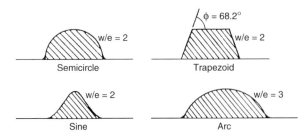

Figure 9.34 Different rib shapes analyzed by Arman and Rabas [1992]. (From Arman and Rabas [1992].)

Table 9.13 Comparison of Mean Nusselt Numbers and Efficiency Index ($e/d_i = 0.02$)

Shape	p/e	Re = 9400		Re = 39,000	
		Nu/Nu_p	η/η_{trap}	Nu/Nu_p	η/η_{trap}
Sine	10	2.17	0.95	2.22	0.96
Sine	20	1.84	0.97	1.88	0.99
Semicircle	10	2.13	0.93	2.21	0.96
Semicircle	20	1.79	0.95	1.83	0.96
Arc	10	1.98	0.86	2.16	0.91
Arc	20	1.71	0.91	1.78	0.94
Trapezoid	10	2.29	1.00	2.32	1.00
Trapezoid	20	1.89	1.00	1.91	1.00

of two-dimensional roughness geometries tested by Hijikata and Mori [1987] and Nunner [1956]. The Nu and f of the Hijikata and Mori arc shaped geometries ($e/d_i = 0.02$, $p/e = 44.4$, and $e/d_i = 0.04$, $p/e = 22.2$) were overpredicted by 17 and 26%, respectively. The Nunner semicircular rib geometry ($e/d_i = 0.04$, $p/e = 20$) Nu and f were overpredicted 5 and 0%, respectively.

They next predicted the effect of the four different rib shapes shown in Figure 9.34 for $e/d_i = 0.02$ and $p/e = 20$ and 40 at $Re_d = 9400$ and 39,000. Table 9.13 shows the enhancement ratios Nu/Nu_p and the efficiency index (η) referred to a plain tube. The table shows that the trapezoidal shape gives the highest Nu/Nu_p and η. The arc shaped rib gives 29% lower Nu/Nu_p and 9% lower η does the trapezoid shaped rib. They also provide plots of the local shear stress and pressure distributions, along with a table of the reattachment lengths.

9.10.3 The Discrete-Element Predictive Model

Taylor et al. [1988] and Taylor and Hodge [1992] developed a seminumerical "discrete-element" method to predict the j and f factors for flow in tubes having a three-dimensional roughness. In this method, one numerically integrates the velocity and temperature profiles across the pipe radius, and accounts for the flow blockage caused by the roughness elements. The drag coefficient on the discrete roughness elements is included using empirical expressions of the form $C_D = a\,Re^n$. The solution to the momentum equation provides the velocity profile and the friction factor. The velocity profile is used in a similar approach to solving the energy equation. Again, the heat transfer from the discrete roughness elements is accounted for by empirical expressions of the form $\text{Nu} = b\,\text{Re}^m$.

The discrete-element roughness model was originally developed by Taylor et al. [1984]. The major physical phenomena that distinguish flow over a rough surface are blockage of the flow, form drag on the roughness elements, and skin friction on the base surface. The heat transfer formulation accounts for the local heat transfer from the roughness elements and from the base surface between the elements. The momentum and energy equations are developed by writing momentum and energy balances on the control volume of thickness δy and length δx shown in Figure 9.35. The momentum equation is

$$0 = (pA_x)\big|_x - (pA_x)\big|_{x+dx} - F_D + (\tau A_y)\big|_y - (\tau A_y)\big|_{y+dy} \tag{9.57}$$

The areas A_x and A_y are given by $A_x = 2\pi r\,\delta y\beta_x$, and $A_y = 2\pi r\,\delta x\,\beta_y$, where β_x and β_y are the fraction of the control surface that is not obstructed by the roughness elements. Taylor et al. [1984, 1988] have shown that $\beta_x = \beta_y = 1 - \pi d_e^2/[4p(1 - 2y/d_i)]$ for a uniform array of elements spaced at p, where d_e is the local element diameter at level y. The drag force is given by

$$F_D = 0.5\rho C_D A_p N_{\text{el}} \tag{9.58}$$

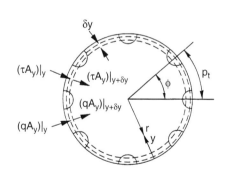

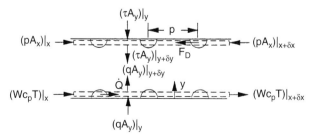

Figure 9.35 Control volume for discrete-element method of Taylor and Hodge [1992]. (From Taylor and Hodge [1992].)

where the drag coefficient on the roughness element (C_D) is obtained from empirical data. The N_{el} is the number of roughness elements contained in the control volume, and A_p is the projected area of the roughness elements to the flow.

The energy equation is given by

$$0 = (Wc_pT)\big|_{x+\delta x} - (Wc_pT)\big|_x + (qA_y)\big|_{y+dy} - (qA_y)\big|_y - h_{el}\delta A_{el} Nu_{el}(T_w - T) \qquad (9.59)$$

The Nusselt number for heat transfer from the element is given by an empirical correlation. Letting $d_e(y)$ be the local diameter of the element at distance y from the surface, the $h_{el}\,\delta A_{el}$ is given by

$$h_{el}\delta A_{el} = \frac{Nu_{el}k}{d_e(y)} \pi d_e(y)\delta y \qquad (9.60)$$

The solution of Equations 9.45 to 9.47 requires a turbulence model for $\in_m$ and $\in_h$ and a roughness model for C_D and Nu_{el}. The turbulence model is not modified to include roughness effects, because the physical effects of the roughness are included explicitly in the differential equations. The Prandtl mixing-length model with Van Driest damping and constant turbulent Prandtl number $Pr_+ = 0.9$ is used. Following Kays and Crawford [1980] the pressure gradient effects on the turbulence are accounted for using a wall parameter. The boundary conditions are applied at the smooth base wall ($y = 0$ at the smooth base surface) and the free stream. At $y = 0$, $u = v = 0$ and $T = T_w$. At $y = d_e/2$, $du/dy = dT/dy = 0$.

Empirical correlations are used for C_D and Nu_{el}. Taylor et al. [1984, 1988] give

$$\log_{10} C_D = -0.125\log_{10} Re_{de} + 3.75 \quad Re_{de} < 6000$$
$$C_D = 0.6 \quad Re_{de} > 6000 \qquad (9.61)$$

and Hosni et al. [1989, 1991] proposed

$$Nu_{de} = 1.7\,Re_{de}^{0.49}\,Pr^{0.4} \quad Re_{de} < 13,800$$
$$Nu_{de} = 0.0605\,Re_{de}^{0.84}\,Pr^{0.4} \quad Re_{de} > 13,800 \qquad (9.62)$$

An iterative solution method is required to provide closure for integration of the momentum and energy equations. After solution for the velocity and temperature profiles, the friction factor and Stanton numbers are obtained from Equations 9.63 and 9.64 below.

$$f = \frac{\beta_{y,w}\mu \dfrac{du}{dy}\Big|_w + \dfrac{\rho}{2p^2}\displaystyle\int_0^e d_e C_D u^2 dy}{\rho u_m^2/2} \qquad (9.63)$$

where $\beta_{y,w} = 1 - \pi d^2/4p$ is the total blockage factor for the elements.

$$\text{St} = \frac{\beta_{y,w} k c_p \left.\frac{dT}{dy}\right|_w + \frac{\pi k}{p^2} \int_0^e \text{Nu}_{de} (T_w - T) dy}{\rho u_m (T_w - T_m)} \qquad (9.64)$$

The Equation 9.61 correlation is based on the friction factor correlation of Žukauskas [1972] for tube banks, reinterpreted as a drag coefficient. The Equation 9.62 correlation for Nu_{de} is based on the Žukauskas [1972] tube bank heat transfer correlation, and modified by Taylor et al. [1984] and Hosni et al. [1989, 1991] to obtain the best prediction of several data sets. The weak point of the analysis is in the accuracy of Equations 9.61 and 9.62. Thus, can these empirical correlations be applied to different element shapes? Although the concept makes sense, the methodology to define C_D and Nu_{de} is relatively weak. A more rationally based method of defining these important parameters would strengthen the method.

Taylor and Hodge [1992] used the discrete-element model to predict the j and f factors for the three-dimensional roughness data of Dipprey and Sabersky [1963], Takahashi et al. [1988], Gowen and Smith [1968], and Cope [1945]. They show quite good ability to predict the j and f vs. Re_d data. The effect of Prandtl number is also well predicted. They note that some inconsistencies may be due to errors in some of the experimental data. Figure 9.36 shows the predictions for the closely packed sand-grain roughened tube A-4 (e/d_i = 0.0488) of Dipprey and Sabersky [1963], which shows very good agreement with the data. Rabas et al. [1993] successfully predicted the performance of helically dimpled tubes of Figure 9.21 using the discrete element method by Taylor and Hodge [1992]. The prediction method was then used to determine the effect of different spherical shapes and spacings. The analysis revealed that significant performance improvements are possible with an increased roughness curvature and a closer roughness spacing. Hosni et al. [1991] and Taylor and Hodge [1992] use the method to predict heat transfer and friction on rough plates in boundary layer flow. Chakroun and Taylor [1992] treat accelerating boundary layer flow over rough plates.

James et al. [1994] extended the discrete-element model to a two-dimensional rib roughness geometry. The same momentum (Equation 9.57) and energy equation (Equation 9.59) as the three-dimensional case is applicable to the two-dimensional case. However, the blockage factors (β_x, β_y), the drag coefficient of rib element (C_D), and the Nusselt number of the rib element must be modeled accordingly. The blockage for three-dimensional roughness is a function of the roughness geometry only. For a rib roughness, however, the significant streamline curvature and the separation and reattachment of the flow over ribs causes the blockage to be a function of the roughness geometry and of the flow field. The flow model used by James et al. [1994] is shown in Figure 9.37, where the separation streamlines are modeled as straight lines. The separated regions upstream and downstream of the ribs were assumed as blockage to the flow. Based on the experimental measurements by Faramarzi and Logan [1991] and Mantle [1966] on a ribbed geometry, James et al. deduced the following blockage factors:

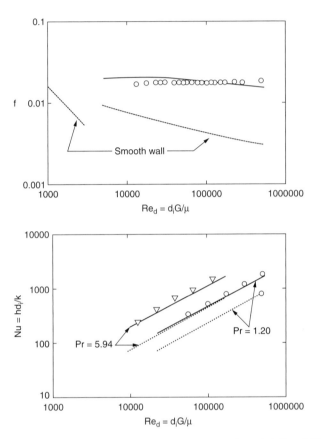

Figure 9.36 Prediction of Dipprey and Sabersky [1963] data on Tube A-4 ($e/d_i = 0.0488$) using the discrete-element method of Taylor and Hodge [1992]. (From Taylor and Hodge [1992].)

$$\beta_x = \beta_y = 1 - \left[w + (\gamma + 1)(e - y) \right] / p \qquad (9.65)$$

where

$$\gamma = \gamma_w (\gamma_L / \gamma_{ss})$$

$$\gamma_{ss} = 9.1,$$

$$\gamma_w = 12.6(w/e) + 11.7 \quad \text{for} \quad w/e \le 2, \quad \gamma_w = 6.5 \quad \text{for} \quad w/e > 2, \qquad (9.66)$$

$$\gamma_L = 4 \quad \text{for} \quad 6.6 \le p/e \le 12, \quad \gamma_L = 0.3667(p/e) - 1.0672 \quad \text{for} \quad 12 \le p/e \le 28$$

The rib element drag coefficient C_D was modeled following the suggestion by Good and Joubert [1968]. They proposed a logarithmic pressure coefficient distribution from the investigation on the bluff body immersed in the turbulent boundary layer.

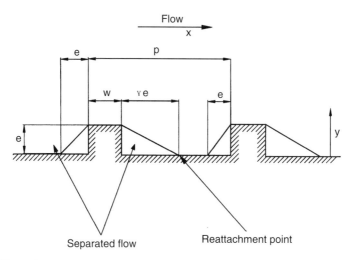

Figure 9.37 The flow model used by James et al. [1994] for repeated rib geometry (From James et al. [1994].)

$$C_D = C_p \left(f_{ref}/2\right)^{0.5} / u_{ref}^2, \quad C_p = C_1 \log\left[(e/2)\mathrm{Re}_d(f_{ref}/2)^{0.5}\right] + C_2$$

$$f_{ref} = 0.625/\left[\log(5.74/\mathrm{Re}_d^{0.9})\right]^2, \quad u_{ref} = 1.244 y^{1/7}$$

(9.67)

In the equations above, u_{ref} and f_{ref} are the reference (smooth tube) values. The C_1 and C_2 are experimentally determined constants, and were determined from the best fit of the Webb et al. [1971] data.

$$C_1 = 2846(p/e)^{-0.5} \quad C_2 = -0.4C_1 \tag{9.68}$$

The Nusselt number of the rib element was modeled by the enhancement ratio (E_r) over the smooth tube value. The E_r was assumed to be 3.0 following the suggestion by Berger and Hau [1979] and Lewis [1975].

$$\mathrm{Nu}_{rib} = 0.023 E_r \, \mathrm{Re}_d^{0.8} \, \mathrm{Pr}^{0.4} \tag{9.69}$$

James et al. [1994] used the model to predict the j and f factors of the repeated rib geometry of Webb et al. [1971], Berger and Hau [1979], Mendes and Mauricio [1987], and Baughn and Roby [1992]. The model predicted the data reasonably well, except for the j and f factors of Mendes and Mauricio [1987], which were underpredicted as much as 30%. The underprediction increased as the rib height increased. The Nusselt number was underpredicted by 10 to 20% for $e/d_i = 0.05$ and 30% for $e/d_i = 0.1$. James et al. [1994] recommended refinement of the model for high e/d_i geometries. Figure 9.38 shows the predictions for $p/e = 10$ tube of Webb et al. [1971]. The figure shows very good agreement with the data. The effect of the

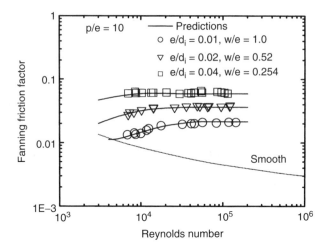

Figure 9.38 Comparison of predicted friction factors with the data of Webb et al. [1971] (From James et al. [1994].)

Prandtl number is also well predicted. It is noted that the James et al. [1994] model is applicable to the repeated rib roughness geometry with a 90° helix angle. Extension of the model to the spirally ribbed geometry should consider the effect of flow swirl.

9.11 CONCLUSIONS

Significant progress has been made in the design and manufacture of roughened surfaces. Innovative manufacturing methods have resulted in the development of tubes having internal roughness (with external enhancement), which provide capital cost savings relative to plain tube designs. Rough surfaces are extensively used for water-side enhancement in air-conditioning condensers and evaporators, and in large steam condensers. Roughness is also used to cool the interior surfaces of gas turbine blades. Table 9.10 shows that the commercially used roughness geometries provide enhancement of 250%. Roughness appears to offer higher performance than other enhancement techniques, with the possible exception of internal fins. However, internal fins are difficult to make in many practical materials of interest. The low-fin, helical-rib geometries of Figure 9.19a through d approach those of helical internal fins and can be made in copper. Engineers are encouraged to seek new engineering applications of tube-side roughness.

Laminar flow heat transfer in rough tubes is strongly affected by bouyancy and entrance effects. The transition to turbulence occurs at smaller Reynolds numbers than the smooth tube. For turbulent flow, roughness has a higher Prandtl number dependency than exists for plain tubes. This means that the heat transfer enhancement ratio (h/h_p) increases with increasing Prandtl number. For Prandtl numbers above, say, 20, a properly designed rough tube may have an efficiency index greater than

1.0. The roughness size must be selected for the Reynolds number range of interest. If the tube is operated above this range, the efficiency index will drop to small values. If the tube is operated below this range, little enhancement will occur.

There are many possible roughness geometries, as suggested by Figure 9.2. At present, the highest performance roughness types are helical ribs and three-dimensional roughness. The three-dimensional roughness appears to offer the highest enhancement level, at a high efficiency index. Recent advances have been made in manufacture of three-dimensional roughness, as shown in Figure 9.18.

Techniques to correlate the performance of roughness have moved well past the empirical correlation method. A semiempirical Reynolds analogy for rough surfaces has been found to be a powerful tool to correlate roughness data. But using data on one roughness geometry, the model may be used to predict the performance of geometrically similar roughness.

Recent advances in hydraulic and thermal modeling of roughness include the discrete-element method of Taylor and Hodge [1992], and numerical solutions, as evidenced by Arman and Rabas [1992], Bhatia and Webb [2001], and others. The discrete-element method has been found to work well for both two- and three-dimensional roughness. Future developments are expected in predictive methods.

Heat transfer surface fouling is a concern for rough surfaces. This subject is discussed in Chapter 10.

NOMENCLATURE

A	Heat transfer surface area (= $\pi d_i L$ for tube), m^2 or ft^2
A_c	Flow cross-sectional area (= $\pi d^2/4$ for tube), m^2 or ft^2
A_p	Projected area of roughness element, m^2 or ft^2
A_r	Aspect ratio of channel having two smooth walls opposite two rough walls (= L_s/L_r), dimensionless
A_x	Cross-sectional flow shown in Figure 9.35, m^2 or ft^2
$B(e^+)$	Correlating function for rough tubes, Equation 9.11, dimensionless
b	Height of rectangular channel, m^2 or ft^2
C_D	Drag coefficient, dimensionless
c_p	Specific heat of fluid at constant pressure, J/kg-K or Btu/lbm-°F
D_h	Hydraulic diameter of flow passages, $4LA_c/A$, m or ft
d_e	Roughness element base diameter, m or ft
d_i	Tube inside diameter (diameter to the base of roughness), m or ft
d_o	Tube outside diameter, m or ft
e	Fin height or roughness height, m or ft
e_{sg}	Equivalent sand-grain roughness, m or ft
e^+	Roughness Reynolds number (= eu^*/ν), dimensionless
F_D	Drag force on roughness element, N or lbf
f	Fanning friction factor, $\Delta p_f d_i/2LG^2$, dimensionless
f_m	Normalized friction factor (Equation 9.5), dimensionless

G	Mass velocity (W/A_c), kg/m²-s or lbm/ft²-s
g	$[f/(2St) - 1]/(f/2)^{1/2} \pm B(e^+)]$, dimensionless
$\bar{g}$	$g\,Pr^{-n}$, dimensionless
g	Acceleration due to gravity, 9.806 m/s² or 32.17 ft/s²
H	Channel height, m or ft
h	Heat transfer coefficient based on A (h for all rough tubes is based on $A_i/L = \pi d_i$), W/m²-K or Btu/hr-ft²-°F
h_o	Shell-side heat transfer coefficient, W/m²-K or Btu/hr-ft²-°F
j	Colburn factor (= $StPr^{2/3}$), dimensionless
K	Thermal conductance (= hA), W/K or Btu/hr-°F
k	Thermal conductivity of fluid, W/m-K or Btu/hr-ft-°F
L	Fluid flow (core) length on one side of the exchanger, m or ft
L_s	Height of smooth wall in rectangular channel having one rough and one smooth wall, m or ft
L_r	Width of rough wall in rectangular channel having one rough and one smooth wall, m or ft
LMTD	Log-mean temperature difference, K or °F
M	Mass of tube material, kg or lbm
N_{el}	Number of elements in control volume (Figure 9.35), dimensionless
Nu	Nusselt number, $Nu_d = hd_i/k$ or hd_o/k $Nu_{el} = h_{el}d_e/k$, dimensionless
Nu_m	Normalized Nusselt number (Equation 9.6), dimensionless
NTU	Number of heat transfer units [= $UA/(Wc_p)_{min}$], dimensionless
n_f	Number of fins in internally finned tube, dimensionless
n_s	Number of starts around the tube circumference for a helical-rib roughness, dimensionless
P	Fluid pumping power, W or hp
Pr	Prandtl number (= $c_p\mu/k$), dimensionless
Pr_e	Effective Prandtl number (v_e/α_e), dimensionless
Pr_t	Turbulent Prandtl number ($\in_m/\in_h$), dimensionless
p	Axial spacing between roughness elements, m or ft
Δp	Fluid static pressure drop, Pa or lbf/ft²
Q	Heat transfer rate, W or Btu/hr
q	Heat flux, W/m² or Btu/hr-ft²
q_o	Heat flux at wall, W/m² or Btu/hr-ft²
Re_{Dh}	Reynolds number based on the hydraulic diameter (= GD_h/μ), dimensionless
Re_d	Reynolds number based on the tube diameter (= Gd/μ), $d = d_i$ for flow inside tube and $d = d_o$ for flow outside tube, dimensionless
Re_{de}	Reynolds number based on the tube diameter (= Gd_e/μ), dimensionless
Re_m	Normalized Reynolds number (Equation 9.4), dimensionless
R_{fi}	Tube-side fouling resistance, m²-K/W or ft²-hr-°F/Btu
St	Stanton number = h/Gc_p, dimensionless
T	Temperature, T_w (wall), T_∞ (free stream), T_{av} (average), K or °F
T^+	Dimensionless temperature, $T/(Q/\rho c_p u^*)$, dimensionless
ΔT	Temperature difference between hot and cold fluids, K or °F

ΔT_i	Temperature difference between hot and cold inlet fluids, K or °F
U	Overall heat transfer coefficient, W/m^2-K or Btu/hr-ft^2-°F
u	Local fluid velocity, m/s or ft/s
u_m	Fluid mean axial velocity at the minimum free flow area, m/s or ft/s
u^*	Friction velocity [$= (\tau_w/\rho)^{1/2}$], m/s or ft/s
u^+	Dimensionless velocity (u/u^*)
V	Heat exchanger total volume, m^3 or ft^3
V_m	Heat exchanger tube material volume, m^3 or ft^3
v	Velocity component normal to wall, m/s or ft/s
W	Fluid mass flow rate ($= \rho u_m A_c$), kg/s or lbm/s, or channel width, m or ft
w	Width of roughness element at base, m or ft
x	Cartesian coordinate along the flow direction, m or ft
x^+	x/D_hRe Pr, dimensionless
y	Coordinate distance normal to wall, m or ft
y_b	Thickness of viscous influenced fluid layer, m or ft
y^+	yu^*/v, dimensionless
y_b^+	$y_b u^*/v$, dimensionless
z	Dimple pitch in helix angle direction, m or ft

Greek Letters

α	Thermal diffusivity ($= k/\rho c_p$), m^2/s or ft^2/s
α	Helix angle relative to tube axis rad or degrees
δ	Boundary layer thickness, m or ft
δ^+	Dimensionless boundary layer thickness ($\delta u^*/v$)
$\in$	Porosity, dimensionless
$\in_m$	Eddy diffusivity for momentum, m^2/s or ft^2/s
$\in_h$	Eddy diffusivity for heat, m^2/s or ft^2/s
η	Efficiency index [$= (h/h_s)/(f/f_s)$], dimensionless
μ	Fluid dynamic viscosity coefficient, Pa-s or lbm/s-ft
v	Kinematic viscosity, m^2/s or ft^2/s
v_e	Effective viscosity in turbulent region ($= v + \in_m$), m^2/s or ft^2/s
ρ	Fluid density, kg/m^3 or lbm/ft^3
τ	Shear stress in fluid, Pa or lbf/ft^2
τ_w	Wall shear stress, Pa or lbf/ft^2
τ_o	Apparent wall shear stress, Pa or lbf/ft^2

Subscripts

el	Refers to roughness elements of Figure 9.35
fd	Fully developed flow
FR	Fully rough condition
m	Average value over flow length
max	Maximum Value
p	Plain tube or surface

s Smooth tube or surface
sg   Sand-grain roughness
tr   Transverse rib roughess
w Evaluated at wall temperature
x Local value
δ Viscous boundary layer thickness

REFERENCES

Achenbach, E., 1977. The effect of surface roughness on the heat transfer from a circular cylinder to the cross flow of air, *Int. J. Heat Mass Transfer*, 20, 359–369.

Almeida, J.A. and Souza-Mendes, P.R., 1992. Local and average transport coefficients for the turbulent flow in internally ribbed tubes, *Exp. Thermal Fluid Sci.*, 5, 513–523.

Arman, B. and Rabas, T.J., 1991. Prediction of the pressure drop in transverse, repeated-rib tubes with numerical modeling, in *Fouling and Enhancement Interactions*, T.J. Rabas and J.M. Chenoweth, Eds., ASME HTD, 164, 93–99.

Arman, B. and Rabas, T.J., 1992. Disruption shape effects on the performance of enhanced tubes with the separation and reattachment mechanism, in *Enhanced Heat Transfer,* M.B. Pate and M.K. Jensen, Eds., ASME Symp. Vol. HTD, 202, 67–76.

Baughn, J.W. and Roby, J., 1992. Enhanced turbulent heat transfer in circular ducts with transverse ribs, in *Enhanced Heat Transfer*, ASME HTD, 202, 9–15.

Becker, B.R. and Rivir, R.B., 1989, Computation of the Flow Field and Heat Transfer in a Rectangular Passage with a Turbulator, ASME Paper 89-GT-30; ASME, New York.

Benodekar, R.W., Goddard, A.J.H., Gosman, A.D., and Issa, R.I., 1985, Numerical prediction of turbulent flow over surface-mounted ribs, *AIAA J.*, 23, 359–366.

Berger, F.P. and Hau, K.-F., 1979. Local mass/heat transfer distribution on surfaces roughened with small square ribs, *Int. J. Heat Mass Transfer*, 22, 1645–1656.

Bhatia, R.S. and Webb, R.L., 2001. Numerical study of turbulent flow and heat transfer in microfin tubes. Part I, model validation, *J. Enhanced Heat Transfer*, 8, 291–304.

Brognaux, L., Webb, R.L., Chamra, L.M. and Chung, B.Y., 1997. Single-phase heat transfer in microfin tubes, *J. Heat Mass Transfer*, 40, 4345–4358.

Burck, E., 1970. The influence of Prandtl number on heat transfer and pressure drop of artificially roughened channels, in *Augmentation of Convective Heat and Mass Transfer*, A.E. Bergles and R. L. Webb, Eds., ASME Symp. Vol., ASME, New York, 27–35.

Burgess, N.K., Oliveira, M.M., and Ligrani, P.M., 2003. Nusselt number behavior on deep dimpled surfaces within a channel, *J. Heat Transfer*, 125, 11–18.

Chakroun, W. and Taylor, R.P., 1992. The effects of modestly strong acceleration on heat transfer in the turbulent rough-wall boundary layer, in *Enhanced Heat Transfer*, M.B. Pate and M.K. Jensen, Eds., ASME Symp. Vol. HTD, 202, 57–66.

Chandra, P.R., Niland, M.E., and Han, J.C., 1997. Effect of rib profiles on turbulent channel flow heat transfer, *J. Turbomachinery*, 119, 373–380.

Chen, H.C. and Patel, V.C., 1988, Near-wall turbulence models for complex flows including separation, *AIAA J.,* 26, 641–648.

Churchill, S.W., 1973. Empirical expressions for the shear stress in turbulent flow in commercial pipe, *AIChE J.*, 19, 375–376.

Chyu, M.K., Yu, Y., Ding, H., Downs, J.P., and Soechting, F.O., 1997. Concavity Enhanced Heat Transfer in an Internal Cooling Passage, ASME Paper 97-GT-437, ASME, New York.

Cope, W.G., 1945. The friction and heat transmission coefficients of rough pipes, *Proc. Inst. Mech. Eng.*, 145, 99–105.

Cuangya, L., Chuanyun, G., Chaosu, W., Jinshu, H., and Cun, J., 1991. Experimental investigation of transitional flow heat transfer of three-dimensional internally finned tubes, in *Advances in Heat*

Transfer Augmentation and Mixed Convection, M.A. Ebadian, D.W. Pepper, and T. Diller, Eds., ASME Symp. Vol., HTD, 169, 45–48.

Dalle-Donne, M. and Meyer, L., 1977. Turbulent convection heat transfer from rough surfaces with two-dimensional rectangular ribs, *Int. J. Heat Mass Transfer*, 20, 583–620.

Dawson, D.A. and Trass, O., 1972. Mass transfer at rough surfaces, *Int. J. Heat Mass Transfer*, 15, 1317–1336.

Dipprey, D.F. and Sabersky, R.H., 1963. Heat and momentum transfer in smooth and rough tubes at various Prandtl numbers, *Int. J. Heat Mass Transfer*, 6, 329–353.

Dong, Y., Huixiong, L., and Tingkuan, C., 2001. Pressure drop, heat transfer and performance of single-phase turbulent flow in spirally corrugated tubes, *Exp. Thermal Fluid Sci.*, 24, 131–138.

Durst, F., Fonti, M., and Obi, S., 1988, Experimental and computational investigation of the two-dimensional channel flow over two fences in tandem, *J. Fluids Eng.*, 110, 48–54.

Edwards, F.J., 1966. The correlation of forced convection heat transfer data from rough surfaces in ducts having different shapes of flow cross section, *Proc. Third Int. Heat Transfer Conf.*, 1, 32–44.

Edwards, F.J. and Sheriff, N., 1961. The heat transfer and friction characteristics of forced convection air flow over a particular type of rough surface, *International Developments in Heat Transfer*, ASME, New York, 415–426.

Ekkad, S.V. and Han, J.C., 1997. Detailed heat transfer distributions in two-pass square channels with rib turbulators, *Int. J. Heat Mass Transfer*, 40, 2525–2537.

Ekkad, S.V., Huang, Y., and Han, J.C., 1998. Detailed heat transfer distributions in two-pass smooth and turbulated square channels with bleed holes, *Int. J. Heat Mass Transfer*, 41, 3781–3791.

Esen, E.B., Obot, N.T., and Rabas, T.J., 1994a. Enhancement: Part I. Heat transfer and pressure drop results for air flow through passages with spirally-shaped roughness, *J. Enhanced Heat Transfer*, 1, 145–156.

Esen, E.B., Obot, N.T., and Rabas, T.J., 1994b. Enhancement: Part II. The role of transition to turbulent flow, *J. Enhanced Heat Transfer*, 1, 157–167.

Faramarzi, J. and Logan, E., 1991. Reattachment length behind single roughness element in turbulent pipe flow, *J. Fluids Eng.*, 113, 712–714.

Farrell, P., Wert, K., and Webb, R.L., 1991. Heat Transfer and Friction Characteristics of Turbulator Radiator Tubes, SAE Technical Paper 910917, SAE International Congress, Detroit, MI.

Fenner, G.W. and Ragi, E., 1979. Enhanced Tube Inner Surface Heat Transfer Device and Method, U.S. patent 4,154,291, May 15.

Fodemski, T.R. and Collins, M.W., 1988. Flow and heat transfer simulations for two- and three-dimensional smooth and ribbed channels, in *Proc. 2nd U.K National Conf on Heat Transfer*, C138/88, University of Stratchlyde, Glasgow, U.K.

Fujita, H., Hajime, Y., and Nagata, C., 1986. The numerical prediction of fully developed turbulent flow and heat transfer in a square duct with two roughened facing walls, *Proc. 8th Int. Heat Transfer Conf.*, 3, 919–924.

Gee, D.L. and Webb, R.L., 1980. Forced convection heat transfer in helically rib-roughened tubes, *Int. J. Heat Mass Transfer*, 23 1127–1136.

Good, M.C. and Joubert , P.N., 1968. The form drag of two-dimensional bluff plates immersed in turbulent boundary layers, *J. Fluid Mech.*, 31, 547–582.

Gowen, R.A. and Smith, J.W., 1968. Turbulent heat transfer from smooth and rough surfaces, *Int. J. Heat Mass Transfer*, 11, 1657–1673.

Groehn, H.G. and Scholz, F., 1976. Heat transfer and pressure drop of in-line tube banks with artificial roughness, in *Heat and Mass Transfer Sourcebook: Fifth All-Union Conf.*, Minsk, Scripta, Washington, D.C., 21–24.

Hall, W.B., 1962. Heat transfer in channels having rough and smooth surfaces, *J. Mech. Eng. Sci.*, 4, 287–291.

Han, J.C., 1984. Heat transfer and friction in channels with two opposite rib-roughened walls, *J. Heat Transfer*, 106, 774–781

Han, J.C., 1988. Heat transfer and friction characteristics in rectangular channels with rib turbulators, *J. Heat Transfer*, 110, 321–328.

Han, J.C. and Zhang, Y.M., 1992. High performance heat transfer ducts with parallel broken and V-shaped broken ribs, *Int. J. Heat Mass Transfer*, 35, 513–523.

Han, J.C., Glicksman, L.R., and Rohsenow, W.M., 1978. An investigation of heat transfer and friction for rib-roughened surfaces, *Int. J. Heat Mass Transfer*, 21, 1143–1156.

Han, J.C., Glicksman, L.R., and Rohsenow, W.M., 1979. An investigation of heat transfer and friction for rib-roughened surfaces, *Int. J. Heat Mass Transfer*, 22, 1587.

Han, J.C., Chandra, P.R., and Lau, S.C., 1988. Local heat/mass transfer distributions around sharp 180 deg turns in two-pass smooth and rib-roughened channels, *J. Heat Transfer*, 110, 91–98.

Han, J.C., Ou, S., Park, J.S., and Lei, C.K., 1989. Augmented heat transfer in rectangular channels of narrow aspect ratios with rib turbulators, *Int. J. Heat Mass Transfer*, 32, 1619–1630.

Han, J.C., Zhang, Y.M., and Lee, C.P., 1991. Augmented heat transfer in square channels with parallel, crossed, and V-shaped angled ribs, *J. Heat Transfer*, 113, 590–596.

Han, J.C., Huang, J.J., and Lee, C.P., 1993. Augmented heat transfer in square channels with wedge-shaped and delta-shaped turbulence promoters *J. Enhanced Heat Transfer*, 1, 37–52.

Han, J.C., Dutta, S., and Ekkad, S.V., 2000. *Gas Turbine Heat Transfer and Cooling Technology*, Taylor & Francis, London.

Hijikata, K. and Mori, Y., 1987. Fundamental study of heat transfer augmentation of tube inside surface by cascade smooth turbulence promoters and its application to energy conversion, *Wärme Stoffüber-trag*, 21, 115–124.

Hijikata, K., Ishiguro, H., and Mori, Y., 1987. Heat transfer augmentation in a pipe flow with smooth cascade turbulence promoters and its application to energy conversion, in *Heat Transfer in High Technology and Power Engineering*, W.J. Yang and Y. Mori, Eds., Hemisphere, New York, 368–397.

Hinze, J.O., 1975. *Turbulence*, 2nd ed., McGraw-Hill, New York.

Hishida, M. and Takase, K., 1987. Heat transfer coefficient of the ribbed surface, *Proc. Third ASME/SME Joint Thermal Eng. Conf.*, 3, 103–110.

Hitachi, 1984. Hitachi High-Performance Heat-Transfer Tubes, Catalog EA-500, Hitachi Cable, Ltd., Tokyo, Japan.

Hosni, M.H., Coleman, H.W., and Taylor, R.P., 1989. Measurement and calculation of surface roughness effects on turbulent flow and heat transfer, Report TFD-89–1, Mechanical and Nuclear Engineering Department, Mississippi State University.

Hosni, M.H., Coleman, H.W., and Taylor, R.P., 1991. Measurements and calculations of rough-wall heat transfer in the turbulent boundary layer, *Int. J. Heat Mass Transfer*, 34, 1067–1081.

Hudina, M., 1979. Evaluation of heat transfer performances of rough surfaces from experimental investigation in annular channels, *Int. J. Heat Mass Transfer*, 22, 1381–1392.

Hung, Y.H., Liou, T.M., and Syang, Y.C., 1987. Heat transfer enhancement of turbulent flow in pipes with an external circular rib, in *Advances in Enhanced Heat Transfer-1987*, M.K. Jensen and V.P. Carey, Eds., ASME Symp. Vol. HTD, 68, 55–64.

Hwang, J.-J., 1998. Heat transfer-friction characteristic comparison in rectangular ducts with slit and solid ribs mounted on one wall, *J. Heat Transfer*, 120, 709–716.

Hwang, J.-J. and Liou, T-M., 1994. Augmented heat transfer in a rectangular channel with permeable ribs mounted on the wall, *J. Heat Transfer*, 116, 912–920.

Hwang, J.-J. and Liou, T.-M., 1995. Heat transfer and friction in a low-aspect-ratio rectangular channel with staggered perforated ribs on two opposite walls, *J. Heat Transfer*, 117, 843–850.

Iacovides, H. and Raisee, M., 1999. Recent progress in the computation of flow and heat transfer in internal cooling passages of turbine blades, *Int. J. Heat Fluid Flow*, 20, 320–328.

Ito, Y., Hozumi, H., Noguchi, K., Oizumi, K., and Seki, K., 1983. Characteristics of newly developed THERMOEXCEL with pin-shaped fins for water cooled coaxial condensers, *Hitachi Cable Rev.*, 2 (August), 45–49.

Jaber, M.H., Webb, R.L., and Stryker, P., 1991. An Experimental Investigation of Enhanced Tubes for Steam Condensers, ASME Paper 91-HT-5, ASME, New York.

James, C.A., Hodge, B.K., and Taylor, R.P., 1994. A validated procedure for the prediction of fully-developed Nusselt numbers and friction factors in tubes with two-dimensional rib roughness, *J. Enhanced Heat Transfer*, 1, 287–304.

Kays, W.M. and Crawford, M.E., 1980. *Convective Heat Transfer*, McGraw-Hill, New York, 174 and 188.

Kukreja, R.T., Lau, S.C., and McMillan, R.D., 1993. Local heat/mass transfer distribution in a square channel with full and V-shaped ribs, *Int. J. Heat Mass Transfer*, 36, 2013–2020.

Kumar, R. and Judd, R.L., 1970. Heat transfer with coiled wire turbulence promoters, *Can. J. Chem. Eng.*, 48, 378–383.

Kuwahara, H., Takahashi, K., Yanagida, T., Nakayama, W., Hzgimoto, S., and Oizumi, K., 1989. Method of Producing a Heat Transfer Tube for Single-Phase Flow, U.S. patent 4,794,775, January 3.

Lewis, M.J., 1974. Roughness functions, the thermohydraulic performance of rough surfaces and the Hall transformation — an overview, *Int. J. Heat Mass Transfer*, 17, 809–814.

Lewis, M.J., 1975. An elementary analysis predicting the momentum and heat transfer characteristics of a hydraulically rough surface, *J. Heat Transfer*, 97, 249–254.

Li, H.M., Ye, K.S., Tan, Y.K., and Den, S.J., 1982. Investigation of tube-side flow visualization, friction factors and heat transfer characteristics of helical-ridging tubes, in *Proceedings 7th International Heat Transfer Conference*, Vol. 3, Hemisphere, Washington, D.C., 75–80.

Liao, Q., Zhu, X., and Xin, M.D., 2000. Augmentation of turbulent convective heat transfer in tubes with three-dimensional internal extended surfaces, *J. Enhanced Heat Transfer*, 7(3), 139–151.

Ligrani, P.M., Mahmood, G.I., Harrison, J.L., Clayton, C.M., and Nelson, D.L., 2001. Flow structure and local Nusselt number variations in a channel with dimples and protrusions on opposite walls, *Int. J. Heat Mass Transfer*, 44, 4413–4425.

Ligrani, P.M., Oliviera, M.M., and Blaskovich, T., 2003. Comparison of heat transfer augmentation techniques, *AIAA J.*, 41(3), 337–362.

Liou, T.M. and Hwang, J.J., 1993. Effect of ridge shapes on turbulent heat transfer and friction in a rectangular channel, *Int. J. Heat Mass Transfer*, 36, 931–940.

Liou, T.-M., Chen, S.-H., and Shih, K.-C., 2002. Numerical simulation of turbulent flow field and heat transfer in a two-dimensional channel with periodic slit ribs, *Int. J. Heat Mass Transfer*, 45, 4493–4505.

Liu, X. and Jensen, M.K., 2001. Geometry effects on turbulent flow and heat transfer in internally finned tubes, *J. Heat Transfer*, 123, 1035–1044.

Mahmood, G.I., 2001. Heat Transfer and Flow Structure from Dimples in an Internal Cooling Passage, Ph.D. thesis, University of Utah, Salt Lake City.

Mahmood, G.I. and Ligrani, P.M., 2002. Heat transfer in a dimpled channel: combined influences of aspect ratio, temperature ratio, Reynolds number, and flow structure, *Int. J. Heat Mass Transfer*, 45, 2011–2020.

Mahmood, G.I., Hill, M.L., Nelson, D.L., Ligrani, P.M., Moon, H.K., and Glezer, B., 2001a. Local heat transfer and flow structure on and above a dimpled surface in a channel, *J. Turbomachinery*, 123, 115–123.

Mahmood, G.I., Sabbagh, M.Z., and Ligrani, P.M., 2001b. Heat transfer in a channel with dimples and protrusions on opposite walls, *J. Thermophys. Heat Transfer*, 15, 275–283.

Mantle, P.L., 1966. A new type of roughened heat transfer surface selected by flow visualization techniques, *Proc. 3rd Int. Heat Transfer Conf.*, 1, 45–55.

Maubach, K., 1972. Rough annulus pressure drop — interpretation of experiments and recalculation for square ribs, *Int. J. Heat Mass Transfer*, 15, 2489–2498.

McLain, C.D., 1975. Process for Preparing Heat Exchanger Tube, U.S. patent 1,906,605, issued to Olin Corp.

Mehta, M.H. and Raja Rao, M., 1979. Heat transfer and friction characteristics of spirally enhanced tubes for horizontal condensers, in *Advances in Enhanced Heat Transfer*, J.M. Chenoweth et al., Eds., ASME Symp. Vol., ASME, New York, 11–22.

Mehta, M.H. and Raja Rao, M., 1988. Analysis and correlation of turbulent flow heat transfer and friction coefficients in spirally corrugated tubes for steam condenser application, *Proc. 1988 Natl. Heat Trans. for Conf.*, HTD-96, 3, 307–312.

Mendes, P.R.S. and Mauricio, M.H.P., 1987. Heat transfer, pressure drop, and enhancement characteristics of the turbulent flow through internally ribbed tubes, in *Convective Transport*, ASME HTD, 82, 15–22.

Metzger, D.E., Fan, C.Z., and Pennington, J.W., 1983. Heat transfer and flow friction characteristics of very rough transverse ribbed surfaces with and without pin fins, *Proc. 1983 ASME-JSME Thermal Eng. Conf.*, 1, 429–436.

Metzger, D.E., Vedula, R.P., and Breen, D.D., 1987. The effect of rib angle and length on convection heat Transfer in rib-roughened triangular ducts, *Proc. 1987 ASME-JSME Thermal Eng. Conf.*, 3, 327–333.

Meyer, L., 1980. Turbulent flow in a plane channel having one or two rough walls, *Int. J. Heat Mass Transfer*, 23, 591–608.

Moon, H.-K., O'Conell, T., and Glezer, B., 1999. Channel Height Effect on Heat Transfer and Friction in a Dimpled Passage, ASME Paper 99-GT-163, ASME, New York.

Nakamura, H. and Tanaka, M., 1973. Cross-Rifled Vapor Generating Tube, U.S. patent 3,734,140, May 22.

Nakayama, W., Takahashi, K., and Daikoku, T., 1983. Spiral ribbing to enhance single-phase heat transfer inside tubes, *Proc. ASME-JSME Thermal Eng. Joint Conf.*, Honolulu, HI, 1, ASME, New York, 503–510.

Newson, I.H. and Hodgson, T.D., 1973. The development of enhanced heat transfer condenser tubing, *Desalination*, 14, 291–323.

Nikuradse, J., 1933. Laws of Flow in Rough Pipes, *VDI Forschungsheft*, pp. 361. [English translation, NACA TM-1292 (1965).]

Nunner, W., 1956. Heat transfer and pressure drop in rough pipes, *VDI-Forschungsheft*, 455, Ser. B, 22, 5–39. [English Translation, AERE Lib./Trans. 786 (1958).]

Obot, N.T., Das, L., and Rabas, T.J., 2001a. Smooth- and enhanced-tube heat transfer and pressure drop. Part I. Effect of Prandtl number with air, water and glycol/water mixtures, in *Proceedings of the Third International Conference on Compact Heat Exchangers and Enhancement Technology for the Process Industries*, R.K. Shah, A.W. Deakin, H. Honda, and T.M. Rudy, Eds., Begell House, New York, 259–264.

Obot, N.T., Das, L., and Rabas, T.J., 2001b. Smooth- and enhanced-tube heat transfer and pressure drop. Part II. The role of transition to turbulent flow, in *Proceedings of the Third International Conference on Compact Heat Exchangers and Enhancement Technology for the Process Industries*, R.K. Shah, A.W. Deakin, H. Honda and T.M. Rudy, Eds., Begell House, New York, 187–198.

Olsson, C.-O. and Sunden, B., 1996. Heat transfer and pressure drop characteristics of ten radiator tubes, *Int. J. Heat Mass Transfer*, 39, 3211–3220.

Olsson, C.-O. and Sunden, B., 1998a. Experimental study of flow and heat transfer in rib-roughened rectangular channels, *Exp. Thermal Fluid Sci.*, 16, 349–365.

Olsson, C.-O. and Sunden, B., 1998b. Thermal and hydraulic performance of a rectangular duct with multiple V-shaped ribs, *J. Heat Transfer*, 121, 1072–1077.

Patankar, S.V., 1990. Numerical prediction of flow and heat transfer in compact heat exchanger passages, in *Compact Heat Exchangers*, R.K. Shah, A.D. Kraus, and D. Metzger, Eds., Hemisphere, Washington, D.C., 191–204.

Petukhov, B.S., 1970. Heat transfer in turbulent pipe flow with variable physical properties, in *Advances in Heat Transfer*, Vol. 6, T.F. Irvine and J.P. Hartnett, Eds., Academic Press, New York, 504–564.

Prakash, C. and Zerkle, R., 1995. Prediction of turbulent flow and heat transfer in a ribbed rectangular duct with and without rotation, *J. Turbomachinery*, 117, 255–264.

Prasad, R.C. and Brown, M.J., 1988. Effectiveness of wire-coil inserts in augmentation of convective heat transfer, in *Experimental Heat Transfer, Fluid Mechanics and Thermodynamics*, R.K. Shah, E.N. Ganic, and K.T. Yang, Eds., Elsevier, New York, 502–509.

Rabas, T.J. and Arman, B., 1992. The influence of the Prandtl number of the thermal performance of tubes with the separation and reattachment enhancement mechanism, in *Enhanced Heat Transfer*, M.B. Pate and M.K. Jensen, Eds., ASME Symp. Vol. HTD, 202, 77–88.

Rabas, T.J., Bergles, A.E., and Moen, D.L., 1988. Heat transfer and pressure drop correlations for spirally grooved (rope) tubes used in surface condensers and multistage flash evaporators, *Augmentation of Heat Transfer in Energy Systems*, ASME Symp. Vol. HTD, 52, 693–704.

Rabas, T.J., Thors, P., Webb, R.L., and Kim, N-H., 1993. Influence of roughness shape and spacing on the performance of three-dimensional helically dimpled tubes, *J. Enhanced Heat Transfer*, 1, 53–64.

Raja Rao, M., 1988. Heat transfer and friction correlations for turbulent flow of water and viscous non-Newtonian fluids in single-start spirally corrugated tubes, *Proc. 1988 Natl. Heat Transfer Conf.* HTD-96, 1, 677–683.

Ravigururajan, T.S. and Bergles, A.E., 1985. General correlations for pressure drop and heat transfer for single-phase turbulent flow in internally ribbed tubes, *Augmentation of Heat Transfer in Energy Systems*, ASME Symp. Vol. HTD, 52, 9–20.

Ravigururajan, T.S. and Bergles, A.E., 1996. Development and verification of general correlations for pressure drop and heat transfer in single-phase turbulent flow in enhanced tubes, *Exp. Thermal Fluid Sci.*, 13, 55–70.

Rigby, D.L., Steinthorsson, E., and Ameri, A.A., 1997. Numerical Prediction of Heat Transfer in a Channel with Ribs and Bleed, ASME Paper 97-GT-431, ASME, New York.

Schlicting, H., 1979. *Boundary-Layer Theory*, 7th ed., McGraw-Hill, New York, 600–620.

Sethumadhavan, R. and Raja Rao, M., 1983. Turbulent flow heat transfer and fluid friction in helical wire coil inserted tubes, *Int. J. Heat Mass Transfer*, 26, 1833–1845.

Sethumadhavan, R. and Raja Rao, M., 1986. Turbulent flow friction and heat transfer characteristics of single- and multi-start spirally enhanced tubes, *J. Heat Transfer*, 108, 55–61.

Sheriff, N. and Gumley, P., 1966. Heat transfer and friction properties of surfaces with discrete roughness, *Int. J. Heat Mass Transfer*, 9, 1297–1320.

Shome, B. and Jensen, M.K., 1996. Experimental investigation of laminar flow and heat transfer in internally finned tubes, *J. Enhanced Heat Transfer*, 4, 53–70.

Smith, J.W. and Gowen, R.A., 1965. Heat transfer efficiency in rough pipes at high Prandtl number, *AIChE J.*, 11, 941–943.

Sumitomo, 1983. *Technical Data of Tred-Fin*, Sumitomo Light Metal Industries, Aichi, Japan.

Takahashi, K., Nakayama, W., and Kuwahara, H., 1988. Enhancement of forced convective heat transfer in tubes having three-dimensional spiral ribs, *Heat Transfer Jpn. Res.*, 17(4), 12–28.

Tan, Y.K. and Xaio, J.W., 1989. Influence of Prandtl number to fluid heat transfer characteristics of spirally corrugated tubes, Presented at the Fourth Asian Congress of Fluid Mechanics, Hong Kong.

Tanasawa, I., Nishio, S., Takano, K., and Tado, M., 1984. Enhancement of forced convection heat transfer in rectangular channel using turbulence promoters, *1983 ASME-JSME Thermal Eng. Joint Conf.*, Vol. 1, Y. Mori and I. Tanasawa, Eds., ASME, New York, 395–402.

Taslim, M.E. and Spring, S.D., 1988. An experimental investigation of heat transfer coefficients and friction factors in passages of different aspect ratios roughened with 45 degree turbulators, *Proc. 1988 Natl. Heat Trans. Conf.*, HTD-96, 1., 661–668.

Taslim, M.E. and Spring, S.D., 1994. Effects of turbulator profile and spacing on heat transfer and friction in a channel, *J. Thermophys. Heat Transfer*, 8, 555–562.

Taylor, R.P. and Hodge, B.K., 1992. Fully-developed heat transfer and friction factor predictions for pipes with 3-dimensional roughness, in *Fundamentals of Forced Convection Heat Transfer*, M.S. Ebadian and P.H. Oosthuizen, Eds., ASME Symp. Vol. HTD, 210, 75–84.

Taylor, R.P., Coleman, H.W., and Hodge, B.K., 1984. A Discrete Element Prediction Approach for Turbulent Flow Over Rough Surfaces, Report TFD-84–1, Department of Mechanical Engineering, Mississippi State University.

Taylor, R.P., Scaggs, W.F., and Coleman, H.W., 1988. Measurement and prediction of the effects of nonuniform surface roughness on turbulent flow friction coefficients, *J. Fluid Eng.*, 110, 380–384.

Thors, P., 2004. Personal communication.

Thors, P., Clevinger, N.R., Campbell, B.J., and Tyler, J.T., 1997. Heat Transfer Tubes and Methods of Fabrication Thereof, U.S. patent 5,697,430, December 16.

Vicente, P.G., Garcia, A., and Viedma, A., 2002a. Experimental study of mixed convection and pressure drop in helically dimpled tubes for laminar and transition flow, *Int. J. Heat Mass Transfer*, 45, 5091–5105.

Vicente, P.G., Garcia, A., and Viedma, A., 2002b. Heat transfer and pressure drop for low Reynolds turbulent flow in helically dimpled tubes, *Int. J. Heat Mass Transfer*, 45, 543–553.

Watkinson, A.P., Miletti, D.L., and Kubanek, G.R., 1975. Heat transfer and pressure drop of forge-fin tubes in laminar oil flow, ASME paper 75-HT-41, presented at the AIChE-ASME Heat Transfer Conference, San Francisco, Aug. 11–13.

Webb, R.L., 1971. A critical evaluation of analytical and Reynolds analogy equations for turbulent heat and mass transfer in smooth tubes, *Wärme Stoffübertrag.*, 4, 197–204.

Webb, R.L., 1979. Toward a common understanding of the performance and selection of roughness for forced convection, *Studies in Heat Transfer: A Festschrift for E.R.G. Eckert*. J.P. Hartnett et al., Eds., Hemisphere Washington, D.C., 257–272.

Webb, R.L., 1982. Performance cost effectiveness and water-side fouling considerations of enhanced tube heat exchangers for boiling service with tube-side water flow, *Heat Transfer Eng.*, 3(3), 84–98.

Webb, R.L., 1991. Advances in shell side boiling of refrigerants, *J. Inst. Refrig.*, 87, 75–86.

Webb, R.L. and Eckert, E.R.G., 1972. Application of rough surfaces to heat exchanger design, *Int. J. Heat Mass Transfer*, 15, 1647–1658.

Webb, R.L. and Robertson, G.F., 1988. Shell-side evaporators and condensers used in the refrigeration industry, in *Heat Transfer Equipment Design*, R.K. Shah, E.C. Subbarao, and R.A. Mashelkar, Eds., Hemisphere, Washington, D.C., 559–570.

Webb, R.L., Eckert, E.R.G., and Goldstein, R.J., 1971. Heat transfer and friction in tubes with repeated-rib roughness, *Int. J. Heat Mass Transfer*, 14, 601–617.

Webb, R.L., Eckert, E.R.G., and Goldstein, R.J., 1972. Generalized heat transfer and friction correlations for tubes with repeated-rib roughness, *Int. J. Heat Mass Transfer*, 15, 180–184.

Webb, R.L., Hui, T.S., and Haman, L., 1984. Enhanced tubes in electric utility steam condensers, in *Heat Transfer in Heat Rejection Systems*, S. Sengupta and Y.F. Mussalli, Eds., Book G00265, HTD, 37, 17–26.

Webb, R.L., Narayanamurthy, R., and Thors, P., 2000. Heat transfer and friction characteristics of internal helical-rib roughness, *J. Heat Transfer*, 122, 134–142.

White, L. and Wilkie, D., 1970. The heat transfer and pressure loss characteristics of some multi-start ribbed surfaces, in *Augmentation of Convective Heat and Mass Transfer*, ASME, New York, 55–62.

Wilkie, D., 1966. Forced convection heat transfer from surfaces roughened by transverse ribs, *Third Int. Heat Transfer Conf.*, 1, 1–19.

Wilkie, D., Cowan, M., Burnett, P., and Burgoyne, T., 1967. Friction factor measurements in a rectangular channel with walls of identical and non-identical roughness, *Int. J. Heat Mass Transfer*, 10, 611–621.

Williams, F., Pirie, M.A.M., and Warburton, C., 1970. Heat transfer from surfaces roughened by ribs, in *Augmentation of Convective Heat and Mass Transfer*, ASME, New York, 55–62.

Withers, J.G., 1980a. Tube-side heat transfer and pressure drop for tubes having helical internal ridging with turbulent/transitional flow of single-phase fluid. Part 1: Single-helix ridging, *Heat Transfer Eng.*, 2(1), 48–58.

Withers, J.G., 1980b. Tube-side heat transfer and pressure drop for tubes having helical internal ridging with turbulent/transitional flow of single-phase fluid. Part 2: Multiple-helix ridging, *Heat Transfer Eng.*, 2(2), 43–50.

Wolfstein, M., 1988. The velocity and temperature distribution of one dimensional flow with turbulence augmentation and pressure gradient, *Int. J. Heat Mass Transfer*, 12, 301–318.

Wolverine, 1984. *Engineering Data Book II*, K.J. Bell and A.C. Mueller, Eds., Wolverine Tube Corp., Decatur, AL.

Ye, Q.Y., Chen, W.L., Marto, P.D., and Tan, Y.K., 1987. Prediction of heat transfer characteristics in spirally fluted tubes by studying the velocity profiles and turbulence structure, in *1987 ASME-JSME Thermal Eng. Joint Conf.*, Vol. 5, P.D. marto and I. Tanasawa, Eds., ASME, New York, 189–194.

Yorkshire, 1982. Heat Exchanger Tubes: Design Data for Horizontal Rope Tubes in Steam Condensers, Technical Memorandum 3, Yorkshire Imperial Metals, Ltd., Leeds, U.K.

Zhang, Y.F., Li, F.Y., and Liang, Z.M., 1991. Heat transfer in spiral-coil-inserted tubes and its application, in *Advances in Heat Transfer Augmentation*, M.A. Ebadian, D.W. Pepper, and T. Diller, Eds., ASME Symp. Vol. HTD, 169, 31–36.

Žukauskas, A.A., 1972. Heat transfer from tubes in crossflow, in *Advances in Heat Transfer*, Vol. 8, J.P. Hartnett and T.F. Irvine, Jr., Eds., Academic Press, New York.

Žukauskas, A.A. and Ulinskas, R.V., 1983. Surface roughness as means of heat transfer augmentation for banks of tubes in crossflow, in *Heat Exchangers: Theory and Practice*, J. Taborek, G.P. Hewitt, and N. Afgan, Eds., Hemisphere, Washington, D.C., 311–321.

Žukauskas, A.A. and Ulinskas, R.V., 1988. *Heat Transfer in Tube Banks in Crossflow*, Hemisphere, New York, 94–118.

FOULING ON ENHANCED SURFACES

10.1 INTRODUCTION

The commercial viability of enhanced heat transfer surfaces is dependent on their long-term fouling characteristics. The problem of heat transfer surface fouling is of concern for both plain and enhanced surfaces. Although fouling is generally regarded as a more serious problem for liquids than for gases, gas-side fouling can be important in certain situations. Section 2.7 describes the importance of fouling in heat exchanger design and explains why fouling is an important concern for enhanced surfaces. A primary concern for enhanced surfaces is their fouling rate, relative to a plain surface, when operated at the same velocity. At present, the state of the art does not allow quantitative prediction of the fouling resistance that will occur on smooth or enhanced surfaces in actual field installations. However, recent and ongoing research is advancing our quantitative understanding of fouling phenomena. Readers interested in general information describing the state of the art on fouling are referred to books by Somerscales and Knudsen [1979], Garrett-Price et al. [1985], Melo et al. [1987], and Bott [1995]. Watkinson [1990, 1991] and Bergles and Somerscales [1995] provide review articles of fouling on enhanced heat transfer surfaces.

Fouling research has resulted in the definition of six different types of fouling that may occur with liquids or gases:

1. Precipitation fouling (scaling) occurs when salts precipitate on a heat transfer surface, if the temperature is sufficiently hot (or cold) to cause supersaturation of the salts at the surface temperature. The salts may have either increasing or

decreasing solubility with increasing temperature. Evaporation of dissolved salts in a liquid beyond the solubility limit is one example. Precipitation of $CaCO_3$ from heated cooling water is an example of inverse solubility.

2. Particulate fouling takes place when suspended solids deposit on the surface. Examples are dirt contained in cooling water, corrosion products formed elsewhere and deposited on the surface, and airborne particulates (such as dust in air or soot in combustion products).

3. Corrosion fouling occurs when the heat transfer surface material reacts with the fluid to yield corrosion deposits.

4. Chemical reaction at the heat transfer surface (the surface material is not a reactant) may yield surface deposits. Examples are polymerization, cracking, and coking of hydrocarbons.

5. Biofouling deposits are formed when biological mechanisms attach and grow on the heat transfer surface. Untreated water cooling systems are susceptible to biofouling.

6. Solidification fouling can occur with either liquids or gases. Water vapor may be condensed from air and frozen on a surface below the freezing point. Similarly, a component such as sodium sulfate may precipitate from a gas and solidify on the heat transfer surface. A layer of ice will form on the surface if the surface temperature is below the freezing point of water. Paraffin waxes in hydrocarbon solutions may deposit as solids on a cooled surface.

Measurement of the fouling resistance is typically performed by measurement of the total thermal resistance ($1/UA$) for the clean and fouled conditions. The fouling resistance (R_f) is obtained by subtraction of the $1/UA$ values for the fouled and clean conditions, respectively, giving

$$\frac{R_f}{A} = \frac{1}{(UA)_f} - \frac{1}{(UA)_c} \tag{10.1}$$

where A is the surface area on which the R_f is based. For tube-side fouling, the UA values for the clean and fouled conditions are defined by Equations 10.2 and 10.3, respectively:

$$\frac{1}{(UA)_c} = \frac{1}{h_i A_i} + \frac{t_w}{k_w A_w} + \frac{1}{h_o A_o} \tag{10.2}$$

$$\frac{1}{(UA)_f} = \left(\frac{1}{h_i} + R_f\right)\frac{1}{A_i} + \frac{t_w}{k_w A_w} + \frac{1}{h_o A_o} \tag{10.3}$$

The inside (h_i) and outside (h_o) heat transfer coefficients must be equal in the dirty and clean tube conditions. Otherwise, the measured fouling resistance, R_f, will be erroneous.

10.2 FOULING FUNDAMENTALS

Fouling is a rate-dependent phenomena. The net fouling rate is the difference between the solids' deposition rate and their removal rate

$$\frac{dm_f}{dt} = \dot{m}_d - \dot{m}_r \tag{10.4}$$

Depending on the magnitudes of the deposition and removal terms, several fouling rate characteristics are possible. Figure 10.1 illustrates the increase of the fouling factor (R_f) with time for different possible fouling situations. Crystallization fouling has an initial delay period (t_d), during which nucleation sites are established; then, the fouling deposit begins to accumulate. A linear growth of the fouling deposit occurs if the removal rate $(\dot{m}_r)$ is negligible or if $\dot{m}_d$ and $\dot{m}_r$ are constant with $\dot{m}_d > \dot{m}_r$. Crystallization, chemical reaction, corrosion, and freezing fouling will show behavior in the linear or falling rate categories. The fouling resistance will attain an asymptotic value if $\dot{m}_d$ is constant and $\dot{m}_r$ eventually attains a constant value. Particulate fouling typically shows asymptotic behavior. Which type of fouling characteristics occurs depends on the fouling mechanism.

Epstein [1983, 1988a, 1988b] provides detailed discussion of the different fouling mechanisms. He also describes models to predict the deposition and removal rates for the different fouling mechanisms. The deposition model is a function of

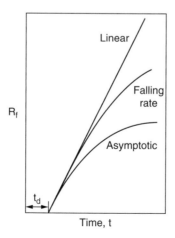

Figure 10.1 Characteristic fouling curves.

the fouling mechanism. The removal rate model depends on the reentrainment rate, which is proportional to the shear stress at the surface.

The heat exchanger operating parameters that may influence the deposition or removal rates are as follows:

1. The bulk fluid temperature will typically increase chemical reaction rates and crystallization deposition rates.
2. Increasing surface temperature will increase chemical reaction rates, or crystallization from inverse solubility salts. Reducing surface temperature will increase solidification fouling.
3. The combination of surface material and fluid will influence corrosion fouling. Copper surfaces act as a biocide to biological fouling. Rough surfaces may promote nucleation-based phenomena.
4. The liquid or gas velocity may influence both the deposition and removal rates. Removal of soft deposits is affected by surface shear stress, which increases with velocity. Increased velocity will increase the mass transfer coefficient and, hence, the deposition rate for diffusion controlled particulate fouling.

10.2.1 Particulate Fouling

To limit the discussion of fouling mechanisms, the focus is placed on particulate fouling, which is important for both gases and liquids. The particulate fouling deposition rate term is given by

$$\dot{m}_d = SK_m(C_b - C_w) \tag{10.5}$$

The deposition process is controlled by the mass transfer coefficient (K_m) and the concentration difference between that in the bulk fluid (C_b) and that in the liquid at the foulant surface (C_w). Typically, $C_w = 0$. The term S is called the sticking probability. This is the probability that a particle transported to the wall will stick to the wall.

The foulant removal rate model was proposed by Taborek et al. [1972] and is assumed to be proportional to surface shear stress (τ_w), foulant deposit thickness (x_f), and inversely proportional to the deposit bond strength factor (ξ). The removal rate is given by

$$\dot{m}_r = \frac{m_f \tau_w}{\xi} = \frac{\rho_f x_f \tau_w}{\xi} \tag{10.6}$$

Substitution of Equations 10.5 and 10.6 in Equation 10.4 using $R_f \equiv x_f/k_f$ and $dm/dt = (\rho_f/k_f) \, dR_f/dt$ gives the differential equation to be solved:

$$\frac{dR_f}{dt} = SK_m C_b - BR_f \tag{10.7}$$

where

$$B = \frac{\rho_f \tau_w k_f}{\xi} \tag{10.8}$$

The solution of Equation 10.7 is

$$R_f = R_f^*(1 - e^{-Bt}) \tag{10.9}$$

where R_f^* is the asymptotic fouling resistance defined by

$$R_f^* = \frac{SK_m C_b}{B} \tag{10.10}$$

Equations 10.7 and 10.8 show that the R_f and R_f^* can be predicted if K_m, τ_w, S, and ξ are known. One must also know the foulant density (ρ_f) and thermal conductivity (k_f). Equation 10.7 was first developed by Kern and Seaton [1959], who assumed that S and ξ are unity.

Papavergos and Hedley [1984] have shown that there are three regimes for particle deposition — the diffusion, inertia, and impaction regimes. Which regime controls the deposition process is determined by the dimensionless particle relaxation time defined by

$$t^+ = \frac{\rho_p d_p^2 (u^*)^2}{18 \mu v} \tag{10.11}$$

Papavergos and Hedley [1984] show that the t^+ values associated with the diffusion, inertia, and impaction regimes are $t^+ < 0.10$ (diffusion), $0.10 \le t^+ \le 10$ (inertia), and $t^+ > 10$ (impaction). Fine particle transport will likely be diffusion controlled. For example, the deposition process will be diffusion controlled $d_p \le 10$ μm for Re = 30,000 in a 15-mm-diameter tube. If the particle transport is diffusion controlled, the mass transfer coefficient may be predicted from heat transfer data on the enhanced surface, using the heat-mass transfer analogy, as reported by Kim and Webb [1991] and Chamra and Webb [1993b] on rough surfaces. If the heat transfer coefficient for the enhanced surface (h) is known, the mass transfer coefficient (K_m) is given by

$$\frac{K_m}{u} \text{Sc}^{2/3} = \frac{h}{\rho u c_p} \text{Pr}^{2/3} \tag{10.12}$$

The Sc $= v/D$ and the diffusion coefficient (D) are defined in the nomenclature at the end of the chapter. Small particles result in high Schmidt numbers. Equation 10.12 shows that a high heat transfer coefficients will result in a high mass transfer coefficient. Hence, enhanced heat transfer surfaces should result in higher foulant deposition rates than occur with plain surfaces at the same operating velocity.

Equation 10.6 shows that the foulant removal rate is directly proportional to the surface shear stress. One may also expect higher surface shear stress for an enhanced surface. The net effect of deposition and removal is defined by the asymptotic fouling resistance given by Equation 10.10. Equation 10.10 shows that $R_f^* \propto \dot{m}_d/\tau_w \propto K_m/\tau_w$. If $K_m/\tau_w > 1$, the enhanced surface will have a higher asymptotic fouling resistance than a plain surface. More discussion of this is provided by Watkinson [1991] and in Section 10.8.

10.3 FOULING OF GASES ON FINNED SURFACES

Marner and Webb [1983] provide a survey and bibliography of gas-side fouling. The bibliography provides 206 references dating from 1970. As noted in Chapter 2, gas-side fouling may not significantly increase the gas-side thermal resistance in finned tube gas-to-liquid heat exchangers. This may occur if the gas-side surface area is much greater than the liquid-side surface area. However, the fouling deposit on the heat exchanger surface increases the gas pressure drop. The increased pressure drop may result in a lower gas flow rate, because of the lower gas flow rate at the balance point on the fan curve. Hence, the heat transfer rate is reduced, because of the lower gas flow rate through the heat exchanger. An example of this is provided by the test data of Bott and Bemrose [1981]. They measured the effects of particulate fouling (3 to 30-μm calcium carbonate dust) on a finned heat exchanger as illustrated in Figure 6.1b. The four-row heat exchanger was composed of spiral wound fins at 2.4-mm pitch on a 25.4-mm diameter tube with 15.9-mm-high fins arranged in a staggered layout. Dust concentrations of 0.65 to 1.5 g/m^3 were injected, which the authors state "is equivalent to desert storm concentration." The heat exchanger was sprayed with a bonding agent prior to starting the fouling tests. This was intended to simulate field conditions and stabilize the foulant layer. Their test 24, taken at 4.8 m/s air frontal velocity (Re = 2400) and 0.78 g/m^3 dust loading resulted in a 67% friction factor increase over the 26-h test period. However, the air-side heat transfer coefficient decreased only 12%. The airflow rate reduced during the test period, because of the increased pressure drop. Although the flow rate reduction was not documented for test 24, they indicate that it may have been on the order of 10%. These data clearly show that the severe fouling caused a relatively small effect in the j factor, but a large increase in pressure drop. The associated increase of fan power is a costly penalty resulting from the fouling. Fouling factor curves were not given by the authors.

Bemrose and Bott [1984] tested finned tubes in one to four tube row configurations with calcium carbonate dust. They used a multiple regression technique to develop a correlation for the fouling resistance. No comparison with plain tube fouling was given.

Gas-side fouling is generally not a problem of concern for air-conditioning and automotive heat exchangers, where heating of the air is involved. This is because ambient air is relatively clean. In an air-cooled refrigerant evaporator, freezing fouling (frost or ice) occurs if the surface temperature is below the freezing point. This results in a thermal resistance, along with plugging of the flow passage. The foulant deposit is removed by melting the frost layer. Heat exchangers used in agricultural equipment are subject to particulate fouling, e.g., dust and chaff of threshed grains. Industrial heat exchangers used in dusty environments are subject to particulate fouling by dust. Such freezing and particulate fouling situations generally cause limitations on the minimum fin pitch.

Zhang et al. [1990] measured accelerated particulate fouling for a plate fin-and-tube automotive heat exchanger containing oval tubes and louvered fins, similar to that illustrated in Figure 6.24. The heat exchanger contained eight rows of tubes (27.6/12-mm major/minor diameter, 8.3-mm fin height, and 2.11-mm fin pitch). The air contained 5 to 12 µm $CaCO_3$ particulates. A foulant "cake" formed on the front surface of the heat exchanger, and fouling rates were higher for the smaller particles, and at higher air velocities. The foulant deposit caused a significant air pressure drop increase. They present fouling rate curves for different particle loadings.

Zhang et al. [1992] investigated the benefits of placing "spoilers" at the upstream face of a four-row bank of circular finned tubes. The spoilers consist of a shaped flow obstruction placed just upstream of each tube in the first row of tubes. They propose that the turbulence generated by the spoiler may reduce fouling. They measured the R_f^* for six different spoiler configurations. They asserted that the largest fouling deposit will occur on the front face of the finned tube heat exchanger. This may be the case for the plate fin-and-tube geometry studied by Zhang et al. [1990], although it may not be true for the circular fin geometry. In the author's opinion, this study was not conclusive in justifying the spoiler concept. Further, the spoilers add to the gas pressure drop.

More severe fouling situations exist for combustion products of coal and fuel oil. Combustion products may also contain vapors, which precipitate upon cooling. Possible fouling mechanisms include particulate, corrosion, and precipitation fouling. Gases that contain SO_2 will form corrosive H_2SO_3 (sulfuric acid) if both SO_2 and water vapor are condensed. Combustion products from glass furnaces contain H_2SO_4, which forms a sticky deposit when combined with condensed water vapor. Two examples, described by Webb et al. [1983], are provided to illustrate fouling resulting from combustion products, for two types of extended surface heat exchangers. Burgmeier and Leung [1981] measured the fouling characteristics of a plate-and-fin geometry using a simulated glass furnace exhaust. The 677°C gases were composed of natural gas combustion products with dilution air, 1600 ppm of 11 to 75-µm particulates, and added contaminants primarily containing SO_2, with SO_3 and HCl. They tested a plain fin geometries (Figure 6.1b) with 7.4-mm fin height with 295 fins/m and an offset-strip fin (OSF) with 3.7-mm fin height with 197 fins/m. They used a traversing air lance for cleaning, which consisted of a series of small

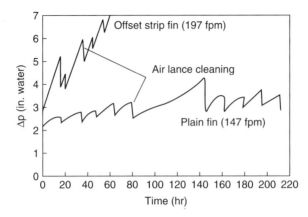

Figure 10.2 Fouling and cleaning characteristics of the plate-and-fin heat exchangers using 677/149°C gas inlet/outlet temperatures, with 1600 ppm particle loading.

holes drilled along the axis of the lance tube. The lance provides a line array of high-velocity air jets, which discharge directly onto the flow passages, at the upstream exchanger face. Figure 10.2 shows the fouling data for the two surface geometries with the air lance operating at 10-hr intervals. The air lance successfully cleaned the plain fin surface, but was ineffective for the OSF.

Roberts and Kubasco [1979] measured fouling on a bank of helical-finned tubes (Figure 6.1b) in the exhaust of a gas turbine. The tubes were 19.05 mm in diameter with 275 fins/m, with approximately 9 mm fin height. The tubes were exposed to 450°C exhaust gases, and the fin temperatures were 150 to 315°C. The deposits were "fluffy" and easily removed by air lancing. The deposit thickness increased as the metal temperature was decreased. Analysis of the deposit suggested a two-stage fouling mechanism. Trace quantities of hydrocarbons condense on the surface forming a thin coating of an adhesive nature, which trapped particulates. A self-cleaning method was evaluated, in which the tube-side coolant flow was stopped, and the hot gases baked away the deposit.

Kindlman and Silvestrini [1979] measured fouling and corrosion on 32-mm-diameter 118 and 197 fins/m finned tubes (Figure 6.1b) in the 200 to 260°C exhaust of a diesel engine, whose acid condensation temperature was 117°C. The test compared fouling on uncooled tubes with that on cooled tubes (104°C wall temperature). No measurable soot accumulation was observed on the uncooled tubes. However, the cooled tubes exhibited severe fouling and plugging.

These three studies show a consistency in several respects: (1) The precipitation fouling is dependent on the wall temperature and the surface-to-gas temperature difference. (2) If vapors do not condense, an easily removed fluffy deposit occurs. (3) Above the acid and water vapor dew points, some vapors may condense. These yield a hard deposit on the tube surface that is difficult to remove by air lancing.

However, it may be washed away with water. (4) Below the acid and water dewpoints, the deposit is sticky and cannot be removed by air lancing, although it may be water washed.

Grillot and Icart [1997] obtained Diesel exhaust fouling data for a circular finned tube heat exchanger. The 5-row heat exchanger was composed of spiral fins at 1.0-mm pitch on a 20-mm-diameter tube with 8.0-mm-high fins arranged in an inline layout. The exhaust gas temperature was controlled to approximately 380°C. Oil was circulated in the tube side to recover heat from the exhaust gas. The average soot size was measured to be 0.67 μm. The fouling resistance showed an asymptotic behavior. The asymptotic fouling resistance increased as the gas velocity decreased and the wall temperature decreased. Correlations for the asymptotic fouling resistance and the initial fouling rate were developed from their data.

It is interesting to speculate on the relative fouling rates of the plate-and-fin and finned-tube exchanger types, and their cleanability. The plate-and-fin geometry has a constant-area flow channel, and no stagnation or separated flow zones within the fin passage. One would expect a relatively uniform fouling deposit on the fins. Air or steam lancing will affect foulant removal as a result of high shear stress at the fin surface. This high velocity should be quite effective over the entire finned surface area. Conversely, finned tubes experience a stagnation/accelerating flow on the upstream half of the tube, and a decelerating flow with flow separation on the downstream half of the tube. Because the surface shear stress is greater on the front half of the tube, smaller fouling deposits are expected on the upstream face. This foulant deposit pattern was observed by Kindlman and Silvestrini [1979] for heat recovery from diesel exhaust. The fouling deposit thickness distribution at the fin tip for the upstream and downstream faces is shown by Figure 10.3. Using air or steam lancing, the smallest local surface shear stresses will occur on the downstream face — where the greatest foulant deposit exists. Thus, the desired cleaning shear stress distribution is opposite to the desired distribution.

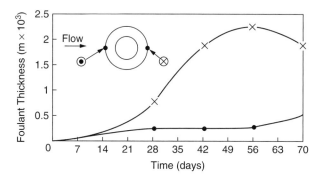

Figure 10.3 Fouling thickness near the fin tip on 31.8-mm-diameter tubes with 118 fins/m and 15.9-mm fin height.

10.4 SHELL-SIDE FOULING OF LIQUIDS

10.4.1 Low Radial Fins

Several papers have been published on fouling of low, integral-finned tubes (Figure 6.1c) in actual heat exchanger service. These long-term studies in field fouling situations typically show that the integral-fin tubes experience equal or less scaling fouling (R_f/A_o) than do plain tubes, and the scale is easier to remove than on plain tubes. Moore [1974] argues that the small axial thermal expansion between adjacent fins acts to break the scale or keep it looser than that on plain tubes. Moore's conclusions are based on actual examination of fouling deposits and cleaning ability using water jets. Webber [1960] conducted laboratory and field studies of precipitation (scaling) fouling on 19-mm-diameter integral-fin tubes (748 fins/m, 1.5 mm). Their single tube laboratory studies indicated that the finned tubes would not foul as readily as plain tubes and that, once dirty, they would be easier to clean. Moore recommends integral-fin tubes in "any dirty service, as long as coking does not occur."

Webber reports several case studies, in which plain and finned tube bundles are compared in service involving scaling fouling. He concludes that the finned tube bundles may be operated for a longer time period, before heat transfer limitations make cleaning necessary.

Katz et al. [1954] compare the performance of plain and finned tube bundles (725 fins/m) having hot No. 5 fuel oil on the shell side and cooling water on the tube side. In the clean condition, the finned tube bundle provided a *UA* value 2.2 times that of the plain tube bundle. After 670 h, the *UA* value of the finned tube bundle was 2.5 times that of the plain tube bundle. These data are also discussed by Watkinson [1990].

10.4.2 Axial Fins and Ribs in Annulus

Sheikholeslami and Watkinson [1986] measured accelerated precipitation fouling from hard water on flow in an annulus. They tested a plain tube and an axially finned tube, as illustrated in Figure 8.17a. The 25°C hard water was heated by tube side steam at 129 kPa. The R_f was based on the nominal tube surface area ($A_o/L = \pi d_o$). Surprisingly, the finned tube (12 fins, 6.0 mm high on a 19-mm-diameter tube) showed a fouling resistance 50 to 60% less than that for a steel plain tube of the same diameter. The deposit on the finned tube was heaviest on the prime surface. The scale thickness decreased with distance from the fin base. This is expected, as high surface temperature promotes scaling and the fin temperature decreases from the base.

Freeman et al. [1990] measured particulate fouling on the outer surface of several 12.7-mm-diameter finned and rough tubes in a double-pipe heat exchanger. Their tests were performed using 0.3-μm Al_2O_3 particles in a nonpolar fluid called X-2, which contained C_6 and C_7 hydrocarbons. They measured the asymptotic

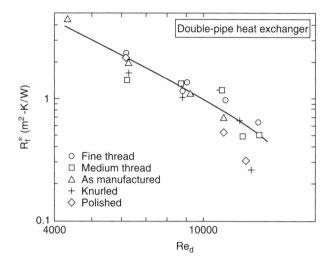

Figure 10.4 Asymptotic particulate fouling resistance for 650 ppm concentration of 0.3 μm Al_2O_3 particles in X-2 fluid tested in double-pipe heat exchanger by Freeman et al. [1990]. (From Freeman et al., 1990.)

fouling resistance for particle concentrations between 700 and 2000 ppm and liquid Reynolds numbers between 4000 and 14,000. Figure 10.4 shows their asymptotic fouling results on three enhanced surfaces and two plain surfaces (one commercially rough and one polished), apparently taken at 650-ppm particulate concentration. The R_f^* of threaded and knurled surfaces is not much different from that of the plain tubes. The thread or knurl depth roughness of the three enhanced surfaces, based on British Standard BS 1134, are fine thread (0.01 mm), medium thread (0.018 mm), and knurled (0.087 mm). These compare to 0.0055 mm-185for the commercial smooth plain tube. The Figure 10.4 data show that $R_f^* \propto Re^{-1.85}$, which compares with $R_f^* \propto Re^{-2.0}$ predicted by theory of Müller-Steinhagen et al. [1988] for adhesion-controlled particulate fouling.

Gomelauri et al. [1992] investigated the effect of roughness geometry on pre-cipitation fouling. Tests were conducted using rib roughness on the inner tube of an annulus using artificially hardened water. The roughness was made on the outer surface of the inner tube by winding a wire onto it. The roughness heights (e) were 0.1, 0.2 and 0.3 mm, with roughness pitch-to height (p/e) values of 12.5 and 25. The smooth geometry fouled much faster than the rough geometry. However, no explanation was provided. The effect of roughness height on the fouling resistance was investigated for $p/e = 12.5$. The fouling resistance was approximately the same irrespective of the roughness height. The effect of the p/e, however, was very prominent, yielding the minimum fouling resistance for $p/e = 12.5$.

10.4.3 Ribs in Rod Bundle

Owen et al. [1987] investigated particulate fouling effects that may occur in gas-cooled nuclear reactors, which use fuel rods having transverse-rib roughness. They observed submicron deposits (0.1 to 0.2 μm diameter) on the fuel rods. They developed a theoretical model to predict the deposition process. The model accounts for the effect of thermophoresis and surface roughness. The model shows that thermophoresis significantly affects the deposition of particles in the 0.1 to 0.7 μm size range.

10.5 FOULING OF LIQUIDS IN INTERNALLY FINNED TUBES

Watkinson et al. [1974] and Watkinson [1975] conducted accelerated scaling experiments on the internally finned and spirally indented tubes listed in Table 10.1. Calcium carbonate was deposited from a solution of $NaHCO_3$, $CaCl_2$, and NaCl containing 400 ppm suspended solids and 3000 ppm dissolved solids. The 330 K hard water was heated by steam condensing at 380 K. The fouling factor was based on the total internal surface area of the tubes (A_{tot}). Figure 10.5 shows the asymptotic fouling resistance R_f^* vs. water velocity for the internally finned tubes. Table 10.1 gives the ratio $R_{f,t}/R_{f,p}$, where $R_{f,t}$ and $R_{f,p}$ are based on the total and the nominal surface areas of the enhanced tubes, respectively. The fouling factor ratio is compared at 1.2 m/s and the same fouling operating conditions. Table 10.1 shows that the tubes have higher fouling resistance than the plain tube, at the same operating conditions. This is somewhat misleading, because the plain and enhanced tube surface areas (A_i/L) are different. As shown by Equation 10.1, the fouling resistance is R_f/A. The ratio of the R_f/A_i values for internal fin 3 and the plain tube is $1.32 \times 0.033/0.082 = 0.53$. Hence, the fouling penalty (R_f/A_i) for the internally finned tube is actually less than for the plain tube!

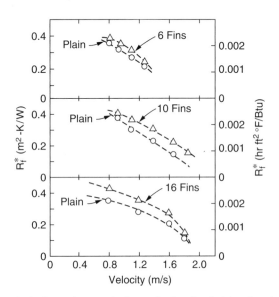

Figure 10.5 Asymptotic fouling resistances for internally finned and plain tubes shown in Figure 10.1, Watkinson et al. [1974]. (From Watkinson [1990].)

Table 10.1 Ratio of Fouling Resistances for Enhanced and Plain Tubes

Geometry	n_f	d_i (mm)	e_i/d_i	A_i/L	A_{tot}/A_n	$R_{f,\pm}/R_{f,p}$
Plain	0	10.4	0.00	0.033	1.00	1.00
Internal fin 1	6	11.8	0.16	0.052	1.42	1.15
Internal fin 2	10	10.4	0.13	0.060	1.85	1.27
Internal fin 3	16	12.7	0.14	0.082	2.46	1.32
Spirally indented	4	17.4	0.23	0.054	1.0	1.15
Spirally indented	4	17.6	0.24	0.055	1.0	2.30

Somerscales et al. [1991] tested two internally finned tubes using 3-μm magnesium oxide foulant at 2500 ppm concentration in water. The tests were performed at two different water velocities, 0.9/1.0 and 1.5 m/s. Heat was supplied by electric heat generation in the tube wall (1.2 m long), and the heat transfer coefficient was measured using a wall thermocouple. The tube geometries tested were as follows: Tube 1 (12.3 mm inside diameter with 10 axial fins 1.57 mm high), Tube 2 (16.2 mm inside diameter with 38 helical fins 0.64 mm high and 27° helix angle).

Data were also obtained using a smooth tube and three rough tubes. The rough tube results are discussed in Section 10.6. Their fouling resistance was based on the internal envelope area ($A_i/L = \pi d_i$). None of the six tubes showed fouling at 1.5 m/s. After 12 h, neither internal fin tube showed much fouling ($R_f \simeq 0.3\text{E-}5$ m²-K/W). However, the helical-finned tube had the smaller R_f. At 1.0 m/s, the helical-finned tube showed initial buildup of high R_f, but it apparently broke away and the R_f settled down to a small value.

Panchal [1989] obtained biofouling data for a spirally indented tube (Figure 8.14c). Tests were conducted using once-through flow of sea water in the test tube. Fouling data were also obtained for the spirally fluted tube illustrated in Figure 8.14a. The experimental results showed that the fouling resistances of the enhanced tubes were comparable to that of the smooth tube. Moreover, the amount of chlorination required to maintain the fouling resistance within an acceptable value was approximately the same both for the enhanced and the smooth tubes.

The fouling resistance based on total internal area may be converted to a resistance based on the nominal internal surface area ($A_n = \pi d_i/L$) using Equation 10.13.

$$R_{f,n} = R_{f,\text{tot}} \left(\frac{A_n}{A_{\text{tot}}} \right)$$

(10.13)

Definition of R_f based on Equation 10.13 allows the R_f to be compared directly with the plain tube R_f for enhanced and plain tubes having the same nominal inside diameter.

10.6 LIQUID FOULING IN ROUGH TUBES

10.6.1 Accelerated Particulate Fouling

Webb and Kim [1989] measured accelerated particulate fouling in water for three different internally ribbed tubes and a plain tube. The purpose of the tests was to compare the fouling rate of the enhanced tubes, relative to a plain tube, and to investigate the use of online brush cleaning of the tubes. The 3 19.05-mm-external-diameter tube geometries installed in the apparatus are the GEWA-SC™ tube (Figure 9.19d), GEWA-SPIN™ tube (same as Figure 9.3a), and the GEWA-TWX™ tube (Figure 9.19f). Ferric oxide powder which is the pigment used in red primer paint, was used as the foulant material. Data were also obtained using 3-μm-diameter aluminum oxide foulant material. Similar results were obtained with both foulants.

Figure 10.6 shows the fouling resistance measured in a 100-h test at 1.8 m/s water velocity with 2000 ppm foulant concentration. The maximum fouling resistance occurred in the SC tube. Very little fouling occurred in the SPIN, TWX, and plain tubes. These results are quite different from those at low velocities (1.2 m/s), where significant fouling occurred in the TWX and SPIN tubes. The 2000 ppm used is much higher than one would expect in actual heat exchange systems. The data cannot be used to predict long-term fouling rates using typical field-quality water.

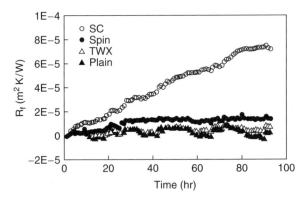

Figure 10.6 Fouling curves for Wieland GEWA tubes for 2000-ppm ferric oxide at 1.8 m/s water velocity. (From Webb and Kim [1989].)

Figure 10.7 Flow-driven nylon brush cleaning system. (Courtesy Water Services of America.)

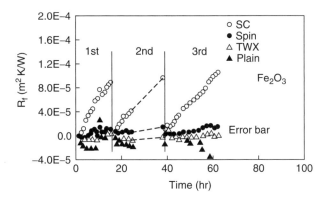

Figure 10.8 Online brush cleaning tests of Wieland GEWA tubes using 1500-ppm ferric oxide with 1.2 m/s water velocity. (From Webb and Kim [1989].)

Flow-driven nylon bristle brushes (Figure 10.7) may be used to clean the tubes during service. This cleaning system is described by Leitner [1980]. Figure 10.8 shows the accelerated fouling test results of Webb and Kim [1989] for operation at 1.2 m/s with 1500 ppm ferric oxide foulant. The figure shows three brushing cycles. The R_f of the SC tube shows a moderately rapid increase, whereas the other tubes show negligible fouling resistance. As shown in Figure 10.8, each brushing cycle returned the fouling resistances back to zero except the SC tube, where a slight residual R_f is observed. This occurred because the brush in the SC tube was slightly undersized.

Leitner [1980] reports fouling tests of an online brush cleaning system used in the condenser tubes of a large refrigeration system. Leitner reports that the brush system maintained the U value of the Turbo-Chil™ tube (Figure 9.3a) within 8% of the clean tube value for cooling tower water.

Webb and Chamra [1991] performed particulate fouling tests of the Figure 12.9c Wieland GEWA-NW™ and the Figure 12.9a Wolverine Korodense™ tubes using 0.3-μm aluminum oxide particles. Figure 10.9 shows the test results for the GEWA-NW (18.0 mm ID), the Korodense (19.7 mm ID), and a plain tube (17.9 mm ID) with 1.8 m/s water velocity and 1500 ppm foulant concentration. Both the GEWA-NW and Korodense tubes foul significantly faster than the plain tube.

Taprogge Corp. has developed an online recirculating sponge ball cleaning system that has been installed in many electric utility plants having plain tubes. This system is described by Keysselitz [1984] and Renfftlen [1991]. Webb and Chamra [1991] tested the ability of flow-driven nylon brushes and sponge balls to clean their tubes. Both the nylon brushes and sponge-ball cleaning system were very effective in cleaning all the enhanced tubes.

Somerscales et al. [1991] used 3-μm magnesium oxide foulant at 2500 ppm concentration in water. They tested the Wolverine Turbo-Chil (Figure 9.3a), the Turbo-B (internal surface of Figure 9.19a), the Korodense (Figure 9.3b), and the two internally finned tubes discussed in Section 10.5. The tests were performed at two different water velocities, 1.0 and 1.5 m/s. Contrary to the tests of Webb and

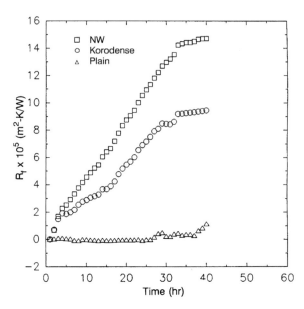

Figure 10.9 Fouling curves for 1500-ppm aluminum oxide (0.3 μm) at 1.8 m/s water velocity. (From Webb and Chamra [1991].)

Chamra [1991] they conclude that the Korodense tube is less susceptible to fouling than a smooth tube. The Wolverine Turbo-Chil tube tested by Somerscales et al. [1991] has the same internal geometry as the Wieland GEWA-SPIN tested by Webb and Kim [1989]. Again, the Somerscales data are below that of Webb and Kim [1989]. One possible reason for the difference of results may be the tube length used; Somerscales et al. [1991] used a relatively short tube length (1.2 m), as compared to the 3.05 m length used by Webb and Kim [1989].

10.6.2 Long-Term Fouling

Boyd et al. [1983] describe operating experience of the Wolverine Korodense tube in steam condensers. Rabas et al. [1990] provide an update of operating experience with the Korodense tube, since publication of the Boyd et al. [1983] paper. Rabas et al. [1991] describe the long-term operating experience of 12 electric utility steam condensers, 9 of which were retubed with Korodense tubes. Although the results show that the Korodense tube fouls faster than plain tubes, a relatively small fouling resistance occurred within 9 months after cleaning. Table 10.2 compares the fouling resistance of plain and Korodense tubes at the 3 plant locations 10 months after cleaning.

No cleaning was performed during the 10-month fouling period. Although Table 10.2 shows that the Korodense tube had a higher fouling rate than the plain tube, large fouling resistances were not attained. The thermal performance of the Korodense tubes remained superior to that of the plain tubes for more than a year without cleaning. In addition, the thermal performance of both the enhanced and plain tubes was restored to the new, clean levels after mechanical brush cleaning. As previously noted, Renfftlen [1991] reports that enhanced tubes can be successfully cleaned using the sponge-ball cleaning system.

Webb and Li [2000] obtained long-term fouling data for cooling tower water flowing inside enhanced tubes. The data were taken using 800-ppm calcium hardness water supplied to an operating chiller/cooling tower system. The water was treated with corrosion inhibitors and biocides, and only precipitation and particulate fouling (P&PF) occurred. Fouling data were measured for 7 different enhanced tubes over a 2500-h period. The tubes tested had internal helical ridges, and the number of internal ridges varied from 10 to 45 with corresponding p/e values varying from 9.88 to 2.81. The tube inner diameter was 15.54 mm, and the water velocity was 1.07 m/s. The geometric details are provided in Table 10.3.

The fouling resistance was based on the nominal (plain tube) area. Figure 10.10 shows the measured fouling resistance for tube 2, which has the largest number of internal ridges ($n_f = 45$). No significant fouling occurred during an induction period of approximately 1350 operating hours. This was also true for the other tubes. After this time, significant fouling increase was observed. The final fouling resistances (R_f) at the end of the cooling season (2500 h) are listed in Table 10.3, where the tubes are listed in order of decreasing R_f values. Table 10.3 shows that the fouling

Table 10.2 Fouling Resistance of Steam Condensers 10 Months after Cleaning (hr-ft^2-°F/Btu)

Plant	Plain	Korodense
Gallatin	0.00035	0.00048
Shawnee	0.0001–0.0002	0.0003–0.0005

Table 10.3 Fouling Resistance vs. Tube Geometry at End of 2500-hr Test Period

Tube	$R_f \times 10^4$ (m^2-K/W)	$R_f \times 10^4$ (Repeat test) (m^2-K/W)	n_f	e (mm)	α (deg)	p/e	h/h_p
2	1.44	1.71	45	0.33	45	2.81	2.32
5	0.95	1.0	40	0.47	35	3.31	2.26
3	0.63	0.92	30	0.40	45	3.50	2.33
6	0.44	0.46	25	0.49	35	5.02	2.08
7	0.42	0.42	25	0.53	25	7.05	1.93
8	0.35	N/A	18	0.55	25	9.77	1.51
4	0.32	0.49	10	0.43	45	9.88	1.74
1	0.28	0.40	NA	NA	NA	NA	1.0

NA = not applicable.

resistance increases as the number of internal ridges increases (or p/e value decreases). Figure 10.11 shows the relation between the fouling resistance and the number of internal ridges. The fouling resistance significantly increases for $n_f > 25$. As the number of internal ridges increases, the axial pitch between ridges decreases and provides a sheltered, low-velocity area between the ridges. The decreased shear

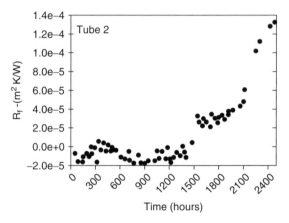

Figure 10.10 Long-term fouling data for cooling tower water for the enhanced tube with ID=15.88 mm, $e = 0.33$mm $p/e = 2.81$, $n_f = 45$, $\alpha = 45°$ at $u = 1.07$m/s. (From Webb and Li [2000]. Reprinted with permission from Elsevier.)

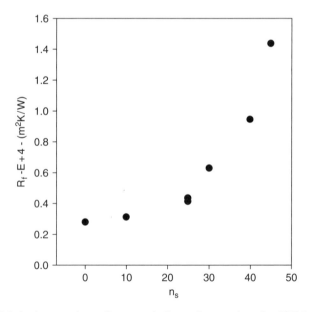

Figure 10.11 Relation between the cooling water fouling resistance taken after 1350 hr operation and number of internal ridges of enhanced tubes. (From Webb and Li [2000]. Reprinted with permission from Elsevier.)

Table 10.4 Enhanced-to-Plain Tube Ratio for Asymptotic Fouling Resistance and Heat Transfer Coefficient

Tube	2	5	3	6	7	8	4	Plain
$x_1 = R_f^*/R_{f,p}^*)_{P\&PF}$	7.43	5.65	3.25	2.68	2.46	2.03	1.93	1.0
$x_2 = R_f^*/R_{f,p}^*)_{part}$	7.28	5.28	3.28	NA	NA	2.00	2.17	1.0
x_1/x_2	1.02	1.08	0.99	NA	NA	1.02	0.90	1.0
h/h_p	2.32	2.26	2.33	2.08	1.93	1.51	1.74	1.0

stress will result in a reduced foulant removal rate, and higher fouling resistance. Li and Webb [2000] sought to predict the long-term P&PF fouling resistance of Webb and Li [2000] using accelerated particulate fouling data taken on the same tubes. Accelerated particulate fouling tests were conducted using 3.0-μm aluminum oxide particles at 1300-ppm concentration. The water velocity was maintained at the same value as the long-term fouling test (1.07 m/s). Table 10.4 lists the ratio of asymptotic fouling resistances of the enhanced tubes relative to the plain tube for the long-term fouling (P&PF) and the accelerated particulate fouling (part). Table 10.4 shows that the ratio $(R_f^*/R_{f,p}^*)_{P\&PF}]/(R_f^*/R_{f,p}^*)_{part}]$ is approximately 1.0. This result suggests that a unique relationship exists between long-term P&PF and accelerated particulate fouling. Webb and Li [2000] suggested that one may infer the relative long-term P&PF performance of enhanced tubes from accelerated particulate fouling data.

10.7 LIQUID FOULING IN PLATE-FIN GEOMETRY

Marsi and Cliffe [1996] obtained water fouling data in plate-fin heat exchangers. Both plain fin (Figure 5.2a) and wavy fin configurations (Figure 5.2c) were tested using aluminum oxide (6 μm) and ferric oxide (3 μm) particles. For both geometries, the fin height was 6.35 mm, the fin pitch was approximately 700 fins/m, and the channel depth was 150 mm. Tests were conducted in the laminar flow regime. The amount of deposit was determined by weighing the fouled test section. The wavy geometry yielded larger deposit weights than the plain fin geometry. For the aluminum oxide fouling, the asymptotic deposit weight (per unit surface area) in the wavy fin was 12% larger than that in the plain fin. Visual inspection indicated the deposits were distributed uniformly both in the plain fin and the wavy fin.

The effect of flow velocity is shown in Figure 10.12a for the plain fin geometry. Figure 10.12a shows that the deposit weight increased with the increasing Reynolds number. Note that the flow is in laminar regime ($400 \leq Re \leq 1800$). For laminar flow conditions, the removal forces are relatively weak, which results in increasing fouling rate with increasing velocities. The wavy fin data are shown in Figure 10.12b. For the wavy fin geometry, the deposit weight increased with increasing flow velocity for $Re < 1500$. At higher Reynolds numbers, the deposit weight

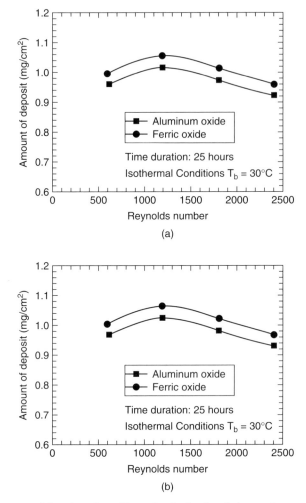

Figure 10.12 Amount of deposit vs. Reynolds number for the plate-fin heat exchanger with 6.35-mm fin height and 1.4-mm fin pitch: (a) plain fin geometry, (b) wavy fin geometry. (From Marsi and Cliffe [1996].)

decreased with increasing velocity. The high degree of turbulence created by the wavy channels at high Reynolds numbers may have broken the deposits resulting in deposit removal.

10.8 CORRELATIONS FOR FOULING IN ROUGH TUBES

Kim and Webb [1990] tested transverse rib-roughened tubes with $0.015 \leq e/d_i \leq 0.030$ and $10 \leq p/e \leq 20$ at $14,000 \leq \mathrm{Re}_d \leq 26,000$ using 0.3-μm aluminum oxide particles at 1500 ppm concentration. The rib cross section was semicircular, with $w/e \simeq 3.0$. They developed an empirical correlation to predict the effect of the roughness parameters and the flow velocity on the asymptotic fouling rate (R_f^*). Their results are correlated by

$$R_f^* \propto (e/d_i)^{-0.3}(p/e)^{0.3} \, \mathrm{Re}_d^{-3.9} \tag{10.14}$$

Equation 10.14 shows that $R_f^* \propto u^{-3.9}$. Hence, increasing velocity will significantly reduce the fouling rate. The velocity dependency is significantly higher than the value predicted by the Kern and Seaton [1959] model $(R_f^* \propto u^{-1.1})$ for plain tubes. Actually, the Kern and Seaton model underpredicts the velocity dependency of R_f. Chamra and Webb [1993b] found that $R_f^* \propto u^{-2.94}$ for plain tubes. Also note that smaller e/d_i gives higher fouling rate for the same p/e.

Chamra and Webb [1993a] determined the effect of particle size, particle distribution, and foulant concentration on the Figure 12.9c Wieland GEWA-NW and the Figure 9.3b Wolverine Korodense tubes. They used 2, 4, and 16 μm diameter with concentrations between 800 and 2000 ppm at 1.2 to 2.4 m/s water velocity. Tests were also performed with a distribution of the three particle sizes. The tests were intended to simulate particulate fouling in the Mississippi River, which contains a range of particle size distributions. Figure 10.13 shows the asymptotic fouling resistance (R_f^*) as a function of particle diameter at 1.8 m/s water velocity and 2000 ppm particle concentration. The figure shows that R_f^* increases as the particle size is reduced. This is because $K_m \propto \mathrm{Sc}^{2/3}$, and $\mathrm{Sc} \propto 1/d_p$. A mixture of containing 33% 2-μm and 67% 4-μm particles has

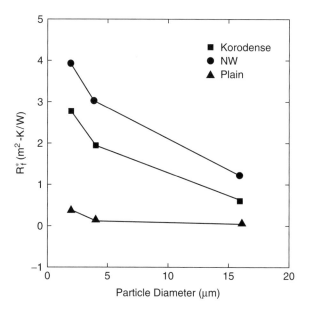

Figure 10.13 Asymptotic fouling resistance as a function of particle diameter for 1.8 m/s water velocity and 2000-ppm particle concentration. (From Chamra and Webb [1993a].)

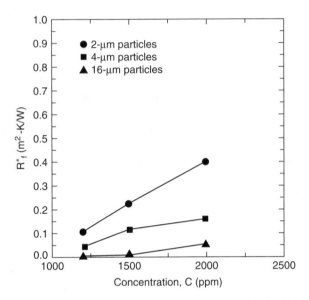

Figure 10.14 Effect of particle concentration on R_f^* for 1.8 m/s water velocity for 2, 4, and 16 μm diameter particles. (From Chamra and Webb [1993a].)

a lower R_f than for 100% 2-μm particles, because the Schmidt number is reduced. Figure 10.14 shows the effect of particle concentration on R_f^* for 1.8 m/s water velocity.

10.9 MODELING OF FOULING IN ENHANCED TUBES

Kim and Webb [1991] and Chamra and Webb [1993b] have worked to develop predictive models for particulate fouling in rough tubes. They used the heat-momentum analogy (Equation 10.12) to predict the mass transfer coefficient (K_m) from the heat transfer coefficient. This is valid, if the particle deposition occurs in the diffusion regime, as discussed by Kim and Webb [1991]. If the particle deposition occurs in the inertia regime, the mass transfer coefficient (K_m) must be corrected to account for particle inertia effects, as discussed by Chamra and Webb [1993b]. The corrected mass transfer coefficient is called the particle deposition coefficient, K_D and is described by Chamra [1992]. The wall shear stress was predicted based on an analytical model proposed by Lewis [1975]. Because the apparent shear stress (τ_a) based on pressure drop includes contributions due to profile drag on the roughness elements, $\tau_w < \tau_a$. The fouling data were curve-fitted to the asymptotic form of Equation 10.9, which provided B (Equation 10.8) and R_f^* (Equation 10.10). With B and R_f^* known they then calculated the sticking probability (S) and the deposit strength factor (ξ) using Equations 10.10 and 10.8, respectively. Chamra and Webb [1993b] show that the ξ and S functions for plain and enhanced tubes (GEWA-NW and Korodense) are reasonably correlated by

$$\frac{S_{\text{enh}}}{S_{\text{ref}}} \propto \tau_w^{-0.721} d_p^{-0.319} C_b^{1.02} (e/D)^{-0.307} \tag{10.15}$$

$$\frac{\xi_{\text{enh}}}{\xi_{\text{ref}}} \propto \tau_w^{-0.435} d_p^{-0.0769} C_b^{0.421} (e/d_i)^{-0.396} \tag{10.16}$$

The S and ξ values are presented relative to those of the smooth tube to avoid the evaluation of the deposit density (ρ_f), and the deposit thermal conductivity (k_f). The reference values (S_{ref}, ξ_{ref}) are for the smooth tube at Re = 24,000, d_p = 2 μm, and C_b = 2000 ppm. The sticking probability decreases as τ_w and d_p increase. The deposit bond strength factor decreases as the Reynolds number and particle diameter increase. The exponents on τ_w, d_p, and C_b agree closely for the three geometries tested. The Equation 10.15 and 10.16 results agree closely with those of Kim and Webb [1991] for repeated-rib tubes. Figure 10.15 and Figure 10.16 show the measured S/S_{ref} and ξ/ξ_{ref} for the Korodense tube at 1500-ppm foulant concentration. Equations 10.15 and 10.16 may be used with Equations 10.6 and 10.10 to predict the fouling rate, relative to a plain tube.

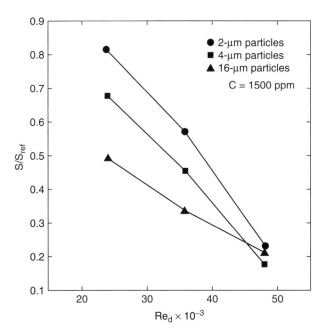

Figure 10.15 Sticking probability curves for the Korodense tube at 1500 ppm foulant concentration. (From Chamra and Webb [1993b].)

The R_f^*, S, and ξ are influenced by the Reynolds number (Re_d). Tables 10.5 and 10.6 compare the Reynolds dependency (exponent on Re_d) of R_f^*, S, and ξ for the referenced models with the test results for the plain and enhanced tubes. Table 10.5 shows that the experimental results are in good agreement with the results of Watkinson and Epstein [1970] for smooth tubes. In addition, Table 10.5 shows that

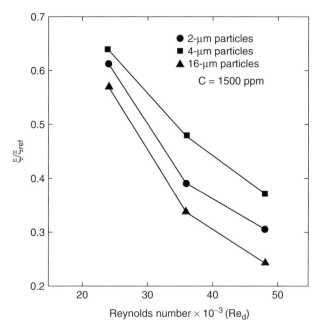

Figure 10.16 Deposit bond strength factor curves for the Korodense tube at 1500-ppm foulant concentration. (From Chamra and Webb [1993b].)

Table 10.5 Comparison of Reynolds Number Exponents with Different Models for a Plain Tube

Model	R_f^*	P	ξ
Kern and Seaton [1959]	−1.10	—	—
Watkinson and Epstein [1970]	−3.00	−2.00	—
Chamra and Webb [1993b]	−2.94	−1.57	−0.95

Table 10.6 Comparison of Reynolds Number Exponents with Different Models for Plain Tube

Model	R_f^*	P	ξ
Kim [1989]	−3.90	−2.20	−0.400
Chamra and Webb [1993b]	−3.93	−1.59	−0.659

the Kern and Seaton [1959] model underpredicts the experimental data. Table 10.6 shows that the experimental results closely match the Kim and Webb [1991] correlation for repeated-rib tubes. The difference between the exponents occurs because the results Chamra and Webb [1993b] are applicable to both the diffusion and inertia regimes.

The detailed model of Chamra [1992] may be used to predict R_f^* for different water velocities, foulant concentrations, particle diameters, or particle size distributions. This model was used to prepare Table 10.7 and Table 10.8. Table 10.7 shows the ratio of the Korodense to smooth R_f^* for $300 \leq C \leq 2000$ ppm at 1.8 m/s (Re_d = 36,000) and $d_p = 2$ μm. Table 10.7 shows that the asymptotic fouling resistance for the GEWA-NW tube is slightly higher than that for the Korodense tube. However, the Korodense tube fouls faster than a plain tube. In addition, both of the asymptotic fouling resistance ratios increase as the concentration decreases. Table 10.8 shows the predicted R_f^* ratio for different particle sizes and size distributions. The concentration and Reynolds number are held constant at 2000 and 36,000 ppm, respectively. The Schmidt number for the mixture is calculated by taking the weighted average of the particle diameter. Table 10.8 shows the asymptotic fouling resistance ratios decreases as the particle diameter increases.

Assuming that ξ, ρ_f, k_f, and S are the same for the plain tube and the enhanced tube and using the heat and mass transfer analogy, Li and Webb [2002] obtained the following expression for the asymptotic fouling resistance ratio:

$$R_f^* / R_{f,p}^* = C \, (\beta \eta)^m \tag{10.17}$$

where $\beta = (A/A_p)/(A_c/A_{c,p})$ is an area ratio and and $\eta = (j/j_p)/(f/f_p)$ is the efficiency ratio. The long-term P&PF data in helical-rib tubes by Webb and Li [2000] were correlated by Li [2003] as follows:

$$R_f^* / R_{f,p}^* = \beta \eta \qquad \text{for} \qquad 5.0 \leq p/e \leq 10.0$$

$$R_f^* / R_{f,p}^* = 8.18\beta\eta - 11.11 \qquad \text{for} \qquad p/e < 5.0 \tag{10.18}$$

The accelerated particulate fouling data by Li and Webb [2000] for the same helical-rib tubes were correlated by the efficiency index alone (Li et al. [2002]).

Table 10.7 Predicted Asymptotic Resistance Ratio for Different Concentrations ($u = 1.8$ m/s, $d_p = 2$ μm)

Concentration, C (ppm)	$(R_f^*)_{KD}/R_{f,s}^*$	$(R_f^*)_{NW}/(R_f^*)_{KD}$
2000	1.470	1.075
1500	1.640	1.114
1200	1.800	1.130
800	2.100	1.170
600	2.340	1.200
300	3.060	1.260

Table 10.8 Predicted Asymptotic Resistance Ratio for Different Particle Sizes ($u = 1.8$ m/s, $C = 2000$ ppm)

Particle Diameter, d_p (μm)	$(R_f^*)_{KD}/R_{f,s}^*$	$(R_f^*)_{NW}/R_{f})_{KD}$
2	2.870	1.670
4	1.940	1.394
16	1.750	1.230
67% 2 and 33% 4	2.210	1.460
33% 2 and 67% 4	2.040	1.230
50% 2 and 50% 4	2.160	1.306

$$R_f^* / R_{f,p}^* = 4.6\,\eta^{1.85} \tag{10.19}$$

10.9.1 Example Problem 10.1

Calculate the ratio of the asymptotic fouling resistance R_f^* of the Korodense tube, relative to a plain tube for particulate fouling of 35°C water in a 19.6-mm ID tube at 1.8 m/s ($Re_d = 36,000$, $Pr = 4.8$) with $C = 800$ ppm using $d_p = 2$ μm particles. How will the results change if the particle size is increased to 4 μm?

The results are obtained from Table 10.7. This table shows that $(R_f^*)_{KD}/R_{f,s}^* = 2.10$. If the particle size is increased to 4 μm, Table 10.8 shows that there is negligible change in the R_f^* ratio.

10.10 FOULING IN PLATE HEAT EXCHANGERS

Plate heat exchangers are widely used in process and food industries because of their compactness and versatility. While reasonably reliable correlations are available for the design of plate heat exchangers (Section 5.6), very limited information is available on the performance under fouling conditions. Generally, it is assumed that plate heat exchangers foul less than comparable shell-and-tube heat exchangers, because of the higher degree of turbulence in the plate type exchanger. Cooper et al. [1980] reports that the fouling resistance of the plate heat exchanger was only one third of the respective shell-and-tube heat exchanger for cooling water fouling. For 0.45 m/s water velocity, the fouling resistance in the plate heat exchanger (with corrugation angle 60° was one fourth of the value recommended by the Tubular Exchanger Manufacturers Association (TEMA).

Müller-Steinhagen and Middis [1989] obtained accelerated particulate fouling data in a plate heat exchager with a corrugation angle 90° (corrugations normal to the main flow). The fluid was a solvent in which aluminum oxide particles of 1.6 μm were added. The flow velocity was varied from 0.06 to 0.75 m/s, and the concentration was varied from 280 to 900 ppm. The asymptotic fouling resistance was inversely proportional to the fluid velocity squared and proportional to the concentration. Bansal and Müller-Steinhagen [1993] obtained crystallization fouling data of calcium sulfate in a plate heat exchanger having a 60° corrugation angle. The scale formation was highly non-uniform with more severe fouling in the outlet part of the plate, where the wall temperature was highest. With increasing flow velocity, both the initial fouling rate and the fouling resistance decreased, which is typical for a reaction-controlled deposition. In a succeeding study, Bansal et al. [1997] investigated the effect of suspended particles on calcium sulfate crystallization fouling in plate heat exchangers. Both crystallizing (calcium sulfate) and non-crystallizing (alumina) particles were added. The calcium sulfate particles acted as nuclei, and significantly increased the crystallization rate. The presence of alumina particles reduced the crystallization rate. Alumina particles attached loosely to the heat transfer surface. Therefore, calcium sulfate crystals growing on these particles were removed easily, which reduced the crystal growth.

Thonon et al. [1999] investigated the effect of corrugation angle on the particulate fouling in plate heat exchangers. The plates had three different corrugation angles (30°, 45°, 60°). Figure 10.17 shows measured fouling resistances. The $CaCO_3$ particles of 7-μm mean diameter were added in water at 4400-ppm concentration. The flow velocity was 0.5 m/s. Figure 10.17 shows that high corrugation angle leads to low fouling resistance. This result is anticipated because the turbulence intensity is stronger (and eventually wall shear stress is increased) as the corrugation angle increases. The strong geometry effect, however, needs to be balanced with actual flow conditions of plate heat exchangers. Generally, plate heat exchangers with low corrugation angle (30°) operate at a higher velocity (1.0 m/s) compared with that (0.5 m/s) of the high corrugation angle (60°) due to pressure drop limitations. Thonon et al. [1999] state that, under these typical conditions, the fouling resistances are similar for both geometries.

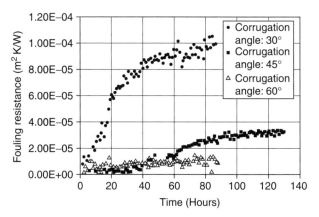

Figure 10.17 Effect of corrugation angle on $CaCO_3$ particulate fouling at 4.4 kg/m^3 concentration in the plate heat exchanger. (From Thonon et al. [1999]. Reprinted with permission from Elsevier.)

10.11 CONCLUSIONS

The fouling data on enhanced tubes appears to show a higher fouling rate than plain tubes. However, cleaning studies using online bristle brushes or sponge balls shows that the tubes can be effectively cleaned. Significant progress is being made in developing correlations to predict the particulate fouling rate of enhanced tubes, relative to plain tubes. The particulate fouling rate is sensitive to the tube geometry and particle size. Small particles result in a higher fouling rate than large particles. Scaling fouling tests on internally finned tubes show that the fouling resistance (R_f/A_i) is less than that of a plain tube. This lower fouling resistance is due to the surface area increase provided by the internally finned tube.

Water chiller condenser tubes now use helically grooved tubes with as many as 40 internal starts. Webb and Li [2000] show that the potential for fouling increases as the number of starts increases. The fouling resistance significantly increases for $n_f > 25$, as shown in Figure 10.11. Webb and Li [2000] obtained long-term fouling data for cooling tower water flowing inside enhanced tubes. They show that long-term fouling of cooling tower water may be related to test results for accelerated particulate fouling.

Recent work has provided data on fouling in plate-type heat exchangers. Thonon et al. [1999] investigated effect of corrugation angle. Figure 10.17 shows that high corrugation angle leads to reduced fouling resistance, because of the high turbulence level.

For a finned-tube, gas-to-liquid heat exchanger, gas-side fouling may be of less concern than liquid-side fouling. This is because the gas-side fouling thermal resistance (R_f/A_o) is apportioned over the total air-side surface area (A_o), as discussed in Section 2.7. However, if the foulant reduces the airflow rate, heat exchanger thermal performance will suffer.

Much less work has been done on gas-side fouling than on liquid-side fouling. Particulate, solidification, and freezing fouling are the major modes of concern. All three fouling modes may cause plugging of the fin channels. Particulate fouling may not be of great concern for finned tube heat exchangers, unless the gases contain very high particulate loading. Gas-side fouling problems are most severe in cooling of combustion products, which may contain vapors that condense and solidify on the fin surface.

NOMENCLATURE

A	Heat transfer surface area ($= \pi d_i/L$), m or ft
A_c	Cross-sectional flow area, m or ft
B	Time constant defined by Equation 10.8 ($= \rho_f \tau_w k_f/\zeta$), s^{-1}

c_p Specific heat, kJ/kg-K or Btu/lbm-°F
C_b Particulate concentration in bulk fluid, ppm
C_w Particulate concentration at wall, ppm
d_e Diameter at tip of fins for a finned tube, m or ft
d_i Plain tube inside diameter, or diameter to the base of internal fins (or flutes) or roughness, m or ft
d_o Tube outside diameter, fin root diameter for a finned tube, minor diameter of rectangular cross-section tube, m or ft
d_p Foulant particle diameter, m or ft
D Diffusion coefficient, ($= K_B T_{abs}/3\pi\mu d_p$), m²/s or ft²/s
e Fin height or roughness height, m or ft
e^+ Roughness Reynolds number ($= eu^*/v$), dimensionless
f Fanning friction factor, $\Delta p_f d/2LG^2$, dimensionless
G Mass velocity–based on the minimum flow area, kg/m²-s or lbm/ft²-s
h Heat transfer coefficient, W/m²-K, or Btu/hr-ft²-°F.
k_w Tube wall thermal conductivity, W/m-K or Btu/hr-ft-°F.
k_f Foulant layer thermal conductivity, W/m-K or Btu/hr-ft-°F.
K_B Boltzmann constant (= 1.38 E-23 J/kg)
K_m Mass transfer coefficient, kg/m²-s or lbm/ft²-s
m_f Foulant mass per unit area attached to wall, kg/m² or lbm/ft²
$\dot{m}_d$ Foulant deposition rate, kg/m²-s or lbm/ft²-s
$\dot{m}_r$ Foulant removal rate, kg/m²-s or lbm/ft²-s
n_f Number of internal ridges or fins, dimensionless
p Axial pitch of surface or roughness elements, m or ft
Pr   Prandtl number (= $c_p\mu/k$), dimensionless
Re_d Reynolds number based on the tube diameter (= Gd/μ), $d = d_i$ for flow inside tube and $d = d_o$ for flow outside tube, dimensionless
R_f Fouling resistance, m²-K/W or ft²-°F-hr/Btu
R_f^* Asymptotic fouling resistance (Equation 10.10), m²-K/W or ft²-°F-hr/Btu
dR_f/dt_o Initial fouling rate, m²-K/kJ or ft²-°F/Btu
S Sticking probability, dimensionless
Sc   Schmidt number (= v/D), dimensionless
t Time, s or hr
t_d Delay time shown on Figure 10.1, s or hr
t^+ Dimensionless particle relaxation time defined by Equation 10.11
T_{abs} Absolute temperature, K or R
u Flow velocity, m/s or ft/s
u^* Friction velocity, m/s or ft/s
U Overall heat transfer coefficient, W/m²-K or Btu/hr-ft²-°F
U_c Clean tube overall heat transfer coefficient, W/m²-K or Btu/hr-ft²-°F
U_f Fouled tube overall heat transfer coefficient, W/m²-K or Btu/hr-ft²-°F
w Width of rib, m or ft
x_f Foulant layer thickness at wall, m or ft

Greek Letters

α Helix angle, degrees

μ Dynamic viscosity, Pa-s or lbm/s-ft

ν Kinematic viscosity, m²/s or ft²/s

ξ Deposit bond strength factor, dimensionless

ρ Fluid density, kg/m³ or lbm/ft³

ρ_p Density of a particle, kg/m³ or lbm/ft³

τ_a Apparent wall shear stress based on pressure drop, N/m² or lbf/ft²

τ_w Wall shear stress, N/m² or lbf/ft²

Subscripts

i Designates inner surface of tube

o Designates outer surface of tube

p Plain tube

part   Particulate fouling

P&PF   Precipitation and particulate fouling

s Smooth surface

w Wall

τ^+ Nondimensional particle relaxation time

REFERENCES

Bansal, B. and Müller-Steinhagen, H., 1993. Crystallization fouling in plate heat exchangers, *J. Heat Transfer*, 115, 584–591.

Bansal, B., Müller-Steinhagen, H., and Chen, X.D., 1997. Effect of suspended particles on crystallization fouling in plate heat exchangers, *J. Heat Transfer*, 119, 568–574.

Bemrose, C.R. and Bott, T.R., 1984. Correlations for gas-side fouling of finned tubes, Institution of Chemical Engineers, Symposium Series 86, presented at First U.K. National Conference on Heat Transfer, University of Leeds. 357–367.

Bergles, A.E. and Sommerscales, E.F.C., 1995. The effect of fouling on enhanced heat transfer equipment, *J. Enhanced Heat Transfer*, 2, 157–166.

Bott, T.R., 1995. *Fouling of Heat Exchangers*, Elsevier, New York.

Bott, T.R. and Bemrose, C.R., 1981. Particulate fouling on the gas-side of finned tube heat exchangers, in *Fouling in Heat Exchange Equipment*, J.M. Chenoweth and M. Impagliazzo, Eds., ASME Symp. Vol. HTD, 17, 83–88.

Boyd, L.W., Hammon, J.C., Littrel, J.J., and Withers, J.G., 1983. Efficiency Improvement at Gallatin Unit 1 with Corrugated Condenser Tubing, ASME Paper 83-JPGC-PWR-4, ASME, New York.

Burgmeier, L. and Leung, S., 1981. Heat Exchanger and Cleaning System Report, AiResearch Mfg. Co. Report 81–17932 on DOE contract EC-77-C-03–11557, August 21.

Chamra, L.M., 1992. A Theoretical and Experimental Study of Particulate Fouling in Enhanced Tubes, Ph.D. thesis, Pennsylvania State University, University Park, August 1992.

Chamra, L.M. and Webb, R.L., 1993a. Effect of particle size and size distribution on particulate fouling in enhanced tubes, *J. Enhanced Heat Transfer*, 65–76.

Chamra, L.M. and Webb, R. L, 1993b. Modeling particulate fouling in enhanced tubes having a particle size distribution, *Int. J. Heat Mass Transfer*, 37(4), 571–579.

Cooper, A., Suiter, J.W., and Usher, J.D., 1980. Cooling water fouling in plate heat exchanger, *Heat Transfer Eng.*, 1(4), 571–579.

Epstein, N., 1983. Fouling of heat exchangers, in *Heat Exchangers: Theory and Practice*, J. Taborek, G.F. Hewitt, and N.H. Afgan, Eds., Hemisphere, New York, 795–815.

Epstein, N., 1988a. General thermal fouling models, in *Fouling Science and Technology, Proceedings of the NATO Advanced Study Institute*, L.F. Melo, T.R. Bott, and C.A. Bernardo, Eds., Kluwer Academic, Hingham, MA, 15–30.

Epstein, N., 1988b. Particulate fouling of heat transfer surfaces: mechanisms and models, in *Fouling Science and Technology, Proceedings of the NATO Advanced Study Institute*, L.F. Melo, T.R. Bott, and C.A. Bernardo, Eds., Kluwer Academic, Hingham, MA, 148–164.

Freeman, W.B., Middis, J., and Müller-Steinhagen, H., 1990. Influence of augmented surfaces and of surface finish on particulate fouling in double pipe heat exchangers, *Chem. Eng. Processing*, 27, 1–11.

Garrett-Price, B.A., Smith, S.A., Watts, R.L., Knudsen, J.G., Marner, W.J., and Suitor J.W. 1985. *Fouling of Heat Exchangers*, Noyes, Park Ridge, NJ.

Gomelauri, V.I., Gruzin, A.N., Magrakvelidze, T.S., and Lekveishvili, N.N., 1992. The effect of two-dimensional artificial roughness on the formation of deposits on heat transfer surfaces, *Thermal Eng.*, 39(8), 439–442.

Grillot, J.M. and Icart, G., 1997. Fouling of a cylindrical probe and a finned tube bundle in a diesel exhaust environment, *Exp. Thermal Fluid Sci.*, 14, 442–454.

Katz, D.L., Knudsen, J.G., Balekjian, G., and Grover, S.S., 1954. Fouling of heat exchangers, *Petroleum Refiner*, 33(4), 123–125.

Kern, D.Q. and Seaton, R.E., 1959. A theoretical analysis of thermal surface fouling, *Br. Chem. Eng.*, 14(3), 258–262.

Keysselitz, J., 1984. Can waterside condenser fouling be controlled operationally? in *Fouling in Heat Exchange Equipment*, J.W. Suitor and A.M. Pritchard, Eds., ASME Symp. Vol. HTD, 35, 105–111.

Kim, N.-H., 1989. A Theoretical and Experimental Study on the Particulate Fouling of Tubes Having Two-Dimensional Roughness, Ph.D. thesis, Department of Mechanical Engineering, Pennsylvania State University, University Park.

Kim, N.-H. and Webb, R.L., 1990. Particulate fouling inside tubes having arc-shaped two-dimensional roughness by a flowing suspension of aluminum oxide in water, *Heat Transfer 1990, Proc. 9th Int. Heat Transfer Conf.*, Jerusalem, Israel, 139–146.

Kim, N.-H., and Webb, R.L., 1991. Particulate fouling in tubes having a two-dimensional roughness geometry, *Int. J. Heat Mass Transfer*, 34, 2727–2738.

Kindlman, L. and Silvestrini, R., 1979. Heat Exchanger Fouling and Corrosion Evaluation, AiResearch Mfg. Co. Report 78–1516(2) on DOE contract DE-AC03–77ET11296, April 30.

Leitner, G.F., 1980. Controlling chiller tube fouling, *ASHRAE J.*, 6(2), 40–43.

Lewis, M.J., 1975. An elementary analysis for predicting the momentum and heat transfer characteristics of hydraulically rough surfaces, *Int. J. Heat Mass Transfer*, 97, 249–254.

Li, W., 2003. The performances of internal helical-rib roughness tubes under fouling conditions: practical cooling tower water fouling and accelerated particulate fouling, *J. Heat Transfer*, 125, 746–748.

Li, W. and Webb, R.L., 2000. Fouling in enhanced tubes using cooling tower water. Part II: Combined particulate and precipitation fouling, *Int. J. Heat Mass Transfer*, 43, 3579–3588.

Li, W. and Webb, R.L., 2002. Fouling characteristics of internal helical rib roughness tubes using low-velocity cooling tower water, *Int. J. Heat Mass Transfer*, 45, 1685–1691.

Li, W., Webb, R.L., and Bergles, A.E., 2002. Particulate fouling of water in internal helical rib roughness tubes, *Heat Transfer 2002, Proc. 12th Int. Heat Transfer Conference*, 4, 435–440.

Marner, W.J. and Webb, R.L., 1983. A bibliography on gas-side fouling, *Proc. ASME-JSME Thermal Engineering Joint Conference*, 1, 559–570.

Marsi, M.A. and Cliffe, K.R., 1996. A study of the deposition of fine particles in compact plate fin heat exchangers, *J. Enhanced Heat Transfer*, 3, 259–272.

Melo, L.F., Bott, T.R., and Bernardo, C.A., 1987. *Fouling Science and Technology, Proceedings of the NATO Advanced Study Institute*, Kluwer Academic, Hingham, MA.

Moore, J., 1974. Fintubes foil fouling for scaling services, *Chem. Processing*, August, 8–10.

Müller Steinhagen, H.M. and Middis, J., 1989. Particulate fouling in plate heat exchangers, *Heat Transfer Eng.*, 10, 30–36.

Müller Steinhagen, H., Reif, F., Epstein, N., and Watkinson, A.P., 1988. Influence of operating conditions on particulate fouling, *Can. J. Chem. Eng.*, 66, 42–50.

Owen, I., El-Cady, A.A., and Cleaver, J.W., 1987. Fine particle fouling of roughened heat transfer surfaces, *Proc. 1987 ASME-JSME Thermal Eng. Joint Conf.*, P.D. Marto and I. Tanasawa, Eds., 3, 95–101.

Panchal, C.B., 1989. Experimental investigation of seawater biofoulng for enhanced surfaces, in *Heat Transfer Equipment Fundamentals, Design, Applications and Operating Problems*, ASME HTD, 108, 231–238.

Papavergos, P.G. and Hedley, A.B., 1984. Particle deposition behavior from turbulent flows. *Chem. Eng. Res. Design*, 62, 275–295.

Rabas, T., Merring, R., Schaefer, R., Lopez-Gomez, R., and Thors, P., 1990. Heat-rate improvements obtained with the use of enhanced tubes in surface condensers, presented at EPRI Condenser Technology Conference, Boston, 1990.

Rabas, T.J., Panchal, C.B., Sasscer, D.S., and Schaefer, R., 1991. Comparison of power-plant condenser cooling-water fouling rates for spirally indented and plain tubes, in *Fouling and Enhancement Interactions*, T.J. Rabas and J.M. Chenoweth, Eds., ASME Symp. Vol. HTD, 164, 29–37.

Renfftlen, R.G., 1991. On-line sponge ball cleaning of enhanced heat transfer tubes, in *Fouling and Enhancement Interactions*, T.J. Rabas and J.M. Chenoweth, Eds., ASME Symp. Vol. HTD, 164, 55–60.

Roberts, P.B. and Kubasco, A.J., 1979. Combined Cycle Steam Generator Gas-Side Fouling Evaluation: Phase 1 Final Report, Solar Turbines International, Report SR79-R-4557–20, July.

Sheikholeslami, P. and Watkinson, A.P., 1986, Scaling of plain and externally finned heat exchanger tubes. *J. Heat Transfer*, 108, 147–152.

Somerscales, E.F.C. and Knudsen, J.G., 1979. *Fouling of Heat Transfer Equipment*, Hemisphere, Washington, D.C.

Somerscales, E.F.C., Ponteduro, A.F., and Bergles, A.E., 1991. Particulate fouling of heat transfer tubes enhanced on their inner surfaces, in *Fouling and Enhancement Interactions*, T.J. Rabas and J.M. Chenoweth, Eds., ASME HTD, 164, 117–128.

Taborek, J., Aoki, T., Ritter, R.B., Palen, J.W., and Knudsen, J.W., 1972. Fouling — the major unresolved problem in heat transfer, *Chem. Eng. Prog.*, 68(2), 59–67; 68(7), 69–78.

Thonon, B., Grandgeorge, S., and Jallut, C., 1999. Effect of geometry and flow conditions on particulate fouling in plate heat exchangers, *Heat Transfer Eng.*, 20, 12–24.

Watkinson, A.P., 1975. Scaling of spirally indented heat exchanger tubes, *J. Heat Transfer*, 108, 147–152.

Watkinson, A.P., 1990. Fouling of augmented heat transfer tubes, *Heat Transfer Eng.*, 11, 57–65.

Watkinson, A.P., 1991. Interactions of enhancement and fouling, in *Fouling and Enhancement Interactions*, T.J. Rabas and J.M. Chenoweth, Eds., ASME Symp. Vol. HTD, 164, 1–7.

Watkinson, A.P. and Epstein, N., 1970. Particulate fouling of sensible heat exchangers, in *Proc. of the 4th Int. Heat Transfer Conf.*, Paper HE 1.6.

Watkinson, A.P., Louis, L., and Brent, R., 1974. Scaling of enhanced heat exchanger tubes, *Can. J. Chem. Eng.*, 52, 558–562.

Webb, R.L. and Chamra, L.M., 1991. On-line cleaning of particulate fouling in enhanced tubes, in *Fouling and Enhancement Interactions*, T.J. Rabas and J.M. Chenoweth, Eds., ASME Symp. Vol. HTD, 164, 47–54.

Webb, R.L. and Kim, N.-H., 1989. Particulate fouling in enhanced tubes, in *Heat Transfer Equipment Fundamentals, Design, Applications, and Operating Problems*, R.K. Shah, Ed., ASME HTD, 108, 315–324.

Webb, R.L. and Li, W., 2000. Fouling in Enhanced Tubes Using Cooling Tower Water, Part II: Long Term Fouling Data, *Int. J. Heat Mass Transfer*, Vol. 43, pp. 3567–3578.

Webb, R.L., Marchiori, D., Durbin, R.E., Wang, Y.-J., and Kulkarni, A.K., 1983. Heat exchangers for secondary heat recovery from glass plants, in *Fouling of Heat Exchange Surface*, R.W. Bryers, Ed., Engineering Foundation, New York, 169–182.

Webber, W.O., 1960. Does fouling rule out using finned tubes in reboilers? *Petroleum Refiner*, 39(3), 183–186.

Zhang, G., Bott, T.R., and Bemrose, C.R., 1990. Finned tube heat exchanger fouling by particles, *Heat Transfer 1990, Proc. 9th Int. Heat Transfer Conf.*, Jerusalem, Israel, 6, 115–120.

Zhang, G., Bott, T.R., and Bemrose, C.R., 1992. Reducing particle deposition in air-cooled heat exchangers, *Heat Transfer Eng.*, 13, 81–87.

POOL BOILING AND THIN FILM EVAPORATION

11.1 INTRODUCTION

Perhaps the most significant advances in enhanced heat transfer technology have been made in special surface geometries that promote high-performance nucleate boiling. These special surface geometries provide an increased heat transfer coefficient. Although the surface area may be increased, it is common to define the boiling coefficient in terms of the projected, or flat plate, surface area. That "roughness" can improve nucleate boiling performance has been known for 60 years, but for nearly 35 years this was regarded as an interesting, but commercially unusable concept because the heat transfer improvement lasted for only a matter of hours before "aging" caused the performance to decay to the plain surface value. In the period 1955 to 1965, fundamental advances were made in understanding the character of nucleation sites and the shape necessary to form stable vapor traps. This understanding provided a basis for industrial research (1960 – 1975) that led to the development of commercially viable enhanced surface geometries. The first of these high-performance geometries was patented in 1968. By 1980, six nucleate boiling surface geometries were commercially available.

This chapter describes the fundamental and applied research, that has brought about this major advance in the field of heat transfer. It also describes available correlations for prediction of the boiling coefficient, as well as current understanding of boiling mechanism. The excellent book by Thome [1990] on enhanced boiling heat transfer is a recommended reference.

The reference list includes a number of patents. Industrial research, which favors publication via patents, has led the way in development of enhanced boiling surfaces. Academic research has been important in explaining the physical mechanism of the enhanced surface.

11.2 EARLY WORK ON ENHANCEMENT (1931 – 1962)

In 1931, Jakob and Fritz investigated the effect of surface finish on nucleate boiling performance, as reported by Jakob [1949]. They boiled water from (1) a sandblasted surface and (2) a surface having a square grid of machined grooves (0.016 mm square with 0.48 mm spacing). The sandblasting provided no more than 15% improvement, which dissipated within a day. The grooved surface initially yielded boiling coefficients about three times higher than those of a smooth surface, but this performance also dissipated after several days.

The performance decay is described as an aging effect. The cavities formed by the roughening were not stable vapor traps and slowly desassed, leaving no vapor to sustain the nucleation process. Sauer's [1935] tests of a grooved tube (0.4 mm square grooves spaced at 1.6 mm) offered no additional promise. Interest in roughness was renewed about 1954, and a sustained effort has advanced the technology to the present level.

Between 1954 and 1962, 3 studies were performed using flat surfaces roughened with emery paper of different coarseness or lapping compound. Berenson [1962] observed nucleate boiling coefficient increases as large as 600% using lapped surfaces (Figure 11.1). The tests were not run long enough to establish the long-term aging performance, although Corty and Foust's [1955] tests of grit-roughened surfaces did show some short-term aging. Kurihari and Myers [1960] boiled water and several organic fluids on copper surfaces roughened with different grades of emery paper. Figure 11.2 shows their acetone results for the different grit sizes. They clearly established that the increased boiling coefficient results from an increased area density of nucleation sites. Their results for five test fluids showed that $h \propto (n_s/A)^{0.43}$. These works demonstrated that boiling enhancement will occur if the site density of stable nucleation sites can be increased. The artificially formed sites must allow incipience to occur at a lower ΔT ($T_w - T_s$) than for naturally occurring nucleation sites. These studies motivated later researchers to investigate other means of making artificial sites that showed incipience at low ΔT and functioned as stable vapor traps.

11.3 SUPPORTING FUNDAMENTAL STUDIES

Research on artificially formed nucleation sites was supported by parallel fundamental studies of the character of nucleation sites, the effect of cavity shape, and theoretical models of the nucleation process.

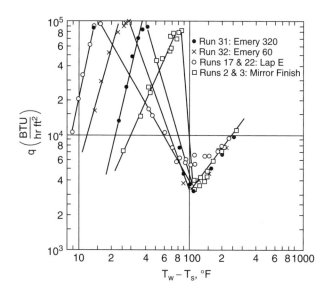

Figure 11.1 Effect of surface roughness for pentane boiling on copper. (From Berenson [1962].)

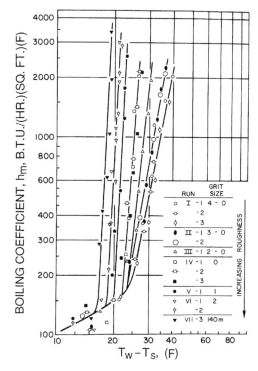

Figure 11.2 Effect of emery paper roughening for acetone boiling on copper. (From Kurihari and Myers [1960].)

Particularly notable are the high-speed photographic studies of Clark et al. [1959], who identified naturally occurring pits and scratches, between 0.008 and 0.08 mm surface width, as active boiling sites for pentane. Griffith and Wallis [1960] showed that the cavity geometry is important in two ways. First, the mouth diameter determines the superheat needed to initiate boiling, and its shape determines its stability once boiling has begun. Figure 11.3a shows the reentrant cavity shape proposed by Griffith and Wallis. Figure 11.3b shows l/r vs. bubble volume. The term l/r is proportional to the liquid superheat required to maintain a vapor nucleus. When the vapor radius of curvature becomes negative (concave curvature of the liquid–vapor interface), the vapor nucleus may still exist in the presence of subcooled liquid. Therefore, a reentrant cavity should be a very stable vapor trap. Benjamin and Westwater [1961] were apparently the first to construct a reentrant cavity and demonstrate its superior performance as a vapor trap. This site remained active after surface subcooling, while the naturally occurring sites became flooded. In a later study Yatabe and Westwater [1966] varied the interior dimensions and shape of reentrant cavities, while maintaining the surface opening diameter at a constant value. These tests showed that the interior shape of a reentrant cavity is not important.

Hsu [1962] developed a model to define the size range of active cavities for incipient boiling on a heated surface. The model assumes that the cavity is of the proper shape to act as a stable vapor trap. Bankoff [1959] developed a dynamic model to predict the ability of a cavity to serve as a stable vapor trap, e.g., one that

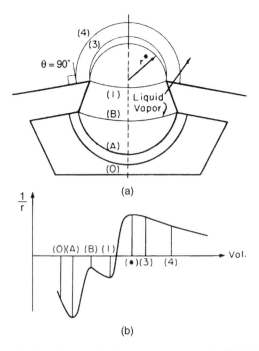

Figure 11.3 (a) States of the liquid–vapor interface in a reentrant cavity. (b) Reciprocal radius (l/r) vs. vapor volume for liquid having a 90° contact angle. (From Griffith and Wallis [1960].)

will not become liquid-filled. The analysis considered triangular and constant width circular cavities and grooves. He showed that grooves are poor vapor traps unless they are steep walled or poorly wetted. Bankoff's suggestion that grooves may function as vapor traps provided motivation for later research on cavities having a reentrant groove shape. Moore and Mesler [1971] showed that high vaporization rates occur across a "microlayer" of liquid at the base of the bubble.

These studies, which provided a strong basis for the later research on artificially formed nucleation sites, almost exclusively focused on nucleation in discrete cavities. However, the state-of-the-art enhanced surfaces generally consist of two-dimensional reentrant grooves or interconnected cavities. Griffith and Wallis's suggestion of the reentrant cavity shape provided a major contribution.

11.4 TECHNIQUES EMPLOYED FOR ENHANCEMENT

A variety of surface treatments have been used to develop enhanced boiling surfaces. Survey articles on such treatments have been published by Webb [1981, 1983, 2004] and Dundin et al. [1990]. This provides an overview of the various techniques that have been investigated. Matijević et al. [1992] compares the enhancement for boiling water using a variety of the enhancement techniques described below. Their highest performance was obtained using a porous coating of sintered spherical particles. Note that they used a very large particle diameter; 3.0 mm was the smallest.

11.4.1 Abrasive Treatment

Because of the severe aging effects, there has been little sustained interest in this abrasive treatment. Chaudri and McDougall [1969] measured the long-term performance (up to 500 hr) of standard abrasive-treated tubes. The abrasive treatment produced parallel scratches 0.025 mm or less in width. They boiled several organic fluids on copper and steel tubes and found only temporary benefits of the surface treatment. After several hundred hours, the treated tubes showed essentially the same performance as the untreated tubes.

A recent experimental study on the effects of emery roughnening and sandblasting for boiling of propane on horizontal copper and steel tubes was published by Luke [1997]. This work attempted to quantify the effect of roughness on boiling, but the paper does not describe whether long-term aging effects were observed. Kang [2000] investigated the effect of sandpaper roughness on boiling of water at 1.0 atm using horizontal and vertical tubes. He tested three tube diameters, 9.7, 19.05, and 25.4 mm. The tubes were unidirectionally roughened with two different grit size sandpapers (No. 800 and No. 3000). The No. 800 paper yielded 60.9 nm rms roughness, and the No. 3000 paper resulted in 15.1 nm rms roughness. He found that neither roughness resulted in noticeable performance improvement for horizontal tubes. However, moderate improvements were observed when the tubes were tested in the vertical orientation. The improvement for the vertical tubes increased with increasing tube L/d ratio. He explains the difference between the horizontal and vertical tube performance to result from increased liquid agitation as the bubbles move up along the vertical tubes.

11.4.2 Open Grooves

Bonilla et al. [1965] formed parallel grooves by dragging a sharp-pointed scriber across a polished copper plate. The boiling coefficient of water was measured for scratch spacings of 1.6, 3.2, 6.4, 12.8, and 25.5 mm. Increased performance was observed for all groove spacings. Surprisingly, they found that the 6.4-mm spacing yielded the highest performance. They concluded that optimum performance occurs with a scratch spacing of 2 to 2.5 bubble diameters. The boiling coefficient was increased 100% at this optimum scratch spacing. Reentrant grooves have shown greater sustained-performance level than that of open grooves. These are discussed later.

11.4.3 Three-Dimensional Cavities

Several tests of three-dimensional cavities have been performed. As previously discussed, Griffith and Wallis [1960] examined fundamental characteristics (incipience and stability) of a single 0.15-mm-diameter conical cavity. They showed that the addition of a nonwetting coating within the cavity allows incipience of water at lower superheat. Such nonwetting phenomena may be beneficial for high-surface-tension fluids, and are discussed later.

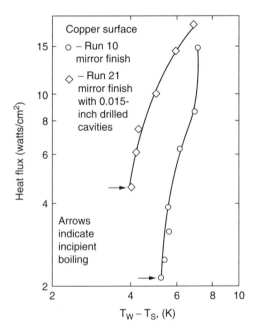

Figure 11.4 Nitrogen boiling at 101 kPa (1 atm) on a 25.4-mm-diameter copper surface. Run 21 surface has 13 circular cavities in the surface.

Marto et al. [1968] measured the boiling performance of nitrogen on a 25.4-mm-diameter copper block. The nucleation sites were formed by pressing cylindrical or conical cavities in the surface. Figure 11.4 shows their results for 0.38 × 0.76 mm deep cavities spaced at 3.7 mm. Significant enhancement was obtained. As previously noted, Griffith and Wallis [1960] have shown that a reentrant cavity shape (Figure 11.3) is preferred over a conical or cylindrical shape; lower superheat is required for incipience, and the reentrant shape provides a more stable vapor trap. Marto and Rohsenow [1966] boiled sodium (a nonwetting liquid) on a 75-mm-diameter block having 12 reentrant cavities. The cavity cross section is shown in Figure 11.5b. They also tested other surface treatments, which included a 100-grit lapped finish (lap *F*), porous welds, a sintered coating, and reentrant cavities (12 cavities in 75-mm-diameter surface). Their results are summarized in Figure 11.5a, which shows the high-performance potential of reentrant cavities and porous coatings.

The formation of discrete cavities is difficult and of limited practicality, which has motivated innovators to develop commercially feasible concepts to produce a high area density of nucleation sites. Reentrant grooves and porous coatings have developed as favored methods.

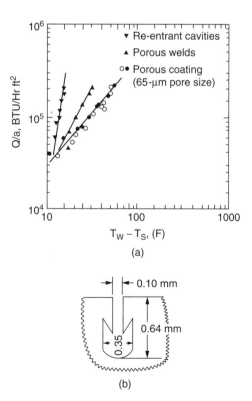

Figure 11.5 (a) Effect of several surface treatments on sodium boiling at 65 mm Hg. (b) Cross section of doubly reentrant cavities tested.

11.4.4 Etched Surfaces

Mechanical roughness may be formed by chemical etching. Vachon et al. [1969] boiled water on a chemically etched stainless steel surface. This showed no better performance than a polished surface. Chu and Moran [1977] described the use of a laser beam to form cylindrical cavities having a reentrant shape (Figure 11.6).

11.4.5 Electroplating

Bliss et al. [1969] electroplated 0.13-mm layers of copper, chrome, nickel, cadmium, tin, and zinc on stainless steel tubes. Some platings increased the heat transfer coefficient in the range of 200 to 300%. However, other coating materials (tin, zinc, and nickel) produced a decreased heat transfer coefficient. They offer no explanation for the observed results, but they do argue that the thermal properties of the plating material do not, by themselves, account for the effect.

   Albertson [1977] shows that electroplating at very high current densities causes the formation of dendrites or nodules on the base surface. This copper porous structure produced a large heat transfer increase for boiling R-12 on a 19-mm-diameter tube. The performance is further enhanced by cold rolling the tube to partially compact the nodules.

11.4.6 Pierced Three-Dimensional Cover Sheets

Ragi [1972] describes a method of forming reentrant cavities using a three-dimensional formed foil sheet, which is brazed to a flat base surface (Figure 11.7). A 0.08-mm-thick foil sheet has a three-dimensional pattern of pyramids whose tops are pierced, thereby forming reentrant cavities. Data are given for R-11 and water with several geometric variations having cavity densities of 10 to 80/cm². Figure 11.7 shows the enhancement obtained with water at 101 kPa and 62 cavities/cm². At 60 kW/m² heat flux, the boiling coefficient was increased a factor of 7.7; the

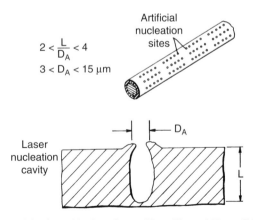

Figure 11.6 Reentrant cavities formed by laser beam. (From Chu and Moran [1977].)

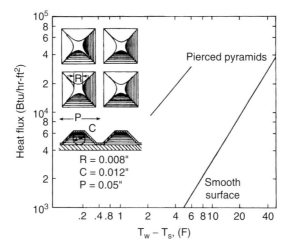

Figure 11.7 Water boiling at 101 kPa (1 atm) on surface having reentrant cavities formed by brazing pierced foil sheet to base surface. (From Ragi [1972].)

corresponding enhancement level with R-11 was 2.5. Greater enhancement levels were measured for increased cavity density, e.g., 180 cavities/cm^2.

Palm [1992] has done additional work using perforated metal foils. He has pierced 0.05-mm-thick flat sheets and investigated a range of hole densities and hole diameters. His R-22 data show earlier incipient boiling, and a higher heat flux than the commercial GEWA-T (Figure 11.12c, later in the chapter) for $(T_w - T_s) <$ 3.5 K. Additional fabrication details are provided in a U.S. patent by Granryd and Palm [1988].

Kim [1996] boiled water and R-113 at 1.0 atm on a horizontal surface, which was covered by a 0.5-mm-thick, perforated copper plate. He varied the perforation hole size, hole pitch, and gap between the plate and the heating surface. Significant enhancement was found and it was especially good for water. The preferred geometry for water was 3.0-mm hole size, 15-mm hole pitch, and 0.3 to 0.5-mm gap. Smaller hole size (2.0 mm) and hole pitch (3.5 mm) with 0.3-mm gap was found for R-113. Kim provides an empirical correlation to define the effect of geometry and fluid properties.

Uma et al. [2000] performed similar experiments using perforated plates in boiling of methanol. They used 2.0-mm hole diameter with hole pitches of 4, 6, and 8 mm with gap dimensions of 0.1, 0.3, and 0.7 mm. Performance generally improved by using the perforated plates. However, transition to film boiling occurred at a lower heat flux than the plain surface (no cover plate). The transition to film boiling occurred earlist at the widest gap dimension.

11.4.7 Attached Wire and Screen Promoters

Hasegawa et al. [1975] boiled water on a polished copper surface on which a screen was placed. The woven screens used in the experiments were 30, 50, 80, and 150 mesh, using either one or two layers of screening. The screens were held down by glass bars laid at the sides of the screens. They were not very effective as enhancement devices. The 150-mesh screen produced the greatest enhancement, about 50% increased boiling coefficient. The screens increased the critical heat flux about 35%. This improvement was generally greater for the larger mesh openings, and one screen layer was better than two layers. Corman and McLaughlin [1976] performed similar experiments, using felt-metal nickel and copper wicking of various thicknesses (0.5 to 3 mm). At low heat flux (60 kW/m^2), the boiling coefficient was improved about 70%, but the slope of the boiling curve was smaller than for the smooth surface. At increasing heat flux, the wick surface becomes inferior; the crossover point occurs at 155 kW/m^2. In spite of the lower slope, the wick surface displayed a higher critical heat flux than the smooth surface. Improvements in the range of 50 to 70% were measured for the thinner wick structures.

Asakavicius et al. [1979] attached 8 to 12 layers of copper screening (0.07-mm open space) and boiled R-113, ethyl alcohol, and water. Enhancements in excess of 100% were observed at low heat fluxes, but little improvement was noted at high heat fluxes.

Danilova and Tikhonov [1996] wrapped mesh copper screen on a horizontal tube and boiled R-113 at 1.0 atm. They used 0.1 and 0.21-mm-diameter wires with one to four mesh layers. Their results showed that the enhancement level decreased with increasing number of layers. For one layer, at fixed heat flux, they found enhancement levels as high as 2.6. Guglielmini et al. [1988] report other R-113 data for boiling on vertical plates using stainless steel mesh spot-welded to the hot surface. Liu et al. [2001] boiled methanol and HFE-7100 on one layer of stainless steel mesh and show that fine mesh is better than coarse mesh. They also provide critical heat flux (CHF) and post-CHF data, which shows that the mesh surfaces have lower CHF than a plain surface.

Tsay et al. [1996] boiled saturated water (unspecified pressure) on smooth and horizontal single layer screen mesh surfaces. The mesh was stainless steel and boiling occurred on a stainless steel surface. They used 16-, 24 and 60-mesh screens and investigated the effect of liquid level (5 to 60 mm) on performance. They found that the 16-mesh screens gave the best improvements and speculated that this was because the enhancement opening in the 16-mesh screen was closest to the bubble departure diameter. They proposed that the greater mesh screens "blocked the bubbles on the surface." Screens provided the greatest enhancement for the smallest liquid level (5 mm).

Figure 11.8a shows a loosely fitting metallic (or nonmetallic) wire-wrapped integral fin tube tested by Webb [1970]. A gap spacing of 0.013 to 0.05 mm provides substantial enhancement. Figure 11.8b, from Schmittle and Starner's [1978] patent, shows that ΔT is reduced approximately 60% by the wire wrap. This patent also discusses an interesting variation applicable to plain tubes (Figure 11.8b). This tube is wrapped with a 0.38-mm stranded nylon cord with 0.25-mm space between wraps. Figure 11.8c shows that the stranded nylon wrap substantially enhances the plain tube, but the wire-wrapped plain tube is ineffective.

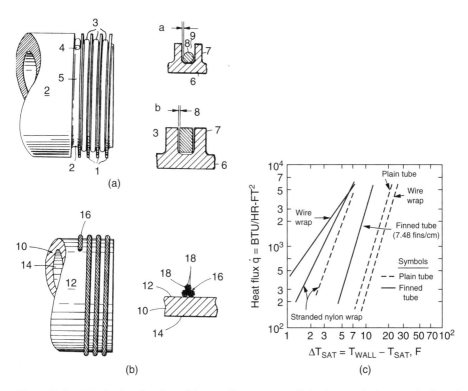

Figure 11.8 (a) Nucleation sites formed by metallic (or nonmetallic) wire wrap in between the fins of an integral fin tubes. (From Webb [1970].) (b) Stranded Nylon cord wrap on smooth tube. (From Schmittle and Starner [1978].) (c) R-11 boiling at 101 kPa (1 atm) on an integral-fin tube (7.48 fins/cm) and enhancement provided by wire or stranded nylon wrap in space between fins. (From Schmittle and Starner [1978].)

11.4.8 Nonwetting Coatings

Griffith and Wallis [1960] were apparently the first to recognize that a thin coating of nonwetting material may improve the nucleating characteristics of water (a high-surface-tension fluid) on a boiling surface. They proposed that a nonwetting coating on the interior walls of a nucleating cavity will enhance its stability as a nucleating site. Thus, a lower temperature level will be needed to deactivate the cavity. The nonwetting coating establishes a larger liquid–vapor interface radius within the cavity, which requires less superheat for its existence. However, they proposed that this coating would not affect the temperature at which a cavity will nucleate, because this is fixed by the cavity mouth radius. They performed experiments on single conical cavities 0.08 mm in diameter, formed by pressing a needle into the boiling surface. They found that it was easier to maintain boiling and obtain reproducible results from the paraffin-coated cavity than from clean ones, and that unwetted cavities are more stable than wetted cavities.

Gaertner [1967] performed further work with artificial nucleation sites, covering the inside surface of the cavities with a nonwetting material (Figure 11.5b). He boiled water on a variety of surfaces having closely spaced nucleation sites formed by needle sharp punches and parallel scratches. The low-surface-energy material was coated over the surface containing the artificial sites, then removed from the flat face by abrasion, leaving a thin film of the material deposited in each cavity. The coated cavity surface boiled at a lower superheat and remained active for a much longer time. The heat transfer coefficient was considerably reduced if the coating was left on the entire surface. For this condition the bubbles spread over the surface and coalesced with bubbles formed at other sites, causing the entire surface to become vapor blanketed.

Hummel [1965] describes a surface coated with small, nonwetted spots (Figure 11.9), which not only improves the boiling stability of water, but also provides a very significant heat transfer enhancement. Young and Hummel [1965] boiled water on a stainless steel surface that had been sprayed with Teflon, producing 30 to 60 spots/cm^2, and 0.25 mm diameter or less. Figure 11.9a shows a dramatic enhancement, with significant nucleation occurring at $\Delta T < 0.5$ K. The theory of this augmentation concept is as follows: The Teflon spot surrounds a naturally occurring or a mechanically prepared nucleation site. When the bubble is in contact with the nonwetted surface, it has a much larger radius of curvature than for a wetted surface. Therefore, lower superheat is required for its existence and initial growth. Once the bubble grows in diameter past the area of the unwetted spot, it quickly reverts to the spherical shape and grows very rapidly because of evaporation of the thin liquid film at the base of the spherical bubble. It is undesirable to have the entire surface Teflon coated, because the vapor would tend to blanket a large area of the surface, thereby reducing heat transfer. Figure 11.9a also shows data for a pitted surface with and without Teflon spots in the pits. The performance of the pitted surface is marginally better than the smooth surface when both have Teflon spots. Apparently, the mechanical pitting did not establish significantly more nucleation sites than those occurring naturally in the stainless steel strip.

The Teflon spotting method should be effective only for surface–liquid combinations that have large contact angles (e.g., water). Bergles et al. [1968] confirmed this in their tests with refrigerants, which have low surface tension and small contact angles. Their results showed that the Teflon spotting method did not favorably affect the boiling performance of the refrigerants.

Vachon et al. [1969] applied thin (0.008 mm) surface coatings of Teflon in an attempt to augment nucleate boiling of water on stainless steel. Gaertner [1967] and Hummel [1965] argue against the merits of a continuous surface coating because of the tendency of the surface to become vapor blanketed. However, Vachon et al. did not observe vapor blanketing. A considerable heat transfer enhancement was observed for the 0.008-mm-thick coating, but a decrease of heat transfer occurred with a 0.04-mm coating because of the insulating effect of the thick coating.

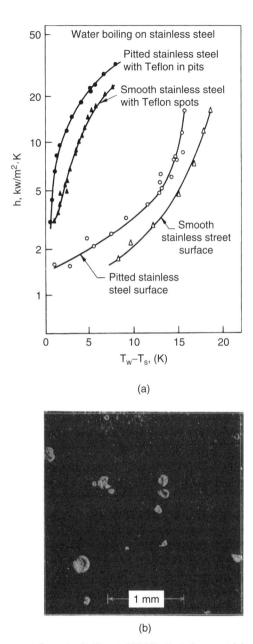

(a)

(b)

Figure 11.9 (a) Enhancement for water boiling at 101 kPa (1 atm) on a stainless steel surface having minute nonwetted spots (30 to 60 spots/cm², 0.25 mm diameter or less). (b) Enlarged photo of Teflon-spotted smooth surface. (From Young and Hummel [1965].)

11.4.9 Oxide and Ceramic Coatings

Zhou and Bier [1997] used plasma spray to apply a 0.2-mm-thick aluminum oxide coating to an 8.0-mm-diameter copper tube. Scanning electron microscope (SEM) images show a porous structure having approximately 2.0 μm particle diameter. Although the oxide coating has low thermal conductivity, they propose that such a coating may be advantageous for corrosive fluids. Pool boiling data were taken for seven fluids, including refrigerants and hydrocarbons. They found that the boiling performance was approximately equal to that for a sandblasted copper tube having the same roughness (2 to 2.8 μm).

Porous oxide coatings can be found on steam boiler tubes. Sridharan et al. [2002] boiled water on tubes removed from the steam generator of pressurized water reactors to investigate the effect of a porous deposit coating resulting from the minerals in the water. They found that the heat transfer performance of the "fouled" tube was the best, followed by the chemically cleaned tube. The performance of a new tube was the worst. The performance of a new tube improved after each test, due to the formation of a porous layer. SEM photographs revealed the presence of significant amount of porous deposits on the surface, which provide ample boiling nucleation sites. Chemical cleaning removes most of the deposits, such that the boiling performance of the cleaned surface is degraded.

Uhle et al. [1998] performed water boiling experiments on the surface of tubes having an artificially created porous coating. It was observed that boiling occurs at lower wall superheats on a porous surface as compared to clean surfaces. If the coating material is conductive and very permeable, the heat transfer is significantly higher than a clean tube, particularly at lower heat fluxes. Uhle et al. also used a modified form of the Kovalev model (see section 11.11.2) and found that it predicted their data reasonably well. From their fouling experiments, Jamialahmadi and Muller-Steinhagen [1993] concluded that the number of nucleation sites increased dramatically, early in time, as the surface fouled. As time progresses, the number of sites diminish as they become covered with the fouling material.

Imadojemu et al. [1995] investigated the effects of surface oxidation on pool boiling on a copper GEWA-SE enhanced boiling tube (Figure 11.12f, later in the chapter). Water and R-11 were boiled at subatmospheric pressure. Oxidation increased the pool boiling heat transfer coefficient for both fluids. The pool boiling heat transfer coefficient increased as much as a factor of 2.4 for water, and 1.5 for the refrigerant. For highly wetting R-11, the boiling heat transfer coefficient increased with increase in the level of oxidation. Contact angles can be used as a rough guide or indicator of the level of surface oxidation or the degree of wettability of fluids on a surface. The contact angles of water and R-11 were shown to decrease with increasing levels of oxidation.

11.4.10 Porous Surfaces

Milton's patents [1968, 1970, 1971] of a porous boiling surface provided a contribution of major practical significance. He describes a sintered porous metallic matrix bonded to the base surface. The coating is approximately 0.25 mm thick and has a void fraction of 50 to 65%. The optimum pore size depends on the fluid properties.

Figure 11.10a based on data of Gottzmann et al. [1973] compares the performance between the High-Flux™ porous surface and a smooth surface. The ΔT required to produce a given heat flux was reduced by more than a factor of 10. The surface showed negligible decay of long-term performance because of aging.

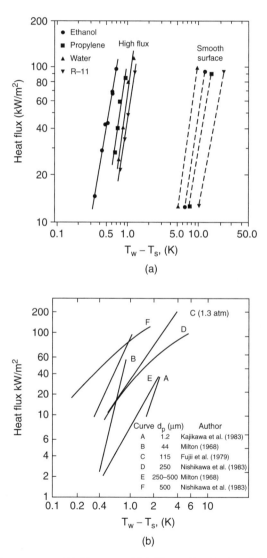

Figure 11.10 (a) Enhancement provided by porous High-Flux surface for three fluids boiling at 101 kPa (1 atm). (From Gottzmann et al. [1971, 1973].) (b) Comparative performance of the Table 11.1 coatings for R-11 boiling at 1.0 atm, except as noted. (From Webb [1983].)

Gottzmann et al. [1971] also found that the critical heat flux for trichlorethylene is 80% higher than for a smooth surface.

Table 11.1 taken from Webb [1983] summarizes four different methods that have been used to make porous coatings. The Milton [1968] patent describes a five-step method for making a sintered particle coating. The particles are mixed into a liquid slurry, which contains a temporary binder. The surface is then coated with this slurry, which is air-dried and then sintered in a baking oven. The sintering process takes place at a temperature slightly below the particle melting point. The sintered method may be disadvantageous if it anneals the base tube. The other methods listed in Table 11.1 are performed at lower temperatures, and the mechanical properties of the base tube are not adversely affected.

Table 11.1 shows that most of the published test data are for copper coatings using R-11 or R-113 as test fluids. The geometric variables of the porous coating are (d_p) the particle size (or size distribution), (2) the particle shape, (3) the coating thickness, and (4) the particle packing arrangement. The packing arrangement establishes the porosity of the matrix and depends on whether the particles are of uniform size and shape or whether a distribution of sizes and shapes is used. Figure 11.10b compares the R-11 boiling performance of the coatings listed in Table 11.1. Table 11.2 gives the geometric characteristics of each coating represented on Figure 11.10b, and are listed in order of increasing particle diameter. The particle diameters are in the range $0.044 < d_p < 1.0$ mm and the coating thickness ranges from 0.25 to 2.0 mm. Experimental error in measuring $T_w - T_s$ may be a major uncertainty, since its value is in the range of 1 K.

Nishikawa et al. [1983] and Nishikawa and Ito [1982] investigated the effect of particle size and coating thickness for R-11 and R-113 boiling on 18-mm-diameter horizontal tubes. They used spherical particles of copper or bronze. The coating porosity ranged from 0.38 to 0.71. In Test 1 using 0.25-mm copper particle diameter (d_p), they varied the coating thickness (δ) from 0.4 to 4 mm. This test showed that

Table 11.1 Porous Coatings Tested

Ref.	Particle Metal	Test Fluid	Coating Structure
O'Neill et al. [1972]	Cu, Ni, Al	R-11 and others	Sintered particles (nonuniform size and shape)
Nishikawa et al. [1983]	Cu, bronze	R-11, R-113, benzene	Sintered spherical particles
Fujii et al. [1979]	Cu	R-11, R-113	Spherical particles bonded by electroplating
Dahl and Erb [1976]	Al	R-113	Metal sprayed powder
Janowski et al. [1978]	Cu	No published data	Electroplated polyurethane foam
Kajikawa et al. [1983]	Cu	R-11	Sintered steel fibers
Yilmaz and Westwater [1981]	Cu	Isopropanol, p-xylene	Metal spray with copper wire
Kartsounes [1975]	Cu	R-12	Metal sprayed coating

Table 11.2 Porous Copper Coatings Shown in Figure 11.10b

Ref.	$d_p(\mu m)$	$\delta(mm)$	δ/d_p	ε
Kajikawa et al. [1983]	12	1.0	8	0.65
Milton [1968]	< 44	0.25	6	
Fujii et al. [1979]	115	0.40	3–4	0.49
Nishikawa et al. [1983]	250	2.0	8	0.66
Milton [1968]	250–500	1.1	3	
Nishikawa et al. [1983]	500	2.0	4	0.42

the maximum performance was obtained for δ/d_p = 4.0. In Test 2, they used 2.0-mm coating thickness (δ) and tested particle diameters of 0.10, 0.25, and 0.50 mm. The coating with δ/d_p = 4 (d_p = 0.25 mm) again yields the highest boiling coefficient. Therefore, for spherical particles with 0.10 < d_p < 0.50 mm, it appears that a coating having δ/d_p = 4 is preferred for maximum performance. Tests 3 and 4 with bronze spherical particles show a much smaller effect of δ/d_p on the boiling coefficient; the δ/d_p = 4 coating yields the highest boiling coefficient only for q > 50 kW/m². Below this heat flux, coatings having 2 < δ/d_p < 6 provided approximately equal performance. Kim and Bergles [1988] also report R-113 data on coatings made with δ/d_p = 4 for particle diameters of 0.059, 0.096, and 0.247 mm.

Considerable data have been published on the sintered coating described by Milton [1968, 1970, 1971]; for example, see O'Neill et al. [1972] and Czikk and O'Neill [1979]. This is made from commercially produced powders, which are not spherical and have a range of particle sizes. About 45% of the particles are between 7 and 44 µm, with the rest smaller. The commercially produced tube is known as High-Flux, and Figure 11.11a shows top and cross section views of the coating.

Figure 11.11c shows a cross section of the spherical copper coating tested by Fujii et al. [1979], which has a nominal particle diameter of 0.115 mm and δ/d_p = 3. The coatings tested by Nishikawa et al. [1983] were also made of spherical, uniform-diameter particles. The pore dimensions and matrix porosity of the Figure 11.11a coating should be different from those for spherical particles of uniform diameter, since the smaller particles will tend to partially fill the voids between the larger particles. This raises a question concerning the important geometric parameters of the coating. Rather than the particle diameter, the pore size (the void space between particles) is probably a key characteristic dimension. Although the thermal conductivity of the particles is important, coatings made of aluminum and copper give comparable performance. Milton and Gottzman [1972] show that low-conductivity nickel particles produced considerable enhancement for boiling oxygen, although the boiling coefficient was only 30% as large as for a copper structure.

Because the independent parameters of the porous structure are the particle diameter (or particle size distribution), particle shape, and the coating thickness, one cannot calculate the pore diameter, unless the particle "stacking arrangement" is known. As seen by examining the porous structures of Figure 11.11, the "pore

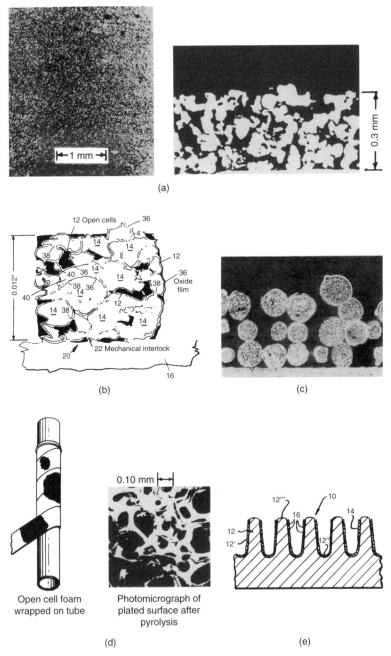

Figure 11.11 (a) Copper-sintered surface (left); cross section of sintered coating (right). (From Gottz-manm et al. [1971, 1973].) (b) Cross section photograph of aluminum flame-sprayed surface. (From Dahl and Erb [1976].) (c) 0.115-mm-diameter spherical particles tested by Fujii et al. [1979]. (d) Porous coating formed by plating copper on polyurethane foam wrapped tube. (From Janowski et al. [1978].) (e) Porous coating on fins formed by electroplating in a bath containing graphite particles. (From Shum [1980].)

size" is not an explicitly measurable dimension. Czikk and O'Neill [1979] used a quantitative scanning microscope to examine cross-sectional slices of the coating in attempt to measure the pore size distribution. This resulted in a statistically defined pore size distribution, which was not wholly realistic. This is because the method could not differentiate between connected and closed pores. A more acceptable method involves use of a capillary rise test using a wetting fluid (near zero contact angle). By observing the height (H) that the wetting liquid rises on a porous coated plate, one may determine the "equivalent" pore size as $r = 2\sigma\rho_{lg}H$. This method is used in the Milton's patents [1968, 1970, 1971] to define the equivalent pore size.

The pores provide a high area density of relatively large cavities, which are believed to function as reentrant nucleation sites. Many of these pores are probably of a reentrant shape, which Griffith and Wallis [1960] have shown to be very stable vapor traps. The pores within the matrix are interconnected so vapor formed in one pore can activate adjacent pores. Gottzmann et al. [1971] theorize that essentially all vaporization occurs within the porous matrix, and the high performance is the result of two factors: (1) the porous structure entraps large radius vapor-liquid interfaces, compared to the very small nuclei in naturally occurring pits and scratches, which considerably reduces the theoretical superheat required for nucleation; and (2) the porous structure provides a much larger surface area for thin film or microlayer evaporation than exists with a flat surface. Thus, high evaporation rates can be accommodated with a very small film ΔT.

The sintering process described by Milton [1968] for copper powder on a copper base tube caused annealing and possible deformation of the base copper tube. Grant [1977] describes an improved metal powder, which allows a lower temperature sintering process (732 to 813°C). Use of a copper alloy base tube prevents full annealing. Further details of the processing conditions for sintering are given by Hausner and Mal [1982].

Hsieh and Weng [1997] and Hsieh and Yang [2001] boiled R-134a on horizontal tubes having flame sprayed, plasma sprayed, and "pitted" coatings. The pitted coatings were formed using a sandblasting technique. The coatings used were aluminum, copper, and molybdenum on a copper substrate. For $q < 10$ kW/m², they found that the pitted coatings provided higher performance than flame or plasma spraying.

Several other methods of forming a porous coating, which do not require high-temperature processing, have been proposed. These methods provide boiling performance comparable to that of the Milton patent [1968]. Fujii et al. [1979] bond several layers of fine copper particles to the base surface by electroplating with copper; this process is accomplished at room temperature. Shum [1980] forms a porous coating by electroplating in a bath containing graphite particles (see Figure 11.11e). The graphite particles are attracted to the surface and bonded by the plating operation. Dahl and Erb [1976] make an aluminum porous coating by flame-spraying the aluminum particles in an oxygen-rich flame atmosphere (see Figure 11.11b). The oxygen-rich flame forms an oxide film on the particle and prevents it from flattening upon impact with the surface. Flame-sprayed copper tends to flatten on impact, providing a less-porous, lower-performance surface. A U.S. patent by Modahl and Lukeroth [1982] describes aluminum flame spray techniques that provide higher enhancement than is obtained using the technique described by Dahl and Erb. [1976]. Sanborn et al. [1982] show that aluminum flame spray on the outer surface of the Figure 11.12f GEWA-T tube provides increased performance.

Liu et al. [1987] formed a porous coating by sintering multiple layers of 185-mesh copper screens to a copper surface. They tested R-113 and water for 1, 2, 3, 4, 5, 6, 7, 9, and 13 mesh layers. The 185-mesh wire screen was formed from 0.046-mm wires, provided 0.091-mm pore diameter, and had area and volumetric porosities of 44 and 56%, respectively. They found that the boiling coefficient increased with number of layers, up to five layers, and decreased for increasing number of layers for both R-113 and water. They provide an empirical, power-law correlation of their data, whose maximum error was ±30%. The correlation assumes that the characteristic dimensions are pore diameter (not varied) and number of layers. They observed interesting hysteresis phenomena, which are discussed in Section 11.7.

Other innovative methods have been developed to make porous surfaces. Janowski et al. [1978] have developed a process for making a porous boiling surface (see Figure 11.11d). A tube is wrapped with an open cell polyurethane foam, which is then copper plated. The copper plating provides structural integrity and good thermal conduction. The polyurethane is removed by pyrolysis (300 to 520° C), which forms additional very small pores within the skeletal structure. The structure has 1.5-mm coating thickness, 97% void volume, 4 pores/mm, and 0.12-mm pore size, augmented by 0.02-mm pores within the skeletal structure. Sachar and Silvestri [1983] used vacuum deposition of aluminum in the presence of an inert gas to form a porous surface coating. Zohler [1990] flame-sprayed two dissimilar metals (e.g., zinc and copper) on a tube, and then etched the tube so that one of the metals was etched out. Voids are left where the metal has been etched away. Chang and You [1997] made a "microporous" surface using a paint containing silver flakes (3 to 10 μm). The pores are substantially smaller (0.1 to 1 μm) than for sintered porous surfaces. The paint contained alcohol, which evaporates and epoxy that bonds the silver flakes. They showed performance improvement over a plain surface for FC-72. O'Conner and You [1995] used a dielectric paint made with diamond particles (8 to 12μm) for electronics cooling. They found that the CHF was increased 109% over that of a plain surface. The performance of the Chang and You [1997] and O'Conner and You [1995] painted coatings is not as good as that made by sintered copper particles.

Metal (or nonmetal) foams may also be used to make porous boiling structures. Metal foaming technology is discussed in Section 5.10. Test results for aluminum and copper foams with FC-72 are presented in Section 19.4.4.

Porous coatings offer two potentially important advantages over enhancement formed by integral roughness. If tube-side enhancement is desired, it may be formed independent of the boiling-side enhancement. If corrosion requirements permit, the porous coating may be made using a less expensive material than required for base tube surface.

11.4.11 Structured Surfaces (Integral Roughness)

State-of-the-art concepts employ cold metal working to form a high area density of reentrant nucleation sites, which are interconnected below the surface. Typically, these are reentrant grooves or tunnels, rather than discrete cavities. Figure 11.12b through f shows five patented boiling surfaces of this type, which have significantly higher performance than the Figure 11.12a standard integral-fin tube. The surfaces were developed for use in refrigeration evaporators, which use shell side boiling on copper tubes. The first enhanced surface (Figure 11.12b) was patented in 1972. All of the enhanced surfaces shown on Figure 11.12 are commercially produced, and they are routinely used in refrigerant evaporators. Pais and Webb [1991] provide a survey of existing data for refrigerants on enhanced boiling surfaces. In single tube pool boiling tests, the heat flux at fixed $(T_w - T_s)$ will be 2-to-4 times greater than that of the Figure 11.12a integral fin tube.

The Figure 11.12b reentrant, single-grooved surface of Webb [1972] consists of integral-fin tubing (1300 fins/m, 0.8 mm fin height) in which the fins are bent to form a reentrant cavity. The groove opening at the surface is a critical dimension, and the recommended gap spacing is 0.038 to 0.089 mm for R-11. The performance is sharply reduced for gap widths outside this range. The preferred gap spacing is expected to be smaller for low-surface-tension fluids. All the enhancements shown in Figure 11.12 are made by deforming the fins on integral fin tubes. Saier et al. [1979] describes the Figure 11.12c tube, which is known as the Wieland GEWA-TW and has 750 fins/m. Figure 11.12f is the Wieland GEWA-SE tube having 1024 fpm. The Figure 11.12d and 11.12e tubes are also made from integral-fin tubes, whose fins are further deformed to provide a three-dimensional structure. Figure 11.12d reported by Fujie et al. [1977] is formed from a low fin tube (1970 fpm),

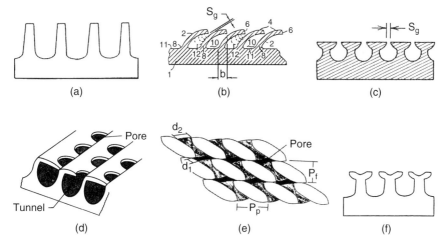

Figure 11.12 Six commercially available enhanced boiling surfaces: (a) integral-fin tube, (b) Trane bent fin, (c) Wieland GEWA-TW, (d) Hitachi Thermoexcel-E, (e) Wolverine Turbo-BII, and (f) Wieland GEWA-SE.

which has small spaced cutouts at the fin tips. These sawtooth fins are bent to a horizontal position to form tunnels having spaced pores at their top. This commercially available tube is called Thermoexcel-E™ and is further discussed by Arai et al. [1977] and Torii et al. [1978]. Figure 11.12e illustrates the Wolverine Turbo-BII, which has 1889 fpm (48 fins/in.). Improvements have been made by Wolverine since the original Turbo-B (1650 fins/m) as described in the *Wolverine Databook II* by Bell and Mueller [1984] and Thors et al. [1997]. The latest version, Turbo-BIII, has 2362 fins/m (60 fins/in) on the outer surface.These tubes also have internal helical grooves, whose dimensions are given in Table 9.9.

Kim and Choi [2001] describe the fundamental surface/subsurface structure of the Figure 11.12 enhancement geometries. Figure 11.13 shows three basic types of finned structures, all of which may be applied to an integral-fin tube. Kim and Choi [2001] state that the Turbo-B type tube (Figure 11.12e) contains subsurface tunnels having "pores with connecting gaps" as shown in Figure 11.13c. One may also consider Figure 11.12b and c as variants of Figure 11.13b and Figure 11.12d as a variant of Figure 11.13a. Chien and Webb [1998a, 1998b, 1998e] provide detailed test results and predictions for the Figure 11.13a geometry, which are given in Section 11.8.4. Kim and Choi [2001] tested the Figure 11.13c surface structure for different pore diameters applied to a tube having 1654 fins/m. The test results are discussed in Section 11.5.

Chien and Chang [2004] made their own structured surfaces having surface pores and subsurface tunnels. The enhanced surfaces were formed in flat plates, which were tested in the vertical orientation using water and HFC-4310. They made the surfaces by cutting 1.0-mm-high × 0.6-mm-pitch fins in the plates. Triangular grooves 0.3 mm deep were cut in the fin tips (0.6 or 1.0 mm pitch) and then the fins were bent to form surface pores of triangular cross section. The resulting surface is much like the Thermoexcel-E surface shown in Figure 11.12d. Figure 11.14a and b shows the fin geometry before and after bending, respectively. Also shown on Figure 11.14c is a photograph of their surface SA-0.6-1.0. The structured surfaces are similar to the surface tested in Chien and Chen [2000], except that the present surfaces have a rectangular fin channel, as compared to the triangular channel of Chien and Chen [2000]. An important element of this work is that it shows how one can make such structured surfaces on flat plates using standard machining practices. In total, they made seven different surfaces having different fin pitches ($p_f = 0.6$ or 1.0 mm) and different pore pitches ($P_p = 0.6$ and 1.0 mm) and heights ($H_t = 0.7$ and 1.0 mm). For water boiling at $T_{sat} = 70°C$, they found that the preferred dimensions were $H_t = 0.7$ mm, $P_p = 1.0$ mm. For HFC-4310, the different pore dimensions had no significant influence. Kim and Choi [2001] also made a tube having subsurface tunnels and surface pores similar to Figure 11.14c. Their test results are presented later in the chapter in Figures 11.21a and 11.22.

Figure 11.15 shows other structured surfaces reported in the literature. Kun and Czikk [1969] were awarded an early patent on a reentrant, cross-grooved surface (Figure 11.15a). A series of fine, closely spaced grooves (1100 to 9000 grooves/m) are cut in the surface, followed by a second set of cross grooves, which are not as deep as the first set. The patent provides data on boiling nitrogen and water for a

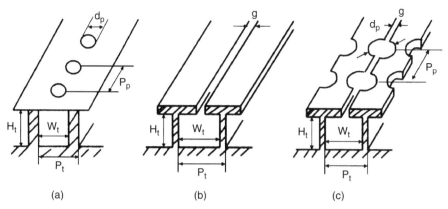

Figure 11.13 Possible basic surface/subsurface structures for the Figure 11.12 enhanced boiling tubes made from an integral-fin tube. (a) Surface pores, (b) surface gaps, (c) surface pores with connecting gaps. (From Kim and Choi [2001].)

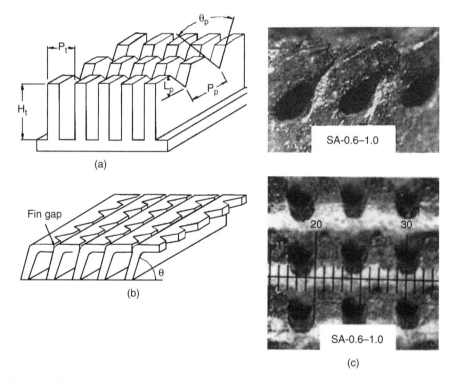

Figure 11.14 Structured surfaces made by Chien and Chang [2003]. (a) Unbent fins, (b) bent fins, (c) cross section through fins of surface SA-0.6-1.0 (0.6-mm fin pitch and pore pitch with 1.0-mm-high fins).

variety of groove densities. Kun and Czikk report that the opening width of the reentrant cavity is a critical dimension, as also reported by Webb [1972]. Maximum

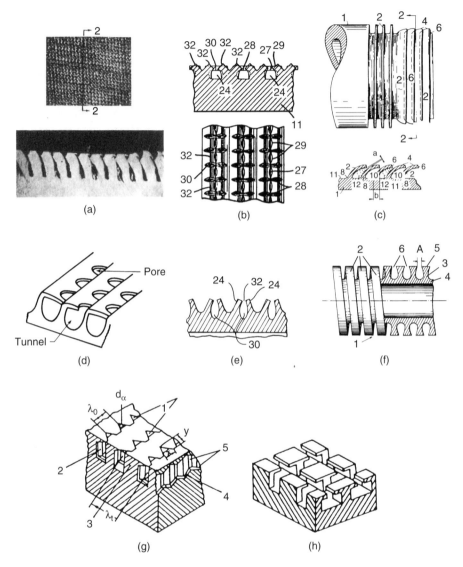

Figure 11.15 Structured boiling surfaces. (a) Cross-grooved surface having reentrant cavities. (From Kun and Czikk [1969].) (b) Deeply knurled Y-shaped fins. (From Szumigala [1971].) (c) Reentrant grooves formed by bending tips of integral fin tube. (From Webb [1972].) (d) Reentrant grooves having minute spaced holes at top of tunnel. (From Fujie et al. [1977].) (e) Reentrant grooves formed by bending of fins of unequal height. (From Brothers and Kallfelz [1979].) (f) Reentrant grooves formed by flattening fin tips of integral fin tube. (From Saier et al. [1979].) (g) JK-2 tubes tested by Zhong et al. [1992]. (h) Geometry developed by Zhang and Dong [1992].

performance is obtained with a restricted opening dimension of 0.0013 mm for nitrogen and 0.006 mm for water.

11.4.12 Combination of Structured and Porous Surfaces

Ma et al. [1986] have found quite high performance with water and methanol boiling on a copper plate having: (1) parallel rectangular grooves covered by fine mesh screens; (2) same as item 1, but with a 40-mesh screen covered by a thin brass, and plate having small pores (0.08 to 0.22 mm); (3) same as item 2, but varying the cross-sectional shape of the grooves (rectangular and vee). They varied the groove size and shape, the screen porosity, and the pore size in the item 2 variant. Their findings and preferred dimensions are as follows:

Variant 1: The best performance was obtained for water and methanol with 0.3-mm-wide, 0.4-mm-deep grooves spaced at 0.6 mm, which were covered with a 185-mesh copper screen (0.046-mm wire diameter). A 40 kW/m^2 heat flux was obtained at 1.0 K with methanol.

Variant 2: Grooved plates (0.60-mm groove pitch) covered by a 40-mesh screen, upon which a plate having 0.08 to 0.22-mm-diameter pores at 277 cm^2 was brazed. The best configuration provided 90 kW/m^2 heat flux at 1.0 K. This surface performance is competitive with the High-Flux surface for ethanol, as reported by Antonelli and O'Neill [1981].

Variant 3: Use of triangular grooves provided higher heat flux than rectangular grooves at heat fluxes above 80 kW/m^2.

The Ma et al. [1986] paper is very interesting and worthy of close reading. Although the use of screening above the grooves (or screening with a plate having pores) provides high performance, it may be relatively expensive. It appears that a standard porous surface (e.g., the High-Flux) may provide competitive performance.

11.4.13 Composite Surfaces

Yang et al. [1996] tested a composite fiber-copper surface formed of 0 to 50% graphite fibers (by volume) with R-113. They found moderate enhancement with the best improvement for 25% graphite fibers. The thermal conductivity of the graphite fibers is three times that of copper. They attributed the improvement to "hot spots" on the graphite tips, which act as nucleation activators. Liang and Yang [1998] describe a very interesting composite surface formed by embedding graphite fibers in copper (or aluminum) substrates. The 8 to 10-μm-diameter graphite fibers are aligned normal to the boiling surface to take advantage of the very high thermal conductivity of graphite (1200 W/m-K). The composite structure contains microsize nucleation sites embedded in the high-thermal-conductivity graphite. Figure 11.16 shows test results for pentane. The figure shows that the highest performance is given by the graphite–copper composite. Note that the enhancement ratio (enhanced to plain-surface) is significantly below that of the porous sintered copper surface shown on Figure 11.10. Liang and Yang speculate that the high graphite thermal conductivity results in higher temperature at the graphite–liquid surface, which allows the very small microcavities in the graphite (0.05 to 0.1μm) to be active.

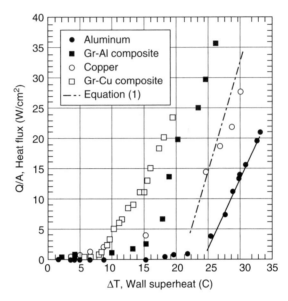

Figure 11.16 Liang and Yang [1998] test results for pentane boiling on a composite surface formed by embedding graphite fibers in aluminum or copper substrates. Also shown are curves for boiling on plain aluminum or copper surfaces.

Further, the poor wettability of the graphite microcavities delay flooding at low ($T_w - T_{sat}$). They also found that significantly lower start-up ($T_w - T_{sat}$) on the composite surfaces than on the plain metal surfaces.

11.5 SINGLE-TUBE POOL BOILING TESTS OF ENHANCED SURFACES

Figure 11.17 shows the single-tube pool boiling performance of four commercially available enhanced surfaces measured by Yilmaz and Westwater [1981] using iso-propyl alcohol. These are the High-Flux (Figure 11.11a), the Thermoexcel-E (Figure 11.12d), the GEWA-T (Figure 11.12c), and the ECR-40 (same as Turbo-B). Figure 11.17 also shows the performance of a copper metal-sprayed surface (CSBS). The tubes were heated by steam condensing on the tube side, and data were taken for saturated isopropyl alcohol boiling at 103 kPa. The data span the nucleate boiling and transition regimes. The highest performance is given by the ECR-40 and High-Flux surfaces. However, at a typical design heat flux of 50 kW/m², the other tubes provide considerable enhancement. The heat flux is defined in terms of the envelope area ($\pi d_o L$), except for the 1200 fins/m integral fin tube (GEWA); the heat flux of this tube is defined in terms of the extended surface area. The enhanced surface tubes exhibit a CHF at least 40% greater than that of the plain tube. If the heat flux of the integral fin tube is defined in terms of the diameter over the fins, the heat flux will be 2.6 times greater. Then its CHF will be approximately equal to that of

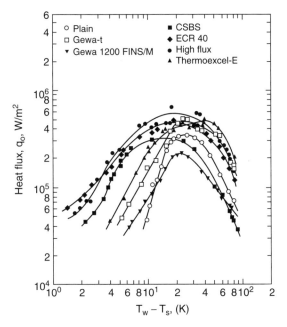

Figure 11.17 Comparative single-tube pool boiling test results for isopropyl alcohol at 101 kPa (1 atm). (From Yilmaz and Westwater [1981].)

the High-Flux tube. More discussion on CHF is presented in Section 11.12. Li et al. [1992] tested a variant of the Thermoexcel-E tube shown in Figure 11.12d. The R-113 boiling coefficient of their best tube, having 0.14-mm pore diameter, was approximately 50% higher than that of the Thermoexcel-E tube.

Figure 11.18 taken from the patent of Thors et al. [1997] shows the progression of boiling performance of old-to-new versions of the generic Turbo-B tube. The boiling data were taken for R-134a at 15°C. The figure shows that the highest performance tube (Turbo-BIII) gives 70% higher performance than the original Turbo-B tube for $(T_w - T_{sat}) = 1.0$ F. Tatara and Payvar [2000] compare their R-134a data on the Turbo-BII tube with test results of Webb and Pais [1992] on the Turbo-B tube. They show that the Turbo-BII has 60 to 90% higher boiling coefficient than the Turbo-B tube data of Webb and Pais.

Webb and Pais [1992] obtained saturated nucleate boiling data for five refrigerants (R-11, R-12, R-22, R-123, and R-134a) on a plain tube and geometries a, c, d, and f shown in Figure 11.12. The copper tubes had 17.5 to 19.1 mm diameter over the fins. Tests were performed at two saturation temperatures, 4.44°C and 26.7°C using refrigerants R-11, R-12, R-22, R-123, and R-134a. The tubes were specially made to have a 9.53-mm smooth inside bore. This was done to facilitate wall thermocouple and electric cartridge heater installation. Figure 11.19 shows the R-22 boiling data at 4.44°C for the tubes tested. Figure 11.20 shows the 4.44°C GEWA-SE data for refrigerants. The complete data set is given by Webb and Pais [1992]. The data for the high-pressure refrigerants (R-12, R-22, and R-134a) are

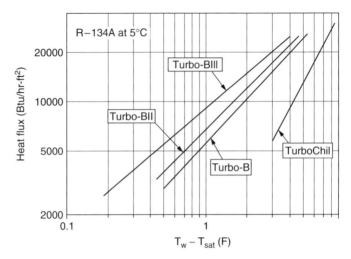

Figure 11.18 Progression of boiling tube performance (R-134a at 5.0°C) from old to new versions of the Turbo-B type tube. Tube I (TurboChil), Tube II (Turbo-B), Tube III (Turbo-BII), Tube IV (Turbo-BIII). (From Thors et al. [1997].)

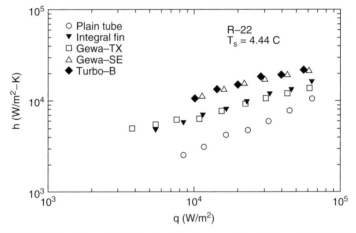

Figure 11.19 R-22 boiling data at 4.44°C for the five tubes tested by Webb and Pais [1992].

higher than that for the low-pressure refrigerants (R-11 and R-123). Note the difference in slope for the high-and low-pressure refrigerants, particularly for the GEWA-SE. At 40 kW/m² heat flux, the boiling coefficient of the GEWA-SE relative to the 26 fins/in. tube is 1.58 for R-12 and 1.38 for R-134a, respectively.

Wang et al. [1998] extended the work of Webb and Pais [1992] to obtain single-tube pool boiling data for R-410A and R-407C on a 748 fins/m integral fin tube, and the Wieland-B enhanced boiling tube (1968 fins/m). They also tested R-22, R-123, and R-134a. For the Wieland-B tube, they found that R-407C gave 80 to 90%

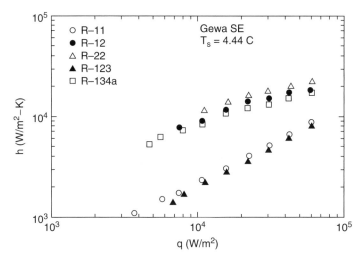

Figure 11.20 Boiling data for five refrigerants boiling at 4.44°C on the GEWA-SE tube. (From Webb and Pais [1992].)

lower boiling coefficient than R-22, while the boiling coefficients of R-410A were nearly equal.

Saidi et al. [1999] provide R-123 data on a 17.1-mm-OD integral-fin tube having 1923-fins/m and 0.91-mm fin height. Compared to the R-123 data of Webb and Pais [1992] on a 1024 fin/m integral-fin tube (1.5-mm fin height), the 1923 fins/m provided 10 to 20% higher boiling coefficient. This improvement is roughly consistent with the surface area increase.

Kim and Choi [2001] provide pool boiling test results on a 1654 fins/m integral-fin tube (1.3-mm fin height) made with surface "pores and gaps" as shown in Figure 11.21a. See the previous discussion of Chien and Chang [2004] concerning Figure 11.14 regarding a similar geometry. The three tube variants were made by cutting 0.9-mm-deep triangular notches in the fin tips and then flattening the fin tips using a rolling process. This formed triangle shaped pores with connecting gaps and "gourd-shaped" tunnels. Three pore/gap geometries (pore diameters of 0.2, 0.23, and 0.27 mm) were tested with R-11, R-123, and R-134a. The test results show that the preferred pore size depends on the refrigerant used. The R-134a test results are shown in Figure 11.22. This figure shows that large pore size provides the highest performance for R-134a, which is a "high-pressure" refrigerant. However, for R-123 (a low-pressure refrigerant) they found that a smaller pore size (0.23 mm) is preferred. They provide empirical correlations to predict the effect for pore diameter. The correlations are a modification of the Cooper [1984] correlation (Equation 2.29) for plain surfaces, with an additional parameter to account for the pore diameter.

Kedzierski [1995] and Srinivasan et al. [2001] provide comparative data on the Turbo-BII) and the sintered High-Flux tubes with R-123 and liquid nitrogen, respectively. Both studies show roughly comparable performance for the two tubes

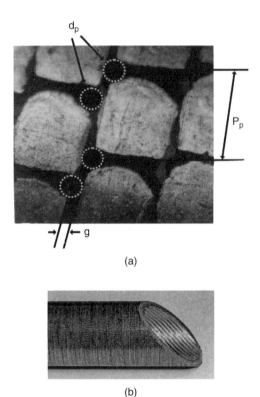

(a)

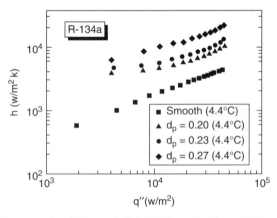

(b)

Figure 11.21 (a) Photograph of 1654 fins/m tube made by Kim and Choi [2001] having 0.2-mm-diameter surface pores and 0.04-mm-gaps. (b) Photograph of commercial Turbo-BIII tube.

Figure 11.22 R-134a test results of Kim and Choi [2001] on the Figure 11.21a type tube for three different pore diameters (0.20, 0.23, and 0.27 mm).

for $q < 20$ kW/m² However, the High-Flux tube has a higher q vs. $(T_w - T_s)$ slope, which results in its giving higher performance at higher heat fluxes.

Several workers have measured the performance of enhanced tubes for boiling of hydrocarbons. Among these are Tarrad and Burnside [1993], who used the High-Flux, Turbo-B, GEWA-T, and a 1024 fins/m integral fin tube for boiling of a pentane/tetradecene mixture. Sokol et al. [1990] boiled propane and propylene on an integral-fin tube. Hübner and Künstler [1997] boiled various hydrocarbons on the Wieland GEWA-T and GEWA-SE tubes. Mertz et al. [2002] boiled isobutane on four variants of the Wieland GEWA-T surface geometry. Kulenovic et al. [2002] report high-speed flow visualization results for propane boiling on the GEWA-SE tube. The visualization results include bubble frequency, bubble departure diameter, and the bubble rise velocity.

11.6 THEORETICAL FUNDAMENTALS

Small bubbles (e.g., $d_b < 1.5$ mm) are periodically grown and released from nucleation sites. After departure of a bubble, the nucleation sites must retain a small vapor embryo to facilitate the next bubble cycle. The cavities must be able to initially entrap vapor that will allow the nucleation process to take place. These phenomena must exist for both plain and enhanced surfaces. Carey [1992] provides an excellent discussion of these fundamental phenomena. This section summarizes the key issues, and then presents a discussion of how the fundamental phenomena of boiling on enhanced surfaces differ from plain surfaces.

11.6.1 Liquid Superheat

The first item to understand is the thermodynamic requirement for a bubble to exist. Consider a static bubble of radius r, which exists in thermal equilibrium with a large volume of liquid. The pressure inside and outside the bubble are p_v and p_l, respectively. The net pressure force, $(p_v - p_l)\pi r^2$ is equal to the surface tension force, $2\pi r\sigma$. Equating the pressure and surface tension forces shows that

$$p_v - p_l = \frac{2\sigma}{r} \tag{11.1}$$

Thermodynamic tables (or an equation of state) give the saturation pressure (p_s) that corresponds to a given saturation temperature (T_s). The $p_s(T_s)$ is for a liquid–vapor interface having zero curvature. For the present case of a vapor bubble of radius r, Rohsenow [1985] shows that the pressure inside the bubble (p_v) is slightly lower than the pressure of a saturated liquid (p_s) at the same temperature. This difference is

$$p_v = p_s - \frac{\rho_v}{\rho_l - \rho_v} \frac{2\sigma}{r} \tag{11.2}$$

If $\rho_v \ll \rho_l, p_v \simeq p_s$.

$$p_s - p_l \simeq \frac{2\sigma}{r} \tag{11.3}$$

It is frequently assumed that $p_v = p_s$ at the liquid temperature. Using this approximation, one may write Equation 11.1 as

$$T_l - T_s = \frac{2\sigma}{mr} \tag{11.4}$$

where the local slope of the saturation curve (p_s vs. T_s) is $dp/dT \equiv m$. Frequently, writers use the Clasius-Clapeyron equation to approximate $dp/dT = \lambda/T_{abs}v_{fg}$. However, here the terminology of $m \equiv dp/dT$ is used for simplicity. Assuming $p_v \equiv p_{sat}$, Equation 11.4 shows that the liquid temperature must be superheated for the vapor bubble of radius r to exist.

Equation 11.4 was written for a vapor bubble (concave liquid shape). If a force balance is written on a liquid droplet (a convex liquid shape) of radius r, the resulting force balance will yield a negative sign on the right-hand side of Equation 11.1. Then, the right-hand side of Equation 11.4 would be negative. In this case, the liquid must be subcooled an amount ($T_s - T_l$) for a convex liquid surface of radius r to exist. Hence, Equation 11.4 may be written in a more general form

$$T_l - T_s = \pm \frac{2\sigma}{mr} \tag{11.5}$$

where the plus and minus signs are for concave and convex liquid curvatures, respectively.

11.6.2 Effect of Cavity Shape and Contact Angle on Superheat

Figure 11.23 shows a liquid–vapor interface growing inside a conical cavity for three different values of liquid contact angle (θ). The interface curvature is larger for increasing contact angles. The liquid–vapor interface is spherical. Its radius (r) depends on the liquid contact angle (θ) the angular opening (2ϕ) of the cavity. The interface radius changes with its location within the cavity. Figure 11.23a shows a highly wetting liquid ($\theta \ll \phi < 90°$), and Figure 11.23c is for a nonwetting liquid ($\theta > \phi + 90°$). Figure 11.23b is for nonwetting inside the cavity ($\theta > \phi + 90°$) but wetting outside the cavity ($\theta < 90°$). Figure 11.23d shows the variation of r_c/r within the conical cavity, for the three wetting characteristics. If $r_c/r < 1$, the required liquid superheat (see Equation 11.5) is less than that required at the cavity mouth. Figure 11.23d shows that a highly wetting surface requires higher superheat inside the cavity than at the mouth. Conversely, a poorly wetting surface requires smaller superheat inside the cavity than at the mouth.

The contact angle is not a thermodynamic property. It depends on the surface cleanliness and microroughness. Oxidized surfaces have smaller contact angles than clean surfaces. For example, water deposited on a waxed surface will exist as beads, but it tends to spread on the surface after the wax has worn away. Chapter 3 of Carey [1992] provides a good discussion of contact angle.

The internal shape of the cavity has a significant influence on the liquid superheat required for a liquid–vapor interface to exist within a cavity. Webb [1983] gives

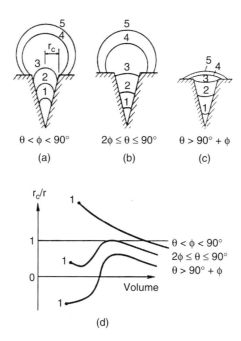

Figure 11.23 Variation of bubble radius within during growth for: (a) $\theta < \phi < 90°$, (b) $2\phi < \theta < 90°$, (c) $\theta > \phi + 90°$, (d) variation of r_c/r with cavity volume. (From Carey [1992].)

equations for the interface radius for the five cavity geometries shown in Figure 11.24. These equations are developed using geometric relations for the cavity shape, as a function of the contact angle (θ) and the cavity angle (ϕ). Depending on θ and ϕ the liquid either wets the cavity (concave liquid interface) or is nonwetting (convex liquid interface). The equations are

1. Conical (Figure 11.24a)

$$\frac{r}{x} = \frac{\sin\phi}{\sin(\theta - \phi \pm \pi/2)} \tag{11.6}$$

The plus sign is used for a wetting fluid ($\theta < \pi/2 + \phi$) and the minus sign for a nonwetting fluid ($\theta > \pi/2 + \phi$)

2. Cylinder (Figure 11.24b)

$$\frac{r}{r_c} = \pm\frac{1}{\cos\theta} \tag{11.7}$$

The cylinder radius is r_c. The plus sign is used for a wetting fluid ($\theta < \pi/2$) and the minus sign for a nonwetting fluid.

3. Reentrant conical cavity (Figure 11.24c and d)

$$\frac{r}{x} = \frac{\sin\phi}{\sin[\pm(\theta+\phi-\pi/2)]} \qquad (11.8)$$

The minus sign is used for a wetting fluid ($\theta < \pi/2 - \phi$) and the plus sign for a nonwetting fluid.

4. Doubly reentrant conical cavity (Figure 11.24e): Same equation as for Figure 11.24c and Figure 11.24d (Equation 11.8).

Equations 11.6 through 11.8 were applied to the cavity shapes in Figure 11.24 for $\theta = 15°$ and $r_c = 0.04$ mm to predict the liquid–vapor interface radius values. Then, from Equation 11.5, we may calculate the $(T_l - T_s)$ values required for a stable liquid–vapor interface. These results are shown in Figure 11.25. Figure 11.25 shows the value $1/r = m(T_l - T_s)/2\sigma$ vs. y/r_c, where r_c is the cavity mouth radius. All of the cavity shapes require the same value of $(T_l - T_s)$ for a vapor bubble to exist at the cavity mouth. However, significantly different values of $(T_l - T_s)$ exist within the cavity. Figure 11.25 shows that cavity a requires the greatest superheat, and that the limiting superheat exists at the bottom of the cavity. The required superheat decreases with vapor volume in cavity c, attaining a maximum at the cavity mouth, and decreases as the bubble grows outside the cavity. The required superheat within the cavity b is intermediate to that cavity a and cavity c. The reentrant cavities (cavities c, d, and e) require the smallest $(T_l - T_s)$ within the cavity. Only the cavities in Figure 11.24d and e, will be stable if a subcooled liquid is exposed to the cavity; this is because the liquid interface takes a convex shape along the conical side walls. The doubly reentrant cavity (Figure 11.24e) allows greater subcooling than the cavity in Figure 11.24d.

Therefore, a reentrant cavity shape, having a high contact angle will be very stable, and if $\theta > 90° - \phi$, it will not flood in the presence of moderate subcooling of the liquid. For very small contact angles (e.g., $\theta = 4°$) a reentrant cavity is susceptible to liquid flooding. However, the doubly reentrant cavity (Figure 11.24e) will establish a convex liquid shape just inside the lower cavity mouth for low-contact-angle fluids and provide a stable interface with a subcooled fluid. Low-surface-tension fluids, such as refrigerants and organics, have small contact angles, e.g., $5°$, and high-surface-tension fluids, such as water, have contact angles, in the range of $70°$.

11.6.3 Entrapment of Vapor in Cavities

Section 11.6.2 has addressed how the radius of the vapor interface changes with its movement within cavities of various shapes. Now, consider the means by which vapor (or air) is initially trapped within a cavity. Boiling systems are typically charged with subcooled liquid. After the system is charged, heat is applied and the subcooled liquid is heated to the boiling state. Some analysis has been done to explain how inert gas (air) can be trapped in cavities, when flooded by subcooled liquid. The fact that an inert gas is trapped in the cavity has implications for the boiling process, which are discussed in Section 11.6.4.

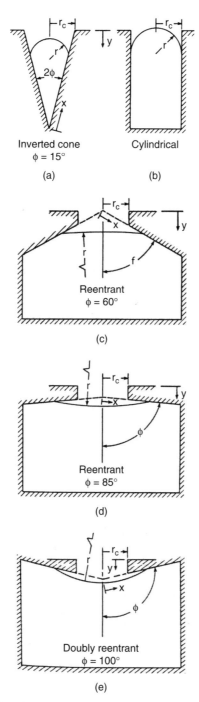

Figure 11.24 Vapor–liquid interface radius in cavities of different shapes for r_c = 0.04 mm and θ = 15°. (From Webb [1983].)

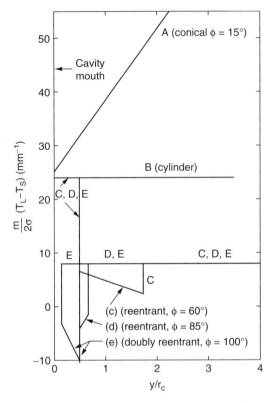

Figure 11.25 Required liquid superheat for stability the Figure 11.24 cavity geometries having $R_c =$ 0.04 mm and $\theta = 15°$. (From Webb [1983].)

When the system is initially charged with liquid, a liquid front advances over the nucleation sites, as illustrated for a conical cavity in Figure 11.26a. The contact angle of the advancing liquid front is θ_a. If the liquid contacts the far wall of the cavity before it contacts the base, gas (air) will be trapped in the cavity. This will occur in Figure 11.26a if $\theta > 2\phi$. As discussed by Carey [1992], the advancing contact angle (θ_a) may be significantly greater than the static contact angle (θ). Hence, the possibility of trapping gas is influenced by the cavity geometry and the magnitude of θ_a.

Lorenz et al. [1974] have provided a simple method to calculate the size of the initially entrapped gas radius. If the cavity geometry and θ_a are known, one may calculate the volume of the trapped gas illustrated in Figure 11.26a. Assuming that this trapped gas volume reverts to the shape illustrated in Figure 11.26b, one may calculate the interface radius, which is shown in Figure 11.27. This may be an oversimplification, because Lorenz et al. assume that the static contact angle applies to both Figure 11.26a and b. As suggested above $\theta_a > \theta$. However, the Lorenz et al. concept is probably useful to explain how a finite liquid–vapor radius can exist in a cavity.

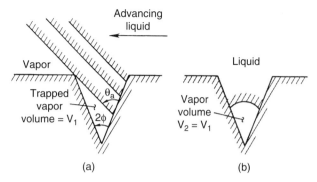

Figure 11.26 Illustration of how a cavity may trap air when the surface is flooded with liquid. (a) Advancing liquid front. (b) Liquid–vapor interface for the Figure 11.26a conditions. (From Carey [1992].)

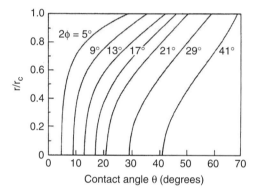

Figure 11.27 Equilibrium radius for the Figure 11.26 conical cavities with different ϕ, θ. (From Lorenz et al. [1974].)

Some boiling systems (e.g., refrigerant systems) are evacuated before charging with liquid. Consider charging an evacuated system with R-22 at 300 K. At 300 K, R-22 will exist as a saturated liquid at 1097 kPa. Thus, R-22 enters the evacuated system as a vapor, and the system pressure increases during the filling process. This vapor will flood the cavities, and may be trapped by the mechanism described by Lorenz et al. [1974]. However, the vapor trapped in the cavity may not contain an inert gas. It is possible that the charged liquid may contain a dissolved gas, which comes out of solution during the process of raising the liquid to boiling temperature. This evolved gas may add to the trapped vapor volume.

11.6.4 Effect of Dissolved Gases

Equations 11.1 through 11.5 assume that the vapor does not contain inert gases. Should the system contain inert gases, the total pressure in the vapor bubble is $p_v + p_G$. Assuming $p_v \simeq p_s$, Equations 11.3 and 11.5 would then be written as

$$p_s - p_l \simeq \frac{2\sigma}{r} - p_G \qquad (11.9)$$

$$T_l - T_s = \pm \frac{1}{m}\left(\frac{2\sigma}{r} - p_G\right) \qquad (11.10)$$

Equation 11.10 shows that a concave vapor interface will require less liquid superheat when inert gas is present in the cavity. When boiling occurs from cavities containing inert gases, each departing bubble will remove a small amount of inert gas from the cavity. With increasing time, a higher $(T_1 - T_s)$ will be required to sustain boiling, because the partial pressure of the inert gas will be reduced.

11.6.5 Nucleation at a Surface Cavity

Consider a bubble growing from a cavity, whose radius at the surface is r_c. Figure 11.28 shows that the cavity is contained in a surface, maintained at temperature T_w. How much liquid superheat must exist at the wall temperature for the bubble of radius r_c to exist? This will define a necessary condition for nucleate boiling to occur at the surface. The liquid superheat at the top of the bubble, $T_l(r) - T_s$, must satisfy Equation 11.4 for $r = r_c$. Assuming steady-state conduction in the liquid, and a linear temperature profile in the liquid, the Fourier heat conduction equation gives

$$\frac{Q}{A} = \frac{k_l[T_w - T_l(r_c)]}{r_c} \qquad (11.11)$$

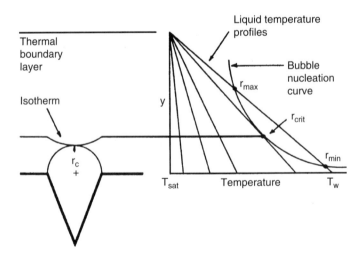

Figure 11.28 Necessary condition for bubble growing from cavity of radius r_c in a heated surface at T_w.

Writing $T_w - T_s = [T_w - T_l(r_c)] + [T_l(r_c) - T_s]$, the first and second terms on the right-hand side may be obtained from Equations 11.11 and 11.4, respectively. The resulting $T_w - T_s$ is

$$T_w - T_s = \frac{qr_c}{k_l} + \frac{2\sigma}{mr_c} \tag{11.12}$$

Hsu [1962] proposed a more refined analytical model to define the wall superheat necessary to support periodic bubble growth at a plain heated surface. The Hsu model assumes transient heat flux to the cold liquid layer that floods the surface after bubble departure.

If the cavity size is known, one may solve Equation 11.12 for the $T_w - T_s$ needed to support nucleation from the cavity at the specified heat flux. A problem in using the above model for plain surfaces is that the size of the natural nucleation sites is typically not known. For an enhanced surface, the bubbles grow from pores in the surface that are typically much larger than the microcavities associated with plain surfaces. Hence, substantially smaller superheat is required for the second term of Equation 11.12. More detail on nucleation on enhanced surfaces is given in Section 11.8.

11.6.6 Bubble Departure Diameter

Bubbles are periodically grown and released from the surface. The bubble departs when buoyancy force exceeds the surface tension force holding the bubble to the surface. Writing a force balance between buoyancy and surface tension forces on a departing bubble, with inertia forces neglected, one obtain

$$d_b = (6r_c \sin\theta)^{1/3} \left(\frac{2\sigma}{(\rho_l - \rho_v)g} \right)^{1/3} \tag{11.13}$$

where r_c is the surface cavity radius. This equation is not normally used for plain surfaces, because the cavity radius is typically not known. However, the surface pore radius is a known dimension for certain types of enhanced surfaces. For plain surfaces, a different form of Equation 11.13 is typically used, and is based on the work of Fritz [1935]. The Fritz equation includes the empirical constant, $c_b = 0.0148$ $\theta°$ and does not use the cavity radius. This equation is

$$d_b = c_b \left(\frac{2\sigma}{(\rho_l - \rho_v)g} \right)^{1/2} \tag{11.14}$$

Other equations have been proposed to calculate the bubble departure diameter. Some of the equations also account for the inertia term in the momentum equation for the growing bubble. These equations are surveyed in Table 6.1 of Carey [1992].

Measured bubble departure diameters are reported for structured surfaces (surface pores and subsurface tunnels) by Chien and Webb [1998c] for methanol, and by Kulevonic et al. [2002] for propane.

11.6.7 Bubble Dynamics

Several studies have shown that the nucleate boiling heat flux is proportional to the nucleation site density (n_s) on the heated surface. For example, the data of Kurihari and Myers [1960] for plain and emery-roughened surfaces show that $q \propto n_s^{0.33}$. Researchers who have modeled boiling on plain surfaces typically regard the bubble density as an empirical parameter in their models. This is the case for the Mikic and Rohsenow [1969] model, which is based on transient heat conduction to liquid during the bubble waiting period. The model assumes that the transient conduction occurs over a surface area twice the bubble departure diameter.

Measured bubble frequencies are reported for structured (surface pores and subsurface tunnels) surfaces by Nakayama [1980a], and Chien and Webb [1998c] for methanol and by Kulevonic et al. [2002] for propane. Haider and Webb [1997] developed an extension of the Mikic and Rohsenow [1969] model and applied it to predicting the external heat flux on a structured surface. The model is discussed in Section 11.9.3. The Haider and Webb model requires knowledge of the bubble frequency, site density, and departure diameter.

11.7 BOILING HYSTERESIS AND ORIENTATION EFFECTS

11.7.1 Hysteresis Effects

The existence of a starting hysteresis effect has been observed by a number of researchers. This phenomenon occurs when heating is initially applied and the data are taken in order of increasing heat flux. As the heat flux is increased from zero, boiling will not commence until a certain surface superheat is attained. Then the surface suddenly "explodes" into nucleate boiling, at which point the surface temperature decreases. It is possible then to traverse up and down the boiling curve, q vs. $(T_w - T_s)$, without significant hysteresis. Apparently, this phenomenon occurs when the nucleation sites contain few, or very small vapor nuclei, and it is necessary to evaporate the trapped liquid to activate the cavities. Bergles and Chyu [1982] have measured hysteresis effects of the High-Flux surface by boiling R-113 on a single electrically heated tube. Figure 11.29 shows test results from this study. The data were taken for different amounts of initial subcooling. This figure shows that the nucleation sites are suddenly activated when $2.5 < \Delta T < 8$ K. In some cases, small boiling patches were observed and full surface activation did not occur until a higher heat flux was applied. They were able to shut off the test section heat input and restart the test from a low heat flux without the occurrence of the initial hysteresis effect. However, if the R-113 was sufficiently subcooled in the shutdown phase, the hysteresis would again occur. The possibility of cavity flooding is more likely for low-surface-tension, wetting fluids (e.g., R-113) than for a less-wetting fluid, such as water. The cavity geometry and the existence of dissolved gas should have a significant effect on hysteresis effects.

Kim and Bergles [1988] extended their evaluation of hysteresis effects on porous boiling surfaces. They boiled R-113 at 1.0 atm on horizontal, 4.7-mm square plates

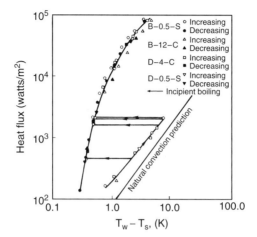

Figure 11.29 Starting hysteresis effects with R-113 boiling at 101 kPa (1 atm) on a horizontal High-Flux tube (Figure 11.11a) measured by Bergles and Chyu [1982]. (From Bergles and Chyu [1982].)

having porous coatings of spherical copper particles. They tested three different coatings having a thickness of four particle diameters, e.g., as in Figure 11.11c. The particle diameters were 0.059, 0.096, and 0.247 mm. A fourth coating had layers of different particle diameters. Incipient boiling occurred at $30 < \Delta T < 40$ K, which is higher than for the High-Flux tube tests of Bergles and Chyu [1982]. The superheat for incipient boiling was independent of the particle size.

Low-surface-tension fluids, such as R-113, are more susceptible to hysteresis effects than are high-surface-tension fluids, such as water. This is because the low-surface-tension fluids typically have smaller contact angles, and are more likely to flood the cavities. Ko et al. [1992] measured hysteresis effects for water and R-113 on four different porous coatings having different particle sizes. Their data, on a given surface, show that R-113 shows significantly higher incipient superheat than does water. They attempted to explain the incipient liquid superheat required for bubble existence by the equation $(T_w - T_s) = 2\sigma(\cos\theta)/r$. They argued that porous structures having small pores will have the largest incipient superheat. Unfortunately, they did not actually define the average pore size of the surfaces they tested. Hence, their conclusions are not well justified. The tests of multiple layers of sintered 185-mesh screen by Liu et al. [1987] showed decreasing hysteresis effects as the number of screen layers increased from one to five. Both R-113 and water showed similar hysteresis effects, although the hysteresis for water was a little less than for R-113.

Ma et al. [1986] note that combination structured/porous surfaces showed substantially less hysteresis with methanol than observed by Bergles and Chyu [1982] for R-113. Bar-Cohen [1992] provides a survey of hysteresis effects and Malyshenko and Styrikovich [1992] observed a peculiar hysteresis phenomenon for boiling on porous coatings made of low-thermal-conductivity particles. They found a significant difference between increasing and decreasing power boiling curves, and attempted to explain the phenomenon.

The starting hysteresis problem involves how to get vapor into the cavities — whether it be a naturally occurring cavity or an appropriately shaped artificial cavity. It appears that the only possible options for forming a vapor (or gas) nucleus in potentially vapor trapping cavities are the following:

1. Gas initially trapped in cavity
2. Dissolution of dissolved gas from the liquid, because of liquid heating
3. Blowing vapor or inert gas into the cavity, from an external source
4. Heterogeneous nucleation

Starting hysteresis in tube bundles is discussed in Section 13.8.2.

11.7.2 Size and Orientation Effects

Rainey and You [2001] investigated the effect of boiling surface size and angular orientation for boiling of FC-72 on flat surfaces. Their tests were performed using plain copper and "microporous painted" surfaces described by Chang and You [1997]. The tests were performed on 3 different surface sizes — 10×10, 20×20, and 50×50 mm. Data were taken for angular orientations between $0°$ (horizontal facing up) and $180°$ (horizontal facing down). Figure 11.30 shows their results for the 20×20 mm sample. Figure 11.30 shows that the performance of the microporous surface is relatively insensitive to both surface size and orientation. However, the

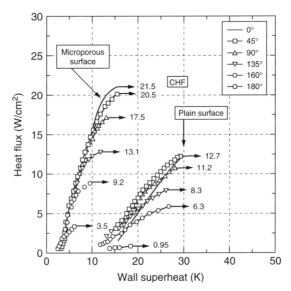

Figure 11.30 Effect of surface size and angular orientation for boiling of FC-72 on 20×20 mm flat plain copper and microporous painted surfaces. (From Rainey and You [2001].)

plain surface performance is strongly dependent on orientation. The performance increased slightly from 0° to 90° and then decreased dramatically from 90° to 180°. For the lower heat fluxes, the performance decreased with increased surface size for 0° to 45° orientation. However, at the higher heat fluxes, the performance increased with surface size for 0° to 45° orientation. The CHF is strongly influenced by orientation for both plain and microporous surfaces. The CHF decreases with inclination angle. For 180°, the CHF is reduced by 90%, relative to that for 0°. The authors concluded that the CHF changed from "hydrodynamically controlled" to "dryout" for angles greater than 90°.

Park and Bergles [1988] also investigated size and orientation effects for boiling of R-113 on plain copper surfaces. They found that the performance was insensitive to heater size.

11.8 BOILING MECHANISM ON ENHANCED SURFACES

11.8.1 Basic Principles Employed

The published literature shows that a complete understanding of the mechanism of boiling on enhanced surfaces does not exist. However, some very good recent progress has been made in theoretical modeling, which is discussed in Section 11.9. As previously discussed, there are two basic types of enhancement geometries, namely, porous and structured surfaces. The structured surfaces are more amenable to analysis and description of the boiling mechanism than are the porous surfaces. This is because the structured surfaces, as illustrated in Figures 11.12, contain well-defined dimensions. However, the porous surfaces of Figure 11.11 are made up of an irregular geometry of the porous structure.

In the author's opinion, four key concepts are employed to attain high performance:

1. The use of a reentrant (or doubly reentrant) tunnel-type cavities is necessary.
2. The subsurface pores are substantially larger than those occurring on natural surfaces. This reduces the superheat requirement, as defined by Equation 11.4.
3. Liquid is supplied via the surface pores to subsurface capillary passages. The liquid is heated as it proceeds through the capillary passages; thus, in the vicinity of an active pore $T_l > T_s$. Thin film evaporation occurs from liquid–vapor interfaces in the subsurface capillaries. When a bubble departs from the surface, the liquid–vapor interfaces in the subsurface capillaries are not subjected to the cold liquid temperature existing above the surface.
4. The subsurface capillaries are interconnected. Vapor produced within the subsurface structure at one site tends to activate other sites, because the subsurface sites are interconnected.

The six structured surfaces illustrated in Figure 11.12 are all made from an integral-fin tube. The Figure 11.12a fins are deformed to form reentrant tunnels in the Figure 11.12b and c geometries. A narrow gap exists at the top of the tunnel. The width of this gap has a strong effect on the boiling performance. A secondary operation is applied to the Figure 11.12d and e surfaces. The initially formed fins have a "sawtooth" configuration, as shown in Figure 12.6. In Figure 11.12d, the fins are bent so that there is no gap between adjacent fins. However, the sawtooth fin shape provides "pores," which communicate between the tunnel interior and the fluid above the surface.

Because both the Figure 11.12b and c and the Figure 11.12d and e geometries provide high performance, it is concluded that the pore structure of the Figure 11.12d geometry is not a necessary geometrical feature. However, if the pore structure is not provided, there is an "optimum" gap between adjacent fins. Ayub and Bergles [1988a, 1988b] boiled R-113 and water on the 748 fins/m (19 fins/in.) Figure 11.12c enhanced surface. They varied the gap spacing s_g at the fin tip. They found that the optimum gap dimension is 0.25 mm. However, Webb [1972] found the "optimum" fin gap is between 0.04 and 0.09 mm for R-11 boiling on the Figure 11.12b geometry.

Nakayama et al. [1980b] investigated the optimum pore diameter for the Figure 11.12d surface. The optimum pore diameter is 0.15 mm for water and 0.07 mm for nitrogen. The optimum pore diameter varies with saturation pressure.

11.8.2 Visualization of Boiling in Subsurface Tunnels

Nakayama et al. [1980a] were the first to perform a visualization study of boiling in a rectangular tunnel, with pores in a very thin copper sheet covering the top surface, as illustrated in Figure 11.31a. Liquid R-11 at 1.0 atm was contained above the cover plate. In one series of experiments, heat was provided from below by a cartridge heater in the base block. In other tests, the lid was electrically heated by dc current and was used as a resistance thermometer. The tunnel heights varied between 0.5 and 1.0 mm, and the width was held fixed at 1.0 mm. The diameters of the pores were uniform, from 0.05 to 0.5 mm in different lids tested. Boiling activity was recorded by a high-speed cine camera or a still camera. Their visual observations are summarized below.

1. In the unheated state, a vapor plug existed in part of the tunnel. The volume of this plug reduced as the ambient temperature was decreased.
2. Upon addition of heat, the vapor region expanded along the tunnel, and the advancing liquid–vapor interface shown in Figure 11.31b was observed. A thin liquid film existed on the walls of the vapor region, with liquid holdup in the corners.
3. At wall superheats greater than 0.6 K, bubbles began to emerge from several pores at a frequency of about 1 every 8 s.
4. Incrementally increasing the wall superheat increased the bubble departure frequency and also the number of active pores.

These characteristics were observed for all test geometries except for the lid with the largest pores (0.5 mm). In this case, expulsion of the liquid from the tunnel

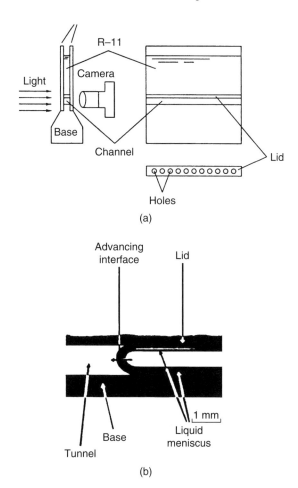

Figure 11.31 (a) Schematic of apparatus used by Nakayama et al. [1980a] to view boiling in tunnels. (b) Liquid–vapor interface observed in tunnel by Nakayama et al. [1980a]. (From Nakayama et al. [1980].)

was never complete, and liquid could be seen sloshing around in the tunnel while occupying roughly 10 to 50% of the tunnel volume.

Arshad and Thome [1983] performed a visualization study with water using an experimental apparatus similar to that of Nakayama et al. [1980a]. They soldered 0.005-mm-thick copper cover plates to heated plate having grooves machined in the top surface. The cover plates contained spaced holes, with diameters between 0.15 and 0.25 mm. The ends of the tunnels were covered with thin glass plates. They tested triangular, rectangular, and circular cross-sectional groove shapes. Their apparatus was designed to allow observation along the axis of the tunnel, as opposed to side viewing method of Nakayama et al. [1980a]. They observed the activity for a triangular groove about 1.0 mm deep with 0.15-mm pore diameter. The following conclusions were drawn from the visualization study for the nucleate boiling mechanism on structured surfaces:

1. Thin film evaporation is the principal heat transport mechanism inside the tunnel.
2. The geometrical shape of the subsurface groove affects the shape and formation of the thin evaporating liquid film.
3. Dryout of the liquid film causes the surface to revert toward the performance of a plain surface.

Because Arshad and Thome viewed the end of the tunnels, their observations were unable to detect a meniscus moving down the tunnel, as was observed by Nakayama et al. [1980a].

Chien and Webb [1998c] performed visualization experiments on a finned tube geometry that closely simulates the actual tube geometry. The experiments viewed a 1575 fins/m, 19.5-mm-diameter copper integral-fin tube having 0.8-mm fin height. A 0.02-mm-thick, transparent polypropylene film was wrapped over the fins. Pores of a fixed diameter were made in the transparent film using a heated needle. The heated needle melted the plastic material and created permanent pores of diameter $(d_p) = 0.23$ mm and pore pitch $(P_p) = 1.5$ mm. In one case, a 0.05-mm copper foil was soldered on the fin tips. Pores of the same diameter and pitch were also formed in the copper foil–wrapped tube. Methanol at 1.0 atm $(T_{sat} = 64.5°C)$ was used as the working fluid and data were taken for $2 < q < 20$ kW/m^2. The tube was tested in either the vertical (or horizontal) orientation using the test cell described by Chien and Webb [1998a, 1998c]. A high-speed photography system was used to photograph the boiling phenomena at speeds between 125 and 1000 frames/s. Still photographs were also taken with magnification, which allowed detailed observation of the liquid–vapor activity inside the tunnels. The transparent plastic film allowed observation of the liquid–vapor conditions inside the tunnels, as shown in Figure 11.32. The vapor and liquid were distinguished by the difference of reflection and the movement of liquid meniscuses. The vapor-filled tunnels are brighter than the liquid-filled tunnels as shown on Figure 11.32, because the reflection from liquid is different from that for the vapor-filled region. For a vapor-filled tunnel, the surface pores appear as shining white spots. However, liquid-filled tunnels are much darker than those on a vapor-filled tunnel, and the surface pores are difficult to see.

For horizontal tube orientation, the copper cover provided moderately higher enhancement than that with the plastic cover. However, the plastic cover provided considerable enhancement compared to the finned tube without a pored cover foil. Factors causing these differences are the higher thermal conductivity of the copper foil than the polypropylene film and the copper foil was deformed into the tunnel forming reentrant shape cavities. Because the boiling curves show the same enhancement trend for $q < 20$ kW/m^2, their boiling mechanism should be the same. When the tube was tested in the vertical orientation, the boiling coefficient is approximately 10 to 20% lower than that of the horizontal tube.

The fins on the tube illustrated in Figure 11.32 were made with a two lead thread. Therefore, alternate tunnels are physically separated. At low heat flux $(q = 2$ kW/m$^2)$, one long vapor region filled two to five internally connected tunnels, as shown on Figure 11.32c. Bubbles emerged from the vapor-filled regions. As shown on the left portion of Figure 11.32c, one set of four vapor-filled tunnels alternate

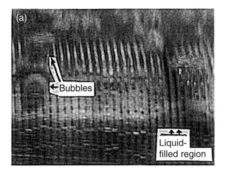

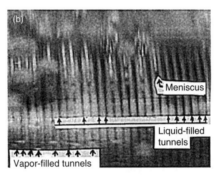

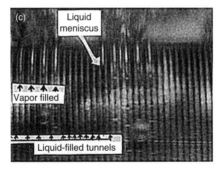

Figure 11.32 Saturated boiling of methanol on a finned tube (1575 fins/m, 0.8-mm fin height) having spaced 0.23-mm-diameter pores in a transparent cover at 20 kW/m², 10 kW/m², and 2 kW/m² heat flux. (From Chien and Webb [1998c]. Reprinted with permission from Elsevier.)

with another set of liquid-filled tunnels. The four vapor-filled tunnels are internally connected and do not connect with the liquid-filled tunnel between two vapor-filled tunnels. Figure 11.33 was prepared based on the visualization observations at low heat flux (5 kW/m²). As shown on Figure 11.33, all the tunnels are internally connected. At low heat flux, part of the tunnel is liquid-filled, as shown by the black regions. Oscillating liquid menisci were observed at the two ends of each vapor region. Bubbles emerge from the vapor-filled regions and the liquid-filled regions are non-active. The overall heat transfer coefficient decreases because the total

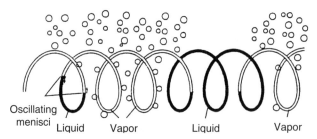

Figure 11.33 Boiling mechanism for a horizontal tube at low heat flux. (From Chien and Webb [1998c]. Reprinted with permission from Elsevier.)

evaporation area decreases. This visualization provides proof that bubbles are formed by evaporation of menisci in vapor-filled tunnels.

11.8.3 Boiling Mechanism in Subsurface Tunnels

Nakayama et al. [1982] described three possible evaporation modes, which may occur in a subsurface tunnel. Conceptual sketches of the three possible evaporation modes in the tunnel are shown in Figure 11.34. Nakayama et al. [1982] state that the "flooded mode" occurs at low heat flux, where most of the tunnel space is occupied by liquid, and an active pore acts as an isolated nucleation site. At higher heat flux, the "suction-evaporation mode," exists, where liquid is sucked in the tunnel space through inactive pores by pumping actions of bubbles growing at active pores. It then spreads along the tunnels and evaporates from menisci at the corners of the tunnel. With further heat flux increase, the "dried-up mode" exists, where the tunnel space is filled with vapor, and vaporization into bubbles takes place outside the tunnels.

Nakayama et al. [1980b] were the first to propose an analytical model to predict boiling performance of the Thermoexcel-E (Figure 11.12d) surface geometry. Their dynamic model described below assumes that the bubble cycle occurs as illustrated by Figure 11.35.

1. Phase I (pressure build-up phase): Evaporation of liquid in the tunnel causes the internal vapor pressure to build up until the vapor nuclei located at the mouth of the pores attain a hemispherical shape, protruding outward into the external liquid pool.
2. Phase II (pressure-reduction phase): At some pores, the vapor nuclei become active and grow into bubbles. Initially, the bubbles grow as a result of the pressure differential across the interface, which reduces the pressure in the tunnel as the vapor flows into the bubbles. They then continue to grow as a result of the inertial force imparted to the surrounding liquid.
3. Phase III (liquid-intake phase): During the inertial stage of bubble growth the pressure inside the tunnel is reduced to less than that of the external liquid pool, and liquid is drawn into the tunnel at inactive pores. At the end of this phase the bubbles depart, all pores are closed by liquid menisci, and the liquid spreads along the angled corners of the tunnel by capillary forces. The cycle then repeats from this point.

The visualization experiments of Chien and Webb [1998c] clearly show that, for saturated boiling, an active tunnel is vapor-filled with liquid menisci in the corners.

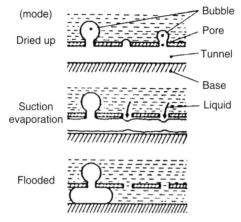

Figure 11.34 Heat transfer modes in tunnels of the Thermoexcel-E surface proposed by Nakayama et al. [1982]. (From Nakayama et al. [1982].)

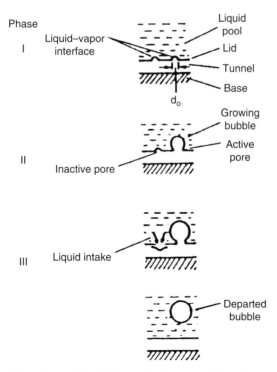

Figure 11.35 Basis of dynamic model for boiling on the Thermoexcel-E surface proposed by Nakayama et al. [1980b]. (From Nakayama et al. [1980b].)

The principal boiling mechanism is evaporation from liquid menisci in the corners of the tunnel. Liquid is sucked into the tunnel when a bubble departs, and the liquid is then pulled into the corners by surface tension force. The observations support the existence of the suction-evaporation boiling model as described by Nakayama et al. [1982].

11.8.4 Chien and Webb Parametric Boiling Studies

These tests were performed to define the effect of the tunnel and pore dimensions on performance. The tests were performed by boiling R-11, R-123, R-134a, or R-22 on a low, integral-fin tube, which was wrapped with a 50-μm copper foil. The pores were made by piercing the copper foil using a needle. Tunnels were created by the copper foil and the fins. The tunnel pitch and the tunnel height are equal to the fin pitch and height, respectively. Pores of specific diameters (d_p) were pierced in the foil at a specific pore pitch (P_p). The tunnel dimensions were varied by using integral-fin tubes of different fin pitch and fin height.

Table 11.3 summarizes the geometric parameters and test ranges investigated by Chien and Webb [1998a, 1998b] using two tunnel base shapes — rectangular and circular. The tunnel dimensions (P_t, H_t, and W_t) are defined in Figure 11.13a. The tube was wrapped with 50 μm copper foil and holes were pierced to provide pore pitches of 0.75, 1.5, or 3.0 mm, and pore diameters of 0.12, 0.18, 0.23, or 0.28 mm. The 50-μm-thick copper sheet was soldered to the fin tips. R-11 or R-123 was used as the working fluid. Chien and Webb [2001] provide additional data for R-134a and R-22.

The first column of Table 11.3 shows the tube code used to describe the test results. The first number is the fins/m and second is the fin height (mm). The second column shows the fin base shape. The surface code in the figure legends indicates the tube code followed by the pore diameter and pore pitch. For example, 1969-0.9-0.23-1.5 means a surface made on a 1969 fins/m and 0.9 mm fin height tube, having 0.23 mm diameter surface pores (d_p) on 1.5 mm pitch (P_p).

The test apparatus is similar to that used by Webb and Pais [1992] and is described in detail by Chien and Webb [1998a, 1998b]. Boiling occurred in a 178-mm-long cylindrical glass vessel with brass end flanges. The wall temperature was measured by 0.25-mm-diameter iron-constantan-sheathed thermocouples located below the tube wall at 180° opposite positions. The design allowed the test section

Table 11.3 Tube Code and Specifications Tested by Chien and Webb [1998a, 1998b]

Tube Code	Fin Base Shape	Fins/m	P_f (mm)	W_t (mm)	H_f (mm)
1378-0.9	Circular	1378	0.73	0.40	0.90
1378-0.5	Circular	1378	0.73	0.40	0.50
1969-0.9	Circular	1969	0.51	0.25	0.95
1575-0.6	Circular	1575	0.64	0.33	0.60
1575r-0.6	Rectangular	1575	0.64	0.33	0.60

to be rotated and wall temperature measured as a function of circumferential angle. The test tubes were 140 mm long, 9.53 mm ID, and 18 to 19.5 mm OD. Heat input was provided by a 9.52-mm-diameter, 129-mm-long, 500-W electric cartridge heater in the copper tube. Data were taken for 26.7°C saturation temperature using procedures described by Chien and Webb [1998a].

Figure 11.36 shows boiling curves for different pore diameters with 1.5 mm fixed pore pitch from Chien and Webb [1998a]. The surfaces with d_p= 0.18, 0.23, and 0.28 mm give high performance at low heat flux. However, all show a maximum, which indicates the dryout heat flux (DHF) and the heat transfer coefficients decrease drastically above their DHF. Above the DHF, evaporation does not occur in the tunnels, and the vapor in the tunnels causes high thermal resistance. The boiling curves are strongly influenced by the pore diameter. As the pore diameter increases, the boiling curves shift upward and to the right. The DHF increases with increase of the pore diameter. The surface having the largest pores (d_p = 0.28 mm) has the greatest heat transfer coefficient at high heat flux and greater DHF. However, the performance of this surface drops rapidly at heat fluxes greater than the DHF. The 0.23 mm pore diameter is the best for $q < 35$ kW/m^2. Figure 11.36 clearly shows that there is a preferred combination of pore diameter and pore pitch for a specified design heat flux.

Figure 11.37 is a composite diagram showing the separate and combined effects of pore diameter (d_p) and pore pitch (P_p) for the 1378 fins/m tube having 0.9-mm fin height. This figure illustrates the separate and combined effects of pore pitch (P_p) and pore diameter (d_p). There are two families of curves on Figure 11.37. The solid lines are for fixed pore pitch (P_p = 1.5 mm) with increasing pore diameter (d_p). The dashed lines show the effect of pore pitch ($0.75 \le P_p \le 4.0$ mm) for fixed pore diameter (d_p= 0.23 mm). Data were obtained only for the d_p= 0.23 mm, P_p = 1.5 and 3.0 mm cases. Based on the data in Chien and Webb [1998a, 1998b], the dashed lines are the authors' qualitative estimates of the effect of pore pitch. Four salient features are shown in Figure 11.37:

1. As shown by the solid curves for fixed P_p, the DHF increases as the pore diameter increases. A lower DHF occurs for the smaller pore diameters, because they cannot supply enough liquid to the tunnel at high heat flux. So, the smaller pores become dry and the DHF occurs at a lower heat flux. Conceivably, a higher DHF would be observed for $d_p > 0.28$ mm. However, the maximum possible pore diameter is limited by the fin pitch, 0.5 mm.
2. For fixed P_p and heat fluxes less than the DHF (e.g., 50% of the DHF), the heat transfer coefficient increases as d_p increases, except for the largest d_p (0.28 mm). The lower heat transfer coefficient for the largest d_p occurs because too much liquid is supplied to the tunnels and floods the intermittent regions. Hence, there is an optimum pore diameter for operation at low-to-moderate heat flux.
3. For fixed d_p (0.23 mm), the DHF decreases as the P_p increases. This is because the tunnel needs greater liquid supply at the higher heat flux. However, the more widely spaced pores are unable to supply the liquid.

4. For low heat flux at fixed d_p (0.23 mm), increasing P_p causes the heat transfer coefficient to increase. At lower heat flux, a smaller liquid supply to the tunnel will prevent tunnel flooding and result in a thin film in the tunnel.

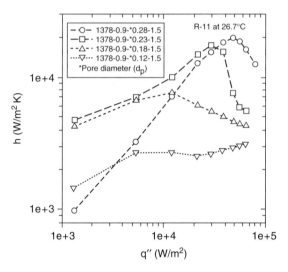

Figure 11.36 Effect of pore diameter for fixed pore pitch (1.5 mm) for 1378 fins/m and 0.9-mm fin height. (From Chien and Webb [1998a].)

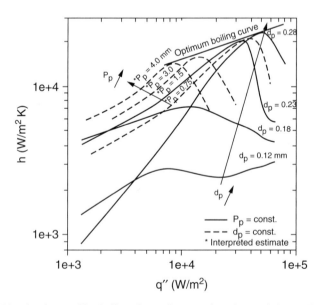

Figure 11.37 Map showing combined effect of pore diameter (d_p) and pore pitch (P_p). (From Chien and Webb [1998b].)

Figure 11.37 clearly shows that there is a preferred combination of pore diameter and pore pitch for a specified design heat flux. The pore pitch/pore diameter combination influences the liquid supply rate to the tunnel via the total open pore area per unit tunnel length ($A_{pt} = \pi d_p^2/4P_p$), which supplies liquid to the tunnel. If the liquid supply rate exceeds the evaporation rate, the performance will be reduced, because of tunnel flooding. For fixed pore pitch, the highest boiling coefficient is obtained by use of the largest pore diameter. However, a smaller pore diameter is preferred for low heat fluxes. A large pore diameter may result in too much liquid supply to the tunnel at low heat flux and result in tunnel flooding.

One may use this information to select the preferred d_p and P_p combination for a specific design heat flux (e.g., an electronic heat sink). However, the heat flux significantly varies over the tube length in refrigerant water chillers. Consider an R-123 water chiller tube bundle, for which one seeks to maximize the boiling coefficient over the tube length. In this example bundle, the heat flux is 48.0 kW/m^2 at the water inlet end and linearly drops to 9.0 kW/m^2 at the water exit end. According to the Figure 11.37 boiling map, one would decrease A_{pt} over the tube length by appropriate selection of the d_p and P_p combination. Table 11.4 shows how one may change d_p and P_p at the heat fluxes that will occur over the tube length. Thus, it is possible to optimize the surface performance for operation within a specific heat flux range (below the DHF). This is equivalent to being able to change the slope of the boiling curve.

The GEWA-TW and bent fin surfaces have continuous open gaps at the fin tips, as opposed to the surface pores of Turbo-B, Turbo-BII, Thermoexcel-E surfaces and the present studied geometry. Although comparable data have not been taken on the continuous fin gap structures, the authors propose that the understanding developed for the pored surfaces can be extended to the continuous fin gap structures. The equivalence is found in the total open area, A_p. Assuming the optimum design for the bent fin surface has the same "total open area per unit length," A_{pt}/L, as for the pored surface, the optimum gap width (s_g) of bent fin is $s_g = \pi d_p^2/4P_p$. Using the optimum combination $d_p = 0.23$ and $P_p = 1.5$ mm for the pored surfaces in the equation for s_g yields $s_g = 0.028$ mm, which is within the range recommended by Webb [1972], who found that the optimum fin gap for the bent fin surface is between 0.004 and 0.09 mm for R-11.

Table 11.4 Selection of Pore Diameter and Pore Pitch vs. Design Heat Flux for R-123 at 5°C

Heat Flux (kW/m^2)	d_p (mm)	P_p (mm)
48	0.28	1.5
35	0.23	0.75
28	0.23	1.5
15	0.23	3
9	0.23	4

11.9 PREDICTIVE METHODS FOR STRUCTURED SURFACES

Many of the Figure 11.12 surfaces have interconnected subsurface tunnels and surface pores (or narrow gaps at the surface). The author proposes that the Figure 11.12 surfaces having surface pores operate by the same mechanism. It is probable that the mechanism of the GEWA-TW and the bent fin geometries having narrow fin gap (s_g) operates by a closely related mechanism. Two key geometric characteristics of the surfaces are the (1) subsurface tunnels and (2) surface pores or fin gaps. These key features may be further defined by the following geometric parameters:

1. Subsurface Tunnels: Tunnel pitch (P_t), tunnel height (H_t), tunnel width (W_t), tunnel base radius (R_b), and tunnel shape
2. Surface Pores (or Gaps): Pore diameter (d_p), and pore pitch (P_p)

Thus, a geometry having surface pores has six geometrical dimensions, plus the tunnel shape. Figure 11.13a illustrates a generic surface having subsurface tunnels and surface pores, which may be defined by the above dimensions. Some of the commercial enhanced tubes do not have round surface pores (e.g., Thermoexcel-E and Turbo-BII). One possibility for definition of the pore diameter is to use the largest circle that can be drawn within the pore (e.g., minor diameter of the elliptical hole for the Figure 11.2e Turbo-BII). A rationally based predictive model must account for the effect of all of the geometric dimensions of the boiling surface.

11.9.1 Empirical Correlations

Empirical correlations typically do not account for the surface and subsurface pore geometry. Rather, they attempt to account for the effect of flux and fluid properties for a specific enhancement geometry. Zhang and Dong [1992] developed a power-law correlation for the Figure 11.15h tube geometry, based on their data for R-113 and ethyl alcohol taken at pressures between 1.0 and 6.0 bar. The dimensionless parameters were selected, based on their speculation of the boiling mechanism. The data were correlated within ± 20%. Unfortunately, they do not give the dimensions of the enhancement geometry. Li et al. [1992] propose a power-law correlation for their 0.10 MPa R-11 and R-113 data on the Figure 11.15g tube. The database supporting the correlation is insufficient for generality.

11.9.2 Nakayama et al. [1980b] Model

Nakayama et al. [1980b] developed an analytically based model to predict boiling performance of the Figure 11.38a surface, which was intended to simulate the Thermoexcel-E enhanced tube (Figure 11.12d). The Figure 11.38a test surface consisted of subsurface tunnels in a flat plate covered with pores, through which vapor bubbles escape into the bulk fluid. Their model is based on the suction-evaporation mode illustrated in Figure 11.34. Nakayama et al. assume that the total heat flux is

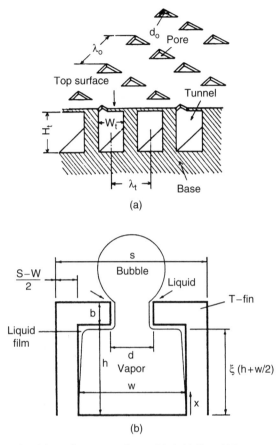

Figure 11.38 Schematic of the surface geometries modeled. (a) From Nakayama et al. [1980b] and (b) from Xin and Chao [1985].

the sum of a latent and a single-phase (external convection) heat flux. This is expressed by

$$q = q_{tun} + q_{ex} \qquad (11.15)$$

The term q_{tun} results from evaporation in the tunnel, and q_{ex} is due to external convection to the bubble after it has emerged from the tunnel. The model for q_{tun} is based on Figure 11.35, and it assumes that wetting occurs only in the corners of the tunnel, the suction-evaporation mode of Figure 11.34. The model for q_{tun} required six empirical constants.

The external convection term is given by the following empirical expression

$$q_{ex} = \left(\frac{T_w - T_s}{c_q} \right)^{5/3} n_s^{1/3} \qquad (11.16)$$

where c_q is an empirical constant that depends upon the fluid and the pressure. The c_q term was derived from experimental data for boiling data on a simulated Thermoexcel-E surface. Hence, one additional empirical constant is required to determine q_{ex}.

Nakayama et al. [1980a] measured q_{lat} heat flux fraction of the total heat flux for boiling on the Figure 11.38a surface and on a plain surface. This was determined using the measured bubble departure diameter and the nucleation site density. Figure 11.39 shows that q_{lat} contributes a much greater fraction of the total for the Figure 11.38a enhanced surface than for a plain surface. Hence, a significant fraction of the total evaporation occurs in the subsurface tunnels. A commercial R-11 refrigerant-flooded evaporator operates at approximately 35 kW/m^2 heat flux.

The major deficiency of the Nakayama et al. [1980a] model is that it requires seven empirical constants. If contact angle is important, it is not directly accounted for in the analysis. Further, the model is limited to the suction-evaporation mode of Figure 11.34.

11.9.3 Chien and Webb [1998e] Model

The previously discussed work of Chien and Webb [1998a, 1998b, 1998c, 1998d] confirm the Nakayama et al. [1980b] suction and evaporation model is correct for saturated boiling. In the Nakayama et al. [1980b] model, the external convection heat flux does not include the bubble frequency and bubble departure diameter. Chien and Webb [1998e] provide an improvement over the Nakayama et al. [1980b] model. The Chien and Webb model has only two empirical constants compared to six in the Nakayama et al. [1980b] model. The Chien and Webb model:

1. Accounts for the temporal evaporation rate variation inside tunnels by analyzing meniscus thickness, bubble departure diameter, and bubble growth
2. Uses the bubble frequency and departure diameter for external the heat flux

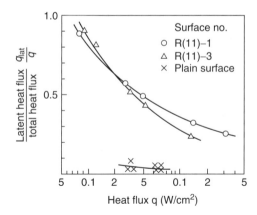

Figure 11.39 Contribution of latent heat transport to the total heat flux on the Figure 11.33a surface. (From Nakayama et al. [1980a].)

As given by Equation. 11.15, the total heat flux is separated into two parts, tunnel heat flux (q_{tun}) and external heat flux (q_{ex}). The heat transfer rate in the tunnel is governed by evaporation of liquid menisci. The external heat flux is contributed by transient conduction and convection caused by the departing bubbles. The complete development of this model is given in Chien and Webb [1998e]. An overview of the model is given here. A bubble cycle includes three periods: waiting period (Δt_w), bubble growth period (Δt_g), and liquid intake period (Δt_e). Figure 11.40 shows the process of evaporation in the tunnel during a boiling cycle. A bubble cycle includes the following three periods:

1. Waiting period (Δt_w): In this period (Figure 11.40a and b), liquid is evaporated in the tunnel. However, the vapor is constrained inside the tunnel by the surface tension on the pore, so the pressure in the tunnel increases with time. Bubble embryos protrude from a pore when $P_v - P_s \equiv \Delta P_{br} = 4\sigma/d_p$, where ΔP_{br} is the breakthrough pressure differential. The radius of the meniscus decreases from $R_{m,i}$ at the beginning of the cycle (Figure 11.40a) to $R_{m,g}$ at the end of this period (Figure 11.40b). During this period, the tunnel is filled with vapor except for the liquid menisci in the upper corners.
2. Bubble growth period (Δt_g): In this period, vapor passes through surface pores and increases the bubble radius as the bubble grows above the pore. Because the liquid in the tunnel continues to evaporate, the meniscus radius decreases from $R_{m,g}$ to $R_{m,e}$. The effective evaporation temperature increases as the meniscus radius decreases. Meniscus evaporation will stop when the elevation of saturation temperature, due to capillary pressure (σ/R) of the meniscus, equals the wall superheat. The meniscus radius, at which its saturation temperature attains the wall temperature is R_{ne}, and is

$$R_{ne} = \frac{\sigma}{\Delta T_{ws}} \left(\frac{dT}{dP} \right) \tag{11.17}$$

3. Depending on the evaporation rate, $R_{m,e} \geq R_{ne}$. At time t_d (Figure 11.40c) bubbles of diameter d_b depart from the surface pores.
4. Liquid intake period (Δt_e): After the bubble departs, the pressure in the tunnel is lower than that of the liquid pool, and liquid flows into the tunnel and is retained in the corners. At the end of the liquid intake period, the radius of the meniscus equals $R_{m,i}$ as shown in Figure 11.40d. The pore diameter and pitch control the amount of the liquid that flows into the tunnel. This period is much shorter than the other two periods and is neglected in calculation of the bubble frequency.

At the end of a bubble cycle, liquid is drawn into the tunnel (see Figure 11.40d) by inertia force and its movement in the s direction is impeded by surface tension force. Figure 11.41 shows a meniscus existing at the top corner of the tunnel. Surface tension force acts to spread the liquid in the x direction. Once the liquid is on the side wall, surface tension will act to pull it into the corners at the fin base. Evaporation

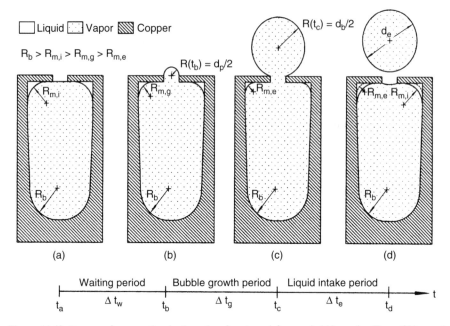

Figure 11.40 Process of evaporation in the subsurface tunnel for one bubble cycle. (From Chien and Webb [1998e].)

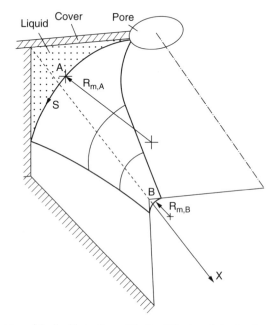

Figure 11.41 Geometry of the liquid meniscus at the top of the tunnel. Surface tension pulls liquid along the tunnel length. (From Chien and Webb [1998e].)

occurs only on liquid menisci in the corners of the tunnel. A one-dimensional model is used to calculate the evaporation rate of this meniscus.

The tunnel heat flux in Equation 11.15 is the product of the latent heat transferred at the menisci during a bubble cycle and the instantaneous bubble frequency, as given by

$$q_{tun} A_{tun} = (Q_m) f \tag{11.18}$$

where Q_m is the latent heat from the menisci in one bubble cycle. Assuming one-dimensional heat conduction in the liquid, the instantaneous heat flux is given by

$$q_{tun} = \frac{k_l}{\delta} (T_w - T_s) \tag{11.19}$$

When the liquid meniscus becomes very thin, the saturation temperature elevates because of the disjoining pressure and capillary pressure (σ/R). So, when the elevated saturation temperature equals T_w, no further evaporation will occur. The total latent heat in the tunnel during one bubble cycle can be calculated by integrating the latent heat transfer rate at the menisci (dQ_m/dt) during a bubble cycle, and is given by

$$Q_{tun} = \int_0^{1/f} \frac{dQ_m}{dt} dt = \int_0^{1/f} \left[\Delta T_{ws} \int_0^{A_m} \frac{k_l}{\delta_m(t,\phi)} dA \right] dt \tag{11.20}$$

where $\delta_m(t, \phi)$ is the local thickness of the liquid meniscus, and A_m is the meniscus surface per unit tunnel length.

The sensible heat flux (q_{ex}) term in Equation 11.15 is due to the external convection induced by bubble agitation. The mechanism is similar to pool boiling on a plain surface in that the bubble evaporates into a superheated liquid layer above the base surface. The key difference between the present situation for q_{ex} and boiling on a plain surface is that the nucleation site density, frequency, and bubble departure diameter are known.

The model used to predict q_{ex} in Equation 11.15 is totally different from the empirical correlation used by Nakayama et al. [1980a]. An analytically based model developed by Haider and Webb [1997] is used to predict the external heat flux. This model is based on an extension of Mikic and Rohsenow [1969], who developed a model based on transient conduction to the superheated liquid layer. The Haider and Webb model includes the effects of transient convection to the liquid, caused by convection in the wake of the departing bubbles. Transient microconvection, rather than transient conduction, is the dominant heat transport mechanism.

The equations for prediction of q_{tun} and q_{ex} both require knowledge of the active nucleation site density (n_s), bubble frequency (f), and bubble departure diameter (d_b). Using a force balance, Nakayama et al. [1980a] used Equation 11.14 to predict the bubble departure diameter. The constant c_b was obtained empirically (0.42 for R-11 and 0.22 for water). Chien and Webb [1998d] formulated the force balance on the bubble using buoyancy and surface tension forces as a function of d_b, d_p, and contact angle (θ), and thus the bubble diameter d_p was predicted without an empirical constant. The bubble frequency is obtained from $1/f = \Delta t_w + \Delta t_g$. The waiting period and the bubble growth period were modeled separately and are given in Chien and Webb [1998e].

The vapor generated in the tunnel is ejected through the surface pores. From an energy balance, $q_{tun} = \lambda V_v \rho_v$, where V_v is the vapor volume. Using a mass balance, one obtains the nucleation site density (n_s) by the following equation:

$$n_s = \frac{q_{tun}}{\rho_v \lambda f (\pi d_b^3)/6} \tag{11.21}$$

The Chien and Webb model predicts the total heat flux (q) for given wall superheat (ΔT) and dimensions (d_b, P_p, P_f, H_t) using the detailed procedure given in Chien and Webb [1998e].

The model was used to predict the data of Chien and Webb [1998a, 1998b, 2001] and for $q = 20\%$ to 100% of the dryout heat flux. The predictions were limited at the low end, because the model assumes the tunnels are vapor filled, which may not be true at low heat flux. These data include R-11, R-123, R-134a, and R-22 pool boiling data on tubular surfaces having a circular fin base for heat fluxes between 10 and 65 kW/m². The range of geometric parameters is shown in Table 11.5. The heat flux was predicted within 20% mean standard deviation.

11.9.4 Ramaswamy et al. [2003] Model

Ramaswamy et al. [2003] developed a variant of the Chien and Webb [1998e] model. The model parallels the Chien and Webb model with some key differences. One difference is in the model for the external heat flux. Whereas Chien and Webb used the Haider and Webb [1997] model having a constant of 6.42, Ramaswamy et al. replaced the constant with a polynomial of $(T_w - T_{sat})$. They predicted their own data within ± 40%. See the paper for details.

11.9.5 Jiang et al. [2001] Model

Jiang et al. [2001] has modified the microheat pipe model of Khrustalev and Faghri [1994] for prediction of boiling performance of structured surfaces. Similar to the suction-evaporation boiling model of Nakayama et al. [1980b] and the Chien and Webb [1998e] model, Jiang et al. [2001] also assumed that evaporation occurs on liquid meniscus in the tunnel corners. The evaporation rate is predicted assuming evaporation from the menisci in the corners of the tunnels. Their analysis of evaporation heat transfer on the menisci is similar to the Chien and Webb [1998e] model. The model for liquid and vapor flow and evaporation of the menisci is taken

Table 11.5 Range of Parameters in the Database

Fluids	d_p (mm)	P_p (mm)	fins/m	W_t (mm)	H_t (mm)
R11 R123	0.12–0.28	0.75–1.5	1378	0.25–0.4	0.5–1.5
R134a R22	0.18–0.28	0.75–1.5	1578, 1968	0.25–0.33	0.6–1.5

from Khrustalev and Faghri [1994], who modeled evaporation in a heat pipe having microgrooves. There are two major differences between the Chien and Webb [1998e] model and the Jiang et al. [2001] model:

1. Chien and Webb [1998e] correlated the initial liquid meniscus radius of a bubble cycle from experimental data of various geometric parameters. Jiang et al. [2001] obtained the initial meniscus radius by solving the momentum equation for the liquid volume during the intake period. Their numerical solution adapted the Khrustalev and Faghri [1994] heat pipe model to account for the transient one-dimensional liquid and vapor flow in the tunnel, and set the boundary condition on the pores using a correlation of others to account for the pressure drop of the liquid flowing through inactive pores.

2. The Chien and Webb [1998e] model accounts for q_{ex} (see Equation 11.15) and the bubble growth process on the active pores. However, Jiang et al. [2001] assumed $q_{ex} = 0$. The model contains an empirical constant (c_f) that was adjusted to fit the nucleation site density data.

Based on the bubble dynamic experiments, Nakayama et al. [1980a] found that the q_{ex}/q increases as heat flux (q) increases, and their experiments showed that q_{ex}/q is greater than 50% for $q > 3.0$ kW/m². The observation of Chien and Webb [1998e] using a high-speed video camera also showed that q_{ex} is an important component and should not be ignored. The Chien and Webb [1998e] model shows that a larger pore diameter results in larger q_{ex}. Hence, a larger optimum pore diameter would likely have been predicted by Jiang et al. [2001] had the q_{ex} been included in their model.

Jiang et al. [2001] validated their model by predicting the Nakayama et al. [1980a] R-11 boiling for 0.1-mm pore diameter. However, it significantly underpredicts the R-11 data for 0.04-mm pore diameter when $q > 1.0$ W/cm². The prediction of nucleation site density and frequency of the Jiang et al. model is in good agreement with that of the Nakayama et al. [1980a, 1982] data. They also performed R-11 boiling tests on a plate having 4.0-mm tunnel pitch and various tunnel hydraulic diameters. Their water data showed that 0.35 mm tunnel hydraulic diameter yields significantly higher performance than 0.29, 0.5, or 0.6 mm diameter. Their simulation is in good agreement with the 0.35-mm tunnel diameter data. However, their simulation also showed that tunnel diameters between 0.28 and 0.4 mm yield the same performance.

Jiang et al. used their model to perform optimization of geometric parameters for R-11, R-134a, R-22, and water. Their predictions showed that the boiling performance has great dependency on pore diameter and tunnel hydraulic diameter. They found that the optimum pore diameter is 0.09 mm for R-11 at 25°C, 0.08 mm for R-134a and 0.07 mm for R-22 at 5°C. The predicted optimum pore diameter for water is 0.15 mm at $q = 25$ W/cm². However, the Nakayama et al. [1980a] water data showed that 0.08-mm-diameter pores performed better than the 0.14-mm pores for heat flux less than 30 W/cm². Readers are encouraged to review the parametric predictions.

It is noted that the predicted preferred pore diameters and pore pitches tend to be significantly smaller than for the Chien and Webb [1998e] model and Kim and Choi [2001] experiments. For example, for R-11, Jiang et al. predict that the preferred pore diameter and pore pitch are 0.09 and 0.5 mm, respectively, vs. about 0.23 mm pore diameter and 1.5 mm pore pitch for the Chien and Webb [1998a, 1998b] data. It is noted that these differences may be somewhat related to the different range of pore pitches and diameters used in the studies. Thus, small pore diameters at close pitch may provide approximately equivalent performance to larger pore diameters at larger pitches. This involves the "total open area" parameter (A_{pt}) described earlier and defined in the nomenclature at the end of the chapter.

11.9.6 Other Models

Several other models, which are less sophisticated than the previously described models, have been proposed. These include the Xin and Chao [1985, 1987] who proposed a model for boiling on a planar (not tubular) GEWA-T (Figure 11.12c) surface. The model, illustrated in Figure 11.38b, assumes steady-state evaporation of a thin liquid film spreading over the inside surface of the tunnel. Wang et al. [1991] attempted to improve the Xin and Chao model. The basic assumptions of the Wang et al. [1991] model are the same as those of Xin and Chao [1985]. This includes Xin and Chao's physically unrealistic assumption of the countercurrent two-phase flow of vapor and liquid through the openings at the top of the tunnel. They concluded with a multiple regression correlation having six empirical constants.

Ayub and Bergles [1987] obtained R-113 and water boiling data at 1.0 atm on the Figure 11.12c GEWA-T tube. They used Equations 11.15 and 11.16 to predict their data. However, this required knowledge of the nucleation site density to calculate q_{ex}. They experimentally determined the nucleation site density on their surface, and expressed it as a polynomial of ($T_w - T_s$). Following the suggestion of Nakayama et al. [1980b], they assumed that q_{tun} may be written

$$q_{lat} = c_1 k (T_w - T_s) \qquad (11.22)$$

They determined the constant c_1 for each fluid at one heat flux, and assumed that it applied at all heat fluxes. Because they did not provide a method to predict n_s or the constant o_1 as a function of fluid properties, the correlation is of very limited value.

Webb and Haider [1992] developed an analytically based model that is applicable to the flooded mode illustrated in Figure 11.34. As noted above, the flooded mode is expected to apply only at very low heat flux, or when significant flooding of the tunnels exists. A dynamic model based on Equation 11.15 was used. The tunnels are two dimensional, having a gap dimension s_g at their mouth.

11.9.7 Evaluation of Models

The models of Chien and Webb [1998e], Jiang et al. [2001], and Ramaswamy et al. [2003] provide state-of-the-art understanding of the boiling process on structured surfaces. These are all improved versions of the suction-evaporation model originally introduced by Nakayama et al. [1980b]. The Chien and Webb model requires only two empirical constants, as compared to the seven empirical constants used in the Nakayama et al. [1980b] model. The three models are all similar in overall structure, but differ in some detailed aspects.

11.10 BOILING MECHANISM ON POROUS SURFACES

11.10.1 O'Neill et al. [1972] Thin Film Concept

O'Neill et al. [1972] theorized that vapor bubbles exist within the pores formed by the void space between the stacked particles. Thin liquid films exist on the surface of the particles. Heat is transferred by conduction through the particle matrix, and then by conduction across the thin liquid film, where evaporation occurs. The pores within the matrix are interconnected so that: (1) liquid can be supplied to the pores, and (2) vapor can pass through the matrix to the free liquid surface. As vapor is generated within a pore, the pressure in the vapor bubble increases. When the pressure is sufficiently high, it overcomes the surface tension retention force, and the vapor is forced through the interconnected pores to the liquid surface. O'Neill et al. assumed that each pore contains a vapor bubble and is therefore an active nucleation site. They outlined an analytical model based on the thin film concept, which is described in Section 11.11.1 In a later publication Czikk and O'Neill [1979] proposed that not all pores are active. They give the following pore classification:

1. Active pores. These are reentrant pores that always contain vapor bubbles.
2. Intermittent pores. These contain vapor as the result of vapor flow generated in interconnected active pores; these are envisioned as larger, non-reentrant pores.
3. Liquid-filled pores. These are small non-reentrant pores, whose opening radius is too small to permit bubbles at the prevailing $T_w - T_s$. They are permanently liquid filled and supply superheated liquid to the active pores.
4. Nonfunctional pores. These are "closed" pores, which do not contain the boiling fluid.

11.10.2 Kovalev et al. [1990] Concept

The static model of O'Neill et al. [1972] is not realistic, because it does not account for the dynamics of fluid flow within the matrix, and it does not include bubble dynamics (bubble departure, frequency, and site density). Kovalev et al. [1990] proposed a more rationally based model, which accounts for (1) evaporation from menisci within the matrix and (2) the dynamics of fluid flow within the porous matrix. Figure 11.42 illustrates the concepts on which their model is based. Figure 11.42a shows that evaporation takes place from stationary menisci, of radius r_m within the matrix of thickness δ. The meniscus radius changes over the matrix thickness. The pressure difference between the vapor and the liquid is $(p_v - p_l) = 2\sigma/r_m$

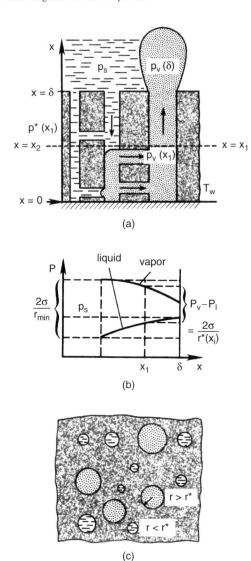

Figure 11.42 (a) Concept of Kovalev et al. [1990] model for porous boiling surface. (b) Distribution of pressure in vapor and liquid channels over matrix depth. (c) Top view of matrix showing vapor- and liquid-filled channels.

at a subsurface meniscus of radius r_m. The temperature of the porous matrix is greatest at the base of the matrix ($x = 0$), and it decreases in the direction of the surface. The local liquid temperature within the matrix (T_l) is equal to the local matrix temperature. Thermodynamic equilibrium requires that T_l must be superheated an amount $(T_l - T_s) = 2\sigma/mr_m$ for a meniscus of radius r_m to exist. Hence, at distance x from the base of the matrix, all pores having a radius $r_m \leq 2\sigma/m(T_l - T_s)$

are assumed to be liquid filled, and pores having $r > r_m$ are assumed to be vapor filled. Because the matrix temperature decreases with distance from its base (x), the value of r_m increases in the direction from the base. Hence, $r_m = r_m(x)$.

Evaporated vapor formed at the menisci within the matrix flows to the surface via the larger vapor-filled, interconnected pores. Bubbles are emitted at the surface from the vapor-filled capillaries. Vapor flow within the matrix is caused by "capillary pressure." Because $(p_v - p_l) = 2\sigma/r_m$ and r_m increases in the direction of the surface, a pressure gradient exists within the matrix. This pressure gradient forces the vapor through the capillaries to the surface. Figure 11.42b shows a pressure gradient in the vapor, due to friction losses. The liquid exists at saturation pressure p_s at the matrix surface. At this point, the vapor pressure is greater than p_s by the amount $2\sigma/r_m(\delta)$. The liquid flow in the small liquid-filled capillaries causes a pressure gradient within the liquid. Some detail of the mathematical model is given in Section 11.11.2.

11.11 PREDICTIVE METHODS FOR POROUS SURFACES

Several models or correlations have been proposed for prediction of porous boiling surfaces. The first model was developed by O'Neill et al. [1972], and is described in detail by Webb [1983]. The O'Neill et al. model assumes thin film evaporation on the particles. A different model was proposed by Kovalev et al. [1990], who assume that evaporation occurs at menisci within the porous structure.

11.11.1 O'Neill et al. [1972] Model

O'Neill et al. [1972] developed an analytical model to predict the heat flux for this envisioned boiling mechanism. The model, described in detail by Webb [1983], assumes that the matrix consists of precisely stacked spherical particles and that each pore contains a vapor bubble. Although the O'Neill et al. [1972] model is a static model, it contains several concepts of fundamental importance and may serve as the starting point for the development of a more sophisticated dynamic model. The model, described in detail by Webb [1983], assumes that $T_w - T_s$ may be written as the sum of two terms:

1. The temperature drop across the thin liquid film between the surface of the particle and the liquid $(T_w - T_l)$ at the surface of the bubble
2. The liquid superheat $(T_l - T_s)$ required for the existence of a vapor bubble within the pore of radius r_b

$$T_w - T_s = (T_w - T_l) + (T_l - T_s) \tag{11.23}$$

The second term of Equation 11.23 is given by Equation 11.4, and the first term is calculated using the Fourier equation for heat conduction across the liquid film of thickness δ. The heat flux q is based on the area of the base surface rather than the total liquid film surface of the matrix (S). This gives

$$(T_w - T_l) = \frac{qA\delta}{k_l S} \qquad (11.24)$$

The key assumptions of the model are as follows:

1. The uniform diameter spherical particles are interconnected with liquid or vapor to flow between them.
2. The pore diameter is defined as the diameter of the largest sphere that may be contained within the void space. The geometric packing arrangement of the spherical particles is known. Hence, the pore diameter is calculable, as illustrated in Figure 11.43 for two possible packing arrangements.
3. Each pore is active and functions as a stable vapor trap.
4. The matrix is a perfect heat conductor; thus $k_m = \infty$, and there is no temperature drop within the coating.

Based on the particle packing geometry and the coating thickness, one may show that

$$T_w - T_s = \frac{\beta_q r_b^2}{k_l} + \frac{2\sigma}{mr_b} \qquad (11.25)$$

where β is a geometry factor given by

$$\beta = \left[1 - (1 - \epsilon)\left(1 + \frac{V_b}{V_T}\right)\right]\left(\frac{V_T}{Sr_b}\right)^2 \qquad (11.26)$$

The geometry factor is calculable for given values of the particle diameter d_p and the particle packing arrangement.

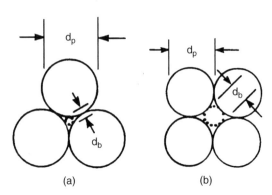

(a)   (b)

Figure 11.43 Definition of d_b and d_p for (a) staggered and (b) inline packings as used in O'Neill [1972] analytical model.

Comparisons of porosity measurements of the High-Flux coating show $0.50 < \in < 0.65$, which approximately agrees with the calculated porosity for an inline packing arrangement ($\in = 0.48$). Predictions by O'Neill et al. [1972] have used $\beta\delta = 1.04$, which is for the inline packing arrangement.

The matrix packings tested by Nishikawa and Ito [1982] and Nishikawa et al. [1983] had $0.38 < \in < 0.71$, which differs substantially from the $\in = 0.48$ associated with the inline packing. It is not possible to perform measurements on actual matrix packings to determine their geometric parameter, $\beta\delta$. Hence, the geometric parameters required for this model are indeterminate for actual packings.

11.11.2 Kovalov et al. [1990] Model

The evaporation concept proposed by Kovalev et al. [1990] was described in Section 11.10.2 and illustrated in Figure 11.42. The pores sizes are distributed between $r_{min} \le r \le r_{max}$. At distance x from the base surface, a meniscus of radius r_m will exist in a pore of radius r. At this x location, pores of $r < r_m$ are liquid-filled, and pores of $r > r_m$ are vapor-filled. The pressure difference across the meniscus interface at location x is given by

$$p_v(x) - p_l(x) = \frac{2\sigma}{r_m(x)} \tag{11.27}$$

The vapor evaporated at a meniscus passes through the interconnected vapor-filled pores and is emitted at the surface of the porous coating. Equation 11.27 is satisfied at each distance x from the base of the coating.

The momentum equation is written for parallel cylindrical capillary channels illustrated in Figure 11.42c. The small channels are liquid-filled, and the large channels are vapor-filled. Prior to describing the model, one must understand how the porosity and permeability for the liquid- and vapor-filled channels are defined. The term $f(r)$ is the pore size distribution function of the porous structure, as shown in Figure 11.44 from Kovalev and Lenkov [1981] for a particular porous coating. Figure 11.44 shows that the dominant pore size, maximum $f(r)$, is approximately 55μm. The term $\phi(r)$ is defined as the integral pore distribution and is shown on Figure 11.44. It is defined by

$$\phi(r) = \int_{r_{min}}^{r} f(r)dr \tag{11.28}$$

The total porosity of the matrix ($\in$) is equal to $\phi(r_{max})$. If all pores between r_{min} and r_m are liquid-filled, the porosity of these liquid filled pores is $\in_l = \phi(r_m)$. Similarly, if all pores having $r > r_m$ are vapor-filled, their porosity is $\in_v = \phi(r_{max}) - \phi(r_m)$.

The frictional pressure drop in the liquid-filled pores is calculated using the Darcy equation,

$$\frac{dp}{dx} = \frac{Gv_l}{K_l} \tag{11.29}$$

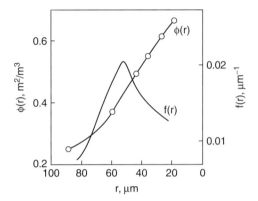

Figure 11.44 Pore size distribution, $f(r)$ and integral distribution function, $\phi(r)$ for a 1.1-mm-thick sintered coating made of 0.1-mm spherical particles. (From Kovalev and Lenkov [1981].)

where the permeability of the liquid-filled pores is K_l. The permeability of the porous matrix for single-phase flow is defined as $K = CF(r_{max})$, where $F(r_{max})$ is defined by

$$F(r_{max}) = \frac{\displaystyle\int_{r_{min}}^{r_{max}} r^2 f(r)dr}{\displaystyle\int_{r_{min}}^{r_{max}} f(r)dr} \tag{11.30}$$

and C is an empirical constant obtained by test data for single-phase flow in the matrix. It is necessary to subdivide the permeability (K) into the permeability for the liquid-filled channels (K_l) and the vapor-filled channels (K_v). The permeability of the matrix having liquid-filled pores of $r_{min} < r < r_m$ is $CK_l = F(r_m)$, where $F(r_m)$ is given by Equation 11.30 with the upper limit of the integral in the numerator set as r_m, rather than r_{max}. The permeability of the matrix for vapor flow is $K_v = K - K_l = K - F(r_m)/C$.

Kovalev et al. [1987, 1990] analytically model the flow and heat transfer within the porous structure by writing momentum and energy equations for the liquid and vapor flow within the liquid- and vapor-filled capillaries. The two-phase momentum equation is given by

$$\frac{dr_m}{dx}\left[\frac{2\sigma}{r_m^2} - G^2 f(r_m)\left(\frac{1}{\rho_l \in_l^3} + \frac{1}{\rho_v \in_v^3}\right)\right]$$

$$= G\frac{dG}{dx}\left(\frac{1}{\rho_v \in_v^2} - \frac{1}{\rho_l \in_l^2}\right) + \frac{v_l G}{K_l} + \frac{v_v G}{K_v} + \frac{2f_v G^2}{\rho_v \in_v^2} \tag{11.31}$$

The pore distribution functions $f(r_m)$, $\phi(r_m)$, and $F(r_m)$ have been previously defined. Solution of the equations requires (1) experimental data for the porous matrix that define the pore distribution functions, (2) the pressure drop characteristic vs. mass velocity for simultaneous liquid and vapor flow in the porous structure containing liquid-filled small pores, and (3) the permeability of the matrix. Kovalev and Ovodkov [1992] describe an experimental technique to determine the Item 2 information. Kovalev et al. [1990] show the measured pressure drop characteristics for a porous matrix.

The energy equation for the porous matrix is

$$k_l \frac{d^2 T_m}{dx^2} = h_v (T_m - T_l) \tag{11.32}$$

where subscript m refers to the matrix properties. The h_v is the volumetric heat transfer coefficient for evaporation from the stationery menisci. The analytical solution for h_v is given by Kovalev et al. [1990]. The momentum and energy equations must be numerically solved over the matrix thickness. The model is simplified if one assumes that the matrix temperature is constant over its thickness. Thus, all menisci are of a constant radius.

Kovalev et al. [1990] shows that the model reasonably predicts the boiling coefficient for water at 1.0 atm boiling on 1.0-mm-thick matrix made of 0.3 to 0.4-mm-diameter particles. It is important to note that their model also predicts the critical heat flux. They provide predictions, which describe the performance of porous structures having different matrix thermal conductivity, permeability, and thickness.

The model appears to be an important contribution to understanding of the boiling mechanism in porous structures. The Kovalev et al. [1990] paper does not provide sufficient detail and definitions necessary to easily understand their analysis method. Polezhaev and Kovalev [1990] provide some additional description of the model.

11.11.3 Nishikawa et al. [1983] Correlation

Nishikawa et al. [1983] developed an empirical correlation of their data on their matrix geometries, which include copper and bronze particles having $0.10 \leq d_p \leq 1.0$ mm and $0.4 \leq \delta \leq 4.0$ mm. Boiling data were taken at 1 atm for R-11, R-113, and benzene. A linear, multiple regression technique was employed, and Figure 11.45 shows the correlated data. Most of the data are correlated within ±30%. The algebraic equation of the correlation is

$$\frac{q\delta}{k_m \Delta T} = 0.001 \left(\frac{\sigma^2 \lambda}{q^2 \delta^2} \right)^{0.0284} \left(\frac{\delta}{d_p} \right)^{0.56} \left(\frac{q_{dp}}{\in \lambda \mu_v} \right)^{0.593} \left(\frac{k_m}{k_l} \right)^{-0.708} \left(\frac{\rho_l}{\rho_v} \right)^{1.67} \tag{11.33}$$

The left side of Equation 11.33 is a Nusselt number based on the coating thickness, and the third term on the right is a Reynolds number based on the total vapor flow in the matrix, using the particle diameter as the characteristic dimension.

Webb [1983] used the O'Neill et al. [1972] and Nishikawa et al. [1983] correlations to predict the boiling coefficient for several porous structures.

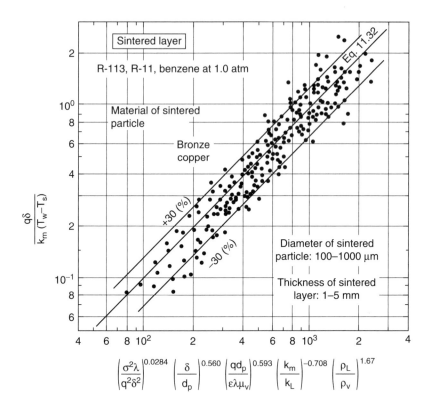

$$\left(\frac{\sigma^2\lambda}{q^2\delta^2}\right)^{0.0284} \left(\frac{\delta}{d_p}\right)^{0.560} \left(\frac{qd_p}{\epsilon\lambda\mu_v}\right)^{0.593} \left(\frac{k_m}{k_L}\right)^{-0.708} \left(\frac{\rho_L}{\rho_v}\right)^{1.67}$$

R-113 and R-11 tests conducted using spherical particles

Test	Mat'l.	d_p (μm)	δ (mm)	δ/d_p
1	Cu	250	0.4, 1, 2, 4	1.6, 4, 8, 16
2	Cu	100, 250[a]		
		500	2	4[a], 8[a], 20
3	Bronze	500	1, 2, 3, 4	2, 4, 6, 8
4	Bronze	250[a], 500[a]		
		750, 1000	2	2, 2.7, 4, 8[a]

[a] Also with R-11.

Figure 11.45 Nishikawa et al. [1983] correlation of their R-11 and R-113 data for coatings made of uniform spherical particles.

11.11.4 Zhang and Zhang [1992] Correlation

This power-law correlation is based on dimensionless parameters assumed to control the boiling process. Zhang and Zhang tested 12 bronze-sintered coatings on a flat surface, which spans $0.11 \leq d_p \leq 0.53$ mm, $0.94 \leq \delta \leq 4.6$ mm, using water, ethyl alcohol, and R-113 at approximately 100 kPa. The data were correlated within ±25%.

11.12 CRITICAL HEAT FLUX

As noted in Section 11.5, enhanced surfaces may have a higher critical heat flux than plain surfaces. The present understanding of critical heat flux in nucleate boiling on a plain, horizontal surface is based on the theory of hydrodynamic instability of the closely spaced vapor jets, as described by Zuber [1958]. These vapor jets carry the latent heat away from the surface. When they break down, the surface can no longer support nucleate boiling. The CHF is given by

$$q_{\text{CHF}} = \rho_v \lambda u_v \left(\frac{A_j}{A} \right) \tag{11.34}$$

where u_v is the velocity of the vapor jet and A_j is the jet cross-sectional flow area. The jet velocity is given by

$$u_v = \left(\frac{2\pi\sigma}{\rho_v \lambda_H} \right) \tag{11.35}$$

where λ_H is the pitch of one-dimensional Taylor instability waves. The λ_H is given by $A_j/A = \pi/16$. The model shows that the CHF is determined by λ_H, which is determined by the balance of surface tension against inertia and buoyancy forces on a plain surface. Lienhard [1987] shows that the CHF on other smooth surface shapes (e.g., a horizontal tube) is influenced by the shape of the surface.

An explanation of higher CHF on an enhanced surface of the same basic shape (e.g., a horizontal plate) would require that the spacing of the vapor jets is different from that on a smooth surface. Polezhaev and Kovalev [1990] have proposed such an explanation for a porous boiling surface. They propose that the vapor jet velocity will still be controlled by Equation 11.35. However, the spacing of the bubble columns is fixed by the spacing of the large pores, from which vapor is emitted. Therefore, the Taylor wavelength used in Equation 11.35 does not control the spacing of these bubble columns. They observed that the spacing of the vapor columns on a porous boiling surface was one tenth that occurring on a plain surface at the same heat flux. As shown by Equation 11.35, the limiting jet velocity, u_v, will be increased if the jet spacing is smaller than λ_H. So, a higher CHF will exist. Polezhaev and Kovalev [1990] show that the term A_j in Equation 11.34 can be written as $\in F$, where $\in$ is the porosity of the matrix, and F is the fraction of the pores that are vapor-filled, and from which the vapor columns are emitted. Based on tests for porous surfaces made from sintered, spherical particles, they provide an empirical correlation for F, given by $F = 0.65\in^{1.28}$. Their resulting correlation for the CHF is

$$q_{\text{CHF}} = 0.52 \in^{2.28} \lambda \left(\frac{\sigma\rho_v}{r_{\text{br}}} \right)^{1/2} \tag{11.36}$$

where r_{br} is defined as the breakdown pressure differential and is approximately 30% larger than the smallest vapor-filled pores. Their correlation shows that q_{CHF} is governed by the porosity of the surface and that $q_{CHF} \propto \in^{2.28}$. In CHF tests of water on various porous coatings, Cieśliński [2002] found little difference between the CHF and that the measured values were in approximate agreement with that for a plain surface. Chang and You [1996] show that their painted microporous surface has 100% higher CHF than a plain surface. They provide an empirical correlation to predict the CHF as a function of orientation for plain and microporous surfaces. As noted in Section 11.7.2, the CHF is strongly influenced by surface orientation. Corresponding work has not been done for the structured boiling surfaces, as illustrated in Figure 11.12. However, it appears that the spacing of the vapor jets would be controlled by the subsurface structure, rather than by the Taylor wavelength.

Ferjančič and Golobič [2002] show that surface treatment (microroughness) can increase the CHF as much as 50%. They worked with flat strips of carbon steel and 302 stainless steel and boiled either water or FC-72. The roughness was accomplished either by sanding the surface with sandpaper, or by acid etching. The roughness formed was in the range of 0.02 to 1.5 μm rms. They did not speculate on the fundamentals that affected this CHF improvement. Law and Dhir [1986] investigated the effects of surface treatment that altered the static contact angle for boiling on a flat vertical plate. They show that the CHF increases as the static contact angle is reduced.

Kandlikar [2001] analytically sought to explain the effect of contact angle on CHF. They show that the dynamic receding contact angle is an important factor that affects the CHF. Specifically, a surface treatment that reduces the receding contact angle will improve the CHF. This fundamental understanding is related to work of Min et al. [2000] discussed in Section 6.13 that relates receding contact angle to surface wettability. With respect to the work of Ferjančič and Golobič [2002], it is known that a microroughness can improve surface wettabilty.

11.13 ENHANCEMENT OF THIN FILM EVAPORATION

Thin film evaporation is the heat transfer mode used in heat pipe evaporators. Much information has been published on evaporator wick design (e.g., Faghri [1995]). Thin film evaporation occurs at the liquid–vapor interface in a heat pipe evaporator. It is desired that nucleation not exist within the liquid film. If nucleate boiling occurs in the wick structure, the thermal performance is reduced. Special surface geometries, such as porous structures and microgrooves, promote high evaporation heat transfer coefficients. This section addresses the heat transfer mechanism of the enhanced geometry structures. Further discussion of thin film evaporation with convection is given in Chapter 13. However, heat pipe design principles are not discussed and readers are referenced to other sources for this, such as Faghri [1995] and Peterson [1994].

The basic approach to enhancement of thin film evaporation is to increase the surface area covered by liquid menisci. Evaporation occurs from the menisci.

Figure 11.46 from Hanlon and Ma [2003] illustrates a porous sintered wick. Liquid menisci exist in the regions between the sintered particles. A meniscus is illustrated on the lower part of Figure 11.46. At any point in the meniscus, the local heat transfer across the meniscus occurs by conduction. The equivalent heat transfer coefficient is given by k_l/δ, where k_l is the liquid thermal conductivity, and δ is the local film thickness. The meniscus is defined to have three regions. Region I is the non-evaporating thin film, Region II is the evaporating thin film, and Region III is the meniscus thin film region. Because δ is thick in region III, little evaporation occurs there. Ma and Peterson [1997] have shown that the dominant fraction of evaporation occurs in Region II and that the active thickness of this region can be

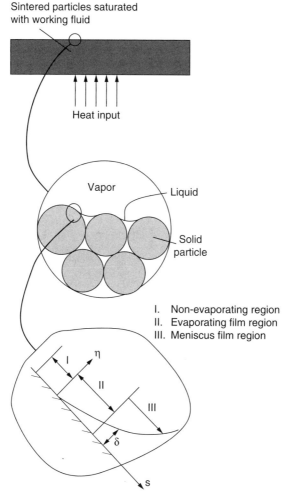

Figure 11.46 Illustration of porous sintered wick, with liquid menisci existing in the regions between the sintered particles. (From Hanlon and Ma [2003].)

as small as 1.0 μm. Hence, the average heat transfer coefficient in Regions II and III is given by

$$\overline{h_{\text{men}}} = \frac{1}{L_{\text{II}}} \int_0^{10^{-6}} \frac{k_l}{\delta} ds \tag{11.37}$$

where s is the projected length along the meniscus measured from the start of the non-evaporating region. Ma and Peterson [1997] and Hanlon and Ma [2003] show that

$$\sigma \frac{d\xi}{ds} - \frac{dp_d}{ds} = \frac{f_1^+ \text{Re}_\delta \mu_l \int_0^s \frac{q_s}{\lambda} ds}{2\delta^3(s)\rho_l} \tag{11.38}$$

where p_d is the disjoining pressure (see Carey [1992]) and ξ is the film curvature given by

$$\xi = \frac{d^2\delta}{ds^2} \bigg/ \left[1 + \left(\frac{d\delta}{ds}\right)^2\right]^{3/2} \tag{11.39}$$

Subject to the boundary conditions $\delta = \delta_0$, $\xi = 0$ and $d\delta/ds = 0$ at $s = 0$. The δ is the thickness of the non-evaporating film given by

$$\delta_0 = \exp\left\{\frac{1}{b}\left[\left(\frac{T_w}{T_v} - 1\right)\frac{\lambda}{RT_w} - \ln a\right]\right\} \tag{11.40}$$

where a and b are constants ($a = 1.5787$ and $b = 0.0243$). If the average heat transfer coefficient is to be referred to the projected base area, then

$$\overline{h}_{\text{proj}} = \overline{h}_{\text{men}}(A_{\text{men}}/A_{\text{proj}}) \tag{11.41}$$

Hence, to obtain a high heat transfer coefficient based on the projected area, one needs a large number of menisci per unit base area. This can be accomplished using a sintered particle geometry or closely spaced microgrooves.

Ma and Peterson [1997] applied the above equations to predict the evaporation coefficient for a channel with parallel microgrooves. They showed that when the groove width was decreased from 600 to 300 μm, the $(T_w - T_s)$ was reduced nearly 50%. Fabrication of microgrooves less than 300 μm in metals becomes difficult to do. However, very small grooves could be made by photolithography. Another approach to provide large A_{men} is to use a porous wick formed by high thermal conductivity spherical particles. The $\overline{h}_{\text{proj}}$ will increase with decreased particle diameter, up to the point that the diameter is less than the characteristic length of the evaporating thin film region (L_{II}). Hence, an optimum particle diameter will exist. Hanlon and Ma [2003] analytically model evaporation and capillary flow in this structure. They also provide experimental data for water evaporating in a sintered

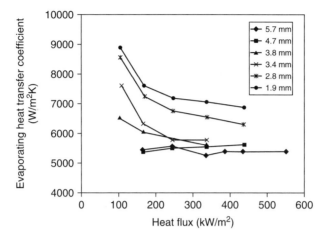

Figure 11.47 Effect of wick thickness on evaporating heat transfer coefficient for water on 100-mesh sintered copper surface. (From Hanlon and Ma [2003].)

copper wick made from 100-mesh particles (43% porosity). Data were taken for 6 wick thickness values between 1.9 and 5.7 mm with heat fluxes as high as 50 W/cm². Included in the data are the evaporation heat transfer coefficient (Figure 11.47) and the heat flux at bubble nucleation in the wick. Above the nucleation heat flux (approximately 10 W/cm²), Figure 11.47 shows that the evaporation coefficient significantly decreases. Their analysis predicts the optimum wick thickness (maximum heat flux) depending on the particle radius, porosity, and unheated flow entry length to the heat applied zone. The optimum wick thickness increases with reduced particle size. Dryout heat flux was predicted (increases with increasing wick thickness) and found to be always below the capillary limit.

Wang and Catton [2001] show that coating the sides of a triangular groove with a thin porous coating will enhance the meniscus evaporation rate. The thin porous film extends the width of the evaporating surface area and also improves the capillary force to maintain the liquid supply and increases the critical heat flux. Their analytical model shows that the evaporation coefficient can be increased three to six times that of a plain triangular groove.

11.14 CONCLUSIONS

This survey has described the evolution of special nucleate boiling surfaces employed on the outer surface of a tube. Two basic types of surface structures exist, namely, porous coatings and reentrant grooves. Rapid advances have been made since 1968, and several high-performance tubes are commercially employed. These have found extensive use in the refrigeration industry and limited use in process applications.

The key to the high performance of the porous and reentrant grooved structures is attributed to three factors: (1) a pore or reentrant cavity within a critical size range,

(2) interconnected cavities, and (3) nucleation sites of a reentrant shape. When the cavities are interconnected, one active cavity can activate adjacent cavities. It appears that the dominant fraction of the vaporization occurs at very thin liquid films within the subsurface structure. The reentrant cavity shape provides a stable vapor trap, which will remain active at very low liquid superheat values.

The porous coated surfaces offer the opportunity for a duplex tube material construction. Then, the boiling surface may be made of less expensive material than required to meet the corrosion characteristics of the tube-side fluid.

Significant progress has been made in analytical models to explain the mechanism of boiling in enhanced surfaces. However, more work is required.

Enhanced tubes may require a substantial liquid superheat for cavity activation. Start-up problems are not observed in refrigeration systems, because the expansion device supplies a liquid–vapor mixture, which impinges on the boiling surfaces and activates the cavities.

Future research and development activity is expected to focus on the following:

1. The development of analytical models and design correlations applicable to a wide range of fluid properties.
2. Application studies to define where such enhanced boiling surfaces may be effectively used.
3. Improved methods of manufacture, which yield lower product cost. The most cost-effective surface geometry is yet to be established. Most of the surfaces produced to date are of copper. Other materials are needed if wide-scale application is to be achieved.
4. Finally, doubly enhanced tubes (inside and outside enhancement) require extensive development effort. The preferred boiling-side enhancement may be dependent on the method of manufacture.

NOMENCLATURE

A	Base area, m^2 or ft^2
A_j	Cross-sectional flow area of vapor jet, m^2 or ft^2
A_{pt}	Total open pore area (= $A_p L/P_p$, where $A_p = \pi d_p^2/4$, P_p/L = pores/length), m^2 or ft^2
c_b	Empirical constant in Equation 11.14
c_p	Specific heat of the liquid, J/kg-K or Btu/lbm-°F
d	Diameter, bubble departure diameter (d_b), pore diameter (d_p), m or ft
f	Bubble formation and departure cycle frequency, 1/s
G	Mass velocity, kg/s-m^2 or lbm/hr-ft^2
g	Acceleration due to gravity, m/s^2 or ft/s^2
h	Depth of the fin gap spacing, m or ft
H_t	Tunnel height, m or ft
h_{nb}	Nucleate boiling heat transfer coefficient, kW/m^2-K or Btu/hr-ft^2-°F
k	Thermal conductivity, W/m-K or Btu/hr-ft-°F

m	Average slope of the P–T curve for the working regime, Pa/K or lbf/ft²-°F
n	Exponent on q vs. $(T_w - T_{sat})$ equation
n_s	Nucleation site density, 1/m² or 1/ft²
p	Pressure, Pa or lbf/ft²
p_{cr}	Critical pressure, Pa or lbf/ft²
p_G	Partial pressure of inert gas, Pa or lbf/ft²
p_{sat}	Pressure of the saturated pool liquid, Pa or lbf/ft²
P_f	Fin pitch, m or ft
P_p	Pore pitch, m or ft
P_t	Tunnel pitch, m or ft
Δp	Pressure differential across a liquid slug, Pa or lbf/ft²
Δp_{max}	Maximum pressure differential across a liquid slug, Pa or lbf/ft²
Q	Heat transfer rate, W or Btu/hr
q_{CHF}	Critical heat flux, W/m² or Btu/hr-ft²
q	Total heat flux, W/m² or Btu/hr-ft²
q_{ex}	Heat flux from the external surface, W/m² or Btu/hr-ft²
q_{tun}	Heat flux from the tunnel surface, W/m² or Btu/hr-ft²
r	Liquid–vapor interface radius, m or ft
r_b	Pore radius used in Equation 11.25, m or ft
r_c	Cavity radius at surface, m or ft
r_m	Meniscus radius shown on Figure 11.42, m or ft
r_t	Radius of subsurface circular tunnel, m or ft
R	Gas constant for saturated water vapors, J/kg-K or lbf-ft/lbm-R
R_m	Meniscus radius, m or ft
s_g	Fin gap spacing, m or ft
S	Surface area of liquid film in porous structure, m² or ft²
t	Time, s
T_l	Temperature at liquid–vapor meniscus, K or °F
T_s	Temperature of saturated pool liquid, K
T_w	Wall temperature, K or °F
ΔT	$T_w - T_s$, K or °F
u	Velocity, u_o (initial velocity of translating meniscus), m/s or ft/s
v	Specific volume, $v_{lv} = v_{lv} - v_l$, m³/kg or ft³/lbm
V	Volume, V_T (total volume of porous surface), V_b (total bubble volume), m³ or ft³
W_t	Tunnel width, m or ft
W_h	Tunnel height, m or ft

Greek Letters

β	Geometry factor defined by Equation 11.25, dimensionless
δ	Liquid film thickness, m or ft
$\in$	Porosity, dimensionless
θ	Liquid contact angle, θ_a(advancing contact angle), θ_r(receding contact angle), degrees

λ Latent heat, J/kg or Btu/lbm

λ_H Pitch of one-dimensional Taylor instability waves, m or ft

v Kinematic viscosity of the liquid, m^2/s or ft^2s

ρ Density, kg/m^3 or lbm/ft^3

σ Surface tension of the liquid, N/m or lbf/ft

τ Time period of the bubble departure and formation cycle, s

τ_w Wall shear stress, N/m^2 or lbf/ft

ϕ Angle between cavity centerline and wall, degrees

Subscripts

CHF   Critical heat flux

l Liquid phase

m Porous matrix

s At saturated condition

t In subsurface tunnel

v Vapor phase

w At tube wall

REFERENCES

Albertson, C.E., 1977. Boiling Heat Transfer Surface and Method, U.S. patent 4,018,264.

Antonelli, R. and O'Neill, P.S., 1981. Design and application considerations heat exchangers with enhanced boiling surfaces, in *Heat Exchanger Sourcebook*, J. Palen, Ed., Hemisphere, Washigton, D.C.

Arai, N., Fukushima, T., Arai, A., Nakajima, T., Fujie, K., and Nakayama, Y., 1977. Heat transfer tubes enhancing boiling and condensation in heat exchangers of a refrigerating machine, *ASHRAE Trans.*, 83(2), 58–70.

Arbelaez, F., Sett, S., and Mahajan, R.L., 2000, An experimental study on pool boiling of saturated FC-72 in highly porous aluminum metal foams, *Proc. 34th National Heat Transfer Conf.*, 759–767.

Arshad, J. and Thome, J.R., 1983. Enhanced boiling surfaces: heat transfer mechanism mixture boiling, *Proc. ASME-JSME Therm. Eng. Joint Conf.*, 1, 191–197.

Asakavicius, J.P., Zukauskav, A.A., Gaigalis, V.A., and Eva, V.K., 1979. Heat transfer from Freon- 113, ethyl alcohol and water with screen wicks, *Heat Transfer- Sov. Res.*, 11(1), 92–100.

Ayub, Z.H., and Bergles, A.E., 1987. Pool boiling from GEWA surfaces in water and R-113, *Wärme Stoffübertrag.*, 21, 209–219.

Ayub, A.H. and Bergles, A.E., 1988a. Pool boiling enhancement of a modified GEWA-T surface in water, *J. Heat Transfer*, 10, 266–267.

Ayub, Z.H. and Bergles, A.E., 1988b. Nucleate pool boiling curve hysteresis for GEWA-T surfaces in saturated R-113, *ASME Proc. of the National Heat Transfer Conference*, 2, ASME Symp. Vol. HTD, 96, 515–521.

Bankoff, S.G., 1959. Entrapment of gas in the spreading of a liquid over a rough surface, *AIChE J.*, 4(1), 24–26.

Bar-Cohen, A., 1992. Hysteresis phenomena at the onset of nucleate boiling, in *Pool and External Flow Boiling*, V.J. Dhir and A.E. Bergles, Eds., ASME, New York, 1–14.

Bell, K.J. and Mueller, A.C. 1984. *Engineering Data Book II*, Wolverine Tube Corp., Decatur, AL.

Benjamin, J.E. and Westwater, J.W., 1961. Bubble growth in nucleate boiling of a binary mixture, *International Developments in Heat Transfer*, ASME, New York, 212–218.

Berenson, P.J., 1962. Experiments on pool boiling heat transfer, *Int. J. Heat Mass Transfer*, 5, 985–999.

Bergles, A.E. and Chyu, M.C., 1982. Characteristics of nucleate pool boiling from porous metallic coatings, *J. Heat Transfer*, 104, 279–285.

Bergles, A.E., Bakhru, N., and Shires, J.W., Jr., 1968. Cooling of high-power density computer components, Massachusetts Institute of Technology, EPL Report 70712–60.

Bliss, F.E., Hsu, S.T., and Crawford, M., 1969. An investigation into the effects of various platings on the film coefficient during nucleate boiling from horizontal tubes, *Int. J. Heat Mass Transfer*, 1, 1061–1072.

Bonilla, C.F., Grady, J.J., and Avery, G.A., 1965. Pool boiling heat transfer from scored surfaces, *Chem. Eng. Prog. Symp. Ser.*, 61(57), 281–288.

Brothers, W.S. and Kallfelz, A.J., 1979. Heat Transfer Surface and Method of Manufacture, U.S. patent 4,159,739.

Carey, V.P., 1992. *Liquid Vapor Phase-Change Phenomena*, Hemisphere, Washington, D.C.

Chang, J.Y. and You, S.M., 1996. Heater orientation effects on pool boiling of micro-porous-enhanced surfaces in saturated FC-72, *J. Heat Transfer*, 118, 937–943.

Chang, J.Y. and You, S.M., 1997. Boiling heat transfer phenomena from microporous and porous surfaces in saturated FC-72, *Int. J. Heat Mass Transfer*, 40, 4437–4447.

Chaudri, I.H. and McDougall, I.R., 1969. Aging studies in nucleate pool boiling of isopropyl acetate and perchlorethylene, *Int. J. Heat Mass Transfer*, 12, 681–688.

Chien, L.-H. and Chang, C.-C., 2003. Enhancement of pool boiling on structured surfaces using HFC-4310 and water, *J. Enhanced Heat Transfer*, 11, 23–44.

Chien, L.-H. and Chang C.-C., 2004. Enhancement of pool boiling on structured surfaces using HFC-4310 and water, *J. Enhanced Heat Transfer*, 11, 23–44.

Chien, L.-H. and Chen, C.-L., 2000. An experimental study of boiling enhancement in a small boiler, in *Proceedings of 34th National Heat Transfer Conference*, Pittsburgh, PA, NHTC 2000-12210.

Chien, L.-H. and Webb, R.L., 1998a. Parametric studies of nucleate pool boiling on structured surfaces, Part I: Effect of tunnel dimensions and comparison of R-11, *J. Heat Transfer*, 120, 1042–1048. Also published in *ASME Proc. National Heat Transfer Conf.*, L. Witte et al., Eds., 4, 129–136, 1996.

Chien, L.-H. and Webb, R.L., 1998b. Parametric studies of nucleate pool boiling on structured surfaces. Part II: Effect of pore diameter and pore pitch, *J. Heat Transfer*, 120, 1048–1054. Also published in *ASME Proc. National Heat Transfer Conf.*, L. Witte et al., Eds., 4, 137–144, 1996.

Chien, L.-H. and Webb, R.L., 1998c. Visualization of pool boiling on enhanced surfaces, *Exp. Fluid Thermal Sci.*, 16, 332–341.

Chien, L.-H. and Webb, R.L., 1998d. Measurement of bubble dynamics on an enhanced boiling surface, *Exp. Fluid Thermal Sci.*, 16, 177–186.

Chien, L.-H. and Webb, R.L., 1998e. A nucleate boiling model for structured enhanced surfaces, *Int. J. Heat Mass Transfer*, 41, 2183–2195.

Chien, L.-H. and Webb, R.L., 2001. Effect of geometry and fluid property parameters on performance of tunnel and pore enhanced boiling surfaces, *J. Enhanced Heat Transfer*, 8, 329–340.

Chu, R.C. and Moran, K.P., 1977. Method for Customizing Nucleate Boiling Heat Transfer from Electronic Units Immersed in Dielectric Coolant, U.S. patent 4,050,507.

Cieśliński, J.T., 2002. Nucleate pool boiling on porous metallic coatings, *Exp. Thermal Fluid Sci.*, 25, 557–564.

Clark, H.B., Strenge, P.H., and Westwater, J.W., 1959. Active sites for nucleate boiling, *Chem. Eng. Prog. Symp. Ser.*, 55(29), 103–110.

Corman, J.C. and McLaughlin, M.H., 1976. Boiling augmentation with structured surfaces, *ASHRAE Trans.*, 82(1), 906–918.

Corty, C. and Foust, A.S., 1955. Surface variables in nucleate boiling, *Chem. Eng. Prog. Symp. Ser.*, L20 51(17), 1–12.

Czikk, A.M. and O'Neill, P.S., 1979. Correlation of nucleate boiling from porous metal films, in *Advances in Enhanced Heat Transfer*, J.M. Chenoweth, J. Kaellis, J.W. Michel, and S. Shenkman, Eds., ASME, New York, 103–113.

Dahl, M.M. and Erb, L.D., 1976. Liquid heat exchanger interface and method, U.S. patent 3,990,862.

Danilova, G.N. and Tikhonov, A.V., 1996. R113 boiling heat transfer modeling on porous metallic matrix surfaces, *Int. J. Heat Fluid Flow*, 17, 45–51.

Dundin, V.A., Danilova, G.N., and Tikhonov, A.V., 1990. Enhanced heat transfer surfaces for shell-and-tube evaporators of refrigerating machines, *Refrig. Mach.*, Ser. XM-7, 1–46 [in Russian].

Faghri, A., 1995. *Heat Pipe Science and Technology*, Taylor & Francis, London.

Ferjančič, K. and Golobič, I., 2002. Surface effects on pool boiling CHF, *Exp. Thermal Fluid Sci.*, 25, 565–571.

Fritz, W., 1935. Berechnung des Maximalvolume von Dampfblasen, *Phys. Z.*, 36, 379–388.

Fujie, K., Nakayama, W., Kuwahara, H., and Kakizakci, K., 1977. Heat Transfer Wall for Boiling Liquids, U.S. patent 4,060,125.

Fujii, M., Nishiyama, E., and Yamanaka, G., 1979. Nucleate pool boiling heat transfer from microporous heating surface, in *Advances in Enhanced Heat Transfer*, J.M. Chenoweth, J. Kaellis, J.W. Michel, and S. Shenkman, Eds., ASME, New York, 45–51.

Fujikake, J., 1980. Heat Transfer Tube for Use in Boiling Type Heat Exchangers and Method of Producing the Same, U.S. patent 4,216,826.

Gaertner, R.F., 1967. Methods and means for increasing the heat transfer coefficient between a wall and boiling liquid, U.S. patent 3,301,314.

Gottzmann, C.F., Wulf, J.B., and O'Neill, P.S., 1971. Theory and application of high performance boiling surfaces to components of absorption cycle air conditioners, *Proc. Conf. Natural Gas Research. Technology*, Session V, Paper 3, Chicago, IL.

Gottzmann, C.F., O'Neill, P.S., and Minton, P.E., 1973. High efficiency heat exchangers, *Chem. Eng. Prog.*, 69(7), 69–75.

Granryd, E. and Palm, B., 1988. Heat Transfer Element, U.S. patent 4,787,441.

Grant, A.C., 1977. Porous Metallic Layer and Formation, U.S. patent 4,064,914.

Griffith, P. and Wallis, J.D., 1960. The role of surface conditions in nucleate boiling, *Chem. Eng. Prog. Symp. Ser.*, 56(49), 49–63.

Guglielmini, G., Misale, M., Schenone, C., Pasquali, C., and Zappaterra, M., 1988, On performances of nucleate boiling enhanced surfaces for cooling of high power electronic devices, *Proc. 22nd Int. Symp. Heat Transfer in Electronic and Microelectronic Equipment*, 589–600.

Haider, S.I. and Webb, R.L., 1997. A transient micro-convection model of nucleate pool boiling, *Int. J. Heat Mass Transfer*, 40, 3675–3688.

Han, C.Y. and Griffith, P., 1965. The mechanism of heat transfer in nucleate pool boiling, Parts I and II, *Int. J. Heat Mass Transfer*, 8, 887–917.

Hanlon, M.A. and Ma, H.B., 2003. Evaporation heat transfer in sintered porous media, *J. Heat Transfer*, 125, 644–652.

Hasegawa, S., Echigo, R., and Irie, S., 1975. Boiling characteristics and burnout phenomena on a heating surface covered with woven screens, *J. Nucl. Sci., Technol.*, 12(2), 722–724.

Hausner, H. and Mal, M.K., 1982. *Handbook of Powder Metallurgy*, 2nd ed., Chemical Publishing Co., New York.

Hsieh, S.-S. and Weng, C.-J., 1997. Nucleate pool boiling from coated surfaces in saturated R-134a and R-407C, *Int. J. Heat Mass Transfer*, 40, 519–532.

Hsieh, S-S. and Yang, T-Y., 2001. Nucleate pool boiling from coated and spirally wrapped tubes in saturated R-134a and R-600a at low and moderate heat flux, *J. Heat Transfer*, 123, 257–270.

Hsu, Y.Y., 1962. On the size range of active nucleation cavities on a heating surface, *Heat Transfer*, 84, 207–216.

Hübner, P. and Künstler, W., 1997. Pool boiling heat transfer at finned tubes: influence of surface roughness and shape of the fins, *Int. J. Refrig.*, 20, 575–582.

Hummel, R.L., 1965. Means for Increasing the Heat Transfer Coefficient between a Wall and Boiling Liquid, U.S. patent 3,207,209.

Imadojemu, H., Hong, K., and Webb, R.L., 1995. Pool boiling of R-11 refrigerant and water on oxidized enhanced tubes, *J. Enhanced Heat Transfer*, 2, 189–198.

Jakob, M., 1949. *Heat Transfer*, John Wiley & Sons, New York, 636–638.

Jamialahmadi, M. and Muller-Steinhagen, H., 1993. Scale formation during nucleate boiling - a review, *Corrosion Rev.* 11(1–2), 25–54.

Janowski, K.R., Shum, M.S., and Bradley, S.A., 1978. Heat Transfer Surface, U.S. patent 4,129,181.

Jiang, Y.Y., Wang, W.C., Wang, D., and Wang, B.X., 2001. Boiling heat transfer on machined porous surfaces with structural optimization, *Int. J. Heat Mass Transfer*, 44, 443–456.

Kajikawa, T., Takazawa, H., and Mizuki, M., 1983. Heat transfer performance of the metal fiber sintered surfaces, *Heat Transfer Eng.*, 4(1), 57–66.

Kandlikar, S.G., 2001. A theoretical model to predict pool boiling CHF incorporating effects of contact angle and orientation, *J. Heat Transfer*, 123, 1071–1079.

Kang, M.-G., 2000. Effect of surface roughness on pool boiling heat transfer, *Int. J. Heat Mass Transfer*, 43, 4073–4085.

Kartsounes, G.T., 1975. A study of surface treatment on pool boiling heat transfer in Refrigerant 12, *ASHRAE Trans.*, 81, 320–326.

Kedzierski, M.A., 1995. Calorimetric and Visual Measurements of R-123 Pool Boiling on Four Enhanced Surfaces, Report NISTIR 5732, November.

Kim, C.J., and Bergles, A.E., 1988. Incipient boiling behavior of porous boiling surfaces used for cooling microelectronic chips, in *Particulate Phenomena and Multiphase Transport*, Vol. 2, T.N. Veziroglu, Ed., Hemisphere, New York, 3–18.

Kim, N.H., 1996. Pool boiling heat transfer enhancement by perforated plates, in *Interfacial Phenomena, Boiling Heat Transfer, Thermal Hydraulics for Advanced Nuclear Reactors*, L. Witte et al., Eds., ASME HTD, 326(4), 145–151.

Kim, N.H. and Choi, K.-K., 2001. Nucleate pool boiling on structured enhanced tubes having pores with connecting gaps, *Int. J. Heat Mass Transfer*, 44, 17–28.

Ko, S.-Y. Liu, L., and Yao, Y.-Q., 1992. Boiling hysteresis on porous metallic coatings, in *Multiphase Flow and Heat Transfer, Second International Symposium*, X.-J. Chen, T.N. Veziroğlu, and C.L. Tien, Eds., Hemisphere, New York, 259–268.

Kovalev, S.A. and Lenkov, V.A., 1981. Mechanism of burnout with boiling on a porous surface, *Thermal Eng.*, 28(4), 201–203.

Kovalev, S.A. and Ovodkov, O.A., 1992. A study of gas-liquid counterflow in porous media, *Exp. Thermal Fluid Sci.* 5, 457–464.

Kovalev, S.A., Solov'yev, S.L., and Ovodkov, O.A., 1987. Liquid boiling on porous surfaces, *Heat Transfer Soviet Research*, 19(3), 109–120.

Kovalev, S.A., Solov'yev, S.L., and Ovodkov, O.A., 1990. Theory of boiling heat transfer on a capillary porous surface, *Proc. 9th Int. Heat Transfer Conf.*, 2, 105–110.

Khrustalev, D. and Faghri, A., 1994. Thermal analysis of a micro heat pipe, *J. Heat Transfer*, 116, 189–198.

Kulenovic, R., Mertz, R., and Groll, M., 2002. High speed flow visualization of pool boiling from structured tubular heat transfer surfaces, *Exp. Thermal Fluid Sci.*, 25, 547–555.

Kun, L.C. and Czikk, A.M., 1969. Surface for Boiling Liquids, U.S. patent 3,454,081. (Reissued August 21, 1979 Re. 30,077.)

Kurihari, H.M. and Myers, J.E., 1960. Effects of superheat and roughness on the boiling coefficients, *AIChE J.*, 6(1), 83–91.

Law, S.P. and Dhir, V.K., 1986. Effect of surface wettability on transition boiling heat transfer from a vertical surface, *Heat Transfer — 1986, Proc. Eighth Intl. Heat Transfer Conf.*, San Francisco, CA, 4, 2031–2036.

Li, Z., Tan, Y., and Wang. S., 1992. Investigation of the heat transfer performance of mechanically made porous surface tubes with ribbed tunnels, *Multiphase Flow and Heat Transfer, Second International Symposium*, Vol. 1, X.J. Chen, T.N. Veziroğlu, and C.L. Tien, Eds., Hemisphere, New York, 700–707.

Liang, H.-S. and Yang, W.-J., 1998. Nucleate pool boiling heat transfer in a highly wetting liquid on micro-graphite-fiber composite surfaces, *Int. J. Heat Mass Transfer*, 41, 1993–2001.

Lienhard, J.H., 1987. *A Heat Transfer Textbook*, 2nd ed., Prentice-Hall, Englewood Cliffs, NJ.

Liu, J.W., Lee, D.J., and Su, A., 2001. Boiling of methanol and HFE-7100 on heated surface covered with a layer of mesh, *Int. J. Heat Mass Transfer*, 44, 241–246.

Liu, X., Ma, T., and Wu, J., 1987. Effects of porous layer thickness of sintered screen surfaces on pool nucleate boiling heat transfer and hysteresis phenomena, *Heat Transfer Science and Technology*, B.-X. Wang, Ed., Hemisphere, New York, 577–583.

Lorenz, J.J., Mikic, B.B., and Rohsenow, W.M., 1974. The effect of surface conditions on nucleate boiling characteristics, *Proc. 5th Intl. Heat Transfer Conf.*, 4, 35–49.

Luke, A., 1997. Pool boiling heat transfer from horizontal tubes with different surface roughness, *Int. J. Refrig.*, 20, 561–574.

Ma, H.B. and Peterson, G.P., 1997. Temperature variation and heat transfer in triangular grooves with an evaporating film, *J. Thermophys. Heat Transfer*, 10, 83–89.

Ma, T., Liu, X., Wu, J., and Li, H., 1986. Effects of geometrical shapes and parameters of reentrant grooves on nucleate pool boiling heat transfer from porous surfaces, *Heat Transfer 1986, Proc. 8th Int. Heat Transfer Conf.*, 4, 2013–2018.

Malyshenko, S.P. and Styrikovich, M.A., 1992. Heat transfer at pool boiling on surfaces with porous coating, in *Multiphase Flow and Heat Transfer, Second International Symposium*, Vol. 1, X.-J. Chen, T.N. Veziroğlu, and C.L. Tien, Eds., Hemisphere, New York, 269–284.

Marto, P.J. and Rohsenow, W.M., 1966. Effects of surface conditions on nucleate pool boiling of sodium, *J. Heat Transfer*, 88, 196–204.

Marto, P.J., Moulson, J.A., and Maynard, M.D., 1968. Nucleate pool boiling of nitrogen with different surface conditions, *J. Heat Transfer*, 90, 437–444.

Matijevic´, M., Djuric´, M., Zavargo, Z., and Novakovic´, M., 1992. Improving heat transfer with pool boiling by covering of heating surface with metallic spheres, *Heat Transfer Engineering*, 13(3), 49–57.

Mertz, R., Kulenovic, R., Chen, Y., and Groll, M., 2002. Pool boiling of butane from enhanced evaporator tubes, *Heat Transfer 2002, Proc. 12th Int. Heat Transfer Conf.*, 3, 629–634.

Mikic, B.B. and Rohsenow, W.M., 1969. A new correlation of pool-boiling data including the effect of heating surface characteristics, *J. Heat Transfer*, 91, 245–250.

Milton, R.M., 1968. Heat Exchange System, U.S. patent 3,384,154.

Milton, R.M., 1970. Heat Exchange System, U.S. patent 3,523,577.

Milton, R.M., 1971. Heat Exchange System with Porous Boiling Layer, U.S. patent 3,587,730.

Milton, R.M. and Gottzmann, C.F., 1972. High efficiency reboilers and condensers, *Chem. Eng. Prog.*, 68(9), 56–61.

Min, J., Webb, R.L., and Bemisderfer, C.H., 2000. Long-term hydraulic performance of dehumidifying heat-exchangers with and without hydrophilic coatings, *Int. J. HVAC&R Res.*, 6, 257.

Mirzamoghadam, A. and Catton, I., 1988. A physical model of the evaporating meniscus, *J. Heat Transfer*, 110, 201–207.

Modahl, R.J. and Lukeroth, V.C., 1982. Heat Transfer Surface for Efficient Boiling of Liquid R-11 and its Equivalents, U.S. patent 4,354,550.

Moghaddam, S., Ohadi, M., and Qi, J., 2000. Pool boiling of water and FC-72 on copper and graphite foams, Paper 35316, *Proc. ASME InterPack '03*, Maui, HI.

Moore, F.D. and Mesler, R.B., 1971. The measurement of rapid surface temperature fluctuations during nucleate boiling of water, *AIChE J.*, 7, 620–624.

Nakayama, W., Daikoku, T., Kuwahara, H., and Nakajima, T., 1980a. Dynamic model of enhanced boiling heat transfer on porous surfaces. Part I: Experimental investigation, *J. Heat Transfer*, 102, 445–450.

Nakayama, W., Daikoku, T., Kuwahara, H., and Nakajima, T., 1980b. Dynamic model of enhanced boiling heat transfer on porous surfaces Part II: Analytical modeling, *J. Heat Transfer*, 102, 451–456.

Nakayama, W., Daikoku, T., and Nakajima, T., 1982. Effects of pore diameters and system pressure on saturated pool nucleate boiling heat transfer from porous surfaces, *J. Heat Transfer*, 104, 286–291.

Nishikawa, K. and Ito, T., 1982. Augmentation of nucleate boiling heat transfer by prepared surfaces, in *Heat Transfer in Energy Problems*, T. Mizushina and W.J. Yang, Eds., Hemisphere, New York, 111–118.

Nishikawa, K., Ito, T., and Tanaka, K., 1983. Augmented heat transfer by nucleate boiling at prepared surfaces, *Proc. 1983 ASME-JSME Thermal Eng. Conf.*, 1, 387–393.

O'Conner, J.P. and You, S.M., 1995. A painting technique to enhance pool boiling heat transfer in saturated FC-72, *J. Heat Transfer*, 117, 387–393.

O'Neill, P.S., Gottzman, C.F., and Terbot, J.W., 1972. Novel heat exchanger increases cascade cycle efficiency for natural gas liquefaction, in *Advances in Cryogenic Engineering*, K.D. Timmerhaus, Ed., Plenum Press, New York, 420–437.

Pais, C. and Webb, R.L., 1991. Literature survey of pool boiling on enhanced surfaces, *ASHRAE Trans.*, 97(1), 79–89.

Palm, B., 1992. Heat transfer enhancement in boiling by aid of perforated metal foils, in *Recent Advances in Heat Transfer*, B. Sunden and A. Zukauskas, Eds., Elsevier Science, New York.

Park, K.A. and Bergles, A.E., 1988. Effects of size of simulated microelectronic chips on boiling and critical heat flux, *J. Heat Transfer*, 110, 728–734.

Peterson, G.P., 1994. *An Introduction to Heat Pipes*, Wiley Interscience, New York.

Polezhaev, Y.U. and Kovalev, S.A., 1990. Modelling heat transfer with boiling on porous structures, *Teploenergetika*, 37(12), 5–9 [in Russian]. Also in *Thermal Eng.*, 37(12), 617–620.

Probstein, R.F., 1989. *Physicochemical Hydrodynamics — An Introduction*, Butterworths, Boston, 280–289.

Ragi, E.G., 1972. Composite Structure for Boiling Liquids and Its Formation, U.S. patent 3,684,007.

Rainey, K.N. and You, S.M., 2001. Effects of heater size and orientation on pool boiling heat transfer from microporous coated surfaces, *Int. J. Heat Mass Transfer*, 44, 2589–2599.

Ramaswamy, C., Joshi, Y., Nakayama, W., and Johnson, W.B., 2003. Semi-analytical model for boiling from enhanced structures, *Int. J. Heat Mass Transfer*, 46, 4257–4269.

Rohsenow, W.M., 1985. Boiling, in *Handbook of Heat Transfer Fundamentals*, McGraw Hill, New York, 12–5.

Sachar, S.S. and Silvestri, V.J., 1983. Porous Film Heat Transfer, U.S. patent 4,381,818.

Saidi, M.H., Ohadi, M., and Souhar, M., 1999. Enhanced pool boiling of R-123 refrigerant on two selected tubes, *Appl. Thermal Eng.*, 19, 885–895.

Saier, M., Kastner, H.W., and Klockler, R., 1979. Y- and T-Finned Tubes and Methods and Apparatus for Their Making, U.S. patent 4,179,911.

Sanborn, D.F., Holman, J.L.M., and Ware, C.D., 1982. Heat Exchange Surface with Porous Coating and Subsurface Cavities, U.S. patent 4,359,086.

Sauer, E.T., 1935. M.S. thesis, Department of Mechanical Engineering, Massachusetts Institute of Technology, Cambridge, MA.

Schmittle, K.V. and Starner, K.E., 1978. Heat Transfer in Pool Boiling, U.S. patent 4,074,753.

Shum, M.S., 1980. Finned Heat Transfer Tube with Porous Boiling Surface and Method for Producing Same, U.S. patent 4,182,412.

Sokol, P., Blein, P., Gorenflo, D., Rott, W., and Schömann, H., 1990. Pool boiling heat transfer from plain and finned tubes to propane and propylene, *Heat Transfer 1990, Proc. Ninth Int Heat Transf. Conf.*, 2, 75–80.

Sridharan, A., Hochreiter, L.E., Cheung, F.B., and Webb, R.L. 2002. Effect of chemical cleaning on steam generator tube performance, *Heat Transfer Eng.*, 23, 38–47.

Srinivasan, V., Augustyniak, J.D., and Lockett, M.J., 2001. Pool boiling experiments with liquid nitrogen on enhanced boiling surfaces, in *Proceedings of the Third International Conference on Compact Heat Exchangers and Enhancement Technology for the Process Industries*, R.K. Shah, A.W. Deakin, H. Honda, and T.M. Rudy, Eds., Begell House, New York, 409–414.

Szumigala, E.T., 1971. Manufacturing Method for Boiling Surfaces, U.S. patent 3,566,514.

Tarrad, A.H. and Burnside, B.M., 1993. Pool boiling tests on plain and enhanced tubes using a wide boiling range mixture, *Exp. Heat Transfer*, 6, 83–96.

Tatara, R.A. and Payvar, P., 2000. Pool boiling of pure R134a from a single turbo-BII-HP tube, *Int. J. Heat Mass Transfer*, 43, 2233–2236.

Thome, J.R., 1990. *Enhanced Boiling Heat Transfer*, Hemisphere, New York.

Thors, P., Clevinger, N.R., Campbell, B.J., and Tyler, J.T., 1997. Heat Transfer Tubes and Methods of Fabrication Thereof, U.S. patent 5,697,430.

Torii, T., Hirasawa, S., Kuwahara, H., Yanagida, T. and Fujii, M., and Ito, T., 1978. The Use of Heat Exchangers with THERMOEXCEL's Tubing in Ocean Thermal Energy Power Plants, ASME Paper 78-WA-HT-65, ASME, New York.

Tsay, J.Y., Yan, Y.Y., and Lin, T.F., 1996, Enhancement of pool boiling heat transfer in a horizontal water layer through roughness and screen coverage, *Heat Mass Transfer*, 32, 17–26.

Uhle, J., Turner C., and Klimas, S., 1998. Boiling heat transfer characteristics of steam generator U-tube fouling deposits, paper ICONE-6421, presented at 6th International Conference on Nuclear Engineering (ASME), May 10–14.

Uma, B.B.K., Rao, M., and Balikrishnan, A.R., 2000. Enhanced pool boililng heat transfer using interference plates, *Proc. NHTC '00, 34th National Heat Transfer Conf.*, 911–929.

Vachon, R.I., Nix, G.H., and Tanger, G.E., 1968. Evaluation of constants for the Rohsenow pool-boiling correlation, *J. Heat Transfer*, 90, 239–247.

Vachon, R.I., Nix, G.H., Tanger, G.E., and Cobb, R.E., 1969. Pool boiling heat transfer from Teflon coated stainless steel, *J. Heat Transfer*, 91, 364–370.

Wang, C.-C., Hsieh, W.-Y., Yang, C-Y., Li, C., and Chang, Y.-J., 1998. Nucleate boiling performance of R-22, R-123, R-134a, R-410A and R-407C on smooth and enhanced tubes, *ASHRAE Trans.*, 104(1), 1314–1321.

Wang, D.Y., Cheng, J.G., and Zhang, H.J., 1991. Pool boiling heat transfer from T-finned tubes at atmospheric and super-atmospheric pressures, in *Phase Change Heat Transfer*, E. Hensel, V.K. Dhir, R. Greif, and J. Fillo, Eds., ASME Symp. Vol. HTD, 159, 143–147.

Wang, J. and Catton, I., 2001. Enhanced evaporation heat transfer in triangular grooves covered with a thin fine porous layer, *Appl. Thermal Eng.*, 21, 1721–1737.

Webb, R.L., 1970. Heat Transfer Surface which Promotes Nucleate Boiling, U.S. patent 3,521,708.

Webb, R.L., 1972. Heat Transfer Surface Having a High Boiling Heat Transfer Coefficient, U.S. patent 3,696,861.

Webb, R.L., 1981. The evolution of enhanced surface geometries for nucleate boiling, *Heat Transfer Eng.*, 2(3–4), 46–69.

Webb, R.L., 1983. Nucleate boiling on porous coated surfaces, *Heat Transfer Eng.*, 4(3–4), 71–82.

Webb, R.L., 2004. Odyssey of the enhanced boiling surface, J. Heat Transfer, 126, 1051–1059.

Webb, R.L. and Chien, L.-H., 1999. Boiling on Structured Surfaces, invited lecture for NATO Advanced Study Institute on Heat Transfer Enhancement, Izmir Turkey, May 1998, in *Heat Transfer Enhancement of Heat Exchangers*, S. Kakaç, A.E. Bergles, F. Mayinger, and H. Yüncü, Eds., Kluwer Academic, Dordreoht, 249–284.

Webb, R.L. and Haider, S.I., 1992. An analytical model for nucleate boiling on enhanced surfaces, in *Pool and External Flow Boiling*, V.J. Dhir and A.E. Bergles, Eds., ASME, New York, 345–360.

Webb, R.L. and Pais, C., 1992. Nucleate boiling data for five refrigerants on plain, integral-fin and enhanced tube geometries, *Int. J. Heat Mass Transfer*, 35, 1893–1904.

Xin, M.D. and Chao, Y.-D., 1985. Analysis and experiment of boiling heat transfer on T-shaped finned surfaces, *AIChE paper* Presented at 23rd Natl. Heat Trans. Conf., Denver, Co.

Xin, M.D., and Chao, Y.-D., 1987. Analysis and experiment of boiling heat transfer on T-shaped finned surfaces, *Chem. Eng. Comm.*, 50, 185–189.

Yang, G.-W., Liang, H-S., and Vrable, D.L., 1996. Nucleate pool boiling on micro graphite-copper composite surfaces, *J. Heat Transfer*, 118(3), 792–796.

Yatabe, J.M. and Westwater, J.W., 1966. Bubble growth rates for ethanol-water and ethanol-isopropanol mixtures, *Chem. Eng. Prog. Symp. Ser.*, 62(64), 17–23.

Yilmaz, S. and Westwater, J.W., 1981. Effect of commercial enhanced surfaces on the boiling heat transfer curve, in *Advances in Enhanced Heat Transfer 1981*, R.L. Webb, T.C. Carnavos, E.F. Park, Jr., and K.M. Hostetler, Eds., ASME Symp. Vol. HTD, 18, 73–92.

Young, R.X. and Hummel, R.L., 1965. Improved nucleate boiling heat transfer, *Chem. Eng. Prog. Symp. Ser.*, 61(59), 264–470.

Zhang, H. and Dong, L., 1992. Analysis and experiment of pool boiling heat transfer from Cit-shaped finned tube above atmospheric pressure, in *Multiphase Flow and Heat Transfer, Second International Symposium*, Vol. 1, X.-J. Chen, T.N. Veziroğlu, and C.L. Tien, Eds., Hemisphere, New York, 384–392.

Zhang, Y. and Zhang, H., 1992. Boiling heat transfer from a thin powder porous layer at low and moderate heat flux, in *Multiphase Flow and Heat Transfer, Second International Symposium*, Vol. 1, X-J. Chen, T.N. Veziroğlu, and C.L. Tien, Eds., Hemisphere, New York, 358–366.

Zhong, L., Tan, Y., and Wang, S., 1992. Investigation of the heat transfer performance of mechanically made porous surface tubes with ribbed channels, in *Multiphase Flow and Heat Transfer, Second International Symposium*, Vol. 1, X.-J. Chen, T.N. Veziroğlu, and C.L. Tien, Eds., Hemisphere, New York, 700–707.

Zhou, X. and Bier, K., 1997. Pool boiling heat transfer from a horizontal tube coated with oxide ceramics, *Int. J. Refrig.*, 20, 552–560.

Zohler, S.R., 1990. Porous Coating for Enhanced Tubes, U.S. patent 4,890,669.

Zuber, N., 1958. On stability of boiling heat transfer, *Trans. ASME*, 80, 711–720.

VAPOR SPACE CONDENSATION

12.1 INTRODUCTION

This chapter is concerned with enhancement of vapor space condensation. Geometries include plates and tubes (horizontal and vertical). If the vapor flows in the direction of the draining condensate film, the interfacial shear stress will enhance condensation. Condensation with significant vapor velocity is called convective condensation, which is discussed in Chapter 14. Because the surface orientation will affect condensate drainage characteristics, one must distinguish between horizontal and vertical tube orientations.

The majority of enhancement techniques of practical interest are limited to the passive types. These include special surface geometries for enhancement of film condensation and nonwetting coatings or additives for promotion of dropwise condensation. Electric field enhancement of film condensation appears to be very promising, and is discussed in Chapter 15.

Condensation will occur on a surface whose temperature is below the vapor saturation temperature. The condensed liquid formed on the surface will exist either as a wetted film or in droplets. Droplets are formed if the condensate does not wet the surface. Although dropwise condensation yields a very high heat transfer coefficient, it cannot be permanently sustained. Dropwise condensation may be promoted by liquid additives or surface coatings that inhibit surface wetting. As the surface slowly oxidizes, the surface will eventually become wetted, and the process will revert to filmwise condensation. Hence, filmwise condensation is currently the more important process.

Study of the literature shows that surface tension effects are an important phenomenon in enhancement of film condensation. The technique classification in Table 1.1 lists rough and extended surfaces separately from surface tension devices. However, surface tension forces are probably the dominant mechanism for enhancement on rough and extended surfaces, in the absence of vapor velocity. The importance of surface tension forces in enhancement of film condensation was first described by Gregorig [1954]. However, its importance in affecting condensation on extended surfaces (e.g., integral-fin tubes) was not recognized until the late 1970s (e.g., Karkhu and Borovkov [1971]). Consequently, the separate classifications of extended surfaces and surface tension devices applied to film condensation is somewhat ambiguous. Here surface tension devices are more precisely defined as concepts that do not increase the surface area of the base surface. However, they may include loosely attached (poor thermal contact) devices, such as wires.

Since 1981, significant advances have been made in understanding the importance of surface tension in enhancement of film condensation on finned surfaces—for plates and horizontal tubes. The key advances involve understanding the role that surface tension force plays in draining the condensate from the fins, and in retaining condensate within the interfin region of finned tubes. This understanding has culminated in the development of analytical models for predicting the condensation rate on horizontal integral-fin tubes and on banks of integral-fin tubes. The models have been validated for both low- and high-surface-tension fluids. In addition, data have been obtained to establish the "row effect" on several high-performance integral-fin tube geometries. High-performance fin geometries have been identified, as well as techniques to model currently used fin geometries. Models have been developed to predict condensate retention and to account for the effect of drainage strips.

12.1.1 Condensation Fundamentals

The Nusselt [1916] analysis provided the foundation for the present understanding of laminar film condensation. By neglecting the convection terms in the energy equation, the thermal resistance across the condensate film, of thickness δ, is given by the Fourier heat conduction equation. Defining the condensation coefficient in terms of $(T_{sat}-T_w)$, one obtains

$$q = h(T_{sat} - T_w) = \frac{k_l(T_{sat} - T_w)}{\delta} \tag{12.1}$$

Hence, $h = k_l/\delta$. This equation is very simple. The difficult part is in determining the film thickness. The Nusselt analysis developed the equation for δ on a vertical, gravity drained plate, as illustrated in Figure 12.1. This analysis, for zero interfacial shear, is given in heat transfer textbooks. Neglecting the inertia terms and assuming $u = u(y)$, the momentum equation is

$$(\rho_l - \rho_v)g + \mu_l \frac{d^2u}{dy^2} = 0 \tag{12.2}$$

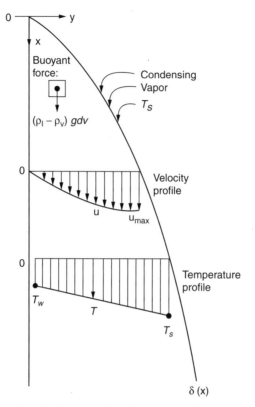

Figure 12.1 Illustration of gravity drained condensation on a vertical plate.

Combining Equation 12.2 with an energy balance on the condensate film, the Nusselt result for $1/\delta$ may be written as

$$\frac{1}{\delta} = \left(\frac{\lambda F_g}{4 \nu k_l x \Delta T_{vs}} \right)^{1/4} \tag{12.3}$$

where $F_g = (\rho_l - \rho_v)g$, is the gravity force per unit volume. Substitution of Equation 12.3 in Equation 12.1 gives the local condensation coefficient at location x from the top of the plate.

$$h = 0.707 \left(\frac{k_l^3 g(\rho_l - \rho_v)\lambda}{\nu_l \Delta T_{vs} x} \right)^{1/4} \tag{12.4}$$

The average condensation coefficient (h) on the plate of length L is obtained by integrating Equation 12.4 over $0 \le x \le L$, to obtain

$$\bar{h} = 0.943 \left(\frac{k_l^3 g(\rho_l - \rho_v)\lambda}{v_l \Delta T_{vs} L} \right)^{1/4} \tag{12.5}$$

Equation 12.5 may be written in terms of the condensate Reynolds number (Re$_l$ = $4\Gamma \mu_l$) as shown in heat transfer textbooks (e.g., Incropera and DeWitt [1990]). The Γ is the condensate mass flow rate per unit wetted perimeter (P_w). The Reynolds number form of Equation 12.5 is

$$\bar{h} = 1.47 \left(\frac{k_l^3 \rho_l (\rho_l - \rho_v)g}{\mu_l^2} \right)^{1/3} \mathrm{Re}_L^{-1/3} \tag{12.6}$$

Equation 12.6 is more convenient for a sizing problem, where the required tube length (L) is unknown.

Interfacial shear stress (τ_i) will cause ripples in the condensate film, thereby yielding a higher condensation coefficient. For $\tau_i = 0$, the condensate film will transition to a turbulent film. Incropera and DeWitt [1990] give equations for rippled and turbulent films. For $\tau_i \gg 0$, transition to a turbulent film may occur at Re$_l$ as small as 300. Collier and Thome [1994] provides equations for these more complex situations.

Nusselt also derived equations for laminar film condensation on inclined plates and horizontal tubes. For an inclined plate, the gravity force component, in the direction of the plate, is $g \sin \theta$, where θ is the plate angle from the vertical. So, the average condensation coefficient on an inclined plate is given by Equation 12.5 with g replaced by $g \sin \theta$. To obtain the equation for horizontal tubes, Nusselt integrated the inclined plate equation over $0 \leq \theta \leq \pi$ to obtain

$$\bar{h} = 0.728 \left(\frac{k^3 g(\rho_1 - \rho_v)\lambda}{v_l \Delta T_{vs} d} \right)^{1/4} \tag{12.7}$$

Nusselt extended his analysis of horizontal tubes to predict the row effect. For a vertical rank of horizontal tubes, each successive tube row will receive condensate generated on the upper tube rows. Nusselt showed that the average condensing coefficient on N-tube rows (h_N) related to the condensing coefficient on a single horizontal tube by the relation

$$\frac{\bar{h}_N}{h_1} = N^{-1/4} \tag{12.8}$$

Equation 12.8 shows that the condensation coefficient on a gravity–drained bank of horizontal tubes should decrease as $N^{-1/4}$. Equation 12.8 assumes no interfacial shear and no mixing of the condensate, which is unrealistic for actual tube bundles.

The row effect in actual tube banks may be substantially less than that predicted by the Nusselt model. This is because of splashing and mixing in the liquid film. More discussion of the row effect is given in Section 12.6.2.

12.1.2 Basic Approaches to Enhanced Film Condensation

As shown by Equation 12.1, the thermal resistance in film condensation is that of conduction across the condensate film. The local film thickness is determined by the force that drains the condensate. Equations 12.3 through 12.8 assume that gravity force drains the condensate film. Other possible drainage forces are surface tension, suction, and centrifugal force. Any technique that yields a reduced film thickness will enhance the film condensation coefficient.

Therefore, a surface geometry that promotes reduced film thickness will provide enhancement. Short, vertical fins on horizontal tubes will have a smaller film thickness than on the base tube, thus providing enhancement.

An alternative to gravity-drained films is the use of surface tension forces for condensate removal. An example is the vertical fluted tube proposed by Gregorig [1954]. This tube has axial fins of a special shape, and is discussed later. In a gravity-drained condenser, the lower portions of the bundle suffer from condensate inundation or flooding. A number of mechanical means may be envisioned to remove the accumulated condensate, which would allow reduced condensate film thickness in the lower portion of the condenser. Similarly, when condensation occurs inside a serpentine tube circuit, one may envision possibilities for condensate removal.

In vapor shear–controlled condensation, high vapor velocity will provide positive effects due to interfacial shear or condensate entrainment. This contributes an exponent in Equation 12.8 less than the 1/4 value predicted by the Nusselt theory, which assumes 0 interfacial shear. In shear-controlled flow, enhancement may be provided by reducing the cross-sectional vapor flow area as condensation proceeds to lower vapor qualities. This is possible for tube-side condensation in a multipass design by reducing the number of tubes in parallel in each succeeding pass. Similar techniques are possible for shell-side condensation in large tube bundles. These concepts are discussed in detail in Chapter 14.

When noncondensibles are present, an additional thermal resistance is introduced in the vapor at the vapor–liquid interface. Mixing of the gas film will substantially reduce this thermal resistance. The maintenance of high vapor velocities, or special surface geometries, which promote a higher heat transfer coefficient in this gas film will substantially alleviate the performance deterioration due to noncondensibles.

Surface roughness may also provide mixing within the condensate film. However, this will not be effective if the film is laminar.

12.2 DROPWISE CONDENSATION

If surface wetting can be prevented, high-performance dropwise condensation will occur. Griffith [1985] reviews recent advances and presents an excellent discussion of the expected performance and practical aspects of applying dropwise condensation to steam condensers. Carey [1992] provides an analytical treatment of the proposed mechanisms of dropwise condensation. Two basic techniques for promoting dropwise condensation exist, namely, nonwetting surface coatings and chemical additives. Because low-surface-tension fluids more easily wet a surface than do high-surface-tension fluids, steam is a much more viable candidate for promotion of dropwise condensation than are the refrigerants or many organics. This is unfortunate, because the condensation coefficients for the second group of fluids are substantially less than that for steam.

Iltscheff [1971] condensed R-22 on several coated horizontal tubes (intended to promote dropwise condensation) and found that the condensation coefficients were below or approximately equal to those predicted by the Nusselt model for film condensation. Iltscheff implies that he observed condensation in the dropwise mode. This raises a question that has plagued many experimental studies of dropwise condensation: Were noncondensible gases present? The presence of noncondensible gases may substantially offset the potential enhancement provided by dropwise condensation.

The majority of the experimentation has been performed with steam. Successful surface coatings include noble metals and plastic (e.g., PTFE) and chemical additives (e.g., oleic acid). Chemical additives may be effective for up to 1000 hr, and then require surface cleaning and re-promotion for further effectiveness. When plastic coatings are used, one must account for the additional thermal resistance of the coating. This may be 60% as large as the resistance associated with film condensation. For a typical steam condenser design, Hanneman [1977] shows that dropwise condensation may provide an order of magnitude smaller steam-side resistance, resulting in a 40% surface area reduction for titanium tubes. If the thermal resistance of a 0.0015-mm-thick PTFE coating is included, the surface area saving is reduced to only 10%. Kim et al. [2001] provide updated information on use of additives. Working with a 12-tube bundle and 2-ethyl-1-hexanol, they obtained a 30% improvement in condensation rate.

More recently (e.g., since 1994), new surface treatment techniques have emerged. Das et al. [2000a] provide a brief review of recent work. They conclude that organic coatings are difficult to maintain unless the coating is relatively thick. Zhao and Burnside [1994] describe an ion implantation method they describe as "activated reactive-magnetron sputtering ion plating" (ARE-MSIP). Koch et al. [1998] describes a plasma-enhanced CVD method of forming "amorphous layers of hydrogenated carbon" (or diamond-like carbon, DLC). Further increase of contact angle can be produced by adding fluorine, silicon, or silicon/oxygen during the CVD process. The coatings show excellent chemical inertness and resistance to acids, alkalis, and solvents. The measured contact angle approached 90° and the coating thickness was 2 to 4 μm. Steam condensation tests on a vertical plate showed an enhancement level 11 times than for film condensation. Leipertz et al. [2002] tested ion implantation–induced dropwise condensation of steam on vertical surfaces of several different metals. They report heat transfer coefficient enhancement up to 17 times that of a plain surface. The dropwise condensation showed long-term stability, with dropwise operation maintained for more than 4 years. Ma et al. [2002] describe the influence of processing conditions for a PTFE ion beam implantation technique on steel and brass alloys. Lixin and Jiehui [1998] electroplated Ni using an electrolyte bath containing small PTFE particles. The resulting composite-plated surface had 92% Ni and 6.8% PTFE (by weight). An SEM photo of the surface showed that the PTFE particles were finely distributed in the surface structure. Their steam condensation tests on a vertical plate showed an enhancement factor of approximately 3.5. They claim that the plated surface has good adhesion and is durable.

Das et al. [2000a, 2000b] investigated self-assembled monolayers (SAMs), which are composed of a single layer of organic molecules adsorbed onto a surface to form a coating. The coating provides strong covalent bonds between alkythiol and the metal substrate. Because it is only a monolayer thick, the coating introduces negligible thermal resistance. They applied the SAM to copper and Cu-Ni tubes and tested horizontal tubes under vacuum steam condensation, which yielded a five-factor enhancement. Relatively large condensate drops remained on the bottom surfaces of the tubes, which reduced the enhancement potential. Drainage strips (see p. 487) would likely be beneficial, but were not investigated. Long-term tests are required to validate the durability of this approach.

Abu-Orabi [1998] briefly surveys models to predict dropwise condensation and developed a model to predict the drop size distribution for small drops that grow by direct condensation. The resistances to heat transfer due to the drop (conduction through the drop, vapor–liquid interfacial resistance, drop curvature), due to the promoter layer, and the sweeping effect of falling drops are incorporated into the model and are also included in calculating the heat transfer rate through a single drop. The total heat flux is calculated from the drop size distributions and the heat transfer rate through a single drop. Drop size distribution, for large drops that grow by coalescence, is included from previously published work of others. The analysis shows that to calculate the heat flux adequately, all the resistances to heat transfer, due to the drop and the promoter layer, must be included. The heat flux will be overestimated if one considers only heat conduction through the drop as the only resistance to heat transfer. The amount of overestimation increases as the temperature difference increases.

Most dropwise condensation studies have been performed on vertical plates or single horizontal tubes. If applied to a tube bundle, the effects may be reduced due to inundation and vapor shear effects, which may cause reversion to film condensation.

12.3 SURVEY OF ENHANCEMENT METHODS

This section discusses film condensation on vertical plates and tubes and on horizontal tubes. Each enhancement technique is separately discussed.

12.3.1 Coated Surfaces

Figure l2.2 shows the cross section of a vertical condensing surface on which nonwetting strips are attached. Theoretical predictions for this geometry have been performed by Cary and Mikic [1973] and Brown and Matin [1971]. Because the condensate film is thinned near the nonwetting strips, high heat transfer coefficients should prevail. Brown and Matin show that the enhancement is dependent on the liquid contact angle and the thermal conductivity of the base surface. Cary and Mikic reason that an additional enhancement mechanism will be present due to a surface tension induced secondary flow. They show that the liquid surface temperature will be reduced near the nonwetting strip, resulting in a surface tension gradient causing the secondary flow (e.g., the Marangoni effect). The predicted results of Cary and Mikic and of Brown and Matin do not appear to be consistent. Accounting for the secondary flows, Cary and Mikic predict much lower enhancement levels than those of Brown and Matin. Cary and Mikic conclude that at best, "modest" enhancement levels can be expected (e.g., 30% or less). Glicksman et al. [1973] measured the effect of nonwetting Teflon strips for steam condensing on a horizontal tube. They helically wrapped 3.2-mm-wide, 0.16-mm-thick tape on a 12.7-mm-diameter copper tube and tested two strip spacings: $p/d_0 = 3$ and 6. The $p/d_0 = 3$ wrap gave a 35% increased condensation coefficient. However, greater enhancement was obtained with a simple axial tape strip along the bottom of the tube. This yielded a 50% enhancement level. Addition of the helical wrap with the bottom axial tape strip did not provide further enhancement.

A U.S. patent by Notaro [1979] describes a coated surface geometry for enhanced film condensation. It consists of an array of small-diameter metal particles bonded to the tube surface. The particles are 0.25 to 1.0 mm high covering 20 to 60% of the tube surface. Condensation occurs on the particle array and drains along the smooth base surface. High condensation rates occur on the convex surfaces of the particles, due to surface tension forces, which maintain very thin condensate films on the particles. Figure 12.3 shows a photograph of the surface and illustrates the thinned condensate films on the particles. The patent provides performance data for several fluids condensing on vertical tubes. For a given particle height, there is an optimum particle spacing. A 6.0-m-long vertical tube having a 50% area density of 0.5-mm-diameter particles yielded a steam condensation coefficient 17 times that predicted by the Nusselt equation for an equal-length smooth tube.

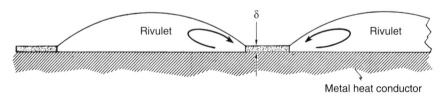

Figure 12.2 Cross section of a vertical condensing surface on which nonwetting strips are attached. (From Brown and Matin [1971].)

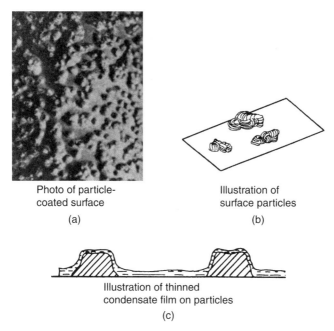

Photo of particle-
coated surface

(a)

Illustration of
surface particles

(b)

Illustration of thinned
condensate film on particles

(c)

Figure 12.3 Enhanced condensation surface formed by small-diameter metal particles bonded to the base surface. (a) Actual surface, (b) particles bonded to surface, (c) thin condensate film on particles and thick condensate film on base surface. (From Notaro [1979].)

A porous coating on the base surface can be a very effective enhancement method for film condensation. Condensate drainage is assisted by capillary flow within the porous coating, resulting in a thinning of the condensate film thickness. Because the temperature drop across a laminar condensate film is the condensation thermal resistance, such capillary-assisted film thinning is effective in reducing the condensate thermal resistance. Consider, a gravity drained laminar condensate film on an uncoated plate, for which the $h = \delta/k_l$. If a porous coating is applied to the plate, the resulting thermal resistance is $h = e/k_{p,e} + \delta_{mod}/k_l$, where the first term is the thermal resistance of the porous film of thickness e and effective thermal conductivity $k_{p,e}$. The δ_{mod} is the reduced film thickness resulting from condensate suction into the porous coating. The enhancement ratio (E_o) is

$$E_0 = \frac{e/k_{p,e} + \delta_{mod}/k_l}{\delta/k_l} \qquad (12.9)$$

where $k_{p,e}$ is the effective thermal conductivity of the liquid saturated porous coating. Ma and Wang [1998] performed analytical analysis of steam condensation on a vertical plate having a thin porous coating. Equation 17 in their analysis provides the solution for δ_{mod}, which is dependent on the suction velocity (V_o) at the porous interface surface. The V_o is given by

$$V_o = \frac{eu_D}{L\varepsilon} \qquad (12.10)$$

where $u_D = K(\rho_l - \rho_v)/\mu_l$, for which u_D is the Darcy velocity (the mean liquid velocity within the porous coating), ε is the coating porosity, and K is its permeability. For a given plate length (L) and fluid properties, the suction velocity $V_o \propto eK\varepsilon$. Enhancement increases with increasing V_o and increased coating thickness—up to a coating thickness of 50 μm. Above 50 μm, the thermal resistance within the porous layer becomes dominant. Wang et al. [2000] obtained condensation data on a "fluted tube" having a thin porous coating.

Renken and Aboyoe [1993a] also provide analysis of condensation on porous surfaces. Renken and Aboyoe [1993b] and Renken and Mueller [1993] obtained steam condensation on porous coated surfaces with coating thickness of 25 to 250 μm and porosity of approximately 50%. Their steam condensation tests showed that the highest enhancement was provided by the thinnest coating. Renken and Raich [1996] extended the measurements to include effects of forced convection steam. Further discussion is given in Section 12.3.3.

12.3.2 Roughness

Medwell and Nicol [1965] and Nicol and Medwell [1966] investigated enhancement due to a closely knurled roughness for a condensate film flowing down a vertical surface. The knurled roughness provides mixing in the condensate film, and hence increases the condensing coefficient. Nicol and Medwell [1966] present a theoretical treatment of the problem. They apply the heat-momentum analogy for roughness developed by Dipprey and Sabersky [1963], which is discussed in Chapter 9. Their theory shows that, the benefits of roughness are characterized by the roughness Reynolds number, $e^+ = eu^*/\nu_l$. For constant film thickness (δ), increasing roughness height increases e^+, which reduces the thermal resistance of the viscous influenced region. This effect continues for larger roughness sizes up to $e^+ = 55$, at which point viscosity no longer influences the thermal resistance. As the condensate film thickness increases, for constant roughness height, the e^+ will decrease, causing viscous effects to influence the thermal resistance. The film may become so thick that the roughness characteristic approaches the hydraulically smooth condition ($e^+ = 5$), where no effective mixing occurs in the roughness zone. Thus, the enhancement produced by a given roughness size is directly related to the condensate film thickness. Their theory shows that the maximum possible enhancement is approximately 100%. The steam condensation tests of Nicol and Medwell [1966] offer good support for the theory. The largest roughness ($e = 0.5$ mm) yielded an enhancement of approximately 90%.

12.3.3 Horizontal Integral-Fin Tubes

Horizontal integral-fin tubing illustrated in Figure 12.4 has found wide commercial acceptance for condensation on horizontal tubes. These tubes are commercially available with 433 to 1575 fins/m (11 to 40 fins/in.). Integral-fin tubes provide substantial performance improvement over plain tubes, particularly for low-surface-tension fluids, which use 748 to 1575 fins/m. This occurs because of the area increase provided by the fins and because of the thin condensate films formed on the short fins. The thin condensate films are primarily due to surface tension effects, which are discussed in detail in Section 12.4. Table 12.1 shows the enhancement ratio determined by Webb et al. [1985] for R-11 condensing on 19-mm-diameter (d_e) commercial integral-fin tubes. The surface area of the integral-fin tubes is based on the envelope diameter over the fins, $A/L = \pi d_e$. The table shows that the enhancement ratio increases with increasing fins/m. The 1378 fins/m tube provides an enhancement ratio of 5.28 compared to a 19-mm-diameter plain tube.

Substantial data have been reported for condensation on integral-fin tubes. Marto [1988] presents an excellent survey of the earlier work. Considerable work has been done since the first edition of this book, and much of the existing data are chronologically listed in Table 12.2. A more recent very good survey is by Browne and Bansal [1999]. A summary of work since 1994 is given below.

Wen et al. [1994] investigated the effect of the fin root radius and found improvements up to 15% for the higher-surface-tension fluids (e.g., water) but little benefit for low-surface-tension fluids (e.g., R-113). Das et al. [1995] measured steam condensation (vacuum and 1.0 atm) on copper, aluminum, CuNi, and stainless steel tubes using 0.5 to 1.5 mm fin heights for 500 and 667 fins/m. Low tube thermal conductivity can significantly decrease tube performance, especially for the higher fins. They found that the Briggs and Rose [1994] model predicted the vacuum data

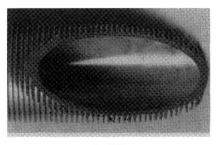

26 FPI

Figure 12.4 Horizontal, integral-fin tube.

Table 12.1 Enhancement Ratio for R-11 Condensing on Integral-Fin Tubes (d_e = 19 mm, T_s = 35°C, ΔT_{vs} = 9.5 K)

Fins/m (m⁻¹)	e_o (mm)	h (mm)	h/h_p (W/m²-K)
748	1.5	8,070	2.64
1024	1.5	11,970	3.91
1378	0.9	16,140	5.28

Table 12.2 Published Data for Condensation on Integral-Fin Tubes

Fins/m	Fluids	Ref.	Comment
608	Methyl chloride, sulfur chloride, R-22, *n*-pentane, propane	Beatty and Katz [1948]	First model for integral-fin tubes
630	R-22, *n*-butane, acetone, water	Katz and Geist [1948]	Earliest published reference on integral fin tubes
396–770	R-12	Henrici [1961]	
748–1024	R-22	Pearson and Withers [1969]	
748–1024	R-22	Takahashi et al. [1979]	
1060–1610	R-11	Carnavos [1980]	
1020–2000	R-113, methanol	Honda et al. [1983]	
95–1000	Water	Yau et al. [1985]	
100–667	Water	Wanniarachchi et al. [1986]	
400–1000	R-113	Masuda and Rose [1987]	
200–1333	Water	Marto et al. [1988]	
1417	R-11	Sukhatme et al. [1990]	
400–667	R-113	Briggs et al. [1992]	
1000	R-113	Wang et al. [1990]	
1026	R-113	Honda et al. [1994]	Effect of fin tip and fin root radius; exp. and pred.
500, 667	R-113, steam, eth. glycol	Briggs and Rose [1994]	Effect of fin root radius
1333, 1575	R-113	Cavallini et al. [1994]	Visualization of drainage on circular integral-fin and spine fin with vapor velocity
787, 1212, 1490	R-152A	Cheng and Tao [1994]	Circular and spine integral fins
667	R-113	Briggs et al. [1995]	Effect of fin height and tube thermal conductivity
1575, 2000	R-11, R-113	Cavallini et al. [1995]	Refrigeration condensation data
500, 664	Steam	Das et al. [1995]	Effect of fin height and tube thermal conductivity
1333	R-123	Honda and Makishi [1995]	Effect of circumferential ribs on fin side; theo. pred.
450–1923	R-113	Briggs and Rose [1995]	Data on 12 commercial integral-fin tubes; includes T- and Y-shaped fins
200–2000	R-11, R-113, steam	Cavallini et al. [1996]	Model for effect of vapor velocity
300–550	Steam	Jaber and Webb [1996]	Effect of fin geometry, fin pitch, and tube thermal conductivity
1024–1640	R-22	Cheng et al. [1996]	Data on six commercial finned tubes
748–2000	R-123	Sreepathi et al. [1996]	Effect of fin pitch and fin height

Table 12.2 (Continued)

Fins/m	Fluids	Ref.	Comment
242–1366	R-12	Gogonin and Kabov [1996]	Data for effect of fin radius
1024, 1654	R-11, R-12	Jung et al. [1999]	Circular fins and Turbo-C
1000	Steam	Liu et al. [1999]	Circular fin tube with wire mesh drainage strip
400	Steam	Das et al. [1999]	Effect of fin height for stainless steel tube
934	R-134a	Kumar et al. [2000]	Circular fins
390–1875	Steam, R-12, R-134A	Kumar et al. [2000]	Circular fins and spine fins
Wire-wrapped tube	Steam	Briggs et al. [2002, 2003]	1–6 mm wire pitch, 0.2, 0.4, 0.75, 1.0 wire diameter
390	Steam	Singh et al. [2001]	Four tubes in a vertical rank
390, 1560	Steam, R-134a	Kumar et al. [2002a]	Circular fins and spine fins; fins on total or only upper or lower half of tube

exp. = experimental; pred. = prediction; theo = theoretical

well, but overpredicted the low-thermal-conductivity stainless steel tube data ($k = 14$ W/m-K) 1.0 atm data as much as 30%. Honda and Makishi [1995] numerically investigated the effect of adding one or two low profiled ribs on the fin flank (1.4-mm fin height) of an integral-fin tube for R-123. They found that 1 rib provided 27% enhancement and 2 ribs provided 54% enhancement. Although this may be a difficult geometry for circular finned tubes, it may be very practical for finned plates. Briggs and Rose [1995] condensed R-113 on 12 commercial integral-fin tubes, including tubes having T and Y cross section fins. They found that the T and Y cross section fins yielded lower performance than the conventional trapezoidal fin cross-sectional shape. They also found that the Rose [1994] model satisfactorily predicted the data for the trapezoidal fin shape.

Condensate Retention. The integral-fin tubes listed in Table 12.1 are routinely used for condensation of low-surface-tension fluids. However, they will not be effective for high-surface-tension fluids, such as water (steam). This is because capillary (surface tension) force retains condensate between the fins on the lower side of the tube, where the condensate thickness is $\delta = e_o$ (fin height). Since $h = k_l/\delta$, the condensation coefficient in this condensate-flooded zone is very small. Figure 12.5 shows that the fins are condensate flooded over an angle β. Rudy and Webb [1983, 1985] and Honda et al. [1983] show that the condensate retention angle for fins of rectangular cross section is given by

$$\beta = \pi c_b = \cos^{-1}\left(1 - \frac{4\sigma}{d_o \rho_l g s}\right) \qquad (12.11)$$

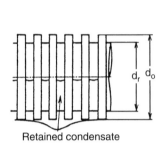 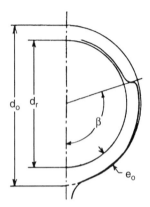

Figure 12.5 Condensate flooding angle (β) on horizontal integral-fin tubes.

Equation 12.11 is based on a force balance between gravity and surface tension forces on the condensate. Equation 12.11 shows that the condensate retention angle (β) increases with increasing surface tension (σ) and with decreasing fin spacing (s). Condensation of steam on a 19-mm-diameter tube having 0.25-mm fin thickness would result in total flooding if the fin density were greater than 1000 fins/m. Jaber and Webb [1993] show that integral-fin tubes for steam condensation should use no more than 630 fins/m (16 fins/in.).

The amount of condensate retained between the fins may be reduced by attaching drainage strips to the bottom of the tube. A drainage strip is simply a thin plate of height H_d that hangs from the bottom of the tube. This alters the balance between surface tension and gravity, which establishes the condensate retention angle. Addition of the drainage plate adds a surface tension force in the downward direction. Because of the added downward surface tension force, less condensate is retained on the tube. Honda et al. [1983] present theoretical relations to predict the condensate retention angle for finned tubes with drainage strips. Their theory shows that if the drainage strip is made of a porous material, the capillary condensate removal force will be increased. Table 12.3 shows the effect of drainage strips for R-113 condensing on an 18.9-mm-diameter tube (diameter over fins) having 2000 fins/m and 1.13-mm fin height, as measured by Honda and Nozu [1987b]. Two drainage strip materials were used. The first was polyvinyl chloride (PVC), and the second was porous nickel having a permeability of 1.7E-9 m^2. For a strip height (H_d) of 12.6 mm, the PVC strip reduced c_b (fraction circumference flooded) by 9%, and the porous strip reduced c_b by 51.6%. The shorter 4.0-mm porous strip reduced c_b by 33%. As shown in Table 12.3, the 14.6-mm porous strip provided a 65% increase in the condensation coefficient, as opposed to only 11% for the PVC strip of the same height. The improvements are solely due to reducing the condensate retention angle. Additional data on integral-fin tubes having drainage strips are provided in Honda and Nozu [1985].

Liu et al. [1999] tested 10 different drainage strip designs for steam condensation data on a horizontal 17-mm-OD tube having 1000 fin/m and 0.75-mm fin height.

Table 12.3 Effect of Drainage Strips for Methanol Condensing on 2000 Fins/m Tube at $(T_s - T_w) = 5$ K

Strip	H_d (mm)	c_b	h (W/m²-K)
None	—	0.62	6200
PVC	12.6	0.57	6900
Porous	12.6	0.32	10,200
Porous	4.0	0.42	7500

The different strips were made using multiple layers of wire mesh. The porosity characteristics of the drainage strips were characterized by the measured minimum meniscus radius and the calculated permeability. They analytically modeled condensate flow in the flooded interfin region of the tube and through the drainage strip. They show that a good porous drainage strip should have small meniscus radius and high permeability. Increased length of the drainage strip is beneficial only in a limited range. They combined the drainage strip with a model for condensation on integral-fin tubes and showed excellent ability to predict the condensation coefficient on 12 different tubes for 3 fluids (steam, R-113, and methanol).

Although significant enhancement is provided by porous strips, it is difficult to envision how such drainage strips may be employed in an actual condenser tube bundle. They would be added after the tube is inserted in the bundle, which is cumbersome. Sufficient vertical clearance between tubes must be provided to accommodate the strip height. These strips would act as barriers to horizontal vapor flow. Hence, the vapor must flow downward within the bundle. Vapor shear would occur on the drainage strips, which may increase the vapor pressure drop. However, the strip may act as a "splitter plate," which decreases the vapor pressure drop.

Yau et al. [1986] have measured the effect of drainage strips on steam condensation. Using an 8.0-mm-high copper drainage strip on a 15.9-mm-OD tube having 667 fins/m, they found a 30% condensation enhancement. They also showed that the same drainage strip provided little benefit for a plain tube.

Advanced Surface Geometries. Integral-fin tubes having a sawtoothed fin shape have been developed, which have higher condensing coefficients than the standard integral-fin tubes. Figure 12.6a, c, and d shows three such sawtooth fin geometries, which are commercially used in refrigerant condensers. Figure 12.4 is the standard integral-fin tube (1024 fins/m, 26 fins/in.), and Figure 12.6b has a Y-shaped fin tip. The geometries of these tubes are given in Table 12.4.

The Figure 12.6d Tred-26D fin has the same height and fin pitch as the Figure 12.4 standard 1024 fins/m (26 fins/in.) integral-fin tube. The notch depth in the Tred-26D fin tip is approximately 40% of the fin height. Test data have been reported for refrigerants condensing on copper tubes for the Figure 12.4 and Figure 12.6 geometries. Figure 12.7a shows the R-11 condensation coefficient data of Webb et al. [1985] for three commercially used integral-fin tubes, which have 748, 1024, and 1378 fins/m. Figure 12.7b shows the R-11 data of Sukhatme et al. [1990] for four

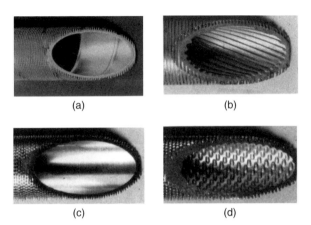

(a)   (b)

(c)   (d)

Figure 12.6 Enhanced condensing tubes. (a) Hitachi Thermoexcel-C, (b) Wieland GEWA-SC, (c) Wolverine Turbo-C, (d) Sumitomo Tred-26D.

Table 12.4 Dimensions of the Figure 12.4 and 12.6 Tubes (All Dimension in mm)

Tube Type	Figure	d_o (mm)	e_o (mm)	Fins/m
Integral-fin	12.4	18.9	1.3	1024
Thermoexcel-C	12.6a	18.9	1.2	1378
GEWA-SC	12.6b	18.9	1.3	1024
Turbo-C	12.6c	18.9	1.1	1575
Tred-26D	12.6d	18.9	1.3	1024

1417 fins/m tubes having fin heights between 0.46 and 1.22 mm. Note that the condensation coefficient is based on the envelope area over the fins, $\pi d_e L$. Figure 12.7a shows that the condensation coefficient increases with decreasing fin pitch. For the same fin pitch, Figure 12.7b shows that the condensation coefficient increases with increasing fin height.

Webb and Murawski [1990] measured the R-11 condensation coefficient for the Figure 12.6b, through d and Figure 12.4 tube geometries, which are shown in Figure 12.8. Figure 12.8 shows that the highest single tube performance is provided by the Turbo-C tube, and the lowest given by the standard 1024 fins/m tube. The condensation coefficient of the standard 1024 fins/m tube is 60% that of the Turbo-C. The second and third best performance is given by the GEWA-SC (80% of Turbo-C) and Tred-26D (72% of Turbo-C) tubes, respectively.

Wang et al. [1990] report the R-113 performance of a unique finned tube having lateral ripples in the fins. The tube had 1000 fins/m with 1.47 mm high fins. Their data show that the condensation coefficient is 30 to 40% higher than that of the Thermoexcel-C tube. They propose that surface tension force pulls the condensate into the valleys of the ripples.

No analytical models have been specifically developed to predict the effect of the notched fin tip geometry. However, surface tension drainage from the sawtooth

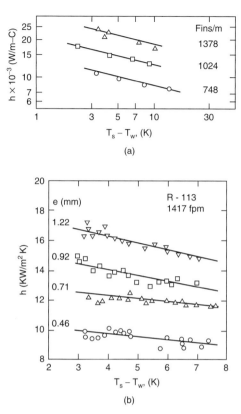

Figure 12.7 (a) R-11 condensation coefficient on standard copper integral-fin tubes (748, 1024, and 1378 fins/m) as reported by Webb et al. [1985], (b) R-11 condensation coefficient on 1417 fins/m copper integral-fin tubes with fin heights between 0.46 and 1.22 mm as reported by Sukhatme et al. [1990].

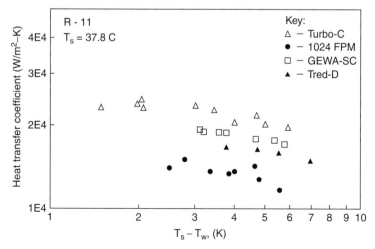

Figure 12.8 R-11 condensation coefficient on copper enhanced horizontal tubes. (From Webb and Murawski [1990].)

fins is expected to account for the higher performance of the fin. A small tip radius establishes a surface tension gradient to affect condensate drainage.

Steam Condensation. Figure 12.9 shows six tube geometries consciously developed for steam condensation. The first four are integral-fin tubes, and the fifth is the attached particle tube shown in Figure 12.3. The sixth tube (Figure 12.9f) is discussed by Wildsmith [1980]; it has a corrugated internal surface ($p = 6.4$ mm, $e_i = 0.7$ mm), and fine triangular threads ($e_o = 0.3$mm, $p_f = 1.0$ mm) on the outer surface. Wildsmith reports that the Figure 12.9f tube provides $UA/U_p A_p = 1.47$ for the same water-side pressure drop. The Figure 12.9 tube dimensions are listed in Table 12.5. Note that they typically have a larger fin pitch than the tubes listed in Table 12.4. Only one of the Table 12.5 tubes is made of copper. Electric utility steam

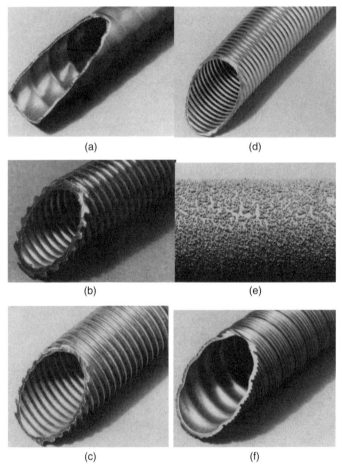

(a)   (d)

(b)   (e)

(c)   (f)

Figure 12.9 Horizontal integral-fin tubes developed for steam condensation by Jaber and Webb [1993]. (a) Wolverine Korodense, (b) copper Wieland 11-NW, (c) copper-nickel Wieland 11-NW, (d) stainless steel Wieland NW-16, (e) UOP attached particle tube, (f) Yorkshire MERT (multiply enhanced roped tube).

Table 12.5 Dimensions of the Figure 12.9 Tubes

Tube Type E_o	Figure	Fins/m	Material	d_o (mm)	e_o (mm)	t_b (mm)	t_t (mm)
Wolverine Korodense	12.9a	None	Cu/Ni	22.2	NA	NA	NA	NA
Wieland NW-11C	12.9b	433	Cu	19.0	1.1	0.9	0.3	2.80
Wieland NW-11C/N	12.9c	433	Cu/Ni	22.2	1.1	0.9	0.3	1.75
Wieland NW-16SS	12.9d	630	S Stl.	18.9	0.3	1.2	0.8	1.70
UOP A/P-50	12.9e	—	Cu/Ni	22.2	0.5	—	—	1.70
Yorkshire MERT	12.9f	1000	Cu/Ni	25.4	0.3	0.3	0.05	2.00 (est)

NA = not applicable.

condensers do not use copper tubes. Rather, they use 90/10 Cu/Ni, stainless steel, or titanium. These materials have much lower thermal conductivity than copper, so fin efficiency becomes a limiting factor, especially for steam condensation. As shown in Table 12.5, quite small fin heights are required. The Table 12.5 tubes were designed to have a thicker fin thickness at the base (t_b) than at the tip (t_t) to increase the fin efficiency. The thermal conductivities of the tube materials are: Cu/Ni (k_w = 45 W/m-K), stainless steel (k_w = 14 W/m-K), and titanium (k_w = 22 W/m-K).

The last column of Table 12.5 shows the outside enhancement ratio for steam condensation at (T_s-T_w) = 2.3 K and T_s = 54°C. The E_o of the copper NW-11C is much higher than the NW-11C/N tube, which has the same fin geometry; the difference is due to the fin efficiency. The 11C/N, 16SS, and A/P-50 tubes all provide condensation enhancement in the range of 1.64 to 1.75.

All the tubes in Tables 12.4 and 12.5, except the standard integral-fin, have tube-side enhancement, which is for the tube-side water flow. Adding the shell-side enhancement results in a reduction of the inside diameter. For tube-side flow at fixed velocity, $\Delta p/L \propto (d_i)^5$. Hence, any reduction of the tube inner diameter will significantly increase the tube-side pressure drop. Efforts to reduce the external fin height will benefit the tube-side pressure drop. Note that the outside fin heights are typically lower in Table 12.5 than in Table 12.4.

Considerable recent work has been done on integral-fin tubes for steam condensation. Among the major works are Jaber and Webb [1993, 1996], Briggs and Rose [1995, 1996], Briggs et al. [1995, 2002], Wen et al. [1994], Das et al. [1995, 1999], Renken and Raich [1996], Liu et al. [1999], Singh et al. [2001], and Kumar et al. [2000, 2002a,b]. Figure 12.10 shows the data of Mitrou [1986] for steam condensing on a 13.7-mm-diameter tube having 472 fins/m with a triangular screw thread geometry with 1.0-mm-high fins. Note the significant decrease of performance as the tube thermal conductivity decreases. Figure 12.11 shows predictions for the effect of fin pitch and material conductivity by Jaber and Webb [1996] for steam condensing at 34°C on a 22.23-mm-OD tube for different tube materials. Above a certain fins/m, the E_o decreases for all tube materials. This occurs because of increased condensate flooding. For copper and admiralty, the maximum E_o occurs at in the range of 475 to 515 fins/m for all materials. The figure clearly shows the significant

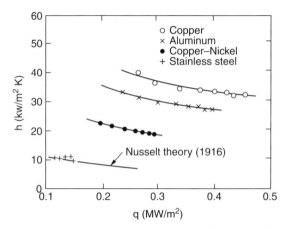

Figure 12.10 Steam condensation on 13.7-mm-diameter tubes having 1.0-mm-high, triangular-shaped fins at 472 fins/m. (From Mitrou [1986].)

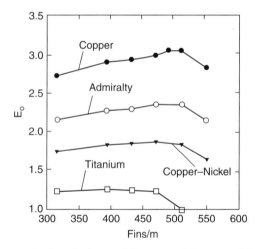

Figure 12.11 Enhancement level vs. fins/m, $e = 1.0$ mm, $t_b = 0.9$ mm, $t_t = 0.2$ mm, $D_o = 22.23$ mm. (From Jaber and Webb [1996].)

effect of material thermal conductivity. Note that fin pitch (below 475 fins/m) has little effect for the lower thermal conductivity materials. Additional predictions provided for smaller fin height (0.5 mm) show that the lower fin height yields higher performance for low conductivity titanium tubes. The predictions are based on a modification to the Adamek and Webb model [1990a] to account for the material thermal conductivity. Das et al. [1999] provide data on stainless steel integral fin tubes (rectangular fins) for 1.5-mm fin spacing and heights between 0.2 and 1.5 mm. Kumar et al. [2002a] provide condensation data on both circular fin (390 fins/m) and spine fins of copper. Singh et al. [2001] provide row-effect data on the circular fin tube tested by Kumar et al. [2002a]. Renken and Aboyoe [1993b] and Renken and

Raich [1996] measured steam condensation on a vertical and inclined porous coated surface, for which the second study investigated vapor velocity effects.

Noncondensible gases can severely reduce the condensation rate, if vapor velocity is low. Abdullah et al. [1995] measured steam and R-113 condensation in the presence of a noncondensible gas on a bank of integral-fin tubes.

Rabas and Taborek [1999] provide a study on the heat-rate improvement that can result from use of the Korodense tube (Figure12.9a) in an operating electric utility condenser. They report that this tube is used in operating condensers in several U.S. and international electric utility locations.

12.3.4 Corrugated Tubes

Figure 12.12 shows three basic types of commercially available corrugated tubes. These tubes also provide tube-side enhancement. Figure 12.12a has axially spaced helical grooves pressed in the outer surface, which form helical ridges in the inner surface. Surface tension forces are responsible for the condensation enhancement: surface tension force pulls the condensate into the outer helical grooves, which act as condensate drainage channels. Thin condensate films exist on the convex profile between the drainage grooves.

Mehta and Rao [1979] condensed steam on the Figure 12.12a tubes for systematically varied groove pitch (p) and the groove depth (e_i) with a constant diameter aluminum tube (19.5-mm outside diameter and 1.8-mm wall thickness). Figure 12.13 shows their results for $p = 6.35$ mm with e_i increasing from 0.13 to 1.0 mm. The enhancement (h/h_p) initially increases with increasing groove depth and attains a maximum of 1.38 at $e_i = 0.35$ mm, after which h/h_s decreases with increasing groove depth. The h/h_p decreases for increasing groove pitch (e_i = constant), as one would expect. Mehta and Rao also measured the water-side heat transfer and friction characteristics for their tubes. These data are reported by Rao [1988].

Withers and Young [1971] show that the Figure 12.12a corrugated tube will provide a 30 to 50% tubing material reduction, compared to a smooth-tube steam condenser designed for equal water-side pressure drop. Their data show that the steam-side enhancement is 35 to 50%, relative to a smooth tube.

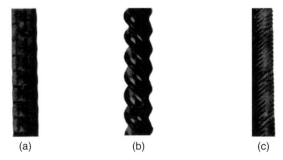

(a)   (b)   (c)

Figure 12.12 Doubly enhanced tube geometries for condensation on horizontal tubes tested by Marto et al. [1979]. (a) Helically corrugated tube, (b) Turbotec tube, (c) corrugated tube formed by rolling a corrugated sheet and steam welding. (From Marto et al. [1979].)

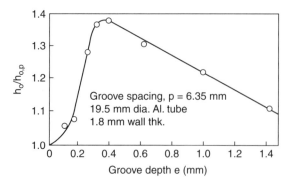

Figure 12.13 Steam condensation data on corrugated tubes reported by Mehta and Rao [1979].

Marto et al. [1979] performed comparative tests of all of the Figure 12.12 geometries with steam condensation (100°C), which included variation of the geometric parameters for each tube type. Their results for p = 9.58 mm, e_o = 0.4 to 0.6 with a 15.9-mm-diameter tube appear to be in conflict with those of Mehta and Rao. Marto et al. found that the condensing coefficient is approximately 10% below the smooth tube value. The low steam-side enhancement may be due to an error in their use of the Wilson plot method to separate the water and steam-side resistances. Only the Figure 12.12a tube provided significant steam-side enhancement, approximately 35%. The dominant enhancement occurred on the water side, where they obtained h/h_p values in the range of 2 to 4 for most tubes. The water side enhancement was accompanied by substantial increased pressure drop. These results imply that such tubing would be of value only if the water side offers the controlling thermal resistance. Further, the steam-side enhancement is marginal, compared to that possible with other surface geometries.

Yorkshire [1982] describes corrugated tubes having different corrugation pitches and depths. Their brochure presents data and a design correlation for steam-side condensing performance as a function of surface geometry. Their data are generally consistent with the Figure 12.13 data of Mehta and Rao [1979].

12.3.5 Surface Tension Drainage

The use of surface tension forces to affect condensate drainage is a very important and effective enhancement technique. We have previously noted enhancement by surface tension forces. In this section, the discussion briefly describes data for various geometries. Section 12.4 provides detailed discussion of the mechanism and theory of surface tension enhancement. Equations are provided in Section 12.4 to select preferred surface shapes. The geometries described in this section are limited to those that do not provide significant increased surface area. Such surface tension drainage may be invoked by loosely attached wires, which have essentially no thermal contact with the base heat transfer surface. In some cases, a surface area increase may be involved, which technically should be discussed under "extended surfaces." However, it is discussed in this section because the enhancement mechanism is caused by surface tension force.

Shah et al. [1999] provide a recent review of surface tension enhancement of film condensation.

Vertical Fluted Tubes. Gregorig [1954] was the first to propose use of surface tension forces to enhance laminar film condensation on a vertical surface. Figure 12.14a illustrates a horizontal cross section through the wall of a vertical fluted tube, and Figure 12.14b is a photograph of a doubly fluted tube. Due to the surface curvature, the liquid pressure in the convex film is greater than that of the vapor. The combination of convex and concave surfaces establishes a surface tension force that draws the condensate from the convex surface into the concave region. A high condensation rate occurs on the convex portions of the fluted surface due to nearly horizontal drainage by the surface tension force. The concave portions serve as vertical, gravity drainage channels. The resulting heat transfer coefficients averaged over the total surface area are substantially higher than for a uniform film thickness on a smooth tube. The size of the flutes should be selected such that the drainage

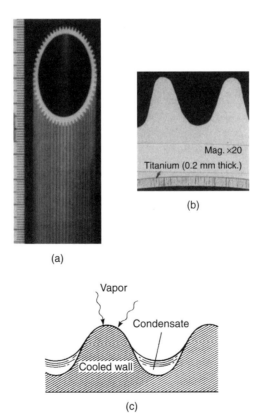

Figure 12.14 Vertical fluted tubes. (a) Cross section of fluted tube, (b) photo of doubly fluted tube, (c) detail of fin cross section. (Courtesy of Sumitomo Light Metal, Ltd.)

channel will be filled to capacity at the bottom of the vertical surface. Therefore, longer tubes would require larger drainage channels. This is discussed by Webb [1979] and by Adamek and Webb [1990a].

Mori et al. [1979] also investigated the effect of placing circular disks (Figure 12.15a) at spaced intervals to remove the condensate flowing down the tube, thereby exposing new condensing surfaces below the disk. Figure 12.15b shows their experimental and predicted results for R-113 condensing on a vertical surface 50 mm high. This figure shows that the 25-mm disk spacing improves the performance of the 50-mm-high surface. Their accompanying theory allows prediction of an optimum disk spacing. Combs and Murphy [1978] also tested the effect of spaced runoff

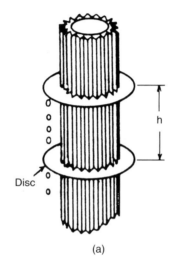

(a)

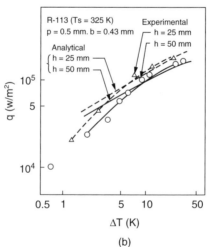

(b)

Figure 12.15 (a) Vertical fluted tube fitted with drainage skirt; (b) predicted and experimental results on tube having drainage skirt. (From Mori et al. [1979].)

disks for ammonia condensation on 1.2-m-long fluted tubes. They show that the disks should be effective, provided that the tube is sufficiently loaded with condensate.

Fluted tubes have received considerable attention for vertical tube condensers used in desalination and are commercially available. Thomas [1968] and Carnavos [1974] give performance data for single and doubly fluted tubes (see Figure 12.14). They show enhancement ratios for h/h_p (total area basis) in the range of 4 to 8 for tube lengths between 0.50 and 0.60 m. Combs and Murphy [1978] report similar enhancement ratios for condensing ammonia on 1.2-m high tubes. The flute geometry of their tubes was not selected based on theoretical relations. In fact, some of the tubes they tested are actually curtain rods they purchased from a department store!

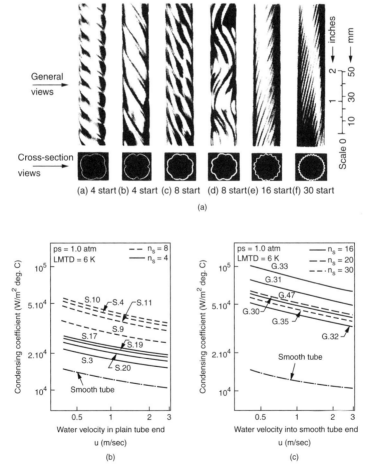

Figure 12.16 (a) Tubes tested by Newson and Hodgson [1973]; (b) condensing coefficient for 4- and 8-start tubes; (c) condensing coefficient for 16-, 20-, and 30-start tubes. (From Newson and Hodgson [1973].)

Newson and Hodgson [1973] tested 32 tubes of the types illustrated in Figure 12.16a. They condensed atmospheric steam on a 1.13-m-long vertical tube against cooling water inside the tubes. Each tube provides inside and outside enhancement, causing a swirl flow to the tube-side coolant. Their tubes have 4, 8, 16, 20, or 30 flutes (n_f). All tubes were made from a 31.8-mm-OD, 0.89-mm wall plain tube. The geometric enhancement parameters are the number of flutes, the groove depth, and the helix angle. They propose that the tubes provide enhancement by draining off the condensate after each gravity-drained, vertical length between the grooves, thus acting as a short vertical tube. Hence, it is proposed that surface tension drainage is the actual condensation enhancement mechanism. Figure 12.16b shows the condensation coefficient for a selected group of the 4 and 8-flute tubes. Figure 12.16c shows the condensation coefficient for a selected group of the 16, 20-, and 30-flute tubes. The tubes having 16, 20, or 30 flutes have a higher condensation coefficient than the 4 or 8-flute tubes. The condensation and water-side heat transfer coefficients were separated using the Wilson plot method. Table 12.6 lists the geometry details, E_o $(=h_o/h_{op})$, and E_i $(=h_i/h_{ip})$, (U/U_p) and the water-side pressure drop ratio $(\Delta p/\Delta p_p)$ at 1.52 m/s water velocity. In contrast to the condensation side enhancement, the lower number of starts and higher helix angles (α) provide the higher E_i values. The relatively large water-side $\Delta p/\Delta p_p$ are partially caused by the reduced cross-sectional flow area of the fluted tubes. The U/U_p values are dependent on the split of thermal resistances between the outside and inside of the tube. At the reference test condition (93°C water temperature, 60°C LMTD, and 1.52 m/s water velocity), 46% of the total thermal resistance of the smooth tube was on the water side.

Axial Wires on Vertical Smooth Tubes. Loosely attached, spaced vertical wires on a vertical surface (Figure 12.17) can also provide surface tension condensation enhancement. If the wire diameter (e) is appreciably larger than the condensate film thickness (δ) and the wires are wetted by the condensate, surface tension force draws the condensate into a rivulet at the wire (region A). This produces film thinning in the space between the wires (region B). The enhancement occurs due to the thinned film in region B, and the condensate drains down along the wires. Thomas [1967] provides experimental data for this enhancement technique. In a second publication, Thomas [1968] shows that square wires are more effective than circular wires of the same height. This is because square wires have a greater condensate carrying capacity. Butizov et al. [1975] and Rifert and Leont'yev [1976] have also worked with loosely attached vertical wires. Their work with steam condensation supports conclusions advanced by Thomas. Using 1.0-mm-diameter circular wires spaced at 9 wire diameters, they measured enhancement levels in the range of 3 to 6 on a 1.3-m-long test section. The enhancement decreases with increasing heat flux because the wires become more quickly loaded with condensate. The Kun and Ragi [1981] patent describes 1.4-mm wires spaced at 2.0 mm on a vertical tube. The vertical wires were held in place by a second set of wires helically wrapped at 28-mm axial pitch. For the same LMTD between the condensing steam and the water-side coolant, the attached wire design increased the heat flux at the bottom of the tube by a factor of 23.

Table 12.6 Test Results of Newson and Hodgson [1973] for Steam Condensation on Vertical Spirally Fluted Tubes

Tube	n_f	e/D_h	α (deg)	E_o	E_i	U/U_p	$\Delta p/\Delta p_p$
S.3	4	0.078	56	1.45	1.79	1.69	4.54
S.17	4	0.068	65	1.90	2.21	1.99	6.97
S.11	8	0.055	20	3.00	1.19	1.61	3.0
S.10	8	0.056	27	3.50	1.26	1.70	3.5
S.4	8	0.050	45	3.20	1.47	1.82	3.2
S.9	8	0.033	56	2.35	1.91	2.02	3.56
G.33	16	0.051	14	6.75	1.23	1.90	1.74
G.31	16	0.070	21	5.00	1.34	1.87	1.91
G.30	16	0.063	27	4.00	1.47	1.93	2.15
G.32	16	0.053	36	6.20	1.66	2.04	2.52
G.35	20	0.048	21	3.60	1.37	1.85	2.03
G.47	30	0.047	14	4.10	1.19	1.61	1.47

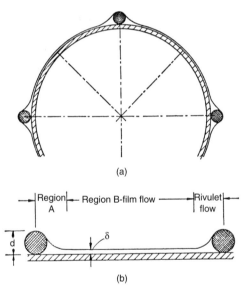

(a)

(b)

Figure 12.17 (a) Vertical tube fitted with loosely attached wires for enhancement of condensation as reported by Thomas [1967]; (b) detail showing condensate drainage. (From Thomas [1967].)

Although the attached wire concept is effective, the fluted tube concept of Figure 12.16 provides enhancement on both sides, is probably cheaper, and is more practical for tube insertion in a tube sheet. The fluted tubes are made with plain tube ends.

Wire Wrap on Horizontal Tubes. Thomas et al. [1979] tested ammonia condensation on a smooth, horizontal tube wrapped with a spaced wire in a helical manner. A 38-mm-diameter aluminum tube was wrapped with 2.4-mm-diameter wire spaced at 9.5 mm. The condenser contained 147 tubes. The measured condensing coefficient was approximately three times that predicted by the Nusselt equation for a smooth tube. Surface tension forces draw the condensate to the base of the wires, which act as condensate runoff channels. This results in a thick condensate film in the region between the wires, thus promoting enhancement. One does not need to use high-thermal-conductivity wires, because the little condensation occurs on the wires. On the basis of this test, it appears that external ridges (wire) are more effective for condensation enhancement than are circumferential grooves.

Other studies of wire-wrapped horizontal tubes are reported by Marto and Wanniarachchi [1984], Fujii et al. [1987], and Marto et al. [1987]. Fujii et al. condensed R-11 and ethanol on an 18-mm-diameter tube and varied the wire pitch (p_f) for wire diameters (e) of 0.1, 0.2, and 0.3 mm. They found maximum enhancement for closely spaced wires, $p_f/e \simeq 2.0$, which provided $h/h_p \simeq 3.5$. For fixed wire pitch, the larger-diameter wires produced higher enhancement. Rifert and Trokoz [1996] provide steam condensation data on 16-mm-OD tubes wrapped with 1.5-mm-diameter wires at 10-mm pitch. They took data for wires wrapped on brass, CuNi, and German silver tubes, whose thermal conductivities were 100, 50, and 25 W/m-K, respectively.

Briggs et al. [2002, 2003] measured steam condensation on 12.7-mm-OD tubes helically wrapped with wire. Data were taken for wire diameters of 0.4, 0.75, and 1.0 mm for 6 wire wrap pitches between 1.0 and 6.0 mm. Briggs et al. [2003] extended the Briggs et al. [2002] data to include 0.2 and 1.0-mm wire diameters. For constant wire pitch, the data show that the enhancement ratio (E_o) for constant (T_{sat}-T_w) increases with decreasing wire diameter. However, for each wire diameter, there is an optimum helix pitch. Their best-performing tube had 0.4-mm wire diameter at approximately 2.0-mm pitch. For wire diameters of 0.75 or more, the E_o decreases for wire pitches below 2.0 mm.

Fujii et al. [1985] formulated the first model for condensation on a wire-wrapped tube. This model was modified later by Briggs et al. [2002, 2003]. Rose [2002] modified the Fujii et al. [1985] model to predict the condensation coefficient on wire wrapped tubes. The model is applicable to any fluid and is given by

$$E_o = \frac{z_o^{-1/4}}{2.827}\left[p\varphi_f - \left(\frac{2e\sigma}{\rho g d_o}\right)^{1/2}\ln\left\{\frac{\varphi_f}{4}+\frac{\pi}{4}\right\}\right] \qquad (12.12)$$

where

$$z_o = 1.5\left(1+\frac{4d_o^2}{s_o}\right)^{-1} \quad s_o = p-4\left(\frac{e\sigma}{2\rho g d_o}\right)^{1/2} \quad \phi_f = \cos^{-1}\left[\left(\frac{16e\sigma}{\rho g d_o p^2}\right)-1\right]$$

Briggs et al. [2004] show that the Briggs et al. [2002] model generally underpredicts the data 10 to 25% for wire diameters 0.4 mm or greater. For wire diameters 0.4 mm or smaller, the Fujii et al. [1985] model does a better job in predicting the data.

Practical problems are associated with this enhancement method for shell-and-tube heat exchangers. It would be necessary to expand the tube end, in order to slip the wire-wrapped tube into the tube sheets.

Surface Tension Effects in Zero Gravity. In a U.S. patent, Staub [1966] describes surface geometries intended to operate in a zero-gravity environment, where gravity forces do not exist to drain the condensate. His geometry consists of an array of circular spine fins having rounded fin tips and a capillary wicking material on the base surface. Surface tension force drains condensate from the rounded spine-fin tip and the condensate is removed from the base surface by the capillary wicking. Such a surface may effectively operate in any orientation, regardless of whether a gravity field is present. No data are presented in the patent.

12.3.6 Electric Fields

Recently, considerable work has been done using electric fields to enhance film condensation. Application of an electric field to a dielectric fluid is termed an electro-hydrodynamic (EHD) effect. Yabe [1991] enhanced R-113 condensation on a vertical plain tube by a factor of 4.5 using an electric field strength of 4 MV/m. However, the enhancement did not exceed that provided by a horizontal Thermoexcel-C tube (Table 12.4). Section 15.4 discusses recent research.

12.4 SURFACE TENSION DRAINED CONDENSATION

12.4.1 Fundamentals

The phenomenon takes advantage of the fact that the pressure difference across a liquid–vapor interface is influenced by the local radius of the interface. By forming an interfacial shape that has a changing radius, one can establish a pressure gradient in the liquid film, which drains condensate from the surface.

To understand the basic principles of the pressure difference across an interface, consider a liquid droplet of radius r. The liquid droplet exists in a saturated vapor environment, whose pressure is p_v. The liquid pressure in the droplet is p_l. Surface tension force acts on the liquid–vapor interface. If the droplet is cut in half, and a force balance is written on the pressure and surface tension forces, one obtains

$$(p_l - p_v)\pi r^2 = 2\pi r\sigma \tag{12.13}$$

By solving for $(p_l - p_v)$,

$$p_l - p_v = \frac{2\sigma}{r} \tag{12.14}$$

which shows that the pressure in the liquid is greater than that in the vapor. The value of $(p_l - p_v)$ increases as the droplet radius, r, is decreased. Next consider a two-dimensional fin having a semi circular shaped fin tip, of constant radius r, illustrated in Figure 12.18. Because of surface tension force, the pressure in the liquid is higher than that of the vapor ($p_v = p_{sat}$). A force balance between pressure and surface tension forces shows that

$$p_l = p_{sat} + \frac{\sigma}{r} \tag{12.15}$$

The ($p_l - p_{sat}$) is equal to σ/r, rather than the $2\sigma/r$ of Equation 12.14, because the droplet considered by Equation 12.15 is a surface of revolution, whereas Figure 12.18 involves a two-dimensional shape. Now, consider the fin profile shown in Figure 12.19. This profile has a small radius at the tip, and the local radius increases with increasing distance from the tip. At the fin base ($\theta_m = \pi/2$), $r = \infty$. As shown by Equation 12.15, the liquid pressure will decrease with increasing distance from the fin tip, because the local radius of the condensate increases. This means that condensate formed on the fin would be drained from the fin tip toward the fin base. The pressure gradient in the liquid film is obtained by differentiating Equation 12.15 and is given by

$$\frac{dp}{ds} = \sigma \frac{d(1/r)}{ds} \tag{12.16}$$

where $dp/ds < 0$. The largest pressure gradient exists at the fin tip (where r is the smallest) and decreases along the fin length. The fin tip radius on Figure 12.18 is constant, and suddenly changes to $r = \infty$ at the tangent point. This radius discontinuity will result in an intense local surface tension gradient at the tangent point. However, gravity force will control after this point.

Figure 12.18 considers a convex liquid profile. Next, consider the concave liquid profile that exists at the base of the fin shown in Figure 12.19. To understand the effects of the concave interface shape, one first considers the forces acting on a vapor bubble, which involves a concave liquid shape. The pressure in the bubble is

Figure 12.18 Fin with constant radius fin tip.

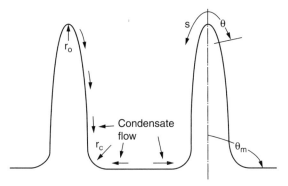

Figure 12.19 Fin with small tip radius with increasing radius from fin tip.

p_v. If the bubble is cut in half, and a force balance is written on the pressure and surface tension forces, one obtains

$$(p_v - p_l)\pi r^2 = 2\pi r \sigma \tag{12.17}$$

By solving for $(p_l - p_v)$,

$$p_l - p_v = -\frac{2\sigma}{r} \tag{12.18}$$

Thus, for a vapor bubble, or a concave interface, the liquid pressure is less than the vapor pressure. Similar to the development for Equation 12.16, the pressure gradient for a two-dimensional concave liquid film is given by

$$\frac{dp}{ds} = -\sigma \frac{d(1/r)}{ds} \tag{12.19}$$

Hence, for a concave interface shape, the liquid pressure decreases in the direction of *decreasing* radius. Consider condensation on the base surface between the fins shown in Figure 12.19. The condensate interface radius at the centerline between the fins is very large as compared to the corner radius r_c at the root of the fin ($r \cong \infty$). Hence, condensate in the root region would be pulled into the corner. Similarly, condensate is pulled from the region near the fin base into the corner.

The larger the pressure gradient, the smaller the condensate film thickness (δ). Assuming that the film is laminar, the local condensation coefficient is given by $h = k_f/\delta$. Thus a strong surface tension–induced pressure gradient will provide a high condensation coefficient. A unique profile shape is not required. Virtually any profile shape that has small tip radius with increasing local radius along the surface (the s direction) will establish a surface tension drainage pressure gradient.

As previously noted, Gregorig [1954] was the first to propose that surface tension can enhance film condensation. His work addressed condensation on a convex surface profile. Later work by Adamek [1981] and Kedzierski and Webb [1990] developed advanced profile shapes. Webb et al. [1982] performed measurements on finned plates having a rectangular fin shape and showed that special fin profile shapes are not required to obtain surface tension drainage. Hence, the prior work on profile shapes can be applied to both vertical and horizontal finned surfaces. Figure 12.20 shows a vertical finned plate consisting of convex surface tension–drained profiles, and concave condensate drainage channels. If the plate is vertical, surface tension force pulls the condensate (generated on the convex profile) horizontally to the drainage channel, where it is drained downward by gravity. Adamek and Webb [1990a] developed equations to predict the condensation coefficient in the flat-sided drainage channels. Both surface tension and gravity forces may be important in the drainage channel. Adamek and Webb [1990b] extended their work to predict condensation on horizontal integral-fin tubes. Honda and Nozu [1987a] have also developed prediction theory for horizontal integral-fin tubes.

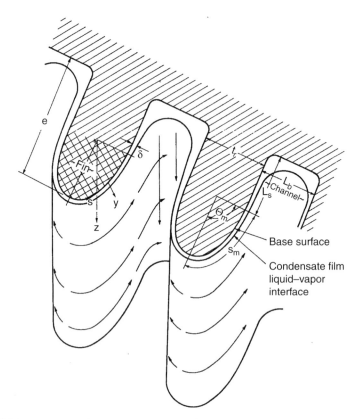

Figure 12.20 Illustration of finned plate with convex fin profiles designed for surface tension–drained condensation, and rectangular condensate drainage channels.

Hence the theoretical equations for surface tension–drained condensation on convex and flat-sided fins, and in the drainage channel, are developed.

A number of quantitative, convex profile shapes have been described. Gregorig [1954] described a profile shape that gives constant film thickness. Smaller film thickness is given by smaller tip radius (r_o). Zener and Lavi [1974] defined a profile shape that gives constant pressure gradient. Consideration of the profile shape described by Zener and Lavi provides important fundamental understanding of the theory of surface tension–drained condensation. Figure 12.21 shows a convex base profile and a thin condensate film on the surface. The coordinate directions along the curved surface and normal to the surface are s and η, respectively. The condensate film (thickness δ) is surface tension drained at velocity $u(\eta)$ in the s direction. The Zener and Lavi profile has radius r_o at $s = 0$, and $r_o = \infty$ at $s = S_m$, where $\theta = \pi/2$. If a profile shape is chosen such that dp/ds = constant, Equation 12.19 shows that

$$\frac{dp}{ds} = \sigma\left(\frac{1/r(S_m) - 1/r(0)}{S_m - 0}\right)$$

$$= -\frac{\sigma}{r_o S_m}$$

(12.20)

Writing the momentum equation in the s direction of Figure 12.21, and assuming $|dp/ds| \gg (\rho_l - \rho_v)g$ and using the boundary layer approximations, the momentum equation in the s direction is

$$-\frac{dp}{ds} + \mu_l \frac{\partial^2 u}{\partial \eta^2} = 0$$

(12.21)

Comparison of Equation 12.21 with the momentum equation for gravity-drained condensation (Equation 12.2) shows that $-dp/ds$ in Equation 12.21 corresponds with the term $F_g = (\rho_l - \rho_v)g$ in Equation 12.2. The solution of Equation 12.2 for gravity-drained condensation is given by Equations 12.3 through 12.5. The solution of the momentum equation for dp/ds = constant may be obtained from Equations 12.3 through 12.5 simply by replacing $(\rho_l - \rho_v)g$ by $\sigma/r_o S_m$, and L by S_m. Thus, the average condensation coefficient over the arc length S_m for the Zener and Lavi profile is obtained from Equation 12.5 as

$$\bar{h} = 0.943\left(\frac{k_l^3 \lambda \sigma}{v_l \Delta T_{vs} r_o S_m^2}\right)^{1/4}$$

(12.22)

The convex profile shape having dp/ds = constant leads to a simple solution for surface tension–drained condensation. Examination of Equation 12.22 shows that a higher condensation coefficient may be obtained simply by reducing the tip radius, r_o.

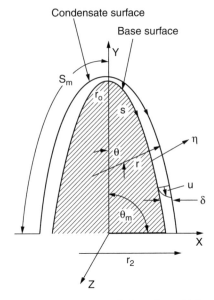

Figure 12.21 Condensation on a profile having small tip radius, with increasing radius along the arc length.

Other profile shapes have been defined, which do not have dp/ds = constant. If $dp/ds \neq$ constant, the local condensate film thickness varies over the profile length. For this case, Adamek [1981] shows that the local film thickness is given by

$$\delta^4(s) = \frac{F_p}{\sigma}\left[\frac{d(1/r)}{ds}\right]^{-4/3}\int\left(\frac{d(1/r)}{ds}\right)^{1/3} ds \qquad (12.23)$$

12.4.2 Adamek [1981] Generalized Analysis

Adamek [1981] defined a family of convex interface profiles that support surface tension drainage. The family includes both the Gregorig [1954] and the Zener and Lavi [1974] profile shapes. The geometrical shape of this family is given by the equation

$$\frac{1}{r} = \left(\frac{\theta_m}{S_m}\right)\left(\frac{\zeta+1}{\zeta}\right)\left[1-\left(\frac{s}{S_m}\right)^\zeta\right] \qquad (12.24)$$

which is valid for $-1 \leq \zeta \leq \infty$. The relation between the tip radius (r_o), the total profile length (S_m), and the angle θ_m is

$$\frac{S_m}{r_o\theta_m} = \frac{\zeta+1}{\zeta} \qquad (12.25)$$

The local pressure gradient in the film is given by applying Equation 12.19 to Equation 12.21. Adamek used Equation 12.24 in Equation 12.23 to obtain the local film thickness, $\delta(s)$. The result is

$$\frac{1}{\delta} = 0.760 \left[\left(\frac{\sigma \theta_m}{F_p \Delta T_{vs}} \right) \left(\frac{(\zeta+1)(\zeta+2)}{S_m^{\zeta+1} s^{2-\zeta}} \right) \right]^{1/4} \tag{12.26}$$

By integrating Equation 12.26 over $0 < s < S_m$, Adamek obtained the average condensate film thickness. Using $h = k/\delta_{av}$, he obtained

$$\bar{h} = 1.712 \frac{k_l}{S_m} \left[\left(\frac{F_p \theta_m S_m}{\Delta T_{vs}} \right) \left(\frac{\zeta+1}{(\zeta+2)^3} \right) \right]^{1/4} \tag{12.27}$$

If $\zeta = 1$ and $\theta_m = \pi/2$, one obtains the result for $dp/ds = $ constant given by Equation 12.22. If $\zeta = 2$, and $\theta_m = \pi/2$, the result is for the $\delta = $ constant case described by Gregorig [1954].

Figure 12.22a shows the average condensation coefficient as a function of arc length (S_m), for $\theta = \pi/2$ and $-0.5 < \zeta < 4$. Figure 12.22b shows the shape of the surface for different values of ζ. The ζ parameter characterizes the aspect ratio of the fin cross section (height/thickness at S_m). The aspect ratio increases as ζ decreases.

In a mathematical sense, the term r in Equations 12.23 and 12.24 is the *radius of curvature*. The term *curvature* (κ) is the rate of change in direction of the local

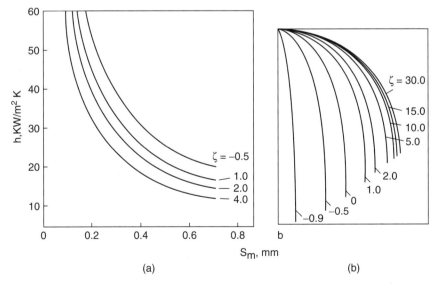

Figure 12.22 (a) Average condensation coefficient vs. S_m; (b) profile shape for different ζ values in Equation 12.24. (From Adamek and Webb [1990b].)

tangent to the curve per unit arc length. The curvature is the inverse of the radius of curvature, $\kappa = 1/r$. So, a curve shape having small local r will have large curvature. All the profiles start with high curvature at the fin tip, which requires a small tip radius. The curvature decreases along the length of the profile, and has zero curvature (infinite radius) at θ_m. For a constant θ_m and S_m, Adamek shows that the maximum h will occur when $\zeta = -0.5$. Inspection of Equation 12.24 shows that when $\zeta \leq 0$, the initial curvature ($\kappa_o = 1/r_o = \infty$). The physical significance of $\kappa_o = \infty$ is that $dp/ds = \infty$ at the fin tip. The practical realization of $r_o = 0$ for $\zeta < 0$ is a physical impossibility—although it is mathematically possible.

Adamek's family of profiles has no special physical significance, other than the fact that the curves promote surface tension drainage from the tip to the fin base. It is possible to identify other curve shapes, and more discussion of alternative fin profiles is presented in a later section.

Adamek's analysis assumes that surface tension dominates gravity force over the entire profile length (e.g., Bo << 1). This condition may not be satisfied over the total arc length, S_m. The Bond number of the Adamek profiles is given by

$$Bo = \frac{\rho_l g S_m^{\zeta+1}}{\sigma \theta_m (\zeta+1) s^{\zeta-1}} \qquad (12.28)$$

For dp/ds = constant, the Bond number is given by

$$Bo = (\rho_l - \rho_v) g \frac{r_o S_m}{\sigma} \qquad (12.29)$$

The Bond number will decrease with increasing direction from the fin tip. At the end of the profile, Bo approaches infinity. This means that the condensate streamline will have a component in the vertical direction.

Kedzierski and Webb [1987] performed experiments to validate the theoretical predictions for the Equation 12.27 profile shapes. They used the electrostatic discharge machining (EDM) method with numerical control of the machining head, to make finned plates with profiles defined to high accuracy. They made finned plates similar to Figure 12.20 having the Figure 12.22b Gregorig and Adamek profiles ($\zeta = 2$ and -0.05, respectively) machined into the face of a 102-mm-long copper test section and tested with R-11 at 1.3 kPa. The theoretical models for the Gregorig and Adamek profiles are based on the curvature of the liquid–vapor interface. Hence, the shape of the base surface is slightly different from the interface shape. The Gregorig and Adamek theories were used to subtract the condensate film thickness from the interface surface to obtain the shape of the base surface. This profile shape was produced by the EDM method. The profiles consist of the convex arc (S_m) and a flat condensate drainage channel of depth L_s width L_b. Because the Gregorig or Adamek theories are applicable only to the convex arc length (S_m), it was necessary to subtract the condensation rate in the drainage channel from the total. The model developed by Adamek and Webb [1990a] allows prediction of the condensation coefficient both on the convex profile and in the drainage channels. Adamek and Webb showed very good agreement between the theoretically predicted and the experimental condensation rates on the finned blocks.

12.4.3 "Practical" Fin Profiles

There are two primary applications for surface tension–drained condensation. The first is for a vertical tube (Figure 12.16) or for horizontal integral-fin tubes (Figure 12.4).

Prior to the Webb et al. [1982] tests on plates having rectangular fins, designers generally did not recognize that surface tension governed the condensate drainage from the fin surfaces on horizontal integral-fin tubes. Figure 12.23 shows the cross section of the fins used on commercial integral-fin tubes. This figure shows that the fins are of a trapezoidal shape and that the sides are flat. Table 12.7 shows the dimensions of the fins, as reported by Webb et al. [1985].

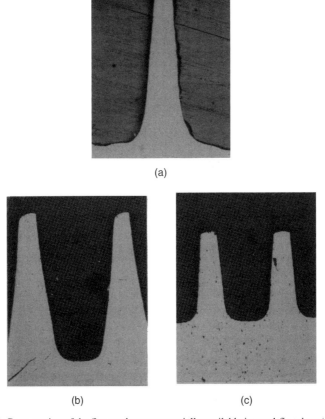

(a)

(b)   (c)

Figure 12.23 Cross section of the fins used on commercially available integral-fin tubes: (a) 748 fins/m, (b) 1024 fins/m, (c) 1378 fins/m.

Table 12.7 Dimensions of Commercially Available Integral-Fin Tubes

Fins/meter (fins/m)	748	1024	1378
Outside diameter d_o (mm)	19.0	19.0	19.0
Area ratio $A_o/(\pi d_o L)$	2.91	3.60	3.18
Fin height e (mm)	1.53	1.53	0.89
Fin thickness at tip t_t (mm)	0.20	0.20	0.20
Fin thickness at base t_b (mm)	0.42	0.52	0.29
Aspect ratio, e/t_b	3.6	2.9	3.1

The basic geometry of the fins is defined by specifying the fin pitch (p_f), the fin height (e), and the fin shape. For the fins in Table 12.7, the shape is defined by the fin thickness at the base (t_b) and the tip (t_t). Generally, one will specify the minimum values of t_b and t_t that are manufacturable. Assume that one wishes to make an integral-fin tube having 748 fins/m with 1.53-mm-high fins. If the Gregorig profile is used, its aspect ratio (fin height/fin base) is 0.75. Hence, the t_b would be 2.04 mm. This thickness is substantially greater than the 0.42 mm value of Table 12.7. Use of greater thickness values will waste material and cause greater condensate retention. For $e = 1.53$, the Gregorig fins are so thick, that it would not be possible to have 748 fins/m. An Adamek $\zeta = -0.86$ profile would provide $e = 1.53$ and $t_b = 0$. The previous discussion concluded that any Adamek profile having $\zeta \leq 0$ is not a practical choice, as it requires a zero-radius liquid–vapor interface at the fin tip. Practical minimum tip radii are probably limited to 0.05 mm. Hence, one quickly concludes that none of the existing theoretical profiles is practical for the design conditions!

Webb and Kedzierski [1990] have defined an alternative set of condensate surface profiles that allow one to independently specify the fin tip radius (r_o), the fin base thickness (t_b), the fin height (e), and the angle (θ_m). For the previously specified parameters, the fin profile shape is a function of the parameter Z, which describes the "fatness" of the profile. The equation describing the fin profiles is presented and discussed in Appendix 12.A. If the fin profile is a continuous function, one may develop an analytical solution for the condensation coefficient, following the same procedure used by Adamek [1981] for his ζ profiles. The local condensate thickness is given by integrating Equation A.1 in Appendix 12.A with respect to the angle θ for the profile of interest. The integration is

$$\delta^4(s) = \frac{F_p}{\sigma}\left[\frac{d(1/r)}{d\theta}\right]^{-4/3}\int_0^{\theta_m}\left(\frac{d(1/r)}{d\theta}\right)^{1/3}ds \tag{12.30}$$

Equation 12.30 may be analytically integrated. Jaber and Webb [1993] give the result of the analytical integration of Equation 12.30. Then, the average condensation coefficient over the profile length, S_m, is given by $h = k/\delta_{av}$, where

$$\delta_{av} = \frac{1}{S_m} \int \delta(s)ds \qquad (12.31)$$

Kedzierski and Webb [1990] have compared the conductance ($\bar{h}S_m$) for $\theta_m = \pi/2$ for several profiles. The results are shown in Table 12.8. For a fin base thickness of $t_b = 0.356$ mm, the value of hS_m of the Webb and Kedzierski profile is within 10% of the Adamek profile, which has zero tip radius. The $t_b = 0.356$-mm Gregorig profile has a small $\bar{h}S_m$ because of its small fin height. If the height of the Gregorig fin is set at 1.45 mm, $\bar{h}S_m$ will increase, but the base thickness will increase to 1.88 mm, which wastes a considerable amount of fin material and limits the fin pitch to considerably larger values than for the $t_b = 0.356$ mm fins.

An additional question of interest arises: How precisely must the convex profiles be machined? Or, what variation in condensation coefficient will occur (for fixed e and t_b) as r_o, or the shape of the profile (Z, θ_m), is varied? These questions were evaluated using the Kedzierski and Webb profiles. Table 12.9 shows the calculated results for a 1.07-mm-fin height (e) and 0.356-mm base thickness (t_b). Lines 1 to 3 show that changing Z from 100 to 10 increases h by 10%. Increasing the tip radius from 0.025 to 0.051 mm decreases h by 3.0% (lines 2, 4), and increasing the tip radius from 0.051 to 0.102 mm decreases h by 7.6% (lines 2, 5). Decrease of θ_m from 90° to 85° causes a 21% decrease (lines 1 and 6). Small r_o and Z provide higher performance. Manufacture of a 0.051-mm fin tip radius is a practical possibility. Moderate variation of the tip radius and profile shape (Z) will have little effect on h. However, it is important to maintain θ_m close to 90°.

12.4.4 Prediction for Trapezoidal Fin Shapes

It is important to be able to predict condensation on trapezoidal (or rectangular) fin shapes. These fins shapes cannot be written in terms of continuous functions as used by Adamek [1981] or by Kedzierski and Webb [1990]. Such fin shapes may have both surface tension and gravity-drained regions. Here, both approximate and quite precise, but also complex, models are described.

Adamek and Webb Model. Figure 12.24a illustrates the Adamek and Webb [1990a] model, which is simple in concept. The model segments the fin into several surface tension–drained or gravity drained zones. The basic concept of the model for condensation on a vertical plate having rectangular fins of pitch p_f, thickness t_f, and height e is illustrated here. The various drainage regions are noted on Figure 12.24a. Linear surface tension gradients are assumed for the surface tension drained zones. The fin tip has a corner radius of r_c. Equation 12.3, which gives the local film thickness for gravity-drained condensation at location x, may be written in the general form:

$$\delta(x) = \left(\frac{F_p x}{F_d} \right)^{1/4} \qquad (12.32)$$

where x is the coordinate in the drainage length and F_d is the drainage force. If the region is gravity drained, $F_d = g(\rho_1 - \rho_v)$, and if surface tension drained, $F_d = \sigma(1/r_1 - 1/r_2)/\Delta x$. The r_1 and r_2 refer to the local film radius at the beginning and end of the region, respectively. Surface tension pulls the condensate from the tip radius r_o, and into the drainage corner, containing a concave film of radius R. Regions that have no change of film radius are gravity drained. Gravity drains regions $L_{1'2'}$, L_{12}, L_{34}, and L_{65} of Figure 12.24a. Table 12.10 defines the drainage length and the drainage force (F_d) for the several drainage regions illustrated in Figure 12.24a. Figure 12.24b shows a detail of the fin tip corner for a rectangular fin, and Figure 12.24c shows a detail of drainage in the corner of a drainage channel, Figure 12.24d shows the curvature ($1/r$) of the condensate film for each of the regions denoted on Figure 12.24a.

Knowledge of the local film thickness permits calculation of the local condensation coefficient, since $h(x) = k_l/\delta(x)$. Regions $L_{1'2'}$, L_{12}, L_{23}, and L_{54} do not start from 0 film thickness, so Equation 12.32 must be modified to account for the film thickness at the beginning of the zone. Consider a region L_{jk} that bounds upstream region L_{ij}, on which condensate flow $\dot{m}_{ij}$ was generated. Adamek and Webb [1990a] show that

$$\delta_k(x) = \left(\frac{3\nu_l \dot{m}_{ij}}{f_{ij}} + \frac{F_p x}{f_{jk}} \right)^{1/4} \tag{12.33}$$

Table 12.8 Comparison of hS_m on Three Profile Types

Profile	Parameter	e (mm)	t_b (mm)	$\bar{h}S_m$ (W/m-s)
Gregorig	$\zeta = 2$	1.45	1.88	8.04
Gregorig	$\zeta = 2$	0.28	0.356	5.31
Adamek	$\zeta = -0.78$	1.45	0.356	9.45
Webb and	$Z = 10$	1.45	0.356	8.45
Kedzierski				
($r_o = 0.025$ mm)				

Table 12.9 Effect of Parameters Z, r_o, and θ_m on $\bar{h}$ ($e = 1.07$ mm, $t_b = 0.356$ mm)

Line	Z	r_o (mm)	θ_m (deg)	S_m (mm)	$\bar{h}$ (W/m²-K)
1	10	0.025	90	1.12	7520
2	50	0.025	90	1.14	7080
3	100	0.025	90	1.14	6790
4	50	0.051	90	1.15	6880
5	50	0.102	90	1.12	6360
6	50	0.051	85	1.11	5940

Equation 12.33 may be appropriately used to calculate the film thickness of regions $L_{1'2'}$, L_{12}, L_{23}, and L_{54}. Readers are referred to Adamek and Webb [1990a, 1990b] for the calculation details. The average condensation coefficient for the surface is given by

$$\bar{h} = \frac{\Sigma h_{ij} l_{ij} \Delta T_{ij}}{L_{ij} \Delta T_{ij}} \tag{12.34}$$

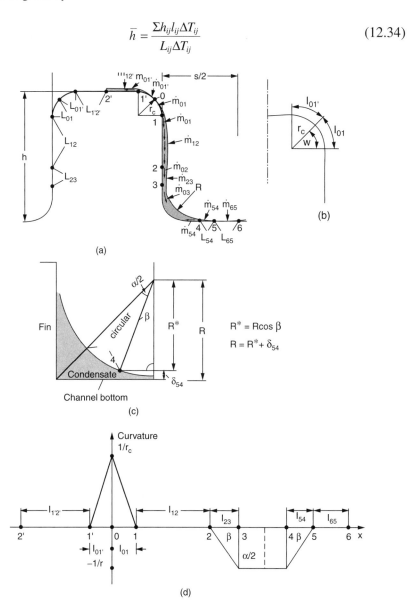

Figure 12.24 (a) Cross section of fin and root illustrating the defined condensate drainage regions (not to scale). (b) Detail of fin tip corner for rectangular fin. (c) Detail of drainage in corner of drainage channel. (d) Curvature of the condensate film in the drainage regions defined on Figure 12.23a. (From Adamek and Webb [1990b].)

Table 12.10 Definition of Drainage Regions for Figure 12.24

Region	x	F_d
1′2′	$(t/2 - r_o)$	ρg
01′	$\pi/2r_o$	$\sigma/r_o l_{01}$
01	$\pi/2r_o$	$\sigma/r_o l_{01}$
12	$e - r_o - l_{23} - R^*$	ρg
23	$4[F_p \Delta T R_c/\sigma]^{1/6}$	$\sigma/R l_{23}$
34	$\pi/4R$	None
54	l_{23}	$\sigma R l_{54}$
65	$p_f - t_f - R - \delta_{54}$	ρg

where the summation refers to all of the L_{ik} drainage zones shown on Figure 12.24a. Adamek and Webb [1990a, 1990b] have validated the theory with good success for vertical finned plates and horizontal finned tubes.

Approximate Models. It is desirable to have a simple model to predict the condensation coefficient on such trapezoidal fins such as those of Figure 12.23. A good approximation is illustrated in Figure 12.25a. A Kiedzierski-Webb K–W profile of tip radius r_c is fitted to the fin sides. The flat top on the fin is of width (t_t-2r_c). The base thickness of the K–W profile is $[t_b - (t_t-2r_c)]$. The condensation coefficient on the fin side is predicted for this K–W profile. The condensation coefficient on the fin tip is predicted using the same procedure as Adamek, and described in Section 12.4.4. Figure 12.25b shows the nature of the approximation at the fin corner. The dashed line part of the K–W profile is shown only for illustration purposes and is not part of the model. Figure 12.25c shows the change of curvature over the fin, where point 0 is the center of the fin tip. Region L_{01} is gravity drained and regions L_{12} and L_{2s} (fin side) are surface tension drained.

An alternative approach is to fit the Adamek [1981] ζ profile having the aspect ratio, $e/(t_b - t_t + 2r_c)$ to the fin side. This is probably not as good an approach as use of the K–W profile, because the Adamek profile results in a zero radius at the fin tip, rather than the actual corner radius, r_c. However, for the same e and t_b, Table 12.8 shows the difference is small.

More precise, but more complex models have been developed for prediction of the condensation coefficient on fins of trapezoidal and rectangular cross section. These models are due to Honda and Nozu [1987a] and Adamek and Webb [1990a, 1990b].

Rose [1994] used simplifying assumptions and dimensional analysis to obtain a model for condensation on integral-fin tubes having rectangular and trapeziod-shaped fins. Their model is presented in terms of the enhancement ratio (E_o) relative to a plain tube of the same root diameter operating at the same ΔT_{vs}. Due to the complex nature of the equations, the reader is referred to equation 47 in the Rose [1994] paper. The model contains two corrections for condensate retention, as compared to the single-term approach of Equation 12.11. The model separately accounts for the fact that some of the condensate is retained in the unflooded zone

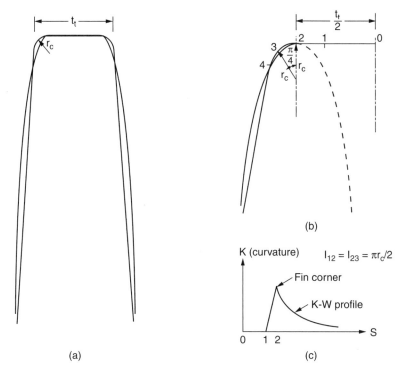

Figure 12.25 (a) Kedzierski–Webb K–W profile fitted to the fin side; (b) approximation at fin tip; (c) change of curvature over fin surface.

in the form of wedges at the root of the fins. They correlated the data of 11 investigators to obtain the 2 empirical constants in the model. This resulted in correlation of almost all the data sets within ±20%. The data included R-113, methanol, glycol, and steam, plus a small several other fluids from Beatty and Katz [1948]. This model assumes an isothermal fin; hence, it does not account for fin efficiency. Briggs and Rose [1994] extended the model of Rose [1994] to account for a temperature gradient in the fin. This Briggs and Rose [1994] model separately calculates condensation on the fin tip and from the flooded and unflooded regions of the fin base and flanks. Using this model, they predicted the preferred fin thickness, height, and spacing dimensions for steam and R-11 with tube thermal conductivities between 80 and 400 W/m-K.

Cavallini et al. [1996] provide a correlation that accounts for the effect of forced vapor convection on condensation on integral-fin tubes having rectangular and trapeziod-shaped fins. This model is discussed in Section 12.6.1. Cavallini et al. used the Briggs and Rose [1994] model to correlate condensation data under low-vapor-velocity conditions. Briggs and Rose [1999] evaluated the ability of several models to predict condensation heat transfer on low-fin tubes.

Belghazi et al. [2002] present a relatively simple model of the relatively complex surface geometry of the Wieland GEWA-C⁺ tube shown in Figure 12.26. They

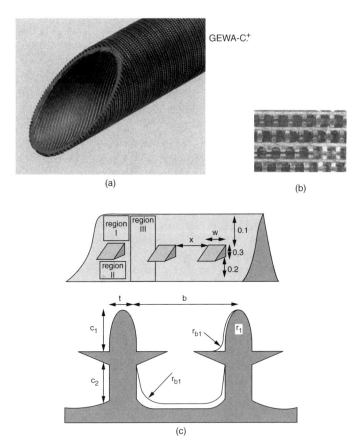

Figure 12.26 GEWA-C+ tube: (a) tube photo, (b) detail of fins, (c) illustration of fin detail. (From Belghazi et al. [2002].)

applied different submodels to the three fin regions illustrated in Figure 12.26c (Regions I, II, and III) plus the condensate drainage channel at the tube outside diameter, between the fins (Region IV). They assumed the Nusselt vertical plate, gravity-drained model for Region II, the Nusselt model for horizontal tubes in Region IV, and a simple linear surface tension drainage model (Equation. 12.29) for Regions I and III. The heat transfer coefficients in each region were weighted by the area to which each model applies. The structure of Equation 12.34 is applied to the several regions and the term c_b predicted using Equation 12.11. The condensation in the flooded region was neglected by neglecting the last term in Equation 12.11. Figure 9 in Belghazi et al. [2002] shows very good ability to predict the measured condensation coefficient of the GEWA-C+ tube.

Srinivasan et al. [2002] proposed an empirical correlation based on a power-law correlation of dimensionless parameters for condensation on horizontal integral-fin tubes. The correlation is applicable to fins of rectangular or trapezoidal cross section and has six empirical constants. The empirical constants were determined by correlating refrigerant and water data of a relatively large database from the open literature. The fluid property–based dimensionless parameters used in the correlation are the surface tension number (Su $= \rho\lambda\sigma p/\mu k\Delta T_{vs}$)and the gravity number (Gy$= \rho^2\lambda g p^3/\mu k\Delta T_{vs}$). The basic structure of the correlation is given by a variant of Equation 12.38, which does not account for fin efficiency:

$$\text{Nu} = (l - c_b)\text{Nu}_u + c_b\text{Nu}_f \tag{12.35}$$

$$\text{Nu}_f = 0.728\frac{t}{p}\left(\frac{pGy}{d_r}\right)^{1/4} \tag{12.36}$$

$$\text{Nu}_u = 15\text{Su}^{0.1968}\left(\frac{t}{d_r}\right)^{0.1062}\left(\frac{e}{p}\right)^{0.1369}\left(\frac{d_r}{p}\right)^{0.7485}\left(\frac{Gy}{Su}\right)^{0.0504} \tag{12.37}$$

Panchal [1994] developed a relatively simple empirical model for condensation on vertical "fluted" tubes, such as illustrated in Figure 12.14. The empirical constants in the model were fitted to the data of Combs and Murphy [1978] on profile geometries approximating a cosine shape. The model does not account for how the profile shape affects performance, as in Equation 12.27. Hence, it is of limited usefulness.

12.5 HORIZONTAL INTEGRAL-FIN TUBE

The first analytical model for horizontal integral-fin tubes was developed by Beatty and Katz [1948]. This model assumes that gravity force drains the condensate from the fins. Later work by Webb et al. [1982] showed that surface tension force, rather than gravity force, drains the condensate from the fins. This observation led the way for application of surface tension–drained models to integral-fin tubes. The first simple surface tension drainage model was proposed by Karkhu and Borovkov [1971]. Rudy and Webb [1983, 1985] showed that the lower part of the fin circumference is flooded by condensate, and that the model should account for this condensate flooding. The first model to incorporate surface tension drainage and condensate retention was developed by Webb et al. [1985]. Later work by Honda and Nozu [1987a] and Adamek and Webb [1990b] provides to the most advanced models for integral-fin tubes. This section will survey these models and advances.

12.5.1 The Beatty and Katz [1948] Model

Beatty and Katz [1948] proposed that heat transfer may be modeled by use of the Nusselt [1916] equations for condensation on horizontal tubes and vertical plates. The model assumes that gravity force drains the condensate from the fins, and that no condensate retention occurs on the lower side of the tube. The area-weighted condensation coefficient (h) is calculated as the average on the finned surface (h_f), and on the tube surface between the fins (h_h) as given by

$$h\eta = h_h \frac{A_r}{A} + \eta_f h_f \frac{A_f}{A} \tag{12.38}$$

where $A = A_f + A_r$. They used Equations 12.5 and 12.7 to calculate h_f and h_h, respectively. Beatty and Katz assumed that the characteristic length (L) needed in Equation 12.5 is given by

$$L = \frac{\pi(d_e^2 - d_o^2)}{4d_e} \tag{12.39}$$

Their test data for six low-surface-tension fluids condensing on several finned tube geometries (433 to 633 fins/m) were predicted within ±10% using Equations 12.38 and 12.39.

The Beatty and Katz [1948] model is fundamentally incorrect, for two reasons: First, it does not account for the condensate-flooded zone illustrated by Figure 12.5. Second, surface tension, rather than gravity, drains the condensate from the fins. Webb et al. [1982] condensed R-12 on the face of a 51-mm-diameter cylinder, in which fins of rectangular cross section were machined (Figure 12.27a). The fins were 1.0 mm high and 0.30 mm thick at 0.84 mm pitch. The finned surface was tested in the three geometric orientations listed in Table 12.11. Figure 12.27b shows the test results, and the predicted condensation coefficient using the Nusselt equation for gravity-drained condensation (Equation 12.5). The different tilt and rotation angles change the characteristic length (L). The characteristic length associated with the several plate orientations is listed in Table 12.11. Figure 12.27b shows the following:

1. The measured condensation coefficient is essentially independent of geometric orientation.
2. The data are underpredicted 70% by Equation 12.5, as applied to a finned plate.

These results show that the condensate drainage from the fin surface is not controlled by gravity. Rather, it is controlled by surface tension force. Thus, the assumption of gravity drainage in the Beatty and Katz [1948] model is invalid. The Figure 12.27 data argue that the theoretical relations for condensation on surface tension–drained fin profiles may be applied to the fins on integral-fin tubes.

12.5.2 Precise Surface Tension–Drained Models

Honda and Nozu [1987a] describe a more complex model, which is applicable to the trapezoidal cross section fins illustrated in Figure 12.23. Figure 12.28 defines the geometry on which their model is based. The condensation coefficient is predicted

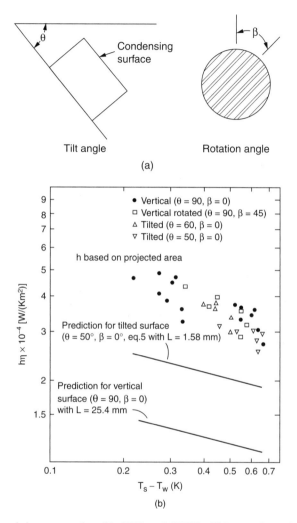

Figure 12.27 Finned plate tests conducted by Webb et al. [1982], which prove that surface tension forces drain the condensate from integral-fins. (a) Illustration of test surface, (b) test data. (From Webb et al. [1982].)

Table 12.11 Summary of Tests Conducted by Webb et al. [1982] on the Figure 12.27 Finned Plate

θ (deg)	β (deg)	L (mm)
90	0	25.4
90	45	25.4
50	0	1.58
60	0	1.60

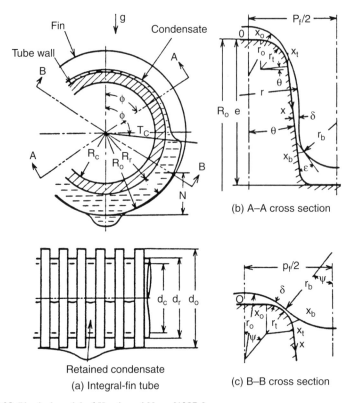

(b) A–A cross section

(a) Integral-fin tube

(c) B–B cross section

Figure 12.28 Physical model of Honda and Nozu [1987a].

for two regions: the u region (unflooded circumferential fraction) and the f region (flooded fraction) shown on Figure 12.28a. These regions are illustrated by Figure 12.28b and c, respectively. The overall structure of their model is

$$h = \frac{(1 - c_b)h_u \eta_u \Delta T_u + c_b h_f \eta_f \Delta T_f}{(1 - c_b)\Delta T_u + c_b \Delta T_f} \tag{12.40}$$

where the first term in the numerator is for the u region, and the second term is for the f region. Equation 12.40 recognizes that the fin efficiency and the $(T_s - T_w) \equiv \Delta T_{vs}$ in the u and f regions are different (e.g., $\Delta T_{vs,u}$ and $\Delta T_{vs,f}$). Equations for calculation of the fin efficiency in the two regions are given by the authors. The heat transfer coefficient is based on the nominal area, $\pi d_e L$. The model predicts the condensation coefficient, based on surface tension drained theory, for the condensate–vapor interface surfaces illustrated in Figure 12.28b and c. Note that this includes the interface in the root area between the fins.

The prediction of h in the u region breaks the interface into four length increments: The fin tip, the corner of the fin tip, and the upper and lower regions of the fin side. A two-region model is used for the f region. Condensation is neglected on the condensate surface in the flooded region, but it is evaluated at the fin tips. The reader should consult their paper for the details of the predictions. The model recognizes that the tube surface temperature is different in the u and f regions, due to circumferential heat conduction in the tube wall thickness.

The values of the wall temperatures $T_{w,u}$ and $T_{w,f}$ were determined by solving the circumferential heat conduction equation in the tube wall, assuming constant heat transfer coefficients in the u and f regions, and neglecting radial conduction. Analytical solutions were obtained, and are given by Honda and Nozu [1987a]. Because the solution depends on the fin efficiency and the heat transfer coefficient, an iterative solution is necessary. Once converged values of $T_{w,u}$ and $T_{w,f}$ have been obtained, the mean tube wall temperature, $T_{w,m} = (T_{w,u} + T_{w,f})/2$ is calculated. The local heat flux is calculated using Equation 12.40 with temperature difference ($T_v - T_{w,m}$). Honda et al. [1987b] have improved the model to account for surface tension drainage in the root region (important for wide fin spacing), and the wall temperature variation between the fin root and the tube wall between the fins. The improved model slightly reduces the error band, relative to the Honda and Nozu [1987a] model.

12.5.3 Approximate Surface Tension–Drained Models

Because of the complexity of the precise models described in Section 12.5.2, it is desirable to have a simpler model for use, where the complex computer models are not available. This section describes a recommended approximate model. A valid model must account for the effect of condensate retention and should also account for surface tension drainage from the fins. In unpublished work, Jaber and Webb developed an unpublished improvement on the model of Webb et al. [1985]. The basic structure of the present model is given by

$$\eta h A = (1 - c_b)[h_h A_r + \eta_{fu} h_{fu} A_f] + c_b \eta_{ff} h_{ff} A_{ft} \qquad (12.41)$$

where $A = A_{fs} + A_{ft} + A_r$ (fin sides, fin tip, and root region), and the finned area is $A_f = A_{fs} + A_{ft}$.

The improvements relative to the Webb et al. [1985] model relate to calculation of the terms h_{fu} and h_f. The first term in square brackets on the right-hand side is for the unflooded fraction of the tube circumference, and the second term is for the condensate-flooded region. The condensate bridged surface fraction is $c_b = \beta/\pi$ and the unflooded fraction is $(1 - c_b)$. The c_b is calculated using Equation 12.11. The condensation coefficients in the unflooded region are h_r (the root tube) and h_{fu} (the fin). The h_{fu} term is composed of condensation contributions on the fin side (h_{fus}) and the fin tip (h_t). The composite h_u is given by

$$h_{fu} l_u = h_{fus} e + h_t t_t \qquad (12.42)$$

The h_{fus} is predicted for a Kedzierski-Webb profile of height e and base thickness $t = t_b - t_t$. This methodology is discussed in Section 12.4.4. The condensation coefficient on the fin tip (h_t) is calculated per Adamek and Webb [1990b] and is described in Section 12.4.4.

If the fin pitch is relatively small (e.g., 1.3 mm) condensation in the root region will be gravity dominated. Then, the condensation coefficient on the base tube (h_h) is calculated from the Nusselt equation for horizontal tubes (Equation 12.7), taking account of the additional condensate drained from the fin profile to the base tube. Equation 12.7 may be written in terms of the condensate Reynolds number as

$$h_h = 1.51 \left(\frac{\mu_l^2 \, \text{Re}_l}{k_l^3 \rho_l^2 g} \right)^{-1/3} \tag{12.43}$$

where the condensate Reynolds number is

$$\text{Re}_l = \frac{4 \dot{m}_r}{\mu_l (p_f - t_b)} \tag{12.44}$$

The term $\dot{m}_r$ in Equation 12.44 is the sum of the condensate formed on the fin ($\dot{m}_f$) and that generated on the base tube between the fins ($\dot{m}_h$). Typically, $\dot{m}_f \gg \dot{m}_h$ so that small error would occur by neglecting $\dot{m}_h$. However, a rigorous solution is possible, and is described by Webb et al. [1985]. For wider fin pitches (e.g., greater than 1.3 mm) one may account for surface tension drainage in the fin root region.

Condensation occurs in the flooded region on the fin tip and on the fin corners. The h_{ff} in Equation 12.41 is given by

$$h_{ff} l_{ff} = h_{ffs} (\pi r_c / 2) + h_t (t_t - \pi r_c / 2) \tag{12.45}$$

where h_{ffs} and h_t are given by k/δ. For h_{ffs}, the δ is calculated using Equation 12.32 with $x = \pi/(4r_o)$ and $F_d = \sigma(1/r_o + 2/s)$. For h_t, the δ is calculated with Equation 12.32 and Table 12.10 using $l_{1'2'}$ and $l_{01'}$.

The fin efficiency for the unflooded region (h_{fu}) is calculated using a standard fin efficiency equation. The fin efficiency in the flooded region (h_{ff}) is calculated as described in Appendix 12.B.

Webb et al. [1985] predicted h_f using an Adamek ζ profile, which fit the aspect ratio of the actual fin cross section (approximately, $\zeta = 0.8$). They neglected condensation on the fin tip, in the flooded region ($h_{ff} = 0$). Their R-11 data for condensation on 748, 1024, and 1378 fins/m tubes were predicted within ±20%. Using the more refined model, Jaber and Webb [1993] predicted the same data within ±13%. The model predicted the condensing coefficient of steam, refrigerant R-11, and refrigerant R-113, for a variety of tubes within ±15%. The tube materials included copper, and the Figure 12.9a to e Cu/Ni, stainless steel finned tubes.

12.5.4 Comparison of Theory and Experiment

Honda and Nozu [1987a] predicted the condensation coefficients for 22 different integral-fin tubes and 12 different fluids using their model and the Webb et al. [1985] model. The test fluids and tube geometries include many of those listed in Table 12.2. The Honda and Nozu model predicts most of the data (including steam) within ±20%. The Rudy and Webb [1985] model also predicted most of the low-surface-tension fluid data within ±20%. The steam data were overpredicted 30 to 40%. The Adamek and Webb [1990b] model was used to predict data for seven different fluids (both low surface tension and steam) on 80 different finned-tube geometries. The data were predicted within ±15%. Hence, the Adamek and Webb [1990b] and the Honda and Nozu models have comparable predictive ability. Sukhatme et al. [1990] tested the ability of the above models to predict his R-11 data for four 1417 fins/m tubes having fin heights between 0.46 and 1.22 mm. They found that the simple Webb et al. [1985] model was as good as the Honda and Nozu [1987a] model for their particular test conditions.

Kumar et al. [2002c] provide an empirically based power law correlation for integral-fin tubes that includes a linear tension force. It was applied to 13 sets of refrigerant and steam data. This is proposed as a simple correlation for design use.

12.6 HORIZONTAL TUBE BANKS

Tube banks are subject to effects of condensate inundation and vapor velocity. Condensate inundation data account for the drainage of condensate from the upper tube rows, without vapor velocity effects. Condensate inundation reduces the condensation coefficient. Vapor velocity generally exerts a shear stress on the liquid film, which acts to thin the film. This will enhance the condensation coefficient, unless the vapor shear opposes gravity. Vapor shear data may be provided for a single horizontal tube, without inundation. An actual condenser tube bundle experiences both inundation and vapor shear effects. Problems involving *both* condensation and vapor shear are more properly classified under "convective condensation," which is discussed in Chapter 14.

Webb [1984] provides a survey of inundation and vapor velocity effects for refrigerant condensers. Recent review papers are provided by Michael et al. [1989] and Browne and Bansal [1999].

12.6.1 Condensation with Vapor Shear

Vapor shear can cause considerable enhancement. Considerable work has been done for vapor shear for both plain and finned tubes. The direction of the vapor flow will affect the results. However, in tube bundles, the vapor direction may be cross-flow, or even radially inward. Most work has been done for downward vapor flow. For downward vapor flow, the vapor shear acts to thin the condensate film, and at sufficiently high vapor velocity, it may entrain the condensate. The authoritative theoretical work on vapor shear effects with plain, horizontal tubes is that of Shek-riladze and Gomealauri [1966] and later works by others based on an extension of this model. Recent work on vapor shear effects on plain tubes is by Briggs et al. [1992] for steam and ethylene glycol, Lee and Rose [1984], Rose [1984] for R-113, McNaught and Cotchin [1989], Michael et al. [1989], and Honda et al. [1986]. Correlations for vapor shear effects are found in most of the above works.

With regard to enhanced condenser tubes, work has been done on both circular integral fin tubes, and spine fin tubes. Correlations exist to predict the effect of vapor shear on circular integral fin tubes as a function of fin geometry and fluid properties. Cavallini et al. [1993, 1996] have measured the effect of vapor velocity for condensation on integral-fin tubes. Figure 12.29 shows the enhancement ratio (with vapor shear/no vapor shear) measured by Cavallini et al. [1993] for R-11 and R-113 on single horizontal circular integral fin tubes. Data are shown for two tubes (1333 and 2000 fins/m) having 0.6 mm fin height. Figure 12.29 shows that enhancement occurs when $Re_v > 70,000$ to $100,000$. Greater enhancement ratio was found for the 2000 fins/m tube than for the 1333 fins/m tube. At the maximum Re_v (about 250,000) the enhancement for the 1333 fins/m tube was 70 to 80%. Cavallini et al. [1996] also shows R-11 data for a 2000 fin/m tube having 0.6 and 1.0 mm fin heights. Approximately equal E_o was shown for the two fin heights. Figure 12.29 also shows the predicted effect of vapor shear for a plain tube. This was calculated using the Honda et al. [1986] model for vapor shear and the Nusselt equation for a single tube without vapor shear. Thus, Figure 12.29 shows that the vapor velocity effect is a little below the predicted value for plain tubes.

Cavallini et al. [1996] developed a correlation to predict the effect of vapor shear. They used an asymptotic correlation given by

$$h^n = h_{st}^n + h_{fc}^n \qquad (12.46)$$

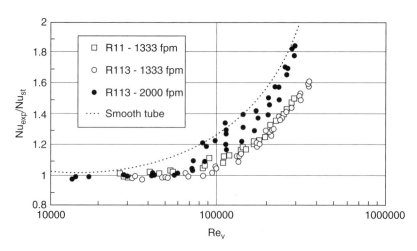

Figure 12.29 Effect of vapor velocity on condensation coefficient for integal-fin tubes. (From Cavallini et al. [1993].)

where h_{st} is for a single tube with no vapor shear, and h_{fc} is for the same tube with vapor shear. Their correlation for h_{fc} was derived by correlating the experimentally determined values of h_{fc}. The correlation for h_{fc} is written in the form of an equivalent Reynolds number and is given by

$$\text{Nu}_{fc} = C\text{Re}_{eq}^{0.8} \text{Pr}^{1/3} \qquad (12.47)$$

The constant was derived from multiple regression analysis of their R-11 and R-113 data and is given by

$$C = 0.03 + 0.166(t/p) + 0.07(e/p) \qquad (12.48)$$

The equivalent Reynolds number is given by

$$\text{Re}_{eq} = \frac{\rho_v u_\infty d_o}{\mu_l}\left(\frac{\rho_l}{\rho_v}\right)^{1/2} \qquad (12.49)$$

They showed that use of Equation 12.46 with the Briggs and Rose [1994] correlation to predict h_{st}, yielded very good prediction of their data R-11 and the R-113 data of Honda et al. [1989, 1991, 1992]. They also made a similar prediction of the steam data of Michael et al. [1989] for 200, 400, and 800 fins/m. Larger deviations were found for the steam data, but they were with in ±30%.

Cavallini et al. [1995] compared the vapor velocity effect for R-113 flowing downward over a 1333 fin/m circular integral-fin tube and the Hitachi Thermoexcel-C (spine fin) tube. Both tubes have approximately equal fin pitch and fin height. At the same Re_v, a smaller enhancement ratio was found for the Thermoexcel-C than for the circular integral-fin tube. The maximum enhancement observed for the Thermoexcel-C tube was 40% at $\text{Re}_v = 550,000$. Based on the observed condensate drainage pattern of Cavallini et al. [1994], they proposed that the different behavior of the two tubes was caused by the different condensate drainage patterns. Hence, one is cautioned that use of Equations 12.47 and 12.48 for spine fin tubes would likely lead to over-prediction of the vapor velocity–induced enhancement.

12.6.2 Condensate Inundation without Vapor Shear

Much work has been done to study condensation on single, horizontal enhanced tubes. However, very few studies have been undertaken to determine the "row effect" exponent for enhanced tubes. The row effect is defined as

$$\frac{\overline{h_N}}{h_l} = N^{-m} \qquad (12.50)$$

where $\overline{h_N}$ is the average condensation coefficient for N tube rows, and h_l is for a single tube. The exponent m determines the row effect. Kern [1958] has recommended

$m = 1/6$ as an approximate value for plain tubes. Katz and Geist [1948] measured the row effect for 6 rows of finned tubes having 590 fins/m and 1.6 mm fin height using R-12, n-butane, acetone, and water. They found $m = 0.04$. Figure 12.30 shows row effect data for steam condensation on plain and enhanced tubes. The plain tube data of Brower [1985] reasonably agrees with the Kern recommendation of $m = 1/6$. Marto and Wanniarachchi [1984] measured the row effect for a wire-wrapped plain 16-mm-diameter tube (1.6-mm wire diameter, 8.0-mm wire pitch) and found $m = 0.025$. Marto [1986] also found $m = 0.04$ for an integral-fin tube (400 fins/m, with 1.0-mm fin spacing and fin height). Singh et al. [2001] provides row-effect data (four tube rank) for steam condensing on the 390 fins/m integral-fin tube, which was tested by Kumar et al. [2002b]. Data were taken at 4 saturation pressures between 112 and 196 kPa. They found the row-effect exponent of Equation 12.50 to be 0.08.

Webb and Murawski [1990] measured the row effect for R-11 condensing on four enhanced tube geometries (1024 standard integral-fin, Tred-26D, Turbo-B, and GEWA-SC). The R-11 was condensed on a vertical rank of five horizontal tubes. The tube geometries are shown in Figure 12.6 and described in Table 12.4. The single tube performance of these tubes was shown on Figure 12.8. Figure 12.31 shows the row effect plotted in the form of the individual tube condensation coefficient vs. condensate Reynolds number leaving the tube. The average condensation coefficient on the Nth tube row is given by

$$\overline{h_N} = a\mathrm{Re}_l^{-n} \qquad (12.51)$$

Figure 12.31 shows that the data for all tube rows of a given tube geometry fall on a single curve, when plotted in the h_N vs. Re_l format. Because the h_N depends only on Re_l, and not on tube rows, Figure 12.31 may be interpreted for any number of tube rows, within the Re_l range of the data. Table 12.12 gives curve fits of straight lines drawn through the h_N vs. Re_l curves of Figure 12.31.

The standard 1024 fins/m tube shows no row effect within the Re_l range tested. Honda et al. [1987a] condensed R-113 on a rank of standard 1024 fins/m tubes, and also found zero row effect. The GEWA-SC tube had the second best single-tube performance, and the second best row effect ($n = 0.22$). The greatest row effect was displayed by the Tred-D tube, having a Reynolds number exponent of $n = 0.58$. Although the Turbo-C tube showed the highest single tube performance, its row effect ($n = 0.51$) is only a little better than that of the Tred-D. These row effect exponents may be compared to that for a plain tube, which has $n = 1/4$, according to the Nusselt theory and given by Holman [1986].

It appears that the superior row effect of the integral-fin tube occurs because the high, solid fins prevent axial spreading of the condensate. Hence, the condensate is channeled over the tube, leaving a region free of condensate between the condensate drainage columns. The poorest row effect was yielded by the geometries having a sawtooth fin shape. It appears that the sawtooth cuts in the fins allow axial spreading of the condensate, which drips from the tube row above.

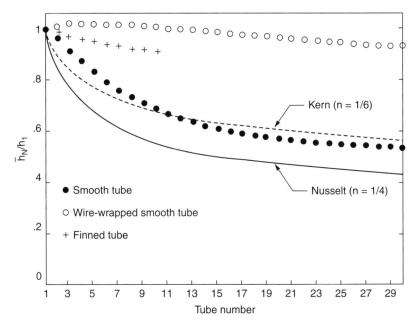

Figure 12.30 Condensation row-effect exponent for steam condensation at 1 bar. (From Marto and Wanniarachchi [1984].)

Table 12.12 Exponent n and Constant a in Equation 12.41

Geometry	$a \times 10^{-3}$	n
1024 fins/m	12.90	0.000
GEWA-SC	54.14	0.220
Turbo-C	257.80	0.507
Tred-D	269.90	0.576

If the condensate Reynolds number leaving the last row of a bank of tubes is known, one may find the average condensation coefficient on the bank of N-tube rows by integrating Equation 12.51. The result is given by

$$\bar{h}_N = \frac{a(\mathrm{Re}_{l,N}^{-n} - \mathrm{Re}_{l,1}^{-n})}{(1-n)(\mathrm{Re}_{l,N} - \mathrm{Re}_{l,1})} \tag{12.52}$$

where $\mathrm{Re}_{l,N}$ and $\mathrm{Re}_{l,1}$ are the condensate Reynolds numbers leaving the Nth and the first tube rows, respectively. An example which shows how the bundle average condensation coefficient may be calculated using Equation 12.52, is given later.

It is interesting to compare the condensate flow rates for steam and R-11 condensing at the same heat flux. For condensation at 38°C, the R-11 condensate Reynolds number will be 25 times greater, and the condensate volume flow rate will

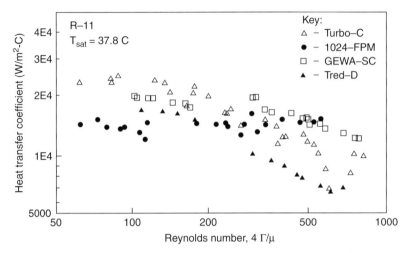

Figure 12.31 Row-effect data for the Figure 12.8 enhanced tube geometries. (From Webb and Murawski [1990].)

be 9.5 times greater. Hence, the condensate flow patterns may be considerably different in R-11 and steam condensers for the same heat flux.

Recent row-effect data for refrigerants are provided by Belghazi et al. [2003] and Honda et al. [2003]. Singh et al. [2001] provide row-effect data for steam. Belghazi et al. [2003] obtained data for 3 columns × 13 rows with R-134a and a R-23/R-134a binary mixture using five different tube geometries. Their tests were performed on integral-fin tubes (11, 19, 26, 32, and 40 fins/in.) and the GEWA-C. tube. They found that the simple Katz and Geist [1948] equation (Equation 12.52 with β = 0.04) predicted their R-134a row effect data quite well. In their tests of four integral-fin tubes and three spine-fin tubes with R-407C, Honda et al. [2003] found that the spine-fin tubes suffered more degradation due to inundation than the integral-fin tubes. The condensation coefficients were smaller for the R-407C (an R-23/R-134a binary mixture) than for R-134a.

12.6.3 Condensate Drainage Pattern

Honda et al. [1987a] have studied the condensate flow pattern for a bank of integral-fin tubes (d_o = 15.8 mm, e = 1.4 mm, 1063 fins/m). Condensate was pumped over a rank of 3 tubes, with vertical tube pitches of 22 or 44 mm using R-113, methanol, and n-propanol. They defined four condensate flow patterns: droplet mode, column mode, column and sheet mode, and sheet mode. Figure 12.32 illustrates the droplet and column modes on plain and 1060 fins/m integral-fin tubes. They analyzed their results to define the flow pattern transitions and the characteristics of the condensate flow. Figure 12.33 shows their flow pattern map for R-113 at 22-mm tube pitch. Assuming that the parameters Γ, σ, ρ, and g are important, dimensional analysis shows that the following dimensionless group is significant:

$$K = \left(\frac{\lambda}{\sigma^{3/4}} \right) \left(\frac{g}{\rho_l} \right)^{1/4} \tag{12.53}$$

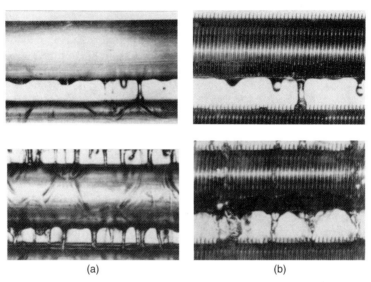

(a)   (b)

Figure 12.32 Flow patterns observed by Honda et al. [1987a] for R-113 condensating at 0.12 MPa on 15.9-diamter plain and 1060 fins/m integral-fin tubes. The top and bottom figures are for the first and 13th rows, respectively. (a) Plain tubes: First row ($Re_l = 19$), 13th row ($Re_l = 150$). (b) 1024 fins/m integral-fin tubes: First row ($Re_l = 50$), 13th row ($Re_l = 550$). (From Honda et al. [1987b]).

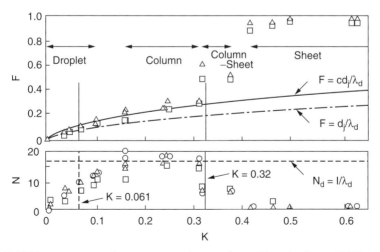

Figure 12.33 Flow pattern map of parameters F and N vs. K for R-113 condensing on 1060 fins/m tube. (From Honda et al [1987b].)

The parameter N in Figure 12.33 is the number of axial droplet release locations along the tube length (160 mm). The parameter F is defined as d_j/λ_d, as illustrated in Figure 12.33. The λ_d parameter is the one-dimensional Taylor instability wavelength, which is defined in the nomenclature as the end of the chapter. As shown by Figure 12.34, the F parameter is the fraction of the tube length that is totally condensate flooded. This fraction of the tube length will have a very small condensation coefficient. They show that $F = cd_j/\lambda_d$, where d_j is the column diameter and $c = 1.5$ for R-12. $F = cd_j/\lambda_d$ is plotted as the upper solid line on the F vs. K curve of Figure 12.33. The authors also present an analytical equation to predict the column diameter, d_j.

As discussed in Section 12.6.1, considerable work has been done to determine the effect of vapor velocity for condensation on plain tubes. Much of this work is summarized by Fujii [1991]. Several studies have been done for integral-fin tubes. Recent work includes Lee and Rose [1984], and Michael et al. [1989] used R-113 vapor flow normal to a single horizontal tube. Honda et al. [1991, 1992] studied R-113 condensing in downward flow over inline and staggered finned tube bundles, respectively. Michael et al. tested downward vapor Reynolds number (Re$_v$ = $d_o u_\infty/\mu_l$ for R-113 and steam. The plain and finned tubes were 19-mm outer diameter. The finned tubes had 1.0-mm fin height and fin thickness, and 3 finned tubes were used (200, 400, and 800 fins/m). They found that vapor velocity benefits both finned and plain tubes. Figure 12.35a shows how the U value is influenced by vapor velocity (u_∞) for 3 m/s water-side velocity. Figure 12.35b shows how the enhancement ratio (E_o) is affected by Re$_v$ for R-12. Figure 12.35b shows that the enhancement ratio tends to decrease with increasing vapor velocity. The reason for the decrease of E_o is that the smooth tube benefits more from vapor velocity than does the finned tube.

Roques and Thome [2003] observed drainage modes for vertical arrays of horizontal plain and enhanced tubes and defined transition thresholds. They observed

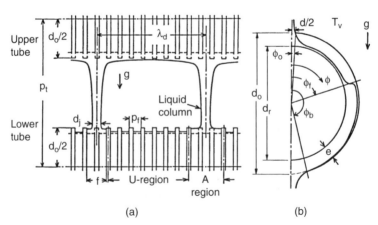

Figure 12.34 Illustration of liquid flow in column region and definition of flow pattern parameters shown on Figure 12.33. (a) Condensate flow in column mode; (b) tube cross section. (From Honda and Nozu [1987c].)

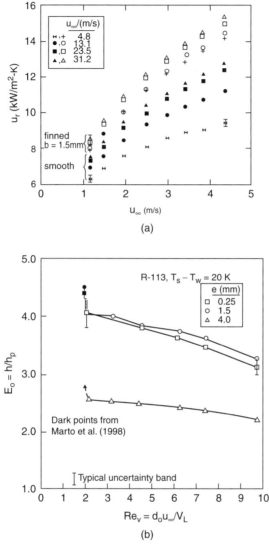

Figure 12.35 (a) Effect of vapor velocity u_∞ u_r for 3-m/s water-side velocity. (b) Variation of the enhancement ratio with the vapor Reynolds number for R-113 condensing on finned tubes having 1.0-mm-thick square fins with fin spacings of 0.25, 1.5, and 4 mm. (From Michael et al. [1989].)

the same drainage modes as previously defined by Honda et al. [1987a]: droplet mode, column mode, column and sheet mode, and sheet mode. Their observations were made for plain tubes, integral-fin tubes (748, 1024, and 1575 fins/m), Thermoexcel-C, and Turbo-BII tubes. Observations were made for drainage of water, 50/50 water/glycol, and glycol. The transitions observed were similar for all fin pitches. However, significantly higher transition thresholds were observed for the lowest fins/m tubes.

12.6.4 Prediction of the Condensation Coefficient

Having developed their analytical model to predict the condensation coefficient on a single finned tube and established the condensate flow pattern for a bank of tubes, Honda and Nozu [1987c] proceeded to predict the condensation coefficient for each tube row. Their predictive model is at present limited to the integral-fin tubes. The basic structure of their model is as follows. The tube length between the draining condensate columns/droplets (λ_d) is divided into the U and A regions illustrated on Figure 12.34. The A region is that receiving condensate from the draining columns. The U region does not receive condensate from the tube above, and essentially acts as if it were in the top tube row. The condensate behavior on the tube circumference is influenced by the condensate retention angle. Hence, the U region is divided into region U_u (unflooded) and U_f (flooded region), where flooding is defined by the condensate retention angle. The A region similarly contains regions A_u (unflooded) and A_f (flooded). Their model assumes that the

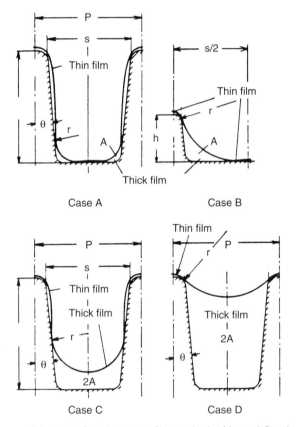

Figure 12.36 Four possible cases of condensate profile on a bank of integral-fin tubes from Honda et al. [1987].

condensate retention angle is the same in the U and A regions. Each of these four regions is modeled as one of the four cases illustrated in Figure 12.36. In the U_f and A_f regions, only case D of Figure 12.36 is possible. Case A, B, or C would typically apply to the U_u and A_u regions, depending on the fin height and condensation rate. Having defined the four tube regions, they applied the model of Honda and Nozu [1987a] to the U_u and U_f regions. The same model was applied to the A_u and A_f regions, with correction for condensate inundation. They used the model to predict the condensation coefficient for a multirow case (6 to 9 rows) of R-12 and 12 rows of steam condensation. The R-12 data were predicted within 12%, and the steam data were 5 to 20% lower than the experimental value. The Adamek and Webb [1990b] model may be equally applied to predicting the condensation coefficient on a bank of integral-fin tubes.

The Honda and Nozu [1987c] model for a bank of tubes is at present limited to integral-fin tubes, which neatly channel the condensate and which have two-dimensional condensate profiles within the fin region. No model exists for tubes, such as the Turbo-C tube (see Figure 12.6), which have sawtooth fin shapes. Although the GEWA-SC tube of Figure 12.6 will have a two-dimensional condensate profile, it may not channel the condensate in the same manner as occurs on the integral-fin tube. Hence, flow pattern studies would be required to characterize this tube. It should be possible to develop a single-tube model, but such work has not yet been done.

12.7 CONCLUSIONS

Surface tension forces are dominant in draining the condensate from the fins on plates and tubes having typical commercial fin geometries. With the geometry of the condensate–vapor interface defined, models exist to predict the condensation coefficient.

Finned horizontal tubes experience condensate retention, which fills the interfin region around a fraction of the tube circumference. Analytical models have been developed to predict the condensate retention angle for typically used integral-fin tubes.

Experiments have been performed to measure the row effect on enhanced horizontal tube geometries. For integral-fin tubes, these experiments show that the row-effect loss is smaller than that of a plain tube. The enhanced tube geometries having a sawtooth fin shape show a greater row-effect loss than do integral-fin tubes.

Work has been done to define the flow patterns that exist on a bank of integral-fin tubes. Honda and Nozu [1987c] have shown excellent success in predicting the condensation coefficient on a bank of integral-fin tubes.

Future research is needed to characterize and predict the condensation coefficient for the more complex surface geometries (e.g., those having a sawtooth fin shape).

APPENDIX 12.A: THE KEDZIERSKI AND WEBB [1990] FIN PROFILE SHAPES

Two key geometric parameters define the basic shape of the profile:

1. The fin height (e) and base thickness (t_b)
2. The fin tip radius (r_o) and the rotation angle, θ_m

Equation A.1 defines the profile shape as a function of e, t_b, r_o, and Z, for $\theta_m = \pi/2$. The term Z is a shape factor parameter.

$$r = c_1 + c_2 e^{z\theta} + c_3\theta \tag{A.1}$$

The constants c_1, c_2, and c_3 are given by Equations A.2, A.3, and A.4.

$$c_1 = r_o - c_2 \tag{A.2}$$

$$c_2 = \frac{0.5t_b(\sin\theta_m - \theta_m\cos\theta_m) - e(\cos\theta_m + \theta_m\sin\theta_m - 1)}{c_4 - [2(1 - \cos\theta_m) - \theta_m\sin\theta_m]}$$

$$+ \frac{r_o[\theta_m\sin\theta_m - 2(1 - \cos\theta_m)]}{c_4 - [2(1 - \cos\theta_m) - \theta_m\sin\theta_m]} \tag{A.3}$$

$$c_3 = \frac{0.5t_b - c_1\sin\theta_m - c_2 Z\exp(Z\theta_m)\cos\theta_m/(Z^2 + 1)}{\cos\theta_m + \theta_m\sin\theta_m - 1}$$

$$- \frac{c_2[\exp(Z\theta_m)\sin\theta_m - Z]/(Z^2 + 1)}{\cos\theta_m + \theta_m\sin\theta_m - 1} \tag{A.4}$$

The constant c_4 in Equation A.3 is given by

$$c_4 = [\exp(Z\theta_m)(1 - \cos\theta_m + Z(\sin\theta_m - \theta_m))]$$
$$+ (Z\theta_m - 1)\cos\theta_m - (Z + \theta_m)\sin\theta_m + 1)/(Z^2 + 1) \tag{A.5}$$

The constants defined by Equations A.2 through A.4 are obtained by applying the following boundary conditions to Equation A1.

$$r = r_o \text{ at } \theta = 0 \tag{A.6}$$

$$t_b = 2\int_0^{\pi/2} r\cos\theta \, d\theta \tag{A.7}$$

$$e = \int_0^{\pi/2} r\sin\theta \, d\theta \tag{A.8}$$

APPENDIX 12.B: FIN EFFICIENCY IN THE FLOODED REGION

In the flooded fraction of the tube, condensation occurs on the fin tip. Because the fins are quite short, a one-dimensional conduction model is acceptable. Assuming no heat transfer from the fin sides, heat conduction in a fin of constant thickness, t, is given by

$$Q = k_w A_b (T_{w,t} - T_{w,r})/h \tag{B.1}$$

where $T_{w,r}$ is the fin base temperature, $T_{w,t}$ is the fin tip temperature, and h is the fin height. The heat transferred by convection from the fin tip is

$$Q = h_t A_t (T_{sat} - T_{w,t}) \tag{B.2}$$

Combining Equations B.1 and B.2 and solving for $T_{w,\pm}$ gives

$$T_{wt} = (T_{w,r} + Nu_t T_{sat})/(Nu_t + 1) \tag{B.3}$$

where $Nu_t = h_{\pm} h/k_w$. Substitution of Equation B.3 in Equation B.2 gives

$$Q = h_t A_t (T_{sat} - T_{w,r})/(Nu_t + 1) \tag{B.4}$$

Let Q_{max} be the heat transfer rate for a fin, having $T_{w,t} = T_{w,r}$, which is

$$Q_{max} = h_{t,\infty} A_t (T_{sat} - T_{w,r}) \tag{B.5}$$

where $h_{t,\infty}$ is the condensation coefficient if $T_{w,t} = T_{w,r}$. The fin efficiency is defined as q/q_{max}, and is the ratio of Equations B.4 and B.5 giving

$$\eta_f = h_t/h_{t,\infty}(Nu_t + 1) \tag{B.6}$$

The fin tip efficiency may be iteratively determined by assuming a fin tip temperature and using Equations B.1 and B.2 to check for convergence of the fin tip temperature.

NOMENCLATURE

A	Heat transfer surface, A_f (fin area), A_r (root region), A_{ft} (fin tip), A_{fs} (fin side), m^2 or ft^2
Bo	Bond number, ratio of gravity force to surface tension force, dimensionless
c_b	Fraction of tube circumference condensate bridged ($= \beta/\pi$), dimensionless
d_j	Column jet diameter (Figure 12.34), m or ft
d_e	Diameter over fins of finned tube, m or ft
d_o	Tube outside diameter, m or ft
d_r	Tube outside diameter at fin root, m or ft
D_h	Hydraulic diameter, m or ft

D_c	Diameter of helix, m or ft
D_m	Mean diameter of fluted tube, m or ft
e	Fin height, roughness height, corrugation depth, or coating thickness, m
e^+	Roughness Reynolds number ($eu*/v_f$), dimensionless
E_i	Tube-side enhancement ratio, compared to plain tube (h_i/h_{ip}), dimensionless
E_o	Outside enhancement ratio, compared to plain tube h_o/h_{op}, dimensionless
F	Parameter on Figure 12.33 (= $d_j/\lambda_d d$), dimensionless
f_{ik}	Force per unit volume within region L_{ik}, N/m³ or lbf/ft³
F_d	Drainage force $F_d = \sigma(1/r_1 - 1/r_2)/\Delta x$ (surface tension drainage), $F_d = g(\rho_l-\rho_v)$ (gravity drainage), N/m³ or lbf/ft³
F_g	Gravity drainage force $(\rho_l-\rho_v)$ g, N/m³ or lbf/ft³
F_p	Property group $4k_1 v\Delta T_{vs}/\lambda$, N or lbf
g	Gravitational constant, m/s² or ft/s²
G	Mass velocity in tube, $G_v = xG$, $G_l = (1 - x)G$, kg/m²-s or lbm/ft²-s
$f\bar{h}$	Average condensation coefficient, W/m²-K or Btu/hr-ft²-°F
h	Heat transfer coefficient, h_l (on top row), h_t (on fin tip), h_h (horizontal plain tube), h_f (on fins), W/m²-K or Btu/hr-ft²-°F
h_N	(On Nth tube row), $\bar{h}_N$ (averge over N tube rows), W/m²-K or Btu/hr-ft²-°F
h_{Nu}	Condensation coefficient calculated from Nusselt's laminar film model, W/m²-K or Btu/hr-ft²-°F
h_s	Heat transfer coefficient on smooth surface, W/m²-K or Btu/hr-ft²-°F
h_w	Water-side heat transfer coefficient, W/m²-K or Btu/hr-ft²-°F
H_d	Height of drainage strip, m or ft
H_L	Helix length per 360° revolution in the Figure 12.16 tubes, m or ft
K	Parameter in Equation 12.53, dimensionless
k_l	Thermal conductivity of condensate, W/m-K or Btu/hr-ft-°F
k_w	Tube thermal conductivity, W/m-K or Btu/hr-ft-°F
l_{ik}	Length between points $x = i$ and $x = k$ (Figure 12.24), m or ft
L_{ik}	Region between points $x = i$ and $x = k$ (Figure 12.24) m or ft
L	Tube or surface length, m or ft
L_b	Width of drainage channel (Figure 12.20), m or ft
L_s	Depth of drainage channel (Figure 12.20), m or ft
LMTD	Log-mean temperature difference, K or °F
m	Row-effect exponent (dimensionless)
$\dot{m}_{ik}$	Condensate flow per unit width rate on region L_{ik} in Figure 12.24, kg/s-m or lbm/s-ft
$\dot{m}$	Condensate flow rate, $\dot{m}_l$ (from tube), $\dot{m}_h$ (on base tube), $\dot{m}_r$ (sum of $\dot{m}_f$ and $\dot{m}_h$), kg/s or lbm/s
N	Number of flutes on tube
N	Number of horizontal tube rows in depth
p	Pressure, p_{sat} (saturation pressure), kPa or lbf/ft²
p	Axial pitch of surface elements, m or ft
P_f	Fin pitch, m or ft

P	Pumping power, W or hp
P_w	Wetted perimeter of inner tube surface, m or ft
Nu	Nusselt number, dimensionless
Pr	Prandtl number, dimensionless
q	Heat flux, W/m² or Btu/hr-ft²
Q	Heat transfer rate, W or Btu/hr
r	Local radius of liquid–vapor interface, m or ft
r_c	Radius at corner of fin tip on trapezoidal fin, m or ft
r_o	Tip radius of convex profile, m or ft
R	Radius of drainage interface at fin root, m
R^*	$R - \delta_{54}$, projection of R on side wall from point 4 of Figure 12.24, m or ft
Re_{eq}	Vapor equivalent Reynolds number Eq. 12.49, dimensionless
Re_l	Condensate Reynolds number $(4\Gamma/\mu_l)$, dimensionless
Re_v	Vapor Reynolds number $(d_o u_\infty v_v)$, dimensionless
s	Spacing between fins $(= p_{ft}-t)$, m or ft
s	Coordinate distance along curved condensing profile, m or ft
S_m	Length of convex profile, m or ft
t	Fin thickness, t_b (at base), $t_\pm$ (at tip), m or ft
T_s	Saturation temperature, also T_{sat}, K or °F
T_w	Wall temperature, K or °F
ΔT	ΔT_u (unflooded region), ΔT_f (flooded region), K or °F
ΔT_{vs}	$T_s - T_w$, K or °F
u	Velocity, u_∞ (vapor velocity), m/s or ft/s
u^*	Frictional shear velocity $(\tau/\rho)^{1/2}$, m/s or ft/s
U	Overall heat transfer coefficient, U_r (based on root diameter), U_p (plain tube), W/m²-K or Btu/hr-ft²-°F
v	Volume of differential control volume, m³ or ft³
V	Volume of heat exchanger tubing material, m³ or ft³

Greek Symbols

α	Angle on Figure 12.24, rad
β	Condensate retention angle (Figure 12.5), rad
Γ	Condensate mass velocity on horizontal tube, $\dot{m}_l/L$, kg/s-m or lbm/s-ft
Γ	Condensate mass velocity leaving vertical tube, $\dot{m}_l/\pi d_o$, kg/s-m or lbm/s-ft
δ	Condensate film thickness, m or ft
ζ	Parameter in Adamek profile shape equation, dimensionless
η_f	Fin efficiency, η_{fu} (unflooded region), η_{ff} (flooded region), dimensionless
θ_m	Rotation angle from fin tip to fin base, rad
λ	Latent heat, J/kg or Btu/lbm
λ_d	One-dimensional Taylor instability wave length, m or ft
π	Constant (3.14159)
μ	Dynamic viscosity μ_l (of liquid), μ_v (of vapor), kg/s-m² or lbm/s-ft²

ρ Density: ρ_l (of liquid), ρ_v (of vapor), kg/m^3 or lbm/ft^3
σ Surface tension, N/m or lbf/ft
τ_i Interfacial shear stress, N/m^2 or lbf/ft^2
τ_w Wall shear stress, N/m^2 or lbf/ft^2

Subscripts

av   Average value
f Flooded region
i Inner surface of tube
m Mean value
o Outer surface of tube
p Plain tube
s Smooth tube
u Unflooded region

Unsubscripted variables refer to enhanced tube.

REFERENCES

Abdullah, R., Cooper, J.R., Briggs, A., and Rose, J.W., 1995. Condensation of steam and R113 on a bank of horizontal tubes in the presence of a noncondensing gas, *Exp. Thermal Fluid Sci*, 11, 298–306.

Abu-Orabi, M., 1998. Modeling of heat transfer in dropwise condensation, *Int. J. Heat Mass Transfer*, 41, 81–87.

Adamek, T.A., 1981. Bestimmung der Kondensationgrossen auf feingewellten Oberflachen zur Auslegun aptimaler Wandprofile, *Wärme Stoffbertrag*, 15, 255–270.

Adamek, T.A. and Webb, R.L., 1990a. Prediction of film condensation on vertical finned plates and tubes—a model for the drainage channel, *Int. J. Heat Mass Transfer*, 33, 1737–1749.

Adamek, T.A. and Webb, R.L., 1990b. Prediction of film condensation on horizontal integral-fin tubes, *Int. J. Heat Mass Transfer*, 33, 1721–1735.

Beatty, K.O., Jr. and Katz, D.L., 1948. Condensation of vapors on outside of finned tubes, *Chem. Eng. Prog.*, 44, 1, 908–914.

Belghazi, M., Bontemps, A., and Marvillet, C., 2002. Condensation heat transfer on enhanced surface tubes: experimental results and predictive theory, *J. Heat Transfer*, 124, 754–761.

Belghazi, M., Bontemps, A., and Marvillet, C., 2003. Experimental study and modelling of heat transfer during condensation of pure fluid and binary mixture on a bundle of horizontal finned tubes, *Int. J. Refrigeration*, 26, 214–223.

Briggs, A. and Rose, J.W., 1994. Effect of fin efficiency on a model for condensation heat transfer on a horizontal, integral-fin tube, *Int. J. Heat Mass Transfer*, 37, 457.

Briggs, A., and Rose, J.W., 1995. Condensation performance of some commercial integral fin tubes with steam and CFC113, *Exp. Heat Transfer*, 8, 2, 131–144.

Briggs, A. and Rose, J.W., 1996. Condensation on low-finned tubes: effects of non-uniform wall temperature and interphase matter transfer, in *Process, Enhanced and Multiphase Heat Transfer*, R.M. Manglik and A.D. Kraus, Eds., Begell House, New York, 455–460.

Briggs, A. and Rose, J.W., 1999. An evaluation of models for condensation heat transfer on low-finned tubes, *J. Enhanced Heat Transfer*, 6, 51–60.

Briggs, A., Huang, X-S., and Rose, J.W., 1995. An experimental investigation of condensation on integral-fin tubes: effect of fin thickness, height and thermal conductivity, *Proc. 30th National Heat Conf.*, HTD-Vol. 308, 21–30.

Briggs, A., Wen, X.L., and Rose, J.W., 1992. Accurate measurements of heat-transfer coefficients for condensation on horizontal integral-fin tubes, *J. Heat Transfer*, 114, 719–726.

Briggs, A., Wang, H.S., and Rose, J.W., 2002. Film condensation of steam on a horizontal wire-wrapped tube, in *Heat Transfer 2002, Proc. 12th Int. Heat Transfer Conf.*, 4, 123–128.

Briggs, A., Wang, H.S., Murase, T., and Rose J.W., 2003. Heat transfer measurements for condensation of steam on a horizontal wire-wrapped tube, *J. Enhanced Heat Transfer*, 10, 355–362.

Briggs, A., Wang, H.S., Murase, T., and Rose, J., 2004. Heat transfer measurements for condensation of steam on a horizontal wire-wrapped tube, *J. Enhanced Heat Transfer*, 10, 355–362.

Brower, S.K., 1985. The Effects of Condensate Inundation on Steam Condensate Heat Transfer in a Tube Bundle, M.S. thesis, Naval Postgraduate School, Monterey, CA.

Brown, C.D. and Matin, S.A., 1971. The effect of finite metal conductivity on the condensation heat transfer to falling water rivulets on vertical heat transfer surfaces, *Heat Transfer*, 93, 69–76.

Browne, M.W. and Bansal, P.K., 1999. An overview of condensation heat transfer on horizontal tube bundles, *Appl. Thermal Eng.*, 19, 565–594.

Butizov, A.I., Rifert, V.G., and Leont'yev, G.G., 1975. Heat transfer in steam condensation on wire-finned vertical surfaces, *Heat Transfer Soviet Res.*, 7, 5, 116–120.

Carey, V.P., 1992. *Liquid Vapor Phase-Change Phenomena*, Hemisphere, Washington, DC.

Carnavos, T.C., 1974. Chapter 17 in *Heat Exchangers: Design and Theory Sourcebook*, N. Afgan and E.U. Schlünder, Eds., McGraw-Hill, New York.

Carnavos, T.C., 1980. An Experimental Study: Condensing R-11 on Augmented Tubes, ASME Paper 80-HT-54, ASME, New York.

Cary, J.D. and Mikic, B.B., 1973. The influence of thermocapillary flow on heat transfer in film condensation, *J. Heat Transfer*, 95, 21–24.

Cavallini, A., Bella, B., Longo, G.A., and Rossetto, L. 1993. Pure vapour condensation of Refrigerant 113 on a horizontal 2000 FPM integral finned tube, in *Condensation and Condenser Design (Proc. Engineering Foundation Conference on Condensation and Condenser Design)*, J. Taborek, J. Rose, and I. Tanasawa, Eds., ASME, New York 357–365.

Cavallini, A., Doretti, L., Longo, G.A., and Rossetto, L., 1994. Flow patterns during condensation of pure refrigerants on enhanced tubes under high vapor velocity, *Proceedings of the 1994 International Refrigeration Conference at Purdue*, D.R. Tree and J.E. Braun, Eds., 311–316.

Cavallini, A., Bella, B., Longo, G.A., and Rossetto, L., 1995. Experimental heat transfer coefficients during condensation of halogenated refrigerants on enhanced tubes, *J. Enhanced Heat Transfer*, 2, 115–126.

Cavallini, A., Doretti, L., and Rossetto, L., 1996. A new model for forced-convection condensation on integral-fin tubes, *J. Heat Transfer*, 118, 689–693.

Cheng, W.Y., Wang, C.C., Hu, R., and Huang, L.W., 1996. Film condensation of HCFC-22 on horizontal enhanced tubes, *Int. Comm. Heat Mass Transfer*, 23(1), 79–90.

Cheng, B. and Tao, W.Q., 1994. Experimental study of R-152a film condensation on single horizontal smooth tube and enhanced tubes, *J. Heat Transfer*, 116(1), 266.

Collier, J.G. and Thome, J.R., 1994. *Convective Boiling and Condensation*, 3rd ed., McGraw-Hill, New York, pp. 430–487.

Combs, S.H. and Murphy, R.W., 1978. Experimental studies of OTEC heat transfer condensation of ammonia on vertical fluted tubes, *Proc 5th OTEC Conf.*, Feb. 20–22, 1978, Miami Beach, FL, Vol. 1, Sect. 6, 111–122.

Das, A.K., Meyer, D.W., Incheck, G.A., Marto, P.J., and Memory, S.B., 1995. Effect of fin height and thermal conductivity on the performance of integral-fin tubes for steam condensation, *Proc. 30th Nat. Heat Transfer Conf.*, V.K. Dhir, Ed., ASME HTD, 308, 111–122.

Das, A.K., Incheck, G.A., and Marto, P.J., 1999. The effect of fin height during steam condensation on a horizontal stainless steel integral-fin tube, *J. Enhanced Heat Transfer*, 6, 237–250.

Das, A.K., Kilty, H.P., Marto, P.J., Andeen, G.B., and Kumar, A., 2000a. The use of an organic self-assembled monolayer coating to promote dropwise condensation of steam on horizontal tubes, *J. Heat Transfer*, 122, 278–286.

Das, A.K., Kilty, H.P., Marto, P.J., Kumar, A., and Andeen, G.B., 2000b. Dropwise condensation of steam on horizontal corrugated tubes using an organic self-assembled monolayer coating, *J. Enhanced Heat Transfer*, 7, 109–124.

Dipprey, D.F. and Sabersky, R.H., 1963. Heat and momentum transfer in smooth and rough tubes at various Prandtl numbers, *Int. J. Heat Mass Transfer*, 6, 329–353.

Fujii, T., 1991, *Theory of Laminar Film Condensation*, Springer-Verlag, New York, 71.

Fujii, T., Wang, W., Kooyama, S., and Shimizu, Y., 1985. Heat transfer enhancement for gravity controlled condensation on a horizontal tube by coiling wires, *Proc. 2nd Int. Symp. on Heat Transfer*, Tsinghua University, Beijing, October 15–18, 1985. Also published in *Heat Transfer Science and Technology*, Hemisphere, Washington D.C., 1987, 773–780.

Fujii, T., Wang, W.C., Koyama, S., and Shimizu, Y., 1987. Heat transfer enhancement for gravity controlled condensation on a horizontal tube by coiling a wire, in *Heat Transfer Science and Technology*, B.X. Wang, Ed., Hemisphere, Washington, D.C., 773–780.

Glicksman, L.A., Mikic, B.B., and Snow, D.F., 1973. Enhancement of film condensation on the outside of horizontal tubes, *AIChE J.*, 19, 636–637.

Gogonin, I.I. and Kabov, O.A. 1996. An experimental study of R-11 and R-12 film condensation on horizontal integral-fin tubes, *J. Enhanced Heat Transfer*, 3, 43–54.

Gregorig, R., 1954. Film condensation on finely rippled surfaces with consideration of surface tension, *Z. Angew. Math. Phys.*, 5, 36–49.

Griffith, P., 1985. Condensation, Part 2: Dropwise condensation, in *Handbook of Heat Transfer Fundamentals*, W.M. Rohsenow, H.P. Hartnett, and E.N. Ganic, Eds., McGraw-Hill, New York, 11.37–11.50.

Hanneman, R.J., 1977. Recent Advances in Dropwise Condensation Theory, ASME paper 77-WA/HT-21, ASME, New York.

Henrici, K., 1961. Kodensation von Frigen 12 and Frigen 22 an glatten und berippten Rohren, Dissertation, TU Karlsruhe.

Holman, J.P., 1986. *Heat Transfer*, 6th ed., McGraw-Hill, New York.

Honda, H., and Makishi, O., 1995. Effect of a circumferential rib on film condensation on a horizontal two-dimensional fin tube, *J. Enhanced Heat Transfer*, 2, 307–315.

Honda, H. and Nozu, S., 1985. Effect of drainage strips on the condensation heat transfer performance of horizontal finned tubes, *Proc. Int. Symp. on Heat Transfer*, Beijing, 455–462.

Honda, H. and Nozu, S., 1987a. A prediction method for heat transfer during film condensation on horizontal low integral-fin tubes, *J. Heat Transfer*, 109, 218–225.

Honda, H. and Nozu, S., 1987b. Effect of drainage strips on the condensation heat transfer performance of horizontal finned tubes, in *Heat Transfer Science and Technology*, B.X. Wang, Ed., Hemisphere, Washington, D.C., 455–462.

Honda, H. and Nozu, S., 1987c. A theoretical model of film condensation in a bundle of horizontal low finned tubes, *J. Heat Transfer*, 111, 525–532.

Honda, H., Nozu, S., and Mitsumori, K., 1983. Augmentation of condensation on horizontal finned tubes by attaching a porous drainage plate, *Proc. ASME-JSME Thermal Eng. Conf.*, Honolulu, 3, 289–295.

Honda, H., Nozu, S., and Uchima, B., 1986. Effect of vapor velocity on film condensaton of R-113 on horizontal tubes in crossflow, *Int. J. Heat Mass Transfer*, 29, 429–438.

Honda, H., Nozu, S., and Takeda, Y., 1987a. Flow characteristics of condensate on a vertical column of horizontal low finned tubes, *Proc. ASME-JSME Thermal Eng. Joint Conf.*, 1, 517–524.

Honda, H., Nozu, S., and Uchima, B., 1987b. A generalized prediction method for heat transfer during film condensation on a horizontal low finned tube, *Proc. ASME-JSME Thermal Eng. Joint Conf.*, 4, 385–392.

Honda, H., Takamatsu, H., and Kim, K., 1994. Condensation of CFC-11 and HCFC-123 in in-line bundles of horizontal finned tubes: effect of fin geometry, *J. Enhanced Heat Transfer* 1, 197–210.

Honda, H., Takata, N., Takamatsu, H., Kim, J.S., and Usami, K., 2003. Effect of fin geometry on condensation in a staggered bundle of horizontal finned tubes, *J. Heat Transfer*, 125, 653–660.

Honda, H., Uchima, B., Nozu, S., Nakata, H., and Torigoe, E., 1989. Film condensation on downward flowing R-113 vapor on in-line bundles of horizontal finned tubes, in *Heat Transfer Equipment Fundamentals, Design, Applications, and Operating Problems*, HTD-vol. 108, ASME, pp. 117–125.

Honda, H., Uchima, B., Nozu, S., Nakata, H., and Torigoe, E., 1991. Film condensation of downward flowing R-113 vapor on in-line bundles of horizontal finned tubes, *J. Heat Transfer*, 113, 479–486.

Honda, H., Uchima, B., Nozu, S., Torigoe, E., and Imai, S. 1992. Film condensation R-113 on staggered bundles of horizontal finned tubes, *J. Heat Transfer*, 114, 442–449.

Iltscheff, S., 1971. Some experiments concerning the attainment of drop condensation with fluorinated refrigerants, *Kaltetech, Klimatisierung*, 23, 237–241.

Incropera, F.P. and DeWitt, D.P., 1990. *Fundamentals of Heat and Mass Transfer*, John Wiley & Sons, New York, 615–619.

Jaber, M.H. and Webb, R.L., 1993. An experimental investigation of enhanced tubes for steam condensers, *Exp. Heat Transfer*, 6, 35–54.

Jaber, M.H. and R.L. Webb, 1996. Steam condensation on horizontal integral-fin tubes of low thermal conductivity, *J. Enhanced Heat Transfer*, 3, 55–71.

Jung, D., Kim, C.B., Cho, S., and Song, K., 1999. Condensation heat transfer coefficients of enhanced tubes with alternative refrigerants for CFC11 and CFC12, *Int. J. Refrig*. 22, 548–557.

Karkhu, V.A. and Borovkov, V.P., 1971. Film condensation of vapor at finely-finned horizontal tubes, *Heat Transfer Sovi. Res.*, 3, *No*. 2, 183–191.

Katz, D.L. and Geist, J.M., 1948. Condensation of six finned tubes in a vertical row, *Trans. ASME*, 70, 907–914.

Kedzierski, M.A. and Webb, R.L., 1987. Experimental measurements of condensation on vertical plates with enhanced fins, in *Boiling and Condensation in Heat Transfer Equipment*, E.G. Ragi, T.M. Rudy, T.J. Rabas, and J.M. Robertson, Eds., ASME Symp. Vol. HTD; 85, 87–95.

Kedzierski, M.A. and Webb, R.L., 1990. Practical fin shapes for surface tension drained condensation, *J. Heat Transfer*, 112, 479–485.

Kern, K.Q., 1958. Mathematical development of loading in horizontal condensers, *AIChE J.*, 4, 157–160.

Kim, K.J., Lefsaker, A.M., Razani, A., and Stone, A., 2001. The effective use of heat transfer additives for steam condensation, *Appl. Thermal Eng.*, 21, 1863–1874.

Koch, G., Zhang, D.C., Grischke, A.L., Trojan, K., and Dimigen, H., 1998. Study on plasma enhanced CVD coated material to promote dropwise condensation of steam, *Int. J. Heat Mass Transfer*, 41, 1899–1906.

Kumar, R., Varma, H.K., Mohanty, B., and Agrawal, K.N., 2000. Condensation of R-134a over single horizontal circular integral-fin tubes with trapezoidal fins, *Heat Transfer Eng.*, 21, 29–39.

Kumar, R., Varma, H.K., Mohanty, B., and Agrawal, K.N., 2002a. Augmenatation of heat transfer during filmwise condensation of steam and R-134A over single horizontal finned tubes, *Int. J. Heat Mass Transfer*, 45, 201–211.

Kumar, R., Varma, H.K., Mohanty, B., and Agrawal, K.N., 2002b. Prediction of heat transfer coefficients during condensation of water and R-134a on single horizontal integral-fin tubes, *Int. J. Refrig.*, 25, 111–126.

Kun, L.C. and Ragi, E.G., 1981. Enhancement for Film Condensation Apparatus, U.S. patent 4,253,519.

Lee, W.C. and Rose, J.W., 1984. Forced convection film condensation on a horizontal tube with and without non-condensing gases, *Int. J. Heat Mass Transfer,* 27, 519–528.

Leipertz, A., Choi, K.H., and Diezel, L., 2002. Dropwise condensation heat transfer on ion implanted metallic surfaces, *Heat Transfer 2002, Proc. 12th Int. Heat Transfer Conf.*, 3, 887–892.

Liu, X., Ma, T., and Zhang, Z., 1999. Investigation of enhancement of steam condensation heat transfer on finned tubes with porous drainage strips, Paper AJTE99–6350, in *Proc. 5th ASME/JSME Joint Thermal Engineering Conf.*, March 15–19, San Diego, CA.

Lixin, C. and Jiehui, Y., 1998. A new treated surface for achieving dropwise condensation, *J. Enhanced Heat Transfer*, 5, 1–8.

Ma, X., Chen, J., Xu, D., Lin, J., Ren, C., and Long, Z., 2002. Influence of processing conditions of polymer film on dropwise condensation heat transfer, *Int. J. Heat Mass Transfer*, 45, 3405–3411.

Ma, X.H. and Wang, B.X., 1998. Suction effect of a vertical coated plain porous layer on film condensation heat transfer enhancement, *Heat Transfer 1998, Proc. 11th Int. Heat Trans. Conf.*, 5, 387–392.

Marto, P.J., 1986. Recent progress in enhancing film condensation heat transfer on horizontal tubes, *Heat Transfer 1986—Proceedings of the Eighth Int. Heat Transfer Conf.*, 1, 161–170.

Marto, P.J., 1988. An evaluation of film condensation on horizontal integral-fin tubes, *J. Heat Transfer*, 110, 1287–1305.

Marto, P.J. and Wanniarachchi, A.S., 1984. The use of wire-wrapped tubing to enhance steam condensation in tube bundles, in *Heat Transfer in Heat Rejection Systems*, S. Sengupta and Y.G. Mussalli, Eds., ASME Symp. Vol. HTD, 37, ASME, New York, 9–16.

Marto, P.J., Reilly, D.J., and Fenner, J.H., 1979. An experimental comparison of enhanced heat transfer condenser tubing, in *Advances in Enhanced Heat Transfer*, J.M. Chenoweth et al., Eds., ASME, New York, 1–10.

Marto, P.J., Mitrou, E., Wanniarachchi, A.S., and Rose, J.W., 1986. Film condensation of steam on horizontal integral finned tubes: effect of fin shape, in *Proc. 8th IHTC*, San Francisco, 1695–1700.

Marto, P.J., Mitrou, E., Wanniarachchi, W., and Katsuta, M., 1987. Film condensation of steam on a horizontal wire wrapped tube, *Proc. ASME-JSME Thermal Eng. Joint Conf.*, 1, 509–516.

Marto, P.J., Zebrowski, D., Wanniarachchi, A.S., and Rose, J.W., 1988. Film condensation of r-113 on horizontal finned tubes, in *Proc. 1988 National Heat Transfer Conference*, Vol. 2, H.R. Jacobs, Ed., ASME Symp. Vol. HTD, 96, 583–592.

Masuda, H. and Rose, J.W., 1987. An experimental study of condensation of refrigerant R113 on low integral-fin tubes, in *Heat Transfer Science and Technology*, Hemisphere, New York, 480–487.

McNaught, J.M. and Cotchin, C.D., 1989. Heat transfer and pressure drop in a shell and tube condenser with plain and low-fin tube bundles, *Chem. Eng. Res. Design*, 67, 127–133.

Medwell, J.O. and Nicol, A.A., 1965. Surface Roughness Effects on Condensing Films, ASME paper 65-HT-43, ASME, New York.

Mehta, H.H. and Rao, M.R., 1979. Heat transfer and frictional characteristics of spirally enhanced tubes for horizontal condensers, in *Advances in Enhanced Heat Transfer*, J.M. Chenoweth et al., Eds., ASME, New York, 11–22.

Michael, A.G., Marto, P.J., Wanniarachchi, A.S., and Rose, J.W., 1989. Effect of vapor velocity during condensation on horizontal smooth and finned tubes, in *Heat Transfer with Change of Phase*, I.S. Habib and R.J. Dallman, Eds., ASME Symp. Vol. HTD, 114, 1–10.

Mitrou, E., 1986, Film Condensation of Steam on Externally Enhanced Finned Tubes, M.S. thesis, Naval Postgraduate School, Monterey, CA.

Mori, Y., Hijikata, H., Hirasawa, S., and Nakayama, W., 1979. Optimized performance of condensers with outside condensing surface, in *Condensation Heat Transfer*, J.M. Chenoweth et al., Eds., ASME Symp. Vol., ASME, New York, 55–62.

Newson, I.H. and Hodgson, T.D., 1973. The development of enhanced heat transfer condenser tubing, *Desalination*, 14, 291–323.

Nicol, A.A. and Medwell, J.O., 1966. The effect of surface roughness on condensing steam, *Can. J. Chem. Eng.*, 66, 3, 170–173.

Notaro, P., 1979. Enhanced Condensation Heat Transfer Device and Method, U.S. patent 4,154,294.

Nusselt, W., 1916. Die Oberflachenkondensation des Wasserdampfes, *Z. Ver. Dtsch. Ing.*, 60, 541–569.

Panchal, C.B., 1994. Generalized correlation for condensation on vertical fluted surfaces, *Heat Transfer Eng.*, 15(4), 19–23.

Pearson, J.F. and Withers, J.G., 1969. New finned tube configuration improves refrigerant condensing, *ASHRAE J.*, pp. 77–82.

Rabas, T.J. and Taborek, J. 1999. Performance, fouling, and cost considerations of enhanced tubes in power-plant condensers, *J. Enhanced Heat Transfer*, 6, 289–315.

Rao, M.R., 1988. Heat transfer and friction correlations for turbulent flow of water and viscous non-Newtonian fluids in single-start spirally corrugated tubes, *Proc. 1988 Natl. Heat Trans. Conf.*, Vol. 1, ASME Symp.Vol. HTD, 96, 677–683.

Renken, K.J. and Aboyoe, M., 1993a. Analysis of film condensation within inclined thin porous-layer coated surface, *Int. J. Heat Fluid Flow*, 14, 48–53.

Renken, K.J. and Aboyoe, M., 1993b. Experiments on film condensation promotion within thin inclined porous coatings, *Int. J. Heat Mass Transfer*, 36, 1347–1355.

Renken, K.J. and Mueller, C.D., 1993. Measurements of enhanced film condensation utilizing a porous metallic coating, *J. Thermophys. Heat Transfer*, 7, 148–152.

Renken, K.J. and Raich, M.R., 1996. Forced convection steam condensation experiments within thin porous coatings, *Int. J. Heat Mass Transfer*, 39, 2937–2945.

Rifert, V.G. and Leont'yev, G.G., 1976. An analysis of heat transfer with steam condensing on a vertical surface with wires to promote heat transfer, *Teploenergetika*, 23(4), 74–80.

Rifert, V.G. and Trokoz, Y.Y., 1996. Surface nonisothermness effect on profiled surface condensation enhancement, *J. Enhanced Heat Transfer*, 8, 1749–1754.

Roques, J.F. and Thome, J.R., 2003. Falling film transitions between droplet, column, and sheet flow modes on a vertical array of horizontal 19 FPI and 40 FPI low-finned tubes, *Heat Transfer Eng.*, 24, 40–45.

Rose, J.W., 1984. Effect of pressure gradient in forced convection film condensation on a horizontal tube, *Int. J. Heat Mass Transfer*, 27, 39–47.

Rose, J.W., 1994. An approximate equation for the vapour-side heat-transfer coefficient for condensation on low-finned tubes, *Int. J. Heat Mass Transfer*, 37, 865–877.

Rose, J.W. 2002. An analysis of laminar film condensation on a horizontal wire-wrapped tube, *Trans. IChemE*, 80 (Part A), 290–294.

Rudy, T.M. and Webb, R.L., 1983. An analytical model to predict condensate retention on horizontal integral-fin tubes, *ASME-JSME Thermal Eng. Joint Conf.*, Vol. 1, ASME, New York, 373–378.

Rudy, T.M. and Webb, R.L., 1985. An analytical model to predict condensate retention on horizontal integral-fin tubes, *J. Heat Transfer*, 107, 361–368.

Shah, R.K., Zhou, S.Q., and Tagavi, K.A., 1999. The role of surface tension in film condensation in extended surface passages, *J. Enhanced Heat Transfer*, 6, 179–216.

Shekriladze, J.G. and Gomealauri, V.I., 1966. Theoretical study of laminar film condensation of a flowing vapor, *Int. J. Heat Mass Transfer*, 9, 581–591.

Singh, S.K., Kumar, R., and Mohanty, B., 2001. Heat transfer during condensation of steam over a vertical grid of horizontal integral-fin copper tubes, *Appl. Thermal Eng.* 21, 717–730.

Sreepathi, L.K., Bapat, S.L., and Sukhatme, S.P., 1996. Heat Transfer during film condensation of R-123 vapour on horizontal integral-fin tubes, *J. Enhanced Heat Transfer*, 3(2), 147.

Srinivasan, P.S., Balasubramanian, R., and Gaitonde, U.N., 2002. Correlation for laminar film condensation over single horizontal integral-fin copper tubes, *Heat Transfer 2002, Proc. 12th Int. Heat Transfer Conf.*, 4, 213–218.

Staub, P.W., 1966. Condensing Heat Transfer Surface Device, U.S. patent 3,289,752.

Sukhatme, S.P., Jagadish, B.S., and Prabhakaran, P., 1990. Film condensation of R-11 vapor on single horizontal enhanced condenser tubes, *J. Heat Transfer*, 112, 229–234.

Takahashi, A., Nosetani, T., and Miyata, K., 1979. Heat Transfer Performance of Enhanced Low Finned Tubes with Spirally Integral Inside Fins, Sumitomo Light Metal Technical Report, Vol. 20, pp. 59–65.

Thomas, A., Lorenz, J.J., Hillis, D.A., Young, D.T., and Sather, N.P., 1979. Performance tests of the 1 MWt shell and tube exchangers for OTEC, *Proc. 6th OTEC Conf.*, Paper lc.

Thomas, D.G., 1967. Enhancement of film condensation rates on vertical tubes by vertical wires, *Ind. Eng. Chem. Fundam.*, 6(1), 97–102.

Thomas, D.G., 1968. Enhancement of film condensation rate on vertical tubes by longitudinal fins, *AIChE J.*, 644–649.

Wang, S.P., Hijikata, K., and Deng, S.J., 1990. Experimental study on condensation heat transfer by various kinds of integral-finned tubes, *Proc. 2nd Intl. Symposium on Condensation and Condensers*, University of Bath, Bath, U.K., 397–406.

Wang, Z.Z., Wei, W.D., and Hong, F., 2000. Experimental study of condensation heat transfer promotion on a fluted tube with thin porous coatings, *Heat Transfer Eng.* 21(4), 46–52.

Wanniarachchi, A.S., Marto, P., and Rose, J., 1986. Filmwise condensation of steam on horizontal tubes with rectangular-shaped fins, in *Multiphase Flow and Heat Transfer*, V.K. Dhir, J.C. Chen, and O.C. Jones, Eds., ASME Symp. Vol. HTD, 47, 93–99.

Webb, R.L., 1979. A generalized procedure for the design and optimization of fluted Gregorig condensing surfaces, *J. Heat Transfer*, 101, 335–339.

Webb, R.L., 1981. The use of enhanced heat transfer surface geometries in condensers, in *Power Condenser Heat Transfer Technology: Computer Modeling, Design, Fouling*, P.J. Marto, and R.H. Nunn, Eds., Hemisphere, Washington, D.C., 287–324.

Webb, R.L., 1984. The effects of vapor velocity and tube bundle geometry on condensation in shell-side refrigeration condensers, *ASHRAE Trans.*, 90 (Part 1B), 39–59.

Webb, R.L. and Kedzierski, M.A., 1990. Practical fin shapes for surface tension drained condensation, *J. Heat Transfer*, 112, 479–485.

Webb, R.L. and Murawski, C.G., 1990. Row effect for R-11 condensation on enhanced tubes, *J. Heat Transfer*, 112, 768–776.

Webb, R.L., Keswani, S.T., and Rudy, T.M., 1982. Investigation of surface tension and gravity effects in film condensation, *Heat Transfer—1982, (Proc. Seventh Intl. Heat Transfer Conf.)*, Vol. 5, Hemisphere, New York, 175–181, .

Webb, R.L., Rudy, T.M., and Kedzierski, M.A., 1985. Prediction of the condensation coefficient on horizontal integral-fin tubes, *J. Heat Transfer*, 107, 369–376.

Wen, X.-L., Briggs, A., and Rose, J.W., 1994. Enhancement of condensation heat transfer on integral-fin tubes using radiused fin-root fillets, *J. Enhanced Heat Transfer*, 1, 211–217.

Wildsmith, G., 1980. Open Discussion section, in *Power Condenser Heat Transfer Technology*, P.J. Marto and R.H. Nunn, Eds., Hemisphere, New York, 463–468.

Withers, J.G. and Young, E.H., 1971. Steam condensing on vertical rows of horizontal corrugated and plain tubes, *Ind. Eng. Chem. Process Design Dev.*, 10(1), 19–30.

Yabe, A., 1991. Active heat transfer enhancement by applying electric fields, *ASME/JSME Thermal Eng. Proc.*, 3, xv-xxiii.

Yau, K.K., Cooper, J.R., and Rose, J.W., 1985. Effect of fin spacing on the performance of horizontal integral-fin condenser tubes, *J. Heat Transfer*, 107, 377–383.

Yau, K.K., Cooper, J.R., and Rose, J.W., 1986. Horizontal plain and low-finned condenser tubes—effect of fin spacing and drainage strips on heat transfer and condensate retention, *J. Heat Transfer*, 108, 946–950.

Yorkshire, 1982. *YIM Heat Exchanger Tubes: Design Data for Horizontal Rope Tubes in Steam Condensers*, Technical Memorandum 3, Yorkshire Imperial Metals, Ltd., Leeds, U.K.

Zener, C. and Lavi, A., 1974. Drainage systems for condensation, *J. Heat Transfer*, 96, 209–205.

Zhao, Q. and Burnside, B.M., 1994. Dropwise condensation of steam on ion implanted condenser surfaces, *Heat Recovery Syst. CHP*, 14, 525–534.

CONVECTIVE VAPORIZATION

13.1 INTRODUCTION

Chapter 11 addressed pool boiling on plates and circular tubes, in which, the generated vapor is removed from the surface by the departing bubbles. However, this is not the case for convective vaporization. For example, for vaporization within a tube the generated vapor is contained within the tube, and it affects the local heat transfer coefficient via the flow pattern.

The combination of total flow rate and local vapor velocity establishes a flow pattern, which changes with vapor quality. Horizontal tubes may behave differently from vertical tubes, because gravity force acts to stratify the liquid. Hence, enhancement requirements for horizontal tubes may be different from those for vertical tubes. Circular tubes are not the only "channel flow" geometry of interest. Brazed aluminum heat exchangers may use flat, extruded aluminum tubes having rectangular flow passages, or they may be of the plate-and-fin construction. Another channel flow geometry of interest is the annulus. Boiling on the outside of tubes in a bundle also involves convective effects. These geometries are also addressed.

Prior to addressing the subject of enhancement, information relating to fundamental understanding of two-phase flow and convective heat transfer is provided.

13.2 FUNDAMENTALS

This section is concerned with fundamentals of vaporization in channels (e.g., in horizontal or vertical plain tubes). Understanding these processes in plain tubes provides a foundation for understanding the performance of enhanced surfaces. Only the aspects of two-phase flow and heat transfer that are particularly relevant to enhanced heat transfer are discussed. For a more detailed discussion of the general topic, the reader should consult a text on the subject area, such as Carey [1992].

When vaporization or condensation occurs in a tube, convective effects occur, which do not exist in pool boiling or vapor condensation, without vapor velocity. The convective effects are associated with the influence of shear stress on the liquid–vapor interface. Further, if gravity force exceeds vapor shear forces, the liquid phase will tend to stratify in horizontal tubes.

As the vapor quality increases in vaporization, the vapor velocity increases, causing increased vapor shear. In condensation, the converse exists. A two-phase flow experiences different "flow patterns" as vapor quality changes along the tube length. The heat transfer coefficient is significantly affected by the flow pattern. Hence, it is important to understand how flow pattern affects the heat transfer coefficient in two-phase heat transfer.

13.2.1 Flow Patterns

Consider vaporization in a tube having subcooled liquid entering and 100% vapor quality leaving. Figure 13.1 and Figure 13.2 illustrate the flow patterns that will exist in vertical and horizontal evaporator tubes, respectively. Gravity leads to stratification in the Figure 13.2 horizontal tube. Stratification effects are more severe at lower vapor qualities, because vapor shear forces are smaller at low vapor quality.

Consider the situation for complete condensation in tubes. Figure 13.3 illustrates the flow patterns for condensation in a horizontal tube. These flow patterns are similar to the reverse of Figure 13.2, with one exception. Vapor may be generated by nucleate boiling at the tube wall, as illustrated in Figure 13.2. Condensation may occur in co-current downward flow. The flow patterns would appear similar to those in Figure 13.1, except vapor bubbles (caused by nucleate boiling) would not exist at the tube wall.

Work has been done to characterize the flow pattern as a function of mass velocity and vapor quality. Figure 13.4 shows the map defined by Hewitt and Roberts [1969] for vertical upward flow. Figure 13.5 show the map defined by Baker [1954] for flow in a horizontal tube. The Taitel and Dukler [1976] map is more accurate, but requires too much space to explain here. See Carey [1992] for details.

The dashed line in Figure 13.4 shows the flow patterns experienced by R-12 evaporating from $0.01 \leq x \leq 0.95$ in a vertical tube. This figure shows that the flow patterns are very sensitive to the mass velocity of the liquid and the vapor. The phase mass velocities are defined as $G_l = G(1 - x)$ and $G_v = Gx$, where G is the total flow

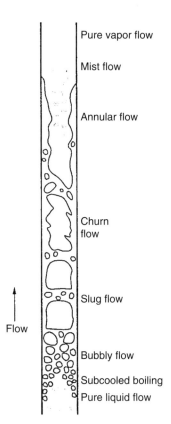

Pure vapor flow

Mist flow

Annular flow

Churn flow

Slug flow

Bubbly flow

Subcooled boiling

Pure liquid flow

Flow

Figure 13.1 Flow patterns for complete vaporization for upward flow in a vertical tube. (From Carey [1992].)

rate divided by the pipe cross-sectional area. The reverse set of flow patterns would be experienced by a condensing flow, which traverses the same vapor qualities.

13.2.2 Convective Vaporization in Tubes

Figure 13.2 shows that three different heat transfer mechanisms can exist, depending on the flow pattern. These are nucleate (subcooled, bubbly, and plug flow), thin film evaporation (annular flow), and single-phase heat transfer to gas in the drywall region, downstream from the annular flow region. The third drywall region may or may not exist, depending on whether the "critical heat flux" has been exceeded in the annular flow region. Chen [1966] proposed that liquid velocity tends to suppress nucleate boiling and that the nucleate boiling contribution is, $h_{nb} = Sh_{nbp}$, where h_{nbp} is for nucleate pool boiling and S is the suppression factor. If nucleate boiling is totally suppressed in the annular film region, heat transfer occurs by thin film evaporation (convective evaporation). If the film flow is laminar, the heat transfer coefficient is given by $h = k_l/\delta$, where δ is the film thickness. In the drywall region,

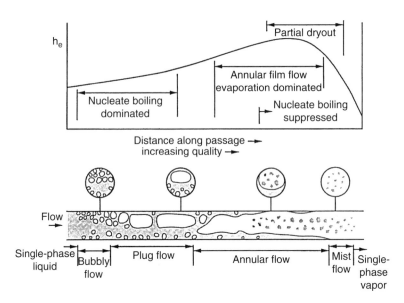

Figure 13.2 Flow patterns for complete vaporization in a horizontal tube. (From Carey [1992].)

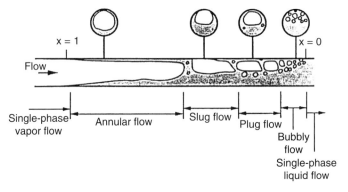

Figure 13.3 Flow patterns for complete condensation in a horizontal tube. (From Carey [1992].)

the heat transfer coefficient is given by the appropriate equation for heat transfer to a gas.

It is possible that both nucleate boiling and convective evaporation heat transfer will exist at the same time. If so, one way to account for both components is via the superposition model, which was proposed by Chen [1966]. This model assumes that the total heat flux (q) is the sum of the nucleate boiling component and the convective evaporation component. This is written as

$$q = q_{nb} + q_{cv} \tag{13.1}$$

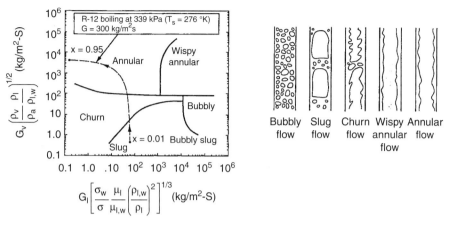

Figure 13.4 Hewitt and Roberts [1969] flow pattern map for vertical co-current upward flow. The figure is annotated to show the flow patterns encountered evaporating by R-12. (Carey [1992].)

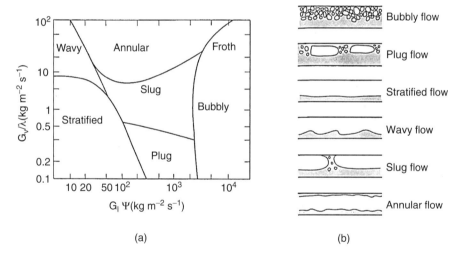

Figure 13.5 (a) Flow pattern map of Baker [1954] for a horizontal tube, (b) defined flow patterns.

The terms q_{nb} and q_{cv} are calculated at the value of wall superheat $[\Delta T_{ws} = (T_w - T_s)]$ that exists in the system. One may write the superposition model in terms of the component heat transfer coefficients by dividing Equation 13.1 by the wall superheat to obtain

$$h = h_{nb} + h_{cv} \tag{13.2}$$

where h_{nb} is the nucleate boiling contribution and h_{cv} is the convective contribution. One calculates $h_{cv} = Fh_l$, where h_l is calculated from an appropriate equation for single-phase flow of the liquid phase alone. For vaporization inside a tube, one may calculate h_l using the Dittus–Boelter equation as given by

$$h_l = \frac{0.023k_l}{d_i} \left(\frac{d_i G (1-x)}{\mu_l} \right)^{0.8} Pr^{1/3} \tag{13.3}$$

Figure 13.6 illustrates the convective vaporization curve (frequently called the flow boiling curve) as a log-log plot of heat flux vs. wall superheat. The constant parameters are mass velocity (G) and vapor quality (x). The convective vaporization curve is asymptotic to q_{cv} at low ΔT_{ws}, and asymptotic to q_{nb} at high ΔT_{ws}. The suppression factor, S, is assumed to be unity in Figure 13.6 for the purpose of illustration. The q_{cv} asymptote has unity slope, because h_{cv} independent of ΔT_{ws}. The objective of a convective vaporization correlation is to predict the heat flux in the curved region between the asymptotes. Steiner and Taborek [1992] have proposed that an asymptotic model gives a better fit of the data in the region between q_{nb} and q_{cv} than does the superposition model (Equation 13.1). This model is written as

$$q^n = q_{nb}^n + q_{cv}^n \tag{13.4}$$

At low heat flux, the vaporization curve is asymptotic to h_{cv} (two-phase convection). At high heat flux, the vaporization curve is asymptotic to h_{nb} (nucleate boiling). The line below and parallel to h_{cv} in Figure 13.6b is that for the liquid phase flowing alone (h_l). The corresponding heat flux and heat transfer coefficient is q_l (liquid phase convection) and h_l, respectively. The ratio, F, is defined as the "two-phase convection multiplier," and is equal to q_{cv}/q_l or h_{cv}/h_l. The curve labeled "superposition" $(n = 1)$ illustrates the superposition model. Using the superposition model, one would derive F from experimental data by the equation

$$F = \frac{q - q_{nb}}{q_l} = \frac{h - Sh_{nbp}}{h_l} \tag{13.5}$$

When $n > 1$, the curved portion of the convective vaporization curve is asymptotic with the limiting straight lines q_{cv} and q_{nb} (Figures 13.6a) and h_{cv} and h_{nb} (Figure 13.6b). One asymptote of the prediction is the line of fully convective vaporization coefficient, where the nucleate boiling contribution is insignificant. The other asymptote is the nucleate boiling coefficient, where the contribution due to convective heat transfer is insignificant. The curved region between the two asymptotes contracts as the value of n increases. As $n \to \infty$, the curved region shrinks to zero length. Thus, the contributions of nucleate boiling and convective evaporation are added, to some extent, in the curved region. If $n = 1$, the total heat flux is the sum of the two components, which is the superposition model.

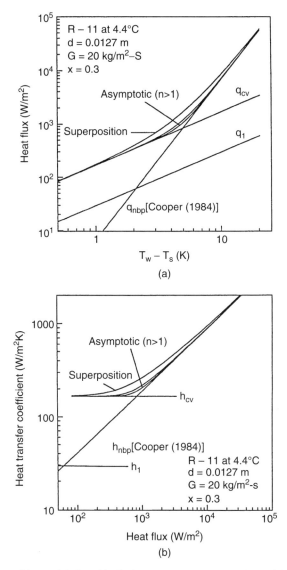

Figure 13.6 Superposition model plotted in the form q vs. ΔT_{ws}. (From Webb and Gupte [1992].)

Enhancement of convective vaporization may be approached by increasing either h_{nbp} or h_{cv} (see Equation 13.2). Using a porous boiling surface on the inner surface of a plain tube will cause $h_{nbp} \gg h_{cv}$, and the tube performance will be "nucleate boiling dominated" up to quite high vapor qualities. In this case, the total heat transfer coefficient will be quite sensitive to heat flux, but relatively insensitive to flow rate. However, if the enhancement consists of internal fins, which should not significantly enhance the nucleate boiling component, the tube performance will be "convection dominated." Here, the total heat transfer coefficient will be sensitive to flow rate, but relatively insensitive to heat flux.

Most tube-side enhancements work to enhance the convective term. Examples are corrugated roughness, internal fins, the "star insert," and twisted tapes. The use of fine grooves at a helix angle on the inner surface may use capillary forces to transport liquid to the upper tube wall. Any approach that thins the film without causing dry spots will provide enhancement of the h_{cv} term in Equation 13.2. Roughness provides enhancement by turbulent mixing of the liquid film. Such a roughness may promote droplet entrainment at moderate vapor qualities, which causes a thinner film on the tube wall.

It is important to recognize that a low heat transfer coefficient will exist if the wall becomes dry. As shown by Figure 13.2 and Figure 13.5, horizontal gravity forces may act to stratify the flow, causing the upper part of the tube to be dry, or intermittently dry.

13.2.3 Two-Phase Pressure Drop

Refrigerant pressure drop affects the local saturation temperature in the bundle. In addition, the shell-side heat transfer coefficient correlation is also directly related to the frictional pressure gradient. The refrigerant pressure drop is given by

$$\Delta p = \Delta p_F + \Delta p_g + \Delta p_a \tag{13.6}$$

The three terms of the right-hand side of Equation 13.6 account for the friction, gravity, and acceleration contributions. For in-tube flow, the friction contribution, Δp_F, is calculated by

$$\Delta p_F = 2 f_f \frac{L}{d_i} G^2 v_l (1-x)^2 \phi_f^2 \tag{13.7}$$

where f_f is the friction factor for liquid flow at mass velocity $G(1-x)$. The term ϕ_f^2 is defined as $\Delta p_F/\Delta p_l$ and is a function of $X_{tt} = (\Delta p_{F,l}/\Delta p_{F,v})^{0.5}$, which is called the *Martinelli parameter*. The Martinelli parameter for turbulent flow of the gas and liquid phases in a circular tube is

$$X_{tt} = \left(\frac{1-x}{x}\right)^{0.9} \left(\frac{\rho_v}{\rho_l}\right)^{0.5} \left(\frac{\mu_l}{\mu_v}\right)^{0.1} \tag{13.8}$$

For flow in tubes, Chisholm [1967] has shown that ϕ_f^2 may be expressed in terms of X_{tt} by

$$\phi_f^2 = 1 + \frac{C}{X_{tt}} + \frac{1}{X_{tt}^2} \tag{13.9}$$

where C is an empirical constant that depends on the flow geometry. Chisholm [1967] shows that use of $C = 20$ in Equation 13.9 is applicable for flow in smooth tubes. Ishihara et al. [1980] proposes $C = 8$ for flow normal to plain tube bundles. Webb et al. [1990] propose that $C = 8$ is also applicable to enhanced tubes, that have a relatively smooth outside diameter. Because no correlations have been developed for the integral-fin geometry, a possible choice is to assume that the Ishihara correlation is also applicable.

Prediction of Δp_g and Δp_a requires knowledge of the void fraction, α, See Carey [1992] or other books on two-phase flow for void fraction correlations.

Tube-side enhancement techniques for certain enhanced geometries may increase the friction component (Δp_F), but they should not affect the gravity or acceleration components of the pressure drop. Hence, a significant increase of Δp_F may not have a significant effect on the total pressure drop. This, of course, depends on what fraction of the total pressure drop is due to Δp_g and Δp_a.

13.2.4 Effect of Flow Orientation on Flow Pattern

For complete vaporization, the flow pattern may be significantly different for the following situations:

1. Vertical tubes with vapor and liquid flowing down: For co-current downflow in a vertical tube, the liquid is confined to the tube wall, except for possibly entrained droplets. In this case, there is essentially no change of flow pattern, except for the amount of entrained liquid. Typically, such evaporators are supplied with a greater liquid rate than is evaporated. The ratio of the inlet-to-evaporated liquid is called the recirculation ratio. Typically, a recirculation ratio of 2 to 4 is used to prevent dryout of the liquid film. This case is called "falling film," or "thin film" evaporation. For plain surfaces, nucleate boiling probably would not occur. However, nucleate boiling may provide a significant contribution if an enhanced nucleate boiling surface is used. Except for enhanced nucleate boiling tubes, convective effects are important, and vapor shear forces may influence the convective contribution.
2. Vertical tubes with vapor and liquid flowing up: This situation experiences the flow patterns shown in Figure 13.1, and convective effects are important. Nucleate boiling may be important at low to moderate vapor qualities. The heating surface may become dry at high vapor qualities.
3. Horizontal tubes with co-current vapor–liquid flow: This situation is quite similar to item 2, above. However, gravity-influenced stratification of the liquid, combined with heat flux, may act to dry the upper tube surface. Hence, an internal enhancement geometry that acts to provide surface wetting may be more important than for vertical tubes.

Refrigerant pressure drop is an important consideration in evaluating tube-side enhancements for vaporization, particularly for fluids with high dT/dp. This concern is addressed in Chapter 4.

13.2.5 Convective Vaporization in Tube Bundles

Webb and Gupte [1992] survey models and correlations for convective vaporization in tubes and tube banks. They give recommended equations for calculation of the suppression factor, S, and the two-phase multiplier, F. The superposition model (Equation 13.1) or the asymptotic model (Equation 13.4) may be used for tube banks. One would use (1) an appropriate equation for pool boiling on horizontal tubes (h_{nbp}), (2) the applicable equation for single-phase heat transfer in tube banks (h_l), and (3) correlations for F and S that are applicable to tube banks. Several F factor correlations have been proposed for tube banks. These are discussed by Webb and Gupte [1992].

The flow pattern that exists in a tube bundle depends on the bundle orientation, and how the bundle is circuited on the tube side. For vertical upflow in a vertical thermosyphon reboiler, one would expect a flow pattern grossly similar to that in a vertical tube. For a horizontal tube bundle, the flow pattern would depend on whether it were a natural circulation reboiler or a flooded refrigerant evaporator. Relatively low vapor qualities (e.g., 10%) exist at the top tube row in a reboiler. However, complete evaporation occurs in a refrigerant evaporator. Polley et al. [1980] provide some description of flow patterns for shell-side vaporization.

13.2.6 Critical Heat Flux

The term *critical heat flux* (CHF) refers to critical conditions that may occur in convective vaporization, at which the heat transfer coefficient precipitously drops to a much lower level. The result of this phenomenon depends on the thermal boundary condition. If a heat flux boundary condition exists, the tube wall temperature will suddenly increase to compensate for the lower heat transfer coefficient. If a temperature boundary condition exists for a two-fluid heat exchanger, the local heat flux will decrease in response to the reduced overall heat transfer coefficient. A destructive condition may exist for a heat flux boundary condition, because the tube wall temperature may increase to the tube melting temperature. Such a condition may exist in a nuclear reactor or a fossil fuel boiler, whose walls are heated by radiant flames. However, this would not be the case in a two-fluid heat exchanger, where the second fluid fixes the tube wall temperature.

A vaporizing flow is subject to CHF at two different flow conditions depending on the heat flux and enthalpy of the flowing fluid. These are explained below:

1. At subcooled or low-vapor-quality conditions, nucleate boiling is the dominant heat transfer mechanism. If the heat flux is high enough at the particular local flow enthalpy, the boiling may suddenly revert to film boiling, for which the heat transfer coefficient drops to a much smaller value. This condition is called *departure from nucleate boiling* (DNB).
2. At moderate or high vapor qualities, the flow pattern is annular with entrained droplets. If the liquid film on the tube wall dries out, the heat transfer coefficient

will precipitously drop. This condition is described as *dryout*. The dryout condition occurs at lower vapor qualities as the mass velocity and local $(T_w - T_s)$ increase.

Both types of CHF conditions may exist in shell-side vaporization as well as tubeside vaporization. Certain enhancement techniques are effective in alleviating both CHF conditions. They are discussed in Section 13.6.

13.3 ENHANCEMENT TECHNIQUES IN TUBES

Enhancement geometries used for pool boiling on the outside of circular tubes are generally not applicable to vaporization inside tubes. This is because the flow pattern and convective effects significantly alter the vaporization process. A variety of enhancement techniques have been investigated. These include twisted-tape inserts, internal fins, and integral roughness. This chapter addresses enhancements applicable to co-current flow in horizontal and vertical tubes.

13.3.1 Internal Fins

In this section, internal fins are defined as having $e/d_i > 0.05$. Microfin tubes have smaller fin height and are discussed in Section 13.4. Work on internally finned tubes was slow to develop, because practical manufacturing techniques to form internal fins did not exist until the late 1960s. Figure 13.7 shows four basic types of internally finned tubes that have been investigated. Figure 13.7a shows axial and helical internal fins made in copper tubes by a swaging process. A 19-mm-diameter tube may have 12 to 30 fins, typically 1.5 to 3 mm in height, and of the same generic type tested by Carnavos, discussed in Section 8.2.2. Such copper-finned tubes are made by a swaging process, which is slow and relatively expensive. Axial internal fins may be made at a higher speed using a hot extrusion process for aluminum. One may add a helix angle by twisting the extruded fins. Figure 13.7b shows internally finned (or grooved) tubes made of steel, which are used in steam power boilers. The fin height is approximately 1.1 mm and is used to increase the departure from nucleate boiling. Figure 13.7c shows an aluminum extended-surface-area tube made by first extruding an aluminum plate with the corrugated pattern, then rolling the strip in a helix, followed by seam welding. This tube has also been made in thin wall titanium. Here, the titanium sheet is corrugated, then rolled in the helical form and welded. Figure 13.7d shows a modern *microfin* tube, which has a small-scale fin structure, and is used for refrigerants. A 9.0-mm tube may have 60 fins, 0.2 to 0.25 mm high at a small helix angle (e.g., 15°). These microfins may be formed in a copper tube at high speed by drawing the tube over a grooved slug.

The earliest work on internal fins was reported by Boling et al. [1953], who worked with refrigerant evaporation. Lavin and Young [1965] were the first to test an integral internal fin. They tested the four tube geometries shown in Figure 13.8 using R-12 and R-22 in horizontal and vertical flow. This very interesting paper

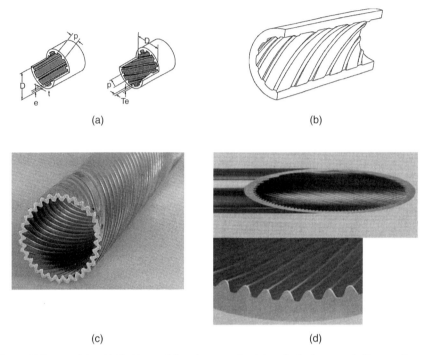

(a) (b)

(c) (d)

Figure 13.7 (a) Axial and helical internal fins, (b) grooved, axial steel tube for power boilers, (c) General Atomics spirally fluted tube, (d) Wieland microfin tube (from Wieland [1991].)

provides data for specific flow regimes (subcooled liquid, subcooled nucleate boiling, annular flow, and mist flow). Kubanek and Miletti [1979] tested R-22 in copper internal fin and aluminum insert tubes (see Figure 8.1c). They measured the average heat transfer coefficient in 0.8 and 2.0-m-long test sections with $\Delta x = 0.7$. Figure 13.9 shows their data, for which the heat transfer coefficient (h) is based on the nominal plain tube area ($A/L = \pi d_i$). Tubes 22, 25, and 30 have either 30 or 32 fins, and are 0.51 to 0.64 mm high. Tube 30 has axial fins, and the fins in tubes 22 and 25 are spiraled. Tube 25, with the tightest fin spiral, provides the highest enhancement — approximately 200% at $G = 100$ kg/m²-s. The aluminum insert device (tube 24c) gave the highest enhancement at low mass velocity, but is below that of the internal fin tubes at high velocity.

Schlunder and Chawla [1969] tested R-11 in 14-mm-diameter tubes having aluminum inserts, of the generic type illustrated by Figure 8.1c, inserted in the tube. The copper tube was "shrink fitted" to the insert to minimize thermal contact resistance. They tested inserts having two, four, or eight legs. These inserts act as full-height fins, and also reduce the hydraulic diameter of the flow passage. They provide empirical correlations for the heat transfer and pressure drop data. Pearson and Young [1970] provide R-22 data on the Figure 13.10 aluminum insert having five legs. Their data are taken for complete evaporation in a four-pass evaporator, with heat transfer coefficients measured for each pass. Wen and Hsieh [1995] tested

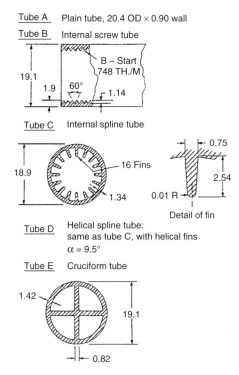

Tube A   Plain tube, 20.4 OD × 0.90 wall

Tube B   Internal screw tube

B – Start
748 TH./M

19.1

1.9   60°   1.14

Tube C   Internal spline tube

18.9   16 Fins

1.34   0.01 R

0.75

2.54

Detail of fin

Tube D   Helical spline tube:
same as tube C, with helical fins
α = 9.5°

Tube E   Cruciform tube

1.42

19.1

0.82

Figure 13.8 Tube geometries tested by Lavin and Young [1965] with R-12 and R-22 in vertical and horizontal flow. (From Lavin and Young [1965].)

two star-insert tubes (similar to Figure 13.10) using R-114, R-22, and R-134a. The test results are discussed in Section 13.3.3.

Panchal et al. [1992] have tested the Figure 13.7c aluminum tube with upflow of R-11. Tests were conducted for subcooled and saturated vaporization. Figure 13.11 shows the saturated vaporization results for exit vapor qualities between 0.4 and 0.9. The heat transfer coefficient is based on the total internal area (1.64 times that of a plain tube of the same inside diameter). The Nusselt and Reynolds numbers are defined as hD_h/k_l and $\mathrm{Re}_{Dh} = D_h G\,(1 - x)/\mu_l$ respectively. Figure 13.11 shows the heat transfer coefficient averaged over $0 < x < x_e$, where x_e is the exit vapor quality. The internal enhancement ($E_{hi} = h/h_p$) is between 1.25 and 1.6. If the enhancement were based on the nominal plain tube area, $2.05 \leq E_{hi} \leq 2.6$. The solid line through the data set is the prediction based on the Chen [1966] convective vaporization model.

Ito and Kimura [1979] and Shinohara and Tobe [1985] describe a tube having very small triangular shaped fins, which has become known as the microfin tube. This tube is used in virtually all currently manufactured refrigerant condensers and evaporators, either air or water cooled. Because of the importance of this tube, it is separately described in Section 13.4 below.

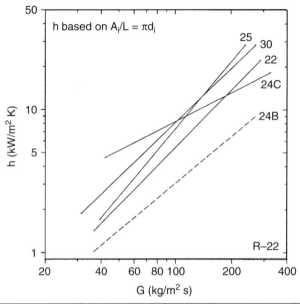

Figure 13.9 R-22 heat transfer vaporization coefficient (h based on $A_i/L = \pi d_i$) data of Kubanek and Miletti [1979] at $T_s = 4.4°C$ in 0.8-m-long tube with $\Delta x = 0.7$. The table defines the tube geometries.

Tube type and number	Number of fins	Tube internal diameter, mm	Tube hydraulic diameter, mm	Fin height, mm	Fin pitch, mm	Wetted area per unit length, $mm^2/m \times 10^{-3}$	Nominal area per unit length, $mm^2/m \times 10^{-3}$	Area ratio	Heated length, m
Plain, 24B		14.4	14.4			45.3	45.3	1.00	0.80
Insert, 24C	5	14.4	4.09		610	90.8	45.2	2.00	0.80
Finned, 22	32	14.7	7.57	0.635	305	87.0	46.2	1.88	0.80
Finned, 25	32	14.7	7.57	0.635	152	87.2	46.2	1.89	0.80
Finned, 30	30	11.9	6.30	0.508	102	68.0	37.4	1.82	0.80

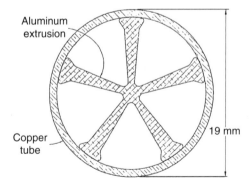

Figure 13.10 Tube having five-leg aluminum insert with shrink fit.

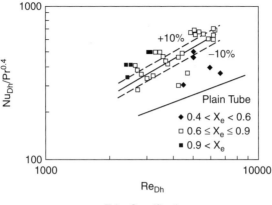

Tube Specifications

Parameter	Value
Tube material	Aluminum, Al 6063
Wall thickness	1.65 mm
Tube flow area	563.2 mm^2
Mean inside diameter	26.8 mm
Inside perimeter	137.54 mm
Outside perimeter	136.42 mm
Equivalent diameter	16.38 mm
Flute angle to tube axis	30°
Flute spacing	2.63 mm
Flute depth	1.54 mm
Effective tube length	4.45 m

Figure 13.11 Convective vaporization coefficient for R-11 (296 to 1010 kPa) in vertical spirally fluted tube. (From Panchal et al. [1992].)

13.3.2 Swirl Flow Devices

Twisted-tape inserts were investigated quite early in the search for enhancement techniques. This is because technology to internally fin or roughen a tube had not yet been developed, and because twisted tapes were easy to make. However, their use is now limited, except for special situations discussed below. The reported test data include water (subcooled and saturated boiling), refrigerants, and liquid metals. The twisted-tape data are typically taken using electrical heating in the tube wall with a loosely fitting tape. The tape geometry is usually described as $y = H/d_i$ where H is the axial length for one half tape revolution. The twist geometry may also be described in terms of the helix angle of the tape, which is $\tan \alpha = \pi d_i/2H = \pi/2y$.

Because the flow pattern changes along the tube length, it is probable that a given enhancement technique may work better for one flow pattern than for others. This is true for the twisted tape. It is particularly effective in the drywall region, where liquid flows as entrained droplets. The tape throws the liquid back on the tube wall. The vapor quality at which the critical heat flux occurs decreases for increasing flow rate and heat flux. Hence, improvements found at high heat fluxes and flow rates would not be expected at low heat fluxes and flow rates. Bergles et al. [1971] provide data on twisted tapes ($\alpha = 10$ and $22°$) to enhance heat transfer in the drywall region for boiling nitrogen at 140 to 170 kPa. Their tests with an electrically heated tube wall showed that the tapes approximately doubled the heat transfer coefficient in the postcritical vapor quality region, relative to a plain tube. However, the tapes did not move the critical vapor quality to a significantly lower value.

Jensen and Bensler [1986] tested R-113 in a vertical upflow using an electrically heated 8.1-mm ID tube. Figure 13.12a shows the effect of vapor quality for $q = 46$

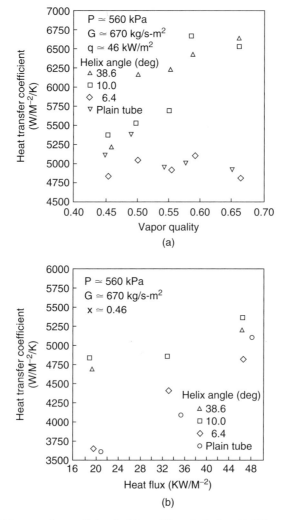

Figure 13.12 R-113 evaporation coefficient in 8.1-mm-ID vertical tubes. (a) Effect of vapor quality for $q = 46$ kW/m², (b) effect of heat flux for $x = 0.45$. (From Jensen and Bensler [1986].)

kW/m^2. The figure shows that h increases with x, and that h increases with increasing tape helix angle (α). Note that the $\alpha = 6.4°$ tape provided equal or smaller heat transfer coefficients than the plain tube. The maximum measured enhancement for the best tape ($\alpha = 38.6°$) is 50%. Figure 13.12b shows the effect of heat flux for $x = 0.46$. Increasing heat flux benefits both the plain and enhanced tubes, apparently because of the nucleate boiling component. The highest performing twisted tape ($\alpha = 38.6°$) benefits least, because of its larger convective contribution. Jensen [1985] reported pressure drop data for the same tapes tested by Jensen and Bensler [1986].

Agrawal et al. [1982, 1986] tested R-12 in a 10-mm ID horizontal tube with electric heating. Agrawal et al. [1986] tested helix angles between 8.8° and 22.7° at 138 kPa with heat fluxes between 8 and 13.6 kW/m^2. Figure 13.13 shows the enhancement ratio, $E_h = h/h_p$ for the low and high heat fluxes. At the lowest heat flux (Figure 13.13a), negligible enhancement was provided for $x < 0.6$. Figure 13.13b shows increasing the heat flux to 13.6 kW/m^2 provided greater enhancement for the

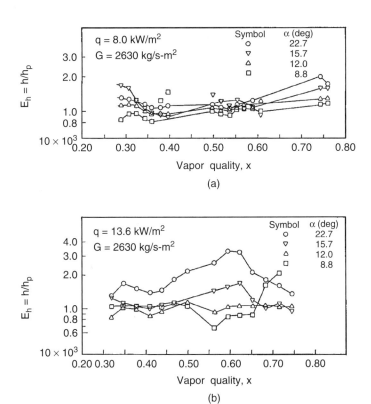

Figure 13.13 Enhancement ratio (h/h_p) for R-12 evaporating in 10-mm-ID horizontal tubes with twisted tapes. (a) Effect of vapor quality for $q = 8.0$ kW/m^2, (b) effect of vapor quality for $q = 13.6$ kW/m^2 (From Agrawal et al. [1986].)

tapes having $\alpha \geq 15.7°$. Surprisingly, the $\alpha > 15.7°$ tapes show poor enhancement for $x > 0.65$. This behavior is not expected, and confirmation is desirable. Note that the highest heat flux in Figure 13.13 is less than the smallest heat flux in Figure 13.12. Hence, convection should play a larger role in Figure 13.13 than in Figure 13.12. Blatt and Alt [1963] tests for R-11 in a 6.4-mm ID tube with condensing steam on the outer surface showed much less sensitivity to heat flux than observed by Agrawal et al.

Kedzierski and Kim [1997] measured refrigerant vaporization and condensation in a 9.64-mm ID tube having a 20.75° twisted tape (0.5 mm thick). Data were taken for five pure refrigerants (R-12, R-22, R-134a, R-152, and R-290) and four refrigerant mixtures. Surprisingly, the data showed a maximum evaporation coefficient at about 0.2 to 0.3 vapor quality and decreased with increasing vapor quality. Data were taken at vapor qualities up to 97%.

Agrawal et al. [1982] measured the adiabatic friction pressure drop characteristics of the twisted-tape tubes. The friction pressure drop of the $y = 3.76$ tube was 1.7 to 3.7 times that of a plain tube. However, they do not define how the pressure drop ratio is influenced by vapor quality. One concludes that the frictional pressure drop ratio substantially exceeds the heat transfer enhancement ratio.

It is concluded that the twisted tape is a fairly low performance enhancement device for refrigerants, as compared to the microfin tube. Further, the pressure drop increases are significant, compared to the heat transfer increases. Its main application appears to be in situations where CHF may exist. Then, the swirl effect may delay the inception of CHF to a higher vapor quality. Jensen [1985] presents a parametric theoretical analysis of R-12 in a tube having a twisted tape for $3 \leq y \leq 21$, with different flow rates and external heat transfer coefficients. For conditions below the CHF, he shows that the R-12 side enhancement is between 20 and 40%. A twist ratio of 3 provides very little benefit over $y = 12$. The Jensen analysis includes calculation of pressure drop and flow power. He concludes that suppression of the CHF is the main attractive feature of a twisted tape insert. More information on the CHF is given in Section 13.6.

13.3.3 Roughness

Among the earliest enhancement devices studied is the wire coil insert shown in Figure 1.2c. These devices typically provide a given enhancement at much higher pressure drop than is obtained by the integral roughness described below. A recent experimental program using wire coil inserts (1.24 mm wire diameter) for R-22 vaporization is described by Varma et al. [1991].

Withers and Habdas [1974] investigated R-12 vaporization at 0°C in 19.1 mm OD (16.3 mm ID) single helix corrugated tubes of the type illustrated in Figure 9.3b. Their data were taken for entering saturated liquid and 100% exit vapor quality, with heat input from high-velocity water in an annulus. A Wilson plot method was used to derive the average vaporization coefficient in the 3.96-m-long test section. The geometry parameters of the corrugated tube are the dimensionless roughness height (e/d_i), and roughness spacing (p/e). Data were obtained for six roughness geometries having $0.4 \leq e \leq 1.6$ mm and $3.2 \leq p \leq 15.9$ mm, and a plain tube. They proposed that the combined effect of e/d_i and p/e may be described by the *severity factor*, which they define as $\phi \equiv (e/d_i)/(p/e)$. Figure 13.14 shows the tube performance as a function of the severity factor. Note that the refrigerant flow rate is not constant on the figure. The R-12 flow rate is determined by the entering water temperature (15.6°C) and the vaporization coefficient. The geometry with the highest vaporization coefficient will have the highest flow rate. The highest performance is provided by $\phi = 0.004$, for which the vaporization coefficient is 2.9 times that of the plain tube ($\phi = 0$) with a 1.8 times higher flow rate. The pressure drop (friction plus momentum) ratio ($\Delta p/\Delta p_p$) and mass flow rate ratio (W/W_p) are also shown in Figure 13.14. At $\phi = 0.004$ $W/W_p = 2.8$, which means that the evaporation rate in the enhanced tube is 2.8 times that in the plain tube. The large $\Delta p/\Delta p_p$ is partially a result of the substantially higher flow rate in the enhanced tube. Withers and Habdas recommend use of $\phi = 0.00234$ ($e = 0.8$ mm and $p = 15.9$ mm) because the R-12 pressure drop is lower. This tube is offered by Wolverine as a commercial product, known as Koro-Chil™.

Shinohara and Tobe [1985] describe an improvement on the corrugated tube tested by Withers and Habdas. This tube is a corrugated microfin tube called TFIN-CR™ and is shown in Figure 13.15d. Table 13.1 compares the dimensions and performance of a 16-mm-OD TFIN-CR tube with the standard corrugated tube (CORG™), a microfin tube (TFIN™ and five-leg star fin insert. The heat transfer coefficients are based on the outer diameter (16 mm). The microgrooves in the TFIN tubes are 0.2 mm and the helix angle is 8°. The 5.7-m-long test section was operated at 2°C. Note that the enhancement level of the TFIN tube is not as high as listed in Table 13.3 later in the chapter. Further, the enhancement ratio of the CORG tube is not as high as measured by Withers and Habdas [1974], although its value of ϕ is higher ($\phi = 0.0042$).

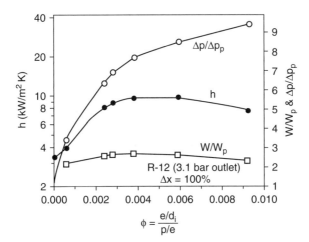

Figure 13.14 R-22 vaporization performance of 19.1-mm-diameter corrugated tubes of Withers and Habdas [1974] plotted vs. severity factor (ϕ).

Akhanda and James [1988] provide data for convective vaporization of sub-cooled water ($p = 11{,}000$ kPa) in a rectangular cross section channel containing 0.5-mm-high ribs with axial pitches between 0.5 and 3.50 mm.

Wen and Hsieh [1995] tested two star-insert tubes (similar to Figure 13.10) and two corrugated tubes (similar to Figure 13.15b) using R-114, R-22, and R-134a. The geometric details are provided in Table 13.2. The test section was a double-pipe configuration 1.5 m long. The heat transfer coefficient was defined in terms of the nominal heat transfer area (see Table 13.2). Figure 13.16 shows the effect of mass flux (G) on the heat transfer enhancement ratio ($h_{TP,a}/h_{TP,s}$). In the low mass flux range (up to 150 kg/m² s), the ratio maintains high values, and decreases rapidly

(a)   (b)   (c)   (d)

Figure 13.15 Evaporator tubes tested by Shinohara and Tobe [1985]: (a) star-fin insert, (b) corrugated tube (CORG), (c) microfin (TFIN), (d) corrugated microfin (TFIN-CR). (From Shinohara and Tobe [1985].)

Table 13.1 Tube Comparison for R-22 Evaporation at $G = 100$ kg/h

Geometry	$e_i$mm	$p_i$mm	$h_i k$ W/m²	Δp kPa	h/h_p	$\Delta p/\Delta p_p$
Plain	NA	NA	1.63	20.6	1.00	1.00
Star insert	NA	NA	2.06	35.8	1.26	1.74
CORG	0.7	8.0	2.15	32.4	1.32	1.57
TFIN	NA	NA	2.19	26.0	1.34	1.26
TFIN-CR	0.7	8.0	2.99	32.4	1.83	1.57

NA = not applicable.

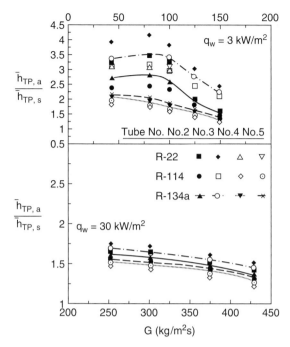

Figure 13.16 Effect of mass flux (G) on the heat transfer enhancement ratio ($h_{Tp,a}/h_{Tp,s}$) for the four tubes tested by Wen and Hsieh [1995] using R-114, R-22, and R-134a. The geometric details are provided in Table 13.2.

Table 13.2 Tubes Tested by Wen and Hsieh [1995] Using R-114, R-22, and R-134a

Tube No.	1	2	3	4	5
Tube configuration	Smooth	Star insert (5 leg)	Star insert (10 leg)	Corrugated	Corrugated
Nominal OD (mm)	16.0	16.0	16.0	16.0	16.0
Nominal ID (mm)	14.0	14.0	14.0	14.0	14.0
Nominal heat transfer area (cm²/m)	440	445	449	471	515
Nominal flow area (cm²)	6.16	5.81	5.60	7.07	6.17
Helix angle (degree)	—	—	—	30	76
Insert thickness or fin height (mm)	—	1.0	0.8	0.7	0.38
Corrugation pitch (mm)	—	—	—	2.6	1.2

with increasing G. For $G > 250$ kg/m² s, the ratios are generally much smaller, and the difference both between the enhanced tubes and between the fluids decreases. The trend is the same both for the star-insert tubes and for the corrugated tubes. This suggests a difference in flow pattern between the enhanced tubes and the smooth tube at low mass flux. The star inserts reduce the hydraulic diameter of the flow passage, and the corrugation may induce some swirl motion of the flow. Wen and Hsieh [1995] also investigated the effect of oil on vaporization in the tubes. Addition of oil reduced the heat transfer coefficients.

13.3.4 Coated Surfaces

High-flux porous coating may be applied to the inner tube surface. Whereas internal-fins, corrugated tubes, and twisted tapes provide enhancement by affecting the convective term of Equation 13.2, the porous internal coating primarily affects the nucleate boiling term. This provides very high performance for convective vapor-ization. The performance of the porous coated tube is quite sensitive to heat flux, because of the large nucleate boiling contribution. Figure 13.17 shows the data of Czikk et al. [1981] for the High-Flux tube with vertical up-flow of oxygen (101 kPa) in an 18.7-mm-diameter tube and is taken from Thome [1990]. The performance is compared with that of a plain tube, whose performance was predicted by Thome [1990] using the Chen [1966] superposition model (Equation 13.2). The exit vapor quality is shown by the numbers in parentheses. At the same heat flux, the $(T_w - T_s)$ of the porous tube is approximately one tenth that of the plain tube. Also shown in Figure 13.17 is the High-flux pool boiling data of Antonelli and O'Neill [1981] at 101 kPa. The convective boiling points are in reasonable agreement with the pool boiling data, which show that the performance of the tube is dominated by the pool boiling term of Equation 13.2. Note that vapor quality and mass velocity have little

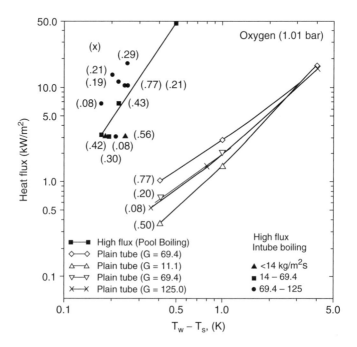

Figure 13.17 Oxygen boiling performance of high-flux and plain tubes. (From Thome [1990].)

effect on the high-flux tube performance, which is another indication that nucleate boiling dominates the tube performance. The very low values of ΔT_{ws} are probably responsible for the data scatter.

The high-flux tube shows an effect of mass velocity for horizontal flow. The ammonia data of Czikk et al. [1981] show increasing performance with mass velocity for the range tested $30 \leq G \leq 450$ kg/m²-s). This is apparently because the mass velocity is not high enough to produce annular flow. Hence, the upper surface of the tube is partially dry. Wetting improves with mass velocity, which explains the mass velocity effect. Although some dryout may exist, the high-flux performance was still ten times better than that of a plain tube.

Ikeuchi et al. [1984] provide R-22 vaporization data in a horizontal porous coated tube. Their tests were performed with 15% inlet vapor quality and exit vapor qualities 70 to 95% or with 5 K superheat. Figure 13.18 shows their data, as reported by Thome [1990]. Note that the data for 5 K superheat are lower than those for incomplete evaporation for both the plain and the high-flux tube. The reduced performance at 5 K superheat is a result of CHF dryout at high vapor quality. These data clearly show the superiority of the high-flux tube at all vapor qualities.

Wadekar [1998] obtained convective vaporization data in a high-flux tube with vertical upflow of R-113. The copper porous coatings used in the high-flux tube had particle sizes less than 150 μm and a coating thickness of 0.2 to 0.3 mm. The test section was 3.0 m long, and consisted of the 1.0-m high-flux tube, followed by a 1.0-m plain tube, and then 1.0-m high-flux tube. The test section was heated by passing direct electric current through the tube wall. Tests were conducted at relatively high heat fluxes to obtain measurable ΔT_{ws}: R-113 was supplied as a subcooled

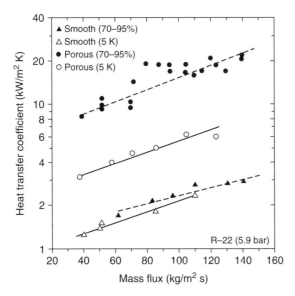

Figure 13.18 R-22 convective vaporization data of Ikeuchi et al. [1984] for porous and plain tubes as reported by Thome [1990].

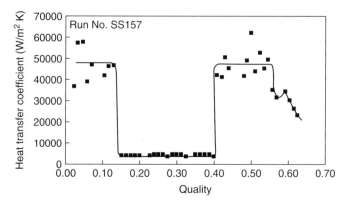

Figure 13.19 Heat transfer coefficients for $G = 282.6$ kg/m²s and $q = 43.2$ kW/m² in high-flux tube tested by Wadekar [1998].

liquid at the test section inlet, and the exit quality was varied depending on the imposed heat flux. Figure 13.19 shows the heat transfer coefficients for $G = 282.6$ kg/m²-s and $q = 43.2$ kW/m². The heat transfer coefficient of the high-flux tube is an order larger than that of the plain tube. At $x = 0.55$, the heat transfer coefficient falls, indicating the CHF. The relatively low CHF quality implies that vapor blanketing, due to high heat flux, is the main cause of CHF. Figure 13.19 shows that the heat transfer coefficients are insensitive to the vapor quality both in the High-Flux tube and in the plain tube, which indicates the dominance of nucleate boiling over convective boiling.

13.3.5 Perforated Foil Inserts

Palm [1990] has tested the perforated foil inserts discussed in Section 11.4.6 for convective vaporization of R-22 in a 15-mm-ID tube. Eight different foil inserts were tested; hole diameters were between 0.10 and 0.22 mm, and the hole densities between 0.5 and 1.5 holes/mm². At 50% vapor quality and heat fluxes below 10 kW/m², the best foil increases the local heat transfer coefficient by 50%. The enhancement is low, relative to the porous coated tube (Section 13.3.4) and the microfin enhancement. Poor performance was obtained at high vapor quality, presumably because the space between the tube wall and the foil was dried out. Conklin and Vineyard [1992] found little performance benefit of a foil insert having 0.5 holes/mm² for R-22.

13.3.6 Porous Media

Topin et al. [1996] measured forced convection vaporization of n-pentane in a vertical container filled with small bronze spheres (140 to 160 μm diameter) for $2 < q < 10$ kW/m² heat flux and vertical upflow of the fluid. They provide temperature profiles in the flow direction, but no heat transfer coefficient data.

13.3.7 Coiled Tubes and Return Bends

Coiled tubes are used in a variety of situations, including chemical reactors, steam generators, agitated vessels, and storage tanks. The centrifugal force imposed on the two-phase flow imposes a secondary flow imposed on the two-phase flow regimes. Liquid droplets are thrown to the outer wall and the liquid film spirals along the tube wall to the inner surface of the coil. This effect causes an increase of the dryout heat flux. A number of publications exist on vaporization in coiled tubes. Jensen and Bergles [1981] provide a literature review.

Crain and Bell [1973] tested 2 helical coils with steam and an electric heated tube wall at approximately 170 kPa with $0.45 \leq x \leq 1.0$ and developed a correlation with an average error of 44%. The correlation, which does not account for the effect of pressure, is

$$\frac{hd_i}{k_l} = 0.0587 \left(\frac{d_i G(1-x)}{\mu_l} \right)^{0.85} \mathrm{Pr}_l^{0.4} \left(\frac{d_i}{D_c} \right)^{0.1} x^{-7.6} \tag{13.10}$$

Equation 13.10 shows that little enhancement is provided by a coil, since the $(d_i/D_e)^{0.1}$ contribution is small. The -7.6 exponent on the vapor quality makes the correlation appear questionable.

Campolunghi et al. [1976] tested a large (1 MW) coil having $d_i = 15.5$ mm and $D_e = 836$ mm with steam over $80 \leq p \leq 170$ bar. Their correlation shows that $h \propto \exp(0.0132\ p)$. Hughes and Olson [1975] measured the DNB and the CHF for R-113 in a planar coil. They found that the CHF is higher on the concave surface than on the convex surface. Jensen and Bergles [1981] measured the DNB and dryout CHF conditions in several electrically heated coils using R-113. They provide correlations for both CHF conditions.

Gu et al. [1989] tested a subcooled fluorocarbon fluid (FC-72) in a single, glass rectangular cross section channel, curved over 180°. The channel radius was 50.8 mm, and the channel cross section was 5.6×27.0 mm. The channel was locally heated on the outer radius by electrical elements 9.5 mm long upstream of the bend, and 135° around the bend. Local heat transfer data were obtained at the location of the two heater elements. They observed higher heat transfer coefficients and higher CHF in the curved channel. They provide a correlation to account for the effect of centrifugal acceleration. Note that the data and correlation do not account for effects that occur on the inner wall of the channel, at which the limiting conditions exist.

13.4 THE MICROFIN TUBE

The tube was first developed by Fujie et al. [1977] of Hitachi Cable, Ltd. and is described by Tatsumi [1982]. An improved Hitachi design is described by Shinohara and Tobe [1985] and by Shinohara et al. [1987]. The version described by Shinohara and Tobe [1985] is close to that now made by tube manufacturers in Japan, Europe, and the U.S. It is also used for condensation of refrigerants, which is discussed in Chapter 14. Details of the microfin geometry are discussed in Section 14.3.1. Figure 13.7d shows Wieland's [1991] version of the microfin tube.

The tube is at present made in diameters between 4 and 16 mm. Figure 13.20 from Torikoshi and Ebisu [1999] shows two common processes used to make the microfin tube. Figure 13.20a illustrates a drawing process, in which a plain tube is pulled over a grooved floating plug with rollers applying pressure on the outer tube surface. Figure 13.20b illustrates the embossing method, in which a flat strip is embossed with the desired pattern, and then the tube is rolled to a circular shape and welded. Figure 13.20b illustrates formation of a herringbone fin pattern. However, one may make single-direction fins, or even "cross-grooved" fins by using a second set of rollers that forms a secondary embossing pattern.

Depending on the manufacturer, the microfin tube provides an R-22 evaporation coefficient 100 to 300% higher than that of a plain tube. Table 13.3, from Yasuda et al. [1990], shows (1) the geometry changes that have been made by Hitachi Cable in their 9.52-mm-OD Thermofin tube, and (2) the resulting performance improvements. The 9.52-mm-OD Thermofin-HEX tube has 60 fins with 0.20 mm height (e), 40° fin apex angle (β), spaced at 2.32 times the fin height (p/e), and at 18° helix angle. The ratio values are stated relative to a plain tube (subscript p). The $A_i/A_{i,p}$ is the total internal surface area ratio, and Wt/Wt_p is the tube weight ratio, per unit tube length. The h/h_p is the evaporation coefficient (based on $A_i/L = \pi d_i$) relative to a plain tube operated at the same mass velocity. Figure 13.21 shows the cross section fin geometries associated with Table 13.3. Material has been added to Table 13.3 that includes recent material from Houfuku et al. [2001] on a 7.0-mm-OD tube. The recent manufacturing development has sought to reduce the material in the fins by reducing the fin apex angle (Houfuku et al. [2001]).

The detailed geometry parameters are defined in Section 14.3, as well as an equation to calculate the surface area ratio, relative to a plain tube. Considerable recent work has been done to reduce the material content of the microfins. This work is summarized in Table 14.5. Table 14.5 includes both vaporization and con-

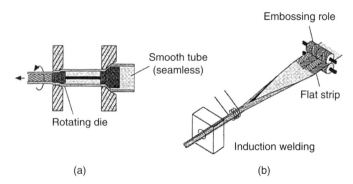

Figure 13.20 Methods used to make microfin tubes: (a) drawing method, (b) embossing method. (From Torikoshi and Ebisu [1999].)

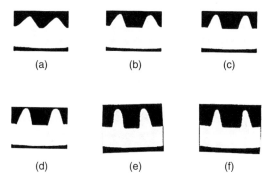

Figure 13.21 Cross sections of Hitachi Thermofin tubes. (a) Thermofin. (b) Thermofin EX. (c) Thermofin HEX. (d) Thermofin HEX-C. (From Yasuda et al. [1990]). (e) TFIN-HG. (f) TFIN-HGL. (From Houfuku et al., 2001 *Hitachi Cable Rev.*, 20. With permission.)

Table 13.3 Chronological Improvements in the Hitachi Thermofin Tube

Year	Geometry	d_o	e	p/e	α	β	n	$A_i/A_{i,p}$	Wt/Wt_p	h/h_p
1977	Orig	9.52	0.15	2.14	25	90	65	1.28	1.22	2.0
1985	EX	9.52	0.20	2.32	18	53	60	1.51	1.19	2.6
1988	HEX	9.52	0.20	2.32	18	40	60	1.60	1.19	3.2
1989	HEX-C	9.52	0.25	2.32	30	40	60	1.73	1.28	3.2
2001	HEX-C	7.0	0.20	2.32	18	40	50			
2001	TFIN-HGL	7.0	0.22	2.03	16	22	54			
2001	HEX-HG	7.0	0.25	2.32	18	15	57			

Sources: From Yasuda [1990] and 7 mm tubes from Houfuku et al. [2001].

densation results. The optimum geometries listed in Table 14.5, are those, that provide the highest heat transfer coefficient.

Research at Hitachi Cable includes publications of Ito and Kimura [1979], Yasuda et al. [1990] for 9.5-mm tube diameter, Hori and Shinohara [2001] and Houfuku et al. [2001] for 7.0-mm tube diameter, and Tsuchida et al. [1993] for 4.0 and 5.0-mm tube diamterers. Early 9.5-mm-diameter microfin tubes, as described by Yasuda et al. [1990], had $e = 0.2$ mm and $\beta = 90°$ (see Figure 14.6). By 1989, Hitachi had increased their fin height to 0.25 mm and β decreased to 40° as reported by Yasuda et al. [1990]. Typically, 60 fins were used in a 9.52-mm-OD tube, as reported by Yasuda et al. [1990], for which $p_n = 0.46$ mm.

Figure 13.21 shows the evolution of cross section fin geometries associated with Table 13.3. The figure shows that since the tube was originally introduced in 1979, the fin height has increased and the fin base thickness reduced. Figure 13.22 shows the R-22 evaporation coefficient of the Table 13.3 tubes as a function of mass velocity. Table 13.3 shows that the evaporating coefficient of the Thermofin HEX-C tube is 3.1 times that of a plain tube at the same mass velocity. The improved versions have higher fins (e) and sharper apex angles (β).

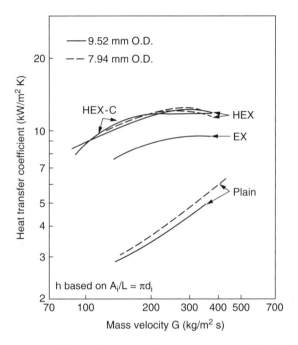

Figure 13.22 Evaporation coefficient (based on $A_i/L = \pi d_i$) vs. mass flux for the Hitachi Thermofin tubes. (From Yasuda et al. [1990].)

13.4.1 Early Work on the Microfin Tube

The first published work on the microfin tube was that of Ito and Kimura [1979]. This section reviews work done between 1979 and 1990. Work done after 1992 is discussed in Section 13.4.2.

In one of the first detailed studies of the microfin tube, Ito and Kimura [1979] investigated the effect of microfin geometry on the convective vaporization of R-22 in 12.7-mm-OD microfin tubes. The test range is provided in Table 14.5. This tube had an included angle β of 90° as shown in Figure 13.21a. Figure 13.23, from Ito and Kimura [1979] shows how the vaporization coefficient of the original Thermofin tube (Figure 13.21a) is influenced by helix angle. This figure shows that the optimum performance occurs at a helix angle of approximately 10°. The vaporization coefficient significantly decreases for helix angles greater than 20°. Then, they varied the fin height from 0.06 to 0.16 mm for $\alpha = 7°$ and $p_n = 0.52$ mm. The heat transfer coefficient increased as the fin height increased. The groove pitch was varied from 0.32 to 0.54 mm for $\alpha = 7°$ and $e = 0.10$ mm. The heat transfer coefficient increased as the groove pitch increased.

A very interesting feature of the tube performance is its low refrigerant pressure drop. Figure 13.24, from Ito and Kimura [1979], shows that the R-22 pressure drop of the Figure 13.21a Thermofin tube is only 10% higher than that of a plain tube. Yasuda et al. [1990] report that the R-22 pressure drop of the 9.52-mm Thermofin

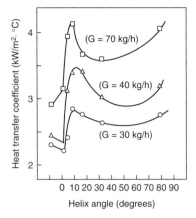

Figure 13.23 Effect of groove angle on evaporation coefficient for the original Hitachi Thermofin tube. (From Ito and Kimura [1979].)

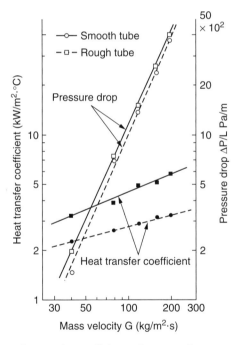

Figure 13.24 Comparison of evaporation coefficient and pressure drop vs. mass velocity for original Hitachi Thermofin tube. (From Ito and Kimura [1979].)

HEX-C tube (Figure 13.21d) is approximately 35% greater than that of a plain tube at 200 kg/s-m^2.

Yoshida et al. [1987] performed a detailed experimental study attempting to explain the enhancement mechanism of the microfin tube. They measured local R-22 heat transfer coefficients on the top, side, and bottom of the tube for a range of vapor qualities and mass velocities. Their measurements were done for the three microfin geometries listed in Table 13.4. Tube C has primary grooves 0.24 mm deep at −30°, and secondary grooves 0.15 mm deep at −15°. Figure 13.25 shows the enhancement ratio, $E_h = h/h_p$ vs. vapor quality for low and high mass velocities. At $G = 300$ kg/m²-s, all 3 tubes give approximately $E_h = 1.5$ for $0.2 \leq x \leq 0.9$. However, at $G = 100$ kg/m²-s, the tubes give much higher enhancement ratios. Tube B with 30° helix angle is better than tube A with 10° helix angle. Figure 13.26 explains why greater enhancement occurs at the lower G than at the higher value. Figure 13.26 shows the local heat transfer coefficients on the top, side, and bottom of tube

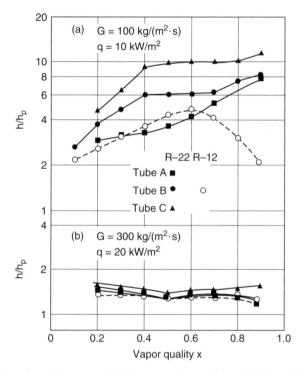

Figure 13.25 Circumferentially averaged R-22 evaporation coefficients in 12-mm-ID microfin tubes described in Table 13.2. (From Yoshida et al. [1987].)

Table 13.4 15.8-mm-OD Microfin Tubes Tested by Yoshida et al. [1987]

Tube	e	α	N	A/A_p
A	0.24	10	60	1.28
B	0.24	30	60	1.35
C	0.24/0.15	30/15	60	1.35

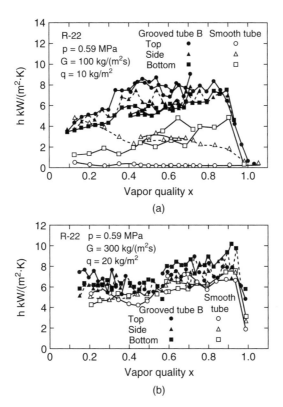

Figure 13.26 Local R-22 evaporation coefficients on top, side, and bottom of 12-mm-ID microfin tubes described in Table 13.4: (a) G = 100 kg/m²-s, (b) G = 300 kg/m²-s. (From Yoshida et al. [1987].)

B and the plain tube. At $G = 300$ kg/m²-s, there is little difference between the h values at the three circumferential locations. However, at $G = 100$ kg/m²-s, the figure shows that much higher h values are obtained on the top and sides of the tube than exist for the plain tube. Yoshida et al. conclude that the narrow grooves carry liquid to the sides and top of the tube by capillary wetting. Thus, thin films are provided around the entire tube circumference. Low heat transfer coefficients exist on the top and sides of the plain tube at low G, because the tube surface is dry.

Although Figure 13.23 shows an optimum helix angle of approximately 8°, there is no uniform agreement on this optimum. The Shinohara et al. [1987] patent shows a relatively flat optimum of 13° for R-22 evaporation in a 9.5-mm-OD tube.

Eckels et al. [1992] provide R-22 evaporation coefficients and pressure drop at 2.0°C for five currently used microfin tube geometries. The reported evaporation coefficient is the average value for 0.10 entering and 0.85 leaving vapor quality. Schlager et al. [1990] tested three 12.7-mm-OD microfin tubes having different helix angles (15°, 18°, and 25°) with R-22. However, their tubes also had different fin heights ($0.15 \leq e \leq 0.3$ mm) and pitch, so they could not define the effect of specific geometry factors on the performance differences. Their test results for all three geometries agree very closely.

Schlager et al. [1988a, 1988b] and Eckels and Pate [1991] report the effect of oil on the performance of R-22, R-134a, and R-12, respectively. Oil slightly reduces the evaporation coefficient for typically used oil concentrations.

13.4.2 Recent Work on the Microfin Tube

Recent work to optimize the microfin geometry for a 7.0-mm-OD tube are reported by Houfuku et al. [2001] using $e = 0.22$ mm, $p_n = 0.37$ mm ($n = 54$), $\beta = 22°$, and $\alpha = 16°$. Included in all of this work were efforts to reduce the material content by reducing the fin base thickness and the fin apex angle. Ishikawa et al. [2002] investigated the number of fins for 7.0-mm tube diameter with $\alpha = 10°$ with $e = 0.21$ to 0.24 mm. They found that 80 fins ($p_n = 0.25$ mm) was the optimum value. Note that this optimum is different from that reported by Houfuku et al. [2001], who found $e = 0.22$ mm, $p_n = 0.37$ mm ($n = 54$), $\beta = 22°$, and $\alpha = 16°$.

Tsuchida et al. [1993] provide extensive data on convective vaporization and condensation of R-22 in microfin tubes with outside diameter ranging from 4.0 to 9.5 mm. Figure 14.7a shows the effect of fin height for different diameter tubes. The evaporation data were taken at 0.5°C and the condensation data at 40°C and show that the evaporation coefficients are moderately higher than the condensation coefficients. For both condensation and evaporation, Figure 14.7a shows that the optimum fin height is about the same for both evaporation and condensation, and that it decreases with decreasing tube diameter. The optimum fin height is 0.13 mm for 4.0-mm diameter, 0.16 mm for 7.0-mm diameter, and further increases to 0.23 mm for 9.52-mm tube diameter. The effect of number of fins is shown in Figure 14.7b and is about the same for both evaporation and condensation. This figure shows that the optimum number of fins decreases as the tube diameter decreases, which is somewhat misleading, because a more relevant parameter is the fin pitch. The fin pitch, p_n, corresponding to the optimum fin number was calculated using $p_n = \pi D_i/n$, and are in the range of 0.26 to 0.34 mm for all tube diameters. The optimum fin pitches are listed in Table 14.5, and the effect of helix angle is shown in Figure 14.7c. The figure shows that the preferred helix is smaller for evaporation than for condensation. The optimum configuration for each diameter tube is listed in Table 14.5.

Many experimental works have addressed measurements of the heat transfer coefficients for vaporization of refrigerants in tubes with different microfin geometries. Reviews on the issue are found in Thome [1996], and Newell and Shah [2001]. Seo and Kim [2000] measured evaporation coefficients in 7.0 and 9.52-mm-OD tubes at −15, −5, and 5° C. They found that both the heat transfer coefficient and pressure gradient increase as the saturation temperature is reduced.

Chamra et al. [1996] investigated the R-22 vaporization characteristics in four two-dimensional microfin tubes and seven three-dimensional cross-grooved microfin tubes. The single groove geometries had 74 to 80 fins in a 15.88-mm-OD tube, 0.35-mm fin height. The corresponding helix angle was varied 15° to 27°. The vaporization coefficient reached a maximum helix angle, and then decreased for higher helix angle. The difference, however, was barely noticeable. This result along with results by Tsuchida et al. [1993] suggest the optimum helix angle decreases as the tube diameter decreases. Oh and Bergles [1998] found that the optimum helix angle changed according to the mass flux.

Houfuku et al. [2001] investigated the effect of total inner surface area on the vaporization coefficient of R-410A in 7.0-mm-OD microfin tubes (TFIN-HEX, TFIN-HG, TFIN-HGL). The geometric dimensions are listed in Table 13.3. The 7% reduction of the tube material was realized for TFIN-HGL compared with TFIN-HG. The vaporization results are shown in Figure 13.27. Figure 13.27 shows the relationship between the inner surface area and the heat transfer coefficient. The vaporization coefficient increases with inner surface area up to 37,000 mm²/m, and then is almost constant afterwards.

Ishikawa et al. [2002] investigated the effect of number of fins on the convective vaporization and condensation of R-22 in 7.0-mm-OD microfin tubes. The number of fins ranged from 55 to 85. The highest heat transfer coefficient was obtained for a tube having 80 fins both for vaporization and condensation. This number of fins is larger than the numbers from Tsuchida et al. [1993] and Houfuku et al. [2001]. The apex angle of Ishikawa et al.'s fin is much smaller (10°) than others (22° and 40°). For small apex angle, the area of the interfin region will increase, which will increase the flow velocity in that region. This is likely to favor larger number of fins. The existence of optimum fin number for the single-phase heat transfer in microfin tubes was evidenced by Bhatia and Webb [2001].

To summarize the geometric effect of the microfin, the optimum number of fins decreases as the tube diameter decreases. Small apex angle increase the optimum

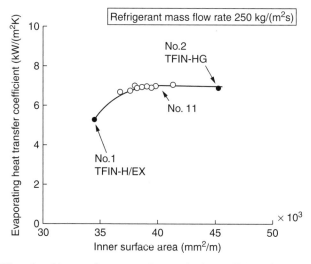

Figure 13.27 Effect of total inner surface area on the vaporization coefficient of R-410A in 7.0-mm-OD microfin tubes (Hitachi TFIN-HEX, TFIN-HG, TFIN-HGL). (From Houfuku et al., 2001. *Hitachi Cable Rev.*, 20. With permission.)

number of fins. The optimum helix angle decreases as the tube diameter increases. Microfin tubes with small apex angle and many inner fins will provide additional enhancement over conventional geometries.

13.4.3 Special Microfin Geometries

Chamra et al. [1996] investigated the R-22 vaporization characteristics in seven 15.88-mm-OD three-dimensional cross-grooved microfin tubes. The tubes were fabricated by embossing a flat strip of copper to form the desired geometry; then the strip was rolled into a round tube and welded. The welded microfin tube is cost competitive compared with the conventional seamless microfin tube. The cross-groove geometries were formed by applying a second set of grooves at the same helix angle, but opposite angular direction as the first set. The depth of the second set of grooves was 40 to 80% of the depth of the first grooves. Photographs of the tubes are shown in Figure 13.28. Data were reported for vaporization at 2.2°C in a 2.44-m-long test section with inlet and exit qualities 0.20 and 0.80, respectively. The highest performance was provided by the cross-grooved tube having 20° helix angle. Its vaporization coefficient was 23% higher than for a 75-fin single-helix tube. The pressure drop was 6% higher than in the 75-fin single-helix tube.

The depth of the second groove had only marginal effect both on the heat transfer and the pressure drop. Kuo and Wang [1996] provide additional data on the cross-grooved tube.

Muzzio et al. [1998] tested a 9.52-mm-OD two-dimensional microfin tube having two different fin heights alternately, 0.23 and 0.16 mm, with an apex angle 40° and a helix angle of 18°. The tube yielded 20% higher heat transfer coefficients for approximately the same pressure drop as the conventional microfin tubes.

Ebisu [1999] tested a 7.0-mm herringbone pattern microfin tube using R-22 and R407C, and compared the performance with conventional two-dimensional microfin tubes. The herringbone tube was fabricated through the embossing and welding process. The strip configuration is illustrated in Figure 13.29. Figure 13.30 shows the measured vaporization coefficients for R-407C. The herringbone tube yielded up to 90% higher vaporization heat transfer coefficient than the single-helix microfin tube. The herringbone tube had higher fin height (e = 0.22 vs. 0.18 mm), smaller

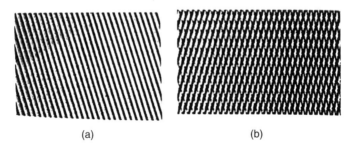

(a)   (b)

Figure 13.28 (a) Two-dimensional, (b) three-dimensional cross-grooved 15.88-mm-OD microfin tubes tested by Chamra et al. [1996].

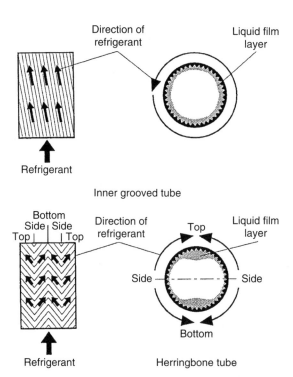

Figure 13.29 Sketches of the groove and wetting pattern of (a) two-dimensional, (b) herringbone microfin tubes tested by Ebisu [1999].

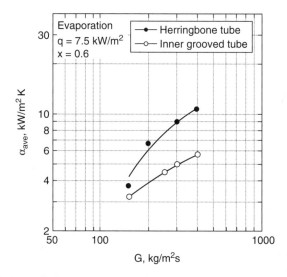

Figure 13.30 R407C vaporization heat transfer coefficient vs. mass flux for tubes illustrated in Figure 13.29. (From Ebisu [1999].)

fin pitch (p = 0.34 vs. 0.41 mm), and smaller helix angle (16° vs. 18°). The enhancement increased as the mass flux increased. The reason was attributed to the capillary flow in the grooves in two different directions, resulting in a film layer on both sides of the tube thinner than at the top and bottom, as illustrated in Figure 13.29. Note that this wetting pattern is believed to be independent of gravity effects. Thus, if the tube were rotated 90° about its axis, the thinner film regions should be at the top and bottom walls. The corresponding pressure drop increase was 30 to 60%, increasing with the mass flux. The effect of tube rotation, however, was not quantified. Goto et al. [2001] provides R-410A vaporization data in 8.0-mm-OD herringbone microfin tubes.

Kim et al. [2001] measured vaporization heat transfer coefficients in an oval microfin tube, and compared the results with the corresponding circular microfin tube. The oval tube cross-section had an aspect ratio of 1:1.5, and was fabricated from a 9.52-mm-OD circular tube with an 18° helix angle. The effect of the orientation angle of the oval tube was also investigated. The oval tube yielded 2 to 15% higher heat transfer coefficients than the circular tube. The pressure drops for both tubes were approximately the same. The variation of the installation angle of the oval tube produced negligible difference in both the heat transfer and the pressure drop.

13.4.4 Microfin Vaporization Data

From the most basic point of view, there are two mechanisms of vaporization active in the microfin tube. These are the convection (h_{cv}) and the nucleate boiling (h_{nb}) components, as described by Equations 13.1 through 13.4. The convection component in vaporization and condensation is governed by the same mechanism. This is discussed in Section 14.3.3. This convection mechanism should apply up to the vapor quality at which dryout occurs. As a test of the validity of this hypothesis, Chamra and Webb [1995] measured local heat transfer coefficients for R-22 at 24.4°C for a cross-grooved microfin tube using hot water heating. Their 15.88-mm-OD tube had 15° helix angle microfins and 0.35-mm fin height that were crosscut (0.17 mm deep) at an opposing angle (–15°) to make three-dimensional fins. Figure 13.31 shows the vaporization heat transfer data compared with the condensation data taken at the same operating condition. Essentially these data illustrate that, in the forced convective region, the evaporation heat transfer coefficient is only a little higher than for condensation. The greatest differences (approximately 10%) occur for vapor qualities below 50%. The difference is attributed to the nucleate boiling contribution in vaporization. At the higher vapor qualities, where convective effects are likely dominant, the evaporation and condensation coefficients are nearly equal. This shows that the mechanisms of evaporation and condensation in the convection-dominated regimes are equal.

Only one study has been reported in attempt to measure the nucleate boiling component. Del Col et al. [2002] measured the condensation and the vaporization coefficients inside a two-dimensional microfin tube (Wolverine DX-75) using R-22. Data were also taken for nucleate boiling on a plain surface and on the a flattened sample of the microfin surface. The pool boiling results are shown in Figure 13.32.

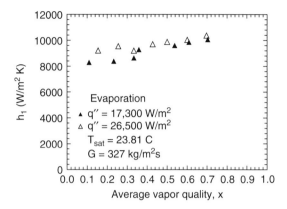

Figure 13.31 Vaporization heat transfer data compared with the condensation data taken at the same operating condition (R-22 at 24.4°C) in the 15.88-mm-OD microfin tube. (From Chamra and Webb [1995].)

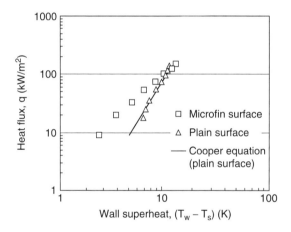

Figure 13.32 Nucleate pool data of the two-dimensional microfin surface (Wolverine DX-75) taken by Del Col et al. [2002] using R-22. (From Del Col [2002].)

The vaporization coefficient was larger than the condensation coefficient. The reason was attributed to the significantly higher nucleate pool boiling coefficient on the microfin surface than on a plain surface. The nucleate boiling contribution will depend on the existing value of $(T_w - T_{sat})$. As $(T_w - T_{sat})$ increases, the nucleate boiling contribution will increase. Coversely, it will decrease at lower values of $(T_w - T_{sat})$. The vaporization heat transfer coefficient was predicted with the asymptotic model (Equation 13.4). They assumed that the convective contributions for condensation and evaporation (q_{cv}) are equal and used the experimental condensation data to calculate q_{cv} and the nucleate boiling curve (Figure 13.32) on the microfin surface to calculate q_{nb}. The q_{nb} was calculated at the existing value of $(T_w - T_{sat})$. It was found that use of $n = 1.3$ in Equation 13.4 yielded the best fit of the evaporation data.

13.5 MINICHANNELS

Flat, extruded aluminum tubes are currently used in automotive evaporators and condensers. Typical geometries used today are in the 0.5 to 1.5 mm hydraulic diameter range. Smaller hydraulic diameter channels are used in electronics cooling. Chapter 18 addresses microchannels, which are defined here to be below 1.0-mm hydraulic diameter. Here minichannels are arbitrarily defined as channels at or below 3.0-mm hydraulic diameter. This section addresses such minichannels. Figure 13.33 illustrates small hydraulic diameter extruded aluminum tubes. The tubes illustrated in the figure have $0.44 \leq D_h \leq 1.56$ mm. It is possible that the use of such channels will be extended beyond automotive air-conditioning applicaitons, e.g., to residential air conditioning. It is further possible to contemplate manufacture of minichannels in different materials, such as copper. Although copper is not amenable to extrusion, it is a simple matter to form such channels in a hollow tube by brazing a corrugated insert.

Kew and Cornwell [1997] report that the effect of confinement becomes significant when the hydraulic diameter is less than $2\sigma/[g\,(\rho_l - \rho_v)]^{1/2}$. For R-22 at 0°C, the corresponding hydraulic diameter is 1.9 mm. The two-phase mass velocity in minichannels, is limited to relatively low values because of pressure drop considerations. This leads to the occurrence of laminar flow and high boiling number ($= q/Gi_{\mathrm{fg}}$). The reduced mass velocity will reduce the convective component. Many of the studies suggest nucleate boiling dominance over forced convection. However, whether nucleate boiling will be dominant depends on the operating heat flux (or wall superheats). Equation 13.1 or 13.4 should also be applicable to minichannels. However, the appropriate correlation for the F factor (Equation 13.5) may be size dependent. Table 13.5 summarizes the reported studies for vaporizaton in minichannels.

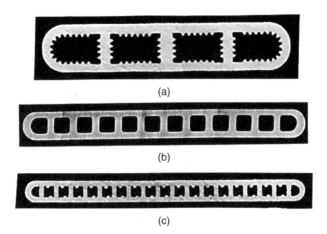

(a)

(b)

(c)

Figure 13.33 Extruded aluminum minichannels: (a) 3×16 mm, $D_h = 1.56$ mm; (b) 1.7×18 mm, $D_h =$ 1.03 mm; (c) 1.35×20 mm, $D_h = 0.44$ mm.

Lazarek and Black [1982] measured convective vaporization in a 3.1-mm round tube using R-113 at relatively high heat flux as shown in Table 13.5. The experiments included vertical cocurrent upflow and downflow configurations. The data showed strong dependence on heat flux, and no dependence on vapor quality. No significant difference between the heat transfer coefficient of upflow and downflow was observed. The following equation successfully correlated their data:

$$Nu_d = 30 \, Re_d^{0.857} \, Bo^{0.714} \tag{13.11}$$

In the above equation, $Nu_d = hd_i/k_l$, $Re_d = G/d_i\mu_l$ and $Bo = q/G\lambda$. Equation 13.11 yields weak dependence of the heat transfer coefficient on mass flux ($h \propto G^{0.153}$).

Wambsganss et al. [1993] obtained R-113 convective vaporization data in a 2.92-mm-diameter tube, also at relatively high heat flux. The data showed dominance of nucleate boiling, because of the relatively high heat fluxes imposed. Tran et al. [1996] extended the study to a low heat flux (down to 3.4 kW/m^2). They evaporated R-12 in a 2.46-mm-diameter tube and in a 1.70×4.06 mm ($D_h = 2.40$ mm) rectangular channel. Figure 13.34 shows the data for the circular tube. The figure shows two distinct regimes based on the wall superheat. For $\Delta T_{ws} > 2.75°C$, the heat transfer

Table 13.5 Reported Convective Vaporization Studies in Minichannels

Ref.	Tube ID (mm)	Fluid	Operating Conditions	Remarks
Lazarek and Black [1982]	3.1	R-113	q: 8.8–90.75 kW/m^2 G: 50–300 kg/m^{2}s	Nucleate boiling dominated
Wambsganss et al. [1993]	2.92	R-113	q: 14–380 kW/m^2 G: 125–750 kg/m^{2}s	Nucleate boiling dominated
Tran et al. [1996]	2.46 (circ.) 2.40 (rect.)	R-12	q: 3.6–129 kW/m^2 G: 44–832 kg/m^{2}s	Nucleate boiling dominated $T > 2.735°C$ Convection dominated $T < 2.75°C$
Yan and Lin [1998]	2.0 (multitube)	R-134a	q: 5–20 kW/m^2 G: 50–200 kg/m^{2}s	Both nucleate boiling and convection effect
Zhao et al. [2000]	0.86 (multichannel)	CO_2	q: 3–23 kW/m^2 G:100–820 kg/m^{2}s	Nucleate boiling dominated
Bao et al. [2000]	1.95	R-11 R-123	q: 5–200 kW/m^{2}s G: 50–1,800 kg/m^{2}s	Nucleate boiling dominated
Lin et al. [2001]	1.0	R-141b	q: 10–1,150 kW/m^2 G: 300–2,000 kg/m^{2}s	Both nucleate boiling and convection effect
Yu et al. [2002]	2.98	Water	q: 50–200 kg/m^{2}s	Nucleate boiling dominated
Fujita et al. [2002]	1.12	R-123	q: 5–20 kW/m^2 G: 50–400 kg/m^{2}s	Nucleate boiling dominated
Pettersen [2003]	0.81 (multichannel)	CO_2	q: 5–20 kW/m^2 G: 190–570 kg/m^{2}s	Nucleate boiling dominated

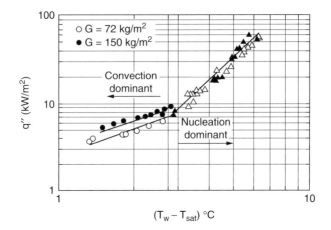

Figure 13.34 R-12 vaporization heat transfer data in a 2.46-mm-ID circular tube obtained by Tran et al. [1996]. Data show two distinct regimes based on the wall superheat: $\Delta T_{ws} > 2.75$ K; nucleation dominant, $\Delta T_{ws} \leq 2.75$ K; convection dominant. (From Tran et al. [1996]. Reprinted with permission from Elsevier.)

coefficients do not change with mass flux, which indicates the nucleate boiling dominance. For $\Delta T_{ws} < 2.75°C$, however, the data clearly show the effect of mass flux, revealing the convection contribution. The forced vaporization data in the rectangular channel showed the same trend. The following correlation was developed, which suggests the dominance of nucleate boiling.

$$h = 8.4 \times 10^{-5} (\mathrm{Bo}^2 \mathrm{We}_l)^{0.3} (\rho_l / \rho_v)^{-0.4} \tag{13.12}$$

where We_l is the Weber number ($= G^2 d_i / \sigma \rho_l$). Tran et al. [1996] compared their data with the Kandlikar [1991] correlation, and the results are shown in Figure 13.35. The predictions were made at three different qualities ($x = 0.3$, 0.5, and 0.8) for two different diameters ($d = 20$ and 2.46 mm). Figure 13.35 shows that the Kandlikar predictions fall well below the data. In addition, the predicted values were very sensitive to vapor quality, while the data do not depend on the quality. The Kandlikar correlation exhibits clear distinction between nucleate and convection-dominant boiling regimes. The transition between them is seen in Figure 13.35 as the abrupt change in slope occurring at $\Delta T_{ws} = 11$ and 12.3 K for the two diameters at $x = 0.5$. The experimental data show this transition at $\Delta T_{ws} = 2.75$ K. This figure shows that nucleate boiling can be dominant in microchannels at small wall superheat, because the convective contribution decreases as the channel size is decreased for a given mass velocity.

The flow visualization study in a minichannel (2.5 × 6.0 mm) by Kasza et al. [1997] provides some insight into the heat transfer enhancement and the nucleate boiling dominance in minichannels. The test fluid was water at $G = 21$ kg/m²-s. Figure 13.36 shows a sketch of the flow visualization results. The flow enters as a subcooled liquid, and when the wall superheat reaches a threshold value, the

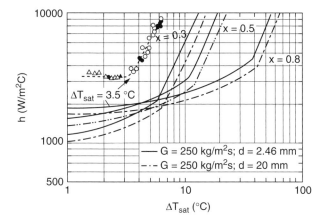

Figure 13.35 R-12 vaporization heat transfer data in a 2.46-mm-ID circular tube compared with predictions (lines) by Kandlikar [1991] correlation. The predictions are at three qualities (x = 0.3, 0.5, and 0.8) for two different diameters (d = 20 and 2.46 mm). (From Tran et al. [1996]. Reprinted with permission from Elsevier.)

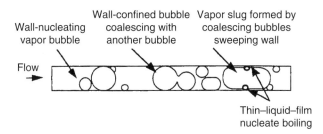

Figure 13.36 Sketch of the generation of vapor from the wall for water flow at G = 21 kg/m²s in a rectangular minichannel (2.5 × 6.0 mm). (From Kasza et al. [1997].)

nucleation sites grow bubbles and soon become the size of the channel cross section and experience confinement. Because the vapor bubbles are confined, they are forced to sweep the walls of the channel as they are conveyed along the channel by the forced flow. As a large vapor bubble sweeps by nucleation sites that are generating bubbles, it coalesces with other bubbles and thus increases the vaporization rate. The large vapor slug resides in the core of the channel forming a thin liquid film between the wall and slug. If the wall superheat is sufficiently high, bubbles are seen to nucleate through the thin liquid layer. The flow visualization study further revealed the increased nucleation frequency and rapid bubble growth dynamics under the large vapor slug. Bao et al. [2000], Yu et al. [2002], and Fujita et al. [2002] also obtained convective vaporization data in minichannels. The test range is provided in Table 13.5. These studies confirmed the nucleate boiling dominance of the convective vaporization in minichannels for the tested heat flux range.

Zhao et al. [2000] obtained convective vaporization data for CO_2 in a flat tube having multiple triangular subchannels 0.86 mm in hydraulic diameter for the test range in Table 13.5. The heat transfer coefficient was nearly constant up to $x = 0.7$, and then for $x > 0.7$ it decreased with increasing vapor quality. Mass flux had a negligible effect on the heat transfer coefficient, while heat flux had a strong effect, which suggests that nucleate boiling is the dominant factor in the convective boiling of CO_2 in minichannels. The heat transfer coefficient increased as the saturation temperature increased. Zhao et al. [2001] compared their convective vaporization CO_2 data with the existing correlations developed for large-diameter tubes. None of the correlations satisfactorily predicted the data. Zhao et al. [2001] modified the Liu and Winterton [1991] correlation by introducing the confinement number to account for the small-diameter effect.

$$h = 1.27 N_{conf}^{0.1} h_{L-W} \qquad (13.13)$$

where $N_{conf} = [\sigma/g(\rho_l - \rho_v)]^{0.5}/D_h$. The above correlation predicted their own data within $\pm$ 20%. Nucleate boiling dominant heat transfer data were also reported by Pettersen [2003] for convective vaporization of CO_2 in a $D_h = 0.81$-mm flat tube having rectangular subchannels.

Yan and Lin [1998] obtained R-134a convective vaporization data in a parallel channels of circular tubes with 2.0-mm inner diameter. At heat fluxes lower than $10\,kW/m^2$, the heat transfer coefficient was almost independent of vapor quality, which is in agreement with other investigations on minichannels. At higher heat fluxes, however, the heat transfer coefficient significantly decreased as the vapor quality increased, possibly due to dryout at the tube wall. Webb and Paek [2003] suggested that these results may be the result of flow mal-distribution in the 20 parallel channels. Pettersen [2003] also reported significant dryout at a high mass flux for the evaporation of CO_2 in a flat multichannel tube. The effect of mass flux on the heat transfer coefficient was evident at low heat fluxes.

13.6 CRITICAL HEAT FLUX (CHF)

The literature review in Chapter 11 showed that structured and porous enhanced surfaces increase the CHF for nucleate pool boiling. Section 11.12 describes a mechanism by which the CHF is increased for a porous boiling surface. Equation 13.1 applies to convective vaporization and states that the total heat flux is the sum of the nucleate boiling and convective contributions.

Hence, increasing the convective component can also lead to an increase of the CHF. As noted in Section 13.2.6, there are two CHF regions, which occur at two different flow conditions. The departure from nucleate boiling (DNB), which occurs at subcooled or low vapor quality and high heat flux, and dryout, which occurs at higher vapor quality.

13.6.1 Twisted Tape

Twisted tape can increase the DNB and the CHF for subcooled boiling. Gambill et al. [1961] performed extensive tests with subcooled boiling of water. At high velocities, the CHF was increased as much as 200%. At lower velocities, the increase was only 20%. The velocity effect is explained by Equation 13.2. At increased velocity, convective contribution supports a higher heat flux at a given $(T_w - T_s)$. Gambill [1963, 1965] also speculates on reasons for the increased CHF. Bergles et al. [1971] investigated dryout CHF for nitrogen (140 to 170 kPa) in a 10-mm-diameter vertical tube having twisted tapes ($y = 4.1$ and 8.5). They show that the tape delays the vapor quality at which dryout occurs. They develop a superposition-based correlation to predict the heat transfer coefficient in the evaporation and the drywall regions. Cumo et al. [1974] also provide data for R-12 in vertical tubes ($y = 4.4$).

Jensen [1984] developed a correlation to predict the CHF for twisted tapes, relative to the CHF in a plain tube. The correlation is

$$\frac{q_{cr}}{q_{cr,p}} = (4.597 + 0.0925y + 0.004154y^2)\left(\frac{\rho_l}{\rho_v}\right) + 0.09012 \ln\left(\frac{a}{g}\right) \qquad (13.14)$$

where $q_{cr,p}$ is the critical heat flux in a plain tube operated at the same inlet flow conditions, and a is the radial acceleration given by

$$a = \left(\frac{2}{d_i}\right)\left(\frac{u_a\pi}{2y}\right)^2 \qquad (13.15)$$

Tong et al. [1994] investigated CHF in small-diameter tubes (2.44 to 6.54 mm ID) with twisted tapes having twist ratios $1.9 \leq y \leq \infty$. Small twist ratios provided significant improvement, with $q_{CHF} \propto y^{-0.343}$. However, for larger twist ratios, the CHF was a weak function of twist ratio, and due to the thermal insulating effect between the twisted tape and heated wall, the CHF values were even lower than those for empty tubes. The gap between the tape and the tube wall was between 0.05 and 0.1 mm. An empirical correlation was proposed based on their data. Additional data are provided by Kabata et al. [1996].

Weisman et al. [1994] proposed a phenomenological model to predict the CHF in tubes containing twisted tapes. The model is an extension of the CHF model developed for empty tubes based on bubble crowding and coalescence in the bubbly layer (Weisman and Pei [1983], Weisman and Illeslamlou [1988]). The model attributes the swirl effect improvement to the increased wall shear stress, which was calculated using the single-phase friction factor correlation of Manglik and Bergles [1992]. A reasonably good agreement was obtained with the data.

13.6.2 Grooved Tubes

Steel grooved tubes of the type shown in Figure 13.7b are used in vertical "wet wall" power boilers. These boilers operate with subcooled or low-vapor-quality water. Swensen et al. [1962] show that the grooved tubes permit operation at higher heat flux and lower mass flow rate than is possible with plain tubes. Watson et al. [1974] also investigated the CHF with water and defined the minimum mass velocity needed to produce swirl flow. In the wet wall boiler, a radiant heat flux is applied to one side of the tube. Kitto and Wiener [1982] investigated the effect of non-uniform circumferential heat flux and tube inclination angle on the CHF. For a peak-to-average heat flux ratio of 1.9, they observed that the CHF was higher than that for a uniform circumferential heat flux. The CHF was 30% smaller for 30° inclination angle (from the vertical), relative to the vertical orientation. Their tests showed that the CHF for the ribbed tube was as much as three times that of a plain tube for the same flow conditions and orientation. Additional data are provided by Chen et al. [1992].

Celata et al. [1994] investigated the effect of a coiled wire insert configuration on the subcooled CHF in 6.0 and 8.0-mm-ID stainless steel tubes. The tests were performed using water at 35 to 50 bar. Both horizontal and vertical upward flow conditions were investigated. The coiled wire consisted of three different wire diameters (0.5, 0.7, 1.0 mm) with various wire pitches (1.5 to 20 mm). The largest CHF increase (50%) was obtained for 1.0-mm wire diameter and 20-mm pitch configuration. The corresponding pressure drop increase was 25%. No observable influence of the channel orientation was detected. Kabata et al. [1996] conducted subcooled water CHF tests in 12-mm-diameter tubes using water at 2 to 11 bar with 7 m/s water velocity. Two different wires (0.5, 1.0, 2.0 mm) with three different pitches (12, 24, 36 mm) were tested. The CHF increased as the wire pitch decreased and the wire diameter increased. The maximum CHF enhancement (90%) was achieved for the 1.0-mm-diameter wire with 12-mm wire pitch. Lan et al. [1997] also provide empirical correlations for the CHF enhancement ratio in tubes with coiled wire inserts.

13.6.3 Corrugated Tubes

Withers and Habdas [1974] evaporated saturated R-12 in 19-mm-diameter horizontal corrugated tubes (Figure 9.3b), as described in Section 13.3.3. The tests investigated the dryout CHF. They found that the dryout condition occurred at much higher vapor quality in the corrugated tubes, relative to a plain tube at the same operating conditions. They were able to obtain 100% evaporation in the corrugated tube at flow rates up to three times that attainable in the plain tube. The supportable heat flux increased in direct proportion to the flow rate increase.

13.6.4 Mesh Inserts

Mergerlin et al. [1974] showed that steel mesh and brush-type inserts increased the CHF by several hundred percent. However, the pressure drop increase was horrendous!

13.7 PREDICTIVE METHODS FOR IN-TUBE FLOW

Predictive equations have been developed for several of the enhancement techniques. These equations are typically based on modifications to vaporization in plain tubes. Schlager et al. [1990] provide a literature survey of correlations for vaporization inside tubes. Although these correlations may contain fluid property groups, their development may have been based on data for only one fluid, and at one pressure. Hence, they cannot be reliably used for other fluids, or even other pressures with the same fluid. Frequently, the modification to account for the enhancement geometry is derived from terms previously developed to account for single-phase flow in that particular geometry. Only heat transfer correlations are presented here. Refer to the references for the accompanying pressure drop correlations. Typically, the modification consists of added empirical terms to account for the effect of the enhancement geometry dimensions. This geometry modifier is written in dimensionless terms, and is derived by linear multiple regression methods. Such methods have no rational basis, and use assumed dimensionless parameters and statistical correlating methods. Further, there is no guarantee that the dimensionless values used are general. In fact, some of the plain tube correlations are based on similar statistical methods. What makes the correlations for the enhanced surfaces riskier is that the enhanced tube correlation contains more geometric variables than does the plain tube correlation.

There are a variety of correlations for vaporization in plain tubes. Thus, the modified correlation developed for the enhanced tube is as strong (or weak) as the base correlation. One of the most popular correlations is the Chen [1966] model defined by Equation 13.2. This model adds a convective term and a nucleate boiling term, which depends on heat flux. An early in-tube vaporization correlation developed by Pierre [1964] for refrigerants is

$$\frac{hd_i}{k_l} = B\left[\left(\frac{\Delta x \lambda}{L}\right)\left(\frac{d_i G}{\mu_l}\right)^2\right]^n \qquad (13.16)$$

when $x \le 0.9$ ($B = 0.0009$, $n = 0.5$) and for $x > 0.9$ ($B = 0.0082$, $n = 0.4$). In contrast to the Chen model, the Pierre correlation does not contain heat flux. If the enhancement is expected to have a significant nucleate boiling contribution, one would not expect that modification of the Pierre correlation would work very well. Conversely, if there is no nucleate boiling contribution (no heat flux dependence), a correlation based on the Pierre correlation would be acceptable. With this introduction, the correlations that have been developed for the various tube-side enhancements are briefly discussed.

13.7.1 High Internal Fins

The Azer and Sivakumar [1984] correlation uses the correlating parameters of the Pierre [1964] equation, multiplied by geometry factors to account for the internal fin geometry. This correlation is

$$\frac{hd_i}{k_l} = B\left[\left(\frac{J\Delta x\lambda}{L}\right)\left(\frac{d_iG}{\mu_l}\right)^2\right]^n (1+0.0024F_1^{3.72}F_2^{-8.88})$$ (13.17)

where $B = 12.24$, $n = 0.146$, and F_1 and F_2 account for the fin geometry. The parameters F_1 and F_2 were used by Carnavos [1980] to correlate single-phase flow data in internally finned tubes and are discussed in Chapter 8. These parameters are defined as $F_1 = A_{fa}/A_{fc}$ (actual flow area/core flow area between fin tips), $F_2 = A_n/A_a$ (plain tube surface area/total surface area). Because the correlation was based only on R-113 data, one should use the correlation with caution.

Schlünder and Chawla [1967] provide an empirical correlation for aluminum inserts (Figure 13.10) based on their R-11 test data. The correlation is not expected to apply to different fluids.

13.7.2 Microfins

Considerable recent work has been done to develop predictive methods for the microfin tube (Figure 13.7d). Cui et al. [1992] developed an empirical correlation of their R-502 data on nine tube geometries. The correlation uses the parameters of the Pierre [1964] plain tube correlation, plus a parameter to account for the microfin geometry. There has been significant recent progress in predictive methods for microfin tubes. Koyama et al. [1996] measured the heat transfer coefficient for vaporization of R-22, R-123, and R-134a in an 8.37-mm-ID microfin tube having 60 fins, 0.168-mm fin height, and 18° helix angle. Koyama used Equation 13.2 to correlate his data. The heat transfer coefficients are based on the total internal surface area, while the Nu and Re are based on the nominal inside diameter (d_i). The h_{cv} term Fh_{lo} is given by

$$F = \frac{h}{h_{lo}} = 1 + \frac{2}{X_{tt}^{0.88}} + \frac{0.8}{X_{tt}^{1.03}}$$ (13.18)

where

$$\frac{h_{lo}d_i}{k_l} = 0.028\,\mathrm{Re}_{lo}^{0.8}\,\mathrm{Pr}^{0.4}$$ (13.19)

The single-phase coefficient (h_{lo}) in Equation 13.18 is based on his correlation discussed in Section 8.4. The nucleate boiling component (h_{nb}) is predicted using a correlation of Takamatsu et al. [1993] for a plain surface and uses the total microfin surface area. The Koyama et al. data [1996] were predicted within +11.9/– 0.8%. The ability of the correlation to predict the data of four other investigators was also examined and found to predict within +18/– 4%.

Thome et al. [1999] extended the Kattan et al. [1998] model for plain tubes to microfin tubes, by assuming annular flow and enhancing the convective term for smooth tube h_{cv} with the factor E_{RB}, and the total heat transfer coefficient with the factor E_{mf}

$$h = E_{mf}[(h_{nb})^3 + (E_{RB}h_{cv})^3]^{1/3} \qquad (13.20)$$

They assumed that the nucleate boiling component (h_{nb}) could be predicted using the Cooper [1984] correlation for pool boiling on a plain surface, augmented by the increase in internal surface area, relative to that of a plain tube. Note that they calculated h_{nb} at the total system heat flux, rather than at the system ($T_w - T_{sat}$). The convective term in h_{cv} is calculated using a turbulent film flow equation, with constants developed from the authors plain tube database. The enhancement factor E_{RB} is that developed by Ravigururajan and Bergles [1985] for single-phase turbulent flow in internally ribbed tubes. Note that a prediction equation for the microfin tube for single-phase turbulent flow has been provided by Brognaux et al. [1997]. The E_{mf} correction factor of Thome et al. [1999] is expressed as a function of mass velocity and obtained using a curve-fitting procedure. The model predicted their R-134a and R-123 data with a mean deviation of 12.5%.

Assuming annular flow, Cavallini et al. [1999] suggested calculating the heat transfer coefficient as the sum of the nucleate boiling, h_{nb}, and of the convective term, h_{cv} components. The nucleate boiling component, h_{nb}, is calculated with the Cooper [1984] equation for plain tubes, expressed as a function of the nucleate boiling heat flux $q_{nb} = h_{nb} (T_s - T_w)$, multiplied by a suppression factor S and a function of the fin tip diameter $F_1 (d_t)$. The S factor depends on the Martinelli parameter. Cavallini et al. [1999] referred to their correlation developed for condensation inside microfin tubes to predict the convective term h_{cv}, and provided the following equation:

$$h_{cv} = (k_1 / d_t) \, \mathrm{Nu}_{cv,sm} \mathrm{Rx}^a (\mathrm{BoFr})^b F_2(d_t) F_3(G) \qquad (13.21)$$

where $\mathrm{Nu}_{cv,sm}$ is the Nusselt number for evaporation in a smooth tube and is equal to the product of all liquid Nusselt number (Nu_{LO}) and the two-phase multiplier (F). The parameter Rx accounts for the effect of the area increase, while the Bond number (Bo) accounts for surface tension effects. The two-phase multiplier (F) is similar to the original one used in their condensation correlation (Cavallini et al. [2000]). The constants in Equation 13.21 were obtained by a nonlinear best-fitting procedure of the local data, for small quality variations in the test section. Cavallini et al. [1999] compared their correlation with those of Koyama et al. [1996] and Thome et al. [1999] to predict the vaporization coefficients from 18 independent sources; these sources contained 643 experimental data points for six different pure refrigerants. The mean absolute deviation for each model was found to be 58% (Koyama et al.), 33% (Thome et al.), and 14% (Cavallini et al.). Note that none of these models are applicable to predict heat transfer in the dryout region. If the dryout vapor quality is known (which is unlikely), one may use an interpolation equation between the dryout point and the all-gas flow heat transfer coefficient, as recommended by Kattan et al. [1998] for horizontal plain tubes.

13.7.3 Twisted Tape Inserts

The Agrawal et al. [1986] correlation is also based on modifying the Pierre [1964] correlation. It is

$$\frac{h}{h_{Pie}} = 0.00188 \, Re_s^{2.23} \, Bo^{1.62} \, y^{-0.357}$$ (13.22)

where h_{Pie} is the value calculated by the Pierre equation (Equation 13.16) for a plain tube. The swirl Reynolds number (Re_s) is defined in Chapter 7, and the term Bo is the boiling number ($q/G\lambda$), which includes a heat flux dependency. The correlation is based on R-12 data for three twisted tapes. As noted in Section 13.3.2, one would not expect the twisted tape to show a heat flux dependency unless it were operated at quite high heat flux.

13.7.4 Corrugated Tubes

No correlations are reported in the literature. However, Withers and Habdas [1974] provide a curve-fit of their R-12 data on two corrugation geometries using the Pierre [1964] correlation, Equation 13.16.

13.7.5 Porous Coatings

No correlations have been developed and tested. However, Ikeuchi et al. [1984] curve-fit their R-22 data in the form of the Pierre [1964] correlation, Equation 13.16. It would be reasonable to predict the porous tube heat transfer coefficient using the Chen [1966] correlation, or a later variant. This approach would add the pool boiling and convective components.

13.8 TUBE BUNDLES

Chapter 11 provided information for pool boiling on enhanced tubes. Because the enhanced tubes are intended for use in tube bundles, it is important to know whether tube bundle data will differ from single-tube pool boiling data. Two effects can be active in a tube bundle, which may alter the performance from single-tube results. These are (1) static head effects and (2) convective effects.

For fluids having a large change of saturation temperature with change of pressure (dT/dp), static liquid head will cause increased saturation temperature in the lower part of a tube bundle, and thus reduce the local driving temperature difference. This is discussed in Chapter 4, and Figure 4.2 shows dT/dp for several fluids. Because of the two-phase mixture in the tube bundle, however, the static head effect will be less than indicated in the nonboiling condition for the same liquid level.

Boiling in tube bundles may span a wide range of flow conditions. In kettle reboilers, this typically involves an entering subcooled liquid and a vapor quality change of 15% or less over the bundle depth. However, a flooded refrigerant evaporator is essentially a once-through device, with 15% entering vapor quality and 100% leaving vapor quality. The mass velocity in a kettle reboiler is higher than that in a flooded refrigerant evaporator.

Early work to simulate boiling in tube bundles was unable to simulate this wide range of flow conditions. Hence, one should be careful to generalize on such simulations. Examples of early laboratory simulations of boiling in tube bundles are Fujita et al. [1986] and Muller [1986]. Muller used saturated R-11 at 101 kPa in a bundle having 18 finned tubes in six rows with three tubes in each row. Fujita et al. [1986] obtained data for R-113 boiling on an 11-tube bundle (7 rows) of 25.4-mm-diameter plain tubes. The effects of tube position in the bundle, distribution of heat flux, and system pressure were studied. They found that the boiling coefficient on the bottom tube row was approximately equal to that for a single tube. However, substantial enhancement was observed for the upper tube rows at moderate heat flux. In the experiments of Muller [1986] and Fujita et al. [1986], the refrigerant entered the boiling vessel as saturated liquid with zero vapor quality. Although these papers give an insight regarding the effect of tube position and intertube spacing on the boiling heat transfer, they involve quite low mass velocity and leaving vapor quality.

As shown by Equation 13.2, the composite heat transfer coefficient will be significantly affected by the use of an enhanced boiling surface. Later studies are discussed that allow a more perceptive understanding of the effect of enhanced tubes in tube bundles.

13.8.1 Convective Effects in Tube Bundles

As noted in Section 13.2.5, the performance of a tube, when used in a tube bundle will depend on the enhancement level for pool boiling. Depending on the pool boiling performance level of the tube, it will show one of the following two characteristics in convective vaporization:

1. *High Pool Boiling Coefficient.* The tube will not be very sensitive to convective effects. However, it will be quite sensitive to heat flux, which indicates that the tube performance is dominated by nucleate boiling.
2. *Low Pool Boiling Coefficient.* This tube will benefit substantially from convective effects. However, its performance in convective vaporization will not be very sensitive to heat flux. This will occur if the performance is convection dominated.

Nakajima and Shiozawa [1975] obtained test data for an R-11 evaporator having 185 19-mm-diameter tubes (748 fins/m, 1.5-mm fin height) with four water passes on the tubeside. Water passes 1 and 2 were in the lower part of the tube bundle and passes 3 and 4 were in the upper part of the bundle. By measuring the water temperature leaving pass 2, they were able to determine the evaporation coefficient for the bottom and top halves of the tube bundle. For 15% inlet vapor quality, the evaporation coefficient in the lower half of the bundle was four to seven times higher than that of a single tube in pool boiling. The evaporation coefficient in the upper half of the bundle was 1.5 times higher than in the lower half of the tube bundle. Convective effects cause the higher evaporation coefficient in the upper half of the bundle.

Webb and Gupte [1992] provide description and discussion of correlations for convective vaporization in tubes and in tube bundles. Gupte and Webb [1994, 1995a, 1995b] also provide convective vaporization data for the Figure 11.12 e and f tube geometries using R-11 and R-123 at operating conditions typical of refrigerant evaporators (5°C). The data were taken on 19.0-mm-diameter tubes on 23.8-mm triangular pitch in a tube bundle simulator, which allowed independent control of heat flux, mass velocity, and vapor quality. Pool boiling data of these geometries for R-22 are shown in Figure 11.19. Figure 11.19 shows that the GEWA-SE tube provides a substantially higher boiling coefficient than the integral-fin tube. The importance of convective effects is demonstrated by comparing the GEWA-SE tube with the lower-performance integral-fin tube.

Figure 13.37a shows the 1024 fins/m integral-fin tube data of Gupte [1994, 1995a] at $G \simeq 13$ kg/m^2-K and that the heat transfer coefficient is a strong function of vapor quality, whereas heat flux has a small effect. This means that the performance is dominated by convection, rather than by nucleate boiling. As the vapor velocity increases, the convective contribution increases because of the increased vapor velocity. Data for different mass velocities (not shown here) show that the integral-fin tube performance is significantly affected by mass velocity.

Figure 13.37b, from Gupte and Webb [1994, 1995b], shows the convective boiling coefficient of the GEWA-SE tube with R-134a at 26.7°C (48 kPa) and mass velocity $G = 24$ kg/m^2-s. In contrast to Figure 13.37a, the figure shows that the GEWA-SE performance is highly sensitive to heat flux, which indicates that the performance is relatively insensitive to convective effects. Examination of the data at different mass velocities (not shown here) shows negligible effect of mass velocity. Note that the GEWA-SE tube maintains its high nucleate boiling–dominated performance up to the highest vapor qualities tested (0.95).

Figure 13.38 shows the ratio of the pool boiling heat transfer coefficient to the convective vaporization coefficient for the two tubes. Figure 13.38a shows that $h_{nbp}/h < 1$ for the integral-fin tube. The h_{nbp}/h ratio decreases with increasing vapor quality, which indicates the importance of convection. The h_{nbp}/h ratio is smaller at low heat flux, where the nucleate boiling contribution is smallest. Figure 13.38b shows that the h_{nbp}/h ratio is approximately 1.0 for the GEWA-SE tube and that the ratio is insensitive to vapor quality. Further, the ratio is relatively insensitive to mass velocity, which indicates that the tube performance is nucleate boiling dominated.

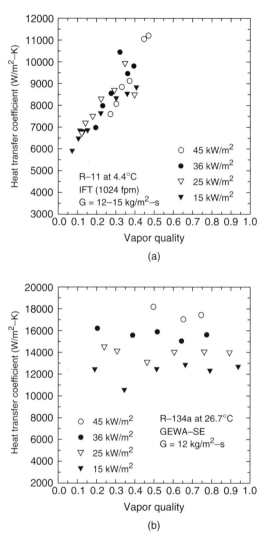

Figure 13.37 Comparison of 19-mm-OD, 1024 fins/m integral-fin and GEWA-SE tubes in a simulated bundle with 23.8-mm equilateral pitch: (a) 1024 fins/m integral-fin tube with R-11 at 4.4°C, (b) GEWA-SE tube with R-134a at 26.7°C.

This example shows that a tube having high pool boiling performance may also be dominated by nucleate boiling in tube bundle applications. However, the performance improvement (relative to integral-fin tubes) in an evaporator tube bundle is not as high as is measured in single tube pool boiling tests. This is because the forced convection enhancement that occurs in the tube bundle is more beneficial to standard integral-fin tubes than it is to the enhanced nucleate boiling tubes.

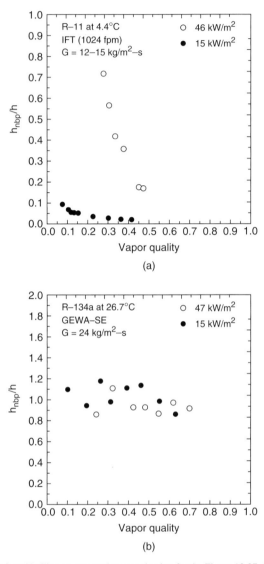

Figure 13.38 Ratio of pool boiling-to-convective vaporization for the Figure 13.37 simulated tube bundle data: (a) 1024 fins/m integral-fin with R-11, (b) GEWA-SE tube with R-134a at 26.7°C.

Webb and Apparao [1990] compare the performance of large flooded refrigerant evaporators having integral-fin and enhanced tubes.

13.8.2 Tube Bundle Convective Vaporization Data

Jensen and Hsu [1987] reported R-113 data for boiling in plain tube bundles. The data were taken on 8-mm-diameter tubes at much higher temperatures (71 to 110°C) and mass velocities (100 to 675 kg/m²-s) than studied by Gupte and Webb [1992]. Because of the high mass flow velocities tested by Jensen and Hsu [1987], the exit vapor qualities were quite low, less, that is, than 0.35. These data showed little effect of vapor quality, possibly because of the low vapor quality range tested. Jensen et al. [1992] also provide tube bundle data on plain, Turbo-B and High-Flux tubes. For p = 206 kPa, 80 kW/m², and G = 217 kg/m²-s, the negligible effect of vapor quality was observed for the High-Flux and Turbo-B tubes, as expected. The plain tube at q = 30 kW/m² inexplicably showed no effect of vapor quality for exit vapor qualities as high as 0.74.

Yilmaz et al. [1981] and Arai et al. [1977] also show that plain or integral-fin tubes yield a substantially higher boiling coefficient in the tube bundle geometry because of the forced convection contribution. However, the boiling coefficient of enhanced tubes is so high that the forced convection effects are not expected to materially enhance their performance in a tube bundle configuration. Because of the higher performance of the plain or integral-fin tubes in the bundle configuration, the designer is cautioned not to expect the same boiling coefficient improvement, as measured in pool boiling.

Figure 13.39 shows the R-11 performance of the High-Flux surface measured in pool boiling and tube bundle tests reported by Czikk et al. [1981]. All surface coatings were made from the same production run. This figure shows that the tube bundle performance is as high as or higher than that measured in pool boiling tests. Arai et al. [1977] report similar results for R-12 boiling on the Thermoexcel-E surface.

Bukin et al. [1982] measured convective vaporization in small tube bundles for 13 different types of porous coatings. Their porous surfaces were made by flame spraying, by metallic deposition, by sintering, and by wrapping the tube with layers of glass or

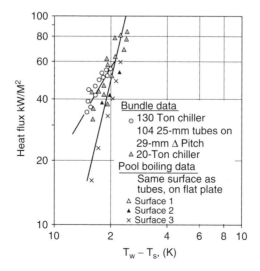

Figure 13.39 R-11 boiling at 39 kPa on the High-Flux surface for pool boiling and for a 106-tube bundle. (From Czikk et al. [1981].)

stainless steel screening. The sintered porous coatings provided the best performance, much better than that of plain or integral-fin tubes. Single tube data were not provided.

Memory et al. [1994] tested the Wolverine Turbo-B with R-113 in a simulated tube bundle having ten tube rows with one to two heated tubes in each row. They found that the average bundle boiling coefficient was greater than for a single tube and attributed the difference to a "bundle effect." The bundle factor is defined as the average coefficient for the bundle/single tube coefficient. The bundle factor varied from 1.61 at q = 1.5 kW/m^2 to 1.22 q = 100 kW/m^2. Note that the existence of a "bundle factor" greater than 1.0 differs from the findings of Gupte and Webb [1994], Arai [1977], and Czikk et al. [1970], who all found a 1.0 bundle factor.

Hsieh et al. [2003] investigated the bundle factor for boiling of R-134a on plain and plasma-coated copper tubes for various bundle layouts. The simulated bundles had 15 tubes in five rows (inline layouts) and ten rows (staggered layouts) for different locations of heated and dummy tubes. Typically only two to six of the tubes were heated and the heated tubes were at the bottom of the bundle. Data were taken for $0.1 < q < 30$ kW/m^2. By investigating different bundle layouts, they also defined a "configuration factor," in addition to the bundle factor. They observed bundle factors as large as 1.5 at q = 30 kW/m^2. Larger bundle factors were observed at lower heat fluxes.

Kim et al. [2002] investigated the effect of pore size of the Turbo-B-type enhanced tube bundles using R-123 and R-134a. This tube is shown in Figure 11.21, where their single-tube pool boiling tests are discussed in Chapter 11.5. Figure 11.21 shows enlarged photographs of the tubes having different pore diameters (d_p = 0.20, 0.23, and 0.27 mm). Tests were conducted for $8 \leq G \leq 26$ kg/m^2-s, $10 \leq q \leq 40$ kW/m^2, $0.1 \leq x \leq 0.9$. Similar to the results by Gupte and Webb [1994, 1995a, 1995b], the convective vaporization coefficients were strongly dependent on the heat flux with negligible dependency on mass flux and quality. Figure 13.40a shows the effect of pore diameter. The optimum pore diameter was different depending on the refrigerant. For R-134a, the heat transfer coefficient increased as the pore diameter increased. For R-123, however, the largest heat transfer coefficient was obtained for d_p = 0.23 mm. The corresponding pool boiling heat transfer coefficients are shown in

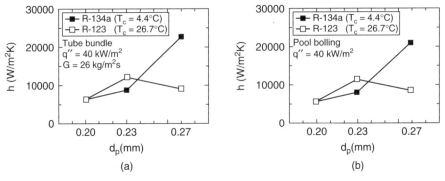

Au: there is no figure 13.53

Figure 13.40 Boiling results on Turbo-B bundle obtained by Kim et al. [2002] for different pore sizes (see Figure 11.21): (a) tube bundle, (b) single-tube pool boiling.

Figure 13.40b, which reveals a similar trend. Figure 13.40b shows that the convective vaporization heat transfer coefficients are approximately the same as the pool boiling heat transfer coefficients. Gupte and Webb [1994, 1995b] also obtained similar results for the GEWA-SE and Turbo-B tube bundles. Kim et al. [2002] successfully correlated the data using the asymptotic model with exponent $n = 1$.

The effect of oil on convective vaporization in an enhanced tube bundle was investigated by Memory et al. [1995]. The effect of oil was different, depending on the tube geometry. For the smooth and finned tube bundles, the heat transfer coefficient increased up to 3% oil concentration. Further increase in oil concentration led to a slight drop-off of the bundle performance. For Turbo-B and High-Flux tube bundles, any addition of oil led to a significant reduction in bundle performance. Tatara and Payvar [1999, 2000a, 2000b] reported the same trend for the integral-fin and enhanced tube bundles. The heat transfer reduction by oil is generally less in the bundle than for single-tube pool boiling (Gan et al. [1993], Marvillet [1989]).

13.8.3 Effect of Spacing between Tubes

Liu and Qiu [2002] have taken data on boiling in tube bundles having small clearance between the tubes. Such clearance can promote high convective velocity within the bundle. Their work was done using pure water (and a 10% saltwater solution) using a 7-row-high, 17-tube bundle. Two tube geometries were used, a plain tube and a "roll-worked" tube similar to Turbo-B (Figure 11.12e). They obtained data on single-tube pool boiling and in a bundle with gap clearances of 0.5, 1, 2, and 4 mm. As expected, when the gap was reduced, significant improvement in performance of the smooth tube boiling coefficient was observed. However, similar benefits were also observed for the enhanced tube. Figure 13.41 shows their results for the enhanced tube bundle. Note that the single-tube performance of the enhanced tube is very high, relative to the single plain tube. For a heat flux below 100 kW/m², this figure clearly shows that significant convection can add to the bundle performance of the enhanced tube for boiling of water. This differs from the findings of Gupte and Webb [1995a, 1995b] for enhanced tubes with 4.6-mm tube clearance, as shown in Figure 13.37.

Liu and Tong [2002] studied the effect of a confining annular space surrounding the tube for boiling on a single horizontal tube. Liquid (water or R-11) was fed into (and out) the tube via a 4.0-mm-wide slot at the top and bottom. Two tube geometries were tested — a plain tube and a "roll-worked" enhanced tube that appears similar to Turbo-B. Tests were also performed on the same tubes in a free liquid space for different annular gaps between 0.4 and 4.0 mm wide. The performance of the smooth tube increased as the annular gap was decreased, up to a point. When the gap became too small, it appears that CHF occurred, because of insufficient liquid supply. This resulted in reduced CHF as the gap dimension was decreased. This was especially evident for the R-11 data. The highest boiling coefficient occurred for a gap dimension of about 0.4 to 0.6 mm for both fluids. For R-11, gap spacing had negligible effect on the boiling coefficient for R-11. However, the CHF decreased with reduced gap dimension, again due to insufficient liquid supply. Because of the high boiling performance of the roll-worked tube, the tube performance was nucleate boiling

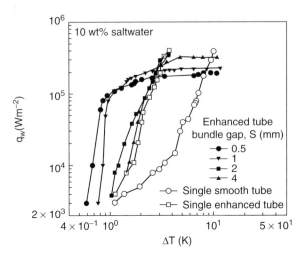

Figure 13.41 Effect of small gap between tubes for a 10% saltwater solution boiling in a horizontal bundle (7 rows) on 18-mm-diameter enhanced tubes. (From Liu and Qiu [2002]. Reprinted with permission from Elsevier.)

dominated and the increased convective effect yielded negligible effect on performance. However, the roll-worked tube showed improved performance with reduced gap dimension for water, up to 0.6-mm gap dimension. Thus, nucleate boiling did not dominate the performance of the roll-worked tube with water.

13.8.4 Convective Vaporization Models

Gupte and Webb [1994, 1995a, 1995b] tried two models (the superposition model and the asymptotic model) to correlate their convective vaporization data on staggered tube banks of 19.2-mm-OD Turbo-B, GEWA SE, and integral-fin tubes with equilateral pitch-to-diameter ratio of 1.25. The models assumed the suppression factor was 1.0 and the single-phase heat transfer coefficient was obtained from the appropriate correlations for tube banks. The Bennet and Chen [1980] two-phase multiplier expression was modified for the corresponding tube bank configuration. The superposition model predicted approximately 80% of the GEWA-SE and Turbo-B data within 20%, but did not predict the integral-fin tube data. All the data were adequately predicted by the asymptotic model with $n = 3$. The asymptotic model with $n = 3$ also adequately correlated the vaporization data of R-113 and R-123 for plain tube bank by Webb and Chien [1994]. However, different values of n were obtained for other investigators. Roser et al. [1999] performed convective vaporization tests of n-pentane in a plain tube bundle of staggered layout with pitch to diameter ratio 1.33. The mass velocity was varied 14 to 44 kg/m²-s, and the heat flux was varied up to 60 kW/m². They correlated their data using the asymptotic model, and obtained $n = 1.5$.

Kim et al. [2002] obtained convective vaporization data for a staggered tube bank of Turbo-B type enhanced tubes with equilateral pitch-to-diameter ratio of

1.25. They obtained the exponent $n = 1.0$. If $n = 1.0$, the asymptotic model has the same form as the superposition model. This is not surprising, because the Turbo-B and GEWA-SE data of Gupte and Webb [1994] were well correlated by the super-position model. The Turbo-B and GEWA-SE yield high pool boiling performance and the convective effect is minor. In such a case, the choice of the exponent n does not have significant influence on the predicted vaporization heat transfer coefficients.

13.8.5 Starting Hysteresis in Tube Bundles

Lewis and Sather [1978] boiled a subcooled ammonia feed from a high-flux tube bundle having 279 tubes. Heat was supplied by warm water on the tube side. The authors observed that a shutdown caused apparent deactivation of the nucleation sites resulting from flooding. When the boiler was restarted after a shutdown period of several hours and operated at the same heat flux (17.4 kW/m²), a smaller overall heat transfer coefficient was observed. About 100 hr of operation were required for the U value to climb from the low initial value ($U_o = 600$ W/m²-K) to the stable value of $U_o = 785$ W/m²-K. It appears that this phenomenon results from the starting hysteresis discussed by Bergles and Chyu [1982]. Apparently, the temperature boundary condition did not provide sufficient liquid superheating to immediately activate all of the nucleation sites. Thus, their activation was a gradual process. Lewis and Sather describe a special start-up process that allows vaporization of the liquid in the flooded pores and hence avoidance of the 100-hr period for site activation. The Bergles and Chyu analysis of the Lewis and Sather data suggests that the boiling coefficient corresponding to the stable $U_o = 785$ W/m²-K is in close agreement to that measured in a single-tube boiling test.

Boiling tests in R-11 and R-12 evaporators by Arai et al. [1977] have not shown the initially low performance on start-up as observed by Lewis and Sather. This results from a different start-up procedure of the refrigeration cycle. In the shutdown condition, the refrigerant liquid is saturated or very close to being saturated. When the compressor is started, the evaporator pressure is lowered, which acts to evaporate the liquid. Because the refrigerant is throttled to the evaporator from the higher condenser pressure, a two-phase mixture enters at the bottom of the tube bundle. This vapor may tend to displace liquid from deactivated cavities. This suggests that there may be some advantage in site activation by bringing a two-phase mixture in at the bottom of the tube bundle.

The author's research has shown that impingement of a two-phase jet on a superheated surface will activate nucleation sites. Czikk et al. [1981] and O'Neill [1981] state that the low initial performance measured by Lewis and Sather has not been noted in process applications of the high-flux boiling surface. O'Neill has observed no advantage in introducing a two-phase feed at the bottom of the tube bundle. The possibility of lower initial performance on start-up may be of little significance in process applications. This equipment is usually brought online and maintained in continuous operation for very long time periods (e.g., months or years). The existence of lower initial performance for the first hundred hours of operation would be of negligible significance. Bergles and Chyu [1982] provide interesting speculations on tube bundle vs. single-tube performance and possible start-up problems due to hysteresis.

Bukin et al. [1982] measured convective vaporization in small tube bundles for 13 different types of porous coatings. Their porous surfaces were made by flame spraying, by metallic deposition, by sintering, and by wrapping the tube with layers of glass or stainless steel screening. The sintered porous coatings provided the best performance, and much better than that of plain or integral-fin tubes. Single tube data were not provided.

13.9 PLATE-FIN HEAT EXCHANGERS

Brazed aluminum, plate-fin heat exchangers are used in cryogenic gas processing operations (air separation and natural gas liquefaction) and for the separation of light hydrocarbons in ethylene plants. Used in an air separation plant, they will condense nitrogen against evaporating liquid oxygen. Figure 13.42 shows a brazed aluminum plate-fin heat exchanger for cryogenic service. Large heat exchangers, such as Figure 13.42, used for gas separation may have as many as ten different process streams.

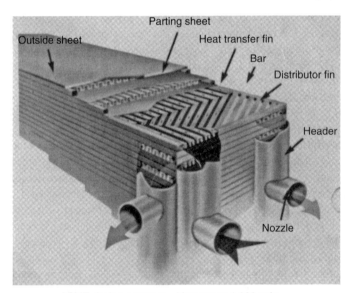

- ASME design pressure—Vacuum to 1,200 psig
- ASME design temperatures—450 to + 150 F
- Cross section—Up to 45" × 72"
- Length—Up to 240"
- Materials—Aluminum alloys 3003, 3004, 5083, 6061

Figure 13.42 Brazed aluminum plate-and-fin heat exchanger for cryogenic service. (Courtesy of Altec, LaCrosse, WI.)

Nunez and Polley [1999] provide a good article on design methodology for such multistream heat exchangers. Typical fin geometries used in this heat exchanger are shown in Figure 5.2. The same fin geometries are used for both single-and two-phase heat transfer.

The same convective vaporization phenomena exist in plate-fin heat exchangers as exist in tube bundles. Equation 13.2 still applies. Typical brazed plate-fin exchangers generally do not use enhanced boiling surfaces. Hence, convective effects will tend to be quite important. References treating convective vaporization in plate fin heat exchangers include Robertson [1980], Carey and Shah [1988], Chapter 13 of Thome [1990], and Chapter 13 of Carey [1992]. Thome [1990] summarizes the available data on plate-fin geometries, and he also summarizes the empirical correlations that have been developed.

Mandrusik and Carey [1989] describe a method to predict the boiling coefficient in a plate-fin exchanger containing the offset-strip fin (OSF) (Figure 5.2d). The prediction is based on use of Equation 13.2. The authors use an accepted plain surface correlation to calculate h_{nb}, and they account for the fin efficiency. The h_{cv} term in Equation 13.2 is predicted using $h_{cv} = Fh_l$ with the Chen model to predict the F factor. The single-phase component (h_l) may be predicted using the correlations for the OSF given in Chapter 6. A correlation for the F factor was developed based on their convective boiling data in a laboratory model of the OSF array using R-113 as the working fluid. They obtained the F factor from their data using Equation 13.5, and correlated it as a function of X_{tt} as is accepted practice. The authors obtained

$$F = \left(1 + \frac{28}{X_{tt}^2}\right)^{0.372}$$

(13.23)

They used the suppression factor of Bennett and Chen [1980]. Their method follows accepted procedures for convective vaporization, as described, for example, by Carey [1992] or by Webb and Gupte [1992]. The reader should be careful about applying the F factor correlation developed for the OSF to other plate-fin surface geometries. The F factor may be geometry sensitive. Work has not been done to develop F factor correlations for other enhanced plate-fin geometries.

The surface geometry used in plate-fin heat exchangers is typically aluminum sheet metal, which should exhibit the nucleate boiling characteristics of a plain surface. A surface geometry that gives high single-phase convective heat transfer will give a high convective evaporation coefficient (see Equation 13.2). The OSF is popular for convective boiling, because of its high single-phase heat transfer performance. It is conceivable that an enhanced boiling surface can be used in plate-fin geometries. In fact, this was the intent of Kun and Czikk [1969], who developed the Figure 11.15a enhanced surface for proprietary use in plate-fin exchangers. It is also feasible to use a porous coating, such as the Figure 11.11a sintered surface. This should have a significant effect on the nucleate boiling term of Equation 13.2.

Thonon et al. [1995] used a method of Cornwell and Scoones [1988] to predict vaporization heat transfer in plate heat exchangers. The two-phase heat transfer coefficient was taken as the greater value of the nucleate boiling term h_{nb} and the convective boiling term h_{cv}. The Gorenflo [1993] correlation was used to obtain h_{nb}. The h_{cv} was calculated using the two-phase multiplier ($h_{cv} = Fh_L$), where h_L is the single-phase heat transfer coefficient of the plate heat exchanger. The F factor equation used is

$$F = 1 + 1.8X_{tt}^{-0.79}$$ (13.24)

Thonon et al. [1995] predicted 85% of their own data within ±30%. However, no detailed information on the plate geometry is provided. They also recommended $C = 8$ in the two-phase pressure drop multiplier equation (Equation 13.9).

Quazia [2001] measured the vaporization heat transfer coefficient for 3 sets of plates having chevron angles 0°, 30°, and 60°. The plate with 0° chevron angle yielded the highest heat transfer coefficient and largest pressure drop. Based on their results, empirical correlations for the heat transfer enhancement ratio and the two-phase pressure drop multiplier were provided. The C values in Equation 13.9 were obtained from a curve-fit of the pressure drop data. The C values were between 2 and 4 and decreased with increasing chevron angle. Yan and Lin [1999] and Hsieh and Lin [2002] provide R-134a and R-410A vaporization data in a plate heat exchanger having 60° chevron angles.

13.10 THIN FILM EVAPORATION

Thin film evaporation is frequently used in the refrigeration, process, desalination, and cryogenic industries. It may be used inside vertical tubes or on the outside of a bundle of horizontal tubes. For a horizontal tube bundle, one uses spray nozzles to distribute the film over the bundle frontal area. Thin film evaporation is particularly valuable for pressure-sensitive fluids. If pool boiling were used for such fluids, the elevation of saturation pressure over the bundle depth decreases the local $(T_w - T_s)$.

Typically, the liquid feed rate is several times the amount evaporated. This is done to prevent de-wetting of the tube surface. The recirculation ratio (R_R) is defined as the total flow rate divided by the rate evaporated. If any fraction of the tube is dry, the tube performance of the dry region is essentially "lost." Heated films are quite susceptible to rupture. When falling films are used in a bundle of tubes, entrainment effects will also act to diminish the liquid film thickness. If $R_R > 1$ a pump must be used to return the unevaporated liquid to the top of the evaporator.

13.10.1 Horizontal Tubes

Chyu et al. [1982] measured the evaporation coefficient on single, horizontal enhanced boiling tubes operated in the thin film evaporation mode. A liquid distributor tube located above the horizontal test tube dripped saturated liquid on the tube. Data for distilled water at 101 kPa were obtained on a plain tube and three enhanced tubes. The enhanced tubes are the Figure 11.11a High-Flux porous coated tube, the Figure 11.12a 1024 fins/m integral-fin tube, and the Figure 11.12c Wieland GEWA-T tube. Figure 13.43 summarizes their results. The operating conditions are defined in Table 13.6. Figure 13.43 shows that plain tubes provide higher performance in thin film evaporation than in pool boiling for $\Delta T_{ws} > 6$ K. Nucleate boiling occurred for $q > 20$ kW/m^2, which accounts for the upward slope of the curve at high heat flux. For a laminar film ($d_o G_l / \mu_l < 1600$), the evaporation coefficient is approximately given by $h = k_l / \delta$, where δ is the film thickness. Hence, the plain tube curve for $q < 20$ kW/m^2 should have unity slope, and the q vs. ΔT_{ws} curve should be higher at lower values of G_l. If the G_l is reduced below a critical value, surface de-wetting will occur, and the evaporation coefficient will decrease. Hence, it is important that the surface remain wetted.

Figure 13.43 shows that the slope of the q vs. ΔT_{ws} curve for the High-Flux surface is the same for thin film evaporation (curve 3) as for pool boiling (curve 4). Nucleate boiling was observed to occur in the film for $q > 8$ kW/m^2. This explains why the thin film and pool boiling curves have the same slope. Curve 3 is taken with increasing power steps. The data for decreasing power steps are approximately the same as curve 4 for pool boiling.

The GEWA-T surface (curve 5) did not show any nucleation in the film for the tested levels of q. Note that the GEWA-T surface gives the highest performance for $q < 40$ kW/m^2. The performance of the 1024 fins/m integral-fin surface was nearly identical to that of the GEWA-T surface. As noted in the discussion of condensation row effect in Chapter 12, the integral-fin tube "channels" the flow, which prevents axial spreading of the film on the surface. Hence, it is unlikely that the 1024 fins/m

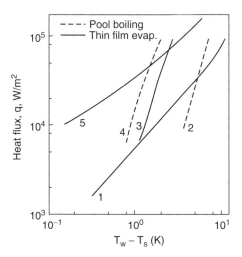

Figure 13.43 Data for thin film evaporation (solid lines) and pool boiling (dashed lines). Film evaporation: curve 1 (plain tube), curve 3 (High-Flux), and curve 5 (GEWA-T and 1024 fins/m integral-fin tube). Pool boiling: curve 2 (plain tube) and curve 4 (High-Flux).

Table 13.6 Operating Conditions for Figure 13.43

Curve	Geometry	G_l (kg/s-m)	Heat Transfer Mode
1	Plain	78	Thin film
2	Plain	0	Pool boiling
3	High-Flux	48	Thin film
4	High-Flux	0	Pool boiling
5	GEWA-T	38	Thin film

tube would provide high performance in a tube bundle, especially at lower values of G_l.

As noted above, nucleation (boiling) may or may not occur in the film. Nucleate boiling would not be expected at normal operating conditions for low-performance surfaces, such as integral-fin tubes. Here, terminology of "evaporation" is used to define cases where boiling does not occur in the film. Owens [1978] and Conti [1978] measured thin film evaporation and thin film boiling of ammonia in the laminar- and turbulent-flow regimes on horizontal tubes. Their tests were done with heat transfer from a two-tube array located below the distributor tube. This allowed determination of the effect of the vertical spacing (H) on the performance. Owens' data are for a 50.8-mm-diameter (d_o) plain tube, which span $120 \leq \mathrm{Re}_1 \leq 10,000$, $5 \leq q \leq 55$ kW/m², and $0.25 \leq H/d_o \leq 2.1$. He provides useful design correlations for the complete range of conditions studied. Conti tested 25.4-mm-diameter integral-fin tubes spaced at $H/d_o = 1.25$. The first tube had 1339 fins/m and 1.3-mm fin height. The second tube had 787 fins/m and 1.5-mm fin height. The fins on the 787 fins/m tube were machined to also provide data for 0.5- and 0.94-mm fin height. The highest performance was provided by the 787 fins/m tube with 1.5 mm fin height. It provided an enhancement (based on $A/L = \pi d_e$) 3.2 times that of a plain tube. The lower fin heights provided smaller enhancement.

Lorenz and Yung [1979] provide a semiempirical model for thin film boiling on plain horizontal tubes. Parken et al. [1990] provide additional data for water on 25.4- and 50.8-mm diameter plain tubes, and they present correlations for the evaporation and boiling regions. Rifert et al. [1992] measured falling film evaporation coefficients for a six-row rank of 38-mm-OD plain tube and longitudinally grooved tubes using water. The groove shapes investigated were triangular and rectangular. Also investigated were spaced longitudinal wires. The heat flux was sufficiently small that nucleate boiling did not occur. The longitudinal grooves produced a heat transfer enhancement factor of 1.4 to 1.9. The enhancement was caused by the partial destruction of the laminar boundary layer in the grooves. Empirical correlations were developed that describe their data. Fujita et al. [1998] provide additional data on R-11 falling film evaporation on a five-tube rank geometry of a plain and three triangular grooved tubes (500 to 2000 fins/m) 25 mm in diameter. The plain tube data were compared to theoretical analysis. The 2000 fins/m grooved tube provided E_o more than five times that of the plain tube.

Moeykens et al. [1995a] measured the R-134a falling film evaporation coefficient on six enhanced tubes; Turbo-CII, Turbo-B, GEWA-SC, GEWA-SE, and two

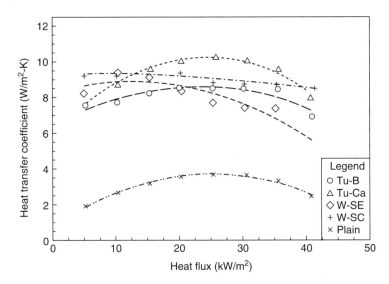

Figure 13.44 R-134a falling film evaporation coefficient of six enhanced tubes (Turbo-CII, Turbo-B, GEWA SC, GEWA SE) and two integral-fin tubes (1575 fpm and 1024 fpm) tested by Moeykens et al. [1995a].

integral-fin tubes (1575 and 1024 fpm).The Turbo-CII and GEWA-SC are enhanced condensation tubes, and the Turbo-B and GEWA-SE are enhanced boiling tubes. The saturated refrigerant was sprayed from nozzles located above the test section. The liquid supply rate was 0.013 kg/m-s, and the overfeed ratio varied from 2.4 to 14.9. Circumferential measurements showed that the dominant fraction of heat transfer occurs on the top half of the tube. Figure 13.44 shows the evaporation heat transfer coefficients of the enhanced tubes at 2°C saturation temperature. The best geometries are the enhanced condensing tubes, which are more effective in promoting thin film evaporation similar to film condensation. The integral-fin tubes (not shown in the figure) yielded slightly lower heat transfer coefficients than for the enhanced tubes. For tests with small fractions (up to 3%) of polyol-ester oils with enhanced condensing tubes, the heat transfer coefficient increased when foaming occurred. No data were presented for the effect of oil with enhanced boiling surfaces. Moeykens et al. [1995b] extended their work to tube bundles comprising of 20 tubes (5 columns and 4 rows) in both square pitch and equilateral triangular pitch layouts. Experiments were conducted using R-134a with liquid overfeed ratios from 1.4 to 7.9. The same tubes as tested in Moeykens et al. [1995a], with the exception of 1024 fpm integral-fin tube, were tested. The heat transfer coefficient decreased as the row number increased. The reduction was much more pronounced (up to 80%) for the GEWA-SC and the integral-fin tube due to the channeling effect of the flow. The heat transfer coefficient increased as the liquid feed rate and the heat flux increased. This dependency was more pronounced for the triangular pitch layout. Bubbles were visible in the thick liquid layer at the bottom of all the tubes, indicating nucleate boiling was occurring in the films. The Turbo-CII yielded the best performance. The falling film evaporation coefficient of the Turbo-CII bundle was 100% larger than the nucleate boiling performance of the Turbo-B bundle.

Moeykens and Pate [1996] extended their earlier tests to investigate the effect of oil on the spray evaporation on enhanced tube bundles. Oil effects were investigated using R-134a and a polyol-ester oil with concentration up to 2.5%. The tube layout was equilateral triangular. The oil promoted foaming and enhanced the heat transfer by retarding the formation of dry patches on the bundles. Increase in the heat transfer coefficient near 100% was obtained for the integral-fin tube bundle. A similar increase was noted for the Turbo-B bundle at a low heat flux (23 kW/m²). The increase, however, dropped off quickly as the heat flux is increased.

Zeng et al. [1998] investigated thin film evaporation of ammonia on a carbon steel integral-fin tube and a corrugated tube. The integral-fin tube had 19.85-mm fin-tip diameter and 1.0-mm fin height. The 19.05-mm-OD corrugated tube had 7.94-mm corrugation pitch. The integral-fin tube increased the performance as much as 2.8 times with respect to the plain tube value at high heat fluxes, but offered little augmentation at low heat fluxes. The corrugated tube performance was approximately the same as that of the smooth tube. The work was extended by Zeng et al. [2000] to tube bundles (3 × 3) having a square pitch layout. Both a plain tube and a 1575 fpm integral-fin tube were tested. The heat transfer coefficient decreased as the row number increased. The reduction was more significant for the integral tube bundle, which is in agreement with the findings of Moeykens et al. [1995b].

Liu and Yi [2002] measured falling film evaporation of water and a water/salt mixture on a vertical three-tube rank of horizontal tubes. They tested a plain tube, a threaded tube (0.4-mm thread height and 0.6-mm pitch), and a roll-worked tube having pyramid-type knurling (0.6 mm high and 0.7 mm pitch). Their roll-worked tube provided enhancement of 2 to 3 over the plain tube and there was no influence of film Reynolds number on performance. The tube performance was little affected by tube pitch. Their single-tube thin film results were also compared to those of Chyu and Bergles [1989].

The drainage pattern on horizontal tubes in a bundle will affect the evaporation rate. The same drainage phenomena occur for thin film evaporation on a bundle of horizontal tubes, as occurs for condensation on a bundle of horizontal tubes. Understanding of this drainage pattern for a variety of enhanced tube geometries is provided in Section 12.6.3.

13.10.2 Vertical Tubes

The correlating parameters for falling film flow (nonboiling) are generally based on the semiempirical correlation developed by Chun and Seban [1971], who studied laminar and turbulent film flow on a vertical plate. They give correlations for the laminar and turbulent regions, and include the effect of Prandtl number. This work was done without vapor shear. Later work by Kosky [1971] defined the effect of co-current vapor shear on the film thickness, and the transition Reynolds number.

Three studies report tests using the enhanced boiling discussed in Chapter 11 for application to refrigeration evaporators: Fagerholm et al. [1985], Nakayama et al. [1982], and Takahashi et al. [1990]. All studies attempted to use R_R close to 1.0. Fagerholm et al. tested a plain tube and three enhanced tubes — Thermoexcel-E (Figure 11.12d), GEWA-T (Figure 11.12c) and the High-Flux (Figure 11.11a) — geometries with R-114 on a single 2.0-m-long vertical tube as the inner tube of an annulus. The enhanced tubes experienced boiling in the film at all heat fluxes tested, $4 \leq q \leq 18$ kW/m². Entrainment caused partial de-wetting when $R_R < 3$. The entrainment induced de-wetting increases with increasing heat flux. The Thermoexcel-E and GEWA-T tubes were more prone to dewetting than was the High-Flux tube. For $R_R > 2$, the heat transfer coefficients on the enhanced tubes was 6 to 12 times that on the plain tube. The High-Flux provided the highest performance.

Nakayama et al. [1982] measured evaporation of R-11 (101 kPa) on a 300-mm-long vertical plate containing the Thermoexcel-E geometry (Figure 11.12d) and a grooved plate (1.1 mm high, 0.4 mm thick, 0.8 mm pitch). The plates were operated over a range of flow rates. The Thermoexcel-E geometry provided higher performance than for pool boiling, which decreased to the plain tube value at $q = 100$ kW/m². The performance remained high until dryout of the liquid. The performance of the finned plate approached that of the Thermoexcel-E plate at $G_l = 0.1$ kg/s-m, but decreased either side of this maximum. The Takahashi et al. [1990] tests were performed using R-22 on the outer surface of a 12-tube bundle, 0.79 m long with water flow in the tubes. The Thermoexcel-E boiling geometry (Figure 11.12d) comprised the outer tube surface, and the inner surface is shown in Figure 9.21b. They consciously sought to work with $R_R \leq 1.2$. Their evaporator was part of a refrigeration system containing a scroll compressor, which could pump the unevaporated refrigerant. Some de-wetting was observed.

Three studies of thin film evaporation on vertical tubes containing axial fins are reported: Rifert et al. [1975], Mailen [1980], and Grimley et al. [1987]. The surface used by both Rifert and Grimley et al. is similar to that in Figure 13.21. Using water, Rifert found 100% enhancement for fins 0.7 mm high at 0.7 mm pitch. However, Mailen's extruded aluminum tube, similar to that in Figure 12.14, showed little performance improvement, relative to a plain tube. Note that the enhancement mechanism of the Figure 12.14 tube is significantly different for condensation than for thin film evaporation. In condensation, surface tension force pulls the condensate into the valley, yielding very thin films, which are beneficial for condensation. However, this is not beneficial for evaporation. Evaporation requires a mechanism to spread the film onto the fin surface. Hence, surface tension force tends to de-wet the Figure 12.14 tube, when used with thin film evaporation.

13.11 CONCLUSIONS

Convective vaporization includes the combined effects of nucleate boiling and convective evaporation. Two heat transfer modes may exist — shell-side boiling (as used in flooded evaporators or kettle reboilers) or thin film evaporation. The relative contributions of nucleate boiling or convective evaporation depend on the surface geometry. The geometry may be enhanced to promote either component.

The enhanced tubes discussed in Chapter 11 are very effective for shell-side boiling in flooded evaporators and kettle reboilers. Their performance is dominated by nucleate boiling. These tubes have also been shown to provide a high nucleate boiling contribution in thin film evaporators. Thus, nucleation occurs in the liquid film, even though it is quite thin. The enhanced boiling tubes generally yield a convective evaporation component typical of that of a plain tube. There is some disagreement concerning whether convection can improve the performance of an enhanced tube.

Enhanced tube geometries used for tube-side convective vaporization generally promote the convective evaporation term, rather than nucleate boiling. However, certain high-performance nucleate boiling surfaces should substantially enhance performance via nucleate boiling. The microfin internal geometry provides very high enhancement for tube-side vaporization of refrigerants. Its application potential for other fluids has yet to be explored.

Geometries beneficial to surface tension drained condensation are generally unfavorable to vaporization. This is because heated films tend to rupture. If the surface cannot be fully wetted, a significant loss of performance potential will exist. However, such tubes can be very effective for thin film vaporization.

NOMENCLATURE

A Inside tube surface area, m² or ft²

A_c Minimum cross-sectional flow area, m² or ft²

Bo   Boiling number ($q/G\lambda$), dimensionless

d_e Diameter over fins of externally finned tube, m or ft

d_i Tube inside diameter, or diameter to the base of internal fins or roughness, m or ft

d_o Tube outside diameter, m or ft

D_c Diameter of helical coil, m or ft

D_h Hydraulic diameter of flow passages, $4LA_c/A$, m or ft

e Fin height, roughness height, or corrugation depth, m or ft

E_h h/h_p at constant Re, dimensionless

f Friction factor, dimensionless

F Two-phase convection multiplier factor, h_{cv}/h_l, dimensionless

g Acceleration due to gravity, m/s² or ft/s²

G Mass velocity, kg/m²-s or lbm/ft²-s

h Heat transfer coefficient: h (total), h_l (liquid phase flowing alone), h_{lo} (total mass rate flowing as a liquid), h_{nbp} (nucleate pool boiling), h_{nb} (= Sh_{nbp}) h_{cv} (two-phase convection, without nucleate boiling), W/m²-K or Btu/hr-ft²-°F

H Length for 180° revolution of twisted tape, m or ft

k Thermal conductivity, W/m-K or Btu/hr-ft

L Flow length, m or ft

p Pressure, Pa or lbf/ft²

p Axial pitch of surface or roughness elements, m or ft

P_n Fin pitch normal to tube axis, m or ft

q Heat flux, W/m²

R_R Recirculation ratio (ratio of feed rate to evaporation rate), dimensionless

Re   Reynolds number, defined where used, dimensionless

S Suppression factor, dimensionless

t Thickness of tube wall fin or twisted tape, m or ft

T Temperature. T_w (wall), T_s (saturation), °C or °F

u Average flow velocity, m/s or ft/s

U Overall heat transfer coefficient, W/m²-K or Btu/hr-ft²-°F

v_l Specific volume of liquid, m³/kg

W Fluid mass flow rate (= GA_c), kg/s or lbm/s

x Local vapor quality, dimensionless

X_{tt} Martinelli parameter [= $(\Delta p_{F,l}/\Delta p_{F,v})^{0.5}$], flow in tubes (Equation 13.8), dimensionless

y Twist ratio [= $H/d_i = \pi/(2 \tan \alpha)$], dimensionless

Greek Symbols

α Void fraction, dimensionless

α Helix angle relative to tube axis [tan $\alpha = \pi/2y$], rad or deg

δ Film thickness, m or ft

Δp Total pressure drop for two-phase flow, Pa, or lbf/ft²

Δp_F Frictional pressure drop for two-phase flow; liquid phase flowing alone ($\Delta p_{F,l}$), two-phase friction (Δp_F), Pa or lbf/ft²

ΔT_{ws} $T_w - T_s$, K or °F

Δx Vapor quality change, dimensionless

η_f Fin efficiency or temperature effectiveness of the fin, dimensionless

λ Latent heat of vaporization, J/kg or Btu/lbm

μ_l Dynamic viscosity: μ_l (of liquid), μ_v (of vapor), $\mu_{l,w}$ (of water), μ_a (of air), Pa-s or lbm/s-ft

v Kinematic viscosity, m/s² or ft/s²

ρ Density: ρ_l (of liquid), ρ_v (of vapor), $\rho_{l,w}$ (of water), ρ_a (of air), kg/m³ or lbm/ft³

σ Surface tension, σ_w (of water), N/m or lbf/ft

ϕ Severity factor, $(e/d_i)/(p/e)$ used in Figure 13.14, dimensionless

ϕ_l^2 Two-phase friction multiplier, $\Delta p_F/\Delta p_{F,l}$ (Equation 13.9), dimensionless

Subscripts

cr   Critical heat flux

cv   Convection

i Inside tube

l,w Of liquid water

l Liquid phase

nb Nucleate boiling
p   Plain tube
v   Vapor phase
w   At tube wall

REFERENCES

Agrawal, K.N., Varma H.K., and Lal, S., 1982. Pressure drop during forced convection boiling of R-12 under swirl flow. *J. Heat Transfer*, 104, 758–762.

Agrawal, K.N., Varma, H.K., and Lal, S., 1986. Heat transfer during forced convection boiling of R-12 under swirl flow, *J. Heat Transfer*, 108, 567–573.

Akhanda, M.A.R. and James, D.D., 1988. An experimental study of the relative effects of transverse and longitudinal ribbing of the heat transfer surface in forced convective boiling, *Two-Phase Heat Exchanger Symposium*, J.B. Pearson and J.B. Kitto, Eds., ASME Symp. 44, 83–90.

Antonelli, R. and O'Neill, P.S., 1981. Design and application considerations for heat exchangers with enhanced boiling surfaces, paper presented at International Conference on Advances in Heat Exchangers, September, Dubrovnik, Yugoslavia.

Arai, N., Fukushima, T., Arai, A., Nakajima, T., Fujie, K., and Nakayama, Y., 1977. Heat transfer tubes enhancing boiling and condensation in heat exchangers of a refrigerating machine, *ASHRAE Trans.*, 83(Part 2), 58–70.

Azer, N.Z. and Sivakumar, V., 1984. Enhancement of saturated boiling heat transfer by internally finned tubes, *ASHRAE Trans.* 90(Part 1A), 58–73.

Baker, O., 1954. Simultaneous flow of oil and gas, *Oil Gas J.*, 53, 185–195.

Bao, Z.Y., Fletcher, D.F., and Haynes, B.S., 2000. Flow boiling heat transfer of Freon 11 and HCFC123 in narrow passages, *Int. J. Heat Mass Transfer*, 43, 3347–3358.

Bennett, D.L. and Chen, J.C., 1980. Forced convective boiling in vertical tubes for saturated pure components and binary mixtures, *AIChE J.*, 26(3), 454–461.

Bergles, A.E. and Chyu, M.C., 1982. Characteristics of nucleate pool boiling from porous metallic coatings, *J. Heat Transfer*, 104, 279–285.

Bergles, A.E., Fuller, W.D., and Hynek, S.J., 1971. Dispersed flow boiling of nitrogen with swirl flow, *Int. J. Heat Mass Transfer*, 14, 1343–1354.

Bhatia, R.S. and Webb, R.L., 2001. Numerical study of turbulent flow and heat transfer in microfin tubes Part 2: Parametric study, *J. Enhanced Heat Transfer*, 8, 305–314.

Blatt, T.A. and Alt, R.R., 1963. The Effects of Twisted-Tape Swirl Generators on the Heat Transfer and Pressure Drop of Boiling Freon-11 and Water, ASME paper 65-WA-42, Asme, New York.

Boling, C., Donovan, W.J., and Decker, A.S., 1953. Heat transfer of evaporating freon with inner-fin tubing, *Refrig. Eng.*, 61, 1338–1340, 1384.

Brognaux, L.J., Webb, R.L., Chamra, L.M., and Chung, B.Y., 1997. Single-phase heat transfer in microfin tubes, *Int. J. Heat Mass Transfer*, 40, 4345–4357.

Bukin, V.G., Danilova, G.N., and Dyundin, V.A., 1982. Heat transfer from freons in a film flowing over bundles of horizontal tubes that carry a porous coating, *Heat Transfer Sov. Res.*, 14(2), 98–103.

Campolunghi, F., Cumo, M., Ferrari, G., and Palazzi, G., 1976. Full scale tests and thermal design of once-through steam generators, AIChE paper presented at 16th Natl. Heat Transfer Conf., St. Louis, MO.

Carey, V.P., 1992. *Liquid-Vapor Phase-Change Phenomena*, Hemisphere, Washington, D.C.

Carey, V.P. and Shah, R.K., 1988. Design of compact and enhanced heat exchangers for liquid-vapor phase-change applications, in *Two-Phase Flow Heat Exchangers: Thermal Hydraulic Fundamentals and Design*, S. Kakaç, A.E. Bergles, and E.O. Fernandes, Eds., NATO ASI Ser. E., Vol. 143, Kluwer Academic, Dordecht. 909–968.

Carnavos, T.C., 1980. Heat transfer performance of internally finned tubes in turbulent flow, *Heat Transfer Eng.*, 4(1), 32–37.

Cavallini, A., Del Col, D., Doretti, L., Longo, G.A., and Rossetto, L., 1999. Refrigerant vaporization inside enhanced tubes: a heat transfer model, *Heat Technol.*, 17(2), 29–36.

Cavallini, A., Del Col, D., Longo, G.A., and Rossetto, L., 2000. Heat transfer and pressure drop during condensation of refrigerants inside horizontal enhanced tubes, *Int. J. Refrig.*, 23, 4–25.

Celata, G.P., Cumo, M., and Mariani, A., 1994. Enhancement of CHF water subcooled flow boiling in tubes using helically coiled wires, *Int. J. Heat Mass Transfer*, 37(1), 53–67.

Chamra, L.M. and Webb, R.L., 1995. Condensation and evaporation in microfin tubes at equal saturation temperatures, *J. Enhanced Heat Transfer*, 2, 219–230.

Chamra, L.M., Webb, R.L., and Randlett, M.R., 1996. Advanced microfin tubes for evaporation, *Int. J. Heat Mass Transfer*, 39, 1827–1838.

Chen, C.C., Loh, J.V., and Westwater, J.W., 1981. Prediction of boiling heat transfer in a compact plate-fin heat exchanger using the improved local technique, *Int. J. Heat Mass Transfer*, 24, 1907–1912.

Chen, J.C., 1966. A correlation for boiling heat transfer to saturated fluids in convective flow, *Indust. Eng. Chem., Process Design Dev.*, 5(3), 322–329.

Chen, T.K., Chen, X-Z., and Chen, X-J., 1992. Boiling heat transfer and frictional pressure drop in internally ribbed tubes, in *Multiphase Flow and Heat Transfer: Second International Symposium*, Vol. 1, X-J. Chen, T.N. Veziroglu, and C.L. Tien, Eds., Hemisphere, New York, 621–629.

Chisholm, D., 1967. A theoretical basis for the Lockhart-Martinelli correlation for two-phase flow, *Int. J. Heat Mass Transfer*, 10, 1767–1777.

Chun, K.R. and Seban, R.A. 1971. Heat transfer to evaporating liquid films, *J. Heat Transfer*, 93, 391–396.

Chyu, M.C. and Bergles, A.E., 1989. Horizontal-tube falling film evaporation with structured surfaces, *J. Heat Transfer*, 111, 518–524.

Chyu, M.C., Bergles, A.E., and Mayinger, F., 1982. Enhancement of horizontal tube spray film evaporators, *Heat Transfer — 1982, Proc. 7th International Heat Transfer Conference*, Vol. 6, Hemisphere, Washington, D.C., 275–280.

Conklin, J.C. and Vineyard, E.A. 1992. Flow boiling enhancement of R-22 and a nonazeotropic mixtue of R-143a and R-124 using perforated foils, *ASHRAE Trans.*, 98(Part 2), 402–410.

Conti, R.J., 1978. Experimental investigation of horizontal tube ammonia film evaporators with small temperature differentials, *Proc. 5th Int. Heat Transfer Conf.*, Vol. 6, Hemisphere, Washington, D.C., 161–180.

Cooper, M.G., 1984. Saturation nucleate, pool boiling — a simple correlation, *Int. Chem. Eng. Symp. Ser.*, No. 86, 785–792.

Cornwell, K. and Scoones, D.S., 1988. Analysis of low quality boiling on plain and low finned tube, *Proc. 2nd UK Heat Transfer Conf.*, ImechE-IChemE, 1, 21–32.

Crain, B., Jr. and Bell, K.J., 1973. Forced convection heat transfer to a two-phase mixture of water and steam in a helical coil, *AIChE Symp. Ser.*, 69(131), 30–36.

Cui, S., Tan, V., and Lu, Y., 1992. Heat transfer and flow resistance of R-502 flow boiling inside horizontal ISF tubes, in *Multiphase Flow and Heat Transfer: Second International Symposium*, Vol. 1, X.-J. Chen, T.N. Veziroglu, and C.L. Tien, Eds., Hemisphere, New York, 662–670.

Cumo, M., Farello, G.E., Ferrari, G., and Palazzi, G., 1974. The influence of twisted tapes in subcritical, once-through vapor generators in counter flow, *J. Heat Transfer*, 96, 365–370.

Czikk, A.M., Gottzmann, C.F., Ragi, E.G., Withers, J.G., and Habdas, E.P., 1970. Performance of advanced heat transfer tubes in refrigerant-flooded coolers, *ASHRAE Trans.*, 76(Part 1), 99–109.

Czikk, A.M., O'Neill, P.S., and Gottzmann, C.F., 1981. Nucleate pool boiling from porous metal films: effect of primary variables, in *Advances in Enhanced Heat Transfer*, R.L. Webb, T.C. Carnavos, E.F. Park, and K.M. Hostetler, Eds., TD Vol. 18, American Society of Mechanical Engineers, New York, 109–222.

Del Col, D., Webb, R.L., and Narayanamurthy, R., 2002. Heat transfer mechanisms for condensation and vaporization inside a microfin tube, *J. Enhanced Heat Transfer*, 9, 25–38.

Ebisu, T., 1999. Evaporation and condensation heat transfer enhancement for alternative refrigerants used in air-conditioning machines, in *Heat Transfer Enhancement of Heat Exchangers*, S. Kakaç, A.E. Bergles, F. Mayinger, and H. Yuncu, Eds., Kluwer Academic, Dordrecht, 579–600.

Eckels, S.J. and Pate, M.B., 1991. In-tube evaporation and condensation of refrigerant-lubricant mixtures of HFC-134a and CFC-12, *ASHRAE Trans.*, 97(Part 2), 62–70.

Eckels, S.J., Pate, M.B., and Bemisderfer, C.H., 1992. Evaporation heat transfer coefficients for R-22 in microfin tubes of different configurations, in *Enhanced Heat Transfer*, M.B. Pate and M.K. Jensen, Eds., ASME Symp. Vol. HTD, 202, 117–126.

Fagerholm, N.-E., Kivioja, K., Ghazanfari, A.-R., and Jarvinen, E., 1985. Using structured surfaces to enhance heat transfer in falling film flow, in *I.I.F.—I.I.R. Commission E2*, Trondheim, Norway, pp. 273–279.

Fujie, K., Itoh, N., Innami, T., Kimura, H., Nakayama, N., and Yanugidi, T., 1977, Heat Transfer Pipe, U.S. patent 4,044,797, assigned to Hitachi Ltd.

Fujita, Y., 1998. Boiling and evaporation of falling film on horizontal tubes and its enhancement on grooved tubes, in *Heat Transfer Enhancement of Heat Exchangers*, S. Kakaç, A.E. Bergles, F. Mayinger, and H. Yuncu, Eds., Kluwer Academic, Dordrecht, 3259–346.

Fujita, Y., Ohta, H., Hidaka, S., and Nishikawa, K., 1986. Nucleate boiling heat transfer on horizontal tubes in bundles, *Proc. 8th Int. Heat Transfer Conf.*, 5, 2131–2136.

Fujita, Y., Yang, Y., and Fujita, N., 2002. Flow boiling heat transfer and pressure drop in uniformly heated small tubes, *Heat Transfer 2002, 12th Int, Heat Transfer Conf.*, 3, 743–748.

Gambill, W.R., 1963. Generalized prediction of burnout heat flux for flowing, subcooled wetting liquids, *Chem. Eng. Prog. Symp. Ser.*, 59(41), 71–87.

Gambill, W.R., 1965. Subcooled swirl-flow boiling and burnout with electrically heated twisted tapes and zero wall flux, *J. Heat Transfer*, 97, 342.

Gambill, W.R., Bundy, R.D., and Wansbrough, R.W., 1961. Heat transfer, burnout, and pressure drop for water in swirl flow tubes with internal twisted tapes, *Chem. Eng. Prog. Symp. Ser.*, 57(3), 127–137.

Gan, Y.P., Chen, Q., Yuan, X.Y., and Tian, S.R., 1993. An experimental study of nucleate boiling heat transfer from flame spraying surface of tube bundle in R-113/R-11-oil mixtures, in *Proceedings of the 3rd Congress on Experimental Heat Transfer, Fluid Mechanics and Thermodynamics*, Honolulu, Hawaii, 1226–1231.

Gorenflo, D., 1993. Pool boiling, in *VDI Heat Atlas*, Section HA, VDI Verlag, Dusseldorf.

Goto, M., Inoue, N., and Ishiwatari, N., 2001. Condensation and evaporation heat transfer of R410A inside internally grooved horizontal tubes, *Int. J. Refrig.*, 24, 628–638.

Grimley, T.A., Mudawwar, I.A., and Incropera, F.P., 1987. Enhancement of boiling heat transfer in falling films, *Proc. 1987 ASME-JSME Thermal Eng. Joint Conf.*, 3, 411–418.

Gu, C.B., Chow, L.C., and Beam, J.E., 1989. Flow boiling in a curved channel, in *Heat Transfer in High Energy/High Heat Flux Applications*, R.J. Goldstein, L.C. Chow, and E.E. Anderson, Eds., ASME Symp. Vol. HTD, 119, 25–32.

Gupte, N.S., 1992. Simulation of Boiling in Flooded Refrigerant Evaporators, Ph.D. thesis, Pennsylvania State University, University Park.

Gupte, N.S. and Webb, R.L. 1992. Convective vaporization of refrigerants in tube banks, *ASHRAE Trans.*, 98(Pt. 2), 411–424.

Gupte, N.S. and Webb, R.L., 1994. Convective vaporization of pure refrigerants in enhanced and integral-fin tube banks, *J. Enhanced Heat Transfer*, 1(4), 351–364.

Gupte, N.S. and Webb, R.L., 1995a. Convective vaporization data for pure refrigerants in tube banks, Part I: Integral-finned tubes, *Int. J. HVAC&R Res.*, 1(1), 35–47.

Gupte, N.S. and Webb, R.L., 1995b. Convective vaporization data for pure refrigerants in tube banks. Part II: Enhanced tubes, *Int. J. HVAC&R Res.*, 1(1), 48–60.

Hewitt, G.F. and Roberts, D.N., 1969. *Studies of Two-Phase Flow Patterns by Simultaneous X-Ray and Flash Photography*, Report AERE-M 2159, Her Majesty's Stationery Office, London.

Hori, M. and Shinohara, Y., 2001. Internal heat transfer characteristics of small diameter thermofin tubes, *Hitachi Cable Rev.*, August, 85–90.

Houfuku, M., Suzuki, Y., and Inui, K., 2001. High performance, lightweight THERMOFIN tubes for air conditioners using alternative refrigerants, *Hitachi Cable Rev.*, August, 97–100.

Hsieh, S.-S., Huang, G.-Z., and Tsai, H.-H., 2003. Nucleate pool boiling characteristics from coated tube bundles in saturated R-134a, *Int. J. Heat Mass Transfer*, 46, 1223–1239.

Hsieh, Y.Y. and Lin, T.F., 2002. Saturated flow boiling heat transfer and pressure drop of refrigerant R-410A in a vertical plate heat exchanger, *Int. J. Heat Mass Transfer*, 45, 1033–1044.

Hughes, T.G. and Olson, D.R., 1975. Critical heat fluxes for curved surfaces during subcooled flow boiling, *Trans. CSME*, 3(3), 122–130.

Ikeuchi, M., Yumikura, T., Fujii, M., and Yamanaka, G., 1984. Heat-transfer characteristics of an internal microporous tube with refrigerant-22 under evaporating conditions, *ASHRAE Trans.*, 90(Part 1A), 196–211.

Ishihara, K., Palen, J.W., and Taborek, J., 1980. Critical review of correlation for predicting two-phase flow pressure drop across tube banks, *Heat Transfer Eng.*, 1, 23–32.

Ishikawa, S., Nagahara, K., and Sukumoda, S., 2002. Heat transfer and pressure drop during evaporation and condensation of HCFC22 in horizontal copper tubes with many inner fins, *J. Enhanced Heat Transfer*, 9, 17–24.

Ito, M. and Kimura, H., 1979. Boiling heat transfer and pressure drop in internal spiral-grooved tubes. *Bull. JSME*, 22(17), 1251–1257.

Jensen, M.K., 1984. A correlation for predicting the critical heat flux condition with twisted-tape swirl generators, *Int. J. Heat Mass Transfer*, 27, 2171–2173.

Jensen, M.K., 1985. An evaluation of the effect of twisted-tape swirl generators in two-phase flow heat exchangers, *Heat Transfer Eng.*, 6(4), 19–30.

Jensen, M.K. and Bensler, H.P., 1986. Saturated forced-convective boiling heat transfer with twisted-tape inserts, *J. Heat Transfer*, 108, 93–99.

Jensen, M.K. and Bergles, A.E. 1981. Critical heat flux in helically coiled tubes, *J. Heat Transfer*, 103, 660–666.

Jensen, M.K. and Hsu, J.T., 1987. A parametric study of boiling heat transfer in a tube bundle, *Proc. 1987 ASME-JSME Thermal Eng. Joint Conf.*, 3, 133–140.

Jensen, M.K., Trewin, R.R., and Bergles, A.E., 1992. Crossflow boiling in enhanced tube bundles, in *Two-Phase Flow in Energy Exchange Systems*, M.S. Sohal and T.J. Rabas, Eds., ASME Symp. Vol. HTD, 220, 11–18.

Kabata, Y., Nakajima, R., and Shioda, K., 1996. Enhancement of critical heat flux for subcooled flow boiling of water in tubes with twisted tape and with a helically coiled wire, *Int. Conf. Nuclear Engi. ASME*, 1(Pt. B), 639–646.

Kandlikar, S.G., 1991. Correlating flow boiling heat transfer data in binary systems, in *Phase Change Heat Transfer*, E. Hensel, V.K. Dhir, R. Grief, and J. Filio, Eds., ASME Symp. Vol. HTD, 159, ASME, New York, 163–170.

Kasza, K.E., Didascalou, T., and Wambsganss, M.W., 1997. Microscale flow visualization of nucleate boiling in small channels: mechanisms influencing heat transfer, in *Proceedings of the International Conference on Compact Heat Exchangers for the Process Industries*, R.K. Shah, K.J. Bell, S. Mochizuki, and V.W. Wadekar, Eds., Begell House, New York, 343–352.

Kattan, N., Thome, J.R., and Favrat, D., 1998. Flow boiling in horizontal tubes. Part 3: Development of a new heat transfer model based on flow pattern, *J. Heat Transfer*, 120, 156–165.

Kedzierski, M.A. and Kim, M.S., 1997. Convective Boiling and Condensation with a Twisted-Tape Insert for R12, R22, R152a, R134a, R290, R32/R134a, R32/R152a, R390/R134a, R134a/R600a, National Institute of Standards and Technology Report NISTR 5905, January 1997.

Kew, P.A. and Cornwell, K., 1997. Correlations for the prediction of boiling heat transfer in small-diameter channels, *Appl. Thermal Eng.*, 17, 705–715.

Kim, M.-H., Shin, J.-S., and Bullard, C.W., 2001. Heat transfer and pressure drop characteristics during R22 evaporation in an oval microfin tube, *J. Heat Transfer*, 123, 301–308.

Kim, N.-H., Cho, J.-P., and Youn, B., 2002. Forced convective boiling of pure refrigerants in a bundle of enhanced tubes having pores and connecting gaps, *Int. J. Heat Mass Transfer*, 45, 2449–2463.

Kitto, J.B. and Wiener, M., 1982. Effects of nonuniform circumferential heating and inclination on critical heat flux in smooth and ribbed bore tubes, *Proc. 7th Int. Heat Transfer Conf.*, 4, 303–308.

Kosky, P.G., 1971. Thin liquid films under simultaneous shear and gravity flows, *Int. J. Heat Mass Transfer*, 14, 1220–1223.

Koyama, S., Yu, J., Momoki, S., Fujii, T., and Honda, H., 1996. Forced convective flow boiling heat transfer of pure refrigerants inside a horizontal microfin tube, in *Proceedings of Convective Flow Boiling, An International Conference*, J.C. Chen, Y. Fujita, F. Mayinger, and R. Nelson, Eds., Taylor & Francis, London, 137–142.

Kubanek, G.R. and Miletti, D.L., 1979. Evaporative heat transfer and pressure drop performance of internally-finned tubes with refrigerant 22, *J. Heat Transfer*, 101, 447–452.

Kun, L.C. and Czikk, A.M., 1969. Surface for Boiling Liquids, U.S. patent 3,454,081 (reissued 1979, Ref. 30,077), assigned to Union Carbide Corp.

Kuo, S. and Wang, C.C., 1996. In-tube evaporation of HCFC-22 in a 9.52 mm microfin/smooth tube, *Int. J. Heat Mass Transfer*, 39, 2559–2569.

Lan, J., Disimile, P.J., and Weisman, J., 1997. Two phase flow patterns and boiling heat transfer in tubes containing helical wire inserts. Part II: Critical heat flux studies, *J. Enhanced Heat Transfer*, 4, 283–296.

Larson, R.L., Quaint, G.W., and Bryan, W.L., 1949. Effects of turbulence promoters in refrigerant evaporator coils, *Refrig. Eng.*, 57, 1193–95.

Lavin, J.G. and Young, E.H., 1965. Heat transfer to evaporating refrigerants in two-phase flow, *AIChE J.*, 11, 1124–1132.

Lazarek, G.M. and Black, S.H., 1982. Evaporative heat transfer, pressure drop and critical heat flux in a small vertical tube with R-113, *Int. J. Heat Mass Transfer*, 25(7), 945–960.

Lewis, L.G. and Sather, N.F., 1978. OTEC Performance Tests on the Union Carbide Flooded-Bundle Evaporator, ANL Report ANL-OTEC-PS-1, Argonne National Lab, Chicago, December.

Lin, S., Kew, P.A., and Cornwell, K., 2001. Two-phase heat transfer to a refrigerant in a 1-mm diameter tube, *Int. J. Refrigeration*, 24, 51–56.

Liu, Z. and Winterton, R.H.S., 1991. A general correlation for saturated and subcooled flow boiling in tubes and annuli, based on a nucleate pool boiling equation, *Int. J. Heat Mass Transfer*, 34, 2759–2766.

Liu, Z.H., and Yi, J., 2002. Falling film evaporation heat transfer of water/salt mixtures from roll-worked enhanced tubes and tube bundle, *Appl. Thermal Eng.*, 22, 83–95.

Liu, Z.-H. and Qiu, Y.-H., 2002. Enhanced boiling heat transfer in restricted spaces of a compact tube bundle with enhanced tubes, *Appl. Thermal Eng.*, 22, 1931–1941.

Liu, Z.H. and Tong, T.F., 2002. Boiling heat transfer of water and R-11 on horizontally smooth and enhanced tubes enclosed by a concentric outer tube with two horizontal slots, *Exp. Heat Transfer*, 15(3), 161–176.

Lorenz, J.J. and Yung, D.T., 1979. A note on combined boiling and evaporation of liquid films on horizontal tubes, *J. Heat Transfer*, 101, 178–180.

Mailen, G.S., 1980. Experimental studies of OTEC heat transfer evaporation of ammonia on vertical smooth and fluted tubes, *Proc. 7th Ocean Energy Conference*, Paper 12.5, pp. 1–10.

Mandrusik, G.D. and Carey, V.P., 1989. Convective boiling in vertical channels with different offset strip fin geometries, *J. Heat Transfer*, 111, 156–165.

Manglik, R.M. and Bergles, A.E., 1992. heat transfer and pressure drop correlations for twisted-tape inserts in isothermal tubes. Part I: Laminar flows, in *Enhanced Heat Transfer*, M.B. Pate and M.K. Jensen, Eds., ASME Symp. HTD-Vol. 202, ASME, New York, 89–98.

Marvillet, C., 1989. Influence of oil on nucleate pool boiling of refrigerants R-12 and R-22 from porous layer tube, *Eurotherm No. 8, Advances in Pool Boiling Heat Transfer*, Paderborn, Germany, 164–168.

Memory, S.B., Chilman, S.V., and Marto, P.J., 1994. Nucleate pool boiling of Turbo-B bundle in R-113, *J. Heat Transfer*, 116, 670–678.

Memory, S.B., Akcasayar, N., Eraydin, H., and Marto, P.J., 1995. Nucleate pool boiling of R-114 and R-114-oil mixtures from smooth and enhanced surfaces, II: Tube bundles, *Int. J. Heat Mass Transfer*, 38, 1363–1376.

Mergerlin, F.E., Murphy, R.W., and Bergles, A.E., 1974. Augmentation of heat transfer by use of mesh and brush inserts. *J. Heat Transfer*, 96, 145–151.

Moeykens, S.A., Huebsch, W.W., and Pate, M.B., 1995a. Heat transfer of R-134a in single-tube spray evaporation including lubricant effects and enhanced surface results, *ASHRAE Trans.*, 101(Part 1), 111–122.

Moeykens, S.A., Newton, B.J., and Pate, M.B., 1995b. Effects of surface enhancement, film feed spray rate, and bundle geometry on spray evaporation heat transfer performance, *ASHRAE Trans.*, 101(Part 2), 408–419.

Moeykens, S.A. and Pate, M.B., 1996. Effect of lubricant on spray evaporation heat transfer performance of R-134a and R-22 in tube bundles, *ASHRAE Trans.*, 102(Part 1), 410–426.

Morita, H., Kito, Y., and Satoh, Y., 1993, Recent improvements in small bore inner grooved copper tube, *Tube Pipe Technol.*, Nov./Dec., 53–57.

Muller, J., 1986. Boiling heat transfer on finned tube bundles — the effect of tube position and intertube spacing, *Proc. 8th Int. Heat Transfer Conf.*, 4, 2111–2116.

Muzio, A., Niro, A., and Arosio, S., 1998. Heat transfer and pressure drop during evaporation and condensation of R22 inside 9.52-mm O.D. microfin tubes of different geometries, *J. Enhanced Heat Transfer*, 5, 39–52.

Nakajima, K. and Shiozawa, A., 1975. An experimental study on the performance of a flooded type evaporator, *Heat Transfer-Jpn. Res.*, 4(3), 49–66.

Nakayama, W., Daikoku, D., and Nakajima, T., 1982. Enhancement of boiling and evaporation on structured surfaces with gravity driven film flow, *Proc. 7th Int. Heat Transfer Conf.*, 6, 409–414.

Newell, T.A. and Shah, R.K., 2001. An assessment of refrigerant heat transfer, pressure drop, and void fraction effects in microfin tubes, *Int. J. HVAC&R Res.*, 7(2), 125–154.

Nunez, M.P. and Polley, G.T., 1999. Methodology for the design of multi-stream plate-fin heat exchangers, in *Recent Advances in Analysis of Heat Transfer for Fin Type Surfaces*, B. Sunden and P.J. Heggs, Eds., Computational Mechanics, Billerica, MA, 277–293.

Oh, S.-Y. and Bergles, A.E., 1998. An experimental study of the effects of the spiral angle on the evaporative heat transfer enhancement in microfin tubes, *ASHRAE Trans.*, 104(Part. 2), 1137–1143.

O'Neill, P.S., 1981. Private communication, February 16. Linde Div., Union Carbide Corp., Tonawanda, NY.

O'Neill, P.S., King, R.C., and Ragi, E.G., 1980. Application of high-performance evaporator tubing in refrigeration systems of large olefins plants, *AIChE Symp. Ser.*, 76(199), 289–300.

Owens, W.L., 1978. Correlation of thin film evaporation heat transfer coefficients for horizontal tubes, *Proc. 5th Ocean Thermal Energy Conversion Conf.*, Miami Beach, FL, 6, 71–89.

Palen, J.W., Yarden, A., and Taborek, J., 1972. Characteristics of boiling outside large-scale horizontal multitube bundles, *AIChE Symp. Ser.*, 68(118), 50–6l.

Palm, B., 1990. Heat Transfer Augmentation in Flow Boiling by Aid of Perforated Metal Foils. ASME paper 90-WA/HT-10, ASME, New York.

Panchal, C.B., France, D.M., and Bell, K.H., 1992. Experimental investigation of single-phase, condensation, and flow boiling heat transfer for a spirally fluted tube, *Heat Transfer Eng.*, 13(1), 43–52.

Parken, W.H., Fletcher, L.S., Sernas, V., and Han, J.C., 1990. Heat transfer through falling evaporation and boiling on horizontal tubes, *J. Heat Transfer*, 112, 744–750.

Pearson, J.F. and Young, E.H., 1970. Simulated performance of refrigerant-22 boiling inside of tubes in a four pass shell and tube heat exchanger, *AIChE Symp. Ser.*, 66(102), 164–173.

Pettersen, J., 2003. Two-phase flow pattern, heat transfer, and pressure drop in microchannel vaporization of CO_2, *ASHRAE Trans.*, 109(Part 1), 523–532.

Pierre, B., 1964. Flow resistance with boiling refrigerants, *ASHRAE J.*, 6(9), 58–65, 6(10), 73–77.

Polley, G.T., Ralston, T., and Grant, I.R., 1980. Forced cross flow boiling in an ideal in-line tube bundle, ASME Paper 80-HT-46, presented at ASME/AIChE Heat Transfer Conference, Orlando, FL.

Quazia, B., 2001. Evaporation heat transfer and pressure drop of HFC-134a inside a plate heat exchanger, in *Proceedings of 2001 ASME International Mechanical Engineering Congress and Exposition*, PID, 6, 115–123.

Ravigururajan, T.S. and Bergles, A.E., 1985. General correlations for pressure drop and heat transfer for single-phase turbulent flow in internally ribbed tubes, in *Augmentation of Heat Transfer in Energy Systems*, ASME HTD-52, ASME, New York, 9–20.

Rifert, V.G., Butuzov, A.I., and Belik, D.N., 1975. Heat transfer in vapor generation in a falling film inside a vertical tube with a finely-finned surface, *Heat Transfer Sov. Res.*, 7(2), 22–25.

Rifert, V.G., Putilin, J.V., and Podberezny, V.L., 1992. Evaporation heat transfer in liquid films flowing down the horizontal smooth and longitudinally-profiled tubes, *ICheME Symp. Ser.*, 129, 1283–1289.

Rifert, V.G., Putilin, J.V., and Podberezny, V.L., 2001. Evaporation heat transfer in liquid films flowing down the horizontal smooth and longitudinally profiled tubes, *J. Enhanced Heat Transfer*, 8, 91–97.

Robertson, J.M., 1980. Review of boiling, condensing and other aspects of two-phase flow in plate fin heat exchangers, in *Compact Heat Exchangers—History, Technological Advances and Mechanical Design Problems*, R.K. Shah, C.F. McDonald, and C.P. Howard, Eds, HTD-Vol. 10, ASME, New York, 17–27.

Roser, R., Thonon, B., and Mercier, P., 1999. Experimental investigations on boiling of *n*-pentane across a horizontal tube bundle, *Int. J. Refrig.*, 22, 536–547.

Schlager, L.M., Pate, M.B., and Bergles, A.E., 1988a. Performance of microfin tubes with refrigerant-22 and oil mixtures, *ASHRAE J.*, November, 17–28.

Schlager, L.M., Bergles, A.E., and Pate, M.B., 1988b. Evaporation and condensation of refrigerant-oil mixtures in a smooth tube and a microfin tube, *ASHRAE Trans.*, 94(Part 1), 149–166.

Schlager, L.M., Pate, M.B., and Bergles, A.E., 1990. Evaporation and condensation heat transfer and pressure drop in horizontal, 12.7-mm microfin tubes with refrigerant 22, *J. Heat Transfer*, 112, 1041–1047.

Schlünder, E.U. and Chawla, J., 1967. Local heat transfer and pressure drop for refrigerants evaporating in horizontal, internally finned tubes, *Proc. Int. Cong. Refrig.*, paper 2.47.

Schlünder, E.U. and Chawla, M., 1969. Ortlicher Warmeubergang und Druckabfall bei der Stromung verdampfender Kaltemittel in innenberippten, waggerechten Rohren, *Kaltetech. Klim.*, 21(5), 136–139.

Seo, K. and Kim, Y., 2000. Evaporation heat transfer and pressure drop of R-22 in 7 and 9.52 mm smooth/microfin tubes, *Int. J. Heat Mass Transfer*, 43, 2869–2882.

Shinohara, Y. and Tobe, M., 1985. Development of an improved thermofin tube, *Hitachi Cable Rev.*, 4, 47–50.

Shinohara, Y., Oizumi, K., Itoh, Y., and Hori, M., 1987. Heat Transfer Tubes with Grooved Inner Surface, U.S. patent 4,658,892, assigned to Hitachi Cable, Ltd.

Shiralkar, B. and Griffith, P., 1969. The Effect of Swirl, Inlet Conditions, Flow Direction and Tube Diameter on the Heat Transfer to Fluids at Supercritical Pressure, ASME paper 69-WA/HT-1, ASME, New York.

Sideman, S. and Levin, A., 1979. Effect of the configuration on heat transfer to gravity driven films evaporating on grooved tubes, *Desalination*, 31, 7–18.

Steiner, D. and Taborek, J., 1992. Flow boiling heat transfer of single components in vertical tubes, *Heat Transfer Eng.*, 13(2), 43–68.

Stephan, K. and Mitrovic, J., 1981. Heat transfer in natural convection boiling of refrigerants and refrigerant-oil mixtures in bundles of T-shaped finned tubes, *Advances in Enhanced Heat Transfer*, HTD, 18, 131–146.

Swenson, H.S., Carver, J.R., and Szoeke, G., 1962. The effects of nucleate boiling versus film boiling on heat transfer in power boiler tubes, *J. Eng. Power*, 84, 365–371.

Taitel, Y. and Dukler, A.E., 1976. A model for predicting flow regime transitions in horizontal and near horizontal gas-liquid flow, *AIChE J.*, 22, 47–55.

Takahashi, K., Daikoku, T., Yasuda, H., Yamashita, T., and Zushi, S., 1990. The evaluation of a falling film evaporator in an R-22 chiller unit, *ASHRAE Trans.*, 96(2), 158–163.

Takamatsu, H., Momoki, S., and Fujii, T., 1993. A correlation for forced boiling heat transfer of pure refrigerants in a horizontal smooth tube, *Int. J. Heat Mass Transfer*, 36, 3351–3360.

Tatara, R.A. and Payvar, P., 1999. Effects of oil on boiling R123 and R134a flowing normal to an integral-finned tube bundle, *ASHRAE Trans.*, 105(Part 1), 478–488.

Tatara, R.A. and Payvar, P., 2000a. Effects of oil on boiling of replacement refrigerants flowing normal to a tube bundle. Part I: R-123, *ASHRAE Trans.*, 106(1), 777–785.

Tatara, R.A. and Payvar, P., 2000b. Effects of oil on boiling of replacement refrigerants flowing normal to a tube bundle. Part II: R-134a, *ASHRAE Trans.*, 106(1), 786–791.

Tatsumi, A., Oizumi, K., Hayashi, M., and Ito, M., 1982. Application of inner groove tubes to air conditioners, *Hitachi Rev.*, 32(1), 55–60.

Thomas, D.G. and Young, G., 1970. Thin film evaporation enhancement by finned surfaces, *Ind. Eng. Chem. Process Design Dev.*, 9, 317–323.

Thome, J.R., 1990. *Enhanced Boiling Heat Transfer*, Hemisphere, New York.

Thome, J.R., 1996. Boiling of new refrigerants: a state-of-the-art review, *Int. J. Refrig.* 19, 435–457.

Thome, J.R., Kattan, N., and Favrat, D., 1999. Evaporation in microfin tubes: a generalized prediction model, in *Convective Flow and Pool Boiling*, Taylor & Francis, London, 239–244.

Thonon, B., Vidil, R., and Marvillet, C., 1995. Recent research and developments in plate heat exchangers, *J. Enhanced Heat Transfer*, 2, 149–156.

Tong, W., Bergles, A.E., and Jensen, M.K., 1994. Critical heat flux and pressure drop of subcooled flow boiling in small diameter tubes with twisted tape inserts, in *Heat Transfer in High Heat Flux*

Systems, A.M. Khounsary, T.W. Simon, R.D. Boyd, and A.J. Ghajar, Eds., ASME Symp. Vol. HTD, 301, 47–58.

Topin, F., Rahli, O., Tadrist, L., and Pantaloni, J., 1996. Experimental study of convective boiling in a porous medium: temperature field analysis, *J. Heat Transfer*, 118, 230–233.

Torikoshi, K. and Ebisu, T., 1999. Japanese advanced technologies of heat exchanger in air-conditioning and refrigeration applications, in *Compact Heat Exchangers and Enhancement Technology for the Process Industries*, R.K. Shah, K.J. Bell, H. Honda, and B. Thonon, Eds., Begell House, New York, 17–24.

Tran, T.N., Wambsganss, M.W., and France, D.M., 1996. Small circular- and rectangular-channel boiling with two refrigerants, *Int. J. Multiphase Flow*, 22, 485–498.

Tsuchida, T., Yasuda, K., Hori, M., and Otarii, T., 1993. Internal heat transfer characteristics and workability of narrow THERMOFIN tubes, *Hitachi Cable Rev.*, 12, 97–100.

Varma, H.K., Agrawal, K.N., and Bansal, M.L., 1991. Heat transfer augmentation by coiled wire turbulence promoters in a horizontal refrigerant-22 evaporator, *ASHRAE Trans.*, 97(Part 1), 359–364.

Wadekar, V.V., 1998. A comparative study of in-tube boiling on plain and high flux coated surfaces, *J. Enhanced Heat Transfer*, 5, 257–264.

Wambsganss, M.W., France, D.M., and Jendrejczyk, J.A., 1993. Boiling heat transfer in a horizontal small-diameter tube, *J. Heat Transfer*, 115, 963–972.

Watson, G.B., Lee, R.A., and Wiener, M., 1974. Critical heat flux in inclined and vertical smooth and ribbed tubes, *Proc. 5th Int. Heat Transfer Conf.*, 4, 275–279.

Webb, R.L. and Apparao, T.R., 1990. Performance of flooded refrigerant evaporators with enhanced tubes, *Heat Transfer Eng.*, 11(2), 30–44.

Webb, R.L. and Chien, L.-H., 1994. Correlation of convective vaporization on banks of plain tubes using refrigerants, *Heat Transfer Eng.*, 15(3), 57.

Webb, R.L. and Gupte, N.S., 1992. A critical review of correlations for convective vaporization in tubes and tube banks, *Heat Transfer Eng.*, 13(3), 58–81.

Webb, R.L. and Paek, J.-W., 2003. Letter to the editors — concerning paper published by Y.-Y. Yan and T.-F. Lin, *Int. J. Heat Mass Transfer*, 46, 1111–1112.

Webb, R.L., Choi, K.-D., and Apparao, T., 1990. A theoretical model to predict the heat duty and pressure drop in flooded refrigerant evaporators, *ASHRAE Trans.*, 95(Part 1), 326–338.

Weisman, J. and Illeslamlou, S., 1988. A phenomenological model for prediction of critical heat flux under highly subcooled conditions, *Fusion Technol.*, 13, 654–659.

Weisman, J. and Pei, B.S., 1983. Prediction of critical heat flux in flow boiling at low qualities, *Int. J. Heat Mass Transfer*, 26, 1463–1478.

Weisman, J., Yang, J.Y., and Usman, S., 1994. A phenomenological model for boiling heat transfer and the critical heat flux in tubes containing twisted tapes, *Int. J. Heat Mass Transfer*, 37, 69–80.

Wen, M.-Y. and Hsieh, S.-S., 1995. Saturated flow boiling heat transfer in internally spirally knurled/ integral-finned tubes, *J. Heat Transfer*, 117, 245–248.

Wieland, 1991. *Ripple Fin Tubes*, Wieland-Werke AG brochure TKI-42e(M)-02.91, Ulm, Germany.

Withers, J.G. and Habdas, E.P., 1974. Heat transfer characteristics of helical corrugated tubes for intube boiling of refrigerant R-12, *AIChE Symp. Ser.*, 70(138), 98–106.

Yan, Y.-Y. and Lin, T.-F., 1998. Evaporation heat transfer and pressure drop of refrigerant R-134a in a small pipe, *Int. J. Heat Mass Transfer*, 41, 4183–4194.

Yan, Y.-Y., and Lin, T.-F. 1999. Evaporation heat transfer and pressure drop of refrigerant R-134a in a plate heat exchanger, *J. Heat Transfer*, 121, 118–127.

Yasuda, K., Ohizumi, K., Hori, M., and Kawamata, O., 1990. Development of condensing Thermofin-HEX-C tube, *Hitachi Cable Rev.*, 9, 27–30.

Yilmaz, S., Palen, J.W., and Taborek, J., 1981. Enhanced boiling surfaces as single tubes and tube bundles, in *Advances in Enhanced Heat Transfer*, R.L. Webb, T.C. Carnavos, E.F. Park, and K.M. Hostetler, Eds., ASME Symp. Vol. HTD-Vol. 18, ASME, New York, 23–24.

Yoshida, S., Matsunaga, T., and Hong, H.P., 1987. Heat transfer to refrigerants in horizontal evaporator tubes with internal, spiral grooves, *Proc. 1987 ASME-JSME Thermal Engineering Joint Conference*, P.J. Marto, Ed., 5, 165–172.

Yu, W., France, D.M., Wambsganss, M.W., and Hull, J.R., 2002. Two-phase pressure drop, boiling heat transfer, and critical heat flux to water in a small-diameter horizontal tube, *Int. J. Multiphase Flow*, 28, 927–941.

Zeng, X., Chyu, M., and Ayub, Z.H., 1998. Ammonia spray evaporation heat transfer performance of single low-fin and corrugated tubes, *ASHRAE Trans.*, 104(Part 1), 185–196.

Zeng, X., Chyu, M.-C., and Ayub, Z.H., 2000. An experimental study of spray evaporation of ammonia in a square-pitch, low-fin tube bundle, *Proceedings of 34th National Heat Transfer Conference*, Pittsburgh, PA, NHTC 2000–12215.

Zhao, Y., Molki, M., Ohadi, M.M., and Dessiatoun, S.V., 2000. Flow Boiling of CO Microchannels, *ASHRAE Trans.*, 106, Pt. 1, 437–445.

Zhao, Y., Molki, M., and Ohadi, M.M., 2001. Predicting flow boiling heat transfer of CO_2 microchannels, Proceedings of the ASME, HTD-Vol. 369–3, Y. Jaluria, Ed., *2001 ASME International Mechanical Engineering Congress and Exposition*, ASME, New York, 195–204.

CONVECTIVE CONDENSATION

14.1 INTRODUCTION

When condensation occurs in a tube, convective effects that do not exist in vapor space condensation occur. The convective effects are associated with the influence of shear stress on the liquid–vapor interface. Further, if gravity force exceeds vapor shear forces, the liquid phase will tend to stratify in horizontal tubes.

The highest vapor shear effects exist near the inlet end of the tube, where vapor velocity is highest. Section 13.2 discusses fundamentals of two-phase flow in tubes, and how heat transfer is influenced by the flow pattern. If the nucleate boiling contribution in convective vaporization were suppressed, e.g., $h_{cv} \gg h_{nb}$, the convective vaporization and convective condensation in a tube would have a great deal in common. This would be particularly true if the wall were wetted in both cases. As noted in Chapter 13, if the film flow is laminar, the heat transfer coefficient is given by $h = k_l/\delta$, where δ is the film thickness. As the vapor quality increases in vaporization, the vapor velocity increases causing increased vapor shear. In condensation, the converse exists. A two-phase flow experiences different flow patterns as vapor quality changes along the tube length, as shown in Figure 13.3. The heat transfer coefficient is significantly affected by the flow pattern. Figure 14.1 shows the flow patterns drawn on the Baker flow map (Figure 13.5) experienced by R-12 condensing in a horizontal tube. This figure shows that the flow patterns are very sensitive to mass velocity. The mass velocity is defined as the total flow rate divided by the pipe cross-sectional area. The reader should review Section 13.2 to understand how flow pattern affects the heat transfer coefficient in two-phase heat transfer.

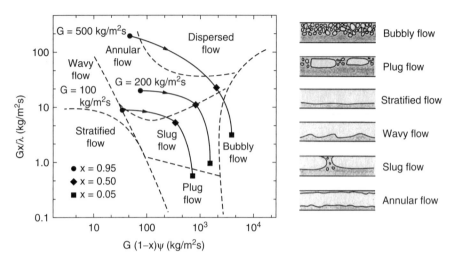

Figure 14.1 Flow patterns for condensation in a horizontal tube drawn on the Baker [1954] flow pattern map. (From Carey [1992].)

The pressure drop in condensation contains the same components as in vaporization. However, the acceleration term, Δp_a, in Equation 13.6 causes pressure recovery in condensation, whereas the same magnitude of Δp_a adds to the pressure drop in vaporization.

Enhancement should also be beneficial for condensing flows that contain noncondensible gases.

14.2 FORCED CONDENSATION INSIDE TUBES

Enhancement techniques used on the outside of vertical tubes are applicable to condensation inside vertical tubes. At low vapor velocities, enhancement requirements for horizontal tubes are different, because gravity force drains the film transverse to the flow direction. Preferred augmentation techniques for horizontal tubes are yet to be established and commercialized. Those that have been tested include twisted-tape inserts, internal fins, and integral roughness. Internal fins have yielded the highest augmentation levels. However, selection of the preferred geometry must consider the tubing cost and pressure drop.

14.2.1 Internally Finned Geometry

Internally Finned Tube. Vrable et al. [1974] and Reisbig [1974] used R-12 in horizontal internally finned tubes. Defining the condensation coefficient in terms of the total surface area, they observed coefficients 20 to 40% greater than the smooth tube value for wet vapor. Accounting for surface area increases of 1.5 to 2.0, the heat conductance hA/L is two to three times the smooth tube value. Reisbig found that his full-height fins provided a smaller hA/L than did smooth tubes when condensing superheated R-l2. No explanation is offered for this unexpected performance.

Royal and Bergles [1978a] condensed steam in four horizontal internally finned copper tubes, three of which had spiraled fins. The inlet steam was saturated at 4 bar and all of the vapor was condensed in the test section. A later work by Luu and Bergles [1979] tested the same tubes with R-113. Their results were generally consistent with the augmentation levels reported by Vrable [1974] and Reisbig [1974] for R-12. The best internal fin geometry provided 20 to 40% higher heat transfer coefficients (total area basis) than the smooth tubes. Figure 14.2 shows their results, with the tube geometries defined in Table 14.1. The value A/A_p is the internal surface area, relative to a plain tube of the same inside diameter.

The heat transfer coefficients of Figure 14.2a are based on the smooth bore area. The figure also contains data for tubes having twisted-tape inserts, to be discussed later. The internally finned tubes provide significant heat transfer improvement, and considerably higher than that provided by the twisted tapes (tubes B and C). The highest performance is provided by Tube G (16 fins). It provides an enhancement ratio of 2.3 relative to its 73% area increase. The figure shows that the area increase is not the controlling factor. Tube D has approximately the same internal area as tube G, but with twice as many fins of half the height. The enhancement provided by tube D is much less than that of tube G. The pressure drop results (Figure 14.2b) show that the twisted tapes cause a much larger pressure drop increase than does the heat transfer enhancement.

Royal and Bergles [1978a] correlated their steam condensation data, by modifying the smooth tube correlation of Akers et al. [1959] to account for geometric parameters. Luu and Bergles [1979] found that the Royal and Bergles [1978a] correlation did not predict their R-113 data well. They developed a modified correlation for the R-113 data, which did not include the steam data. Because of the large fluid property differences of steam and R-113, they conclude that different fin geometries may be preferred for water and refrigerants.

Said and Azer [1983] tested the four internal fin geometries listed in Table 14.2 using R-113. The fins are typically higher and have larger helix angles than those tested by Luu and Bergles [1979]. Because the two investigators performed their tests at different saturation pressures, the results cannot be directly compared. Said

Table 14.1 Geometries Tested by Royal and Bergles [1978a]

Code	d_i (mm)	Geometry	e (mm)	n	α (deg)	A/A_p
A	15.9	Smooth	None	None	0	1
B	15.9	Twisted-tape	None	None	18.7	1
C	15.9	Twisted-tape	None	None	9.3	1
D	15.9	Internal fin	0.60	32	2.95	1.70
E	12.8	Internal fin	1.74	6	5.25	1.44
F	12.8	Internal fin	1.63	6	0	1.44
G	15.9	Internal fin	1.45	16	3.22	1.73

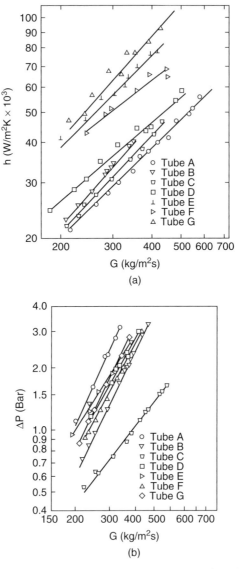

Figure 14.2 Test results of Royal and Bergles [1978a] for steam condensation (4 bar) in horizontal tubes containing internal fins and twisted tapes. (a) Heat transfer coefficient based on plain tube area, (b) pressure drop. Geometries defined in Table 14.1. (From Royal and Bergles [1978a].)

and Azer found that the Luu and Bergles [1979] R-113 heat transfer correlation overpredicted their R-113 data. Hence, Said and Azer developed their own empirical correlation. Unfortunately, they did not include the Luu and Bergles [1979] R-113 data. Kaushik and Azer [1988] developed a new empirical power law correlation for internally finned tubes. This correlation predicts the data of Luu and Bergles [1979], Royal and Bergles [1978a], and Said and Azer [1983] with only modest

Table 14.2 Internal Fin Geometries Tested by Said and Azer [1983]

Code	d_i (mm)	Fin height (mm)	No. of fins	Helix Angle (deg)	A/A_p
2	13.84	1.58	10	0	1.50
3	17.15	1.80	16	9.7	1.89
4	19.87	1.98	16	12.3	1.76
5	25.38	2.13	16	19.4	1.64

success. Some points are under- or overpredicted as much as 100%. The conclusion is that satisfactory correlations to predict the effect of fluid properties and fin geometry on condensation do not exist. Kaushik and Azer [1990] also developed an empirical power law correlation for pressure drop in internally finned tubes. The correlation predicted 68% of the data points of three investigators within ±40%. Kaushik and Azer [1989] and Sur and Azer [1991] propose analytically based models to predict the condensation coefficient and the pressure drop in internally finned tubes, respectively. The models were not sufficiently validated against existing data to justify recommendation of their use. More work is needed on correlations and theoretically based models.

The selection of internally enhanced tubes for two-phase applications should account for the effect of pressure drop on the saturation temperature of the condensing or evaporating fluid. Increased pressure drop, for fixed inlet pressure, will reduce the available LMTD, as discussed in Chapter 4. Royal and Bergles [1978b] present a performance evaluation analysis of the Royal and Bergles [1978a] test geometries. However, they do not include the effect of pressure drop in their performance evaluation analysis.

Ito and Kimura [1979] and Shinohara and Tobe [1985] describe a tube having very small triangular shaped fins, which has become known as the microfin tube. This tube is used in virtually all refrigerant condensers, either air or water cooled. Because of the importance of this tube, it is separately described in Section 14.3.

Smit and Meyer [2002] obtained R22/R-142b condensation data in 9.53-mm-OD internally finned, twisted-tape insert and microfin tubes for $100 < G < 800$ kg/m²s. The internally finned tube had 6 fins 0.6 mm high, 0.2 mm thick, and a spiral angle of 18°. The twisted tape was 0.3 mm thick and had a twist ratio of 3.0. The microfin tube had 60 fins of 0.20-mm fin height, 51° apex angle, and 18° helix angle. The microfin tube yielded the highest heat transfer coefficient and the lowest pressure drop. The heat transfer coefficient enhancements (relative to a plain tube) were 46, 87, and 113% using twisted tapes, internal fins, and microfins, respectively. The pressure drop increases were 81, 148, and 38% for twisted tape, internal fins, and microfins, respectively.

Wire Loop Finned Annulus. Honda et al. [1988] have investigated a different type of internally finned channel than the continuous internal fins previously discussed. Their work is applicable to tube-in-tube condensers, which are used in small, water-cooled refrigeration condensers. Condensation occurs in an annulus, with water flow

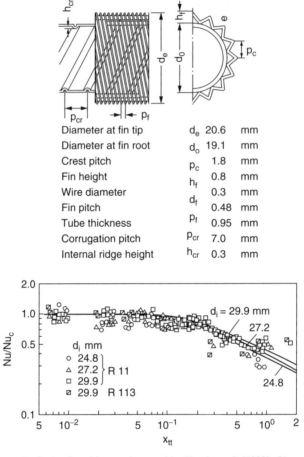

Diameter at fin tip	d_e	20.6 mm
Diameter at fin root	d_o	19.1 mm
Crest pitch	p_c	1.8 mm
Fin height	h_f	0.8 mm
Wire diameter	d_f	0.3 mm
Fin pitch	p_f	0.48 mm
Tube thickness		0.95 mm
Corrugation pitch	p_{cr}	7.0 mm
Internal ridge height	h_{cr}	0.3 mm

Figure 14.3 (a) Detail of wire-finned inner tube tested by Honda et al. [1988], (b) correlation of data vs. X_{tt}. (From Honda et al. [1988].)

in the inner corrugated tube. Their annulus corrugated copper wire fins were wrapped on the outer surface of a 19.1-mm-diameter tube, and were soldered to the tube. Figure 14.3a shows the wire fin geometry, which provided an area enhancement of 3.04. Three outer tube diameters were tested: 24.8, 27.2, and 29.9 mm. The local R-113 condensation coefficient was enhanced a factor of 2 to 13, relative to a plain inner tube at mass velocities between 52 and 201 kg/m²-s, respectively. Nozu et al. [1995] extended the study by including an upstream U bend. The U bend increased the heat transfer and pressure drop of the downstream straight section.

14.2.2 Twisted-Tape Inserts

The previously discussed work of Royal and Bergles [1978a, 1978b] included the evaluation of two twisted-tape geometries for steam. Luu and Bergles [1979] also tested the same twisted tapes with R-113. These works show that the performance of twisted tapes is distinctly poorer than that of internally finned tubes (see Figure 14.2). They provide only 30% heat transfer enhancement, but exhibit pressure drops equal to or higher than those of internally finned tubes. Data on R-113 by Said and Azer [1983] generally confirm this conclusion.

Lin et al. [1980] tested the static mixer insert device shown in Figure 7.1c using R-113. This consists of successive axial increments of twisted-tape segments. The insert device successively rotates the flow 180° clockwise, then 180° counterclockwise. The leading edge of each successive element is at 90° to the trailing end of the upstream element. Lin et al. tested elements approximately 38 mm long in 6.35-, 12.7-, and 19.1-mm-diameter tubes. Significantly higher enhancement is obtained with the static mixer than with a twisted tape. The enhancement was approximately 80% for all tube sizes. However, the pressure drop increased by a factor of four.

Kedzierski and Kim [1997] measured refrigerant vaporization and condensation in a 9.64-mm-ID tube having a 20.75° twisted tape (0.5 mm thick). Data were taken for five pure refrigerants (R-12, R-22, R-134a, R-152, and R-290) and four refrigerant mixtures. The data showed decreasing condensation coefficient with decreasing vapor quality. Data were taken at vapor qualities up to 97%, and empirical correlations are given for the data.

Several correlations have been developed for condensation in tubes having twisted tapes. The correlations are based on modifying correlations for condensation in plain, circular tubes. Royal and Bergles [1978a] developed a correlation for their steam data, based on a modification of the smooth tube correlation of Akers et al. [1959]. The modification consisted of defining the Reynolds number in terms of the condensate properties, using the hydraulic diameter, and calculating the velocity as done in the Lopina and Bergles [1969] correlation for single-phase flow. The velocity used is the resultant of the average axial velocity and tangential velocity at the wall, assuming that the tape induces solid body rotation. The correlation assumed that the tape acted as a fin, with no contact resistance, which is probably optimistic. Luu and Bergles [1980] abandoned the Royal and Bergles [1978a] correlation for their R-113 data. Rather, they chose to use a modification of the Boyko and Kruzhilin [1967] plain tube correlation. The Reynolds number was defined by the hydraulic diameter and the average swirling velocity (rather than that at the wall). Luu and Bergles [1980] showed that most of their R-113 data were correlated within ±30%. However, Said and Azer [1983] state that the Luu and Bergles correlation underpredicted their R-113 data by 0 to 30%. They developed their own heat transfer correlation based on a modification of the Shah [1979] plain tube correlation. They also developed a pressure drop correlation, based on modifying plain tube correlations.

14.2.3 Roughness

The work of Nicol and Medwell on the effect of roughness for vapor space condensation (Section 12.3.2) provided motivation for investigating roughness for forced condensation inside tubes. Luu and Bergles [1979] found that two-dimensional repeated-rib roughness ($e/d = 0.013$, $p/e = 9.7$, and $e/d = 0.021$, $p/e = 20.8$) increased the R-113 condensing coefficient more than 100%. Both roughness sizes provided approximately equal augmentation levels. For the same augmentation level, roughness may be of greater interest than the relatively high internal fins discussed in Section 14.2.1, because roughness requires less tube material.

In a U.S. patent, Fenner and Ragi [1979] describe a method for applying a "sand-grain" type roughness inside tubes. The favored roughness consists of a single layer of metal particles bonded to the surface, and covering approximately 50% of the projected surface. The spaced metal particles provide extended surface for high vapor qualities and cause turbulence in the condensate film at lower vapor qualities. Data are presented for R-12 condensing in a 14.5-mm-diameter tube having a 50% area density of $e/d_i = 0.031$ particles. This tube yielded enhancement levels (smooth tube area basis) of 2.4 for low exit qualities (0.25 to 0.60) and 4.0 for high exit qualities. The accompanying pressure drops were only 68 and 105% larger than the corresponding smooth tube values. It would appear that this roughness geometry can provide the same enhancement as the Section 14.2.1 internally finned tubes with smaller pressure drop and using less additive material content.

Shinohara and Tobe [1985] provide R-22 condensation data on the tubes shown in Figure 13.15. Table 14.3 compares the dimensions and performance of a 16-mm-OD TFIN-CR tube with a plain and several other geometries for the standard Figure 13.15b corrugated tube (CORG). The geometry parameters of the corrugated tube are the roughness height (e) and roughness spacing (p). The table also contains performance data on a microfin tube (TFIN) and five-leg star fin insert (Figure 13.10). The heat transfer coefficients are based on the outer diameter (16 mm). The microgrooves in the TFIN tubes are 0.2 mm and the helix angle is 8°. The 5.7-m-long test section was operated at 35°C. Note that the enhancement level of the TFIN tube is not as high as that listed in Table 14.4.

Hinton et al. [1995] obtained R-22 condensation data in 15.8-mm-OD corrugated tube from a test loop model of a heat pump system. The corrugated tube had 0.8 mm indentation at 7.1-mm pitch. The mass flux was estimated from the measured heating capacity. At 275 to 300 kg/m²s mass flux, the average heat transfer coefficient of the corrugated tube was 40% higher than that of the smooth tube and the pressure drop increase was 23%.

Table 14.3 Tube Comparison for R-22 Condensation at $G = 100$ kg/hr

Geometry	e_i (mm)	p (mm)	h (kW/m²)	Δp (kPa)	h/h_p	$\Delta p/\Delta p_p$
Plain	NA	NA	1.28	7.80	1.00	1.00
Star insert	NA	NA	2.62	15.7	2.04	2.01
CORG	0.7	8.0	2.03	14.2	1.59	1.82
TFIN	Micfin	NA	2.06	7.80	1.61	1.00
TFIN-CR	0.7	8.0	3.29	14.2	2.57	1.82

NA = not applicable.

Table 14.4 Chronological Improvements in the Hitachi Thermofin Tube

Year	Geometry	d_o	e	p/e	a	b	n	A/A_p	Wt/Wt_p	h/h_p
1977	Original	9.52	0.15	2.14	25	90	65	1.28	1.22	1.8
1985	EX	9.52	0.20	2.32	18	53	60	1.51	1.19	2.4
1988	HEX	9.52	0.20	2.32	18	40	60	1.60	1.19	2.5
1989	HEX-C	9.52	0.25	2.32	30	40	60	1.73	1.28	3.1
2001	HEX-C	7.0	0.20	2.32	18	40	50			
2001	TFIN-HGL	7.0	0.22	2.03	16	22	54			
2001	HEX-HG	7.0	0.25	2.32	18	15	57			

14.2.4 Wire Coil Inserts

Wang [1987] investigated wire coil inserts for tube-side condensation with stratified flow. He developed a theoretical model to predict the condensation coefficient. The model assumes that surface tension force at the base of the wire pulls the condensate to the base of the wire, where it is gravity-drained to the liquid level in the tube. This is the same concept described by Thomas [1967] and illustrated in Figure 12.17. They show that

$$\frac{h}{h_p} = \frac{0.75(1+F_1)^{1/4}(p-e)}{p} \tag{14.1}$$

where the parameter $F_1 = 4\sigma d_i / \rho_l g(p - e)^2 r_s$. The F_1 parameter defines the strength of the surface tension force, relative to gravity force. The r_s is the radius of the condensate film at the base of the wire. Data were obtained with R-12 condensing in an 8.4-mm-ID tube using two different wire coil inserts, $e = 0.5$ mm, $p = 2.6$ mm, and $e = 0.3$ mm, $p = 1.5$ mm. Near the exit of the tube, where the flow was stratified, they obtained 40% enhancement. They also ran tests with the wire coil insert in the full circuit length, and obtained an average enhancement of approximately 35%, with $\Delta p / \Delta p_p \cong 10$. At high vapor qualities, the wire coil acts as a roughness. Equation 14.1 correlated their data within ±20%.

Agrawal et al. [1998] investigated R-22 condensation in 12.7-mm-ID tubes with wire coil inserts. The wire coil geometry had $0.65 \leq e \leq 1.5$ mm and $6.5 \leq p \leq 13.0$ mm. For the same coil pitch, the thickest wire yielded the highest heat transfer enhancement at high vapor qualities. For low vapor qualities, however, the maximum enhancement was obtained from the thinnest wire. The effect of wire pitch was not conclusive, although lower pitch generally provided larger enhancement at low qualities. The enhancement ratio generally increased as the vapor quality increased, with a maximum of 100%. No pressure drop data are provided.

14.2.5 Coiled Tubes and Return Bends

Brdlik and Kakabaev [1964] and Miroploskii and Kurbanmukhamedov [1975] have studied steam condensation in helical tube coils. Brdlik and Kakabaev [1964] show that the local condensation coefficient is given by

$$\frac{hd_i}{k_l} = \frac{388}{1-x}\left(\frac{d_i G}{\mu_l}\right)^{0.15}\left(\frac{D_c}{d_i}\right)^{-0.54} \tag{14.2}$$

where d_i is the tube diameter and D_c is the coil diameter. Traviss and Rohsenow [1973] measured the local condensation coefficients downstream from a return bend for R-12 condensing in a 0.80-mm-diameter tube. The effect of the return bend is negligible when averaged over a length of 90 tube diameters.

14.3 MICROFIN TUBE

This tube was introduced in Chapter 13, and is illustrated in Figure 13.7d. Virtually all air-cooled residential air conditioners use this tube. This includes central air conditioners and window units. It is also frequently used in automotive refrigerant condensers, which use aluminum tubes. Some manufacturers use the same tube for enhancement of both evaporation and condensation. Hence, the descriptive information presented in Section 13.4 is equally applicable to the condensation version and will not be repeated here. The tube is made in diameters between 4 and 16 mm. The microfin tube provides a R-22 condensation coefficient approximately 100% higher than that of a plain tube. Table 14.4 is a repeat of Table 13.3, except the last column shows the condensing coefficient relative to a plain tube (h/h_p) operated at the same mass velocity. Figure 13.21, from Yasuda et al. [1990], shows the cross section fin geometries associated with Table 14.4.

The condensation tube (HEX-C) differs from the evaporator tube (HEX) only in the helix angle. Figure 14.4 shows the R-22 condensing coefficients of the 9.52-mm-diameter Table 14.4 tubes as a function of mass velocity. Table 14.4 shows that the condensing coefficient of the HEX-C tube is 3.1 times that of a plain tube at the same mass velocity. The improved versions have higher fins (e) and a smaller fin included angle (β). Work by Shinorara and Tobe [1985] for the Thermofin-EX tube shows that the condensing coefficient gradually increases with helix angle (α) from 7°, up to 30°. The preferred helix angle for condensation is 30°.

The pressure drop for condensation should be slightly less than for vaporization, as the momentum contribution reduces pressure drop. The pressure drop for evaporation was shown in Figure 13.24.

In addition to vapor shear, the author proposes that surface tension forces (see Chapter 12) are influential. Surface tension pulls the condensate from the fin tips into the drainage channel at the base of the fins. No correlations have been published to predict the performance of the tube as a function of the flow, fluid properties, and geometric variables.

Khanpara et al. [1987] measured R-113 evaporation in 9.5-mm-OD microfin tubes made by eight different manufacturers. There were significant differences in the cross section fin profiles, as shown in Figure 14.5. All tubes, except tube 6, had 60 to 70 fins, 0.15 to 0.19-mm fin height, and 20 to 25° helix angle. Tube 6 had

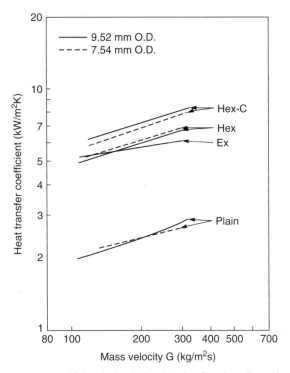

Figure 14.4 R-22 condensing coefficient in the Table 14.4 microfin tubes. (From Yasuda et al. [1990].)

0.1-mm fin height and 9° helix angle. At $x = 0.5$, the enhancement ratio of the eight tubes varied from 1.6 to 2.2 at $G = 200$ kg/m²-s, and from 1.8 to 3.3 at $G = 575$ kg/m²-s. The Shinohara et al. [1987] patent shows that the R-22 condensation coefficient in a 9.5-mm-OD tube increases approximately 20% as the helix angle increases from 10° to 35°. Schlager et al. [1990c] tested three 12.7-mm-OD microfin tubes having different helix angles 15°, 18°, and 25° with R-22. However, their tubes also had different fin heights ($0.15 \leq e_i \leq 0.3$ mm) and pitch, so they could not define the effect of specific geometry factors on the performance differences. The highest condensation coefficient was provided by the tube having the highest fin height and 18° helix angle.

Schlager et al. [1988, 1989] and Eckels and Pate [1991] report the effect of oil on the performance of R-22 and of R-134a and R-12, respectively. Oil slightly reduces the condensation coefficient for typically used oil concentrations.

Much experimental work has addressed measurement of the heat transfer coefficients for condensation of refrigerants in tubes with different microfin geometries. Reviews on the issue are found in Cavallini et al. [2000, 2002, 2003], Newell and Shah [2001], Webb [1999], and Liebenberg et al. [2000]. Most experimental results reveal significant heat transfer enhancement (1.5 to 2.0) with minor pressure drop increase (1.1 to 1.3). The discussion here focuses on efforts at optimization of fin

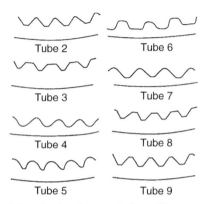

Figure 14.5 Cross sections of the microfin tubes tested. (From Khanpara et al. [1987].)

geometry. Recent findings on condensation in advanced microfin geometries are also provided.

14.3.1 Microfin Geometry Details

The basic algebraic dimensions that define the single-groove microfin geometry are shown in Figure 14.6:

1. The rib layout parameters are defined by the rib height (e), the rib pitch normal to the fins (p_f) and the rib helix angle (α). Alternately, one may define the rib pitch normal to the tube axis as $p_n = p_f / \cos \alpha$. The number of fins (n_f) is given by $n_f = \pi d_i / p_n = \pi d_i \cos \alpha / p_f$.
2. The rib shape parameters are defined by the fin base thickness (t_b), and the apex angle of the fin (β).

The fin tip may be either a circular arc or have a flat top. It appears that tubes made by pulling a seamless tube over a plug will have a curved fin tip. However, tubes made by embossing a flat strip and seam welding will tend to have a relatively flat fin tip. For microfins having a flat fin tip, the area, relative to that of a plain tube of the same diameter to the root of the fins (d_i), is given by

$$A/A_p = 1 + (\sec \beta / 2 - \tan \beta / 2) 2e / p_f \qquad (14.3)$$

Note that p_f is independent of helix angle, but p_n is dependent on helix angle. Because $n_f \propto d_i / p_f$, the number of fins typically increases with increasing tube diameter. For a constant tube diameter, decrease of the fin pitch (p_f) will increase the number of fins, which provides increased A/A_p.

Thus, the principal dimensions are e, p_f, t_b, α, and β. Alternately, one may replace p_f by $n_f = \pi d_i \cos \alpha / p_f$. Tube manufacturers typically offer only one helix angle (approximately 18°) for both condensation and evaporation. There is little difference

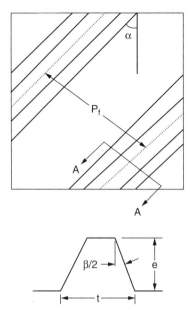

Figure 14.6 Geometric parameters of the microfin surface geometry.

in the geometry of microfin tubes offered by most manufacturers. For a 9.5-mm-diameter tube, typical values are e = 0.20 to 0.25 mm, p_f = 0.46 mm, and α = 18°.

14.3.2 Optimization of Internal Geometry

Considerable recent work has been done to reduce the material content of the microfins. A major focus has involved reducing the fin base thickness (t_b) and the apex angle (β), and investigating fin pitch. Note that A/A_p is not influenced by the fin base thickness. Work to optimize the microfin geometry is summarized in Table 14.5 and includes both vaporization and condensation results. The optimum geometries listed in Table 14.5 are those that provide the highest heat transfer coefficient. Only the condensation data are discussed in this section. The vaporization data are discussed in Section 13.4. Tsuchida et al. [1993] provide extensive data on convective vaporization and condensation of R-22 in microfin tubes with outside diameters ranging from 4.0 to 9.5 mm. Figure 14.7a shows the effect of fin height for different diameter tubes. For both condensation and evaporation, Figure 14.7a shows that the optimum fin height decreases with decreasing tube diameter. The optimum fin height is 0.13 mm for 4.0 diameter, 0.16 mm for 7.0 diameter, and further increases to 0.23 mm for 9.52 mm tube diameter. The effect of number of fins is shown in Figure 14.7b. This figure shows that the optimum number of fins decreases as the tube diameter decreases. This is somewhat misleading, because a more relevant parameter is the fin pitch (p_f) normal to the fins. The p_f corresponding to the optimum number of fins was calculated using $p_f = \pi d_i \cos \alpha / n_f$ and are in the range of 0.26 to 0.32 mm for all tube diameters. The effect of helix angle is shown in Figure 14.7c. The figure

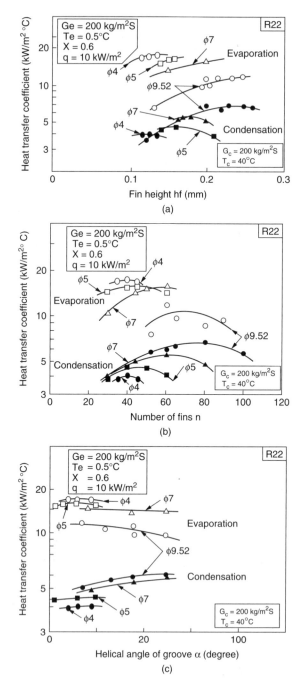

Figure 14.7 Effect of fin height, number of fins, and helix angle on R-22 evaporation and condensation in 4, 5, 7, and 9.52 mm diameter tubes. (a) Effect of fin height, (b) effect of number of fins, (c) effect of helix angle. (From Tsuchida, 1993 *Hitachi Cable Rev.*, 12. With permission.)

Table 14.5 Optimum Microfin Configuration

Ref.	Geometry	Test Conditions	Optimum Configuration (Vaporization)	Optimum Configuration (Condensation)
Ito and Kimura [1979]	$d_o = 12.7$ $d_i = 11.2$	R-22 $40 < G < 203$	$\alpha = 7°$ $p_f = 0.5{-}1.0$ $e = 0.2$	
Yasuda et al. [1990]	$d_o = 9.52, 7.94$ $d_i = 8.8, 7.22$ $\alpha = 40°$	R-22 $100 < G < 300$ $q = 10 \text{ kW/m}^2$	$\alpha = 18°$ $d_o = 9.52$ $n_f = 60$ $p_f = 0.44$	$\alpha = 30°$ $d_o = 9.52,$ $n_f = 50$ $p_f = 0.48$
Morita et al. [1993]	$d_o = 9.52, 7.00, 4.0$ $d_i = 8.92\ 6.4,\ 3.4$ $\alpha = 18°$	R-22	$d_o = 7.0$ $n_f = 50$ $d_o = 4.0$ $n_f = 34$	$d_o = 9.52,$ $n_f = 50$ $d_o = 4.0,$ $n_f = 34$
Tsuchida et al. [1993]	$d_o = 9.52$ $(d_i = 8.92)$ $\beta = 40°$ $d_o = 7.0$ $(d_i = 6.4)$ $\beta = 40°$ $d_o = 5.0$ $(d_i = 4.2)$ $\beta = 40°$ $d_o = 4.0$ $(d_i = 3.3)$ $\beta = 40°$	R-22 $q = 10 \text{ kW/m}^2$ $150 < G < 600$	$\alpha = 7°$ $e = 0.25$ $n_f = 70\ (p_f = 0.40)$ $\alpha = 18°$ $e = 0.2$ $n_f = 60\ (p_f = 0.32)$ $\alpha = 6°$ $e = 0.15$ $n_f = 45\ (p_f = 0.29)$ $\alpha = 6°$ $e = 0.13$ $n_f = 40\ (p_f = 0.26)$	$\alpha = 25°$ $e = 0.25$ $n_f = 80\ (p_f = 0.32)$ $\alpha = 25°$ $e = 0.18$ $n_f = 60\ (p_f = 0.30)$ $\alpha = 10°$ $e = 0.15$ $n_f = 45\ (p_f = 0.29)$ $\alpha = 10°$ $e = 0.13$ $n_f = 40\ (p_f = 0.26)$
Chamra et al. [1996a, 1996b]	$d_o = 15.88$ $e = 0.35$ $p = 0.58$ $\alpha = 30°$	R-22 $72 < G < 289$	$\alpha = 20°$	$\alpha = 27°$
Houfuku et al. [2001]	$d_o = 7$ $(d_i = 6.4)$	R-410A $G = 250$	$\alpha = 16°, \beta = 22°$ $n_f = 54\ (p_f = 0.36)$ $e = 0.22$	$\alpha = 16°, \beta = 22°$ $n_f = 54\ (p_f = 0.36)$ $e = 0.22$
Ishikawa et al. [2002]	$d_o = 7$ $(d_i = 6.4)$ $\alpha = 16°, \beta = 10°$ $e = 0.24$	R-22 $160 < G < 320$	$n_f = 80$ $(p_f = 0.24)$	$n_f = 80$ $(p_f = 0.24)$

Units: d, e, p_f (mm), α, β (deg), G (kg/m²-s), q (kW/m²).

shows that the condensation coefficient increases with increasing helix angle. However, the effect of helix angle is nearly insignificant for 4- and 5-mm-diameter tubes. The optimum configuration for each diameter tube is listed in Table 14.5.

Research at Hitachi Cable includes publications of Ito and Kimura [1979] and, Yasuda et al. [1990] for 9.5-mm-tube OD, Houfuku et al. [2001] for 7.0-mm tube

OD, and Tsuchida et al. [1993] for 4.0- and 5.0-mm tube OD. Early 9.5-mm-diameter microfin tubes, as described by Yasuda et al. [1990], had $e = 0.2$ mm and $\beta = 90°$ (see Figure 14.6). By 1989, Hitachi had increased their fin height to 0.25 mm and decreased β to 40° as reported by Yasuda et al. [1990]. Typically 60 fins were used in a 9.52-mm-OD tube, as reported by Yasuda et al. [1990], for which $p_f = 0.47$ mm. Recent work to optimize the microfin geometry for a 7.0-mm-OD tube reported by Houfuku et al. [2001] use $e = 0.22$ mm, $p_f = 0.37$ mm ($n_f = 54$), $\beta = 22°$, and $\alpha = 16°$. Included in all of this work were efforts to reduce the material content by reducing the fin base thickness and the fin apex angle. Ishikawa et al. [2002] investigated the number of fins for 7.0-mm-tube diameter with $\alpha = 10°$ with $e = 0.21$ to 0.24 mm. They found that 80 fins ($p_f = 0.26$ mm) was the optimum value. Note that this optimum is different from that reported by Houfuku et al. [2001], who found $e = 0.22$ mm, $p_f = 0.36$ mm ($n_f = 54$), $\beta = 22°$, and $\alpha = 16°$.

Chamra et al. [1996a] investigated the R-22 condensation characteristics in four two-dimensional microfin tubes and seven three-dimensional cross-grooved microfin tubes, 15.88 mm OD. These are the same tubes as those used for the vaporization test [1996b]. The vaporization results are discussed in Section 13.4. The single groove geometries had 74 to 80 fins and 0.35-mm fin height. The helix angle was varied 15° to 27°. Figure 14.8 shows the effect of helix angle. The condensation coefficient increases as the helix angle increases, which is in agreement with Tsuchida et al. [1993].

Houfuku et al. [2001] investigated the effect of total inner surface area on the condensation coefficient of R-410A in 7.0-mm-OD microfin tubes. The vaporization counterpart is provided in Section 13.4. The geometric dimensions are listed in Table 14.5. Figure 14.9 shows the relationship between interfin area and heat transfer coefficient. The condensation coefficient is well correlated with interfin area.

Ishikawa et al. [2002] investigated the effect of number of fins [or fin pitch] on the condensation of R-22 in 7.0-mm-OD microfin tubes. The vaporization in these tubes is discussed in Section 13.4. The number of fins ranged from 55 to 85. The highest heat transfer coefficient was obtained for a tube having 80 fins ($p_f = 0.24$ mm). The enhancement is higher than measured by Tsuchida et al. [1993] and Houfuku et al. [2001]. The small apex angle of Ishikawa et al. fin may be responsible.

14.3.3 Condensation Mechanism in Microfin Tubes

Prior to discussing models to predict the condensation coefficient in microfin tubes, it is appropriate to understand the fundamentals of the heat transfer mechanism and its relation to single-phase heat transfer. Moser et al. [1998] developed a theoretically based predictive model for condensation in plain tubes, which they describe as the equivalent Reynolds number concept. The model is also applicable to microfin tubes. The model assumes annular flow and is based on the analogy between heat and momentum transfer. It further assumes that surface tension force does not contribute to condensate drainage from the fins. The possibility of surface tension contribution is discussed in a later section. The model is based on the following concepts (or assumptions):

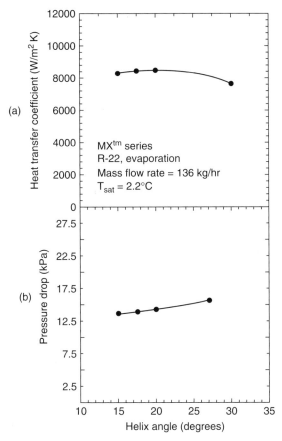

Figure 14.8 Variation of R-22 condensation coefficient in single-grooved tubes with helix angle. (a) Condensation coefficient, (b) pressure drop. (From Chamra et al. [1996a].)

1. The condensation coefficient is directly related to the heat transfer coefficient for single-phase flow in the same geometry. The linkage is the equivalent single-phase Reynolds number (Re_{eq}). At the Re_{eq}, the single-phase flow will have the same wall shear stress as for the condensing flow.

2. Assuming the equivalent single-phase flow is turbulent, one may use the Petukhov equation (Equation 2.22) to calculate the single-phase heat transfer coefficient (h_{sp}).

The condensing coefficient (h) is related to the calculated single-phase heat transfer coefficient by the following analytically based equation given by Moser et al. [1998]:

$$\frac{h}{h_{sp}} = \frac{T_b - T_w}{T_s - T_w} = F_{eq} \tag{14.4}$$

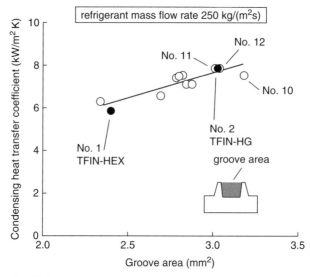

Figure 14.9 Relationship between groove area and condensing coefficient. (From Houfuku et al. [2001].)

Moser et al. [1998] show that

$$F_{eq} = 1.31(R^+)^{C_1} \operatorname{Re}_l^{C_2} \operatorname{Pr}_l^{-0.185} \tag{14.5}$$

where $R^+ = 0.0994\operatorname{Re}_{eq}^{-7/8}$, $C_1 = 0.126\operatorname{Pr}_l^{-0.448}$ and $C_2 = -0.113\operatorname{Pr}_l^{-0.563}$. The F_{eq} correction factor is relatively small for typical conditions. The $0.95 \le F_{eq} \le 1.15$ for $\operatorname{Pr}_l = 3.0$ and $30{,}000 < \operatorname{Re}_{eq} < 200{,}000$, and $\operatorname{Re}_l > 10{,}000$.

Moser et al. [1998] specify that the h_{sp} be calculated using the Petukhov equation (Equation 2.19) at an equivalent Reynolds number (Re_{eq}) defined by a force balance on the two-phase and the equivalent single-phase flow. To obtain the same wall shear stress as for the condensing flow, the single-phase Re_{eq} is given by:

$$\operatorname{Re}_{eq} = \left[\frac{1}{2} \frac{d^3 \rho_l}{f_l \mu_l^2} \left(\frac{dp}{dz} \right)_f \right]^{1/2} \tag{14.6}$$

where $f = 0.079\operatorname{Re}_{eq}^{-0.25}$. The $(dp/dz)_f$ is the frictional pressure gradient for the two-phase (condensing) flow. This may be calculated using experimental data, or predicted using an appropriate correlation for the frictional pressure gradient. Moser et al. [1998] used the Friedel [1979] correlation to predict the two-phase friction pressure gradient for round tubes having of 3.14 to 20 mm ID. Local and average condensation data were predicted within ±14%. Webb et al. [1998] used the model to predict the condensing coefficient for a variety of tube geometries and refrigerants. Webb [1999] also used the model to predict the condensation coefficient in microfin tubes having various helix angles. The condensation data were generally predicted within ±20%.

This model states that a tube having high single-phase performance will also give a high condensation or evaporation coefficient. Bhatia and Webb [2000] performed a parametric study of single-phase flow in microfin tubes (see Section 8.4) to define how variation of the geometry parameters will affect the single-phase turbulent heat transfer coefficient in a microfin tube. They show that as e/p_f increases, the fluid velocity is slowed in the interfin region. This will reduce the wall shear stress, and thus the heat transfer coefficient. Although the surface area will increase with decreasing p_f, the reduced heat transfer coefficient will result in an optimum p_f (or number of fins).

The condensation model neglected the possibility that surface tension force may act to drain condensate from the fin tips. If surface tension force is to be active, the microfins must not be submerged in the annular liquid film. For a 14.88-mm-diameter microfin tube, the cross-sectional area of the interfin region is only 6.25% of the total flow area. Thus, if the void fraction is less than 0.935 (1.0 − 0.0625), the tips of the microfins may protrude through the annular film. Using the Zivi [1964] equation, one can calculate the vapor quality (x_{fill}) that will result in 0.935 (1.0 − 0.0625) void fraction. This calculation shows that $x = 0.70$ at $\alpha = 0.935$ for R-22 at $T_{sat} = 25°C$. That means the surface tension effect should be negligible for $x < 0.70$. For the same fin pitch and height, the value of x_{fill} will be smaller as the tube diameter is reduced, because the cross-sectional area of the interfin region will become a larger fraction of the total flow area. Figure 14.10 shows the calculated value of x_{fill} for several tube diameters. The lower curve on Figure 14.10 is for a small-hydraulic-diameter (1.56 mm) flat extruded aluminum tube used in automotive condensers (Figure 13.33a); this tube is subject to surface tension drainage at considerably lower vapor quality than is the large-diameter microfin tube. Models for small-diameter microchannel tubes are discussed in a later section.

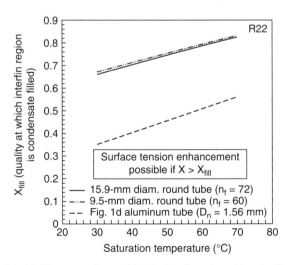

Figure 14.10 Predicted R-22 vapor quality at which the interfin region of the microfin tube will be filled (x_{fill}).

Liquid entrainment in the vapor core can also result in underprediction of the equivalent Reynolds number model. If entrainment exists, the underprediction should be more obvious for high mass velocities with high vapor qualities.

14.3.4 Convective Condensation in Special Microfin Geometries

Chamra et al. [1996a] investigated R-22 vaporization in seven three-dimensional cross-grooved microfin tubes. The same tubes were used for vaporization tests (Chamra et al. [1996b]) and are discussed in Section 13.4.3. The depth of the second set of grooves was varied 20 to 80% of the depth of the first grooves. Data were reported for condensation at 24°C in a 2.44-m-long test section with inlet and exit qualities 0.80 and 0.20, respectively. The condensation coefficient increased as the depth of the second groove increased. The highest performance was provided by the cross-grooved tube having 80% depth of the first groove and 27° helix angle (MCG-27). The condensation coefficient was 51% higher than the 15°, single helix tube (MX-15). The pressure drop, however, was about 77% higher at 140 kg/m²s. Goto et al. [2003], Ferreira et al. [2003], Tang et al. [2000a] provide additional data on refrigerant condensation in cross-grooved tubes. Xin et al. [1998] provide steam condensation data in cross-grooved tubes.

Muzzio et al. [1998] tested a 2-dimensional microfin tube having 2 different fin heights alternately, 0.23 and 0.16 mm, with an apex angle of 40° and helix angle of 18°. The vaporization data are discussed in Section 13.4.3. The tube yielded slightly higher condensation heat transfer coefficients and pressure drops compared with the conventional microfin tubes.

Ebisu [1999] tested a 7.0-mm herringbone pattern microfin tube using R-22 and R-407C and compared the performance with conventional two-dimensional microfin tubes. The vaporization results are discussed in Section 13.4.3 and the strip configurations are illustrated in Figure 13.21. Figure 14.11 shows the measured condensation coefficients for R-407C. The herringbone tube yielded up to 200% higher vaporization heat transfer coefficient than the single helix microfin tube. The herringbone tube had higher fin height (e = 0.22 mm vs. 0.18 mm), smaller fin pitch (p_n = 0.34 mm vs. 0.41 mm), and smaller helix angle (16° vs. 18°). The enhancement increased as the mass flux increased.

The effect of the microfin geometry on the R-410a condensation heat transfer coefficient in 7.0-mm-OD herringbone microfin tubes was investigated by Miyara et al. [2003]. They tested 5 herringbone microfin tubes with fin heights varying from 0.16 to 0.22 mm and helix angles varying from 8° to 28°. The heat transfer coefficient increased with fin height up to 0.18 mm. At higher fin heights, the heat transfer coefficients remain approximately the same as those of the 0.18-mm fin height. Pressure drop increased as the fin height increased. Tubes with larger helix angle yielded higher heat transfer coefficient and pressure drop. The heat transfer enhancement was noticeable at high mass flux (G = 400 kg/m²s) yielding 2 to 4 times higher heat transfer coefficient compared with the helical microfin tube having 0.21-mm fin height and 18° helix angle. Miyara et al. [2003] also investigated the effect of tube rotation. Tests were conducted with the tube rotated 90° about the axis. The

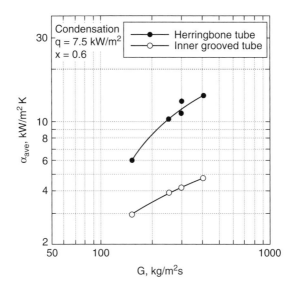

Figure 14.11 Condensation improvement offered by herringbone pattern for R-407C. (From Ebisu [1999].)

effect of rotation on the heat transfer coefficient and the pressure drop was negligible. Goto et al. [2001] provides additional R-410A condensation data in 8.0-mm-OD herringbone microfin tubes.

Graham et al. [1999] tested a 9.52-mm-OD axial microfin tube over a mass flux range 75 to 450 kg/m²s using R-134a. The tube had 60 microfins with 0.18-mm fin height. A helical microfin tube with the same geometry (n_f = 60 and e = 0.18 mm) and 18° helix angle was also tested. At 75 kg/m²s, the heat transfer coefficient of the axial microfin tube was smaller than that of the helical microfin tube. For mass fluxes higher than 150 kg/m²s, however, the axial microfin tube outperformed the helical microfin tube. The difference was more pronounced in the low-quality region. Both tubes yielded approximately the same pressure drops. Tang et al. [2000a] tested 9.52-mm-OD axial microfin tube having 72 fins of 0.2-mm height and 15° apex angle using R-22, R-410a, and R-134a. The heat transfer coefficients are compared with those of the 18° helical microfin tube having 60 fins of 0.2 mm height and 40° apex angle. The axial microfin tube yielded 5 to 10% higher heat transfer coefficients. Chiang [1993] also showed 10 to 20% higher heat transfer coefficients for the axial microfin tube than the 18° helical microfin tube of the same fin height.

14.4 FLAT TUBE AUTOMOTIVE CONDENSERS

Automotive air conditioners frequently use a flat tube as illustrated in Figure 14.12a. Use of a flat tube, rather than a round tube, provides lower air-side pressure drop. A flat tube presents less projected frontal area to the air stream, and hence will reduce the air-side pressure drop. The tubes contain membrane webs between the flat surfaces for pressure containment. Initially, these tubes contained a smooth inner surface as shown in Figure 14.12b. Based on the success of the round microfin tube, it is logical to apply microfins inside a flat extruded aluminum tube. Such an extruded aluminum tube with axial microfin grooves is shown in Figure 14.12c. Ohara [1983] describes an extruded tube with a corrugated insert brazed to the top and bottom walls of the extruded tube passages (Figure 14.12d). Rather than using an extruded tube, Figure 14.12e shows a welded tube with corrugated insert brazed to the tube wall. Figure 13.33 shows cross sections of more recent extruded flat aluminum microfin tubes. Figure 13.33 shows tubes having hydraulic diameters as small as 0.44 mm. These tubes are also used in automotive condensers.

If surface tension forces promote condensation from the fins for round microfin tubes, the same phenomena should occur for the Figure 14.12c aluminum tube with microgrooves. This is illustrated in Figure 14.12f. The role of surface tension force is different in the Figure 14.12c and e tubes. This is because the Figure 14.12d and e

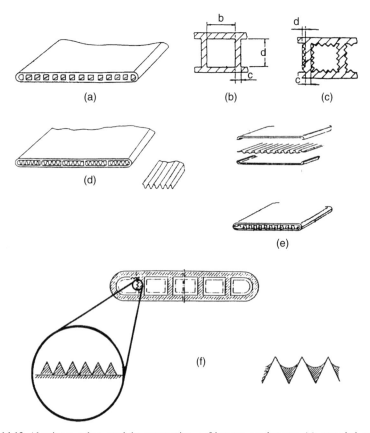

Figure 14.12 Aluminum tubes used in automotive, refrigerant condensers: (a) extruded tube with membrane partitions, (b) detail of part a tube, (c) detail of tube having microgrooves, (d) extruded tube with corrugated insert, (e) tube made by brazing a corrugated strip in a tube formed from flat strip and seam welded, (f) illustration of surface tension–drained condensation on the tips of the microgrooves.

tubes do not have sharp fin tips, on which a surface tension pressure gradient may act. However, surface tension may act to pull the condensate into the acute corners, where the corrugated insert joins with the base tube.

The automotive industry has used small-diameter, flat tubes in radiators for many years. Use of flat tubes for condensers is a recent trend that began in the late 1970s using extruded aluminum tubes with membrane webs. The condenser operates at a much higher pressure than does a radiator. Hence, the membrane webs illustrated in the Figure 14.12 tubes are required for an automotive condenser.

The flat tubes are used with air-side fins. If the minor diameter of a flat tube is reduced, the air-side pressure drop will be reduced. This has been known for many years, and trends in both the refrigeration and automotive industries have been toward smaller tube diameters. When the tube diameter is reduced, more refrigerant circuits are required. Figure 14.13 shows two methods of circuiting automotive refrigerant condensers. Figure 14.13a uses a one-pass, serpentine construction, with the refrigerant supplied directly to the flat tube. Figure 14.13b from Hoshino et al. [1991] shows such a condenser designed for three refrigerant passes, as illustrated in Figure 14.13c. The condenser has inlet and outlet headers, which consist of round aluminum tubes. Figure 14.13d shows the method of dividing the flow at the manifold headers to provide multipasses on the refrigerant side. Figure 14.13c shows that the number of tubes in parallel is reduced with each subsequent pass, which maintains high vapor velocity in each pass. A key advantage of the parallel flow, multipass design is that it allows use of a thinner tube (smaller minor tube diameter), because only part of the total refrigerant flow rate is carried by each tube. Use of a thinner tube results in lower air-side pressure drop. The extruded aluminum tube used in automotive condensers is typically 1.0-to-1.5mm minor tube diameter. The 180° bends required by the Figure 14.13a serpentine arrangement impose a minimum fin height requirement, to satisfy the minimum allowable bend radius. This limitation does not exist in the Figure 14.13b design.

14.4.1 Condensation Data for Flat, Extruded Tubes

Yang and Webb [1996] measured R-12 condensation heat transfer in flat extruded aluminum tubes. The tube outside dimensions are 16 × 3 mm. Two internal geometries were tested: one with plain inner surface and the other with microfins, 0.2 mm high. Data were presented for vapor qualities from 12 to 97%, mass velocity from 400 to 1400 kg/m²s, and heat flux from 4 to 12 kW/m². The saturation temperature was 65°C. The overall heat transfer coefficient was measured for water-to-refrigerant heat transfer, and the modified Wilson plot was used to determine the heat transfer coefficient for water-side flow in the annulus. The condensation coefficient was defined in terms of the total internal surface area. Webb and Yang [1995] extended the work to include R-134a. They found that the Shah [1979] correlation, which is widely used for the prediction of the refrigerant condensation coefficient in larger-diameter tubes, significantly overpredicted the data. However, the Akers et al. [1959] correlation agreed well with the plain tube data. The tube hydraulic diameter was used in these correlations. Akers et al. proposed that the condensation coefficient is

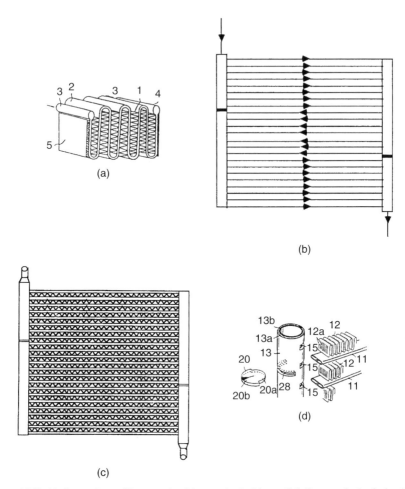

Figure 14.13 (a) Serpentine refrigerant circuiting method, (b) parallel flow method of circuiting, (c) illustration of Figure 14.13b condenser having three refrigerant circuits, (d) separator plate brazed into header for multipass circuiting.

equal to that for an equivalent all-liquid flow having the same wall shear stress as that of the condensing flow. Akers et al. proposed that the mass velocity for the equivalent all-liquid flow that gives the same wall shear stress as the condensing flow is given by

$$G_{eq} = G\left[(1-x) + x(\rho_l / \rho_v)^{1/2}\right] \tag{14.7}$$

The equivalent all-liquid Reynolds number corresponding to G_{eq} is given by $Re_{eq} = G_{eq}D_h/\mu_l$. Assuming Reynolds analogy, the heat transfer coefficient for this equivalent all-liquid flow should be that same as the annular film condensing flow. The resulting Akers correlation is

$$\frac{hd}{k_l} = 0.0265 \, \text{Re}_{eq}^{0.8} \, \text{Pr}_l^{1/3} \tag{14.8}$$

Akers et al. [1959] specified the validity of Equation 14.8 to be $\text{Re}_{eq} > 50,000$. However, Yang and Webb [1996] found that it predicted their data well down to their lowest Reynolds numbers. Figure 14.14 shows that, at the same Re_{eq}, the condensation coefficient for the microfin tube is greater than that for the plain tube for vapor qualities greater than 0.5. Yang and Webb [1996] proposed that the surface tension force is effective in enhancing the condensation coefficient for vapor qualities larger than 0.5. The surface tension–induced enhancement was particularly strong at lower mass velocities, where the convective condensation component is reduced. Figure 14.15 from Yang and Webb [1997] shows a conceptual drawing of the condensate film on the fin surface at high and low vapor qualities. For vapor qualities greater than 0.5, Figure 14.15a shows that the condensate film is sufficiently thin where that part of the fin height penetrates into the vapor region. Thus, it is not flooded by condensate, which will occur at lower vapor qualities, as shown in Figure 14.15b. A surface tension–induced pressure gradient acts to drain condensate from the small-radius fin tip into the concave channel between the fins. The subcooled heat transfer coefficient for both geometries were well predicted using the Petukhov equation (Eq. 2.19) with hydraulic diameter. Webb and Ermis [2001] extended the study to hydraulic diameters down to 0.44 mm. Kim et al. [2000, 2003] obtained R-22 and R-410a condensation data for similar tube geometries. The studies of Webb and Ermis [2001] and Kim et al. [2000, 2003] generally confirmed the findings by Yang and Webb [1996].

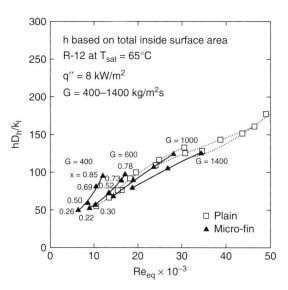

Figure 14.14 Comparison of condensation Nusselt number in flat 3 × 16 mm plain and microfin tubes based on equivalent Reynolds number. (From Yang and Webb [1996].)

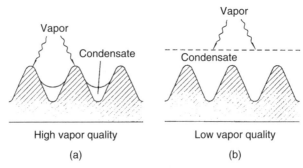

Figure 14.15 Illustration of condensate film on microfins: (a) $x > 0.5$, for which the fin tip penetrates into the vapor region, (b) condensate flooded fins. (From Yang and Webb [1997].)

Koyama et al. [2003] obtained R-134a condensation data in 1.11- and 0.80-mm-hydraulic-diameter aluminum extruded tubes. They found that the Haraguchi et al. [1994] correlation developed for large-diameter tubes highly overpredicted their data.

Wang et al. [2002] obtained R-134a condensation data in a 1.46-mm-hydraulic-diameter extruded aluminum tube. A flow visualization study was also conducted using R-134a for the simulated machined test section with the viewing window on top of the test section. For mass flux from 75 to 350 kg/m²s, slug, wavy/slug, wavy, wavy/annular, and annular flow patterns were observed. Comparison with the Breber et al. [1980] flow regime map revealed that the transition from stratified to annular flow occurred at a lower superficial gas velocity (0.24 m/s) compared with the original recommendation (0.46 m/s). The heat transfer data in the annular region were well predicted by the Akers et al. [1959] correlation. Most of the data (83.7%) were predicted within ±25%. For the heat transfer data in the stratified region, Jaster and Kosky [1976] correlation predicted 80% of the data within ±20%.

14.4.2 Other Predictive Methods of Condensation in Flat Tubes

The condensation heat transfer studies in small-hydraulic-diameter aluminum extruded tubes (Yang and Webb [1996], Webb and Yang [1995], Kim et al. [2000, 2003], Koyama et al. [2003]) show that the Shah [1979] correlation significantly overpredicts the data. The Shah correlation has been evaluated by Moser et al. [1998], who showed that the Shah correlation is valid for condensation in plain round copper tubes with diameters as low as 3.14 mm for refrigerants condensing at saturation temperatures below 40°C. The evaluation included refrigerants R-11, R-12, R-113, R-22, R-125, R-134a, and R-410A). As reported in Webb [1999], Zhang [1998] has further evaluated the Shah correlation using R-134a at 40°C and 65°C in plain round copper tubes of 6.2 and 3.2 mm, and in $D_h = 2.13$ mm aluminum extruded tube. For the 3.2-mm-ID round tube and the aluminum extruded tube, the Shah correlation increasingly failed as the saturation temperature increases above 40°C. The Shah correlation is given by

$$\mathrm{Nu} = \mathrm{Nu}_l \left[1 + a \left(\frac{p}{p_{\mathrm{cr}}} \right)^{-b} \left(\frac{1}{1-x} \right)^{0.76} \right] \tag{14.9}$$

where $a = 3.8$ and $b = 0.38$ and Nu_l is for all liquid flow. The Shah database includes data for $0.001 \leq p/p_{\mathrm{cr}} \leq 0.44$. All of the data were in the range $0.10 \leq p/p_{\mathrm{cr}} \leq 0.44$, except for the R-11 data, which had $p/p_{\mathrm{cr}} \leq 0.001$. Using the original Shah database with the R-11 data excluded, Zhang [1998] found that changing the exponent on p/p_{cr} from 0.38 to 0.8 and using $a = 2.35$ substantially improved the correlation of the data in the Shah database. Further, the 3.2-mm-ID round tube and the aluminum extruded tube condensation data measured by Zhang [1998] were well predicted. This modified Shah correlation works very well for $0.10 < p/p_{\mathrm{cr}} < 0.80$.

Many condensation studies in small-hydraulic-diameter extruded aluminum tubes reveal that the Akers et al. [1959] correlation reasonably predicts the data. However, the Akers equivalent Reynolds definition is theoretically flawed, as discussed by Moser et al. [1998]. Although the Akers correlation seems to work well for tube diameters below approximately 4.0 mm, it increasingly underpredicts as the tube diameter increases above 4.0 mm.

Webb et al. [1998] used the Moser et al. [1998] model to predict the R-134a condensation coefficient in the 3.2-mm-ID round tube and the $D_h = 2.13$ mm aluminum extruded tube. The saturation temperatures were 40°C and 65°C. The $\mathrm{Re}_{\mathrm{eq}}$ was predicted using measured pressure drop, and the single-phase heat transfer coefficient was calculated using the Petukhov equation. Local and average condensation data were predicted within ±15%. Zhang and Webb [2001] found that when the Friedel [1979] correlation was used for the pressure gradient, the data were overpredicted by 20 to 50%. This suggests that the Friedel equation does not accurately predict the pressure drop in small-diameter tubes. Zhang and Webb [2001] provide a correlation for two-phase pressure drop in small-diameter tubes.

Kim et al. [2003] compared the R-22 and R-410A condensation data in a $D_h = 1.41$ mm aluminum extruded tubes with the Moser et al. [1998] model using Zhang and Webb [2001] two-phase pressure drop correlation. The prediction was reasonably good (±30%).

Koyama et al. [2003] compared the R-134a data in 1.11- and 0.80-mm-hydraulic-diameter aluminum extruded tubes with the correlation of Moser et al. [1998]. Good agreement (±30%) was obtained except for the low mass fluxes. They modified the large-diameter tube model (Haraguchi et al. [1994]) by incorporationg the small-diameter frictional pressure drop correlation of Mishima and Hibiki [1995].

Wang et al. [2002], noting that annular and stratified flows are possible for the condensation in aluminum extruded tubes, proposed the following additive model:

$$\mathrm{Nu} = f_{\mathrm{ann}} \mathrm{Nu}_{\mathrm{ann}} + (1 - f_{\mathrm{ann}}) \mathrm{Nu}_{\mathrm{str}} \tag{14.10}$$

where f_{ann} is the fractional length of the tube experiencing the annular flow, $\mathrm{Nu}_{\mathrm{ann}}$ and $\mathrm{Nu}_{\mathrm{str}}$ are the Nusselt number for the annular and stratified regions, respectively.

The Nu_{ann} was modeled following the procedure by Traviss et al. [1973] with the friction multiplier and the dimensionless boundary layer profile empirically modified for the tested tube geometry. The Nu_{str} was modeled using the Chato [1962] correlation for film condensation and the Dittus–Boelter equation for liquid flow. The Zivi [1964] void fraction model was used for the stratification height. The transition from stratified to annular flow was assumed to occur at the vapor superficial velocity $j_v^* = 0.2$ m/s based on the flow visualization results. The additive model predicted their data successfully (79.2% of the data within ±10%).

Yang and Webb [1997] developed a semiempirical model for condensation inside flat extruded aluminum tubes having axial microfins. The key dimensions of the fins include fin pitch (p_f), fin height (e), tip radius (r_b), and apex angle (β). Figure 14.15 illustrates the cross section of the liquid film between two microfins. At low vapor quality, the microfins are flooded by the condensate, so surface tension force cannot be active. When the fin tips are not flooded, both vapor shear and surface tension forces are important. For the microfin geometry tested by Yang and Webb [1996], Yang and Webb [1997] found that fin tips would be flooded with condensate if $x < 0.5$, when the Zivi [1964] void fraction model was used. The resulting heat transfer coefficients averaged over the total surface area can be obtained from

$$h = h_u \frac{A_u}{A} + h_f \frac{A_f}{A} \qquad (14.11)$$

where A is the total heat transfer surface area, A_u is the unflooded area, A_f is the flooded area, and h_u and h_f are the heat transfer coefficients in the unflooded region and flooded region, respectively. Yang and Webb [1997] provides an expression for the flooded fraction (A_f/A), which depends on the microfin tube geometry and the void fraction. The expression was obtained assuming a circular liquid–vapor interface as illustrated in Figure 14.15. The Akers et al. [1959] correlation was used for h_f, based on the study of Yang and Webb [1996]. For the unflooded region, both vapor shear and the surface tension forces are active. Figure 14.15 illustrates the vapor flow parallel to axial microfins. The total wall shear stress (τ_w) consists of that by vapor shear (τ_z) and that by surface tension (τ_s). Solving the momentum equation and consideration of the heat-momentum transfer analogy in the liquid film resulted in the following expression for the h_u.

$$h_u^2 = h_{sh}^2 + h_{st}^2 \qquad (14.12)$$

$$h_{sh} = \tau_z \frac{k_l}{D_h} Re_{eq} Pr_l^{1/3} \frac{\rho_l}{G_{eq}^2} \qquad (14.13)$$

$$h_{st} = C \frac{\tau_z - \tau_i}{dp/dz} k_l \frac{d(1/r)}{ds} \frac{Re_{eq} Pr_l^{1/3}}{We} \qquad (14.14)$$

Here, $h_{\rm sh}$ and $h_{\rm st}$ denote the vapor shear and the surface tension contributions, respectively. Yang and Webb [1997] calculated the vapor shear component using the Akers et al. [1959] correlation, and the surface tension component using a linear surface tension drainage model. The terms τ_s and dp/dz were obtained from the measured single-phase and two-phase pressure drop data. An empirical constant $C = 0.0703$ was obtained from the best fit of experimental data. The model predicted 95% of Yang and Webb [1996] microfin data within $\pm16\%$. Kim et al. [2003] compared the R-22 and R-410A condensation data in a $D_h = 1.56$ mm aluminum extruded microfin tube with the Yang and Webb [1997] model modified with Moser et al. [1998] model for the calculation of vapor shear component. The prediction was reasonably good ($\pm30\%$).

14.5 PLATE-TYPE HEAT EXCHANGERS

Figure 14.16 shows the plates used in common plate-type heat exchangers. Although the plate-type exchanger was developed for liquid–liquid applications, it is now frequently used for condensation and vaporization. Figure 14.16a shows the cross section of a plate heat exchanger developed for condensation, with heat rejection to a single-phase coolant. Condensation occurs in the "A" channels, with the coolant in the "B" channels. The recommended corrugation pitch (p) and height (e) are $1 \le p \le 2$ mm, and $0.3 \le e \le 0.6$ mm, respectively. Surface tension pulls the condensate into the crevices formed by the intersection of the corrugation and the adjacent flat plate. This maintains a thin film on the flat plate region between corrugations. Figure 14.16b shows the overall heat transfer coefficient for ammonia and R-114 as a function of the corrugation pitch (p). The figure shows that the optimum corrugation pitch is approximately 1.0 mm for ammonia and 1.5 mm for R-114. The condensing coefficient significantly decreases for pitches greater than 3.0 mm. This is because there are fewer channels to carry the condensate, so thicker condensate film exists on the flat plate regions.

14.6 NONCONDENSIBLE GASES

When noncondensibles are present, an additional thermal resistance is introduced in the gas at the vapor–liquid interface. Mixing in the gas film will substantially reduce this thermal resistance. The maintenance of high vapor velocities, or special surface geometries, which promote a higher heat transfer coefficient in this gas film will substantially alleviate the performance deterioration due to noncondensibles.

Enhancement should be beneficial when noncondensibles are present. This possibility was investigated by Chang and Spencer [1971], who condensed R-12 in a vertical annulus geometry. Noncondensibles cause a mass transfer resistance at the liquid–vapor interface, and depress the liquid–vapor interface temperature. The O-rings (diameter unspecified) were spaced at 25-m axial pitch on the inner annulus condensing surface. Mixing of this gas boundary layer by the O-ring roughness

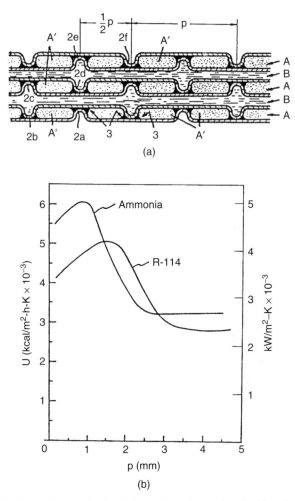

Figure 14.16 Plate heat exchanger developed for condensation against liquid coolant by Uehara and Sumitomo [1985]. (a) End view showing plate configuration with condensing vapor (A) and coolant (B) streams, (b) measured overall heat transfer coefficient for condensation of ammonia and R-114.

should reduce this resistance, increase the interface temperature, and increase the condensation coefficient. Figure 14.17 shows the dimensionless condensing coefficient of R-12 with noncondensible gas in the annulus with and without O-rings. As shown in Figure 14.17, the O-rings cause considerable enhancement at low Reynolds (Re_d) number, but not at high Reynolds numbers. The entering vapor velocity increases with increasing Re_d. The lack of enhancement at high Re_d is believed to be due to surface roll waves, which naturally occur because of an interaction between the light noncondensible layer and the heavy condensing vapor. As the molecular weight of the noncondensibles approaches that of the condensing vapor, enhancement due to the O-rings decreases. Chang and Spencer found negligible enhancement for condensation of pure R-12.

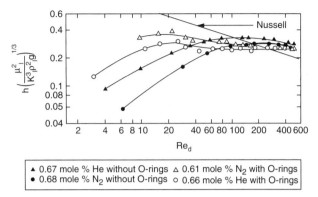

Figure 14.17 Effect of spaced ring roughness on condensation of an R-12/air mixture for vertical flow in an annulus. (From Chang and Spencer [1971].)

The enhancement concept discussed in Section 16.2.2 is directly applicable to the noncondensible gas problem. This concept (Figure 16.2) consists of a transverse-rib roughness, which is displaced from the tube wall and mixes the vapor boundary layer. This is probably a better approach for dealing with noncondensibles than the wall-attached Chang and Spencer O-rings.

14.7 PREDICTIVE METHODS FOR CIRCULAR TUBES

Predictive equations have been developed for several of the enhancement techniques. These equations are based on modifications to condensation in plain tubes. Schlager et al. [1990a] provide a literature survey of correlations for condensation inside tubes. The remarks and warnings concerning predictive equations for vaporization in tubes given in Section 13.7 also apply here. Only heat transfer correlations are presented here. Refer to the references for the accompanying pressure drop correlations.

14.7.1 High Internal Fins

The first correlation for internal fins was developed by Vrable et al. [1974] based on their R-12 data. They proposed an empirical correlation of their R-12 data for one internal fin geometry.

The Royal and Bergles [1978a] correlation of their steam data in four internally finned tubes is based on modifying the Akers et al. [1959] correlation for smooth tubes. The correlation is

$$\frac{hD_h}{k} = 0.0265 \left(\frac{G_{eq}D_h}{\mu_l} \right)^{0.8} Pr_l^{0.33} \left[160 \left(\frac{e^2}{sD_h} \right)^{1.91} + 1 \right] \qquad (14.15)$$

The modification is the term in square brackets, which includes the fin geometry parameters and is empirical. Because the data were taken only for steam at 1.0 bar, there is no guarantee that it will apply to other pressures or fluids. It is likely that surface tension drainage effects occur on the fin tips and fin base, as discussed in Chapter 12, which are not accounted for in Equation 14.15. Note that the equation does not include fin efficiency.

Luu and Bergles [1980] found that Equation 14.15 did not predict their R-113 data very well. They proceeded to develop another correlation for the R-113 data, which did not include the steam data. This correlation was based on an empirical correction to the Boyko and Kruzhilin [1967] correlation for smooth tubes. This correlation should have the same limitations as for the steam correlation.

Kaushik and Azer [1988] developed a correlation based on steam, R-113, and R-11 data. The correlated data included the steam data of Royal and Bergles [1978a], the R-113 data of Luu and Bergles [1980] and of Said and Azer [1983], and the R-11 data of Venkatesh [1984]. The correlation is a strictly empirical power-law regression type given by

$$\frac{hd_i}{k_l} = C\left(\frac{G_{eq}d_i}{\mu_l}\right)^{0.507}\left(\frac{\Delta x d_i}{L}\right)^{0.198}\left(\frac{p}{p_{cr}}\right)^{-0.14}F_1^{n_1}F_2^{n_2} \qquad (14.16)$$

where F_1 and F_2 are geometric parameters (see the nomenclature at the end of the chapter) used by Carnavos [1980] to correlate his single-phase flow data. For $F_1 \le$ 1.4, $n_1 = 0.874$ and $n_2 = -0.814$. For $F_2 > 1.4$, $n_1 = 4.742$ and $n_2 = 0.0$. Again, this empirical correlation does not account for probable surface tension drainage effects or fin efficiency. However, it is probably the most general of those presented. The correlation predicted 71% of the data points within $\pm 30\%$.

14.7.2 Wire Loop Internal Fins

Honda et al. [1988] developed a rationally based correlation for the friction and heat transfer in an annulus containing wire loop fins discussed in Section 14.2.1. The heat transfer correlation is shown in Figure 14.3b, which shows the heat transfer results for R-11 and R-113 plotted in the form Nu/Nu_c vs. the Martinelli parameter, X_{tt}. The Nusselt number is defined in terms of the wire diameter, $Nu_c = he/k_l$. The normalizing parameter in Figure 14.3b is defined as

$$Nu_c = (Nu_s^4 + Nu_v^4)^{1/4} \qquad (14.17)$$

where Nu_s and Nu_v are calculated by correlations for the surface tension and vapor shear controlled regimes, respectively. These correlations are given by Honda et al. [1988]. Figure 14.3b shows that the vapor shear controlled regime prevails for $X_{tt} \le$ 0.1. In the surface tension–controlled regime, condensation occurs on the wire fins, and is pulled to the fin root, and then it is gravity-drained from the tube. In the vapor shear–controlled regime, the model treats condensation on the wire fins similar to shear controlled condensation normal to a cylinder. This is a very comprehensive model, and is well justified by theoretical reasoning.

Kaushik and Azer [1989] worked to develop an analytically based model for condensation in internally finned tubes. This model includes surface tension drainage effects. This rationally based model predicted the steam data of Royal and Bergles [1978a] within ±30%. However, it overpredicted the R-113 data of Said and Azer [1983]. Sur and Azer [1991] developed an analytical model to predict the pressure drop in internally finned tubes.

14.7.3 Twisted Tapes

The Royal and Bergles [1978a] correlation for steam condensation in tubes is based on modifying the Akers et al. [1959] correlation for smooth tubes. The correlation assumes no thermal contact resistance of the tape at the tube wall. The correlation is

$$\frac{hD_h}{k_l} = 0.0265 \left(\frac{F_t G_{eq} D_h}{\mu_l} \right)^{0.8} \mathrm{Pr}_l^{0.4} F_{tt} \tag{14.18}$$

The F_t in the Reynolds term accounts for the velocity of swirling flow at the wall, and is given by

$$F_t = (1 + \tan^2 \alpha)^{1/2} = \left[1 + (\pi / 2y)^2 \right]^{1/2} \tag{14.19}$$

The term F_{tt} accounts for the fin efficiency of the tape and is given by

$$F_{tt} = \frac{(\pi d_i - 2t_t + \eta_f d_i)}{\pi d_i} \tag{14.20}$$

where η_f is the fin efficiency. Their steam data for two tapes were correlated within ±30%. Because Royal and Bergles [1978a] state that their tape did not have good thermal contact with the tube wall, their assumption of zero contact resistance in the correlation raises questions concerning validity of the correlation. It is doubtful that typical twisted tapes will have perfect thermal contact with the tube wall. Assuming no heat transfer from the tape, one would set $\eta_f = 0$ in Equation 14.20.

Luu and Bergles [1980] correlate their R-113 data using a modification of the Boyko and Kruzhilin [1967] smooth tube correlation. The modification is similar to that of Royal and Bergles [1978a], except they assume that only 50% of the tape thickness is in contact with the wall. Again, most of the data were correlated within ±30%.

Said and Azer [1983] found that the Luu and Bergles [1980] correlation under-predicted their R-113 data. Said and Azer [1983] proposed a different correlation, based on the steam data of Royal and Bergles [1978a], and the R-113 data of Luu and Bergles [1980] and of Said and Azer [1983]. This empirical correlation is

$$\frac{hd_i}{k_l} = 0.023\,\mathrm{Re}_{d,l}^{0.8}\,\mathrm{Pr}_l^{0.4}(0.665+1.86p_r^{-0.38})\left[1+0.019\left(\frac{\mathrm{Re}_{d,l}}{y}\right)^{0.26}\right] \qquad (14.21)$$

Equation 14.21 is a modification of the Dittus–Boelter equation for turbulent, single-phase heat transfer in tubes. The first modifying parameter in parentheses accounts for pressure, and the term in square brackets accounts for the tape twist. The Reynolds number is based on the nominal tube diameter, rather than on the hydraulic diameter; it is defined by $\mathrm{Re}_d = d_iG\mu_l$. Equation 14.21 predicted all three data sets within $\pm 30\%$. Note that the correlation makes no assumptions concerning the fin efficiency of the tape. This correlation is recommended because of its ability to predict data for several fluids, and because an assumption concerning the fin efficiency is not required.

14.7.4 Roughness

Luu and Bergles [1980] provide an interesting correlation of their helical rib roughness R-113 data. The correlation is based on the single-phase roughness correlation of Webb et al. [1971] discussed in Chapter 9. They correlated their data on two roughness geometries within $\pm 30\%$. The reader should consult Luu and Bergles [1980] for use of the correlations.

14.7.5 Microfins

Cavallini et al. [2002] provides an extensive literature survey and critical discussion of existing correlations to predict the condensation coefficient in microfin tubes. Cavallini et al. [1995] developed one of the first successful correlations for microfin tubes. This is based on approximately 300 data points for seven refrigerants condensing in microfin, low-fin ($e/d_i > 0.04$), and cross-grooved microfin tubes. Nearly all of the data points were for the average condensing coefficient with high entering and low-leaving vapor quality (e.g., 80 to 20%), and the average vapor quality is typically 0.50 to 0.60. The vast majority of the data points are for the single-grooved microfin tube. Most of the tubes tested were commercially produced single-groove microfin tubes, which do not span a wide range of dimensional variables. The correlation is an extension of the Cavallini and Zecchin [1974] correlation for plain tubes, which is

$$\mathrm{Nu}=0.05\,\mathrm{Re}_{eq}^{0.8}\,\mathrm{Pr}_l^{1/3} \qquad (14.22)$$

where the equivalent Reynolds number Re_{eq} is calculated as specified by Akers et al. [1959] and per Equation 14.7. The Cavallini et al. [1995] microfin correlation adds an empirical geometry factor (total/nominal surface area at fin tips), and the (Bo Fr) group to account for surface tension effects (Bond number) and stratification (Fr). Their correlation for single and cross-grooved tubes is

Table 14.6 Exponents of Equation 14.23

Exponent	Low-Fins ($e/d_i > 0.04$)	Microfins ($e/d_i < 0.04$)	Cross Grooved
s	1.4	2	2.1
t	-0.08	-0.26	-0.26

$$\text{Nu}=0.05\,\text{Re}_{eq}^{0.8}\,\text{Pr}_{l}^{1/3}(\text{Bo}\cdot\text{Fr})^{t}\,\text{Rx}^{s} \tag{14.23}$$

where Bo is the the Bond number, Fr is the Froude number, and the diameter in Nu and Re is the diameter between fin tips (d_t), which is somewhat unconventional. Others typically use the diameter to the base of the fins (d_i). The Rx parameter accounts for the microfin geometry and is defined as Rx $= A_{tot}/A_{Dt}$, where A_{tot} is the total internal surface area, $A_{Dt}/L = \pi d_t$. The exponent (s) on Rx depends on the basic microgroove geometry, and is given in Table 14.6. The Re$_{eq}$ is defined for the equivalent all-liquid flow using Equation 14.7.

Note that the exponents on the microfin and cross-grooved tubes are nearly equal. The Bond number is the ratio of gravity and surface tension forces, and is defined as $gp_l \pi ed_t/8\sigma n_f$. Equation 14.23 suggests that the Nu will increase with increasing surface tension. Because the range of surface tension values for the test fluids is very small, one should be careful in drawing this conclusion. Equation 14.23 correlated the data generally within $\pm20\%$.

Cavallini et al. [1999] also developed a model to predict pressure drop in microfin tubes. The model is a modified plain tube correlation based on the plain tube correlations of Friedel [1979] and Sardesi et al. [1982]. The model uses a two-phase multiplier (ϕ_o^2) given in the reference, and the associated single-phase friction factors (f_{vo} and f_{lo}) are determined using an equivalent friction factor based on the Moody friction factor vs. Reynolds number chart for flow in commercially rough pipes. They developed an empirical correlation to define the "equivalent Moody chart roughness" (e_m/d_t) of the microfin tubes in their database. The e_m/d_t was obtained from regression analysis and is given by

$$\frac{e_m}{d_t} = A\frac{e/d_t}{0.1+\cos\alpha} \tag{14.24}$$

where A is an empirical constant, e is microfin height, d_t is the fin-tip diameter, and α is the helix angle. Two different values were found for the constant $A = 0.18$ for condensation and $A = 0.30$ for evaporation. The following equation developed by Colebrook [1939] may be used to calculate the Moody friction factor

$$\frac{1}{2f^{1/2}} = 1.74 - 2\ln\left(\frac{2e_m}{d} + \frac{18.7}{2\,\text{Re}\,f^{1/2}}\right) \tag{14.25}$$

Note that Equation 14.25 does not account for apex angle or the fin pitch. Cavallini et al. [1999] show that Equation 14.25 predicts all of the data within $\pm20\%$.

Webb [1999] applied the equivalent Reynolds number model of Moser et al. [1998] to predict the condensing coefficient in 14.8-mm-ID round microfin tubes. R-22 condensation at 27°C and single-phase data were taken on 5 microfin tubes, whose helix angle varied from 0° to 34°. The condensation and single-phase data for most of the tubes are reported by Chamra et al. [1996a] and Brognaux et al. [1997], and are discussed in Chapter 14 and Chapter 9, respectively. The Re_{eq} was predicted using measured pressure drop. More than 90% of the data are predicted within ±15%. Note that this model does not account for the possible effect of surface tension–induced condensate drainage from the fin tips. The cross section of the interfin region is only 6.25% of the total flow area for the 14.88-mm-diameter microfin tube. Using the Zivi [1964] equation, Webb [1999] calculated the vapor quality (x_{fill}) that will result in 0.935 void fraction. Below this void fraction, fins will be flooded if the liquid–vapor interface is smooth. This calculation showed that $x = 0.70$ at 0.935 void fraction for R-22 at $T_{sat} = 25$°C. That means that the surface tension effect should be negligible for $x < 0.7$. Webb [1999] also conducted a similar evaluation of the vapor quality at which the interfin space will be condensate filled for other microfin geometries and tube diameters. Figure 14.10 shows the predicted vapor quality (x_{fill}) at which the interfin region of the microfin will be filled. The figure shows the calculated results for R-22 condensing in three different microfin geometries, and spans 25 to 75°C saturation temperature. The void fraction was calculated using the Zivi's [1964] equation. As saturation temperature increases, x_{fill} increases. Note the much smaller values of x_{fill} for the smaller-hydraulic-diameter flat extruded aluminum tube.

Tang et al. [2000b] investigated the ability of existing correlations to predict condensation of R-22, R-134a, and R-410A in smooth tubes and found that all overpredicted their R-410A data. They developed a new correlation for smooth tubes based on a modification of the Shah [1979] correlation. This correlation predicted 173 data points for the 3 refrigerants within ±15%. However, it underpredicted the data of Dobson and Chato [1998] by about 7%. They continued to develop a correlation for axial and helical single-groove and cross-groove microfin tubes, which was based on a modification of their smooth tube correlation. This correlation predicted 596 data points for the 3 refrigerants with the following mean deviations: helical (6.0%), axial (9.3%), and cross-groove (6.5%). They also give a correlation to predict the single-phase heat transfer coefficient in the same three geometries.

Nozu and Honda [2000] developed a complex numerical model that predicts the condensation coefficient in an annular flow regime of microfin tubes. The model assumes that all the condensate flows through the grooves with a smooth liquid–vapor interface. The vapor shear term was added to the surface tension to account for the effect of vapor shear on convective condensation. The condensate film was segmented into thick and thin film regions. In the thin film region, the condensate is assumed to be drained by the combined surface tension and vapor shear forces. In the thick film region, the condensate is assumed driven by the vapor shear force. A similar procedure as the external condensation on integral fin (see Honda and Nozu [1987] in Chapter 12) was taken to formulate and solve the problem. Readers are encouraged to refer to Chapter 12 for Honda and Nozu's [1987] analysis on external condensation on integral fin tubes.

A stratified flow model was later developed by Honda et al. [2002]. The height of the stratified condensate was estimated from Taitel and Dukler's [1976] analysis on the friction of the stratified flow. The effect of shear force was assumed negligible. For the upper part of the tube exposed to the vapor flow, laminar film condensation due to the combined effects of gravity and surface tension forces was assumed. For the lower part of the tube exposed to the condensate flow, the heat transfer coefficient was estimated from Carnavos's [1980] single-phase forced convection correlation discussed in Chapter 8. Again, a similar procedure as the external condensation on integral fin (Honda and Nozu [1987]) was taken to formulate and solve the problem. The theoretical predictions by the stratified flow model and the previous annular flow model (Nozu and Honda [2000]) were compared with the experimental data for five tubes and five refrigerants taken by the authors. The mass flux range was from 100 to 300 kg/m²s. The stratified model successfully predicted the data—even to the highest vapor quality ($x = 0.9$), where the flow is estimated to be annular. The annular flow model highly underpredicted the low-quality data. Wang et al. [2002] refined the stratified flow model by accounting for the curvature of the stratified condensate due to the surface tension force. No criteria for the usage of the two models were provided. Instead, they recommend that one use the higher of the predictions from the two models. Their model predicted the condensation data of four refrigerants in seven microfin tubes with 14.3% rms error (Wang and Honda [2003]).

Additional correlations have been provided by Yu and Koyama [1998] and Goto et al. [2003]. Yu and Koyama [1998] applied the Haraguchi et al. [1994] smooth tube correlation to microfin tubes with minor modification of inner surface area increase. More refinement of the model was made by Goto et al. [2003] by adopting the two-phase multiplier obtained from the pressure drop tests in microfin tubes. Miyara et al. [2000] provide a modified version of Yu and Koyama [1998] correlation applicable to herringbone microfin tubes.

14.8 CONCLUSIONS

Clearly, the microfin tube provides the highest heat transfer performance and the lowest pressure drop of the competing internal enhancements. Considerable work has been done to develop predictive models for condensation in the microfin tube, as a function of surface geometry and fluid properties. Also, much work has been done to identify optimized microfin geometries for tube diameters between 4 and 15 mm. It is a very cost-effective enhancement concept and is routinely used in air-conditioning and refrigeration condensers.

The twisted tape is a relatively low enhancement device that may be of value only to increase capacity of an existing system. Rationally based semiempirical correlations have been developed for the twisted tape. However, the correlations for the high internal fins discussed in Section 14.2.1 are largely empirical, and do not seem to predict data other than the developers' data very well.

NOMENCLATURE

A Heat transfer surface, m^2 or ft^2

A_c Cross-sectional flow area, m^2 or ft^2

d Tube diameter, d_i (inside diameter, or diameter to the base of enhancement), d_{im} (internal diameter if enhancement material is returned to the tube wall), d_o (outside diameter), d_t (diameter between tips of internal fins), m or ft

e Fin height, e_m (Equation 14.24), m or ft

D_c Diameter of helix, m or ft

D_h Tube hydraulic diameter, m or ft

F_1 Geometric parameter used in Equation 14.16, $[(d_i/d_{im})(1-2e/d_i]^2$, dimensionless

F_2 Geometric parameter used in Equation 14.16, d_iD_h/d^2_{im}, dimensionless

e Roughness height, corrugation depth, or wire diameter, m or ft

G Mass velocity in tube, G_v (of vapor component), G_l (of liquid component), G_{eq} (Equation 14.7), kg/m^2-s or lbm/ft^{2}s

h Heat transfer coefficient, W/m^{2}K or Btu/hr-ft^2-°F

h_{cv} Convective component of two-phase heat transfer coefficient, W/m^2-K or Btu/hr-ft^2-°F

h_{nb} Nucleate boiling component for convective vaporization, W/m^2-K or Btu/hr-ft^2-°F

h_s Heat transfer coefficient on smooth surface, W/m^2-K or Btu/hr-ft^2-°F

H Axial distance for one fin (or flute, or twisted-tape) 180° revolution, m or ft

j^*_v Vapor superficial velocity, $[(xG)/(gd\rho_v(\rho_l-\rho_v))^{0.5}]$, m/s or ft/s

L Length of vertical condensing surface; tube length, m or ft

LMTD   Log-mean temperature difference, K or °F

n_f Number of flutes or internal fins in tube, none

N Number of horizontal tube rows in depth, none

Nu   Nusselt number, hd/k_l, dimensionless

Nu$_c$ Nusselt number defined by Equation 14.17, dimensionless

Nu$_e$ Nusselt number based on wire diameter, he/k_l, dimensionless

p_f Rib or fin pitch normal to the fins, m or ft

p_n Rib or fin pitch normal to the tube axis, m or ft

p Pitch of enhancement surface elements, m or ft

p Fluid pressure, kPa or lbf/ft^2

p_{cr} Critical fluid pressure, kPa or lbf/ft^2

p_r Reduced pressure, p/p_{cr}, dimensionless

Δp Pressure drop, kPa or lbf/ft^2

P Pumping power, W or hp

Pr   Prandtl number, dimensionless

q Heat flux, W/m^2 or Btu/hr-ft^2

Q Heat transfer rate, W or Btu/hr

Re	Reynolds number, $\mathrm{Re}_d\,(=Gd_i/\mu_l)$, $\mathrm{Re}_{\mathrm{Dh}}\,(=GD_h/\mu_l)$, $\mathrm{Re}_{d,l}\,[=G(1-x)d_i/\mu_l]$, $\mathrm{Re}_{\mathrm{eq}}$ (defined by Equation 14.6 or 14.7), dimensionless
Re_L	Condensate Reynolds number, $4\Gamma/\mu_l$, dimensionless
s	Spacing between fins, m or ft
t	Fin thickness, fin base thickness (t_b), fin tip thickness (t_t), m or ft
T	Temperature, saturation (T_s), wall temperature (T_w), mixed fluid temperature (T_b), K or °F
U	Overall heat transfer coefficient, W/m²-K or Btu/hr-ft²-°F
u^*	Frictional shear velocity $(\tau/\rho)^{1/2}$, m/s or ft/s
Wt	Tube weight, kg or lbm
x	Vapor quality, x_{av} (average in tube), x_{fill} (vapor quality at which microfins are condensate covered), m or ft
X_{tt}	Martinelli parameter, defined by Equation 13.8, dimensionless
y	Twist ratio $= H/d_i = \pi/(2\tan\alpha)$, dimensionless

Greek Symbols

α	Helix angle, measured from tube axis $(\alpha=\tan^{-1}\pi d_i/H)$, degrees
β	Included angle of fin cross section, degrees
δ	Condensate film thickness, m or ft
η_f	Fin efficiency, dimensionless
μ	Dynamic viscosity μ_l (of liquid), μ_v (of vapor), kg/m-s or lbm/ft-s
ρ	Density: ρ_l (of liquid), ρ_v (of vapor), kg/m³ or lbm/ft³
τ_o	Wall shear stress based on pressure drop, N/m² or lbf/ft²
ϕ	Severity factor, $(e/d_i)/(p/e)$, dimensionless

Subscripts

i	Designates inner surface of tube
o	Designates outer surface of tube
p	Plain tube
s	Smooth surface

Unsubscripted variables refer to enhanced tube.

REFERENCES

Agrawal, K.N., Kumar, A., Akhavan Behabadi, M.A., and Varma, H.K., 1998. Heat transfer augmentation by coiled wire inserts during forced convection condensation of R-22 inside horizontal tubes, *Int. J. Multiphase Flow*, 24, 635–650.

Akers, W.W., Deans, H.A., and Crosser, O.K., 1959. Condensing heat transfer within horizontal tubes, *Chem. Eng. Prog. Symp. Ser.*, 55(29), 1711–1176.

Baker, O., 1954. Simultaneous flow of oil and gas, *Oil Gas J.*, 53, 185–195.

Bhatia, R.S. and Webb, R.L., 2001. Numerical study of turbulent flow and heat transfer in microfin tubes. Part 2: Parametric study, *J. Enhanced Heat Transfer*, 8, 305–314.

Boyko, L.D. and Kruzhilin, G.N., 1967. Heat transfer and hydraulic resistance during condensation of steam in a horizontal tube and in a bundle of tubes, *Int. J. Heat and Mass Transfer*, 10, 361–373.

Brdlik, P.M. and Kakabaev, A., 1964. An experimental investigation of the condensation of steam in coils, *Int. Chem. Eng.*, 2, 216–239.

Breber, G., Palen, J.W., and Taborek, J. 1980. Prediction of horizontal tubeside condensation of pure componenets using flow regime criteria, *J. Heat Transfer*, 102, 471–481.

Brognaux, L., Webb, R.L., Chamra, L.M., and Chung, B.K., 1997. Single-phase heat transfer in microfin tubes, *Int. J. Heat Mass Transfer*, 40, 4345–4358.

Carey, V.P., 1992. *Liquid Vapor Phase-Change Phenomena*, Hemisphere, Washington, D.C.

Carnavos, T.C., 1980. Heat transfer performance of internally finned tubes in turbulent flow, *Heat Transfer Eng.*, 4(1), 32–37.

Cavallini, A. and Zecchin, R., 1974. Dimensionless correlation for heat transfer in forced convection condensation, in *Proc. of 15th Unione Italiana di Termofluodinamica National Heat Transfer Conf.*, Torino, Italy, 19–20 June 1974, I, 521–531.

Cavallini, A., Doretti, L., Klammsteiner, N., Longo, G.A., and Rossetto, L., 1995. A new model for forced convection condensation on integral-fin tubes, *Proc. 30th National Heat Transfer Conference*, V. K. Dhir, Ed., HTD, 308, 87–94.

Cavallini, A., Del Col, D., Doretti, L., Longo, G.A., and Lossetto, L., 1997. Pressure drop during condensation and vaporization of refrigerants inside enhanced tubes, *Heat Technol.*, 15(1), 3–10.

Cavallini, A., Del Col, D., Doretti, L., Longo, G.A., and Rossetto, L., 1999. A new computational procedure for heat transfer and pressure drop during refrigerant condensation inside enhanced tubes, *J. Enhanced Heat Transfer*, 6, 441–456.

Cavallini, A., Del Col, D., Longo, G.A., and Rossetto, L., 2000. Heat transfer and pressure drop during condensation of refrigerants inside horizontal enhanced tubes, *Int. J. Refrig.*, 23, 4–25.

Cavallini, A., Censi, G., Del Col, D., Doretti, L., Longo, A.G., and Rossetto, L., 2002. Condensation heat transfer and pressure drop inside channels for AC/HP application, *Heat Transfer 2002, Proc. 12th Int. Heat Transfer Conf.*, 1, 171–186.

Cavallini, A., Censi, G., Del Col., D., Doretti, L., Longo, G.A., Rossetto, L., and Zilio, C., 2003. Condensation inside and outside smooth and enhanced tubes—a review of recent research, *J. Refrig.*, 26, 373–392.

Chamra, L.M., Webb, R.L., and Randlett, M.R., 1996a. Advanced microfin tubes for condensation, *Int. J. Heat Mass Transfer*, 39, 1839–1846.

Chamra, L.M., Webb, R.L., and Randlett, M.R., 1996b. Advanced microfin tubes for evaporation, *Int. J. Heat Mass Transfer*, 39, 1827–1838.

Chang, K.I., and Spencer, D.L., 1971. Effect of regularly spaced surface ridges on film condensation heat transfer coefficients for condensation in the presence of noncondensible gas, *Int. J. Heat Mass Transfer*, 14, 502–505.

Chato, J.C., 1962. Laminar condensation inside horizontal and inclined tubes, *ASHRAE J.*, February, 52–60.

Chiang, R., 1993. Heat transfer and pressure drop during evaporation and condensation of Refrigerant-22 in 7.5 mm and 10 mm diameter axial and helical grooved tubes, *AIChE Heat Transfer Sympo.*, Atlanta, 89(295), 205–210.

Colebrook, F., 1939. Turbulent flow in pipes with particular reference to the transition region between smooth and rough pipe laws, *J. Inst. Civ. Eng.*, 4, 14–25.

Dobson, M.K. and Chato, J.C., 1998. Condensation in smooth tubes, *J. Heat Transfer*, 120, 193–213.

Ebisu, T., 1999. Evaporation and condensation heat transfer enhancement for alternative refrigerants used in air-conditioning machines, in *Heat Transfer Enhancement of Heat Exchangers*, S. Kakaç, A.E. Bergles, F. Mayinger, and H. Yuncu, Eds., Kluwer Academic, Dordrecht, 579–600.

Eckels, S.J. and Pate, M.B., 1991. In-tube evaporation and condensation of refrigerant-lubricant mixtures of HFC-134a and CFC-12, *ASHRAE Trans.*, 97(2), 62–70.

Fenner, G.W. and Ragi, E., 1979. Enhanced Tube Inner Surface Heat Transfer Device and Method, U.S. patent 4,154,293.

Ferreira, I., Newell, T.A., Chato, J.C., and Nan, X., 2003. R404A condensing under forced flow conditions inside smooth, microfin and cross-hatched horizontal tubes, *Int. J. Refrig.*, 26, 433–441.

Friedel, L., 1979. Improved friction pressure drop correlations for horizontal and vertical two phase pipe flow, Paper E2, European Two Phase Flow Group Meeting, Ispra, Italy.

Goto, M., Inoue, N., and Ishiwatari, N., 2001. Condensation and evaporation heat transfer of R410A inside internally grooved horizontal tubes, *Int. J. Refrig.*, 24, 628–638.

Goto, M., Inoue, N., and Yonemoto, R., 2003. Condensation heat transfer of R410A inside internally grooved horizontal tubes, *Int. J. Refrig.*, 26, 410–416.

Graham, D., Chato, J.C., and Newell, T.A., 1999. Heat transfer and pressure drop during condensation of refrigerant 134a in an axially grooved tube, *Int. J. Heat Mass Transfer*, 42, 1935–1944.

Haraguchi, H., Koyama, S., and Fujii, T., 1994. Condensation of refrigerants HCFC22, HFC134a and HCFC123 in a horizontal smooth tube (2nd Report, Proposal of Empirical Expressions for the Local Heat Transfer Coefficients), *Trans. JSME (B)*, 60(574), 245–252 [in Japanese].

Hinton, D.L., Conklin, J.C., and Vineyard, E.A., 1995. Condensation of refrigerants flowing inside smooth and corrugated tubes, *Proc. ASME/JSME Thermal Eng. Joint Conference*, L.S. Fletcher and T. Aihara, Eds., 2, 439–446.

Honda, H. and Nozu, S., 1987. A prediction method for heat transfer during film condensation on horizontal low integral-fin tubes, *J. Heat Transfer*, 109, 218–225.

Honda, H., Nozu, S., Matsuoka, Y, and Aomi, T., 1988. Condensation of refrigerants R-11 and R-113 in horizontal annuli with an enhanced inner tube, *Proc. 1st World Conf. on Experimental Heat Transfer, Fluid Mechanics and Thermodynamics*, R.K. Shah, E.N. Ganic, and K.T. Yang, Eds., Elsevier Science, New York, 1069–1076.

Honda, H., Wang, H., and Nozu, S., 2002. A theoretical study of film condensation in horizontal microfin tubes, *J. Heat Transfer*, 124, 94–101.

Hori, M. and Shinohara, Y., 2001. Internal heat transfer characteristics of small diameter thermofin tubes, *Hitachi Cable Rev.*, August, 85–90.

Hoshino, R., Sasaki, H., and Yasutake, K., 1991. Condenser for Use in Car Cooling System, U. S. Patent 5,025,855, assigned to Showa Aluminum Co., Japan.

Houfuku, M., Suzuki, Y., and Inui, K., 2001. High-performance, lightweight thermofin tubes for air conditioners using alternative refrigerants, *Hitachi Cable Rev.*, August, 97–100.

Ishikawa, S., Nagahara, K., and Sukumoda, S., 2002. Heat transfer and pressure drop during evaporation and condensation of HCFC22 in horizontal copper tubes with many inner fins, *J. Enhanced Heat Transfer*, 9(1), 17–24.

Ito, M. and Kimura, H., 1979. Boiling heat transfer and pressure drop in internal spiral-grooved tubes. *Bull. JSME*, 22(171), 1251–1257.

Ito, M., Kimura, H., and Senshu, T., 1977. Development of high efficiency air-cooled heat exchangers, *Hitachi Cable Rev.*, 26, 323–326.

Jaster, H. and Kosky, P. G., 1976. Condensation in a mixed flow regime, *Int. J. Heat Mass Transfer*, 19, 95–99.

Kaushik, N. and Azer, N.Z., 1988. A general heat transfer correlation for condensation inside internally finned tubes, *ASHRAE Trans.*, 94(Part 2), 261–279.

Kaushik, N. and Azer, N.Z., 1989. An analytical heat transfer prediction model for condensation inside longitudinally internally finned tubes, *ASHRAE Trans.*, 95(Part 2), 516–523.

Kaushik, N. and Azer, N.Z., 1990. A general pressure drop correlation for condensation inside internally finned tubes, *ASHRAE Trans.*, 96(Part 1), 242–255.

Kedzierski, M.A. and Kim, M.S., 1997. Convective Boiling and Condensation with a Twisted Tape Insert for R12, R22, R152a, R134a, R290, R32/R134a, R32/R152a, R390/R134a, R134a/R600a, National Institute of Standards and Technology Report NISTR 5905, January.

Khanpara, J.C., Pate, M.B., and Bergles, A.E., 1987. Local evaporation heat transfer in a smooth tube and a microfin tube using refrigerants 22 and 113, in *Boiling and Condensation in Heat Transfer Equipment*, E.G. Ragi, Ed., ASME Symp. 85, 31–39.

Kim, N.H., Cho, J.-P., and Kim, J.-O., 2000. R-22 condensation in flat aluminum multi-channel tubes, *J. Enhanced Heat Transfer*, 7, 427–438.

Kim, N.H., Cho, J.-P., Kim, J.-O., and Youn, B., 2003. Condensation heat transfer of and R-410A in flat aluminum multi-channel tubes with or without microfins, *Int. J. Refrig.*, 26, 830–839.

Koyama, S., Kuwahara, K., Nakashita, K., and Yamamoto, K., 2003. An experimental study on condensation of refrigerant R-134a in a multiport extruded tube, *Int. J. Refrigeration*, 26, 425–432.

Liebenberg, L., Bergles, A.E., and Meyer, J.P., 2000. A review of refrigerant condensation in horizontal microfin tubes, in *Proceedings of the ASME Advanced Energy Systems Division*, AES-Vol. 40, S. Garimella, M. Von Spakovsky, and S. Somasundaram, Eds., pp. 155–168.

Lin, S.T., Azer, N.S., and Fan, L.T., 1980. Heat transfer and pressure drop during condensation inside horizontal tubes with static mixer inserts, *ASHRAE Trans.*, 86(Part 2), 649–651.

Lopina, R.F., and Bergles, A.E., 1969. Heat transfer and pressure drop in tape-generated swirl flow of single-phase water, *J. Heat Transfer*, 91, 434–442.

Luu, M. and Bergles, A.E., 1979. Experimental study of the augmentation of the in-tube condensation of R-113, *ASHRAE Trans.*, 85(Part 2), 132–146.

Luu, M. and Bergles, A.E., 1980. Enhancement of horizontal in-tube condensation of R-113, *ASHRAE Trans.*, 85(Part 2), 293–312.

Miroploskii, Z.L. and Kurbanmukhamedov, A., 1975. Heat transfer with condensation of steam within coils, *Thermal Eng.*, 5, 111–114.

Mishima, K. and Hibiki, T. 1995. Effect of inner diameter on some characteristics of air-water two-phase flow in capillary tubes, *Trans. JSME(B)*, 61(589), 99–106 [In Japanese].

Miyara, A., Nonaka, K., and Taniguchi, M., 2000. Condensation heat transfer and flow pattern inside a herringbone-type microfin tube, *Int. J. Refrig.*, 23, 141–152.

Miyara, A., Otsubo, Y., Ohtsuka, S., and Mizuta, Y., 2003. Effects of fin shape on condensation in herringbone microfin tubes, *Int. J. Refrig.*, 26, 417–424.

Morita, H., Kito, Y., and Satoh, Y., 1993. Recent improvements in small bore inner grooved copper tube, *Tube Pipe Technol.*, Nov./Dec., 53–57.

Moser, K., Webb, R.L., and Na, B., 1998. A new equivalent Reynolds number model for condensation in smooth tubes, *J. Heat Transfer*, 120, 410–417.

Muzzio, A., Niro, A., and Arosio, S., 1998. Heat transfer and pressure drop during evaporation and condensation of R22 inside 9.52-mm OD microfin tubes of different geometries, *J. Enhanced Heat Transfer*, 5, 39–52.

Newell, T.A. and Shah, R.K., 2001. An assessment of refrigerant heat transfer, pressure drop, and void fraction effects in microfin tubes, *Int. J. Heating Ventilation Air Conditioning Refrigerating Res.*, 7(2), 125–154.

Nozu, S. and Honda, H., 2000. Condensation of refrigerants in horizontal, spirally grooved microfin tubes: numerical analysis of heat transfer in the annular flow regime, *J. Heat Transfer*, 122, 80–91.

Nozu, S., Honda, H., and Nishida, S., 1995. Condensation of a zeotropic CFC114-CFC113 refrigerant mixture in the annulus of a double-tube coil with an enhanced inner tube, *Exp. Thermal Fluid Sci.*, 11, 364–371.

Ohara, 1983. Heat Exchanger, Japanese patent 58-221390, assigned to Nippondenso Co.

Reisbig, R.L., 1974. Condensing heat transfer augmentation inside splined tubes, presented as *AIAA/ASME Thermophysi. Conf.*, Boston, paper 74-HT-7.

Royal, J.H. and Bergles, A.E., 1978a. Augmentation of horizontal in-tube condensation by means of twisted-tape inserts and internally finned tubes, *J. Heat Transfer*, 100, 17–24.

Royal, J.H. and Bergles, A.E., 1978b. Pressure drop and performance evaluation of augmented tube condensation, *Proc. 6th Int. Heat Trans. Conf.*, Toronto, 2, 459–464.

Said, S.A., and Azer, N.Z., 1983. Heat transfer and pressure drop during condensation inside horizontal tubes with twisted tapes, *ASHRAE Trans.*, 89(Part 1), 114–134.

Sardesi, R.G., Owen, R.G., and Pulling, D.J., 1982. Pressure drop for condensation of a pure vapor in downflow in a vertical tube, *Proc. 8th Int. Heat Transfer Conf.*, Munich, 139–145.

Schlager, L.M., Pate, M.B., and Bergles, A.E., 1988. Performance of microfin tubes with refrigerant-22 and oil mixtures. *ASHRAE J.*, November, 17–28.

Schlager, L.M., Pate, M.B., and Bergles, A.E., 1989. A comparison of 150 and 300 sus oil effects on refrigerant evaporation and condensation in a smooth tube and a microfin tube, *ASHRAE Trans.*, 95(Part 1), 387–397.

Schlager, L.M., Pate, M.B., and Bergles, A.E., 1990a. Performance predictions of refrigerant-oil mixtures in smooth and internally finned tubes. Part 1: Literature review, *ASHRAE Trans.*, 96(Part 1), 160–169.

Schlager, L.M., Pate, M.B., and Bergles, A.E., 1990b. Performance predictions of refrigerant-oil mixtures in smooth and internally finned tubes. Part 2: Design predictions, *ASHRAE Trans.*, 96(Part 1), 170–182.

Schlager, L.M., Pate, M.B., and Bergles, A.E., 1990c. Evaporation and condensation heat transfer and pressure drop in horizontal, 12.7-mm microfin tubes with refrigerant 22, *J. Heat Transfer*, 112, 1041–1047.

Shah, M.M., 1979. A general correlation for heat transfer during film condensation inside pipes, *Int. J. Heat Mass Transfer*, 22, 547–556.

Shinohara, Y. and Tobe, M., 1985. Development of an improved Thermofin tube, *Hitachi Cable Rev.* 4, 47–50.

Shinohara, Y., Oizumi, K., Itoh, Y., and Hori, M., 1987. Heat Transfer Tubes with Grooved Inner Surface, U.S. patent 4,658,892.

Smit, F.J. and Meyer, J.P., 2002. R-22 and zeotropic R-22/R-142b mixture condensation in microfin, high-fin, and twisted tape insert tubes, *J. Heat Transfer*, 124, 912–921.

Sur, B., and Azer, N.Z., 1991. An analytical pressure drop prediction model for condensation inside longitudinally internally finned tubes, *ASHRAE Trans.*, 97(Part 2), 54–61.

Taitel, Y. and Dukler, A.E., 1976. A model for predicting flow regime transitions in horizontal and near-horizontal gas-liquid flow, *AIChE J.*, 22, 47–55.

Tang, L., Ohadi, M.M., and Johnson, A.T., 2000a. Flow condensation in smooth and microfin tubes with HCFC-22, HFC-134a and HFC-410A refrigerants. Part 1: Experimental results, *J. Enhanced Heat Transfer*, 7, 289–310.

Tang, L., Ohadi, M.M., and Johnson, A.T., 2000b. Flow condensation in smooth and microfin tubes with HCFC-22, HFC-134a and HFC-410 refrigerants. Part 2: Design equations, *J. Enhanced Heat Transfer*, 7, 311–326.

Thomas, D.G., 1967. Enhancement of film condensation rates on vertical tubes by vertical wires, *Ind. Eng. Chem. Fund.*, 6, 97–l02.

Traviss, D.P. and Rohsenow, W.M., 1973. The influence of return bends on the downstream pressure drop and condensation in tubes, *ASHRAE Trans.*, 79(Part l), 129–137.

Traviss, D.P., Rohsenow, W.M., and Baron, A.B., 1973. Forced convection condensation in tubes: a heat transfer correlation for condenser design, *ASHRAE Trans.*, 79(Part. 1), 157–165.

Tsuchida, T., Yasuda, K., Hori, M., and Otani, T., 1993. Internal heat transfer characteristics and workability of narrow Thermofin tubes, *Hitachi Cable Rev.,* 12, 97–100.

Uehara, H. and Sumitomo, H., 1985. Condenser, U. S. patent 4,492,268, assigned to Iisaka Works, Ltd.

Venkatesh, K., 1984. Augmentation of Condensation Heat Transfer of R-11 by Internally Finned Tubes, M.S. thesis, Department of Mechanical Engineering, Kansas State University, Manhatlan.

Vrable, D.A., Yang, W.J., and Clark, J.A., 1974. Condensation of refrigerant-12 inside horizontal tubes with internal axial fins, *Heat Transfer 1974, 5th Int. Hear Transfer Conf.*, 3, 250–254.

Wang, H.S. and Honda, H., 2003. Condensation of refrigerants in horizontal microfin tubes: comparison of prediction methods for heat transfer, *Int. J. Refrig.*, 26, 452–460.

Wang, H.S., Honda, H., and Nozu, S., 2002. Modified theoretical models of film condensation in horizontal microfin tubes, *Int. J. Heat Mass Transfer*, 45, 1513–1523.

Wang, W., 1987. The enhancement of condensation heat transfer for stratified flow in a horizontal tube with inserted coil, in *Heat Transfer Science and Technology*, B.-X. Wang, Ed., Hemisphere, New York, 805–811.

Wang, W.-W., Radcliff, T.D., and Christensen, R.N., 2002. A condensation heat transfer correlation for millimeter-scale tubing with flow regime transition, *Exp. Thermal Fluid Sci.*, 26, 473–485.

Webb, R. L., Eckert, E.R.G., and Goldstein, R.J., 1971. Heat transfer and friction in tubes with repeated-rib roughness, *Int. J. Heat Mass Transfer*, 14, 601–617.

Webb, R.L., Zhang, M., and Narayanamurthy, R., 1998. Condensation heat transfer in small diameter tubes, *Heat Transfer 1998, Proc. 11th Int. Heat Transfer Conf.*, Kyongju, Korea, 6, 403–408.

Webb, R.L. and Yang, C.-Y., 1995. A comparison of R-12 and R-134A condensation inside small extruded aluminum plain and microfin tubes, in *1995 Vehicle Thermal Management Systems Conference Proceedings*, SAE, Warrendate, PA, 77–85.

Webb, R.L., 1999. Prediction of condensation and evaporation in microfin and micro- channel tubes, in *Heat Transfer Enhancement of Heat Exchangers*, S. Kakaç, A.E. Bergles, F. Mayinger, and H. Yunci, Eds., Kluwer Academic Dordrecht, 529–550.

Webb, R.L. and Ermis, K., 2001. Effect of hydraulic diameter on condensation of r-134a in flat, extruded aluminum tubes, *J. Enhanced Heat Transfer*, 8, 77–90.

Xin, M.D., Wang, Z.J., and Liao, Q., 1998. Condensation for steam in the horizontal tubes with three-dimensional inner microfins, *Heat Transfer 1998, Proc. 11th Int. Heat Transfer Conf.*, 6, 469–472.

Yang, C.Y. and Webb, R.L., 1996. Condensation of R-12 in small hydraulic diameter extruded aluminum tubes with and without microfins, *Int. J. Heat Mass Transfer*, 39, 791–800.

Yang, C.Y. and Webb, R.L. 1997. A predictive model for condensation in small hydraulic diameter tubes having axial microfins, *J. Heat Transfer*, 119, 776–782.

Yasuda, K., Ohizumi, K., Hori, M., and Kawamata, O., 1990. Development of condensing Thermofin-HEX-C tube, *Hitachi Cable Rev.* 9, 27–30.

Yu, J. and Koyama, S., 1998, Condensation heat transfer of pure refrigerants in microfin tubes, *Proceedings of 1998 International Refrigeration Conference at Purdue*, pp. 325–330.

Zhang, M., 1998. A New Equivalent Reynolds Number Model for Vapor Shear-Controlled Condensation inside Smooth and Microfin Tubes, Ph.D. thesis, Pennsylvania State University, University Park.

Zhang, M.H. and Webb, R.L., 2001. Correlation of two-phase friction for refrigerants in small-diameter tubes, *Exp. Thermal Fluid Sci.*, 25, 131–139.

Zivi, S.M., 1964. Estimation of steady state steam void fraction by means of the principle of minimum entropy production, *J. Heat Transfer*, 86, 247–252.

ENHANCEMENT USING ELECTRIC FIELDS

15.1 INTRODUCTION

Considerable work has been done using electric fields to enhance heat transfer. Ohadi [1991] provides a survey of the technology in lay terms. Yabe et al. [1987] and Yabe [1991] provide an excellent discussion of the fundamentals of electrohydrodynamic (EHD) enhancement. Yabe [1991] provides particular focus on enhancement of condensation and boiling. The application of an electric field to a dielectric fluid will impose a body force (F_e) on the fluid. This body force adds a term to the momentum equation and influences the fluid motion. It is important that the fluid be a dielectric—that is, not conduct an electric current. The body force (F_e) used in the Navier–Stokes equations is described by the change of Helmholtz free energy for virtual work with the energy stored in the fluid by the electric fluid. As described by Panofsky and Phillips [1962], application of electric field strength E to a constant-temperature dielectric fluid of permittivity ϵ and density ρ results in F_e given by

$$F_e = \rho_c E - 0.5 E^2 \nabla \epsilon + 0.5 \nabla \left[\rho E^2 \left(\frac{\partial \epsilon}{\partial \rho} \right) \right] \tag{15.1}$$

665

The electric field strength will vary with distance from the electrode, and depends on the electrode design. The physical significance of the three terms in Equation 15.1 are as follows:

1. The first term is the "electrophoretic" (Coulomb) force acting on the net free charge of electric field space charge density, ρ_c. The direction of the force depends on the relative polarities of the free charges and the electric field.
2. The second term is the "di-electrophoretic" force produced by the spatial change of permittivity, ε (also called the dielectric constant). In two-phase flow, this force arises from the difference in permittivity of the vapor and liquid phases.
3. The third term is called the "electrostriction" force and is caused by inhomogeneity of the electric field strength. This force is dependent on inhomogeneities in the electric field strength and may be likened to an "electrical pressure."

The electric current is given by

$$i = \rho_c u + \sigma_e E \qquad (15.2)$$

where u is the fluid velocity and σ_e is the electrical conductivity of the fluid. In steady state, without convection, the electric charge density is given by

$$\rho_c = -\frac{\varepsilon}{\sigma_e} E \nabla \sigma_e \qquad (15.3)$$

Thus, the electric charges are generated by the gradient to the electrical conductivity. Because the electrical conductivity of liquids is temperature dependent, Equation 15.3 shows that temperature gradients in the fluid generates the electric charge. This makes the Coulomb force (first term of Equation 15.1) an active force.

The momentum and energy equations are coupled through the temperature dependence of the permittivity and thermal conductivity of the fluid. Cooper [1992] notes that analytical solution of the EHD coupled momentum and energy equations is not possible, except for the simplest configurations. Rather, data are correlated, based on the predicted electric field strength distribution at the heat transfer surface on the basis of uniform electrical fluid properties.

The sum of the second and third terms of Equation 15.1 gives the force exerted on dielectric fluids. For polarization of dielectric fluids, the net charge is zero. However, the force on the polarized charges formed in the stronger field zone is greater than the force on the charges in the weaker field zone. Thus, the resultant force (the sum of the forces exerted on each polarized charge) moves fluid elements to the stronger electric field region. Yabe [1991] notes that the effect of the magnetic field generated by the current is negligible compared with the pressure generated by the electric field.

Table 15.1 from Ohadi [1991] summarizes enhancement levels obtained by various investigators for natural convection, laminar flow forced convection, boiling, and condensation.

Table 15.1 EHD Heat Transfer Enhancement in Heat Exchangers

Ref.	Maximum Enhancement (%)	Test Fluid	Wall/Electrode Configuration	Process
Fernandez and Poulter [1987]	2300	Transformer oil	Tube-wire	Forced convection
Ohadi et al. [1991a]	320	Air	Tube-wire/rod	Forced convection
Levy [1964]	140	Silicone oil	Tube-wire	Forced convection
Yabe and Maki [1988]	10000	R-113 (4% oil)	Plate-ring	Natural convection
Cooper [1990]	1300	R-113 (10% oil)	Tube-wire mesh	Pool boiling
Uemura et al. [1990]	1400	R-113	Plate-wire mesh	Film boiling
Bologa et al. [1987]	2000	Diethylether, R-113, hexane	Plate-plate	Film condensation
Sunada et al. [1991]	400; 600	R-113; R-123	Vertical wall-plate	Condensation
Ohadi et al. [1992]	480	R-123	Tube-wire	Boiling

15.2 ELECTRODE DESIGN AND PLACEMENT

The design and location of the electrodes is dependent on whether the flow is single phase or two phase, as well as on the thermophysical and electrical properties of the fluid. The electrodes must be positioned such that the electric field force aids the hydrodynamic forces, particularly at the heat transfer surface. For separated two-phase flows, the permittivity of the liquid and vapors will be distinctly different. Hence $\nabla \in$ exhibits a singularity at a liquid–vapor interface. Jones [1978] shows that the fluid component having the highest $\in$ will move to the region of highest field strength, E. A liquid drop will move toward the electrode, and a bubble will move toward the region of lower field strength.

Placement of the electrodes is different for tube-side and shell-side heat transfer enhancement. Consider heat transfer to a fluid flowing inside a tube. Figure 15.1 shows a wire electrode in the center of a tube. If the fluid inside the tube has constant permittivity ($\in$), the field strength distribution for this configuration is given by

$$E = \frac{V}{r} \left[\ln(R_2 / R_1) \right]^{-1} \tag{15.4}$$

where V is the applied voltage, r is the radial distance from the tube axis, and R_1 and R_2 are defined in Figure 15.1. The field strength decays with radial distance from the central electrode wire. If a liquid–gas mixture flows in the tube, the liquid phase will have the highest permittivity. Hence, the liquid will move toward the central electrode, leaving vapor at the tube wall. This will enhance condensation, but not evaporation, which requires liquid at the tube wall. Condensation enhancement would occur, because the electrode pulls condensate away from the condensing surface. Yabe et al. [1992] used a perforated sheet wrapped in a cylindrical form

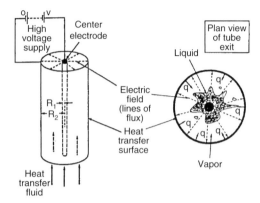

Figure 15.1 Wire electrode in center of tube with two-phase flow. (From Cooper [1992].)

(5 mm diameter with 1- to 2-mm-diameter holes) as the electrode for enhancement of tube-side convective evaporation of refrigerants. Figure 15.2 shows electrode designs used by Cooper [1992] for enhancement of shell-side boiling in a tube bundle. Figure 15.3 illustrates electrode designs used by Yabe et al. [1987], for enhancement of condensation on the outer surface of vertical tubes. The Figure 15.3 electrodes produce a non-uniform electric field. Both ac and dc voltages provide enhancement.

The enhancement effect will be increased by increasing the applied voltage. The electrical breakdown strength of the liquid and vapor must be known and should not be exceeded. As noted by Cooper [1992], careful design of the solid insulation system is required. Poor design may result in electrical breakdown, which would make the EHD enhancement system useless. The insulation material must be compatible with the heat transfer media. Inert dielectrics, such as PTFE and proprietary epoxy composites, are candidates. Cooper [1992] notes that high electric stress exists near small radii of curvature on solids, so electric breakdown may occur at these locations. Electrode and insulation surfaces should be rounded, and conducting rods of large radius should be used to reduce the local electrical field strength. Design aspects of the insulation system are discussed in texts on power transformers and switch-gear, such as Kuffel and Zaengl [1984].

Voltages up to 25 kV are typically used in EHD enhancement. Although high voltages are used, the electric current is quite small (e.g., 1 mA) and thus the power consumption is small. To place these voltages in perspective, automotive spark plugs and electronic air cleaners operate at approximate voltages of 25 and 10 kV, respectively. The power is supplied by a relatively low kVA power supply consisting of a high-voltage transformer operating at very small current. A device, such as an automotive spark plug, may be modified to bring the high-voltage supply conductor through the heat exchanger wall.

The use of very small currents minimizes physical hazards. All high-voltage surfaces are completely contained within the heat exchanger, which is electrically

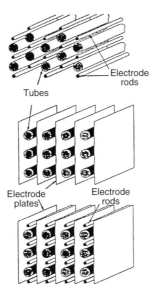

Figure 15.2 Electrode designs used by Cooper [1992] for enhancement of shell-side boiling in a tube bundle. (From Cooper [1992].)

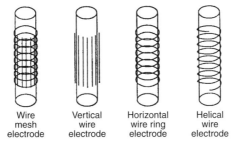

| Wire mesh electrode | Vertical wire electrode | Horizontal wire ring electrode | Helical wire electrode |

Figure 15.3 Electrode designs used by Yabe [1987], for enhancement of condensation on the outer surface of vertical tubes. (From Yabe [1987].)

insulated from the shell and the tubes. Provided that adequate insulation is used, no danger of electric shock should exist.

15.3 SINGLE-PHASE FLUIDS

For single-phase fluids, the permittivity (ε) is approximately constant and the second and third terms of Equation 15.1 are negligible. Thus, for single-phase flows, the first term of Equation 15.1 can cause EHD-induced fluid motion known as the ionic (or corona) wind effect. Ions are produced close to the surface of a wire anode. The Coulomb force on the ions causes them to move to the cathode surface. Interaction of the corona wind with the main flow produces mixing by secondary flows.

The greatest enhancement occurs for natural convection and low-Reynolds-number laminar flows. As the main flow velocity increases into the turbulent regime, the secondary flow speed is negligible compared to the strength of the turbulent eddies. Table 15.1 identifies several studies for natural convection and laminar flow forced convection.

15.3.1 Enhancement on Gas Flow

Data for forced convection enhancement are illustrated in Figure 15.4 by the data of Ohadi et al. [1991a]. They used single and double axial (two axial wires spaced at 4.0 mm) electrodes to measure enhancement of airflow in a 32-mm-ID, 305-mm-long tube. Figure 15.4 shows that the two-electrode design provided a modestly higher enhancement than did the single-electrode design. A maximum enhancement of 3.5 was obtained at 7.5 kV using the two-electrode system at $Re_d = 3000$. Note that the two-electrode design provided only 20% enhancement at $Re_d = 10,000$. For 7.75 kV, the current flow was 385 µA, which results in 2.8 W power expenditure. Hence, the power consumption is minimal. Ohadi et al. [1991b] used axial wire electrodes in a double-pipe heat exchanger with airflow in both streams. They provide test results for EHD enhancement of the tube side, the shell side, and both fluids. Their results show almost total decay of the enhancement for $Re_d > 4000$. However, for shell-side and tube-side Reynolds numbers of 500 and 2000, respectively, the overall heat transfer coefficient was enhanced by a factor of four. Ohadi et al. [1994] extended the work to a shell and tube heat exchanger consisting of seven tubes. The Reynolds number was varied from 1000 to 6000 for the tube and the

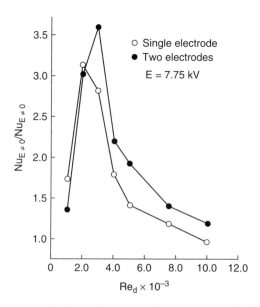

Figure 15.4 Enhancement provided by axial single and double wire electrodes in a 38.4-mm-diameter tube. (From Ohadi et al. [1991b].)

shell sides. A maximum enhancement of 110% was obtained at the lowest Reynolds numbers (1000 for both sides) when both sides were excited. The enhancement decreased as the Reynolds number increased. Nelson et al. [1991] studied EHD enhancement of airflow in a circular tube using a central, axial wire electrode. Their results show that the friction factor increase is generally comparable with the heat transfer enhancement.

Blanford et al. [1995] applied the EHD technique to enhance the air-side heat transfer of a one-row finned-tube heat exchanger. The heat exchanger consisted of two 19.0-mm-OD copper tubes with 50.0-mm tube pitch and copper fins spaced 2.54-mm apart. The fins were soldered to the tubes. Four stainless steel wire electrodes were installed between the fins in the flow direction. The air-side Reynolds number (based on hydraulic diameter) was varied between 1000 and 3000. A maximum heat transfer enhancement of 3.3 was obtained at the lowest Reynolds number and at the lowest air to tube-side fluid temperature (20°C). The enhancement decreased as the Reynolds number increased, or the temperature difference increased. The effect of an electric field on frost formation was also investigated at $Re_{Dh} = 1000$. The amount of frost increased with application of electric current due to increased convective currents by corona wind. At a very low electric current (less than 20 μA), however, slight decrease of frost formation was observed. Under the influence of an electric field, the solidifying water vapor molecules align themselves in the direction of an electric field, resulting in long and thin crystals. The long and thin structures are weak, and can be easily removed by airflow or by electric force (Molki et al. [2000a]). Munakata et al. [1993] and Molki et al. [2000a] report reduced frost formation by EHD under natural convection environment.

15.3.2 Enhancement on Liquid Flow

Fernandez and Poulter [1987] worked with transformer oil flowing on the annulus side of a double pipe heat exchanger. The electric field was applied across the annular spacing. Although the heat transfer coefficient was enhanced a factor of 20 (Table 15.1), the oil pressure drop was increased by a factor of only 3.0. The effects of oil properties on the EHD enhancement of forced convective heat transfer of oil were studied by Paschkewitz and Pratt [2000]. Three different oils (PAO, ECO-C, Beta) with viscosities ranging from 6.6×10^{-6} m^2/s to 108×10^{-6} m^2/s, and electrical conductivities ranging from 2.7×10^{10} to 6.25×10^{13} Siemens/m were tested. Oils with low viscosity and low electrical conductivity (PAO) yielded the greatest enhancement for a given power input.

Working in a 20-mm-high rectangular channel, Ishiguro et al. [1991] placed axial wire electrodes spaced at 20-mm pitch, 5 mm from the wall. A mixture consisting of 96% R-113 and 4% ethanol was used as the working fluid. The authors sought to disturb the controlling thermal resistance in the boundary layer at the wall. As expected, the width-wise distribution of the local Nu occurred at the position closest to the wire electrodes. A maximum Nu enhancement of 23 occurred for laminar flow. A ninefold enhancement was measured at $Re_{Dh} = 9100$ for 10-kV applied voltage. At $Re_{Dh} = 27,300$, the enhancement at 10 kV was 2.8.

15.3.3 Numerical Studies

Numerical studies on forced convection enhancement in a channel by EHD have been conducted by Hasegawa et al. [1999] for turbulent airflow, and by Molki et al. [2000b] for laminar air flow. The Navier–Stokes equation with the electric force term, the conservation equation of electric current, the electric potential equation, and the energy equation were simultaneously solved. The turbulence was modeled using the large eddy simulation method. The numerical results clearly showed corona-induced secondary flow in the cross section of the channel. Comparison with the experimental results, however, revealed marginal success.

15.4 CONDENSATION

15.4.1 Fundamental Understanding

Early work was performed by Velkoff and Miller [1965], who used a screen-type electrode, which enhanced R-113 condensation on a vertical plate by 150%. Recent work by Yabe [1991] and Sunada et al. [1991] provides fundamental understanding of the concept. Working with Yabe, Sunada et al. [1991] condensed R-113 and R-123 on a vertical plate with electric field strengths between 5 and 8 MV/m. The electrode consisted of a transparent glass plate coated with a 500-Å-thick, electrically conducting InO_3 film. Figure 15.5 shows the local condensation coefficient vs. field strength for R-113 condensation at 48°C. The electrode was placed 6.0 mm from the surface, and the condensing plate was cooled by silicone oil. No enhancement was observed until the electrode voltage was increased to 4 MV/m (24 kV electrode

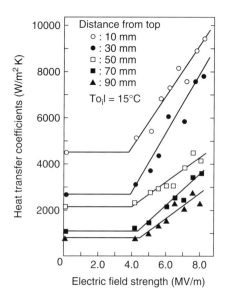

Figure 15.5 Heat transfer enhancement vs. electric field strength for R-113 condensing at 48°C on a vertical plate. (From Sunada et al. [1991].)

voltage). The test results for no electric field are shown by the data points for zero field strength. The average enhancement on the lower part of the plate exceeds 300% for 8 MV/m.

Above 4 MV/m, the character of the condensate film radically changed, as illustrated in Figure 15.6. The condensate formed in surface drops, similar to those observed in dropwise condensation. Yabe calls this "pseudo-dropwise" condensation. R-113 is not amenable to dropwise condensation without an electric field, because its low surface tension causes surface wetting. The high enhancement results from the thin film thickness between the drops.

Use of a screen electrode, as shown in Figure 15.3, provides a second enhancement phenomenon known as *liquid jetting*, which is illustrated in Figure 15.7. Spaced electrodes near the surface cause a jet flow of condensate away from the surface (EHD liquid extraction), which thins the condensate film. This removed 95% of the condensate from the surface and resulted in a 2.8 factor enhancement level.

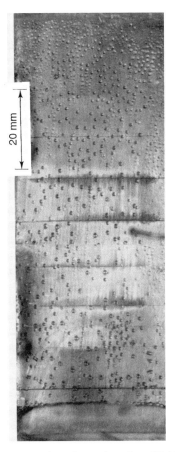

Figure 15.6 Photograph of EHD pseudo-dropwise condensation of R-113 at 48.7°C on a vertical plate observed by Sunada et al. [1991]. Electric field strength (7.5 MV/m), condensing temperature (48.7°C). (From Sunada et al. [1991].)

Finally, Yabe [1991] combined the liquid jetting and pseudo-dropwise phenomena in condensation on a vertical tube. This was done using a helical wire electrode (extraction mode) and a perforated, curved plate (pseudo-dropwise mode). The R-113 condensation coefficient was increased by a factor of 4.5.

Yamashita et al. [1991] proceeded to develop a vertical tube condenser containing 102 19-mm-diameter tubes, 1.4 m long. Tests were performed condensing perfluorohexane C_6F_{14}. The Figure 15.3a electrode design was used, which promoted both liquid extraction and pseudo-dropwise condensation. The authors obtained a maximum heat transfer enhancement ratio of approximately 6.0, relative to that for zero electric field (for the same heat flux). Tests were also performed with R-114. For a total condensation rate of 60 kW, the electric power consumption by the electric fields at 18 kV voltage was less than 1.4 W! Figure 15.8 compares test results on

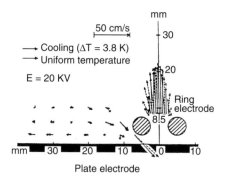

Figure 15.7 EHD liquid jet phenomenon. (From Yabe [1991].)

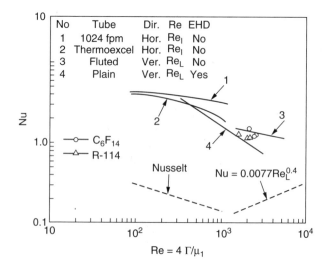

Figure 15.8 Comparison of test results on 102-tube vertical condenser with test results on single horizontal mechanically enhanced tubes. (From Yamashita et al. [1991].)

the 102 tube vertical condenser with data for single horizontal and vertical, mechanically enhanced tubes. For a condensate film Reynolds numbers greater than 1000, the EHD enhancement is higher than that for vertical fluted tubes. The dashed lines at the bottom of the figure show the laminar and turbulent equations for film condensation on a vertical plate. Typically, a vertical tube will operate with higher Re_L ($4\Gamma_L/\mu_1$) than the Re_l ($4\Gamma/\mu_1$) for a horizontal tube, so the comparison is somewhat inconsistent. The vertical EHD enhanced tube operates at a higher film Reynolds number than do the horizontal mechanically enhanced tubes. Although EHD phenomena offer interesting possibilities, they have not shown enhancement significantly superior to that of high-performance horizontal enhanced tubes, e.g., the Figure 12.6a Thermoexcel-C. The data do not suggest that the vertical EHD enhanced tube will be superior to horizontal mechanically enhanced tubes for tube bundle applications.

15.4.2 Vapor Space Condensation

Wawzyniak and Seyed-Yagoobi [1996] conducted EHD-enhanced R-113 condensation tests on a vertical enhanced tube (Turbo-CII, shown in Figure 12.6) and on a smooth tube. The Figure 15.3d helical wire electrode was placed with a 1.6-mm gap from the tube. The heat transfer enhancement by EHD was larger for the enhanced tube—approximately 100% for the Turbo-CII and less than 100% for the smooth tube. The reason was attributed to the non-uniform electric fields caused by the fins of the enhanced tube. Cheung et al. [1999] investigated the effect of electrode–tube gap width of the Figure 15.3d electrode using R-134a on smooth tubes. Both vertical and horizontal configurations were studied. The gap width was varied from 0.8 to 3.2 mm. For both configurations, the largest (3.2 mm) gap yielded the highest heat transfer enhancement and with the least amount of electrical power to the electrode.

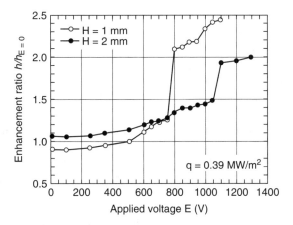

Figure 15.9 Heat transfer enhancement ratio vs. applied voltage for condensing steam on 16-mm-OD, 1.0-mm integral-fin tube. A 1.0-mm-diameter stainless wire electrode placed under the tube at a distance $H = 1$ and 2 mm. (From Chu et al. [2001].)

Da Silva et al. [2001] investigated the effect of polarity on the EHD enhanced condensation of R-134a on horizontal enhanced tubes. The Figure 15.3d helical wire electrode was placed with 3.0-mm gap from the tube. An enhancement ratio of 3.5 was obtained when a positive polarity field was applied compared with 1.5 for the negative polarity case. The condensate extraction from the tube, which was seen clearly for the positive polarity case, was not readily observed for the negative polarity case.

Chu et al. [2001] investigated the film condensation enhancement of steam on a 16-mm-OD, 1.0-mm-pitch integral fin tube (shown in Figure 12.4). A 1.0-mm-diameter stainless-wire electrode was placed under the finned tube parallel to the tube axis. Figure 15.9 shows the heat transfer enhancement ratio by EHD vs. applied voltage

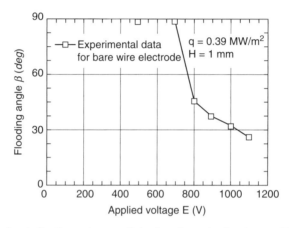

Figure 15.10 Condensate flooding angle vs. applied voltage for condensing steam on 16-mm-OD, 1.0-mm integral-fin tube. A 1.0-mm-diameter stainless wire electrode placed under the tube at a distance $H = 1$ mm. (From Chu et al. [2001].)

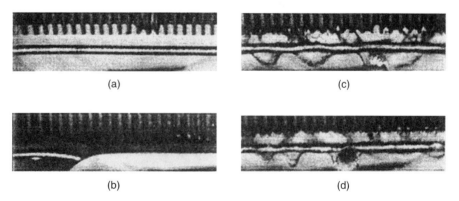

Figure 15.11 Photographs showing the effect of applied voltage for condensing steam on 16-mm-OD, 1.0-mm integral-fin tube. A 1.0-mm-diameter stainless wire electrode placed under the tube at a distance $H = 1$ mm: (a) $E = 0$ V, $q = 0$ W/m^2, (b) $E = 0$ V, $q = 390$ W/m^2, (c) $E = 765$ V, $q = 390$ W/m^2, (d) $E = 900$ V, $q = 390$ W/m^2. (From Chu et al. [2001].)

for two different electrode distances (H). The figure shows that the enhancement ratio jumps to a higher value above a threshold voltage. The threshold voltage is higher for larger H. The maximum enhancement ratio of 2.5 was obtained. Figure 15.10 shows the measured flooding angles and Figure 15.11 shows corresponding flow patterns for $H = 1.0$ mm. For zero voltage (Figure 15.11b), the tube and the wire electrode are bridged by the condensate film, and the interfin spaces are completely flooded (90° flooding angle). When the applied voltage is above the threshold value (765 V), the bridging film becomes liquid columns, and the flooding angle decreases to 45° (Figure 15.11c). With further increase of the voltage, the pitch of the liquid columns decreases, and the flooding angle decreases yielding higher enhancement ratio (Figure 15.11d). Chu et al. [2001] also provides an analytic model that predicts the EHD condensation on integral-fin tubes.

15.4.3 In-Tube Condensation

The EHD technique was applied to in-tube condensation by Singh et al. [1997] and Gidwani et al. [2002]. Singh et al. [1997] investigated EHD enhancement of R-134a condensation in a 12.7-mm-OD smooth tube and in a microfin tube of the same outside diameter for mass flux from 50 to 300 kg/m²-s. Six different electrodes having various electrode diameters and spacings were tested. The optimum electrode (maximum heat transfer enhancement) was the one placed coaxially with a sufficiently large electrode gap. For the smooth tube, up to 6.4-fold heat transfer enhancement was obtained at the lowest mass flux and quality investigated. The enhancement decreased as the mass flux and the quality increased. The pressure drop increase was more pronounced than the heat transfer increase. For the microfin tube, the heat transfer enhancement was much less than that of the smooth tube. A maximum value of 1.8 was obtained. They state that the electric field strength is high at the tip of the microfin, and the condensate is not easily pulled to the electrode as occurs in the smooth tube. For the smooth tube, the electric potential is the highest at the electrode, which pulls the condensate to the electrode, leaving thin condensate film at the tube wall. Gidwani et al. [2002] extended the study to a corrugated tube. Similar to the microfin tube, the enhacement ratio of the corrugated tube was less than that of the smooth tube.

15.4.4 Falling Film Evaporation

Yamashita and Yabe [1997] applied the EHD technique to enhance R-123 falling film evaporation from a vertical smooth tube. In addition to the conventional electrode geometries shown in Figure 15.4, punched electrodes shown in Figure 15.12 were tested. The electrodes were made from rolled plates with punched slits of 2.0 mm width and 30 mm length. Electrode III has slit configuration similar to Electrode I, except that the slits are arranged offset in the vertical direction. The gap between the tube and the electrode was 3.05 mm. Generally, the punched electrodes yielded higher enhancements than conventional electrodes. Different from the condensation case, there existed an optimum applied voltage, at which the heat transfer coefficients were maximized. They state that above the optimum value, EHD causes dryout of the film. Figure 15.13 compares the measured optimum Nusselt numbers

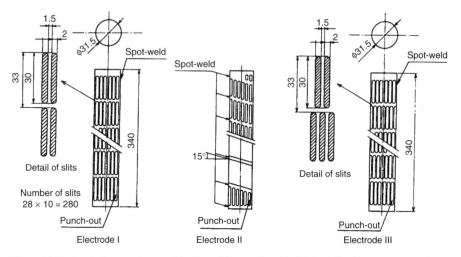

Figure 15.12 Punched electrodes tested by Yamashita and Yabe [1997] for falling film evaporation from vertical tubes. (From Yamashita and Yabe [1997]. Reprinted with permission of ASME.)

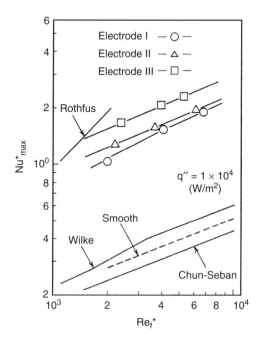

Figure 15.13 Optimum Nusselt number vs. film Reynolds number, obtained with punched electrodes of Figure 15.12. (From Yamashita and Yabe [1997]. Reprinted with permission of ASME.)

vs. film Reynolds numbers for the three punched electrodes. Data were taken at q = 10,000 W/m^2. Figure 15.13 shows that the maximum sixfold enhancement was obtained for Electrode III.

Darabi et al. [2000a] investigated the effect of EHD on falling film evaporation of vertical enhanced tubes (748 fpm integral fin tube and Turbo-BIII) using R-134a. Turbo-BIII is shown in Figure 11.12e. A punched electrode was used for Turbo-BIII and a wire electrode wrapped with nylon strand was used for the integral fin tube. Up to 80% enhancement was obtained for the smooth tube, 110% for Turbo-BIII, and 30% for the integral fin tube. The enhancement decreased with increasing heat flux. Darabi et al. [2000b] extended the study to the horizontal configuration. Tested tubes included: smooth, 748-fpm integral-fin, Turbo-BIII. Enhancement up to 280% was obtained with the smooth tube, 90% with the Turbo-BIII tube, and 60% with the integral-fin tube.

15.4.5 Correlations

As discussed by Yabe [1991], the mechanism of condensation enhancement EHD consists of liquid extraction and the formation of small droplets on the condensing surface. Thus, a successful model should consider both terms. Recently, Al-Ahmadi and Al-Dadah [2002] performed a comprehensive review on available correlations for EHD enhanced condensation heat transfer. Comparison of the correlations with available data was not satisfactory. They proposed a new correlation given by

$$\text{Nu}_E = \frac{h\lambda^*}{k_l} = A\left[\frac{V^{1.75}h'_{fg}(\lambda^*/l)^{4.5}}{c_{pl}\Delta T}\right]^n \tag{15.5}$$

Here, $V = \varepsilon_g E_g^2 l/\sigma$ and is the ratio of the EHD force to the surface tension force, and indicates the effect of the liquid extraction. For λ^*, which is the most unstable wavelength, and denotes the effect of the condensate droplets, Choi and Reynolds' [1965] expression was adopted.

$$\lambda^* = \frac{2\pi}{\dfrac{3}{4\sigma}\left(1-\dfrac{\varepsilon_g}{\varepsilon_l}\right)^2 \varepsilon_g E_g^2} \tag{15.6}$$

In Equation 15.5, l is the gap between electrodes, and h'_{fg} ($= h_{fg} + 0.86c_{pl}\Delta T$) is the modified latent heat. The A and n of Equation 15.5 were obtained from a best-fit of the data. Different values of A and n are provided depending on geometric configurations (horizontal and vertical, inside and outside).

15.5 BOILING

15.5.1 Fundamental Understanding

EHD also provides dramatic enhancement of nucleate boiling. Figure 15.14 shows Yabe's [1991] test results on a plain horizontal tube using a 90%/10% mixture (by weight) of R-11 and ethanol. The Figure 15.3d helical wire electrode design was used with the electrodes placed 5 mm from the boiling surface. The performance is also compared to that of the High-Flux porous surfaces and the Thermoexcel-E mechanically enhanced surface geometries (see Chapter 11) both operating without EHD enhancement. The performance increases with increasing applied voltage, and is as high as or higher than provided by the mechanically enhanced surfaces. Although it is unclear in the Yabe [1991] publication, it appears that the porous surface and the Thermoexcel-E data are for pure R-11.

The proposed mechanism for EHD enhancement on a plain surface is as follows. The EHD body force pushes the vapor bubbles away from the electrode. The greatest force occurs on the part of the bubble closest to the electrode. The bubble is pushed toward the heating surface (away from the electrode), and the radial components of the body force cause violent movement of the bubbles on the heating surface. This enhances thin film evaporation at the base of the bubble, and promotes breakup of large bubbles into a greater number of smaller-diameter bubbles. This theory was substantiated by Ogata and Yabe [1991], who observed the effect of an electric field on a single air bubble in a silicone oil/ethyl alcohol mixture. They explain that the

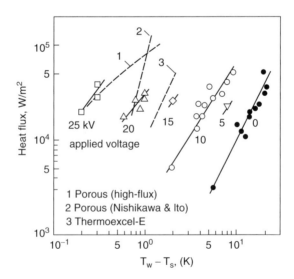

Figure 15.14 Yabe's [1991] test results for EHD boiling enhancement on a plain horizontal tube using a 90/10% mixture (by weight) of R-114 and ethanol. Also shown are data for boiling on mechanically enhanced tubes without EHD enhancement.

bubble breakup can be explained by the Taylor instability in the electric field on the liquid–gas interface, whose shape is deformed by the field.

Boiling mixtures have a smaller heat transfer coefficient than do pure fluids. Yabe theorized that the violent mixing at the heater surface should be beneficial for enhancement of mixture boiling. Figure 15.14 supports this proposal. This mechanism is supported by the visualization studies of Ohadi et al. [1992], who boiled oil–refrigerant mixtures using R-11 and R-123. The high-speed films showed that EHD force causes the number of bubbles to increase and causes the diameter of the bubbles to decrease. The bubble dynamics measurements by Kweon and Kim [2000] from R-113 pool boiling on a single wire support the preceding observation. The nucleation site density and the bubble frequency increased as the electric field density increased. The bubble diameter, however, decreased as the electric field density increased.

Cooper [1992] states that the electrode should be designed to promote a high degree of non-uniformity of the electric field at the surface. Figure 15.15 shows Cooper's [1990] results on a standard integral-fin tube using R-114. A non-uniform electric field was generated by a co-axial electrode. Figure 15.15 shows that significant hysteresis exists with no electric field (0 kV). The 0 kV data are for increasing (curve A) and decreasing (curve B) heat flux, respectively. However, application of the electric field totally eliminated hysteresis. Cooper argues that EHD forces activate nucleation sites at lower superheat than are required without an electric field. After

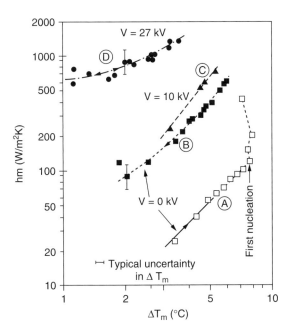

Figure 15.15 R-114 boiling at 21.5°C on a 1428 fins/m horizontal, integral-fin tube tested by Cooper [1990]. (From Cooper [1990].)

activation, the forces provided by the electric field further enhance nucleate boiling. This is shown by comparing the decreasing heat flux data at 0 kV (curve B), with the EHD enhanced data. The elimination of the boiling hysteresis under EHD was also reported by Oh and Kwak [2000] and Zaghdoudi and Lallemand [2001]. Zaghdoudi and Lallemand [2001] state that the electric field modifies the surface tension so that the fluid has nonwetting behavior. The disappearance of the hysteresis was attributed to the increase of the apparent surface tension with the electric field.

The above results have been corroborated by Damianidis et al. [1992], who tested plain, integral-fin, and a 1420 fins/m, sawtooth finned tube (designed for condensation enhancement) using R-114. The electric field increased the boiling coefficient for all three tube geometries. They also provide data on a nine-tube horizontal tube bundle.

15.5.2 Pool Boiling

Papar et al. [1993] investigated the effect of electrode geometry on EHD enhanced boiling of R-123/oil mixtures on a horizontal smooth tube. The electrode geometry included wire mesh, straight wire, and helical wire (Figure 15.3 a, b, and d). The mesh electrode performed better than the other geometries yielding up to a fivefold enhancement. The oil degraded the enhancement and increased the electric power consumption. Singh et al. [1995a] extended the study to 748 and 1575 fpm integral fin tubes. An 11.5-fold enhancement was obtained for the 1575 fpm integral fin tube when a mesh electrode was used. Yan et al. [1996] investigated the EHD effect on the pool boiling of high-performance boiling tubes Thermoexel-HE and GEWA-T (surface geometry shown in Figure 11.12). A mesh electrode was used to enhance R-114 boiling. Approximately 75% enhancement was obtained for the Thermoexel-HE. The enhancement was much smaller for GEWA-T. The boiling hysteresis was completely eliminated with the application of electric field. Darabi et al. [2000c] provide additional data on R-134a boiling of integral-fin tubes and Turbo-BIII.

Zaghdoudi and Lallemand [1999] investigated the effect of electric field polarity on nucleate pool boiling of *n*-pentane on a horizontal plate. The negative polarity configuration yielded a larger enhancement than the positive polarity configuration. They state that, for the positive polarity, the dielectrophoretic force (second term in Equation 15.1) is directed from the heated surface to the liquid. For the negative polarity, the dielectrophoretic force is directed from the liquid to the surface. The electrostriction force (third term in Equation 15.1) always acts to the surface independent of the polarity. Thus, stronger wall-bound electric force is manifested for the negative polarity compared with the positive polarity.

15.5.3 Convective Vaporization

Yabe et al. [1992] studied convective vaporization of R-123 and R-134a inside a circular tube. A three-factor enhancement was obtained at 33 kg/m²-s over a wide range of vapor qualities. The enhancement ratio decreased as the mass flux increased. More dramatic enhancement is obtained in nucleate pool boiling. Singh et al. [1994,

1995b] provide additional data on EHD enhancement of R-123 and R-134a in a circular tube. A similar level of enhancement to that measured by Yabe et al. [1992] was obtained. Bryan and Seyed-Yagoobi [2001] conducted R-134a boiling in a smooth tube. Interestingly, they noticed that, at high vapor qualities, the heat transfer coefficient was reduced with application of EHD. This phenomenon was more prominent at low mass fluxes and high heat fluxes. The electric force, which thins the liquid film by extracting the liquid from the wall, was suspected to cause an early dryout. The reduction of heat transfer coefficient at high qualities by EHD was also reported by Norris et al. [1999].

Ogata et al. [1992] tested R-123 boiling on a horizontal bundle of 50 plain tubes using six axial wire electrodes spaced at 3.0 mm from each tube. They show that the performance of the tube bundle is competitive with that of a single tube. An economic analysis of EHD enhancement is also provided. Cheung et al. [1995] conducted EHD enhanced boiling of R-134a in a bundle of seven integral fin tubes having 1575 fpm. Six straight wire electrodes were spaced at 3.5 mm from each tube. A threefold increase in the overall heat transfer coefficient was obtained at 5 kW/m². The enhancement decreased as the heat flux increased. When a circular mesh electrode (Figure 15.3c) was used, the enhancement increased to fivefold (Cheung et al. [1997]). Karayiannis [1998] investigated the EHD effect for R-123 and R-11 boiling on a horizontal tube bundle of five tubes. Fourteen straight electrodes were placed around the tubes. A 9.3-fold enhancement was obtained for R-123 at 5 kW/m². The enhancement decreased significantly as the heat flux increased. The enhancement with R-11 was only marginal. Karayiannis [1998] attributed the reason to the longer electric relaxation time of R-11 than the bubble detachment period. If the relaxation time is long, the electric forces generated on the bubbles are too weak to affect the bubble behavior significantly.

15.5.4 Critical Heat Flux

Zaghdoudi and Lallemand [2001] investigated the effect of EHD on CHF from a horizontal plate. For *n*-pentane and R-113, the increase in CHF with the electric field was 17 and 23%, respectively. For R-123, however, a threefold increase was obtained. They state that the electric field has an effect of decreasing the most critical wave length (refer to Equation 11.35), which leads to an increase of CHF. Yabe [1991] reported the 20% increase of CHF of R-113 when a uniform electric field of 20 kV/cm is applied.

15.5.5 Correlations

Cooper [1990] developed a correlation for EHD enhanced pool boiling heat transfer, extending the Rohsenow [1952] model.

$$h_E = a h_o \text{Ne}^{0.165} \text{Re}_o^b \qquad (15.7)$$

where h_o is the heat transfer coefficient obtained using the Rohsenow correlation, Ne is an electrical influence number, and Re_o is the boiling Reynolds number. The definitions of Ne and Re_o are provided in Cooper's paper. The constants a and b were obtained from the best fit of the available experimental data as $a = 0.3$ and $b = -0.16$.

15.6 CONCLUSIONS

EHD enhancement of boiling and condensing processes appears to offer exciting new possibilities. Data taken to date suggest that the performance of both plain and mechanically enhanced tube geometries is increased.

However, the case is not yet proved that EHD enhancement will lead to higher enhancement than may be obtained from well-designed mechanically enhanced surface geometries. More data are required using the same fluids and operating conditions for the EHD and mechanical enhancements.

NOMENCLATURE

d_i	Plain tube inside diameter, or diameter to the base of internal fins or roughness, m or ft
d_o	Tube outside diameter, fin root diameter for a finned tube, m or ft
E	Electric field strength, V/m or V/ft
F_e	Electric field body force, N or lbf
h	Heat transfer coefficient, W/m²-K or Btu/hr-ft²-°F
H	Distance between an electrode and a tube, m or ft
i	Electric current, amperes
L	Tube length, m or ft
Nu	Nusselt number, dimensionless
p	Pressure, Pa or lbf/ft²
R	Radius of tube or electrode, m or ft
Re_d	Reynolds number based on the tube diameter $= Gd/\mu$, $d = d_i$ for flow inside tube and $d = d_o$ for flow outside tube, dimensionless
Re_{Dh}	Reynolds number based on the hydraulic diameter $= GD_h/\mu$, dimensionless
Re_l	Condensate Reynolds number $(4\Gamma/\mu_l)$, dimensionless
Re_L	Condensate Reynolds number leaving vertical tube $(4\Gamma_L/\mu_l)$, dimensionless
r	Radial coordinate, m or ft
T	Temperature, °C or °F
t	Time, s
u	Fluid velocity, m/s or ft/s
V	Electric potential, V
W	Condensate flow rate leaving tube, kg/s or lbm/s

Greek Symbols

β Condensate flooding angle (defined in Figure 12.5), degree

Γ Condensate flow rate, per unit tube length, leaving horizontal tube, kg/s-m or lbm/s-ft

Γ_L Condensate flow rate, per unit plate width, leaving vertical plate, kg/s-m or lbm/s-ft

∇ Gradient, dimensionless

ε Fluid permittivity (dielectric constant of fluid), dimensionless

μ Fluid viscosity, kg/m-s or lbm/ft-s

ρ Fluid density, kg/m^3 or lbm/ft^3

ρ_c Electric field space charge density, C/m^3 or C/ft^3

σ_e Electrical conductivity, A/V-m or A/V-ft

Subscripts

g Vapor

l Liquid

REFERENCES

Al-Ahmadi, A. and Al-Dadah, R.K., 2002. A new set of correlations for EHD condensation heat transfer of tubular systems, *Appl. Thermal Eng.*, 22, 1981–2001.

Blanford, M.D., Ohadi, M.M., and Dessiatoun, S.V., 1995. Compound air-side heat transfer enhancement in a cross-flow refrigerant-to-air heat exchanger, *ASHRAE Trans.*, 101 (Part 2), 1049–1054.

Bologa, M.K., Savin, I.K., and Didovsky, A.B., 1987. Electric-field-induced enhancement of vapor condensation heat transfer in the presence of a non-condensible gas, *Int. J. Heat Mass Transfer*, 30, 1577–1585.

Bryan, J.E. and Seyed-Yagoobi, Y., 2001. Influence of flow regime, heat flux, and mass flux on electro-hydrodynamically enhanced convective boiling, *J. Heat Transfer*, 123, 355–367.

Cheung, K., Ohadi, M.M., and Singh, A., 1997. EHD-enhanced boiling coefficients and visualization of R-134a over enhanced tubes, *J. Heat Transfer*, 119, 332–338.

Cheung, K., Ohadi, M.M., and Dessiatoun, S.V., 1999. EHD-assisted external condensation of R-134a on smooth horizontal and vertical tubes, *Int. J. Heat Mass Transfer*, 42, 1747–1755.

Cheung, K.H., Ohadi, M.M., and Dessiatoun, S., 1995, Compound enhancement of boiling heat transfer of R-134a in a tube bundle, *ASHRAE Trans. Symp.*, 101, (Part 1), 1009–1019.

Choi, H.Y. and Reynolds, J.M., 1965. Study of Electrostatic Fields on Condensing Heat Transfer, Air Force Dynamics Laboratory Report AFFDL-TR-65-51, Wright-Patterson Air Force Base, Ohio.

Chu, R.C., Nishio, S., and Tanasawa, I., 2001. Enhancement of condensation heat transfer on a finned tube using an electric field: experiment and modeling analysis on enhancement of heat transfer using a bare wire electrode, *J. Enhanced Heat Transfer*, 8, 99–118.

Cooper, P., 1990. EHD enhancement of nucleate boiling, *J. Heat Transfer*, 112, 458–464.

Cooper, P., 1992. Practical design aspects of EHD heat transfer enhancement in evaporators, *ASHRAE Trans.*, 98, (Part 2), 445–454.

Da Silva, L., Molki, M., and Ohadi, M.M., 2001. Effect of polarity on electrohydrodynamic enhancement of R-134a condensation on enhanced tubes, in *Proceedings of 35th National Heat Transfer Conference*, Anaheim, CA, NHTC 2001–20067.

Damianidis, C., Karayinnis, T., Al-Dadah, R.K., James, R.W., Collins, M.W., and Allen, P.H.G., 1992. EHD boiling enhancement in shell-and-tube evaporators and its application to refrigeration plants, *ASHRAE Trans.*, 98, (Part 2), 462–473.

Darabi, J., Ohadi, M.M., and Dessiatoun, S.V., 2000a, Falling film and spray evaporation enhancement using an applied electric field, *J. Heat Transfer*, 122, 741–748.

Darabi, J., Ohadi, M.M., and Dessiatoun, S.V., 2000b. Augmentation of thin falling-film evaporation on horizontal tubes using an applied electric field, *J. Heat Transfer*, 122, 391–398.

Darabi, J., Ohadi, M.M., and Dessiatoun, S.V., 2000c. Compound augmentation of pool boiling on three selected commercial tubes, *J. Enhanced Heat Transfer*, 7, 347.

Fernandez, J. and Poulter, R., 1987. Radial mass flow in electrohydrodynamically-enhanced forced heat transfer in tubes, *Int. J. Heat Mass Transfer*, 30, 2125–2136.

Gidwani, A., Molki, M., and Ohadi, M.M., 2002. EHD-enhanced condensation of alternative refrigerants in smooth and corrugated tubes, *Int. J. HVAC&R Res.*, 8(3), 219–238.

Hasegawa, M., Yabe, A., and Nariai, H., 1999. Turbulent generation and mechanism analysis of forced convection heat transfer enhancement by applying electric fields in the restricted region near the wall, *Proc. 5th ASME/JSME Thermal Eng. Joint Conf.*, Paper ATJE99-6380.

Ishiguro, H., Nagata, S., Yabe, A., and Nariai, 1991. Augmentation of forced-convection heat transfer by applying electric fields to disturb flow near a wall, *Heat Transfer Science and Technology*, B.-X. Wang, Ed., Hemisphere, New York, 25–31.

Jones, T.B., 1978. Electohydrodynamically enhanced heat transfer in liquids—a review,. *Adv. Enhanced Heat Transfer*, 14, 107–148.

Karayiannis, T.G., 1998. EHD boiling heat transfer enhancement of R123 and R11 on a tube bundle, *Appl. Thermal Eng.*, 18, 809–817.

Kuffel, E. and Zaengl, W.S., 1984. *High-Voltage Engineering*, Pergamon, Oxford.

Kweon, Y.C. and Kim, M.H., 2000. Experimental study on nucleate boiling enhancement and bubble dynamic behavior in saturated pool boiling using a nonuniform DC electric field, *Int. J. Multiphase Flow*, 26, 1351–1368.

Levy, E., 1964. Effects of Electrostatic Fields on Forced Convection Heat Transfer, M.S. thesis, Massachusetts Institute of Technology, Cambridge, MA.

Molki, M., Ohadi, M.M., and Bloshteyn, M., 2000a. Frost reduction under intermittent electric field, *Proceedings of 34th National Heat Transfer Conference*, Pittsburgh, PA, NHTC 2000-12052.

Molki, M., Ohadi, M.M., Baumgarten, B., Hasegawa, M., and Yabe, A., 2000b. Heat transfer enhancement of airflow in a channel using corona discharge, *J. Enhanced Heat Transfer*, 7, 411.

Munakata, T., Yabe, A., and Tanasawa, I., 1993. Effect of electric fields on frosting phenomenon, *The 6th International Symposium on Transport Phenomena in Thermal Engineering*, pp. 381–386.

Nelson, D.A., Ohadi, M.M., Zia, S., and Whipple, R.L., 1991. Electrostatic effects on heat transfer and pressure drop in cylindrical geometries, *Heat Transfer Science and Technology*, B.-X. Wang, Ed., Hemisphere, New York, 33–39.

Norris, C.E., Cotton, J.S., Shoukri, M., Chang, J.-S., and Smith-Pollard, T., 1999. Electrohydrodynamic effects on flow redistribution and convective boiling in horizontal concentric tubes under high inlet quality conditions, *ASHRAE Trans.*, 105 (Part 1), 222–236.

Ogata, J. and Yabe, A., 1991. Augmentation of nucleate boiling heat transfer by applying electric fields: EHD behavior of boiling bubble, *Heat Transfer Science and Technology*, B.-X. Wang, Ed., Hemisphere, New York, 41–46.

Ogata, J., Iwafuji, Y., Shimada, Y., and Yamaziki, T., 1992. Boiling heat transfer enhancement in tube-bundle evaporator utilizing electric field effects, *ASHRAE Trans.*, 98 (Part 2), 435–444.

Oh, S.-D. and Kwak, H.-Y., 2000. A study of bubble behavior and boiling heat transfer enhancement under electric field, *Heat Transfer Eng.*, 21(4), 33–45.

Ohadi, M., Faani, M., Papar, R., Radermacher, R., and Ng, T., 1992. EHD heat transfer enhancement of shell-side boiling heat transfer coefficients of R-123/oil mixture, *ASHRAE Trans.*, 98 (Part 2), 427–434.

Ohadi, M., Sharaf, N. and Nelson, D., 1991b. Electrohydrodynamic enhancement of heat transfer in a shell-and-tube heat exchanger, *Exp. Heat Transfer*, 4(1), 19–39.

Ohadi, M.M., 1991. Heat transfer enhancement in heat exchangers, *ASHRAE J.*, 33(12), 42–50.

Ohadi, M.M., Nelson, D.A., and Zia, S., 1991a. Heat transfer enhancement of laminar and turbulent pipe flows via corona discharge, *Int. J. Heat Mass Transfer*, 34, 1175–1187.

Ohadi, M.M., Li, S.S., and Dessiatoun, S., 1994. Electrostatic heat transfer enhancement in a tube bundle gas-to-gas heat exchanger, *J. Enhanced Heat Transfer*, 1, 327.

Panofsky, W. and Phillips, M., 1962. *Classical Electricity and Magnetism*, 2nd ed., Addison-Wesley, Reading, MA, pp. 107–116.

Papar, R.A., Ohadi, M.M., Kumar, A., and Ansari, A.I., 1993. Effect of electrode geometry on EHD-enhanced boiling of R-123/oil mixture, *ASHRAE Trans.*, 99 (Part 1), 1237–1243.

Paschkewitz, J.S. and Pratt, D.M., 2000. The influence of fluid properties on electrohydrodynamic heat transfer enhancement in liquids under viscous and electrically dominated flow conditions, *Exp. Thermal Fluid Sci.*, 21, 187–197.

Rohsenow, W.M., 1952. A method of correlating heat transfer data for surface boiling of liquids, *ASME Trans.*, 74, 969–976.

Singh, A., Ohadi, M.M., Dessiatoun, S., and Chu, W., 1994. In-tube boiling heat transfer enhancement of R-123 using the EHD technique, *ASHRAE Trans.*, 100 (Part 2), 818–825.

Singh, A., Ohadi, M.M., and Dessiatoun, S., 1995a. EHD-enhanced boiling of R-123 over commercially available enhanced tubes, *J. Heat Transfer*, 117, 1070–1073.

Singh, A., Ohadi, M.M., Dessiatoun, S., and Chu, W., 1995b. In-tube boiling heat transfer coefficients of R-123 and their enhancement using the EHD technique, *J. Enhanced Heat Transfer*, 2, 209.

Singh, A., Ohadi, M., and Dessiatoun, S., 1997. EHD enhancement of in-tube condensation heat transfer of alternate refrigerant, R-134a, *ASHRAE Trans.*, 103(1), Symp. PH-97-10.

Sunada, K., Yabe, A., Taketani, T., and Yoshizawa, Y., 1991. Experimental study of EHD pseudo-dropwise condensation, *Proc. Third ASME/JSME Thermal Eng. Conf.*, 3, 47–53.

Uemura, M., Nishiio, S., and Tanasawa, I., 1990. Enhancement of pool boiling heat transfer by static electric field, *Proc. of the Ninth International Heat Transfer Conference*, Jerusalem, Israel, 4, 69–74.

Velkoff, H.R. and Miller, T.H., 1965. Condensation of vapor on a vertical plate with a transverse electrostatic field, *J. Heat Transfer*, 87, 197–201.

Wawzyniak, M. and Seyed-Yagoobi, J., 1996. Experimental study of electrohydrodynamically augmented condensation heat transfer on a smooth and an enhanced tube, *J. Heat Transfer*, 118, 499–501.

Yabe, A., Taketani, T., Maki, H., Takahashi, K., and Nakadai, Y., 1992. Experimental study of electro-hydrodynamically (EHD) enhanced evaporator for nonazeotropic mixtures, *ASHRAE Trans.*, 98 (Part 2), 455–461.

Yabe, A., Taketani, T., Kikuchi, K., Mori, Y., and Hijikata, K., 1987. Augmentation of condensation heat transfer around vertical cooled tubes provided with helical wire electrodes by applying nonuniform electric fields, in *Heat Transfer Science and Technology*, Bu.-X. Wang, Ed., Hemisphere, Washington, D.C., 812–819.

Yabe, A., 1991. Active heat transfer enhancement by applying electric fields, *Proc. Third ASME/JSME Thermal Eng. Conf.*, 3, xv–xxiii.

Yabe, A. and Maki, N., 1988. Augmentation of convective and boiling heat transfer by applying an electro-hydrodynamical liquid jet, *Int. J. Heat Mass Transfer*, 31, 407–417.

Yamashita, K. and Yabe, A., 1997. Electrohydrodynamic enhancement of falling film evaporation heat transfer and its long-term effect on heat exchangers, *J. Heat Transfer*, 119, 339–347.

Yamashita, K., Kumagai, M., Sekita, S., Yabe, A., Taketani, T., and Kikuchi, K., 1991. Heat transfer characteristics of an EHD condenser, *Proc. Third ASME/JSME Joint Thermal Eng. Conf.*, Reno, Nevada, 61–67.

Yan, Y.Y., Neve, R.S., and Allen, P.H.G., 1996. EHD effects on nucleate boiling at passively enhanced surfaces, *Exp. Heat Transfer*, 9(3), 195–212.

Zaghdoudi, M.C. and Lallemand, M., 1999. Analysis of the polarity influence on nucleate pool boiling under a DC electric field, *J. Heat Transfer*, 121, 856.

Zaghdoudi, M.C. and Lallemand, M., 2001. Nucleate pool boiling under DC electric field, *Exp. Heat Transfer*, 14(3), 157–180.

SIMULTANEOUS HEAT AND MASS TRANSFER

16.1 INTRODUCTION

There are a number of convective heat transfer processes that also involve convective mass transfer. Typically, these processes involve two-phase heat transfer involving either condensation or evaporation of mixtures, one of which may be inert at the process conditions. Examples of evaporation processes include vaporization of binary fluids, air humidification, and desorption of gases from liquid mixtures. Heat/mass transfer in a cooling tower is an excellent example of air humidification. Examples of condensation processes include condensation of mixtures or condensation with noncondensible gases, air dehumidification, and absorption of vapor in a liquid film. The cooling of moist air involves both heat and mass transfer. These problems involve simultaneous heat and mass transfer. A mass transfer resistance exists, because the active component must diffuse through the possibly inert component. Enhancement techniques offer excellent possibilities for performance improvement.

The processes differ depending whether the mass transfer resistance exists in the gas phase, the liquid phase, or both. Important examples of each type of problem are presented in the following sections.

Many applicable processes are within the domain of chemical engineering heat transfer, which involve separation of mixtures. This chapter does not intimately delve into this complex area of transport phenomena of mixtures. Rather, the intent is to illustrate how enhancement may be applied to such processes.

16.2 MASS TRANSFER RESISTANCE IN THE GAS PHASE

This class of problems involves either condensation onto a liquid film or evaporation from a liquid film. In many problems, the inert component is air, although it may be any inert component.

16.2.1 Condensation with Noncondensible Gases

Figure 16.1 shows the cross section of a tube wall with a condensing vapor on one side and cooling water on the other side. The curves in the condensing region show the temperature and pressure distributions. The dashed line profiles indicate conditions that would exist if a noncondensible gas were not present. The solid line shows the situation with noncondensible gas. For a pure vapor, there is no temperature drop between the bulk vapor and the liquid–vapor interface, where condensation occurs; thus, the interface temperature (T_i) is equal to the saturation temperature (T_s) at the system pressure. With noncondensible gas present, the value of T_i is reduced, yielding a smaller value of $(T_i - T_s)$. The $(T_i - T_s)$ is reduced for two reasons:

1. Because the condensing vapor exists at its partial pressure, its saturation temperature is reduced.
2. Because the condensing vapor must diffuse through the vapor–gas mixture, the vapor partial pressure at the interface is less than that of the bulk mixture. This reduces the saturation temperature at the interface.

The first reduction of $(T_i - T_s)$ is directly related to the volume concentration of the noncondensible gas. However, the second reduction of $(T_i - T_s)$ is inversely proportional to the convective heat transfer coefficient of the gas–vapor mixture. High vapor velocity or enhancement will reduce the temperature drop across the gas boundary layer.

Because the vapor is saturated at the liquid–vapor interface, the vapor in the bulk mixture at T_{vb} is superheated. Because $T_{vb} > T_i$, there is a sensible heat flow though the gas to the liquid–vapor interface, where condensation occurs. Neglecting subcooling of the condensate, the heat transfer rate to the coolant is the sum of the sensible and latent portions. The heat transfer rate from the vapor mixture to the interface is given by

$$q = h_g \alpha_s (T_{vb} - T_i) + K_p i_{gv} (p_{vb} - p_{vi}) \tag{16.1}$$

The first term of Equation 16.1 is for sensible heat transfer from the vapor, and the second term is the latent heat transfer. The α_s term is the Ackermann correction factor, which accounts for the effect of the mass flux on the sensible heat transfer rate. The gas-phase mass transfer coefficient (K_p) may be obtained from the heat-mass transfer analogy,

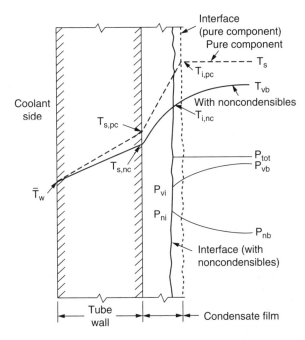

Figure 16.1 Temperature and pressure profiles for condensation of pure vapor (solid lines) and with noncondensible gas (dashed line). (From Webb et al. [1980].)

$$\frac{h_g}{c_p} \Pr^{2/3} = K_p p M \mathrm{Sc}^{2/3} \tag{16.2}$$

where h_g is for single-phase flow of vapor in the geometry of interest. Use of Equation 16.2 to obtain K_p assumes that h_g for the condensation situation is the same as for all gas flow. Roughness on the liquid film may act to slightly enhance h_g for the condensation situation.

Equation 16.2 shows that K_p is directly proportional to the gas-phase heat transfer coefficient (h_g). Hence, techniques applicable to enhancement of the gas-phase heat transfer coefficient will also be effective in enhancing the mass transfer coefficient. Note that the enhancement is required at the gas–liquid interface, rather than at the pipe wall. Therefore, selection of the enhancement should account for the effect of the liquid film thickness.

Section 14.5 describes an enhancement technique that increased the h_g term for shell-side condensation in an axial flow condenser. This consisted of a two-dimensional rib roughness on the tube outer surface. The roughness increased the gas-phase heat transfer coefficient, and, hence, the mass transfer coefficient. Any enhancement technique that is effective for convective heat transfer to gases should be beneficial to convective condensation with noncondensible gases.

Webb [1991] discusses the heat-mass transfer analogy (Equation 16.2) and gives a standard nomenclature for mass transfer processes. The nomenclature of Webb [1991] is used here.

16.2.2 Evaporation into Air

Evaporation from a water film into an air stream is an important process and is used in several engineering applications. Perhaps the most important is the cooling tower. Other applications include the humidifier, evaporative fluid coolers, and evaporative condensers. The heat transfer process involves evaporation from a water film into a flowing air stream. The controlling mass transfer resistance is in the gas boundary layer. Here, how enhancement may be applied to enhance the gas-phase mass transfer coefficient is discussed.

Cooling Tower Packings. Cooling tower packings are good candidates for enhancement. In this case, water evaporated at the liquid film surface is transferred by convective heat and mass transfer to the moist air, which flows over the surface, Figure 1.17 shows typical film packings used in a cooling tower. Both packings provide enhancement for the gas flow. Equation 16.1 is applicable to the cooling tower problem, as discussed by Webb [1988]. However, the latent heat term is typically written in terms of the specific humidity (W). Then, Equation 16.1 becomes

$$q = h_g \alpha_s (T_{vb} - T_i) + K_w i_{gv} (W_b - W_i) \tag{16.3}$$

Equation 16.3 is frequently expressed in terms of the enthalpy driving potential, which involves use of the heat-mass transfer analogy to express h_g in terms of K_w. The relation between K_w, K_p, and h_g is given by Webb [1991] and is

$$K_W = pM(1-y)^2 K_p = \frac{h_g}{c_p}(1-y)^2 \left(\frac{\text{Pr}}{\text{Sc}}\right)^{2/3} \tag{16.4}$$

Equation 16.4 shows that a surface geometry that provides high h_g will also provide a high mass transfer coefficient. Substitution of h_g in terms of K_w into Equation 16.3, along with use of approximations, allows one to combine the sensible and latent terms. The result is a driving potential defined in terms of the moist air enthalpy (i). This derivation is shown by Webb [1988]. Thus,

$$q = K_W(i_b - i_i) \tag{16.5}$$

PEC Example 16.1. The airflow through the Figure 1.17a ceramic cooling tower packing is similar to that which occurs in the offset strip-fin (OSF) geometry (Figure 5.3). Hence, one may use correlations for the OSF to estimate the mass transfer coefficient in the ceramic cooling tower packing. Assume that air flows at 3.8 m/s and 32°C in the ceramic cooling tower packing. Predict the mass transfer coefficient using the heat-mass transfer analogy. The ceramic block is 150 mm high, and has 45.0-mm square cells with 6.4-mm membrane thickness. Using the heat-mass transfer analogy, the mass transfer coefficient is $K_w \simeq h_g/c_p (Pr/Sc)^{2/3}$, where h_g is the heat transfer coefficient for airflow in the packing. The Reynolds number, based on the cell length, is given by

$$\mathrm{Re}_L = \frac{L_p u}{v} = \frac{0.15 \times 3.8}{16.2E\text{-}6} = 35{,}185 \tag{16.6}$$

Using the OSF correlation (Equation 5.10), which uses Re_{Dh} (= 10,555) gives a j factor of 0.0028. Then, $h_g = j \, (\rho u c_p \, P_r^{-2/3}) = 23.9$ W/m²–K. Using the heat-mass transfer analogy (Equation 16.4), one obtains

$$K_W = \frac{h_g}{c_p}\left(\frac{Pr}{Sc}\right)^{2/3} = \frac{23.9}{1007}\left(\frac{0.7}{0.6}\right)^{2/3} = 0.0263 \, \text{kg/s-m}^2 \tag{16.7}$$

Enhancement for Evaporation into an Air Stream. Webb and Perez-Blanco [1986] used gas-phase enhancement for a process involving evaporation of water into an air–steam mixture flowing inside a vertical tube. Figure 16.2 shows a water film draining down the inside surface of a round vertical tube, with moist air in counterflow. They placed a $p/e = 10$ transverse-rib roughness at the edge of the gas boundary layer to provide enhancement of h_g. The roughness was displaced from

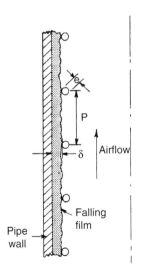

Figure 16.2 Transverse-rib concept applied to enhance mass transfer in the vapor boundary layer. (From Webb and Perez-Blanco [1986].)

the wall a distance equal to the calculated liquid film thickness, so that the ribs were not contained in the liquid film. The heat transfer coefficient of the rib roughness was calculated using the correlation of Webb et al. [1970] described in Section 9.4.1. The wire of diameter e was chosen to operate at roughness Reynolds number (Equation 9.9) of $eu^*v = 20$, as recommended in Section 9.7.3. They obtained a 38% enhancement of the gas-phase heat transfer and mass transfer coefficients.

16.2.3 Dehumidifying Finned-Tube Heat Exchangers

Finned-tube heat exchangers used for cooling of air will experience moisture condensation on the fin surface, if the fin surface temperature is below the dewpoint temperature. This is typical for refrigerant evaporators. Equation 16.1 (or Equation 16.3 or 16.5) applies to this process. Hence, a surface geometry having a high h_g should also enhance the mass transfer coefficient and provide high performance under wet conditions. Moisture condensation on the fin surface provides a naturally occurring enhancement.

Senshu et al. [1981] measured the performance of the Figure 6.7 convex-louver-finned tube geometry, under dehumidifying conditions for which moisture condensation occurred on the fin surface. Their tests were run using 5°C water flow in the tubes. The Figure 6.7 surface geometry provides high enhancement for dry airflow. The heat transfer rate for the wetted surface condition was then predicted in terms of the enthalpy driving potential (e.g., Equation 16.5). In this case the mass transfer is from the bulk air to the interface, so the equation is written as

$$q = K_W(i_b - i_i) \tag{16.8}$$

The authors used their measured h_g (obtained without moisture condensation) and the heat-mass transfer analogy relation (Equation 16.4) to obtain K_w. For their steam–air mixture, they used the acceptable approximation $(1-y)^2 (Pr/Sc)^{2/3} \simeq 1.0$. Thus, $K_w = h_g/c_p$. The predicted K_w was compared with the experimental value obtained using Equation 16.8. Equation 16.8 requires evaluation of the moist air enthalpy at the fin surface temperature. Senshu et al. [1981] developed an equation for this purpose. They showed that their K_w obtained from h_g/c_p was within ±5% of measured value. The good prediction of K_w using the heat-mass transfer analogy shows that the draining condensate did not bridge the fin louvers or substantially alter the airflow pattern over the louvers.

The above theory is applicable, provided that the surface is fully wetted. Or, it is applicable to the portion of the surface that is wetted. Because the air must be cooled to the dewpoint temperature, it is probable that a fraction of the entering air region of the heat exchanger will not be wetted.

The j and f factors of finned-tube heat exchangers under dehumidifying conditions have been studied by several investigators. The data reduction issue of wet surface heat transfer is addressed in Section 6.13. The log mean enthalpy method (LMED) is recommended. Wang et al. [1997] tested nine staggered plain finned tube heat exchangers with $1.82 \leq p_f \leq 3.20$ mm and $2 \leq N \leq 6$. The j factors were reduced using the LMED method. The j factors under wet conditions were comparable with those under the dry condition. Similar to the dry case, the effect of fin spacing on the j factor was negligible. The row effect, however, was less pronounced compared with the dry case. The reason was attributed to the retained condensate on the lower part of the tube, which enhances the downstream mixing. The friction factors of wet coils were much greater than those of the dry coils. Similar conclusions were drawn for fin-and-tube heat exchangers having louvered fins (Wang et al. [2000]). The wet surface pressure drop can be significantly reduced by applying a hydrophilic coating on the fin surface. This issue is addressed in Section 6.13.

16.2.4 Water Film Enhancement of Finned-Tube Exchanger

In Section 16.2.3, heat was transferred from the air stream to the fin surface, with condensation occurring at the liquid–vapor interface. The reverse situation (evaporation from a wetted surface into the air stream) is of interest here. The concept involves spraying water on the finned-tube heat exchanger. The intent is to cause the water to flow as a film on the fin surface, with water evaporation from the film surface. Unless good surface wetting is achieved, the concept will not work well. Tests have been done by spraying water on operating heat exchangers. Performance data are obtained for both the dry and the wetted conditions. Quantification of the degree of surface wetting and drainage characteristics of the water film typically has not been investigated. Among the earliest work to measure the effect of water spray on heat exchanger performance was by Yang and Clark [1975], who sprayed a water mist on the face of an automotive radiator. They found only a 10% increase of performance. Apparently, the water did not wet the fins.

Hudina and Sommer [1988] used mist cooling to enhance two different plate-finned tube heat exchangers (Figure 6.1a) having plain fins. The 1.04 m^2 frontal area heat exchangers were tested in a wind tunnel using upstream water spray nozzles. The heat exchanger dimensional data are listed in Table 16.1. The dimensions differ only in the tube pitch.

Figure 16.3 shows their heat transfer and pressure drop test results for fixed water spray rate, tube-side flow rate (7.3 kg/s), 50% relative humidity, 293 K water spray temperature, and inlet air temperatures between 298 and 303 K. Figure 16.3a

Table 16.1 Geometries Tested by Hudina and Sommer [1988]

Geometry Code	Fs	Fo
Tube diameter (d_o)	17.3	17.3
Transverse tube pitch (S_t)	60	57
Longitudinal tube pitch (S_l)	25	45
Number of tube rows	6	6
Fin pitch	0.028	0.028
Fin thickness	0.30	0.30

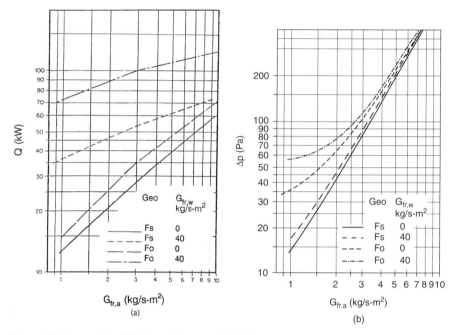

Figure 16.3 Test results of Hudina and Sommer [1988] for spray enhancement of the Table 16.1 finned tube heat exchangers. (a) Heat transfer enhancement, (b) pressure drop enhancement. (From Hudina and Sommer [1988].)

shows that the enhancement level decreases with increasing air mass velocity. Figure 16.3a shows an enhancement level of 5.0 was obtained for the Fo geometry at 1.0 kg/m^2-s air mass velocity. Figure 16.3b shows that the pressure drop increase is less than 15% for the Fs geometry. However, a much larger pressure drop enhancement occurs for the Fo geometry at $G_a < 3$ kg/m^2-s. Note that the Fo geometry was operated at a 2.6 times higher water spray rate than for the Fs geometry. Figure 16.4 shows the effect of water spray rate on the heat transfer enhancement level. The enhancement level increased with increasing water spray rate.

Associated bench-scale experiments by Sommer [1984] showed that a uniform, thin film does not exist over the fin surface. At low ITD ($T_{wl} - T_{al}$), Sommer showed that the sprayed water is collected on the fins near the air inlet in the form of large droplets, which eventually bridge the fin spacing. The water will drain downward by gravity. As the ITD (or air velocity) is increased, water bridging will be less, but the atomized water may pass through the heat exchanger as fine entrained droplets. Because of its high contact angle, water will not wet the fin surface unless the surface is moderately oxidized or specially treated.

Kreid et al. [1983b] tested three different fin-and-tube geometries (Figure 16.5) in a wind tunnel. The dimensions of the aluminum fin heat exchangers are given in Table 16.2, and were 0.61 m wide and 1.8 m high. The heat exchangers were oriented with the fins in the vertical direction to facilitate water drainage. The water was supplied by what is called the deluge method. This involves flowing the water onto

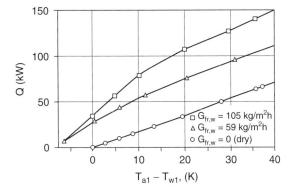

Figure 16.4 Effect of water spray rate on heat transfer enhancement for Fs finned-tube geometry of Table 16.1 tested by Hudina and Sommer [1988]. (From Hudina and Sommer [1988].)

Table 16.2 Geometries Tested by Kreid et al. [1983b]

	Geometry		
	A	B	C
$A_o/A_{fr}N$ (m²/m²)	13.88	27.05	14.70
A_c/A_{fr} (m²/m²)	0.495	0.534	0.509
S_t (mm)	25.0	38.0	30.93
S_l (mm)	60.0	38.0	35.7
D_h (mm)	3.87	2.96	3.18
d_o (mm)	18.0	16.0	16.0
t_f (mm)	0.43	0.22	0.20
Fins/m (m⁻¹)	354	393	315
N	6	3	4
d_i (mm)	20.2	13.4	13.4

the heat exchanger from a supply manifold above the heat exchanger. They used a water deluge rate approximately ten times the amount evaporated. Surface B provided the highest performance for a fixed air pressure drop. The heat transfer rate of sprayed surface B was two to five times that for operation without water spray. The water did not flow as a film on surfaces A and C; rather it broke up into droplets, which significantly increased the air pressure drop. They did not precisely determine to what degree the water wetted the entire fin surface as a film. Kreid et al. [1978, 1983a] developed an analytical model, based on use of Equation 16.3. The model, which assumes that the surface is fully wetted, moderately predicted the results under wet conditions. These studies show that, with a good mechanism for wetting, the heat transfer performance of a finned tube heat exchanger can be significantly increased. The Kreid et al. [1983b] tests were part of a program described by Johnson et al. [1981] to develop a water-enhanced dry cooling tower.

Bentley et al. [1978] performed a bench-scale laboratory investigation of a vertical plate having spaced, gravity-drained water channels. They did not attempt to wet the entire surface. Rather, they sought to enhance the dry surface performance by evaporation from the surface of spaced, discrete water channels. The water channels occupied about 5% of the total area. Their tests showed that the heat transfer by evaporation varied between 20 and 40% of the total heat transfer. Their aluminum plate provided enhancement for dry heat transfer by use of repeated-rib roughness, whose channels also drained water. However, they obtained only 20% improvement in dry heat transfer over a plain surface at the same fan power. This dry enhancement level is quite small, relative to currently used dry tower surfaces. They did not address practical means to supply and distribute the water to finned-tube banks.

Successful evaporative enhancement of a dry cooling tower fin surface requires that a significant fraction of the surface be wetted, and that the dry air heat transfer coefficient be enhanced over that for a plain surface. Further, the surface must be continuously wetted during the evaporative enhancement period, or mineral deposits and corrosion will occur. The excess water drained from the surface should prevent accumulation of corrosion forming deposits, provided that de-wetting of the surface does not occur. The particular challenge is how to obtain a high air-side heat transfer coefficient, and to continuously wet a significant fraction of the surface. To maintain a continuously wetted surface, it is necessary to flow an amount of water in excess of that evaporated. Not all enhanced fin geometries are amenable to surface wetting

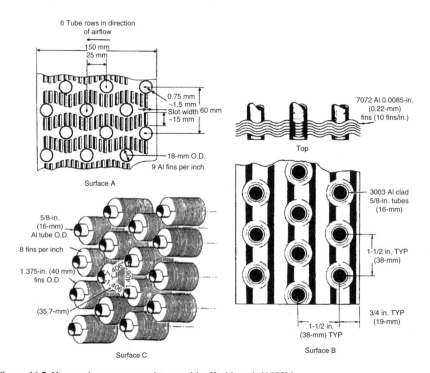

Figure 16.5 Heat exchanger geometries tested by Kreid et al. [1983b].

and water drainage. The Figure 16.5 surface A interrupted strip fins would prevent lateral spreading of a water film, which enters from the front in a spray. The concave channel elements of the Figure 16.5 surface B wavy fins should provide natural drainage channels for a water film introduced at the top of the heat exchanger. In this case, the entire fin surface would not be wetted. The water–air interface should exist in the concave fin channels.

Commercial coatings exist that reduce the contact angle of water on a surface and that promote surface wetting. Use of such coatings will reduce the amount of excess water that must be applied to the surface.

16.3 CONTROLLING RESISTANCE IN LIQUID PHASE

Examples of this problem are the absorption process, and the condensation of mixtures. In the absorption process, a vapor (e.g., steam) is condensed at the surface of a liquid film, and the condensed liquid is absorbed in a binary solution (e.g., ammonia–water). If the condensing vapor is a pure component, there is no mass transfer resistance in the vapor phase. The mass transfer resistance exists in the liquid film. Enhancement is needed in the liquid film, which has both heat and mass transfer resistances. One should determine whether the limiting resistance is heat or mass transfer. In the present problem illustrated, the mass transfer resistance is expected to be the limiting resistance. The primary mass transfer resistance exists at the liquid–vapor interface, not at the tube wall. Hence, reduction of the mass transfer resistance requires mixing of the liquid film at the liquid–vapor interface. Mixing of the liquid film at the wall will be beneficial to the heat transfer resistance. An extended surface would benefit the thermal resistance, but probably not the mass transfer resistance.

This situation also applies to problems involving boiling of mixtures. Mixtures have a lower heat transfer coefficient than do pure components, because of the added mass transfer resistance in the liquid phase.

Specially configured tubes for use in the LiBr absorption process were tested by Miller and Perez-Blanco [1993]. The tests were conducted using 62 wt% LiBr concentration, 10 mm Hg absolute pressure with a single tube in the vertical orientation. The tubes tested were three spine-fin tubes, one spirally indented tube, one spirally fluted tube, and one ribbed tube. Photographs of the tubes are shown in Figure 16.6. The three spine fin tubes had fin pitches of 6.4, 4.8, and 3.2 mm. Each tube had 26 fins per turn, and the fin height was 3.2 mm. The fins were integral to the 21.3-mm-OD stainless steel tube wall. The 22.2-mm-OD spirally indented tube had four start 45° angled grooves and was made of carbon steel. The spirally fluted tube had 24 starts and 22° angled grooves of 0.64 mm depth. The tube was made by machining a 19.1 mm stainless steel tube. The ribbed tube had 1.5-mm rib height and 6.4-mm rib pitch and was machined from a 19.1-mm-diameter stainless steel tube. All tube lengths were 1.53 m. The cooling water inlet temperature was controlled to 46°C flowing at 6.309E-5 m³/s and was in counterflow to the falling film. Figure 16.7 shows the mass of absorbed water vapor plotted against falling film mass flow rate. The mass absorbed by the 6.4-mm fin pitch spine fin tube is 125% larger than the mass absorbed by the smooth tube. Note that the outside diameter of the spine fin tube and the spirally indented tube is 11.5 and 16.2% larger than that of the smooth tube (OD = 19.1 mm). The absorbed mass of the grooved tube is 75% larger than that of the smooth tube, followed by the spirally indented tube (approximately 50% larger), and then the spirally fluted tube (slightly better). Figure 16.7 shows that, for the enhanced tubes, the absorbed mass increases as the falling film mass flow rate increases. However, the absorbed mass tends to level off for the smooth tube with film flow rates above 0.025 kg/s. This suggests that enhanced tubes are more effective for higher falling film flow rates. The performance of spine-fin tubes having 3.2- and 4.8-mm fin pitch were lower than that for 6.4-mm fin pitch. Flow visualization results revealed thicker films as the fin pitch decreased.

Yoon et al. [2002] obtained LiBr absorption heat transfer coefficients in a horizontal staggered bundle (6 × 8) of enhanced tubes. Three tube geometries—smooth, hydrophilic coated, and axially fluted were tested. The effect of additive (normal octyl alcohol) was also investigated. All the tubes had 15.88 mm OD. The hydrophilic tube was made by plasma coating on the smooth tube. The axially fluted

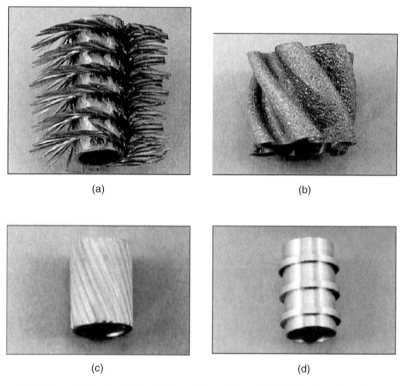

(a)

(b)

(c)

(d)

Figure 16.6 Enhanced tubes tested by Miller and Perez-Blanco [1993] for absorption of water vapor into LiBr solution: (a) spine fin tube, (b) spirally indented tube, (c) spirally fluted tube, (d) ribbed tube. (From Miller and Perez-Blanco [1993].)

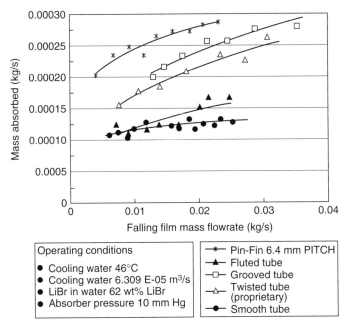

Figure 16.7 Water vapor mass absorbed vs. falling film flow rate of LiBr solution for the enhanced tubes of Figure 16.6. (From Miller and Perez-Blanco [1993].)

tube had 11 circumferential corrugations. Tests were conducted at 60 wt% LiBr concentration, 7 mm Hg absolute pressure. The 32°C cooling water was supplied at 1.0 m/s. Figure 16.8 shows the heat transfer coefficient against solution flow rate. Without the additive, the heat transfer coefficients of the hydrophilic tube are 10 to 35% higher than those of the smooth tube. The enhancement for the axially fluted tube is 5 to 25%. With the addition of 3500 ppm surfactant, the heat transfer coefficient of the smooth tube increased by 35 to 90%. The increase was less for the enhanced tubes; 40 to 70% for the axially fluted tube, and 30 to 50% for the hydrophilic tube. In addition, the heat transfer coefficient of the axially fluted tube is larger than that of the hydrophilic tube. The enhancement by Marangoni convection by the additive is believed to be more effective for the axially fluted tube. Hoffmann et al. [1996] report 20 to 40% increase of the heat transfer coefficient for a knurled tube over the smooth tube for LiBr absorption on a horizontal tube bundle (1 × 24). Addition of additives (1-octanol and 2-ethyl-1-hexanol) increased the smooth tube bundle heat transfer coefficient by 60 to 140%. The increase was less (55 to 85%) for the knurled tube. Isshiki et al. [1991] report 30 to 80% heat transfer enhancement for the constant curvature surface (CCS) tubes. They claim that the solution spreads more evenly on the CCS tube than on the smooth tube, eliminating dry spots.

An active enhancement test using vibrating screen was conducted by Tsuda and Perez-Blanco [2001] for steam absorption on vertical falling film of LiBr–water solution. The screen had a wire diameter of 0.2 mm and an open area of 70%, and

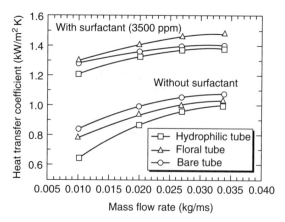

Figure 16.8 Heat transfer coefficient vs. LiBr solution flow rate for enhanced tubes tested by Yoon et al. [2002]. (From Yoon et al. [2002].)

was installed at the vapor–solution interface. The tests varied the screen-vibrating amplitude from 0.2 to 1.0 mm, and the frequency from 20 to 100 Hz. Generally, the mass transfer enhancement increased as the amplitude increased, or the frequency increased. At 1.0 mm and 60 Hz frequency, approximately twofold enhancement was obtained.

16.4 SIGNIFICANT RESISTANCE IN BOTH PHASES

If the condensing vapor of Section 16.3 contained a noncondensible gas, mass transfer resistance would exist in both phases. Mass transfer resistance will exist in both phases for the boiling or condensation of mixtures. The previously discussed approaches to reducing the mass transfer resistance for noncondensible gases are applicable. Hence, mixing is required in both the liquid and vapor phases.

Distillation columns involve two-phase flow and separation of mixtures in either tray towers or packed columns. The packing in Figure 1.17b is an enhanced mass transfer device for packed towers. It causes a tortuous flow path for the gas phase and causes gas-phase mixing. Thin liquid films exist on the packing.

16.5 CONCLUSIONS

Significant potential exists for heat transfer enhancement in simultaneous heat-mass transfer processes. One must first determine which phase requires enhancement—the gas-phase or the liquid phase, or possibly both. The controlling mass transfer resistance is in the gas phase for processes involving humidification or dehumidification of steam–air mixtures, and condensation with noncondensible gases. Use of the heat-mass transfer analogy provides a powerful tool to evaluate enhanced heat transfer surfaces for application to gas-phase mass transfer.

Typical problems involve heat and mass transfer between a gas phase and a liquid film. Cooling towers currently use enhanced surface geometries to reduce the gas-phase resistance.

Significant potential exists for water film enhancement in fin-and-tube heat exchangers used for heat rejection. For evaporation of the liquid film, it is very important that the surface be fully wetted. Surface geometries that promote, or allow, full surface wetting are not easy to obtain, and problems in surface wetting (and corrosion) have prevented their successful commercial exploitation. Commercial coatings that promote surface wetting of a water film are available and offer potential for this application.

Applications involving mass transfer in both the liquid and the gas phases are yet to be seriously addressed. Further developments in water film–enhanced heat transfer to gases are expected.

NOMENCLATURE

a Ackermann correction factor ($q_{lat}c_p/i_{gv}h_g$), dimensionless

A Heat transfer surface area on one side of a direct transfer-type exchanger, m^2 or ft^2

A_c Flow cross-sectional area in minimum flow area, m^2 or ft^2

A_{fr} Airflow frontal area, m^2 or ft^2

A_o Total air-side heat transfer surface area, m^2 or ft^2

c_p Specific heat of fluid at constant pressure, J/kg-K or Btu/lbm-°F

d_o Tube outside diameter, fin root diameter for a finned tube, m or ft

D Diffusion coefficient, m^2/s or ft^2/s

D_h Hydraulic diameter of flow passages, $4LA_c/A$, m, or ft

e Fin height or roughness height, m or ft

G Mass velocity based on the minimum flow area, G_a (air), G_w (water), kg/m^2-s or lbm/ft^2-s

G_{fr} Mass velocity based on flow frontal area, $G_{fr,a}$ (air), $G_{fr,w}$ (water), kg/m^2-s or lbm/ft^2-s

h_g Heat transfer coefficient, W/m^2-K or Btu/hr-ft^2-°F

i Enthalpy of steam–air mixture, kJ/kg or Btu/lbm

i_{gv} Enthalpy of saturated vapor, kJ/kg or Btu/lbm

ITD   Temperature difference between entering fluid temperatures, K or °F

j Colburn factor = $StPr^{2/3}$, dimensionless

K_p Mass transfer coefficient based on partial pressure driving potential, kg/s-m^2 or lbm/s-ft^2

K_w Mass transfer coefficient based on specific humidity or enthalpy driving potential, or kg/s-m^2 or lbm/s-ft^2

L_p Strip flow length of OSF or louver pitch of louver fin, m or ft

L Flow length, m or ft

M Molecular weight of mixture, kg/kmol or lbm/kmol

N Number of tube rows in the flow direction, or number of tubes in heat exchanger, dimensionless

p Pressure, Pa or lbf/ft^2

p Axial spacing between roughness elements, m or ft

q Heat flux, W/m^2 or Btu/hr-ft^2-°F

q_{lat} Latent component of heat transfer, W/m^2 or Btu/hr-ft^2-°F

Q Heat transfer rate in the exchanger W or Btu/hr

Pr   Prandtl number = $c_p\mu/k$, dimensionless

Re_{Dh} Reynolds number based on the hydraulic diameter = GD_h/μ, dimensionless

Sc   Schmidt number, ν/D, dimensionless

S_l Longitudinal tube pitch, m or ft

S_l Transverse tube pitch, m or ft

St   Stanton number = h/Gc_p, dimensionless

T Temperature, T_s (saturated), T_w (wall), T_{vb} (of bulk vapor), T_i (interface), T_{av} (average), T_{el} (entering air), T_{wl} (entering water) K or °F

u Local fluid velocity, m/s or ft/s

W Fluid mass flow rate = $\rho u_m A_c$, kg/s or lbm/s

W Specific humidity, dimensionless

y Mass fraction of diffusing component in vapor, dimensionless

y Coordinate distance normal to wall, m or ft

Greek Letters

α $(1 - e^{-1})/a$, dimensionless

δ Liquid film thickness, m or ft

μ Fluid dynamic viscosity coefficient, Pa-s or lbm/s-ft

ν Kinematic viscosity, m^2/s or ft^2/s

ρ Fluid density, kg/m^3 or lbm/ft^3

Subscripts

al Entering air

b In bulk vapor

i At liquid–gas interface

v Vapor

wl Entering water

REFERENCES

Bentley, J.M., Snyder, T.K., Glicksman, L.R., and Rohsenow, W.M., 1978. An experimental study of a unique wet/dry surface for cooling towers, *J. Heat Transfer* 100, 520–526.

Hoffmann, L., Greiter, I., Wagner, A., Weiss, V., and Alefeld, G., 1996. Experimental investigation of heat transfer in a horizontal tube falling film absorber with aqueous solutions of LiBr with and without surfactants, *Int. J. Refrig.*, 19(5) 331–341.

Hudina, M. and Sommer, A. 1988. Heat transfer and pressure drop measurements on tube and fin heat exchangers, *Proc. 1st World Conf. on Experimental Heat Transfer, Fluid Mechanics and Thermodynamics*, R.K. Shah, E.N. Ganic, and K.T. Yang, Eds., Elsevier Science, New York, 1393–1400.

Isshiki, N., Ogawa, K., Sasaki, N., and Funato, Y., 1991. R & D of constant curvature surface (CCS) tubes for absorption heat exchangers, *Proceedings of Absorption Heat Pump Conference*, Tokyo, Japan, 377–382.

Johnson, B.M., Bartz, J.A., Alleman, R.T., Fricke, H.D., Price, R.E., and McIlroy, K. 1981. Development of an Advanced Concept of Dry/Wet Cooling for Power Plants, Battelle Pacific Northwest Laboratories Report BN-SA-1296. Also presented at the American Power Conf., April 27–29, Chicago, IL.

Kreid, D.K., Johnson, B.M., and Faletti, D.W., 1978. Approximate Analysis of Heat Transfer from the Surface of a Wet Finned Heat Exchanger, ASME paper 78-HT-26, ASME, New York.

Kreid, D.K., Hauser, S.G., and Johnson, B.M., 1983a. Investigation of combined heat and mass transfer from a wet heat exchanger. Part 1: Analytical formulation, *Proc. ASME-JSME Joint Thermal Eng. Conf.*, 1, 517–524.

Kreid, D.K., Hauser, S.G., and Johnson, B.M., 1983b. Investigation of combined heat and mass transfer from a wet heat exchanger. Part 2: Experimental results, *Proc. ASME-JSME Joint Thermal Eng. Conf.*, 1, 525–534.

Miller, W.A. and Perez-Blanco, H., 1993. Vertical tube aqueous LiBr falling film absorption using advanced surfaces, *International Absorption Heat Pump Conference*, AES, 31, 185–202.

Senshu, T., Hatada, T., and Ishibane, K., 1981. Heat and mass transfer performance of air coolers under wet conditions, *ASHRAE Trans.*, 87(Part 2), 109–115.

Sommer, A., 1984. Wasserverteilung aufuden Lamellen bespruhter FORGO-GLATT-Wurmetauscher, Beobachtungen und einem Plexiglas-Aluminum Modell, Report EIR-TM-23– 84–09, Wurenlingen.

Tsuda, H. and Perez-Blanco, H., 2001. An experimental study of a vibrating screen as means of absorption enhancement, *Int. J. Heat Mass Transfer*, 44, 4087–4094.

Wang, C.-C., Hsieh, Y.-C. and Lin, Y.-T., 1997. Performance of plate finned tube heat exchangers under dehumidifying conditions, *J. Heat Transfer*, 119, 109–117.

Wang, C.-C., Lin, Y.-T. and Lee, C.-J., 2000. Heat and momentum transfer for compact louvered fin-and-tube heat exchangers in wet conditions, *Int. J. Heat Mass Transfer*, 43, 3443–3452.

Webb, R.L., 1988. A critical evaluation of cooling tower design methodology, in *Heat Transfer Equipment Design*, R.K. Shah, E.C. Subbarao, and R.A. Mashelkar, Eds., Hemisphere, Washington, D.C., 547–558.

Webb, R.L., 1991. Standard nomenclature for mass transfer processes, *ASHRAE Trans.*, 97(Part 2), 114–118.

Webb, R.L. and Perez-Blanco, H., 1986. Enhancement of combined heat and mass transfer in a vertical tube heat and mass exchanger, *J. Heat Transfer*, 108, 70–75.

Webb, R.L., Eckert, E.R.G., and Goldstein, R.J., 1970. Heat transfer and friction in tubes with repeated-rib roughness, *Int. J. Heat Mass Transfer*, 14, 601–617.

Webb, R.L., Wanniarachchi, A.S., and Rudy, T.M., 1980. The effect of noncondensible gases on the performance of an R-11 centrifugal water chiller condenser, *ASHRAE Trans.*, 86(Part 2), 170–184.

Yang, W.-J., and Clark, D.W., 1975. Spray cooling of air-cooled compact heat exchangers, *Int. J. Heat Mass Transfer*, 18, 311–317.

Yoon, J.I., Kim, E., Choi, K.H., and Seol, W.S., 2002. Heat transfer enhancement with a surfactant on horizontal tube bundles of an absorber, *Int. J. Heat Mass Transfer*, 45, 735–741.

ADDITIVES FOR GASES AND LIQUIDS

17.1 INTRODUCTION

Additives for liquids include solid particles or gas bubbles in single-phase flows and liquid trace additives for boiling systems. Additives for gases are liquid droplets or solid particles, either dilute phase (gas-solid suspensions) or dense phase (packed beds and fluidized beds).

17.2 ADDITIVES FOR SINGLE-PHASE LIQUIDS

17.2.1 Solid Particles

Kofanov [1964] performed a very detailed study of solid particle additives, for flow in a circular tube, which spanned $4000 \leq \mathrm{Re}_d \leq 200{,}000$. Figure 17.1a shows his heat transfer correlation, which includes 18 data sets from five authors. The data shown in Figure 17.1a are as follows:

1. Kofanov [1964]: 1, pure water; 2, water–chalk; 3, water–coal; 4, water–sand; 5, water–aluminum; 6, water–iron
2. Orr and Dallavalle [1954]: 7, water–clay; 8, water–copper; 9, water–glass; 10, water–graphite; 11, water–aluminum; 12, ethylene glycol–graphite; 13, water–aluminum; 14, water
3. Bonilla et al. [1953]: 15, water–chalk

4. Salamone and Newman [1955]: 16, water–copper; 17, water–sand
5. Miller and Moulton [1956]: 18, kerosene–graphite

The correlation shown in Figure 17.1 is given by

$$\mathrm{Nu}_d = 0.026\,\mathrm{Re}_d^{0.8}\,\mathrm{Pr}^{0.4}\,F_p \tag{17.1}$$

where F_p is a property group defined by

$$F_p = \left(\frac{x_v}{1-x_v}\right)^{0.15}\left(\frac{\rho}{\rho_p}\right)^{0.15}\left(\frac{c_p}{c_{p,p}}\right)^{0.15}\left(\frac{d_i}{d_p}\right)^{0.02} \tag{17.2}$$

The fluid properties used in the Nusselt, Reynolds, and Prandtl numbers are those of the solid–liquid suspension. The properties of the solid–liquid mixture (ρ_m, $c_{p,m}$, μ_m, and k_m) are calculated as follows. The specific heat and the thermal

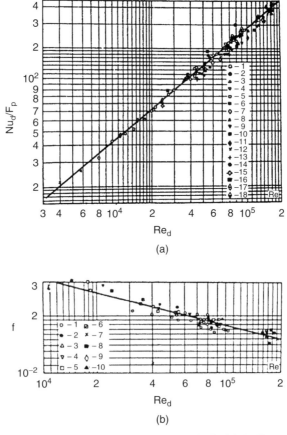

Figure 17.1 Correlations of Kofanov [1964] for solid particles in liquid. (a) Heat transfer correlation, (b) friction factor correlation. See text for data sources. (From Kofanov [1964].)

conductivity are calculated on the basis of particle mass fraction (x_m). The suspension density is calculated on the basis of the particle volume fraction (x_v). The suspension viscosity is calculated using the volume fraction by the equation

$$\mu_m = \mu(1 + 2.5x_v + 7.17x_v^2 + x_v^3) \tag{17.3}$$

where μ is the viscosity of the liquid in which the particles are suspended. Kofanov obtained good correlation of the friction factor of the flowing suspension using the Blasius equation

$$f = 0.079\,\mathrm{Re}_d^{-0.25} \tag{17.4}$$

where Re_d is defined as specified above. The correlated data are shown in Figure 17.1b. Watkins et al. [1976] tested plastic beads in laminar oil flow. They observed a maximum enhancement of 40%.

One should be aware that suspensions of very small, low-thermal-conductivity particles can cause fouling at low velocity. This possibility should exist for particles, such as clay or chalk. Such fouling would be caused by particulate fouling, which is discussed in Chapter 10.

Recently, significant interest has grown on using nanosized metallic particles to increase the thermal conductivity of the flowing medium. Current nanoscience technology can produce powders with average particle sizes up to 10 nm using gas-phase condensation or chemical synthesis technique (Choi [1995]). Nanosized particles are so small that they can remain dispersed in solutions for a long time (many weeks) as described by Eastman et al. [1997]. Some particles, however, tend to agglomerate, and settle rapidly. To prevent agglomeration, auxiliary activators or dispersants are used. Selection of suitable activators or dispersants depends on the properties of solutions and particles. Xuan and Li [2000] obtained stabilized suspensions of nanosized copper particles in water using oleic acid as dispersant. Laurate salt was used for oil. Thermal conductivity of the fluid is significantly improved with metallic nanoparticle addition. Eastman et al. [1997] report that the thermal conductivity of aqueous solution containing nanosize (36 nm) CuO particles increased 60% over that of pure water at 5% volume concentration. Xuan and Li [2000] obtained approximately the same amount of enhancement of thermal conductivity for 100-nm copper particles suspended in water. The enhancement in oil was approximately half of the enhancement in water (30%, at 5% volume concentration).

The literature reveals very limited work on heat transfer characteristics of nanoparticle solutions. Pak and Cho [1998] obtained heat transfer data of nanosize metallic oxide particles, γ-Al$_2$O$_3$ and TiO$_2$, with mean diameters of 13 and 27 nm, respectively, flowing in water for $10{,}000 \leq \mathrm{Re}_d \leq 100{,}000$. To disperse the particles homogeneously, they controlled the pH of the water (3 for γ-Al$_2$O$_3$ and 10 for TiO$_2$) by adding hydrochloric acid or sodium hydroxide. No attempt was made to measure the thermal conductivity of the solutions. The experimentally measured viscosity was used for the calculation of Reynolds number. The Nusselt number increased as the particle volume concentration increased. At the same Reynolds number, they obtained approximately 75% higher Nusselt number with 2.78% volume concentration of γ-Al_2O_3, relative to that for pure water flow. The increase was smaller for the TiO_2 particles. The Nusselt number increase was attributed to the change of viscosity due to addition of particles. When compared at the same flow velocity, however, the heat transfer coefficient of the dispersed flow was smaller than that of the water flow. The friction data were well correlated by the Blasius friction correlation. Li and Xuan [2002] report the Nusselt number of nanosize copper particle–water flow in a circular tube for $800 \le Re_d \le 25,000$. The particle diameter was approximately 100 nm, and the volume concentration was varied up to 1.2%. Compared to pure water flow, approximately 24% higher heat transfer coefficient was obtained with 1.0% concentration. No pressure drop data were reported.

17.2.2 PEC Example 17.1

Calculate the enhancement provided by 15% volume fraction of 0.05-mm-diameter glass beads in water flowing in a 20-mm diameter tube at 2.0 m/s and 20°C.

Solution. First calculate the heat transfer coefficient for pure water (h_o). The Reynolds number is 39,840. Using the Dittus–Boelter equation, $h_o = 7192$. Next calculate the properties of the suspension and use Equation 17.1 to calculate h. $F_p = 0.6517$, $Re_{d,m} = 30,600$, $Pr_m = 5.14$, and $h = 4102$ result. Hence, the enhancement ratio is $h/h_o = 4102/7192 = 0.57$. The suspension of glass beads actually reduces the heat transfer coefficient by 43%, compared to pure water flow. Note that the F_p term significantly contributes to the reduction. Further, the Prandtl number of the suspension is 5.14, as compared to 6.99 for water.

Examination of Equations 17.1 and 17.2 shows that a high particle concentration and high values of $\rho_p/\rho c_{p,p}/c_p$, and k_p/k are required to obtain $h/h_o > 1$. Hence, the conclusion is that typical solid–liquid suspensions will give little enhancement, if any.

17.2.3 Gas Bubbles

Bubbling a gas through a stationary liquid simulates the conditions that occur in nucleate boiling on a surface. Enhancement will occur because of the liquid agitation on the surface, caused by the vapor bubbles. Tamari and Nishikawa [1976] injected air into water (or ethylene glycol) heated on a vertical plate. The air was injected at the base of the plate. They measured enhancement levels up to 400%. Kenning and Kao [1972] injected air into turbulent flow of water in a tube, and they measured 50% enhancement level.

17.2.4 Suspensions in Dilute Polymer and Surfactant Solutions

Considerable work has been done on flowing suspensions that reduce fluid friction. Long fibers dampen fluid turbulence, and reduce the turbulent energy dissipation.

For example, Lee et al. [1974] used asbestos fibers at 200 and 800 ppm in an aqueous dilute polymer solution (150 ppm Separan AP-30). For the asbestos fibers in water, no drag reduction was achieved. The asbestos fiber–polymer solution gave 50 to 64% drag reduction in the turbulent regime.

Moyls and Sabersky [1975] investigated the effects of dilute suspensions of asbestos fibers (300 ppm) in an aqueous solution containing a polymer (50 ppm Polyox). For $10,000 < Re_d < 200,000$, the Nusselt number of the polymer solution (without fibers) was only 20% as high as that of pure water. When the asbestos fibers were added, heat transfer was improved. However, the heat transfer coefficient was far below that of pure water. Hence, the conclusion is that both suspensions in dilute polymer solutions reduce both heat transfer and fluid friction.

Fossa and Tagliafico [1995] tested a dilute polymer–water solution (0 to 40 ppm of Polyox WSR 301) in an annular passage with ribs machined on the outer surface of the inner tube with $1600 \leq Re_{Dh} \leq 9600$. The inner surface of the outer tube was 11-mm ID and had a smooth surface. The ribs ($e = 1.0$ mm, $p = 4.0$ mm) were formed on the 8.0-mm inner tube. The heat transfer and friction characteristics of the ribbed geometry were quite different from that of the smooth geometry. For the smooth geometry, significant reduction of Nu/Nu_o (compared with pure water flow) and f/f_o were obtained at 5.0 ppm polymer concentration. Further addition of the polymer did not reduce the ratios. Both Nu/Nu_o and f/f_o decreased as the Reynolds number increased. For the ribbed geometry, however, Nu/Nu_o was almost independent of the Reynolds number. In addition, the f/f_o curve yielded a minimum at 5.0 ppm. The heat transfer performance of the ribbed geometry was compared with that of the smooth geometry at the same pumping power (FG-1a of Table 3.1). Both geometries yielded approximately the same heat transfer.

Surfactants are also used to reduce the pressure drop. Pressure drop reduction typically accompanies the heat transfer reduction. Qi et al. [2001] tested two surfactant additives (2300 ppm Ethoquad T13-50 and 1500 ppm SPE98330) in a smooth tube, and in a four-start spiral indented tube (Figure 8.14c) for $10,000 \leq Re_d \leq 50,000$. The solution temperature changed from 50 to 70°C. For the smooth tube, 40 to 80% pressure drop reduction was obtained with addition of the surfactants. The reduction increased as the Reynolds number increased, or the solution temperature decreased. For the spirally indented tube, however, the pressure drop increased with addition of the surfactant additives. For Ethoquad T13-50, which is cationic, the pressure drop increased up to three times depending on the solution temperature. For SPE98330, which is zwitterionic, the pressure drop increase was smaller (less than two times). Heat transfer tests were conducted for the spirally indented tube. The addition of surfactants reduced the heat transfer coefficient up to 50%.

Sato et al. [1999] and Kumada et al. [2002] investigated the effect of turbulence promoters (fences, sawtooth plates, porous plates, vortex generators) in channel flow of an aqueous solution containing a surfactant (150 ppm CTAC). A single row of promoters was installed at the frontal part of the rectangular channel (20×250 mm). Generally, the pressure drop increase was larger than the heat transfer increase. The best performance was obtained for the vortex generator geometry. Li et al. [2001] proposed a novel method of increasing the heat transfer of surfactant flow by installing fine mesh structures at the frontal part of the interested area. The mesh induces high shear stress, and breaks the rodlike microcellular structure of the surfactant. The rodlike microcellular structure is known to reduce the turbulence intensity in the near-wall region, and to decrease the heat transfer. They tested three mesh configurations in channel flow of aqueous surfactant solution (25 ppm CTAC). Using a mesh having 0.07 mm diameter and 0.098 mm opening width, they could obtain heat transfer coefficients close to those of pure water flow.

17.3 ADDITIVES FOR SINGLE-PHASE GASES

17.3.1 Solid Additives

Two situations exist, where solid additives are quite important. The first is flow of gas–solid suspensions in ducts. A number of papers have been written on this subject over the last 30 years. This discussion focuses on heat transfer enhancement, and heat transfer correlations for suspended particles in ducts. The principal mechanism for enhancement is increased capacity of the flowing media at a given Re_d. The heat transferred from the heated surface to the particle is usually assumed to be negligible.

The second situation is fluidized beds, which is a high-interest topic. This involves heat transfer between a bundle of horizontal (or vertical) tubes and a fluidized gas–solids media. No attempt is made to survey the extensive publications on fluidized beds, which are reviewed by Gabor and Botterill [1985]. In this discussion attention is directed toward enhancement of gas-to-wall heat transfer, rather than toward understanding the gas-to-particle heat transfer mechanism.

Typical particulate sizes are in the range of 20 to 600 μm, and the solid–gas particle loading ratio (G/G_o) spans $1 < G/G_o < 15$. Figure 17.2 from Bergles et al. [1976] shows the enhancement ratio (h/h_o) for various gas–solid suspensions flowing inside a tube. Enhancement ratios (h/h_o) as high as 3.5 are seen. However, several of the suspensions give very low enhancement.

Several empirical correlations have been developed for gas–solid suspensions flowing inside tubes, none of which appear to be totally satisfactory. Furchi et al. [1988] present recent experimental information and correlations for the heat transfer coefficient of gas–solid suspensions. Figure 17.3 shows the correlated data of Furchi et al. [1988] and three other investigators. The correlation shown in the figure was developed by Sadek [1972]. The dimensional correlation [$(d_i, d_p$ (mm), C_p (m^{-3})] may be written in terms of the enhancement ratio (h/h_o):

$$h/h_o = 1 + 0.20(C_p d_p^2 d_i)^{1.19} \tag{17.5}$$

Figure 17.3 shows that Equation 17.5 does not do a very good job correlating the data, which spans $17 < d_i < 102$ mm, $20 < d_p < 600$ μm, $4000 < Re_d < 80,000$, and $0 < G_m/G_g < 300$. The correlation shows that h/h_o increases with particle concentration, and particle diameter. Notice that the correlation does not include the thermal conductivity of the particles, which may not be valid for high-thermal-conductivity

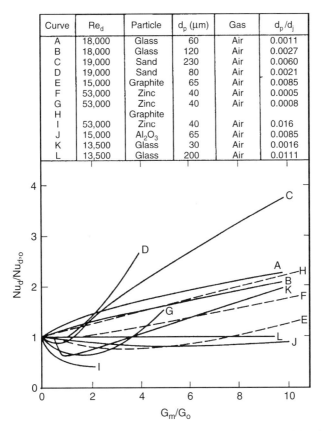

Curve	Re_d	Particle	d_p (μm)	Gas	d_p/d_j
A	18,000	Glass	60	Air	0.0011
B	18,000	Glass	120	Air	0.0027
C	19,000	Sand	230	Air	0.0060
D	19,000	Sand	80	Air	0.0021
E	15,000	Graphite	65	Air	0.0085
F	53,000	Zinc	40	Air	0.0005
G	53,000	Zinc	40	Air	0.0008
H		Graphite			
I	53,000	Zinc	40	Air	0.016
J	15,000	Al_2O_3	65	Air	0.0085
K	13,500	Glass	30	Air	0.0016
L	13,500	Glass	200	Air	0.0111

Figure 17.2 Enhancement ratio h/h_o vs. G/G_o loading ratio for various solid particle mixtures. (From Bergles et al. [1976].)

particles. Contrary to Equation 17.5, the Furchi et al. [1988] data show that h/h_o increases with decreasing particle size.

Gabor and Botterill [1985] give heat transfer correlations for horizontal bundles of plain and finned tubes. Several studies have been performed to measure the advantages offered by finned and rough surfaces in fluidized beds. Use of an enhanced surface is considered to be compound enhancement, because two enhancement techniques are combined. Petrie et al. [1968] used finned tubes in a horizontal fluidized bed, and examined the effect of fin pitch for plain tubes, and for 9.5-mm-high fins. For 199 fins/m, they measured a gas-side enhancement ratio ($E_{ho} = \eta hA/h_s A_s$) of 1.7, compared to an area increase of 6.8. Because the heat transfer coefficient in the fluidized bed is so high, it operates at low fin efficiency, which reduces the enhancement level. Bartel and Genetti [1973] measured the heat transfer coefficients for horizontal bundles of steel tubes having radial, plain aluminum fins. They tested 15.9-mm-diameter tubes having 0, 3.18, 7.1, 12.8, and 16.6 mm fin height. Krause and Peters [1983] also tested steel segmented finned tubes in a fluidized bed. Their tubes had $d_o = 19.2$ mm, 300 fins/m, and 0.76-mm fin thickness.

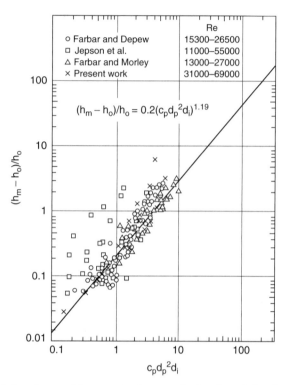

Figure 17.3 Correlation of gas–solids suspension heat transfer data for turbulent flow in tubes. (From Furchi et al. [1988].)

Table 17.1 Segmented Finned-Tube Performance in Horizontal Fluidized-Bed Bundle ($G/G_{mt} \cong 1.2$)

	$d_p = 0.21$ mm			$d_p = 0.43$ mm		
e_o (mm)	η_f	h/h_s	$hA/h_s A_s$	η_f	h/h_s	$hA/h_s A_s$
4.76	0.78	0.88	3.74	0.81	0.89	3.80
8.33	0.65	0.72	4.93	0.64	0.81	5.57
11.11	0.66	0.34	3.03	0.68	0.65	5.82

Three fin heights were tested (4.76, 8.33, and 11.11 mm) using 0.21- and then 0.43-μm-diameter particles. Table 17.1 summarizes their experimental results, which are presented as ratios, relative to the plain tube bundle. The highest enhancement ratio ($hA/h_s A_s$) of 5.82 was obtained with the largest particle size and the highest fins. However, the lowest fins gave the best performance for the smaller particle size.

Chen and Withers [1978] tested vertical 19-mm-diameter integral-fin tubes (see Figure 6.1c) in a vertical 140-mm-diameter tube using glass particles ($d_p = 0.13$, 0.25, and 0.6 mm). Table 17.2 shows their results for $d_p = 0.25$ mm, presented as ratios, relative to the plain tube. Table 17.2 shows that the highest $\eta h/h_s$ was obtained

Table 17.2 Performance of Vertical Integral-Fin Tubes in Fluidized Bed (Glass Particles, $d_p = 0.25$ mm)

E_{ho}	Fins/m	e_o (mm)	A/A_s	$\eta h/h_s$
0.60	197	3.18	2.22	0.70
1.50	354	1.57	1.88	1.00
2.30	433	3.18	3.87	0.90
1.80	748	3.06	3.06	0.80

from the 354 fins/m tube having 1.57-mm fin height. Further, their data showed that the E_{ho} does not increase with decreasing particle size. Rather, they found that the maximum E_{ho} occurred when $3 < s/d_p < 4$, where s is the fin spacing.

Grenwal and Saxena [1979] investigated tubes having a closely spaced V-thread or knurled roughness in a horizontal fluidized bed. The roughness height was 1.07 mm or less. The best roughness provided 40% higher heat transfer coefficient than did the plain tube.

17.3.2 Liquid Additives

Liquid additives generally refer to adding water droplets to the air stream. The water wets the heat transfer surface, providing evaporation from the water film surface into the air stream. Enhancement provided by an evaporating water film on a heat transfer surface is discussed in Section 16.2.3. Very high enhancement can be provided, if the surface can be fully wetted. Using a water spray onto a wedge, Thomas and Sunderland [1970] measured an enhancement (E_{ho}) of 20 by adding 5% water to the air stream.

Moderate enhancement can also be obtained without wetting the surface. Under this condition, the upstream water mist would cool the incoming air to its wet-bulb temperature, and the water droplets would exchange heat with the air in transit through the heat exchanger. Greater enhancement would be obtained for water mist temperature below the air temperature. Bhatti and Savery [1975] analyze the problem of suspended water droplets in a boundary layer flow over a flat plate. Kosky [1976] and Nishikawa and Takase [1979] experimentally and analytically study the problem of a mist flow normal to a circular cylinder.

17.4 ADDITIVES FOR BOILING

As discussed by Thome [1990], the pool boiling heat transfer coefficient of a mixture is less than that of either pure component. In spite of this understanding, work has been done using minute amounts of a volatile additive, which show that the boiling coefficient may be modestly increased. Examples of this poorly under-stood phenomenon are provided by Lowery and Westwater [1959] and by Gannett and Williams [1971].

The presence of surfactant or polymer additives at low concentration in water has been found to significantly affect the nucleate boiling heat transfer coefficient by decreasing the surface tension (surfactant) or increasing the viscosity (polymer) of the solution. The contact angle is also influenced by additives. The enhancement level depends on the type and the concentration of the additives. Several mechanisms have been proposed to explain the observed enhancement. These proposed mechanisms include Marangoni convection, increased bubble frequency, reduced bubble coalescence, etc. A recent review on nucleate boiling enhancement by additives is provided by Wasekar and Manglik [1999].

Typical results for aqueous solutions by Tzan and Yang [1990] are shown in Figure 17.4. They obtained nucleate pool boiling data of aqueous SLS (surfactant) solutions on a 3.35-mm cylindrical heater. The heat transfer coefficient increased as the concentration increased. As much as 160% increase in the heat transfer coefficient, over that of pure water, was obtained. However, above 1000 ppm, reduced heat transfer enhancement is observed. Similar observations were reported by Wasekar and Manglik [2000] for boiling of aqueous SLS solutions on a 22.2-mm heater, and by Ammerman and You [1996] for boiling of aqueous SDS solutions on a 0.39-mm platinum wire. Yang et al. [2002], however, reported no noticeable enhancement for boiling of aqueous Tritron SP-190 and SP-175 solutions on a 5.3-mm heater. They measured the surface tension and the contact angle of the solutions. Both the surface tension and the contact angle decreased as the concentration increased. They postulated that nucleate boiling is enhanced by reduction of surface tension, and deteriorates by the reduction of contact angle. They explained that the opposing effect of the surface tension and the contact angle by additives resulted in negligible heat transfer enhancement by Tritron SP-190 and SP-175. The previous studies by Tzan and Yang [1990], Wasekar and Manglik [2000], and Ammerman and You [1996] do not report the contact angles of their solutions. However, Wu et al. [1998] report

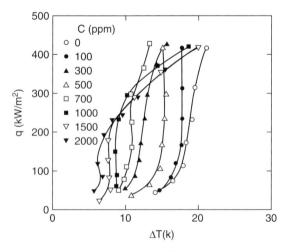

Figure 17.4 Pool boiling curves of aqueous solution of SLS on a 3.35-mm stainless heater. (From Tzan and Yang [1990].)

that the contact angle data of SDS aqueous solutions are almost independent of concentration. Wen and Wang [2002] successfully predicted their experimental data and those by Wu et al. [1998] by incorporating the measured contact angle to the Mikic–Rohsenow [1969] pool boiling model.

The effect of polymer additives is also anomalous. Kotchaphakdee and Williams [1970] report enhancement of nucleate boiling heat transfer by adding HEC-H and PA-10 additives in water. Paul and Abdel-Khalik [1983] and Wang and Hartnett [1992], however, report deterioration of boiling heat transfer performance with aqueous polymeric solutions.

Wu and Yang [1992] studied the effect of surfactant additives on boiling incipience and bubble dynamics on a 5.3-mm-OD stainless steel tube using a high-speed camera. SLS was used as the surfactant additive with concentrations ranging from 0 to 500 ppm. The incipient superheat was found to decrease significantly by the addition of surfactant. The bubble growth period increased slightly, while the waiting period decreased drastically, resulting in increased bubble frequency. The bubble size was considerably reduced with addition of the surfactant as illustrated in Figure 17.5. The photographs in Figure 17.5 were taken by Ammerman and You [1996] for the pool boiling of SDS solution on a 0.39-mm platinum wire. Note that the bubble size decreases as the concentration increases. Wu et al. [1998] provide bubble dynamics data for SDS and Triton-X aqueous solutions. No convincing conclusion was drawn for bubbling period and departure diameter due to the large scatter of data.

Further, it appears that a trace additive may also provide a significant increase of the critical heat flux (CHF) in pool boiling, as shown by van Wijk et al. [1956] and by van Stralen [1959]. van Stralen [1959] measured a 240% increase of the CHF for 3 to 5% concentration of 1-penthanol in water. The optimum concentration is a function of boiling pressure. Bergles and Scarola [1966] found that addition of 1-penthanol reduced the CHF for subcooled boiling.

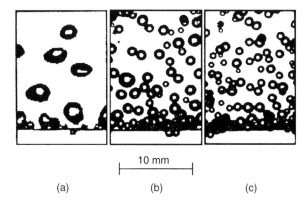

10 mm

(a)   (b)   (c)

Figure 17.5 Flow visualization photographs illustrating the effect of surfactant on the bubble generation from 0.39-mm platinum wire in SDS aqueous solution: (a) 0 ppm, (b) 500 ppm, (c) 1000 ppm. (From Ammerman and You [1996].)

17.5 ADDITIVES FOR CONDENSATION AND ABSORPTION

The objective of such additives is to promote dropwise condensation. Typically, the heat transfer coefficient for dropwise condensation is much higher than that for filmwise condensation. Additives that establish the existence of dropwise condensation are called dropwise condensation "promoters." Griffith [1985] surveys the technology of dropwise condensation and discusses various promoters that have been found for steam (water) condensation. A satisfactory promoter must cause the surface to become hydrophobic to the condensate. This means that the contact angle of the condensate must be increased to approach 90°. Hence, the condensate will not spread on the surface but will exist as discrete droplets. The droplets agglomerate and run off the surface. This is possible for high-surface-tension fluids, such as water. However, low-surface-tension fluids (e.g., organics, alcohols, and refrigerants) have such a low surface tension, that no promoter has been found that will make the surface hydrophobic to the condensate.

Table 2 of Griffith [1985] lists various promoters that work for steam condensation. However, contamination of the surface or depletion of the promoter will cause cessation of dropwise condensation. The desired promoter is also a function of the base surface material. To date, practical concerns have prevented the commercial use of such promoters for steam condensation. Tanasawa [1978] discusses practical concerns for commercial use of dropwise condensation promoters. A recent study of using additives to promote dropwise condensation is reported by Kim et al. [2001].

The only known promoters that are indefinitely durable are certain coatings on the base surface. These are gold and Teflon. Gold has obvious cost problems. Although Teflon effectively promotes dropwise condensation, the coating also adds a significant thermal resistance that substantially diminishes the benefits of the coating. Depew and Reisbig [1964] found that a 1.27-µm-thick Teflon coating, which promoted dropwise condensation, will increase the condensing coefficient 100% over the filmwise condensation coefficient on a 12.7-mm-diameter horizontal tube.

Commercial LiBr absorption chillers typically use additives to improve the absorber performance. As discussed in Section 16.3, the primary mass transfer resistance exists at the liquid–vapor interface during absorption process. The additives enhance mixing at the liquid–vapor interface by Marangoni convection. The Marangoni effect is an interfacial disturbance resulting from surface tension gradients. The additives also improve the surface wettability. The most widely used additive is octyl-alcohol (2-ethyl-1-hexanol). Rush et al. [1991] and Kim et al. [1996] discuss various additives for LiBr absorption system. Rush [1968] reports approximately 20% cooling capacity increase by adding additives in absorption chillers. A recent review on the effects of additives on the absorption process is provided by Ziegler and Grossman [1996].

17.6 CONCLUSIONS

Typical solid–liquid suspensions will provide little, if any, enhancement. However, significant enhancement can be obtained from solid–gas suspensions. The significant enhancement provided by solid–gas suspensions provides justification for the fluidized bed heat exchanger. Finned surfaces can be used in the fluidized bed to further increase the heat transfer coefficient.

Aqueous, dilute polymer additives will reduce both the heat transfer coefficient and friction factor. Use of a rough surface will recover some of the heat transfer reduction, but at the expense of increased friction.

Adding a minute amount of a volatile fluid may moderately enhance the nucleate boiling coefficient. However, this phenomenon is poorly understood, since mixtures typically reduce the heat transfer coefficient. Addition of a surfactant can significantly increases the nucleate boiling heat transfer of water. The effect of polymer on nucleate boiling heat transfer is anomalous.

Additives can promote high-performance dropwise condensation for water or other high-surface-tension fluids. However, long-term surface contamination, or depletion of the additive will likely reduce the performance. Additives that promote dropwise condensation for low-surface-tension fluids have not been found.

NOMENCLATURE

C_p	Particle volume concentration, m^{-3} or ft^{-3}
d_i	Tube inside diameter, m or ft
d_p	Particle diameter, m or ft
e	Rib height, m or ft
e_o	External fin height, m or ft
E_{ho}	Heat transfer enhancement ratio (hA/h_sA_s) relative to plain surface, dimensionless
G	Mass velocity, kg/s-m^2 or lbm/s-ft^2
G_{mf}	Minimum fluidization mass velocity, kg/s-m^2 or lbm/s-ft^2
h	Heat transfer coefficient, W/m^2-K or Btu/ft^2-°F
k	Thermal conductivity, W/m-K or Btu/hr-ft-°F
Nu_d	Nusselt number based on tube diameter, dimensionless
p	Rib pitch, m or ft
Pr	Prandtl number, dimensionless
Re_d	Reynolds number based on tube diameter, dimensionless
Re_{Dh}	Reynolds number based on hydraulic diameter, dimensionless
s	Fin spacing, m or ft
x_m	Particle-fluid mass fraction ($= x_v\rho_m/\rho_p$), dimensionless
x_v	Particle-fluid volume fraction, dimensionless

Greek Symbols

η_f Fin efficiency, dimensionless

μ Viscosity, kg/s-m or lbm/s-ft

ρ Density, kg/m^3 or lbm/ft^3

Subscripts

m Solids–gas or solids–liquid mixture

o Flow without solids

p Particles

s Smooth or plain surface

REFERENCES

Ammerman, C.N. and You, S.M., 1996. Determination of the boiling enhancement mechanism caused by surfactant addition to water, *J. Heat Transfer*, 118, 429–435.

Bartel, W.J. and Genetti, W.E., 1973. Heat transfer from a horizontal bundle of bare and finned tubes in an air fluidized bed, *AIChE Symp. Ser*, 69(128), 85–92.

Bergles, A.E. and Scarola, L.S., 1966. Effect of a volatile additive on the critical heat flux for surface boiling of water in tubes, *Chem. Eng. Sci.*, 21, 721–723.

Bergles, A.E., Junkhan, G.H., and Hagge, J.K., 1976. Advanced Cooling Systems for Agricultural and Industrial Machines, SAE paper 751183; Warrendale, PA.

Bhatti, M.S. and Savery, S.W., 1975. Augmentation of heat transfer in a laminar external gas boundary layer by the vaporization of suspended droplets, *J. Heat Transfer*, 97, 179–184.

Bonilla, C.F., Cervi, A., Jr., Colven, T.J., Jr., and Wang, S.J., 1953. Heat transfer to slurries in pipe, chalk, and water in turbulent flow, *AIChE Symp. Ser.*, 49(5), 127–134.

Chen, J.C. and Withers, J.G., 1978. An experimental study of heat transfer from plain and finned tubes in fluidized beds, *AIChE Symp. Ser.*, 74(174), 327–333.

Choi, S.U.-S., 1995. Enhancing thermal conductivity of fluids with nanoparticles, in *Developments and Applications Non-Newtonian Flows*, ASME FED-231, ASME, New York, 99–103.

Depew, C.A. and Kramer, T.J., 1973. Heat transfer to flowing gas-solid mixtures, in *Advances in Heat Transfer*, Vol. 9, J.P. Hartnett and T.F. Irvine, Eds., Academic Press, New York, 113–180.

Depew, C.A. and Reisbig, R.L., 1964. Vapor condensation on a horizontal tube using Teflon to promote dropwise condensation, *Ind. Eng. Chem. Process Design Dev.*, 11, 365–369.

Eastman, J.A., Choi, U.S., Li, S., Thompson, L.J., and Lee, S., 1997. Enhanced thermal conductivity through the development of nanofluids, in *Nanophase Nanocomposite Mater.*, II, 3–11.

Fossa, M. and Tagliafico, L.A., 1995. Experimental heat transfer of drag-reducing polymer solutions in enhanced surface heat exchangers, *Exp. Thermal Fluid Sci.*, 10, 221–228.

Furchi, J.C.L., Goldstein, L., Lombardi, G., and Mohseni, M. 1988. Heat transfer coefficients in flowing gas-solid suspensions, *AIChE Symp. Ser.*, 84(263), 26–30.

Gabor, J.D. and Botterill, J.S.M., 1985. Heat transfer in fluidized and packed beds, in *Handbook of Heat Transfer Fundamentals*, McGraw-Hill, New York, chap. 6.

Gannett, H.J., Jr. and Williams, M.C., 1971. Pool boiling in dilute nonaqueous polymer solutions, *Int. J. Heat Mass Transfer*, 11, 1001–1005.

Grenwal, N.S. and Saxena, S.C., 1979. Effect of surface roughness on heat transfer from horizontal immersed tubes in a fluidized bed, *J. Heat Transfer*, 101, 397–403.

Griffith, P., 1985. Condensation. Part 2: Dropwise condensation, in *Handbook of Heat Transfer Applications*, McGraw-Hill, New York, chap. 11.

Herold, K., Radermacher, R., and Klein, S.A., 1996. *Absorption Chillers and Heat Pumps*, CRC Press, Boca Raton, FL.

Kenning, D.B.R. and Kao, Y.S., 1972. Convective heat transfer to water containing bubbles: enhancement not dependent on thermocapillarity, *Int. J. Heat Mass Transfer*, 15, 1709–1718.

Kim, K.J., Kulankara, S., Herold, K., and Miller, C., 1996. Heat transfer additives for use in high temperature applications, in *Proceedings of International Absorption Heat Pump Conference*, Montreal, Canada, 1, 89–97.

Kim, K.J., Lefsaker, A.M., Razani, A. and Stone, A., 2001. The effective use of heat transfer additives for steam condensation, *Appl. Thermal Eng.*, 21, 1863–1874.

Kofanov, V.I., 1964. Heat transfer and hydraulic resistance in flowing liquid suspensions in piping, *Int. Chem. Eng.*, 4(3), 426–430.

Kosky, P.G., 1976. Heat transfer to saturated mist flowing normally to a heated cylinder, *Int. J. Heat Mass Transfer*, 19, 539–543.

Kotchaphakdee, P. and Williams, M.C., 1970. Enhancement of nucleate pool boiling with polymeric additives, *Int. J. Heat Mass Transfer*, 13, 835–848.

Krause, W.B. and Peters, A.R., 1983. Heat transfer from horizontal serrated finned tubes in an air-fluidized bed of uniformly sized particles, *J. Heat Transfer*, 105, 319–324.

Kumada, M., Chu, R., and Sato, K. 2002. Heat transfer enhancement and flow characteristics of drag-reducing surfactant aqueous solutions using the turbulent promoter, *Heat Transfer 2002, Proc. 12th Int. Heat Transfer Conf.*, 4, 129–134.

Lee, W.K., Vaseleski, R.C., and Metzner, A.B., 1974. Turbulent drag reduction in polymeric solutions containing suspended fibers, *AIChE J.*, 20, 128–133.

Li, P., Kawaguchi, Y., Daisaka, H., Yabe, A., Hishida, K., and Maeda, M., 2001. Heat transfer enhancement to the drag-reducing flow of surfactant solution in two-dimensional channel with mesh-screen inserts at the inlet, *J. Heat Transfer*, 123, 779–789.

Li, Q. and Xuan, Y., 2002. Convective heat transfer performance of fluids with nano-particles, in *Proc. 12th Int. Heat Transfer Conf.*, 1, 483–488.

Lowery, A.J., Jr. and Westwater, J.W., 1959. Heat transfer to boiling methanol—effect of added agents, *Ind. Eng. Chem.*, 49, 1445–1448.

Mikic, B.B. and Rohsenow, W.M., 1969. A new correlation of pool boiling data including the effect of heating surface characteristics, *J. Heat Transfer*, 83, 245–250.

Miller, A.P. and Moulton, R.W., 1956. Heat transfer to liquid-solid suspensions in turbulent flow in pipes, *Trend Eng.*, April, 15–21.

Moyls, A.L. and Sabersky, R.H., 1975. Heat transfer to dilute asbestos dispersions in smooth and rough tubes, *Lett. Heat Mass Transfer*, 2, 293–302.

Nishikawa, N. and Takase, H., 1979. Effects of particle size and temperature difference on mist flow over a heated circular cylinder, *J. Heat Transfer*, 101, 705–711.

Orr, C. and Dallavalle, J.M., 1954. Heat transfer properties of liquid-solid suspensions, *Chem. Eng Prog. Symp. Ser.*, 50(9), 29–45.

Pak, B.C. and Cho, Y.I., 1998. Hydrodynamic and heat transfer study of dispersed fluids with submicron metallic oxide particles, *Exp. Heat Transfer*, 11, 151–170.

Paul, D.D. and Abdel-Khalik, S.I., 1983. Nucleate boiling in drag reducing polymer solutions, *J. Rheol.*, 27(1), 59–76.

Petrie, J.C., Freeby, J.A., and Buckham, J.A., 1968. In-bed heat exchangers, *Chem. Eng. Prog.*, July, 45–51.

Qi, Y., Kawaguchi, Y., Lin, Z., Ewing, M., Christensen, R.N., and Zakin, J.L., 2001. Enhanced heat transfer of drag reducing surfactant solutions with fluted tube-in-tube heat exchanger, *Int. J. Heat Mass Transfer*, 44, 1495–1505.

Rush, W.F., 1968. Field testing of additives, in *Symposium on Absorption Air Conditioning*, American Gas Association, Chicago, IL.

Rush, W., Wurum, J., and Perez-Blanco, H., 1991. A brief review of additives for absorption enhancement, *Absorption Heat Pump Conf.*, Tokyo, Japan, 91, 183–187.

Sadek, S.E., 1972. Heat transfer to air-solids suspensions in turbulent flow, *Ind. Eng. Chem. Process Design Dev.*, 11, 133–135.

Salamone, J.J. and Newman, M., 1955. Heat transfer design characteristics—water suspensions of solids, *Ind. Eng. Chem.*, 47(2), 283–288.

Sato, K., Mimatsu, J., and Kumada, M., 1999. Drag reduction and heat transfer augmentation of surfactant additives in two dimensional channel flow, *Proc. 5th ASME/JSME Thermal Eng. Joint Conf.*, Paper ATJE99-6452.

Tamira, M. and Nishikawa, K., 1976. The stirring effect of bubbles upon the heat transfer to liquids, *Heat Transfer Jpn. Res.*, 5(2), 31–44.

Tanasawa, I., 1978. Dropwise condensation: the way to practical applications, *Proc. 6th Int. Heat Transfer Conf.*, 6, 393–405.

Thomas, W.C. and Sunderland, J.E., 1970. Heat transfer between a plane surface and air containing water droplets, *Ind. Eng. Chem. Fundam.*, 9, 368–374.

Thome, J.R., 1990. *Enhanced Boiling Heat Transfer*, Hemisphere, Washington, D.C.

Tzan, Y.L. and Yang, Y.M., 1990. Experimental study of surfactant effects on pool boiling heat transfer, *J. Heat Transfer*, 112, 207–212.

van Stralen, S.J.D., 1959. Heat transfer to boiling binary liquid mixtures, *Br. Chem. Eng.*, 4 (Part I), 8–17; 4 (Part II), 78–82.

van Wijk, W.R., Vos, A.S., and van Stralen, S.J.D., 1956. Heat transfer to boiling binary liquid mixtures, *Chem. Eng. Sci.*, 5, 68–80.

Wang, T.A.A. and Hartnett, J.P., 1992. Influence of surfactants on pool boiling of aqueous polyacrylamide solutions, *Wärme Stoffubertrag.*, 27, 245–248.

Wasekar, V.M. and Manglik, R.M., 1999. A review of enhanced heat transfer in nucleate pool boiling of aqueous surfactant and polymeric solutions, *J. Enhanced Heat Transfer*, 6, 135–150.

Wasekar, V.M. and Manglik, R.M., 2000. Pool boiling heat transfer in aqueous solutions of an anionic surfactant, *J. Heat Transfer*, 122, 708–715.

Watkins, R.W., Robertson, C.R., and Acrivos, A., 1976. Entrance region heat transfer in flowing suspensions, *Int. J. Heat Mass Transfer*, 19, 693–695.

Wen, D.S. and Wang, B.X., 2002. Effects of surface wettability on nucleate pool boiling heat transfer for surfactant solutions, *Int. J. Heat Mass Transfer*, 45, 1739–1747.

Wu, W.-T. and Yang, Y.-M., 1992. Enhanced boiling heat transfer by surfactant additives, in *Proceedings of the Engineering Foundation Conference on Pool and External Flow Boiling*, V.K. Dhir and A.E. Bergles, Eds., Santa Barbara, CA, 361–366.

Wu, W.-T., Yang, Y.-M., and Maa, J.-R., 1998. Nucleate pool boiling enhancement by means of surfactant additives, *Exp. Thermal Fluid Sci.*, 18, 195–209.

Xuan, Y. and Li, Q., 2000. Heat transfer enhancement of nanofluids, *Int. J. Heat Mass Transfer*, 21, 58–64.

Yang, Y.-M., Lin, C.-Y., Liu, M.-H., and Maa, J.-R., 2002. Lower limit of the possible nucleate pool boiling enhancement by surfactant addition to water, *J. Enhanced Heat Transfer*, 9, 153–160.

Ziegler, F. and Grossman, G., 1996. Heat transfer enhancement by additives, *Int. J. Refrig.*, 19, 301–309.

MICROCHANNELS

18.1 INTRODUCTION

Heat transfer in microchannels is a relatively new research interest and was not covered in the first edition. The diameter (or hydraulic diameter) associated with the term "microchannels" is ambiguous. As used here, the term is defined to mean diameters below 500 μm. Typical applications may involve diameters in the range of approximately 10 to 200 μm. Fabrication methods to form microchannels are discussed by Grande [2002]. This discussion includes channels as small as 0.1 μm hydraulic diameter. This chapter addresses fundamentals of heat transfer and friction in microchannels for single- and two-phase flows. Two recent conferences provide significant information on this subject. These are Celata et al. [2000] and Kandlikar et al. [2003].

The concept of microchannel cooling was first proposed by Tuckermann and Pease [1981] for cooling of electronic chips. This concept has gained importance in the past decade because of the advent of Micro-Electro-Mechanical Systems (MEMS) and its application to cooling of electronic devices. Active cooling using single-phase fluid flow in micron-sized channels is of interest for cooling applications such as electronic equipment, microreactors, microcombustors and micro-heat pumps. Cooling of electronic chips using multiple, parallel micron-sized channels is of considerable present interest. For example, Koo et al. [2000] studied two-phase cooling for a 25-mm-square heat sink having either 150 or 200 μm channel depth, as illustrated in Figure 18.1. Other studies are those of Jiang et al. [2001] and Perret et al. [2000]. These three studies are discussed in Chapter 19.

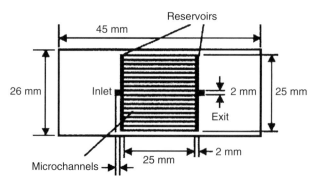

Figure 18.1 Microchannel heat sink studied by Koo et al. [2001]. (From Jiang et al., 2001. *Proc. 17th IEEE Semi-Therm Symp.*, March 20–22, 2001. With permission.)

It is best to approach this understanding by examining data taken in single channels. However, much data have been published on flow in multiple microchannels connected in parallel. Experimental work on flow in parallel microchannels introduces the possibility of flow maldistribution, which complicates the test results. Hence, flow in parallel microchannels and flow in single channels are addressed separately. An important application of flow in parallel microchannels is in cooling of electronic equipment. Such applications of microchannels are addressed in Chapter 19.

Applications may require either single- or two-phase flow. Two-phase flow models for friction and heat transfer typically require single-phase flow characteristics. Hence, analysis of single phase fluid flow in microchannels is the first step for design and performance prediction of the above-mentioned devices. Most of the microchannel applications envisage the use of fluid flow in multiple, parallel channels. A key question in microchannel research is to determine if the heat transfer and friction laws for macrosize channels also apply to microchannels.

Research on microchannel flow has been performed using two types of test sections — single channels, and multiple, parallel channels. Such tests have examined both friction and heat transfer, particularly in single-phase flow. Data for both laminar and turbulent single-phase flow have been reported, but most of the published data are for laminar flow. Study of the literature shows that some of the reported data for multiple, parallel channels does not agree well with that for single microchannels. Understanding of applicable laws for friction and heat transfer in microchannels is best approached by examining work on single-phrase friction in single channels, and then extending this to examine single-phase heat transfer in single channels. Then, one next seeks to understand how friction and heat transfer in multiple, parallel microchannels differs from that in single channels. This assessment will include the possibility that flow maldistribution in multiple channels may exist. If flow maldistribution is probable, it is necessary to identify manifold design guidelines to alleviate this problem. Obot [2000] provides an excellent review of the published friction and heat transfer data in single and multiple channels.

18.2 FRICTION IN SINGLE MICROCHANNELS

For fully developed laminar flow in single macrochannels, the friction factor is governed by the equation:

$$f = \text{Po}/\text{Re} \tag{18.1}$$

where Po is the Poiseuille number, which depends on the channel shape. In the turbulent regime the Darcy friction factor is approximated by the Blasius equation:

$$f = 0.316\,\text{Re}^{-0.25} \tag{18.2}$$

Table 18.1 lists a number of microchannel studies for single-channel flow. Wu and Little [1983] were among the first to experimentally investigate the fluid behavior in single microchannels. They used channels of trapezoidal cross section with hydraulic diameter in the range of 50 μm. The ratio of the perpendicular height of the channel to its base width was approximately 1:4 with 55° included angle. The test channels were etched on silicon or glass and were classified as smooth or rough depending on the relative roughness (e) of the channel surfaces, as compared to the channel dimensions (D). Figure 18.2 shows the variation of Darcy friction factor with Re, with gas flowing in the channel. For a 1:4 rectangular channel, which corresponds to the nominal channel geometry, Po ≅ 73, which is 14% higher than for a round channel (Po ≅ 64). Figure 18.2 shows that the data for smooth channels S(1) and S(3) fall between the theoretical values for a round channel and for a 1:4 rectangular channel. Further, the transition Re occurs at approximately 2000, which is the same as in macroflows. In the turbulent regime, the data for S(1) and S(3) agree very well with Equation 18.2. The rough channel S(2) has $0.2 < e/D_h < 0.3$ and the friction factor is higher than that for a smooth channel in both the laminar and turbulent regimes, as expected.

Choi et al. [1991] studied gas flow in five single, circular microtubes having diameters of 3.0, 6.9, 9.7, 53, and 81.2 μm. Figure 18.3 shows that friction factor agrees well with the theoretical values for all tested diameters. Further, transition occurred at approximately 2000 Reynolds number. Yu et al. [1995] measured the friction factor for liquids and gases in single circular microtubes with 19 μm ≤ D_h ≤ 100 μm and also found good agreement with the theoretical values in the laminar as well as the turbulent regime. Furthermore, the transition was found to occur at a Reynolds number of 2000. Investigations of gas flow in a single trapezoidal microtube by Harley et al. [1995] with hydraulic diameter in the range of 19 μm also show similar results.

Experimental investigation of pressure drop in single channels conducted by Pfund et al. [1998] also show excellent agreement for the laminar flow friction factor in channels with hydraulic diameter in the range of 100 to 500 μm for Reynolds numbers less than 1500. Kim et al. [1998] also report an excellent agreement of the measured friction factor with the theoretical values for straight channels with hydraulic diameter in the range of 200 to 250 μm. They report a distinct transition to turbulence at a Reynolds number of 2500.

Table 18.1 Microchannel Studies for Single-Channel Flow

Ref.	Fluid	Channel Shape	D_h (μm)	H/w	Data Taken
Wu and Little [1983]	Nitrogen	Trapezoid	45.5–83.1	0.019–0.05	f vs. Re lam/turb
Wu and Little [1984]	Nitrogen	Trapezoid	134–164	0.17–0.29	Nu vs. Re Re > 3000
Pfahler et al. [1990]	n-Propanol	Rectangular	1.6, 1.34	0.008, 0.017	—
Choi et al. [1991]	Nitrogen	Cylindrical	3.0–81.2	—	Nu vs. Re lam/turb
Yu et al. [1995]	Nitrogen, water	Cylindrical	19.2–102	—	Nu vs. Re lam/turb
Harley et al. [1995]	Nitrogen, helium, argon	Trapezoid	1.0–35.9	0.0053–0.161	—
Adams et al. [1998]	Water	Cylindrical	760, 1090	—	Nu vs. Re Re > 3000
Adams et al. [1999]	Water	Cylindrical	1130	—	Nu vs. Re Re > 3000

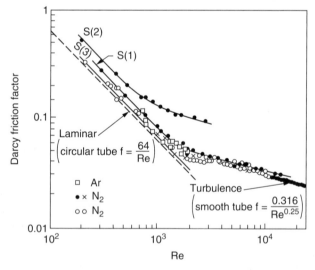

Figure 18.2 Variation of f with Re by Wu and Little [1983]: S(1), S(3): smooth trapezoidal channels; S(2) - rough channel. (From Wu and Little [1983]. Reprinted with permission from Elsevier.)

The above experimental investigations show that friction in single microchannels as small as 3.0 μm diameter agree well with the macroscale friction laws. Further, the transition occurs at a Re of approximately 2000. Based on the above, one can conclude that the fluid flow characteristics in single microchannels agree well with the flow characteristics in macrochannels. This conclusion is supported by Obot [2000].

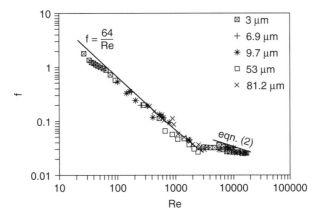

Figure 18.3 Variation of f with Re by Choi et al. [1991].

18.3 FRICTION IN A SINGLE CHANNEL VS. MULTICHANNELS

Because the single-channel data conform to the macroscale laws (within some permissible deviation), it is expected that the multichannel data should follow the same trend, unless additional phenomena exist in the multichannel flow. This section addresses multichannel data.

Peng and co-workers have published a number of related papers on heat transfer and friction in multiple channels. An example is Peng et al. [1994a] who measured friction with water in 50-mm-long multiple parallel rectangular channels with hydraulic diameters in the range of 150 to 300 μm for a range of aspect ratios. The test sections had four to eight channels in parallel. Table 18.2 lists the width (W), depth (H), aspect ratio (H/W), and the hydraulic diameter (D_h) of the tested channels.

The friction data were reduced after subtracting the inlet and exit losses. The water experimental results were reported to be highly sensitive to the hydraulic diameter and the aspect ratio of the channels. The data for different test sections show significant differences. Test sections 1, 2, 4, 5, and 6 have an apparent friction factor much higher than the theoretical value, and test sections 3 and 7 display an apparent friction factor lower than the theoretical value. Detailed test results are shown here for two test sections. Figure 18.4a shows the friction factor vs. Re for the D_h = 150 μm test section with $400 \le \mathrm{Re_{Dh}} \le 850$. Figure 18.4b shows the friction factor vs. Re for the D_h = 267 μm test section. The apparent friction factors for the two channels are significantly different from the theoretical laminar values. At Re $\cong$ 120, Figure 18.4a shows three different friction values for three different test runs, which indicates that the data are not repeatable. However, Peng et al. [1994a] speculate that this unexpected result is caused by very early transition to turbulence. They propose that transition to turbulence occurs at Reynolds numbers as low as 200 to 700, and that the transition Reynolds number decreases as the D_h decreases. Peng and Peterson [1996] report friction and heat transfer data for 12 test

Table 18.2 Multiple Parallel Rectangular Channels Tested by Peng et al. [1994a]

Test	w, mm	H, mm	L, mm	D_h, mm	H/w
1	0.4	0.3	50	0.34	0.75
2	0.3	0.3	50	0.3	1
3	0.4	0.2	50	0.27	0.5
4	0.3	0.2	50	0.24	0.67
5	0.2	0.2	50	0.2	1
6	0.3	0.1	50	0.15	0.33
7	0.2	0.1	50	0.13	0.5

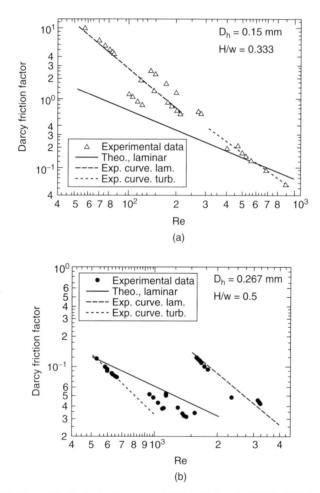

Figure 18.4 Variation of f with Re by Peng et al. [1994a]. (a) Test 6 of Table 18.2, D_h = 0.15, H/W = 0.333. (b) Test 3 of Table 18.2, D_h = 0.267, H/W = 0.50. (From Peng et al. [1994]. With permission from Elsevier.)

sections with varying channel dimensions and $133 \leq D_h \leq 343$ μm; these results appear to include the data of Peng et al. [1994a]. The number of channels in parallel varied from 4 to 8 with 45-mm channel length. The authors report findings similar to those reported by Peng et al. [1994a].

Cuta et al. [1996] measured the friction for single- and two-phase flow in parallel microchannels (aspect ratio 4:1) with $D_h \cong 450$ μm. The test section had 54 channels in parallel with a length to-diameter ratio of 50. The inlet and exit losses were reported to be negligible in comparison with the friction in the channel and hence were not taken into consideration for the apparent friction factor. Figure 18.5 shows the variation of their single-phase data, as well as considerable scatter in their data. The single-phase friction factor is much lower than the theoretical value.

Using water flow, Gui and Scaringe [1995] tested three test sections with each having 12 trapezoidal channels in parallel. The hydraulic diameter of the test sections was 338 to 388 μm and the channel height varied from 217 to 270 μm. The entrance and exit losses were taken into consideration for the computation of friction factor. Their friction factor agreed reasonably with the theoretical fully developed laminar equation up to Reynolds number of 1500, at which transition occurred. However, the friction factor in the transition region showed considerable scatter.

Kawano et al. [1998] worked with an array of 110 rectangular channels (57×180 μm and 57×370 μm) and found that the friction factor agreed reasonably with the theoretical predictions in the laminar regime for low Reynolds numbers (Re < 300). The entrance and exit losses were not taken into consideration for the reduction of friction factor. However, the experiments were conducted at very low Reynolds numbers.

The multichannel test results provide inconsistent results, relative to one another. Part of the reason for the differences in the multiple channel data may be due to different methods used to measure pressure drop in multichannel arrays. Kawano et al. [1998] and Gui and Scaringe [1995] measured the pressure drop across the inlet and exit of the manifolds. However Peng et al. [1994a] had pressure taps on the sides of the manifold. Measurement of the pressure drop at specific sections of the manifold requires the assumption that the pressure in the manifold is constant.

Although some experimental investigations result in friction factors almost five times higher (or lower) than the theoretical predictions (Peng et al. [1994a], Cuta et al. [1996]), some others have obtained results in agreement with the macroscale laws (Gui and Scaringe [1995], Kawano et al. [1998]). The investigations in multiple microchannels appear to yield results that are experiment/test section design specific. Because the L/D_h of the channels were quite large, it does not appear that hydraulic entrance length issues are important. Sobhan and Garimella [2000] provide an excellent detailed discussion of the published microchannel data and conclude that the published data and proposed correlations are not reliable.

The experimental data were reduced assuming equal flow distribution in all of the parallel channels. A probable reason for the discrepancy with single-channel data is flow maldistribution in the parallel channels. Flow maldistribution will cause varying flow across the different channels, and thus the Reynolds number for the individual channels will be different. Further, the pressure in the inlet and exit

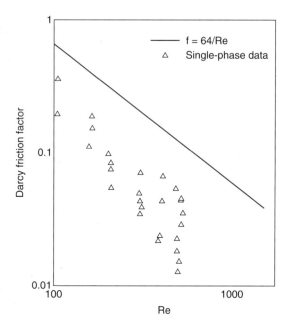

Figure 18.5 Variation of f with Re by Cuta et al. [1996].

manifolds will not be constant. Hence the location of the pressure taps, and the means to account for inlet/exit losses will also contribute to the observed differences. Figure 18.6 shows an exploded view of the test section similar to those used in the tests of Peng et al. [1994a]. This is typical of the test sections used in the multichannel experiments discussed above. The flow maldistribution in these channels can be twofold:

1. Passage-to-passage flow maldistribution, caused by differences in channel sizes
2. Gross maldistribution, associated with the manifold design

Passage-to-passage flow maldistribution will exist if the size of the various channels differ. One may show that for laminar flow with constant pressure drop, $W \propto (A_c)^2$, or for circular channels, $W \propto d^4$. A change in the channel cross-sectional dimension along the flow length will cause passage-to-passage maldistribution. It will be extremely difficult to control manufacturing tolerances in microchannel manufacture to prevent such maldistribution. For example, a variation of only 2.0 μm in the size of a 20-μm channel would result in ±45% difference in the channel mass flow rate. Passage-to-passage flow maldistribution is of little concern in macroscale flows, because manufacturing tolerance will have a negligible effect on passage-to-passage maldistribution because of the large passages. In the microscale size range, the effect of these factors becomes increasingly important. Gross maldistribution can be reduced by proper manifold design. Manifold design is discussed in a later section.

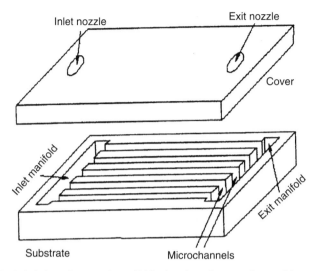

Figure 18.6 Exploded view of a normal manifold microchannel test section used in most of the multi-channel experimental investigations.

18.4 SINGLE-PHASE HEAT TRANSFER IN MICROCHANNELS

Experimental determination of the friction factor is simple, as compared to the Nusselt number. The friction factor depends only on the measurement of the flow rate and the pressure drop. The additional difficulties include measurement of the heat flux (which may involve a non-uniform heat flux), determination of the wall temperature, and effect of variation of the fluid properties with temperature. The experimental setup, test section design, and the assumptions made for data reduction also influence the results obtained. Also of concern is the driving temperature difference used for the data reduction. If heat exchange occurs between two fluids, one must separate the two thermal resistances. Significant errors may be introduced here. Further, if the experimental values are compared with theoretical solutions, an uncertain thermal boundary condition (e.g., constant heat flux vs. constant wall temperature) can affect the comparison.

18.4.1 Single Channel Flow

Choi et al. [1991] investigated heat transfer in circular microtubes with hydraulic diameters of 9.7, 53.0, and 81.2 μm. Heat was transferred between a hot gas flowing normal to the single test tube. They determined the average heat transfer coefficient over the tube length, and their data reduction assumed infinite thermal conductance on the outside of the tube. They report a strong dependence of the Nusselt number on the Reynolds number even in the laminar regime. The authors also report very low values of Nusselt numbers at lower Re values in the laminar regime. In the turbulent regime, the authors report Nu values 70 to 100% larger than those predicted

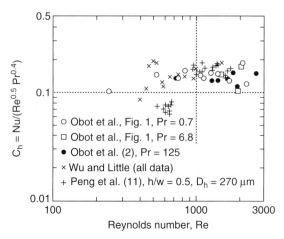

Figure 18.7 Comparison of macro- and microchannel results for single channel flow. (From Obot [2000].)

by the Dittus–Boelter equation. Yu et al. [1995] have reported a similar underprediction of their turbulent flow data by the Dittus–Boelter equation for tubes with 19 to 100 μm hydraulic diameter. They found that the Nu is 40 to 75% greater than predicted by the Dittus–Boelter equation, with the difference increasing at increasing Re. They did not measure heat transfer in the laminar regime. In summary, Choi et al. [1991] and Yu et al. [1995] show similar disparities in the turbulent regime: 50 to 100% higher Nu than the predicted value.

Obot [2000] provides a different assessment of the above data. He first presents friction and heat transfer data for a single macrochannel (13.39-mm-ID smooth tube) taken at Pr = 0.7 and 6.8. The heat transfer data in the laminar regime show that

$$\frac{Nu}{Re^{1/2} Pr^{0.4}} = 0.14 \tag{18.3}$$

and increases with Re in the transition and turbulent regime. He then proceeded to compare Nu data measured by others vs. their data on the 13.39-mm-ID smooth tube plotted in the form of $Nu/Re^{1/2} Pr^{0.4}$ vs. Re. Figure 18.7 shows the macrochannel data, plus that of Wu and Little [1984] and Peng et al. [1994b] for multichannels. The figure shows very good agreement among the macrochannel and microchannel data, and with Equation 18.3. Obot [2000] also concludes that the effect of channel aspect ratio is small.

The apparent disparity in the heat transfer data in single microchannels shown by some investigators may be due to a number of factors — among them is the experimental method and the accuracy with which the data were measured. Obot [2000] provides three important conclusions in regard to microchannels:

1. There is no strong evidence to support that transition will at occur at Re < 1000, as shown by Peng et al. [1994a]. It is possible that the effects shown by Peng et al. [1994a] were due to flow maldistribution in their multichannels.

2. Both macro- and microchannel laminar flow data show $Nu/Re^{1/2}Pr^{0.4} = 0.14$ in laminar flow and the effect of channel shape is small.
3. It appears that macrochannel correlations can be used to predict friction and heat transfer data in microchannels
4. Reported deviation of single-channel microchannel data may be partially due to experimental method and measurement accuracy.

18.4.2 Heat Transfer in Multiple Microchannels

The previously noted flow maldistribution problems associated with multichannel flows will also influence the heat transfer characteristics in these channels. Thus, if flow distribution problems are responsible for disagreement with the macroscale friction laws, one would expect the same concerns for the multichannel heat transfer data.

Several studies have been published for heat transfer in multichannel flow. Among these are Cuta et al. [1996], Gui and Scaringe [1995], and a series of papers by Peng and co-workers (e.g., Peng et al. [1994b]). The geometries tested by Peng et al. [1994b] are described in Table 18.2. Their friction data were previously discussed in this chapter. Because the reported friction data of Peng et al. [1994b] did not agree with macrochannel data, one would not expect the multichannel heat transfer data to do so. The only difference between the single channels and the multichannels is the flow distribution in these multichannels, and hence significant differences between single and multichannel flow is likely due to flow maldistribution. Any such maldistribution should influence both friction and heat transfer.

Cuta et al. [1996] measured heat transfer coefficients in the multiple parallel rectangular channel array of 54 microchannels (D_h = 450 mm, aspect ratio 4:1, L/D_h ratio ~ 50). They found that $Nu \propto Re^{0.6}$ in the laminar regime. Gui and Scaringe [1995] worked with trapezoid-shaped channels of hydraulic diameters between 338 and 388 µm. They also found that $Nu \propto Re^{0.6}$ in the laminar regime. Note that this is reasonably consistent with the findings of Obot [2000] for macrochannels, who shows that $Nu \propto Re^{0.5}$. The reader should also review the previously referenced detailed comparative study of published data by Sobhan and Garimella [2000].

Wang and Peng [1994] computed the heat transfer coefficient on the basis of the fluid inlet temperature of the fluid instead of the mean bulk fluid temperature. This affected their comparison with macrochannel correlations, which base the heat transfer coefficient on the log-mean temperature difference. Although Wang and Peng [1994] proposed a correlation of the laminar and turbulent region heat transfer data, their data correlation may have been strongly influenced by the specific manifold geometry used and that significant flow maldistribution may have existed in their channels.

18.5 MANIFOLD SELECTION AND DESIGN

A poor manifold design can cause gross maldistribution among the various channels of the test section. For the Figure 18.6 design (called a "normal manifold") used in most of the referenced experimental investigations, the fluid enters as a jet at the inlet nozzle and tries to diffuse in the manifold to reach the end channels. Most of the entering high-velocity flow passes through the central channels causing flow maldistribution. Normal flow manifolds are known to give poor flow distribution. Other possible manifold designs are given in Figure 18.8. Figure 18.8a shows a "parallel-flow" manifold and Figure 18.8b shows a "reverse-flow" manifold. In parallel-flow and reverse-flow manifolds, the maximum flow occurs through the first and the last channels, respectively.

18.5.1 Single-Phase Flow

When the fluid flows through the entrance manifold, friction acts to reduce the pressure, and fluid deceleration acts to increase the fluid pressure. However, in the exit manifold, both friction and fluid acceleration decreases the fluid pressure. The local pressure difference between the entrance and exit manifolds determines the flow in each branch tube. Datta and Majumdar [1980] investigated parallel-flow and reverse-flow manifolds both analytically and experimentally. They showed that the reverse-flow manifold yields better flow distribution than the parallel-flow manifold.

One of the earliest and the most extensive studies of flow distribution in manifolds was that of Bajura and Jones [1976]. They developed an analytical model to define the single-phase flow distribution in the lateral branches of dividing and

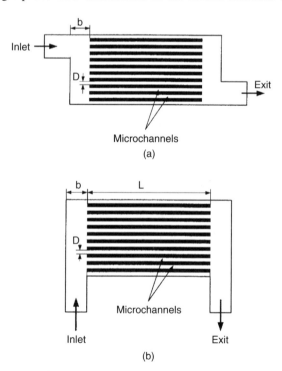

Figure 18.8 Manifold design: (a) parallel-flow manifold, (b) reverse-flow manifold.

combining headers for parallel and reverse manifold systems. The analytical model was solved for a number of cases, assuming continuous outflow along the length of the dividing header. Thus, the lateral channels were treated as porous ducts. Experimental measurements of the branch flows were done for both parallel and reverse manifold designs. The analytical model was able to predict the single-phase flow distribution in the manifolds quite well.

Datta and Majumdar [1980] presented a numerical solution to the Bajura and Jones equations using a finite difference scheme. Unlike Bajura and Jones [1976], they formulated the governing equations for discrete branch points. The momentum and continuity equations in the header and the integral form of momentum equation in the lateral channels were solved iteratively. This method is relatively simple, compared with that of Bajura and Jones [1976] and is well suited for numerical computation. They compared their predicted results with the experimental results of Bajura and Jones [1976] and showed reasonably good agreement.

An improved reverse manifold design is the "oblique reverse flow" manifold shown in Figure 18.9a and is recommended by Mueller [1985]. This design maintains more uniform pressure in the entrance and exit manifolds. Because the pressure drop of all branch channels is more uniform, the flow distribution is improved. Figure 18.9b shows a multiple microchannel test section having oblique parallel-flow manifolds, manufactured by IMM (Mainz, Germany).

To minimize flow maldistribution, it is desirable that the cross-sectional area of the manifold (A_{man}) be large, relative to the channel flow area. Based on the analysis of Datta and Majumdar [1980], Mueller [1985] recommends that the cross-sectional area of the manifold should be at least as large as the sum of the channel cross sections. Thus, the manifold-to-total branch tube area defined as $A_R = A_{man}/NA_c \geq 1.0$. Three other parameters influence the flow distribution:

1. The ratio of the inlet nozzle cross-sectional area to the channel header area (A_{noz}/A_{man})
2. The location and direction of the inlet nozzle relative to the channels
3. The ratio of the channel pitch to the channel diameter (or width) (P/D)

A summary of guidelines for reducing flow maldistribution in parallel channels is as follows:

1. The ratio of the manifold cross-sectional flow area to total branch cross-sectional flow area should be at least 1.0. Thus, assuming round tubes in Figures 18.8a and 18.8b, $D^2/Nd^2 \geq 1.0$.
2. For fixed header and branch tube sizes, the Figure 18.8b reverse-flow manifold is better than the Figure 18.8a parallel-flow manifold. The Figure 18.9a oblique reverse-flow manifold is the best, and the Figure 18.6 normal-flow manifold is the worst.

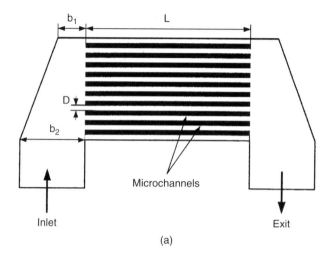

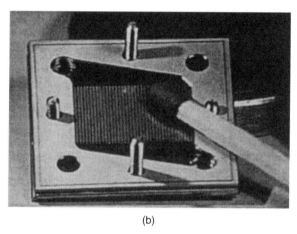

(b)

Figure 18.9 Improved manifold design: (a) reverse oblique flow manifold test section, (b) oblique parallel-flow manifold test section. (Courtesy of IMM, Mainz-Germany.)

18.5.2 Two-Phase Flow

If two-phase flow occurs in parallel microchannels, additional flow maldistribution problems are possible. Two-phase flow in the manifolds is subject to phase separation caused by inertia and gravity forces. If the manifolds are horizontal, gravity forces are minimized. However, if the microchannel array is oriented in the vertical direction, gravity force will contribute to flow separation. Fundamental understanding of two-phase flow distribution is found in the handbook chapter of Azzopardi [1994]. A recent paper by Kim et al. [2003] provides a review of published information on two-phase flow distribution in manifolds and means to improving it. Webb and Chung [2004] discuss problems associated with two-phase flow distribution. They also provide a review of recent patent technology that describes innovative methods for obtaining improved flow distribution. Hrnjak [2004] provides a detailed flow visualization study for a brazed aluminum minichannel evaporator having horizontal headers. Basic options to improve flow distribution are discussed. He also provides a map of flow patterns in the header.

18.6 NUMERICAL SIMULATION OF FLOW IN MANIFOLDS

A two-dimensional numerical simulation was performed by Rao [2000] and further reported by Webb [2003] to evaluate the influence of the manifold type and design on the flow distribution in parallel microchannels. Laminar flow was studied for several manifold designs using Fluent (Ver. 5). Two of the manifold geometries studied are shown in Figure 18.8: reverse-flow manifold in Figure 18.8b and the oblique reverse manifold Figure 18.9a. Each test section modeled had 17 channels and a length of 20 mm. The fluid entered the manifold through a nozzle in the plane of the flow. The mesh had 20 grid points over the channel width and 50 grid points along the length of the channel. Because the grid was structured, the fine mesh was retained in the inlet and the exit manifolds. The velocity in the inlet pipe was chosen to maintain laminar flow throughout the test section. The analysis was performed for a nominal channel Reynolds (assuming uniform flow distribution) number of 245 or 490 where $D_h = 2D$. The chosen channel dimensions are moderately larger than the dimensions of the microchannels reviewed. The slightly larger channels were chosen to provide a uniform computational mesh in the channel and to maintain continuity of the grid aspect ratio in the channels and the manifold. Two calculation series were performed and these are described below.

The Series 1 calculations were performed using a header area ratio less than recommended. The calculations used $A_{noz}/A_{man} = 1.0$, and $A_{man}/NA_c = 5/(17 \times 1) = 0.294$ (as compared to the recommended value of 1.0 or greater). The analysis was performed for 1.0-mm-wide channels with $p/D = 2.0$ and Re = 245. Referring to Figure 18.8, $b_1 = b_2 = 5.0$ mm for the normal manifold (similar to Figure 18.6), and $b_1 = 7.0$, $b_2 = 3.0$ for the Figure 18.8b, reverse-flow manifold. For the reverse oblique manifold (similar to Figure 8.9a), the A_{man}/NA_c dimension is referred to the average flow width in the manifold. Figure 18.10 shows the mass flow rate in the individual channels. The figure shows that significant flow maldistribution exists in both of the designs. The flow rate variation in the reverse-flow manifold geometry is 3.5:1, as compared to 6.5:1 in the normal-flow manifold geometry. A key problem with the normal manifold is the high velocity at the inlet nozzle, which feeds a high flow rate to the central channels and starves the end channels in the array. The flow distribution in the normal manifold would be improved by use of a larger inlet nozzle. This figure clearly shows the effect that the manifold geometry can have on flow distribution.

The Series 2 calculations were performed for 0.5-mm-wide channels at Re = 490 and used $p/D = 3.0$ and $A_{noz}/A_{man} = 1.0$. The normal-flow manifold was 8.5-mm wide giving $A_{man}/NA_c = 8.5/(17 \times 0.5) = 1.0$. The width of the oblique reverse flow changes from 8.5 to 3.64 mm at the two ends, which gives an average A_{man}/NA_c of 0.71.

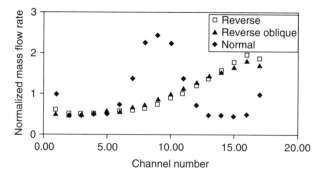

Figure 18.10 Comparison of flow distribution in normal, reverse, and oblique reverse flow manifolds (Re = 245) for p/D = 2.0, A_{man}/NA_c = 0.294, and average A_{man}/NA_c = 0.294 (reverse oblique flow). (From Webb, 2003. *Proc. IPACK03*, July 6–11, 2003. With permission of IEEE.)

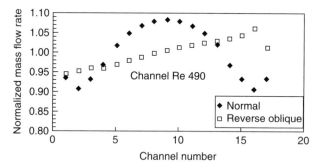

Figure 18.11 Comparison of flow distribution in normal- and oblique reverse-flow manifolds (Re = 490) for p/D = 3.0, A_{man}/NA_c = 1.0, and average A_{man}/NA_c = 0.71 (reverse oblique flow). (From Webb, 2003. *Proc. IPACK03*, July 6–11, 2003. With permission of IEEE.)

Figure 18.11 shows the flow distribution for these two designs. The normal-flow manifold has a ±10% variation, while the reverse-flow manifold is ±5% non-uniformity. If a smaller inlet nozzle were used in the normal manifold, the flow distribution would have been poorer. Note that the average width of the reverse manifold is 29% smaller than that of the normal-flow manifold. The Series 1 and 2 calculations clearly show the importance of good manifold design.

For a given manifold size, the oblique reverse-flow manifold yields significantly better flow distribution than does the normal-flow manifold. Further, the flow distribution for the reverse-flow manifold is less sensitive to inlet nozzle size. Use of the normal-flow manifold should be avoided, if possible.

18.7 TWO-PHASE HEAT TRANSFER IN MICROCHANNELS

Less work has been done on two-phase flow in microchannels than for single-phase flow, especially in channels having hydraulic diameters less than 1.0 mm. However, it is expected that Equation 13.4 will apply. Because of the small channel size, a small liquid-phase Reynolds number will exist, so it is possible that the convective term (h_{cv}) may be small compared to the nucleate boiling term (h_{nb}). The nucleate boiling term will not be significantly affected by channel size. However, it will be strongly affected by the magnitude of the driving temperature difference. Hence, one would expect the boiling coefficient to increase with increasing heat flux. High heat flux would be expected in electronic cooling applications. For plain internal channels, one may use the Cooper correlation given by Equation 2.32. If the channels contain some form of enhancement, an "enhancement factor" would multiply the h_{nb} prediced by the Cooper equation.

Palm [2000] provides a review of work on heat transfer in microchannels, including two-phase heat transfer. Recent data on convective vaporization in microchannels include work on single and multiple channels. Yan and Lin [1998] measured two-phase vaporization data for R-134a in 2.0-mm-OD circular tubes. The test section had 28 parallel tubes, 200 mm long, in a plane array having 100 mm heated length. This geometry is possibly susceptible to flow maldistribution. The authors provide no detailed description of the inlet and exit manifold designs. The heat flux was varied between 5 and 20 kW/m². At low vapor qualities, the evaporation coefficient increases with increasing heat flux. However, at higher vapor qualities, one sees small heat flux dependence and higher some dependence on mass velocity. For $G = 100$ kg/s-m² and $x = 0.2$, the evaporation coefficient is approximately five times that predicted by the popular Shah [1982] correlation developed for tube diameters greater than 8.0 mm. However, at $x = 0.8$ the data are approximately 25% above the Shah correlation prediction. They also provide friction data. Wang et al. [2001] have published a correlation for friction pressure drop that applies to small-diameter tubes. Yan and Lin [1998] provide a correlation of their heat transfer data that is based on a variant of the Kandlikar [1990] correlation for forced convection vaporization inside tubes. The correlation is limited to the laminar flow regime because the all-liquid Reynolds numbers of the data points are less than 2000. Webb and Paek [2003] found that the Yan and Lin correlations show extremely poor ability to predict the Yan and Lin data. In their response to this discussion, Yan and Lin [2003] provided a revised correlation that does a good job of predicting their data.

Steinke and Kandlikar [2003] report single- and two-phase heat transfer data, plus flow visualization for water flowing in six parallel, rectangular microchannels of 207 μm hydraulic diameter. Their flow pattern observations show that the same flow patterns observed in macrochannels were also observed in their microchannel tests. They conclude that theory applicable for convective vaporization in macrochannels is also applicable to microchannels. Their laminar region single-phase heat transfer and friction data showed good agreement with theoretical predictions. The two-phase data spanned an operating range of $157 < G$ (kg/m²-s) < 1782 and at relatively high heat fluxes [$55 < q$ (kW/m²) < 898]. Three data sets are shown in Figure 18.12. Figure 18.12a shows all of their data plotted as q vs. ΔT_{sat}. For $\Delta T_{sat} = 5$ K, one may remove 20 to 70 W/cm² for the range of mass velocities. For $G = 366$ kg/m² K, Figure 18.12b shows that the heat transfer coefficient is independent of heat flux, so it is convection dominated. Surprisingly, the data show decreasing

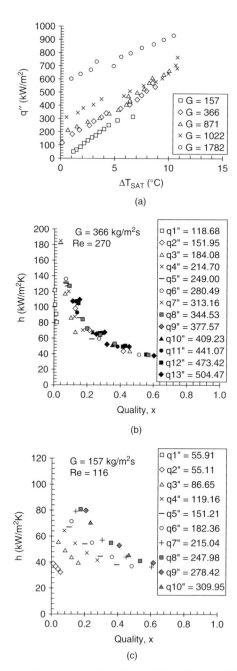

Figure 18.12 Convective vaporization data of Steinke and Kandlikar [2003] for water at 120°C in parallel, rectangular microchannels of 207 μm hydraulic diameter (a) Heat flux vs. ΔT_{sat}; (b) h vs. vapor quality at b = 366 kg/m²-s; (c) h vs. vapor quality at G = 157 kg/m²-s, heat flux (g) units are kW/m². (From Steinke and Kandlikar, 2003. Proc. First Int'l Conf. Microchannels and Minichannels, Apr. 24–25. With permission of IEEE.)

heat transfer coefficient with increasing vapor quality. Figure 18.12c shows a plot similar to Figure 18.12b, but for the lowest mass velocity (157 kg/m²-s). These data are not convection dominated and show a considerable nucleate boiling component.

Serizawa et al. [2002] performed flow visualization for air–water in 20 to 100-μm tube and steam–water in a 50-μm circular tube. They found that the flow patterns agree with the general trends of the Mandhane flow pattern map for macrochannels. Further, the measured void fraction (α) is in reasonable agreement with Armand and Treschev [1946] void fraction correlation for macrochannels. This is $\alpha = 0.833 \, \beta$, where β is the gas flow volume flow fraction.

Kandlikar and Balasubramanian [2003] used the data of Kandlikar et al. [2003] to extend the Kandlikar [1990] convective vaporization correlation to allow prediction of the heat transfer coefficient as a function of heat flux, mass flux, and vapor quality. The correlation was extended to all-liquid Reynolds numbers to the 50 to 500 range. Kandlikar [2003] provides discussion of the mechanisms of convective vaporization in microchannels.

Peng and Wang [1994, 2000] have published data for convective vaporization in multiple channels using a test section similar to the one used for single-phase flow (e.g., Peng et al. [1994b]). Their channels had a glass cover to allow flow visualization. They observed no bubbles in the microchannels even at high flux. However, they did observe streams of bubbles in the exit manifold. Their proposed explanation for this perplexing behavior is that there is a required critical minimum "evaporating space" required for bubble nucleation. They called boiling without visible nucleation "fictitious boiling." For this phenomenon, they propose that "internal evaporation and bubble growth may not have been realized, or there may exist countless microbubbles within the liquid that cannot be visualized by ordinary means." These results appear to be contrary to the findings of Steinke and Kandlikar [2003].

An important issue for convective vaporization in microchannels is the vapor quality at which dryout occurs. Hapke et al. [2000] measured the wall temperature vs. axial position for convective vaporization of *n*-heptane and water in single microchannels 1000 μm wide and 300 μm (or 700 μm) high. Their tests were performed for $20 \leq q$ (kW/m²) ≤ 350 and $25 \leq G$ (kg/m²-s) ≤ 350. They observed that dryout (x_{dry}) was linearly decreased with increasing boiling number (Bo = q/Wi_{fg}). For Bo = 0.0015, dryout occurred at $x = 0.5$ and decreased to 0.3 at Bo = 0.0035.

18.8 CONCLUSIONS

The fluid flow behavior in single microchannels reasonably follows the same hydraulic resistance laws as macroscale flows. Hence, the heat transfer characteristics for liquid flow in single microchannels is also expected to reasonably follow the macroscale laws, provided care is taken to account for the entrance length effects and viscous dissipation. The suppression of turbulence in microchannels is a plausible cause for increased heat transfer in the turbulent regime and should be investigated further.

Flow maldistribution is believed to be a major cause for the anomalous flow behavior in multiple microchannels. Gross maldistribution can be reduced by proper manifold design. Passage-to-passage maldistribution cannot be completely eliminated and depends on the manufacturing tolerances. Reasonably good single-phase flow distribution can be attained for single phase flow, if careful attention is placed on manifold sizing and design. The oblique reverse manifold is best and the uniform flow manifold is worst. Gravity-induced phase separation will cause additional problems in flow distribution for two-phase flow.

NOMENCLATURE

A_c	Cross-sectional area of tube or a single branch channel, m^2 or ft^2
A_{man}	Cross section of manifold at inlet, m^2 or ft^2
A_{noz}	Cross section of inlet or exit flow nozzle, m^2 or ft^2
b	Channel width, m or ft
Bo	Boiling number (= q/Wi_{fg}), dimensionless
c_p	Specific heat, kJ/kg-K or Btu/lbm-°F
D_h	Hydraulic diameter, m or ft
D	Channel diameter, or channel width (two-dimensional channels), m or ft
f	Darcy friction factor, dimensionless
G	Mass velocity, kg/m^2-s or lbm/ft^2-s
h	Heat transfer coefficient, W/m^2-K or Btu/hr-ft^2-°F
H	Channel height, used in Table 18.1, m or ft
ℓ	Mean free path, m or ft
L	Channel length, m or ft
L_c	Characteristic length scale, m or ft
k	Thermal conductivity, W/m-K or Btu/hr-ft-°F
N	Number of channels
Nu	Nusselt number (= hD_h/k), dimensionless
p	Pitch of branch channels, m or ft
Δp	Pressure drop, Pa or lbf/ft^2
Po	Poiseuille number (f Re), dimensionless
Re	Reynolds number [$D_h G/\mu$], dimensionless
ΔT_{sat}	Temperature difference between wall and saturated fluid, K or °F
W	Fluid mass flow rate, kg/s or lbm/s
w	Spacing between branch channels, m or ft
x	Vapor quality, dryout vapor quality (x_{dry}), dimensionless

Greek Symbols

ρ	Density, kg/m^3 or lbm/ft^3
μ	Dynamic viscosity, Pa-s or lbm/s-ft
ν	Kinematic viscosity, m^2/s or ft^2/s

Subscripts

dh   Hydraulic diameter
nb   Nucleate boiling
cv   Convection

REFERENCES

Adams, T.M., Abdel-Khalik, S.I., Jeter, S.M., and Qureshi, Z.H., 1998. An experimental investigation of single phase forced convection in micro-channels, *Int. J. Heat Mass Transfer*, 41, 851–857.

Adams, T.M., Dowling, M.F., Abdel-Khalik, S.I., and Jeter, S.M., 1999. Applicability of traditional turbulent single-phase forced convection correlations to non-circular microchannels, *Int. J. Heat Mass Transfer*, 42, 4411–4415.

Arkilic, E.B., Schmidt, M.A., and Breur, K.S., 1997. Gaseous slip flow in long microchannels, *J. Microelectromech. Syst.*, 6(2), 167–177.

Armand, A.A. and Treschev, G., 1946. *Izv. Vses. Teplotek. Inst.*, 1, 16–23.

Azzopardi, B.J., 1994. Two-phase flows in junctions, in *Encyclopedia of Fluid Mechanics*, Vol. 3, Gulf Publishing, Houston, TX, Chap. 25.

Bailey, D.K., Ameel, T.A., Warrington, R.O., Jr., and Savoie, T.I., 1995. Single phase forced convection heat transfer in microgeometries — a review, in *Proceedings of the 30th Intersociety Energy Conservation Engineering Conference*, ASME, New York, 301–310.

Bajura, R.A. and Jones, E.H. Jr., 1976. Flow distribution manifolds, *J. Fluids Eng.*, 98, 654–666.

Beskok, A., Karniadakis, G.E., and Trimmer, W., 1996. Rarefaction and compressibility effects in gas microflows, *J. Fluids Eng.*, 118, 448–456.

Celata, G.P., Carey, V.P., Groll, M., Tanasawa, I., and Zummo, G., Eds., 2000, *Heat Transfer and Transport Phenomena in Microscale, Proc. Int. Conf on Heat Transfer and Transport Phenomena in Microscale*, Begell House, New York.

Choi, S.B., Barron, R.F., and Warrington, R.O, 1991. Fluid flow and heat transfer in microtubes, *Micromechanical Sensors, Actuators and Systems*, ASME DSC, 32, 123–134.

Cuta, J.M., McDonald, C.E., and Shekarriz, A., 1996. Forced convection heat transfer in parallel channel array microchannel heat exchanger, *Advances in Energy Efficiency Heat/Mass Transfer Enhancement*, ASME HTD, 338, 17–23.

Datta, A.B. and Majumdar A.K., 1980. Flow distribution in parallel and reverse manifolds, *Int. J. Heat Fluid Flow*, 100, 253–262.

Grande, W.J., 2002. Fabrication technologies for advanced heat transfer applications, in *First Int. Conf. on Microchannels and Minichannels*, April 24–25, Rochester, NY, ASME, New York, 215–222.

Gui, F. and Scaringe, R.P., 1995. Enhanced heat transfer in the entrance region of microchannels, in *Proceedings of the 30th Intersociety Energy Conservation Engineering Conference*, ASME, New York, 289–295.

Hapke, I., Boye, H., and Schmidt, J., 2000. Flow boiling of water and heptane in microchannels, in *Heat Transfer and Transport Phenomena in Microscale, Proc. Int. Conf. on Heat Transfer and Transport Phenomena in Microscale,* G.P. Celata, V.P. Carey, M. Groll, I. Tanasawa, and G. Zummo, Eds., Begell House, New York, 222–228.

Harley, J.C., Huang,Y., Bau, H.H., and Zemel, J.N., 1995. Gas flow in micro-channels, *J. Fluid Mech.*, 284, 257–274.

Hrnjak, P., 2004. Developing adiabatic two phase flow in headers — distribution issue in parallel flow microchannel heat exchangers, *Heat Transfer Eng.*, 25, 61–68.

Jiang, L., Koo, J.M., Zeng, S., Mikkelsen, J.C., Zhang, L., Zhou, P., Santiago, J.G., Kenny, T.W., Goodson, K.E., Maveety, J.G., and Tran, Q.A., 2001. Two-phase microchannel heat sinks for an electrokinetic VLSI chip cooling system, in *Proc. of the 17th IEEE SEMI-THERM Symposium*, March 20–22, 2001, San Jose, CA, 153–157.

Kandlikar, S.G., 1990. A general correlation for saturated two-phase flow boiling heat transfer inside horizontal and vertical tubes, *J. Heat Transfer*, 112, 219–228.

Kandlikar, S.G., 2003. Heat transfer mechanisms during flow boiling in microchannels, *First Int. Conf. on Microchannels and Minichannels*, April 24–25, Rochester, NY, ASME, New York, 33–46.

Kandlikar, S.G. and Balasubramanian, P., 2003. Extending the applicability of the flow boiling correlation to low Reynolds number flows in microchannels, *First Int. Conf. on Microchannels and Minichannels*, April 24–25, Rochester, NY, ASME, New York, 603–608.

Kandlikar, S.G., Celata, G.P., Nishio, S., Stephan, P., and Thonon, B., Eds., 2003. *First Int. Conf. on Microchannels and Minichannels*, April 24–25, Rochester, NY, ASME, New York.

Kavehpour, H.P. and Faghri, M., 1997. Effects of compressibility and rarefaction on gaseous flows in microchannels, *Numerical Heat Transfer A*, 32, 677–696.

Kawano, K., Minakami, K., Iwasaki, H., and Ishizuka, M., 1998. Micro channel heat exchanger for cooling electrical equipment, *Proc. ASME Int. Mech. Eng. Cong. Exposition*, ASME HTD, 361(3), 173–180.

Kim, J.-S., Im, Y.B., Kim, J.-H., and Lee, K.-T., 2003. Two-phase flow distribution in compact heat exchanger header, *Proc. 1st Int. Conf. Microchannels and Minichannels*, Kandilikar, S.G., Ed., Rochester, NY, ASME, pp. 513–518.

Kim, M., Yi, M., Zhong, J., Bau, H.H., Hu, H., and Ananthasuresh, G.K., 1998. The fabrication of flow conduits in ceramic tapes and the measurement of fluid flow through these conduits, *Micro-Electro-Mechanical Systems (MEMS)*, ASME DSC, 66, 171–177.

Koo, J.M., Jiang, L., Zhang, L., Zhou, P., Banerjee, S.S., Kenny, T.W., Santiago, J.G., and Goodson, K.E., 2000. Modeling of Two-Phase Microchannel Heat Sinks for VLSI Chips, *Proc. of the 14th Annual IEEE International MEMS-01 Conference*, Interlaken, Switzerland, 422–426.

Mala, G.M., Li, D., and Dale, J.D., 1997. Heat transfer and fluid flow in microchannels, *Int. J. Heat Mass Transfer*, 40, 3079–3088.

Mueller, A.C., 1985. Compact heat exchangers, in *Handbook of Heat Transfer Applications*, 2nd ed., McGraw-Hill, New York, 4.174–4.312.

Obot, N.T., 2000. Toward a better understanding of friction and heat/mass transfer in microchannels — a literature review, *Proc. Int. Conf on Heat Transfer and Transport Phenomena in Microscale*, G.P. Celata et al., Eds., Banff, Canada, October 15–20, 2000, Begell House, New York, 72–79.

Palm, B., 2000. Heat transfer in microchannels, *in Proc. Int. Conf. on Heat Transfer and Transport Phenomena in Microscale*, G.P. Celeta et al., Eds., Banff, Canada, October 15–20, 2000, Begell House, New York, 54–64.

Peng, X.F. and Peterson, G.P., 1995. The effect of thermofluid and geometrical parameters on convection of liquids through rectangular microchannels, *Int. J. Heat Mass Transfer*, 38, 755–758.

Peng, X.F. and Peterson G.P., 1996. Convective heat transfer and flow friction for water flow in microchannel structures, *Int. J. Heat Mass Transfer*, 39, 2599–2608.

Peng, X.F. and Wang, B.X., 1994. Liquid flow and heat transfer in microchannels with/without phase change, Heat Transfer 1994 Vol. 1, *Proc. 10th Int. Heat Transfer Conf.*, Brighton, England, Taylor and Francis, pp. 159–178.

Peng, X.F., Liu, D., Lee, D.J., Yan, Y., and Wang, B.X., 2000. Cluster dynamics and fictitious boiling in microchannels, *Int. J. Heat Mass Transfer*, 43, 4259–4265.

Peng, X.F., Peterson, G.P., and Wang B.X., 1994a. Friction flow characteristics of water flowing through microchannels, *Exp. Heat Transfer*, 7(4), 265–283.

Peng, X.F., Peterson, G.P., and Wang B.X., 1994b. Heat transfer characteristics of water flowing through rectangular microchannels, *Exp. Heat Transfer*, 7(4), 249–264.

Perret, C., Boussey, J., Schaeffer, C., and Coyaud, M., 2000. Analytic modeling, optimization, and realization of cooling devices in silicon technology, *IEEE Trans. Components Packaging Technol.*, 23(4), 665–672.

Pfahler, J., Harley, J., Bau, H., and Zemel, J., 1990. Liquid transport in micron and submicron channels, *Sensors Actuators*, A21–A23, 431–434.

Pfund, D., Shekarriz, A., Popescu, A., and Welty, J.R., 1998. Pressure drop measurements in a microchannel, in *Micro-Electro-Mechanical Systems (MEMS)*, ASME DSC, 66, 193–198.

Rao, P., 2000. Effects of Flow Mal-Distribution in Parallel Micro-Channels, M.S. thesis, Department of Mechanical Engineering, Pennsylvania, State University, University Park.

Ravigururajan, T.S., Cuta, J., McDonald, C.E., and Drost, M.K., 1996. Single-phase flow thermal performance characteristics of a parallel micro-channel heat exchanger, *Proc. National Heat Transfer Conference*, ASME HTD, 329, 157–165.

Serizawa, A., Feng, Z., and Kawara, Z., 2002. Two-phase flow in microchannels, *Exp. Thermal Fluid Sci.*, 26, 703–714.

Shah, M.M., 1982. Chart correlation for saturated boiling heat transfer: equation and further study, *ASHRAE Trans.*, 88, 185–196.

Sobhan, C.B. and Garimella, S.V., 2000. A comparative analysis of studies on heat transfer and fluid flow in microchannels, in *Heat Transfer and Transport Phenomena in Microscale, Proc. Int. Conf on Heat Transfer and Transport Phenomena in Microscale*, G.P. Celata, V.P. Carey, M. Groll, I. Tanasawa, and G. Zummo, Eds., Begell House, New York, 80–92.

Steinke, M.E. and Kandlikar, S.G., 2003. Flow boiling and pressure drop in parallel flow microchannels, *Proc. First Int. Conf. on Microchannels and Minichannels*, April 24–25, Rochester, NY, 567–579.

Tuckermann, D.B. and Pease, R.F., 1981. High performance heat sinks for VSLI, *IEEE Electron Device Lett.*, EDL -2, 126–129.

Wang, B.X. and Peng, X.F., 1994. Experimental investigation of liquid forced convection heat transfer through microchannels, *Int. J. Heat Mass Transfer*, 37(Suppl. 1), 73–82.

Wang, C.-C., Chiang, S.-K., Chang, Y.-J., and Chung, T.-W., 2001. Two-phase flow resistance of refrigerants R-22, R-410A and R-407C in small diameter tubes, *Trans. IChemE*, 79(Part A), 553–560.

Webb, R.L., 2003. Effect of manifold design on flow distribution in parallel microchannels, paper 35251, *Proc. IPACK03, International Electronic Packaging Technical Conference*, July 6–11, 2003, Maui, HI.

Webb, R.L. and Chung, K., 2004. Two-phase flow distribution in tubes of parallel flow heat exchangers, *Heat Transfer Eng.*, 26, 3–18.

Webb, R.L. and Paek, J.-W., 2003. Letter to the Editors — Concerning paper published by Y.-Y. Yan and T.-F. Lin, *Int. J. Heat Mass Transfer*, 46, 1111–1112. (Also see Y.-Y. Yan and T.-F. Lin, 2003. Reply to Prof. R.L. Webb's and Dr. J.W. Paek's comments, *Int. J. Heat Mass Transfer*, 46, 1112–1113.)

Wu, P. and Little, W.A., 1983. Measurement of friction factors for the flow of gases in very fine channels used for microminiature Joule-Thomson refrigerators, *Cryogenics*, 25, 273–277.

Wu, P. and Little, W.A., 1984. Measurement of the heat transfer characteristics of gas flow in fine channel heat exchangers used for microminiature refrigerators, *Cryogenics*, 24, 415–420.

Yan, Y.-Y. and Lin, T.-F., 1998. Evaporation heat transfer and pressure drop of refrigerant 134a in a small pipe, *Int. J. Heat Mass Transfer*, 41, 4183–4194.

Yan, Y.-Y. and Lin, T.-F., 2003. Reply to Prof. R.L. Webb's and Dr. J.W. Paek's comments, Int. J. Heat Mass Transfer, 46, 1112–1113.

Yu, D., Warrington, R., Barron, R., and Ameel, T., 1995. An experimental investigation of fluid flow and heat transfer in microtubes, *ASME/JSME Joint Thermal Eng. Conf.*, 1, 523–530.

ELECTRONIC COOLING HEAT TRANSFER

19.1 INTRODUCTION

The most popular device used for first-generation desktop computer cooling is the "active" heat sink. The currently used version consists of a small aluminum heat sink on which a small fan (e.g., 60 mm diameter) is mounted. Such a heat sink of 60×80 mm plan area has been found adequate to remove up to approximately 100 W. The device has evolved since it was first used with the Pentium processor. The major changes have been in the physical size of the heat sink and fan, and in use of a copper "heat spreader" to spread the heat from the ever decreasing Central Processor Unit (CPU) size. As a result of smaller CPU size and increased power, the heat flux at the CPU has significantly increased. This has resulted in great improvements of thermal interface material (TIM).

As CPU power increases and die size decreases, it appears that limits will exist on the possible power dissipation of CPU-mounted heat sinks. The factors that will limit the possible heat dissipation are the heat sink plan area and height, and the fan size, speed, and allowable noise. When this limit is reached, it will be necessary to identify a new concept for heat removal.

The author believes it is appropriate to introduce a new terminology for heat removal concepts. Rather than using terms, such as "active" heat sink, the following definitions are proposed:

1. Direct heat removal (DirHR) for an ambient heat sink that is directly attached to the hot source

2. Indirect heat removal (IndHR) for an ambient heat sink that is remote from the hot source, and uses a "working fluid" to transport heat from the hot source to the heat sink

The term "ambient heat sink" (AmbHS) is defined to mean the final heat rejection device that rejects heat to the ambient, via air or water cooling. Use of IndHR establishes the need for a second heat sink located at the hot source. This is called the "hot source heat sink" (HS-HS). For several years, notebook computers have used a heat pipe to accomplish indirect heat removal.

Either water or air may be used as for heat rejection to ambient. Although water cooling is considered an option for ambient heat removal, there is strong interest in maintaining heat rejection to ambient air as long as is possible. This chapter deals only with heat rejection to ambient air. Thus, DirHR devices with air cooling are defined as first-generation devices. IndHR devices with heat rejection to ambient air are defined as second-generation devices. Refrigeration-enhanced cooling has also been under investigation, and is used in some servers. In terms of definitions used here, refrigerated systems with heat rejection to air are called third-generation technology. IndHR devices that reject heat to water are defined as fourth-generation technology. Refrigerated IndHR devices that reject ambient heat to water are termed fifth-generation technology.

A key advantage of DirHR with air cooling is that the device is orientation insensitive. The importance of this advantage has been taken for granted until recent work on high-power second-generation devices. A variety of second-generation devices have been proposed for future-generation desktop computers and servers, which are the key focus of this chapter. One concern of some of the devices under current consideration is their orientation sensitivity. Designers of heat pipe systems for notebook computers are already aware of the orientation sensitivity of heat pipes. A device that is sensitive to orientation may not be as desirable as one that is orientation insensitive. Hence, this concern is addressed in the discussion of second-generation IndHR devices.

19.2 COMPONENT THERMAL RESISTANCES

The heat transfer rate is typically defined in terms of the driving temperature difference between the hot surface (T_{hot}) and the inlet air temperature ($T_{air,in}$). For a heat dissipation of Q (W), the total thermal resistance is

$$R_{tot} = \frac{T_{hot} - T_{air,in}}{Q} \tag{19.1}$$

For a DirHR device, the R_{tot} typically consists of three basic components in series: interface (R_{int}), spreading (R_{sp}), and the heat sink (R_{amb}). The thermal resistance associated with the ambient heat sink is called R_{amb} and is a convection resistance.

If an IndHR heat sink is located remote from the hot source, a "working fluid" must be used to transfer the heat from the hot surface to the convection heat sink. Two means of heat transport are possible:

1. A heat pipe, which is typically used in notebook computers
2. Convection via a single phase or two-phase fluid

Use of a working fluid in the IndHR device will involve additional convection thermal resistances. Consider, for example, the device illustrated in Figure 19.1, which boils a fluid at the hot source and condenses it at the ambient heat sink. Following are associated thermal resistances of the indirect system:

1. $R_{H,cv}$, which is the convection resistance at the hot source
2. R_{transp}, which is the thermal resistance, associated with transporting the working fluid to the ambient heat sink
3. $R_{C,cv}$, which is the convection resistance, associated with transferring heat from the working fluid to the ambient heat sink

The sum of these three additional thermal resistances is defined as R_{ind} and is the additional resistance introduced by use of a working fluid with an IndHR device.

Notebook computers may use a heat pipe to transfer heat from the hot surface to the heat sink. The heat pipe has the same three additional thermal resistances. Thus, the thermal resistance associated with pressure drop of the vapor between the boiler and the condenser is R_{transp}. The working fluids used in an IndHR devices may be gases, single-phase liquids, and two-phase fluids. Specific devices that may be

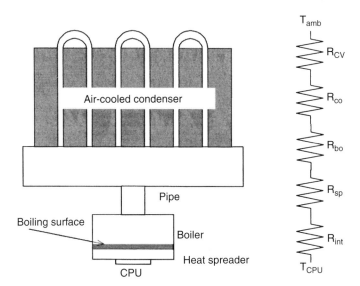

Figure 19.1 Schematic of remotely located thermo-syphon heat rejection device.

used to remove heat at the hot source ($R_{H,cv}$) and transfer it to the ambient heat sink, where $R_{C,cv}$ exists are addressed in a later section.

19.3 LIMITS ON DIRECT HEAT REMOVAL WITH AIR-COOLING

It is important to define probable limits (Watts heat rejection) that may be attained by DirHR devices. It is expected that DirHR will be used as long as possible, because of their low cost. For the same Watts heat rejection, second-generation IndHR are expected to cost more than DirHR devices.

Figure 19.2 illustrates a DirHR air-cooled heat sink that is designed to attach directly to the CPU of a desktop computer. The plan area of computer heat sinks is limited by the space allocated on the circuit board. Further limitations exist in the height of the fan–heat sink combination. The mounted fan provides airflow though the finned array, as shown in Figure 19.2. The airflow through the heat sink may be either "impinging" flow (Figure 19.3a) or "duct flow" (Figure 19.3b). The Figure 19.2 heat sink must operate at the balance point on the fan curve. Because of the design geometry constraints, it is probable that the airflow frontal area will be fixed (for a given flow orientation). For a given heat sink/fan combination, it is important to determine whether higher performance will be obtained by duct flow or impinging flow. This question was studied by Saini and Webb [2002b] for plain fins. They fixed the plan area at 60 × 60 mm and the total heat sink/fan height at 50 mm. They performed analysis to determine the optimum fin pitch and fin thickness for each flow geometry using a 60 × 60 mm fan operating at 4700 rpm. They found that the minimum R_{cv} will occur for the highest allowable fin height. They also extended the analysis to consider increased airflow rate and a larger heat sink size 80 × 80 mm. Their results are shown in Figure 19.4. For the base design (60 × 60 mm heat sink and 60 mm fan operating at 4700 rpm), Figure 19.4 shows the optimum geometry in impinging flow has 23% lower convection resistance than the optimum duct flow geometry. When the heat sink base area was increased to 80 × 80 mm or the fan

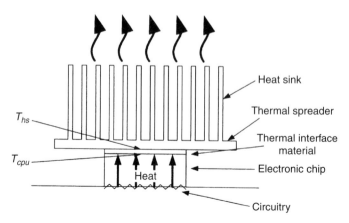

Figure 19.2 Air-cooled heat sink with fan.

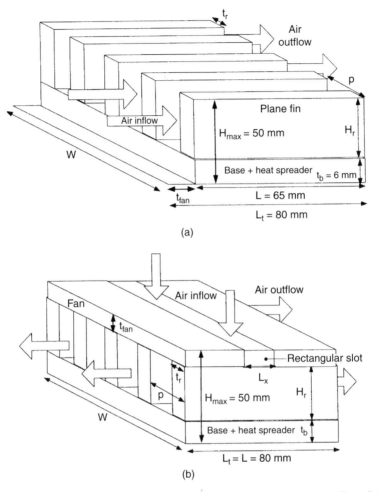

Figure 19.3 (a) Duct flow through heat sink, (b) impinging flow through heat sink. (From Saini and Webb, 2002a. *IEEE Trans. Components and Packaging Technologies.* With permission of IEEE.)

speed increased, the impinging flow geometry maintained superiority over the duct flow geometry. Figure 19.4 raises an interesting question: What is the smallest convection resistance that can be achieved with a DirHR heat sink? For the 80×80 mm heat sink, the impinging flow geometry yields $0.185 < R_{cv} < 0.154$ K/W, depending on the air flow rate.

Saini and Webb [2002a] also investigated the potential advantages of using an "enhanced" fin geometry, such as pin fins or offset strip fins (OSF). The tests showed that in impinging flow, a pin-fin geometry will have higher R_{cv} than the optimum plain fin geometry of the same base area (60×60 mm base area), both using the same fan. They also investigated the possible performance benefits of the OSF in duct flow. The optimum geometry yielded an R_{cv} 9% higher than minimum R_{cv} for

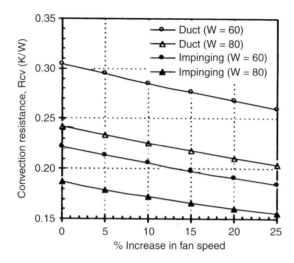

Figure 19.4 Comparison convection resistance for duct flow and impinging flow in a 60 × 60 mm heat sink with 60 mm fan, plus effect of fan speed increase. (From Saini and Webb, 2002a. *IEEE Trans. Components and Packaging Technologies*. With permission of IEEE.)

the plane fin geometry. The principal reason for this is that the high pressure drop of the OSF geometry yields reduced flow rate at the fan balance point. Although the tested pin-fin geometry may not be the optimum, it is doubtful that an optimized geometry would show better performance than the optimum plane fin geometry.

Kim and Webb [2003] have performed further analysis of enhanced vs. plain fins for duct flow using the same geometry restrictions of Saini and Webb [2002b]. They found that optimized round pin fins in both staggered and inline arrangement provide higher convective thermal resistance than the optimized plain-fin heat sink. This is primarily because of the increased pressure drop, which results in a reduced airflow rate at the balance point on the fan curve. Hence, the reduced airflow rate results in lower heat dissipation.

It is known that enhanced fin geometries, such as pin fins and OSF are practically used in various engineering applications. However, such applications typically allow some freedom of frontal area, which was not allowed in this work. As noted above, the poor performance of the pin fins and short louver pitch OSF is primarily caused by the reduced airflow rate at the fan balance point. If the frontal area were sufficiently increased, the pressure drop would be reduced and it would be possible to have the same balance point on the fan curve. Then, the enhanced fin geometries should show significant improvement over the plain fins. Chapter 3 provides quantitative performance evaluation criteria to compare the performance of enhanced vs. plain fin geometries for different operating conditions and constraints. However, Kim and Webb [2003] show that significantly larger frontal areas would likely be needed than can be accommodated within computer design layout requirements.

Based on the analysis of Saini and Webb [2002b], it appears that impinging flow with plain fins will yield the lowest possible R_{amb}. And, it appears that the 0.15

K/W may be a practical limit for R_{cv}, within current design limitations. Setting values for R_{int} and R_{sp} allows one to predict the limiting heat dissipation. Assuming the hot source is 16 mm square and R_{int} = 0.0157 K-in.^{2}/W the interface resistance is 0.04 K/W. With 80 × 60 mm base area case and 4.0-mm-thick copper heat spreader, the R_{sp} = 0.142 K/W for impinging flow. Substituting these values in Equation 19.1 gives R_{tot} = 0.336 K/W. Using T_{hot} = 70°C and T_{amb} = 35°C, one would predict the limiting heat dissipation to be 104 W. Thus, it appears that the DirHR devices will be limited to heat rejection values close to 100 W.

Clearly, the highest performance for any heat sink geometry will be attained if all of the airflow from the fan passes through the heat sink. Any design factors that allow airflow to bypass the heat sink fins will reduce the heat sink performance.

19.3.1 PEC Example 19.1 Enhanced Fin Geometry Heat Sink

The CPU in desktop computers is cooled by a heat sink, on which a 60-mm-diameter fan is mounted to provide "duct flow" through the fin passages. There is no air bypass. A typical heat sink made of aluminum alloy Al 6063 (thermal conductivity 215 W/m-K) consists of plane fins 50 mm high, 60 mm wide, and 65 mm long, with fins on a 3.0-mm pitch, and 1.0-mm thickness. The airflow through the heat sink is determined by the balance point on the fan curve. For the plain fin case, the air friction power is 0.1 W. Of interest here is whether the OSF will provide greater heat rejection than the plain fin geometry for the same airflow frontal area and airflow depth. Use the OSF with four louvers over the 65-mm flow depth, 3.0-mm fin pitch, and 1.0-mm fin thickness. Calculate the heat dissipation (Q) assuming that the OSF geometry will operate with the same flow power as the plane fin heat sink. Use 35°C for ITD and air properties at 35°C. Would you expect improvement if you used a smaller louver pitch? Discuss.

Solution. The heat transfer rate is given by $Q = c_{air} \, \varepsilon$ ITD and the $\varepsilon = 1 - e^{NTU}$, where NTU = hA/c_{air}. Use the Nomenclature of Chapter 2. To simplify the problem, inlet and exit losses were ignored. For the heat transfer coefficient and pressure drop of the plain fin and OSF geometry, the analysis was performed using the plain fin correlations of Saini and Webb [2002b] and the OSF correlations of Manglik and Bergles [1990]. Table 19.1 shows the calculated results for the plain fin and OSF heat sinks at constant-flow power and constant ITD. In Table 19.1, L_p, V_{air}, and Δp are the louver pitch of the fin, volume flow rate of air, and air pressure drop, respectively. Table 19.1 shows that calculations were made for different louver pitches (all having 65 mm airflow depth). The surface area of all geometries is 0.1326 m^2.

Discussion. Table 19.1 shows that the volumetric airflow rate through the OSF geometry is significantly decreased for all louver pitches. For the largest L_p, the flow rate is reduced 28% and the airflow reduction is increases as L_p is reduced. This is because of the higher pressure drop of the OSF geometry operating at fixed frontal area. Although the OSF has higher heat transfer coefficients at the reduced flow rate, the effectiveness (ε) increase caused by the reduced flow rate results in lower

Table 19.1 Performance of Plain Fin and OSF Heat Sinks (For ITD = 35°C and Fan Air Friction Power = 0.1W)

Geometry	L_p (mm)	V_{air} (m³/s × 10³)	Δp (Pa)	h (W/m²K)	NTU	∈	Q (W)
Plain fin	65	6.22	16.07	56.23	1.048	0.649	161.6
OSF	16.3	4.85	20.59	58.04	1.386	0.75	145.7
OSF	13	4.64	21.52	59.26	1.48	0.772	143.5
OSF	10.8	4.46	22.39	60.26	1.565	0.791	141.3
OSF	8.13	4.17	23.98	61.86	1.722	0.821	136.9
OSF	6.5	4.04	24.73	62.53	1.795	0.833	134.8

heat transfer rate (Q). The situation worsens as the L_p is reduced for the OSF geometry. If it were possible to maintain the same total surface area and increase the airflow frontal area (PEC VG-1 in Table 3.1), one may show that the louver fin would give higher performance than the plain fin. Note that the present analysis uses PEC FG-2a of Table 3.1

19.4 SECOND GENERATION IndHR DEVICES FOR HEAT REMOVAL AT HOT SOURCE

Figure 19.5 from Webb [2004] shows a tree of various indirect concepts that may be used to remove heat at the hot source surface. The first-generation DirHR method has limited heat rejection capability. The second-generation IndHR methods are expected to provide greater heat removal capability and are of specific interest here. The heat removed from the hot source surface by a working fluid is transferred to the remote ambient air-cooled heat sink. Excluding gases, the working fluid may be a single-phase liquid or a two-phase fluid. A key concern is whether the methods are orientation sensitive or insensitive. Figure 19.5 is annotated to indicate whether the concepts are expected to be orientation sensitive or insensitive. The Figure 19.5 concepts are separated into the basic categories, "single-phase" and "two-phase," which refers to the working fluid. Discussion of the ambient heat sink used to cool the working fluid is presented later.

19.4.1 Single-Phase Fluids

This concept uses a pump to force a single-phase fluid through multiple, parallel microchannels at the hot source. The fluid must be subcooled and it must be pressurized to remain subcooled. The heated single-phase liquid must be cooled in the ambient heat exchanger. The heated liquid will be cooled in microchannels in the ambient heat exchanger. Discussion of the ambient heat sink is presented later. Because the system operates under forced convection, it is orientation insensitive. This concept is discussed further in a later section.

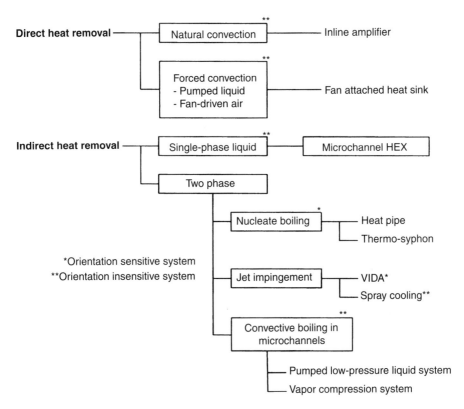

Figure 19.5 Tree of methods of removing heat from hot source surface.

19.4.2 Two-Phase Fluids

Higher performance is expected by using a two-phase working fluid than is possible with a single-phase liquid. For two-phase heat removal at the hot source, the working fluid must be condensed at the ambient air-cooled heat sink. Then, the condensed liquid is returned to the hot source. As previously noted, orientation sensitivity is of concern. It is possible that the process at the hot device may be orientation insensitive, but that the condensation and liquid return process may still be sensitive to orientation. Hence, one must consider both the boiling and the condensing devices to determine orientation sensitivity. If the condensate is returned to the hot source by gravity, the device will be orientation sensitive. In this case, the boiler must be located at a lower elevation than the condenser. This limitation may be overcome by using a forced convection concept. Note that a forced convection concept will require a pump to transport the fluid between the hot source and the heat sink. Pool boiling systems, which do not use a pump, are orientation sensitive. Heat pipes have height limits to their orientation sensitivity. Jet impingement devices may or may not be orientation sensitive at the hot surface, but how the condensate is returned to the pump sump will determine their orientation sensitivity. These several concepts are discussed below.

Table 19.2 Properties and Operating Conditions of Candidate Working Fluids

Refrigerant	M	λ @ 1 atm (kg/kJ)	P_{cr} (MPa)	P (MPa) (Temp = 70°C)	Figure of Merit ($\times 10^{-9}$)	GWP
R-134a	102	178	4.06	2.162	8.37	1300
FC-72		88.47			6.80	9000
HFE-7100	250		2.23	0.127	4.49	90
HFE-7200	264	125.61	2.07	0.1		55
Water	19.02	2445	22.12	0.031	203	0

Several factors will affect the choice of an acceptable working fluid. Material compatibility prevents use of aluminum with water. However, copper is compatible with water. Flammability concerns will tend to prevent use of propane, or iso-butane. Table 19.2 shows some possible candidate fluids. Environmental concerns eliminate fluids having ozone-depletion potential (ODP). Recently introduced concerns include the global warming potential (GWP). Note that FC-72 has very high GWP. Although HFE-7100/7200 have significantly lower GWP than R-134a, they also have significantly lower heat transfer coefficients for the same operating conditions. Interest in using dielectric fluids (e.g., FC-72 and HFE-7100/7200) with silicon heat transfer surfaces would allow the silicon structure to be fabricated on the back surface of the actual electronic chip. However, as is shown below, such dielectric fluids on silicon have lower performance than water on copper. The trade-off is use of higher-performance heat transfer fluids and materials, which require a thermal joint. The thermal conductivity of copper is approximately three times that of silicon.

19.4.3 Heat Pipe

The heat pipe is a well-known concept and is routinely used in notebook computers. It has moderate orientation sensitivity. The capacity will decrease as the evaporator is elevated above the condenser.

19.4.4 Nucleate Boiling

Much work has been done on development of high-performance nucleate boiling surfaces. Much of the work discussed in Chapter 11 on enhanced nucleate boiling is directly applicable to cooling microelectronics. Sintered, porous boiling surfaces described in Chapter 11 were used by Webb and Yamauchi [2002] for their thermosyphon device. It is even possible to apply the the high-performance "structured" surfaces, which were developed for the outer surface of tubes, to small flat surfaces. A method to make structured surfaces having surface pores and subsurface tunnels is described by Chien and Chang [2002]. Wire screen mesh is another possibility, although mesh surfaces do not give performance as high as sintered surfaces. Honda and Wei [2004] review recent advances in boiling dielectric liquids on surface microstructures. They also compare the boiling performance, critical heat flux, and incipience superheat for such surface microstructures. They describe work done on surfaces made by sandblasting, an etched SiO_2 layer, CVD of SiO_2, dendriditic structures, laser-drilled holes, microchannels, microfins, alumina particle spraying, painted silver flakes, diamond particles, micropin fins, etc.

Foamed materials (aluminum, copper, and graphite) have also been used for boiling of FC-72. Arbelaez et al. [2000] boiled FC-72 on a variety of aluminum foam structures. The porosity varied from 90 to 98% and the pore size between 5 and 40 pores/in. (PPI). Significant enhancement was obtained. Lower porosity and higher PPI structures yielded the best performance. Moghaddam et al. [2000] boiled water and FC-72 on copper and graphite foams, which have much higher effective thermal conductivity than do aluminum foams. They found that the copper foam (80 PPI) yielded the best performance with FC-72, and it was significantly better than for the graphite foam. The measured performance for the copper foam was much higher than for the best aluminum foam tested by Arbelaez et al. [2000]. Composite structures made with graphite have also been tested. Liang and Yang [1998] tested graphite–copper and graphite–aluminum fiber composite structures (50% volume concentration) with pentane. The pentane boiling performance was significantly higher than for the foamed structures using FC-72. Further discussion of foams and composite structures is given in Section 11.4.10. Further discussion of foamed metal heat sinks is presented in Section 5.10, including data reported by Bhattacharya and Mahajan [2000].

Ramaswamy et al. [2003] describe a "stacked matrix" of small channels for application to cooing microelectronics, which is shown in Figure 19.6b. The matrix is made by stacking the 10-mm-square single-layer structure shown in Figure 19.6a. The single layer structure contains an array of parallel rectangular channels cut in opposite faces of the substrate. The channels in the top surface are cut 90° to those in the bottom surface. Because the depth of each set of channels is more than half the thickness of the substrate, square openings (pores) exist at the intersection of the channels in the top and bottom surfaces. The pore size is defined as the diameter of a circle that may be inscribed inside the square pore. They investigated the effect of channel width, channel height, and channel pitch (pore pitch) on boiling performance of FC-72. Tests were made on single-layer silicon and copper structures, and on multilayer (three and six layers) copper structures. For a multilayer structure, the heat is conducted upward through the matrix, so good thermal joints between layers and high material thermal conductivity are important. The channel dimensions investigated were 260 and 550 μm channel height, 90 to 320 μm channel width, and 500 to 2100 μm channel pitch. The 90 to 320 μm channel widths resulted in corresponding pore diameters. Data were taken on the single-layer silicon surfaces between 4 and 40 W/cm² and for the copper surfaces between 3 and 80 W/cm². The vapor was condensed in a reflux condenser. The highest performance was obtained for the largest pore diameter and smallest pore pitch. However, the effect of pore size was negligible at the high heat flux. Performance improved by reducing the pore pitch, which resulted in increasing the pore density. For the same geometry, the copper structures showed higher performance, because of the higher thermal conductivity of copper. For the copper multilayer surfaces, performance improved as the number

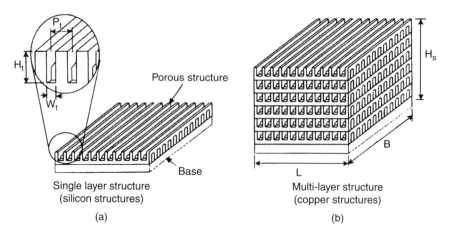

Single layer structure
(silicon structures)

(a)

Multi-layer structure
(copper structures)

(b)

Figure 19.6 The 10-mm-square microfabricated boiling surface made by Ramaswamy et al. [2003] and tested with FC-72.

of layers increased. The best boiling performance of the single-layer copper surface with FC-72 is about 70% below that shown by Webb and Yamauchi [2002] for R-134a boiling on a sintered copper surface used in their thermo-syphon device.

Honda et al. [2002] boiled FC-72 on a microfabricated structure silicon structure having either a microroughness (32 μm rms) or "micropin fins" (50 μm square and 60 m high). The micropin fins increased the surface area by a 2.2 factor. Data were taken for saturated and subcooled boiling. The test results for saturated boiling are shown in Figure 19.7. Figure 19.7 includes test results for O'Connor and You [1995] who used an aluminum particle paint, which shows the best performance. The micropin fin surface (EPF) is second best and is slightly below the micropin fin surface of Mudawar and Anderson [1989]. The rough flat surface E (32 μm rms) shows better performance than the smooth surface S having 2.3 μm rms. Note that the enhanced surface data show a 1.8 to 2.3 factor enhancement of critical heat flux (CHF) compared to the smooth surface S.

If water is used, the boiling surface must especially configured to boil water at very low system pressure. Figure 19.8 from Chien and Chang [2005] shows how the boiling coefficient is influenced by particle size, for 0.5 mm coating thickness. The highest boiling coefficient is provided by the largest particle size (150 μm, surface P1). Chien and Chang [2002] show that larger particle sizes give even a higher boiling coefficient than the P1 surface of Figure 19.8. Their B1 surface (247 μm average particle diameter) at T_{sat} = 70°C provides three times the boiling coefficient of the 150 μm particle diameter P1 surface. Note that the boiling coefficient for surface P1 on Figure 19.8 is about three times higher than for the Figure 19.7 enhanced surfaces using FC-72.

Arik and Bar-Cohen [2001] measured the boiling performance of HFE-7200 on a vertical ATC 2.6 chip package. Their data were for subcooled boiling and include CHF data. To compare the nucleate boiling performance of different working fluids, the pool boiling heat transfer coefficient (h) for saturated boiling at 70°C on a plain

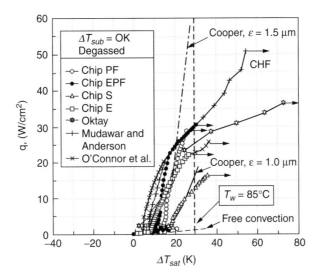

Figure 19.7 FC-72 boiling performance on smooth and enhanced surfaces from Honda et al. [2002]. Chip PF (micropin fin), Chip EPF (rough, 32 μm rms), Chip S (smooth, 2.3 μm rms), Chip E (rough, 25 μm rms), Oktay [1982], Mudawar and Anderson [1989] (micropin fin), O'Connor et al. [1995] (aluminum particle painted).

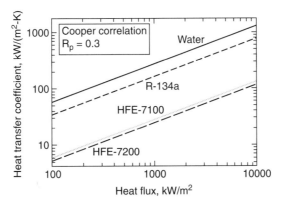

Figure 19.8 Boiling performance of water at T_{sat} = 65°C on 0.5-mm-thick porous boiling surfaces. P-1 (100 to 200 μm), P-2 (50 to 100 μm), P-3 (30 to 50 μm). (From Chien and Chen [2005]. With permission.)

surface using the Cooper correlation (Equation 2.32) with the R_p roughness factor equal to 0.3 has been estimated. Figure 19.9 shows the predicted boiling coefficient for several candidate fluids used to cool microelectronics. The figure shows that water has the highest boiling coefficient, followed by R-134a. The boiling coefficient of HFE-7200 is much less than that of R-134a. If water is used as the working fluid, copper components are required. Water, the most environmentally acceptable working fluid, cannot be used in aluminum, because it causes pinhole corrosion. Although

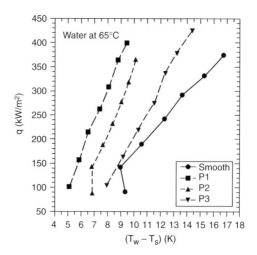

Figure 19.9 Nucleate boiling heat transfer coefficient for several candidate working fluids predicted by the Cooper [1984] correlation. (From Webb and Yamauchi, 2002. *THERMES 2002* and *IEEE Trans. Components and Materials Packaging*, 2002. With permission of IEEE.)

water is an environmentally acceptable fluid, a principal concern is the problem that may result from freezing. Hence, special design is required, or an ethylene glycol–water mixture would be acceptable to protect from freezing.

Boiling usually occurs in a confined space (or narrow channel) in electronic applications. Typically, boiling will occur on a small horizontal or vertical surface having an opposite confining wall only a few millimeters away from the boiling surface. Several studies have been done to establish the effect of confinement on boiling performance. Misale et al. [1999] present data on boiling of FC-72 on a finned surface in a narrow channel that is oriented in the horizontal direction. They conclude that a narrow channel width does not affect performance for the vertical orientation. However, it causes a drastic performance reduction for the horizontal orientation. Chien and Chen [2000, 2001] performed similar studies, for enhanced boiling surfaces in the vertical orientation. They tested gap widths of 0.5 and 1.0 mm, and compared the effect of the narrow gap with boiling on the same unconfined surfaces. The boiling performance improved as the gap width reduced from 2.0 to 0.5 mm. However, the CHF decreased as the gap was reduced.

19.4.5 Forced Convection Vaporization

As shown in Equation 13.2, there are "forced convection" and "nucleate boiling" components in forced convection vaporization. If vaporization occurs in microchannels, it is probable that the forced convection component will be small, because of the low Reynolds number. The nucleate boiling component will be dependent on the value of $(T_w - T_{sat})$. Hence, to obtain a high heat transfer coefficient, it is likely necessary that nucleate boiling be dominant. It is important that the exit vapor quality be controlled to prevent dryout.

19.4.6 Spray Cooling

Thin film evaporation produced by droplet spray cooling has been shown to provide higher heat transfer coefficients, and higher CHF than provided by nucleate boiling. Thin film evaporation is discussed in Section 13.8. In this approach, one uses a pump and atomizing nozzles to spray fine droplets on the hot surface, where they evaporate as a thin film. The vapor must be condensed in an attached condenser. The system pressure is established by the condensing temperature. The condenser may be either air or water cooled, depending on the system requirements. Typically, a higher mass flow rate is sprayed on the surface than is evaporated. This will allow attainment of a higher CHF. Because only part of the liquid flow rate is evaporated, the remaining liquid must gravity-drain to a sump for recirculation by the pump. Thus, the concept is not totally orientation insensitive. Although use of a subcooled liquid will increase the CHF, the fluid will be only slightly subcooled in practical heat rejection systems having a condenser. Although the primary heat transfer mechanism is evaporation of the thin liquid film at the liquid–vapor surface, some nucleation may also occur at the surface.

19.5 DISCUSSION OF ADVANCED HEAT REMOVAL CONCEPTS

19.5.1 Jet Impingement/Spray Cooling Devices

For two-phase applications, both jet impingement and spray cooling involve spraying a liquid on the hot surface, which spreads as a thin film and is evaporated. However, spray cooling specifically means that the liquid impinges on the surface as small droplets. Most high-performance work has involved spray cooling.

Work has been done for single nozzles or on plates having closely spaced nozzles of very small diameter. It is very important to have a uniform spray distribution over the heated surface area, or non-uniform surface temperature can result. Results have been reported by Pais et al. [1992], Marcos et al. [2002], and Xia [2002] for water. If the working fluid is water, the evaporation will occur at very low absolute pressure for a hot surface temperature in the range of 70°C. Figure 19.10 shows two design concepts that have been tested with water. Figure 19.10a illustrates the single-nozzle concept used by Marcos et al. [2002] to cool a 1.0 cm² heat source. Figure 19.10b shows the Xia [2002] multiple-nozzle design with atomization assisted by a vibrating plate driven by a piezoelectric actuator. The outlet nozzles were approximately 100 μm diameter with 160 to 240-μm nozzle pitch. When the piezoelectric plate vibration reaches a certain frequency (in the vicinity of 5.0 kHz), the microliquid streams are broken into microdrops.

Figure 19.11 shows the performance measured by Xia [2002] for water boiling on a 1.0 cm² electrically heated flat plate. The data were taken for 1.6 m/s jet velocity

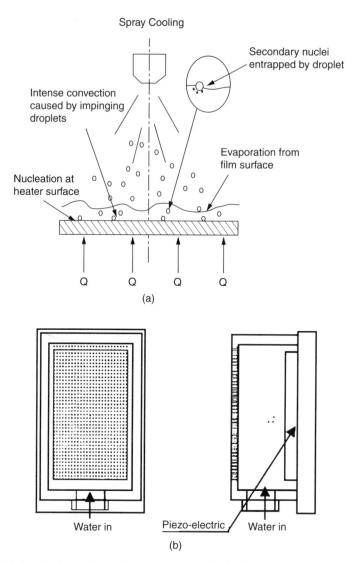

Spray Cooling

Secondary nuclei entrapped by droplet

Intense convection caused by impinging droplets

Evaporation from film surface

Nucleation at heater surface

Q   Q   Q   Q

(a)

Water in

Piezo-electric   Water in

(b)

Figure 19.10 Two impinging jet nozzle design concepts: (a) Single nozzle concept used to cool a 1.0 cm² heat source. (From Marcos et al., 2002 *18th IEEE SEMI-THERM Symp.* With permission of IEEE.) (b) Multiple nozzle design with atomization assisted by a vibrating plate driven by a piezoelectric actuator. (From Xia *18th IEEE SEMI-THERM Symp.* With permission of IEEE.)

and 8.0 mm spacing between the hot surface and the nozzle plate. The different nozzle diameters and pitches (A, B, C, D) made little difference, except at the lowest heat fluxes. This figure shows that heat fluxes as high as 900 W/cm² can be supported. The "temperature rise" on the abscissa is the surface temperature above the saturation temperature. Thus, at 400 W/cm² heat flux, the boiling coefficient is approximately

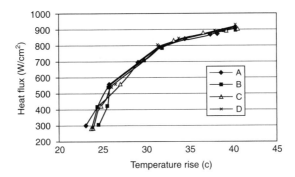

Figure 19.11 Boiling heat flux vs. superheat for four different micronozzle plates. (From Xia, [2002] *18th IEEE SEMI-THERM Symp.* With permission of IEEE.)

16.6 W/cm^2-K, or 166 kW/m²-K. Comparing this with the nucleate boiling data of Chien and Chang [2002] for a porous surface, one sees that the nucleate boiling coefficient on the porous surface at 400 W/cm^2 is significantly lower (4.2 W/cm^2-K). Hence, spray cooling provides much higher heat transfer coefficients than were measured for nucleate boiling of water on a porous surface. Further, jet impingement provides a significant increase of the CHF. It is possible that a heat spreader will not be required for jet impingement, because of the high heat transfer coefficient. At 400 W/cm^2, the thermal resistance at the 1.0 cm^2 hot surface is 0.06K/W.

Although the atomized spray on the hot surface is insensitive to orientation, the spray rate is several times higher than the evaporation rate, so a liquid sump must exist from which the excess liquid is pumped. If total system orientation insensitivity is desired, it will be necessary to have a design that allows the pump suction line to automatically relocate to the lowest point in the system. Such concepts, which allow the pump suction line to automatically locate via gravity force to the lowest point in the system, are possible.

The previous works addressed jet impingement on smooth, flat surfaces. Murthy et al. [2001] investigated the benefits of using closely pitched microjets on a three-dimensional structured surface with HFE-7200. They show that the thin film can wet a three-dimensional shape, which was intended to lead to higher heat flux (based on projected base surface area) at a given wall superheat. However, this was found to be effective only in the subcooled region. They also experimented with different nozzle orifice shapes. Data taken at the highest heat flux (30 W/cm^2) show a heat transfer coefficient of 1.2 W/cm^2-K, which is quite low compared to spray cooling with water (Figure 19.11) or for nucleate boiling of water (Figure 19.8). As shown on Figure 19.9, HFE-7200 yields considerably lower boiling performance than water.

Heffington et al. [2001] used a piezoelectric transducer to vibrate a plate at about 2.5 kHz, which produces a shower of small-diameter drops on the boiling surface. A multihole orifice plate (1.6-mm-diameter holes) just above the driver assists in forming the small drops. These drops impinge on the hot surface and evaporate as a thin film. This concept is similar to that of Xia [2002], although the

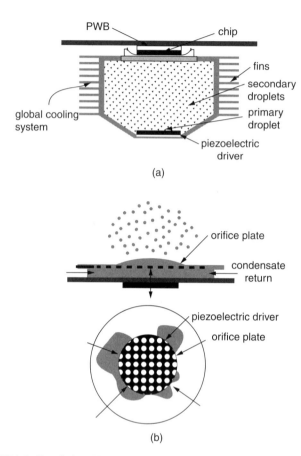

Figure 19.12 VIDA boiling device: (a) cross-section schematic with air-cooling fins on outer cylinder walls, (b) illustration of droplets formed at multiple orifice plate. (From Heffington, 2001. *Proc. IPACK '01*. With permission of IEEE.)

nozzle diameter used by Xia was much smaller (approximately 100 μm). The Heffington et al. concept is called vibration-induced droplet atomization (VIDA), and is illustrated in Figure 19.12. The vapor can be taken to a remote condenser, or can condense on the side walls of the cell, if the outside of the cell is cooled by convection. Tests were performed using water at 100°C, and at lower temperatures. Their measured CHF was 109 W/cm² for 100°C heater temperature. They do not provide data that allows calculation of the boiling coefficient. As tested by Heffington et al. [2001], the hot surface must be above the driver to allow the overspray to gravity-drain to the driver. Another concern is the significant noise generated by the piezoelectric transducer.

It is noted that none of the devices discussed in this section used forced convection vaporization. Thus, in all, the vapor was condensed in a reflux condenser, which makes them orientation sensitive. It is probable that these concepts could be adapted to a forced convection concept.

19.5.2 Single-Phase Microchannel Cooling

Realistic applications would use laminar, rather than turbulent, liquid flow in micro-channels. The very small hydraulic diameter would require very high fluid velocity to obtain turbulent conditions. Further, the pressure drop of turbulent flow would be extremely high, which would require high pump cost.

19.5.3 Two-Phase Microchannel Cooling

Use of two-phase heat transfer will significantly reduce the required flow rate and pumping power, compared to single-phase liquid flow. An example of two-phase microchannel cooling is that of Jiang et al. [2001]. Their concept used 40 parallel, 100-μm-square microchannels machined in a 0.5-mm-thick silicon substrate, which is illustrated in Figure 19.13a. They provide both single- and two-phase test results for 26°C inlet water flow supplied by a mechanical pump. Figure 19.13b shows the wall temperature vs. heat input for the electric heated test section operated at two water flow rates (2 and 5 ml/min). For single-phase flow, the higher flow rate yields the lower wall temperature, as expected. Figure 19.13b is annotated to show the power input at which two-phase flow exists in the microchannels. They observed a periodic oscillation of wall temperature rise and static pressure for two-phase flow. The wall temperature oscillation was as high as 60 K, as shown on Figure 19.13b. The thermal resistance values (K-cm²/W) calculated from the Figure 19.13b data are much higher that for nucleate boiling of water reported by Webb and Yamauchi [2002]. The measured pressure two-phase flow drop was seven to ten times higher than for single-phase flow. It is possible that flow maldistribution in the 40-channel heat sink may be partially responsible for the poor performance. Further comment on two-phase instabilities is given below.

Another etched silicon wafer water-cooled heat sink is described by Perret et al. [2000]. Water flowed in a 20-mm-square wafer having 230-mm-wide rectangular microchannels. For 0.5 l/min water flow rate and 100 W heat input, the measured thermal resistance was quite high—approximately 0.85 K-cm²/W. The performance of the Perret et al. [2000] microchannel heat sink is considerably better than that of Jiang et al. [2001]. Note that the microchannel concept requires use of a high-pressure micropump. The pump adds considerably to the cooling system cost. However, the system will be orientation insensitive. If the microchannel concept were used in a refrigerated server application, use of a high-pressure micropump would not be required.

Most studies of two-phase microchannel flow have typically shown instabilities and uneven temperature distribution in the hot surface. Examples of these studies are Hestroni et al. [2002] and Zhang et al. [2002]. It is possible that the flow instabilities are in part caused by flow maldistribution in the parallel channels. Flow from the pump is discharged into a manifold, which feeds the parallel flow channels. It is very difficult to achieve uniform flow distribution in the microchannels, and if some channels have less flow than others, local regions of overheating will exist. However, two-phase flow instability can also occur in single channels. Considerable

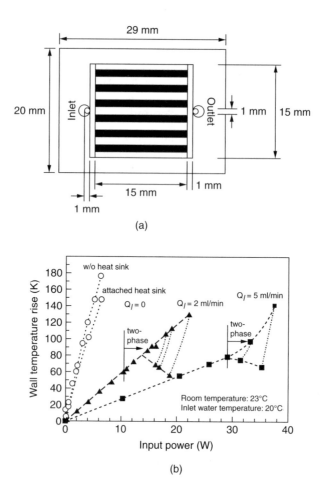

Figure 19.13 (a) Schematic of a microchannel heat sink, (b) wall temperature rise vs. input power. (From Koo et al., 2000. *Proc. 14th Annual IEEE) Intl. MEMS-01 Conf.* With permission of IEEE.)

work has been done on design of manifold flow distributors for single-phase flow. Some of this work is discussed in Section 18.6.5.

Chapter 18 discusses recent measurements by Steinke and Kandlikar [2003] for forced convection vaporization of water in six parallel, rectangular microchannels of 207 μm hydraulic diameter. These data show considerably higher performance than those discussed above. The reader is referred to this discussion in Section 18.7.

For two-phase applications, it is not practical to design for complete evaporation. This is because dryout will occur before 100% vapor quality is attained, resulting in significant surface overheating. In a practical system, flow instabilities and uneven heating cannot be accepted. More work is needed to define the conditions that affect the dryout condition, the possibility of avoiding flow instabilities, and the manifold design needed to ensure uniform flow distribution.

19.5.4 Enhanced Two-Phase Forced Convection Cooling

This envisions that a high-performance nucleate boiling enhancement will be used for the boiling surface. Then, the process will be nucleate boiling dominated. This concept differs from the enhanced nucleate pool boiling concept in that the condenser may also use forced convection condensation, so the entire system will be insensitive to orientation. See Section 13.3.4 for forced convection vaporization in macrosized channels with nucleate boiling enhancement.

19.6 REMOTE HEAT-EXCHANGERS FOR IndHR

The remote heat exchanger is the final heat sink, at which heat is rejected from the system. On Earth, this will involve sensible heat rejection to air and water. However, it is possible to have "air cooling with evaporative assist." In this system, a water film is sprayed on the cooling surface and is evaporated into the air stream. This is called an evaporative condenser, or evaporative fluid cooler, depending on whether the working fluid is a vapor that is condensed or a liquid that is cooled. Some discussion of this enhancement is found in Section 16.2.2. When working in air-borne or space applications, one may also consider rejecting heat to an expendable fluid that is evaporated or radiated to space. These are beyond the scope of the present book.

Designers of electronic cooling systems have been speculating for some time about when the limits to air-cooling will be reached, forcing use of water-cooled systems. Use of air cooling is preferred, and it is desired to use this concept as long as possible. The author believes that power dissipation requirements in the 100 to 300 W per hot source are within the domain of air cooling. However, it is necessary to consider design technology different from the DirHR devices that have been commonly used in the past. By using a "thermal bus system," one may duct the vapor generated at multiple heat sources to a central air-cooled condenser. The IndHR is ideal for use of such advanced air-cooling technology.

19.6.1 Air-Cooled Ambient Heat Exchangers

More design freedom is possible for an IndHR ambient heat exchanger than is possible for the DirHR type. For example, it is possible to locate the IndHR ambient heat exchanger in the back wall of the computer case and supply the airflow via a dedicated fan. This will ensure that all of the airflow from the fan passes through the ambient heat sink. Very little work has been reported for high-performance ambient heat sinks. Table 19.3 lists four design options for the air-cooled ambient heat exchanger. It is assumed that heat is transferred from a two-phase condensing vapor (or a liquid) to air. The means of liquid return to the boiler is listed as "forced convection" or "gravity return." The forced convection concepts, which use one pump, are insensitive to orientation.

Option 1 assumes the two-phase (or single-phase) fluid flows in a flat plate containing parallel microchannels, such as shown Figure 19.13. It would probably

Table 19.3 Air-Cooled Ambient Heat Sink Concepts for Indirect Heat Removal

Design Option	Liquid Return	Liquid (or Two-Phase) Side	Air-Side
1	Forced convection	Microchannels	Flat plate with heat spreader and air fins
2	Forced convection	Round tubes with serpentine circuit	Plate fins
3	Gravity	Flat tubes with parallel circuiting	Serpentine fins
4	Forced convection	Flat tubes with serpentine circuiting	Serpentine fins

be necessary to use an air-cooled heat sink similar to the design used in the current DirHR systems. To achieve heat rejection goals of 100 to 300 W, one would need a heat sink having a quite large plan area. This would require spreading the heat over a large area. Such a design is not believed to be a practical approach to the present goal.

Option 2 would use a "finned tube" air-cooled heat exchanger, such as used for condensers in small residential air conditioning. The tubes are joined by return bends to form "serpentine" refrigerant circuiting. The preferred tube diameter would be chosen to meet capacity requirements and may be in the range of 6 to 9 mm diameter. The design would also be applicable to the condenser section of a heat pipe.

Option 3 uses a design typical of that in automotive refrigerant condensers. Figure 19.14 illustrates the air-cooled condenser design concept. This concept uses flat tubes that provide little obstruction to the airflow. It provides higher air heat transfer coefficients and lower air pressure drop than that of Option 2. If all of the tubes are in parallel, the liquid return to the boiler would be gravity dominated, and the device would be orientation sensitive.

Option 4 is the same as Option 3, except that the tube side has serpentine circuiting. This will make the liquid return forced convection dominant and the device will be orientation insensitive. One would use flat tubes having small internal passages yielding very small hydraulic diameter.

A prototype of the Option 3 concept was built and tested by Webb and Yamauchi [2002] for use with their thermo-syphon IndHR. This prototype is shown in Figure 19.14 and was designed to reject 100 W. The flat tubes of the aluminum design (see cross section photograph in Figure 19.14) uses flat tubes of 3×16 mm cross section and having internal membranes that permit high internal fluid pressure. The air-side fins have high-performance louver fins. The use of flat tubes yields air pressure drop considerably below that of typical 9.5-mm-diameter round tubes used in residential air-cooled refrigerant condensers. It is important that a dedicated fan supply air to this heat exchanger, and that all of the air from the fan passes through the heat exchanger. A typical fin height is approximately 8.0 mm, and a small louver pitch (e.g., 1.0 mm) is used to obtain high air-side heat transfer coefficients. The heat exchanger may be made either of aluminum or copper. Figure 19.14 shows aluminum and copper heat exchangers made by Webb and Yamauchi [2002]. Joining of the

Figure 19.14 Air-cooled thermo-syphon and tube cross section. (a) Aluminum, (b) Copper-brass (From Webb and Yamauchi, 2002. *THERMES 2002* and *IEEE Trans. Components and Materials Packaging*, 2002. With permission of IEEE.)

tubes, headers, and fins is done by brazing. The preferred design will have a high air-side heat transfer coefficient and the smallest possible friction factor, which will result in high airflow rate at the fan/heat exchanger balance point. Copper fins offer performance advantages over aluminum fins, because the high thermal conductivity copper fins may be made thinner (e.g., 25 μm) and will yield a higher airflow rate at the balance point than an aluminum design.

Figure 19.15 compares the heat transfer and pressure drop performance of the Figure 19.14a aluminum flat tube design (21.5 fins/in., 16 mm deep) with that of a round tube heat exchanger (9.5-mm tube diameter and plain fins) designed to give the same air-side thermal performance. The equal thermal performance round tube design has two rows, is 44.0 mm deep, and has 15.7 fins/in. The flat tube/louver

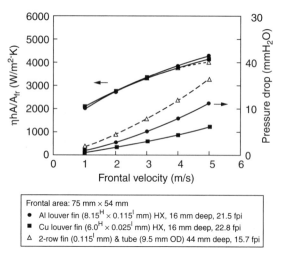

Figure 19.15 Comparison of flat tube and round tube heat exchanger performance. (From Webb and Yamauchi, 2002. *THERMES 2002* and *IEEE Trans. Components and Materials Packaging*, 2002. With permission of IEEE.)

fin heat exchanger performance was predicted using the correlation of Webb et al. [1995], and the round tube design was predicted by the Wang [2000] correlation. Both heat exchangers have the same finned frontal area (75×54 mm). The air-side heat transfer performance is defined by $\eta hA/A_{fr}$, where η is the fin efficiency, and A_{fr} is the heat exchanger frontal area. For the same $\eta hA/A_{fr}$ as the flat tube design the round tube design has 100% higher air pressure drop than that of the flat tube design. Also, the airflow depth of the round tube design is 1.75 times greater.

As previously stated, the Figure 19.14b heat exchanger is made of copper construction for use with water working fluid. The thickness of copper fins is only 25 vs. 115 μm for aluminum fins. For the same $\eta hA/A_{fr}$, Figure 19.15 shows that the flat tube pressure drop of the copper heat exchanger is 36% lower than that of the aluminum flat tube heat exchanger. For 2.0 m/s air frontal velocity and 97×70 mm^2 frontal area, the air-side R value of the copper heat exchanger is 0.05 K/W.

19.6.2 Condensing Surfaces

Vapor condenses in the tubes of the air-cooled heat exchanger. Surface tension force may be used to enhance the condensation coefficient. This technology uses small microgrooves to remove condensate from the fin tips and is well developed within the refrigeration industry. This technology is discussed in Chapter 14. The extruded aluminum tube used in the Figure 19.14a heat exchanger has 0.2-mm-high microfins. If the tube is made of copper or brass, one may use a brazed, corrugated insert. Condensate will be pulled into the corners, which thins the condensate film on the flat surfaces. The author has work in progress to make a flat copper tube having small internal hydraulic diameter, which is similar to the Figure 19.14a aluminum tube.

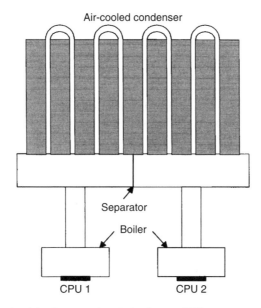

Figure 19.16 Schematic of the thermo-syphon device for two-CPU systems.

19.6.3 Design for Multiple Heat Sources

If multiple heat sources exist, as in a server, it is possible to design the remote air-cooled heat exchanger so that the vapor generated at each heat source is routed to one ambient air-cooled heat exchanger for heat rejection. Figure 19.16 is a schematic of the concept for two-CPU systems. To ensure proper return of the working fluid to each boiler, the air-cooled heat condenser is divided into multiple zones—one for each boiler. This is done using a separator disk in the headers to create a partitioned condenser, as illustrated in Figure 14.13d.

19.7 SYSTEM PERFORMANCE FOR THE IndHR SYSTEM

Consider the IndHR system designed to use water as the working fluid with jet impingement cooling of the hot surface operating at a heat flux of 200 W/cm² on a 1.0 cm² surface area. The vapor generated from this 200-W heat load is condensed in an air cooled ambient heat exchanger of the Figure 19.14 design having 6800 mm² frontal area (82 mm square). The total thermal resistance is composed of the boiling resistance at the hot source ($R_{H,cv}$), the condensing resistance ($R_{C,cond}$), and the air-side convection resistance ($R_{C,cv}$). At q = 200 W/cm², Figure 19.11 indicates $R_{H,cv}$ = 0.10 K/W. Based on the tests of Webb and Yamauchi [2002], the $R_{C,cond}$ = 0.03 K/W, and $R_{C,cv}$ = 0.05 K/W. The sum of these thermal resistances is 0.18 K/W. Note that a heat spreader is not required for jet impingement. System installation would also involve a spreading resistance (R_{sp}), which may be as low as 0.015 K/W.

19.8 CONCLUSIONS

Direct heat removal devices typically used for desktop computers are expected to be limited to approximately 100 W heat rejection, assuming a plan area no larger than 80 × 80 mm. Heat removal for more than 100 W will require new technology. An indirect heat removal concept using a working fluid will likely meet the higher heat rejection requirements. The working fluid may be either single or two phase. All of these concepts, except a thermo-syphon, will require use of a pump. Whether orientation insensitivity is required will likely dictate the working fluid system design of choice.

Several boiling concepts are available for heat removal from the hot source. The simplest and lowest cost is nucleate pool boiling. The highest performance and highest heat flux will be provided by pump actuated jet impingement. Hot source heat removal by two-phase flow in microchannels is yet to be established as a viable system, because of uneven heat removal at the hot source. The indirect heat removal system will require a high-performance ambient heat exchanger using a dedicated fan. Little work has been done to identify such concepts. Webb and Yamauchi [2002] have described a high-performance indirect ambient heat exchanger that uses a two-phase working fluid.

NOMENCLATURE

A — Total heat transfer surface area, m² or ft²

A_{fr} — Frontal heat exchanger area, m² or ft²

h — Heat transfer coefficient, W/m²-K or Btu/hr-ft²°F

ITD — $T_h - T_{air,in}$ (K or F)

M — Molecular weight, kg/mol

Q — Heat dissipation or thermal power, W or Btu/hr

q — Heat flux, W/m² Btu/hr-ft²

p — Pressure, Pa or lbf/ft²

P_{cr} — Critical pressure, Pa or lbf/ft²

P_r — p/p_{cr} dimensionless

R_{bo} — Boiling thermal resistance, K/W or °F-hr/Btu

R_{co} — Condensing thermal resistance, K/W or °F-hr/Btu

$R_{H,cv}$ — Convection thermal resistance, K/W or °F-hr/Btu

R_{int} — Interface thermal resistance, K/W or °F-hr/Btu

R_{sp} — Spreading thermal resistance, K/W or °F-hr/Btu

R_{tot} — Total thermal resistance, K/W or °F-hr/Btu

R_{transp} — Thermal resistance, associated with transporting the working fluid to the ambient heat sink, K/W or °F-hr/Btu

$T_{air,in}$ — Inlet air temperature, °C or °F

T_{hot} — Hot source surface temperature, °C or °F

T_{sat} — Saturation temperature, °C or °F

T_w — Surface temperature, °C or °F

Greek Symbols

η   Fin efficiency, dimensionless
λ   Heat of vaporization, kJ/kg or Btu/lbm
Δp Pressure drop, mm-H_2O or in-H_2O
ε   Thermal effectiveness, dimensionless

REFERENCES

Arbelaez, F., Sett, S. and Mahajan, R.L., 2000. An experimental study on pool boiling of saturated FC-72 in highly porous aluminum metal foams, *Proc. 34th Nat. Heat Transfer Conf.*, 759–767.

Arik, M. and Bar-Cohen, A., 2001. Ebullient cooling of integrated circuits by NOVEC fluids, *Proc. of IPACK'01*, July 8–13, 2001, Kauai, HI, paper IPACK2001-15515.

Chien, L.-H. and Chang, C.-C., 2002. Experimental study of evaporation resistances on porous surfaces in flat heat pipes, *Proc. ITherm 2002, 8th Intersociety Conf. on Thermal and Thermomechanical Phenomena in Electronic Systems*, May 30–June 1, San Diego, CA, 236–242.

Chien, L.-H. and Chen, C.-L., 2000. An Experimental Study of Boiling Enhancement in a Small Boiler, *Proc. ASME NHTC'00, 34th Nat. Heat Transfer Conf.*, 903–909.

Chien, L.-H. and Chen, C.-L., 2001. Experiments of boiling on cross-grooved surfaces in a confined space, *Proc. ASME NHTC'01, 35th Nat. Heat Transfer Conf.*, 175–182.

Chein, L.-H. and Chen, C.-L. 2005. Boiling of enhanced surfaces at high heat fluxes in a small boiler Proc. Inter Pack05

Cooper, M.G., 1984. Saturation nucleate, pool boiling—a simple correlation, *Int. J. Heat Mass Transfer*, 23, 73–87.

Ghui, C.-D., Joshi, Y.K., and Nakayama, W., 2001. Visualization study of pool boiling from transparent enhanced structures, *Proc. ASME Nat. Heat Transfer Conf. '01, 35th Nat. Heat Transfer Conf.*, 697–704.

Heffington, S.N., 2001. Vibration-induced droplet atomization heat transfer cell for cooling of microelectronic components, *Proc. IPACK'01*, paper IPACK2001-15596.

Hestroni, G., Mosyak, A., Segal, Z., and Zisking, G., 2002. A uniform heat sink for cooling of electronic devices, *Int. J. Heat Mass Transfer*, 45, 3275–3286.

Honda, H. and Wei, J.J., 2004. Enhanced boiling heat transfer from electronic components by use of surface microstructures, *Exp. Thermal Fluid Sci.*, 28, 159–169.

Honda, H., Takamastu, H., and Wei, J.J., 2002. Enhanced boiling of FC-72 on silicon chips with micro-pin-fins and submicron-scale roughness, *J. Heat Transfer*, 124, 383–390.

Jiang, L., Koo, J.M., Zeng, S., Mikkelsen, J.C., Zhang, L., Zhou, P., Santiago, J.G., Kenny, T.W., Goodson, K.E., Maveety, J.G., and Tran, Q.A., 2001. Two-phase microchannel heat sinks for an electrokinetic VLSI chip cooling system, *Proc. of the 17th IEEE SEMI-THERM Symposium*, March 20–22, 2001, San Jose, CA, 153–157.

Kim, S.-Y. and Webb, R.L., 2003. Thermal performance analysis of fan-heat sinks for CPU cooling, paper IMECE2003-42173, presented at IMECE meeting, Washington, D.C., November.

Liang, H.-S. and Yang, W.-J., 1998. Nucleate pool boiling heat transfer in a highly wetting liquid on micro-graphite-fiber composite surfaces, *Int. J. Heat Mass Transfer*, 41, 1993–2001.

Manglik, R.M. and Bergles, A.E., 1990. The thermal-hydraulic design of the rectangular offset-strip-fin compact heat exchanger, in *Compact Heat Exchangers*, R.K. Shah, A.D. Kraus, and D. Metzger, Eds., Hemisphere, Washington, D.C., 123–150.

Marcos, A., Chow, L.C., Du, J., Lei, S., Rini, D.P., and Lindauer, J. J., 2002. Spray cooling at low system pressure, *18th IEEE SEMI-THERM Symposium*, 169–175.

Misale, M., Guglielmini, G., Frogheri, M., and Bergles, A.E., 1999. Influence of confinement on FC-72 pool boiling from a finned surface, in *Heat Transfer Enhancement of Heat Exchangers*, S. Kakaç et al., Eds., Kluwer Academic, Dordrecht, 515–522.

Moghaddam, S., Ohadi, M., and Qi, J., 2000. Pool boiling of water and FC-72 on copper and graphite foams, Paper 35316, *Proc. ASME InterPack '03*, Maui, HI.

Mudawar, I. and Anderson, T.M., 1989. High heat flux electronic cooling by means of pool boiling. Part II: Optimization of enhanced geometry, *Heat Transfer in Electronics*, R.K. Shah, Ed., ASME HTD, 111, 35–50.

Murthy, J.Y., Amon, C.H., Gabriel, K., Kumta, P., and Yao, S.C., 2001. MEMS-based thermal management of electronics using spray impingement, *Proc. IPACK'01*, July 8–13, 2001, Kauai, HI, Paper IPACK2001-15567.

Nakayama, W., Nakajima, T., and Hirasawa, S., 1984. Heat Sink Studs Having Enhanced Boiling Surfaces for Cooling Microelectronic Components, ASME paper 84-WA/HT-89, ASME, New York.

O'Conner, J.P. and You, S.M., 1995. A painting technique to enhance pool boiling heat transfer in saturated FC-72, *J. Heat Transfer*, 117, 387–393.

Pais, M.R., Chow, L.C. and Mahefkey, E.T., 1992. Surface roughness and its effects on the heat transfer mechanism in spray cooling, *J. Heat Transfer*, 114, 211–219.

Perret, C., Boussey, J., Schaeffer, C., and Coyaud, M., 2000. Analytic modeling, optimization, and realization of cooling devices in silicon technology, *IEEE Trans. Components Packaging Technol.*, 23(4), 665–672.

Ramaswamy, C., Joshi, Y., Nakayama, W., and Johnson, W.B., 2003. Effects of varying geometrical parameters on boiling from microfabricated enhanced structures, *J. Heat Transfer*, 125, 103–109.

Saini, M. and Webb, R.L., 2002a. Heat rejection limits of air cooled plane fin heat for computer cooling, *Proc. ITherm 2002, 8th Intersociety Conf. on Thermal and Thermomechanical Phenomena in Electronic Systems*, May 30–June 1, San Diego, CA, 1–8.

Saini, M. and Webb, R.L., 2002b. Validation of models for air cooled plane fin heat sinks used in computer cooling, *Proc. ITherm 2002, 8th Intersociety Conf. on Thermal and Thermomechanical Phenomena in Electronic Systems*, May 30–June 1, San Diego, CA, 243–250.

Wang, C.C., 2000. Recent progress on the air-side performance of fin-and-tube heat exchangers, *Int. J. Heat Exchanger*, 1, 49–76.

Webb, R.L., 2003. Effect of manifold design on flow distribution in parallel micro-channels, Paper 35251, *Proc. ASME InterPack '03*, Maui, HI, July.

Webb, R.L., 2004. Next generation devices for electronic cooling, *J. Heat Transfer*, 126, 2–10.

Webb, R.L. and Yamauchi, S., 2002. Remote heat sink concept for high power heat rejection, in *THERMES 2002, Thermal Challenges in Next Generation Electronic Systems*, Y. Joshi and S. Garimella, Eds., Millpress, Rotterdam, 201–210. Also published in *IEEE Trans. Components Materials Packaging*, 25, 608–614, 2002.

Webb, R.L., Chang, Y., and Wang, C., 1995. Heat transfer and friction correlations for the louver fin geometry, in *1995 Vehicle Thermal Management Systems Conference Proceedings*, Society of Automotive Engineers, Warrendale, PA, 533–541.

Xia, C., 2002. Spray/jet cooling for heat flux high to 1 kW/cm^2, *18th IEEE SEMI-THERM Symp.*, 159–163.

Zhang, L., Wang, E., Koo, J., Jiang, L., Goodson, K., Santiago, J., and Kenny, Y., 2002. Enhanced nucleate boiling in microchannels, *Proc. 15th Int. Conf. on Micro Electro Mechanical Systems MEMS 2002*, Las Vegas, NV, 89–92.

PROBLEM SUPPLEMENT

1.1 Benefits of Enhancement. (a) An existing four-row finned-tube refrigerant evaporator has 551 fins/m on 12.7-mm-diameter tubes located on 31.8-mm equilateral triangular pitch. The fins are plain and the tube inside surface is smooth. The air- and refrigerant-side heat transfer coefficients are 400 and 3000 W/m²-K, respectively. The heat exchanger has $A_o/A_{fr}N$ = 21.6 and $A_i/A_{fr}N$ = 1.19, where N is the number of tube rows. If the Figure 1.3e air-side surface geometry is employed, the air-side heat transfer coefficient will be increased 80%. To obtain the same UA value, how many tube rows would be required? [Ans. $0.83N$]

(b) It is also possible to enhance the refrigerant-side using the Figure 1.15 microfin tube, which will provide 120% increased refrigerant heat transfer coefficient. Would use of the tube-side enhancement, retaining the plain fins on the air side, provide comparable size reduction benefits to those obtained in part a? [Ans. $0.61N$]

3.1 Simple Single-Phase PEC. A smooth tube has f_s = 0.079 Re$^{-0.25}$ and j_s = $f_s/2$. An internally roughened tube provides j = 0.079 Re$^{-0.25}$ and has f = 0.11 Re$^{-0.2}$. Assuming all of the thermal resistance is on the tube side, calculate G/G_s, A/A_s and h/h_s using Case VG-1 for Re$_s$ = 20,000. What is A/A_s? [Ans. G/G_s = 0.937, A/A_s = 0.525, h/h_s = 1.90]

Now consider Case FN-1. Outline the procedure you would use to obtain G/G_s and A/A_s. Will A/A_s be greater or smaller than for Case VG-1? Why?

3.2 Complex Single-Phase PEC. Reconsider the rough tube of Problem 3.1 for a case in which $h_{i,s}$ = 4000 kW/m²-K and 75% of the total thermal resistance is on the inside of the smooth tube. For simplicity, assume B_s = B = 1 and ignore the metal wall and fouling resistances. Calculate G/G_s and A/A_s for Case VG-1. Will A/A_s be greater or smaller than for Case VG-1? Why? [Ans. Smaller]

4.1 Two-Phase PEC. (a) Consider steam condensing in an annulus against water inside the tubes. Water enters the tube at 1.0 kg/s and 300 K. Steam enters the annulus at 330 K for which dp/dT = 22.94 kPa/K. The water-side heat transfer coefficient is 10,000 W/m^2-K. If plain tubes are used, the steam condensation coefficient is 5000 W/m^2-K and the steam pressure drop is 50 kPa.

Because the steam side has the controlling thermal resistance, a tube having steam-side enhancement is to be applied. This tube provides 2.5 times higher condensing coefficient, and the steam pressure gradient is also increased 2.5 times. Calculate the tube length reduction for $Q/Q_s = 1$ with the same operating conditions. For simplicity, neglect the effect of the steam pressure drop. [Ans. $L/L_s = 0.6$]

(b) If you use $L/L_s = 0.6$, and account for the effect of steam pressure drop, what would be the resulting value of Q/Q_s ? [Ans. $Q/Q_s = 0.98$]

5.1 Compare Louver and OSF Fin Geometries. Compare the performance of the offset strip fin (OSF) with the louver fin using PEC FN-2. Both fin geometries have 472 fins/m, 7.62 mm fin height, 2.03 mm louver pitch, 0.10 fin thickness, and $D_h = 3.18$ mm. The louver geometry has 26° louver angle and $L_L/H = 0.8$. Compare for airflow at 26.7°C (80°F) with 8.94 m/s (29.3 ft/s) air frontal velocity. Calculate the j and f factors using the Davenport correlations (Equations. 5.12 through 5.14) for the louver fin, and the Manglik and Bergles power law correlations (Equations 5.9 and 5.10) for the OSF geometry. Calculate the A_{osf}/A_{louv} and P_{osf}/P_{louv} where A is the fin surface area. [Ans. $A_{osf}/A_{louv} = 1.03$ and $P_{osf}/P_{louv} = 0.66$]

5.2 Effect of Strip Length on OSF Performance. (a) Consider an OSF having 787 fins/m, 0.025 mm fin thickness, and 9.52 mm fin height. Air enters the fin array at 5.1 m/s frontal velocity ($\nu = 15.9 \times 10^{-6}$ m^{-2}/s, $\rho = 1.16$ kg/m^3). Compare the performance of 3.18 mm ($\frac{1}{8}$ in.) and 12.7 mm ($\frac{1}{2}$ in.) strip lengths using the simple Kays model to predict j and f. Let subscripts 1 and 2 refer to the 3.18 and 12.7 mm strip lengths, respectively. Compute j and f for each surface. [Ans. $j_1 = 0.929 \times 10^{-2}$, $j_2 = 1.86 \times 10^{-2}$, $f_1 = 2.74 \times 10^{-2}$, $f_2 = 7.24 \times 10^{-2}$]

(b) Calculate $(j/f)_2/(j/f)_1$ and the fraction of the friction factor due to form drag. [Ans. $(j/f)_2/(j/f)_1 = 1.32$; for $L_p = 3.18$ mm, 48.6% is form drag]

(c) If $hA = $ constant, calculate A_2/A_1 and P_2/P_1 [Ans. $A_2/A_1 = 0.5$ and $P_2/P_1 = 1.32$] Draw conclusions regarding the performance and acceptability of the two surface geometries.

5.3 Predict j and f of OSF. Use the Manglik and Bergles correlation to predict the performance of the OSF geometry, whose j and f vs. Reynolds number characteristic is shown in figure 10-58 of Kays and London [1984]. The Kays and London geometry has $L_p = 2.8$ mm, $t = 0.10$ mm, $\alpha = 0.1837$, $\delta = 0.036$, $\gamma = 0.111$, and $D_{h,KL} = 1.495$ mm. The j and f vs. Re characteristic is the same as that of the OSF geometry shown on Figure 5.4.

Predict the j and f values for air flow at 27°C with Re = 500 and 3000 and compare with the experimental values. At Re = 500, Figure 5.4 yields $j = 0.0207$ and $f = 0.088$, and at Re = 3000, $j = 0.0095$ and $f = 0.0398$.

5.4 Comparison of Joshi and Webb Model with Numerical Prediction. We wish to use the analytical model developed by Joshi and Webb to predict the heat transfer and friction characteristics (j and f) of the OSF array. Test the ability of the model to predict the numerical solution for f of Patankar and Prakash (*Int. J. Heat Mass Transfer*, 24, 1801–1810, 1981) for $\alpha = 0$, $t/s = 0.111$ and 0.25 with $L_p/s = 1.0$ at Re$_{Dh} = 500$. For $\alpha = 1$, you may use the conversion Re$_s$/Re$_{Dh} = (1 + t/L_p)$.

The Patankar and Prakash prediction for f gives $f = 0.111$ for $t/s = 0.111$ and $f = 0.096$ for $t/s = 0.25$. [Ans. $f = 0.105$ for $t/s = 0.110$, $f = 0.096$ for $t/s = 0.25$]

5.5 Simplified Joshi and Webb Model of OSF. We will use a simplified version of the model of Joshi and Webb [1987] to predict the performance of the OSF geometry, whose j and f vs. Reynolds number characteristic is shown in figure 10-58 of Kays and London [1984]. The Kays and London definition of Reynolds number is $\mathrm{Re} = D_{h,\mathrm{KL}}G/\nu$, where $D_{h,\mathrm{KL}} = 2s/(1+\alpha+\delta)$ with G based on flow area $A_c = s(b - t)$. The Kays and London geometry has $L_p = 2.8$ mm, $t = 0.10$ mm, $\alpha = 0.1837$, $\delta = 0.036$, $\gamma = 0.111$, and $D_{h,\mathrm{KL}} = 1.495$ mm. The j and f vs. Re characteristic is the same as that of the OSF geometry shown on Figure 5.4.

Predict the j and f values for airflow at 27°C with Re = 500 and 2000 and compare with the experimental values. Use 100% fin efficiency. At Re = 500, Figure 5.4 yields $j = 0.0207$ and $f = 0.088$, and at Re = 2000, $j = 0.011$ and $f = 0.044$. For the simplified Joshi and Webb model, we will use the following modifications relative to the equations given in Table 5.1:

1. Predict Nu_p and f_p using the Pohlhausen model for laminar flow over a flat plate with $\mathrm{Re}_{\mathrm{Lp}}$ based on G with flow area $A_c = s(b - t)$. Thus, $\mathrm{Re}_{\mathrm{LP}} = \mathrm{Re}L_p/D_h$.
2. Assume $F_{h\alpha}$ and $F_{fa} = 1.0$.
3. Use the equations in Table 5.1 for f_e, C_D, and Nu_e.

Compare your predicted j and f with the experimental values. [Ans. At Re = 500, predicted $j = 0.0149$ and $f = 0.0437$] Also compare your predicted values for j_p and f_p with those from the complete Joshi and Webb model. [Ans. At Re = 500, the Joshi and Webb model gives $j_p = 0.0267$ and $f_p = 0.08667$]

5.6 Performance of OSF Variants. Figures 5.4, 6.7, and 6.20 show the performance of three closely related heat transfer enhancement concepts. At abscissa values corresponding to the same air velocity, the heat transfer enhancement ratio (j/j_p), relative to the plain fin geometry (j_p) for the geometries are Figure 5.4 $(j/j_p = 2.5)$, Figure 6.7 $(j/j_p = 1.8)$, and Figure 6.20 $(j/j_p = 1.4)$. Explain why the enhancement ratio successively decreases for Figures 6.7 and 6.20. Assume that the strip width is the same for all the enhancements.

5.7 Determine Enhancement Potential. Consider an automotive radiator, such as that illustrated in Figure 5.10. One problem associated with the flat tube design is that the flat sides of the tubes tend to "balloon out" at increased operating pressure. The strength of the tubes would be increased if two-dimensional grooves of depth e were spaced at a pitch of, say, 10 to 15 e. Such a groove geometry would also provide heat transfer enchancement for the water flow inside the tubes.

The water flow in the tubes has Re = 10,000 and $h_i = 10,000$ kW/m²-K. On the air side, $h_o = 250$ kW/m²-K and the surface efficiency is $\eta = 0.87$. The heat transfer surface areas on the water and air sides are $A_o/A_{fr} = 27$ and $A_i/A_{fr} = 3.1$ ft²/ft² frontal area. The enchanced tube has $h_i = 21,500$ kW/m²-K.

Your employer would be interested in this idea if it will provide a moderate radiator size and weight reduction, in addition to increased tube strength. Assume that the VG-1 criterion applies to the water side. Perform a quick preliminary analysis and provide a recommendation on the merits of this proposal.

6.1 Circular Finned Tubes. Predict j and f for the five circular finned tube bank geometries taken from figure 10-89 of Kays and London [1984] shown in Problem Figure P6.1. All geometries have 19.66 mm diameter tubes, 37.16 mm fin diameter, 356 fins/m, and 0.305 mm fin thickness. The j and f vs. Re_{Dh} curves are fit by the equations $j = C_j(Re_{Dh}/1000)^{-m}$ and $f = C_f(Re_{Dh}/1000)^n$, where the constants and exponents are given in the table. The term β is the total surface area per unit heat exchanger volume. Assume 27°C (80°F) air at 3.05 m/s (600 ft/min) approaches the tube bank. Use the Briggs and young correlation (Equation 6.14) for heat transfer, and the Robinson and Briggs correlation (Equation 6.15) for pressure drop. Compare the predicted values with the experimental values shown on the graph. Draw conclusions regarding the ability of the correlations to predict the data. [Ans. The last two lines in the table shows the ratio of the predicted-to-experimental values]

Item	a	b	c	d	e
S_d (mm)	48.51	51.05	56.39	40.13	42.93
D_h (mm)	5.131	8.179	13.59	5.334	6.426
$\sigma = A_c/A_{fr}$	0.455	0.572	0.688	0.537*	0.572
$\beta = A/V$ (m²/m³)	354	279	203	443	354
C_j	0.00925	0.0105	0.0128	0.0090	0.0099
m	0.387	0.343	0.373	0.362	0.362
C_f	0.0365	0.0520	0.0715	0.0400	0.0506
n	0.316	0.278	0.264	0.309	0.275
j_p/j_{exp}	1.09	1.08	1.09	1.10	1.07
f_p/f_{exp}	1.06	1.17	1.34	1.72	1.19

* σ is based on the diagonal dimension.

6.2 PEC Analysis of Circular Finned Tubes. Which of the five tube-bank layouts described in Problem Figure 6.1 would you choose for a heat exchanger application? Assume fixed heat duty (Q), inlet temperature difference (ΔT_i), and pumping power (P). Use geometry a as the reference surface, with 27°C (80°F) air at 3.05 m/s (600 ft/min) approaching the tube bank. Consider only the air-side performance in your PEC analysis.

(a) Use the PEC VG-1 shown in Table 3.1 to justify your choice. Also calculate $A_{ft}/A_{fr,r}$, A/A_r, and L/L_r, where L is the air-flow depth. [Ans. See table below]

	a	b	c	d	e
A/A_r	1.0	1.08	1.18	1.06	1.09
$A_{ft}/A_{fr,r}$	1.0	0.90	0.83	0.89	0.92
L/L_r	1.0	1.54	2.54	1.09	1.23

(b) Calculate A/A_r for geometry d using PEC case FN-1 of Table 3.1. Assume counterflow performance with $C_{min}/C_{max} = 0$ and reference geometry a designed for $\varepsilon = 0.6$ (heat exchanger effectiveness). Compare the calculated A/A_r with the result obtained in part a. Explain why the results for case FN-1 are greatly different from the value calculated for case VG-1. *Hint:* You must account for the effect of reduced flow rate on the required hA. [Ans. $A/A_r = 2.60$]

7.1 Twisted-Tape with Laminar Flow. Consider heat transfer at constant heat flux to fully developed laminar flow inside a plain tube at Re = 1000 for two cases: Water at 60°C (140°F) with Pr 3 and oil at 100°C (212°F) with Pr = 276. The tube contains a twisted-tape insert with having $y = 2.5$ and $t = 0$. Calculate h/h_p and f/f_p for each fluid, where subscript p refers to a plain tube with no tape insert. Comment on the enhancement ratio (h/h_p) and the efficiency index, $(h/h_p)/(f/f_p)$ for each case. For the plain tube, Nu = 4.36 and $f = 16/$Re. For the twisted tape, use the Hong and Bergles [1976] correlation for Nu, and the Manglik and Bergles [1992a] correlation for the friction factor. [Ans. $Nu_p = 4.36$, Nu = 23.4 (water), Nu = 113.8 (oil), $f_p = 0.016$, $f/f_p = 6.17$] Would h/h_p increase or decrease if the oil were cooled?

7.2 Manglik and Bergles Correlation. Repeat the twisted-tape heat transfer calculation for Example Problem 7.1 using the Manglik and Bergles correlation (Equation 7.28), which is for constant wall temperature. Assume 0.6 of the total resistance is on the tube side. From Example Prob. 7.1, $\Delta T_{lm} = 112.2$ K. Also use the Manglik and Bergles friction correlation (Equation 7.17) for the friction factor and compare it to Equation 7.14 used in Example Problem 7.1. [Ans. Nu = 35.6 W/m²-K vs. 43.6 from Example Problem 7.1]

8.1 Laminar Flow in Internal-Fin Tubes. Evaluate the heat transfer enhancement for laminar flow in internally finned tubes, based on the numerical results shown in Table 8.1. Assume a constant wall temperature boundary condition, and that the fin height is limited to $e/d_i = 0.1$. What is the preferred number of fins to obtain the highest enhancement level? What is the magnitude of h/h_p for the chosen geometry? What is the efficiency index? [Ans. $n_f = 8$, $\eta = 0.68$]

Compare your results with the experimental data of Marner and Bergles [1985] for $n_f = 12$ and $e/d_i = 0.084$ as shown on Figure 7.9. Comment on the observed differences.

8.2 Entrance Length for Laminar Flow. Oil having the properties of Problem 8.5 flows on the tube side of an oil cooler having internally finned tubes ($d_i = 17$ mm, $e/d_i = 0.10$, $n_f = 16$) at $Re_d = 1000$. The tube is 1.5 m long giving $L/d_i = 88$. Assuming constant wall temperature, what is the thermal entrance length for this flow? [Ans. 13.2 m]

Would you expect the fully developed flow solutions shown in Table 8.1 to be applicable for estimating the average heat transfer coefficient in this tube?

8.3 Internally Finned Tubes. If the hydraulic diameter concept is used to predict the j and f factors for turbulent flow in an internally finned tube, would you expect to over- or underpredict the data? Provide a rational explanation for your answer.

8.4 Carnavos Correlation. The data Carnavos used to develop his heat transfer correlation (Equation 8.5) for internally finned tubes spanned helix angles of $0 < \alpha < 30°$. Does it seem reasonable that the correlation may be used for helix angles greater than 30°? Would you expect any changes to the flow structure that may limit the applicability of the correlation?

8.5 Internal-Fin Tube Heat Exchanger. Evaluate the benefits of an internal-fin tube design, which provides the same heat duty, tube-side pressure drop, and flow rate as a smooth tube design (Case VG-1 of Table 3.1). The heat exchanger has 19.1 mm (3/4 in.) outside diameter, 0.89 mm (0.035 in.) wall copper tubes on a triangular pitch with 25.4 mm (1.0 in.) tube pitch. The smooth tube exchanger has 136 tubes, 305 mm (12.2 in.) diameter shell (D_s) with four tube-side passes. Oil flows on the tube side at 3.05 m/s (10 ft/s) and is cooled from 98.9 to 87.8°C (210 to 190°F). Cooling water flows on the segmentally baffled shell side. The water enters at 32.2°C (90°F) and exits at 54.4°C (130°F). The shell-side heat transfer coefficient is $h_o = 4825$ W/m²-K (850 Btu/hr-ft²-°F). Assume $R_f = 0$ on both the tube and shell sides. The oil properties at 93°C (200°F) are $\rho = 865$ kg/m³ (54 lbm/ft³), $c_p = 2.13$ kJ/kg-K (0.51 Btu/lbm-°F), $\nu = 4.3$E-6 m²/s (4.6 × 10⁻⁵ ft²/s), $k = 0.128$ W/m-K (0.074 Btu/hr-ft-°F), and Pr = 62.

First, evaluate the potential for internal-fin tubes (IFT) by calculating R_i/R_{tot} (internal/total thermal resistance of the smooth-tube exchanger). Will reduction of the tube-side resistance be beneficial? [$R_i/R_{tot} = 0.83$]

Select a candidate IFT geometry based on your study of Figure 8.13. Do you concur that 16 fins, 1.5 mm high × 0.50 mm thick, with 30° helix angle is a reasonable choice?

Calculate the h and f vs. Re characteristics of the IFT using the Carnavos [1980] correlation (Equations 8.5 and 8.6). Convert the h and f based on the total area to values based on the nominal area using Equations 8.9 and 8.10 for use in the PEC equations.

After calculating A/A_s, calculate L/L_s (H-X length ratio), and V/V_s (tube material volume). Draw conclusions regarding the value of IFT for this application. [Ans. $\beta_s = 0.207$, $G/G_s = 0.908$, $A/A_s = 1.04$, $N/N_s = 1.174$, $L/L_s = 0.435$]

9.1 Sand-Grain Roughness. A heat exchanger uses 15.9 mm (0.625 in.) inside diameter tubes with 27°C (80°F) water having $\nu = 0.982$ E-6 m²/s (9.29 E-6 ft²/s) at 2.44 m/s (8 ft/s) velocity. What is the largest roughness size (e) that will act as hydraulically smooth? *Hint:* Figure 9.10b shows that the hydraulically smooth line is approached at $\log_{10}$ e⁺ = 0.7. [Ans. $e = 0.037$ mm]

What heat transfer enhancement (h/h_s) would you expect at this value of e? [Ans. None]

9.2 Smooth Tube Turbulent Friction Factor. The law of the wall for smooth surfaces is $u⁺ = 2.5 \ln y⁺ + 5.5$. An expression for the friction factor may be obtained by integrating this equation across the radius of a circular pipe to obtain

$$\frac{\bar{u}}{u^*} = \sqrt{\frac{2}{f}} = 2.5 \ln\left(\frac{Re_d}{2}\sqrt{\frac{f}{2}}\right) + 1.75$$

Compare the prediction factor for $Re_d = 50,000$ with that predicted by Equation 9.22.

9.3 Rough Rectangular Channel. (a) Consider a 2.0×1.0 mm rectangular channel having a sand-grain wall roughness. How would you use Equation 9.8 to calculate the friction factor for the fully rough condition? (b) Assume that the 1.0 mm walls are smooth, and that air flows at 10 m/s and 20°C. Predict the pressure gradient by applying Equation 9.8 to the rough walls and the universal velocity distribution ($u^+ = 2.5 \ln y^+ + 5.5$) to the smooth walls.

9.4 Geometric Similarity. Assume that you have tested a 25-mm-ID (d_i), helically ribbed rough surface having $e = 0.5$ mm, $p = 10$ mm, and $\alpha = 45°$, and put the friction results in the form $(2/f)^{1/2} = -2.5 \ln(2e/d_i) - 3.75 + B(e^+)$. For which of the tube geometries in the table below would this friction correlation apply? Briefly explain your reasoning.

No.	d_i (mm)	e (mm)	p (mm)	α (deg)
1	25	0.25	5	45
2	25	0.25	10	45
3	19	0.50	10	45
4	19	0.50	10	25
5	19	0.60	5	45

9.5 Two-Dimensional Rib Roughness. Two-dimensional rib roughness ($p/e = 10$, $\alpha = 90°$) is to be used to enhance heat transfer on the tube side of a water chiller evaporator, in which water flows at 3.05 m/s (10 ft/s) and 7.2°C (45°F) in 16.0 mm (0.63 in.) inside diameter tubes. The evaporator tubes presently use a smooth tube inner surface. The heat conductance on the outer tube surface is $h_o A_o/L = 865$ W/m-K (500 Btu/h-ft-°F) and the fouling resistance is zero.

The objective is to select an appropriate roughness height (e) and to satisfy the VG-1 PEC of Table 3.1. As suggested on page 328, select $e^+ = 14$ as the rough tube design criterion. Obtain $\overline{g}$ (e^+) = 10.8 and $B(e^+) = 4.5$ at $e^+ = 14$ from Figures 9.14 and 9.15, respectively. Because the required G/G_s is unknown, assume $u = 2.6$ m/s (8.5 ft/s) tube-side velocity. Use Equation 9.40 to solve for the roughness height (e) and the friction factor (f). Then use Equation 9.21 to solve for St. [Ans. $e/d_i = 0.00563$, $f = 0.014$, St = 1.716E-3]

Calculate L/L_s and P/P_s for $Q/Q_s = 1$. [Ans. $L/L_s = 0.706$, $P/P_s = 1.06$]. Because $P/P_s > 1$, G/G_s must be reduced to meet the $P/P_s = 1$ constraint. Estimate the G/G_s required to meet the $P/P_s = 1$ constraint. [Ans. $G/G_s \cong 0.833$ and $u = 2.54$ m/s]

9.6 Optimum Internal Roughness. Before deciding to use the internal-fin tube design evaluated in Problem 8.5, it may be advisable to consider an internally roughened tube. Calculate the material savings for the VG-1 case offered by two-dimensional rib roughness ($\alpha = 90$, $p/e = 10$, $t/e = 0.5$) assuming $R_{fi} = 0$. Following the guidance of Webb and Eckert [1972] discussed on page 328, select $e^+ = 20$ as the design criterion. Figures 9.14 and 9.15 yield $\overline{g}$ (20) = 11.0 and $B(20) = 4.2$. Since f and St are functions of e/D and Re, the PEC equation for A/A_s implicitly contains two independent variables, e/d_i and Re. Therefore, a range of possible e/d_i exist. Use the iterative solution procedure outlined in Webb and Eckert [1972].

The solution will converge on the "optimum" e/d_i and the G/G_s needed to satisfy the VG-1 PEC equation. [Ans. $e/d_i = 0.0176$, $G/G_s = 0.81$, $A/A_s = 0.535$, $N/N_s = 1.234$, $L/L_s = 0.434$]

9.7 Design a Product Line of Rough Tubes. Your company plans to manufacture tubes having $p/e = 10$, transverse-rib roughness ($\alpha = 90°$). The tubes will be produced in 19.1 mm ($\frac{3}{4}$ in.) outside diameter with 0.64 mm (0.028 in.) wall thickness. A market survey indicates the product line should be applicable to $2000 < \text{Re} < 80,000$ and $1 < \text{Pr} < 100$. Your job is to select the number of different roughness sizes needed to support the desired operating range, and to select the specific roughness dimensions, p and e. As suggested on page 328 use $14 < e^+ < 35$ as the operating range that you will apply to each tube. As shown in Figure 9.15 the relationship between e^+ and $B(e^+)$ is given below:

e^+	10	14	20	35	60	80	100
$B(e^+)$	5.1	4.5	4.0	3.2	3.0	3.0	3.0

Start the analysis by selecting the e/d_i size you will use for $\text{Re} = 80,000$. Solve Equation 9.56 for the required e/d_i. Over what Reynolds number range will the e/D size calculated be applied? [Ans. $e/d_i = 0.00485$, $35,600 < \text{Re} < 80,000$]. Now repeat the previous procedure for the next lower Reynolds number range. [Ans. $e/d_i = 0.0092$, $16,340 < \text{Re} < 35,600$]

It is necessary to provide curves of St and f vs. Re and Pr for each of the selected rough tubes. Assume that you will prepare curves of St and f vs. Re for Pr = 1, 10, 100. How will you develop these graphs?

11.1 Superheat Required for Bubble Existence. Consider a vapor bubble of radius $R = 0.75$ mm in a uniformly superheated liquid. Calculate the superheat required $(T_f - T_s)$ for existence of the bubble at the following conditions: (a) Saturated water 300, 100, and 10°C. [Ans. 0.38, 0.043, 2.4 K] (b) Saturated R-11 at 7°C [Ans. 0.026 K] The surface tension of water at 10, 100, and 300°C is 0.0742, 0.05878, and 0.01439 N/m, respectively. The R-11 surface tension at 7°C is 0.021 N/m.

11.2 Effect of Cavity Shape on Required Superheat. Consider the cavity shapes shown in Figure 11.24a, b, and d. All have the same cavity radius (r_c) and depth, $H = 4r_c$. Perform the following analysis for liquid contact angles (θ) of 15° and 90°: (a) Will any of the cavities trap air when the surface is flooded with the liquids having the two contact angles? [Ans. $\theta = 15°$ (no, no, yes), $\theta = 90°$ (yes, yes, yes)] (b) Sketch the shape of the liquid–vapor interface within the cavities for the two liquids. (c) Draw a graph of $(T_f - T_s)m/2\sigma$ vs. y/r for the cavities. (d) At what position in the cavity (y/r_c) is the greatest superheat required for existence of the vapor–liquid interface? (e) Will any of the cavities allow any subcooling of the liquid? [Yes, cavity C]

11.3 Effect of Cavity Shape on Incipient Boiling. (a) You plan to develop an enhanced surface geometry for boiling R-11 at 4.4°C (40°F) on 19.1 mm ($\frac{3}{4}$ in.) diameter copper tubes. The R-11 properties at 4.4°C are $v_1 = 0.381\text{E-}6$ m^2/s, $\sigma = 0.0205$ N/m, $\text{Pr}_l = 5.4$, $k_l = 0.0935$ W/m-K, $m = dp/dT = 1994$ Pa/K, $\beta = 1.5\text{E-}3$ K^{-1}

(volume expansion coefficient). Assume the boiling will occur in conical cavities (Figure 11.24a) having $r_c = 0.025$ mm (0.001 in.), and $2\phi = 30°$. Calculate the liquid superheat $(T_l - T_s)$ required for a bubble of radius r_c to exist. [Ans. 0.80 K]

(b) Calculate $T_w - T_s$ required for incipient boiling. For this condition, required $T_l - T_s$ at a distance $y = 1.5r_c$ above the heating surface to equal $2\sigma/mr_c$. Assume that the temperature profile in the liquid is linear. Heat will be transferred from the surface by natural convection at the incipient boiling condition. For natural convection, assume $\mathrm{Nu}_d = 0.5 \, (\mathrm{Gr}_d\mathrm{Pr})^{1/4}$, where Nu and Gr use the tube diameter for their characteristic dimension. The thermal boundary layer thickness is calculated by $\delta = k_l/h$. Calculate the temperature at the top of the bubble assuming a linear temperature profile in the liquid. [Ans. 0.86 K]

11.4 Effect of Cavity Shape. (a) Continuing with Problem 11.3, consider the effect of the cavity shape and contact angle on the value of $T_l - T_s$ required for existence of the vapor nucleus. The contact angle (θ) of R-11 on copper is approximately 10° (or less). If a conical cavity having $2\phi = 30°$ is employed, what value of $T_l - T_s$ is required for a nucleus to exist at $x/l = 1/4$? Compare your calculated $T_l - T_s$ with the $T_w - T_s$ of part b of Problem 11.3. [Ans. 3.2 K]

(b) Assume a reentrant cavity (Figure 11.24c) having $r_c = 0.025$ mm (0.001 in.), $2\phi = 120$ and $y_l = 2r_c$. What $T_l - T_s$ is required at $x/H = 0.75$ (H = cavity depth)? [Ans. 0.11 K]

11.5 Enhanced Boiling Surface. Now, consider an enhanced R-11 boiling surface, such as Figure 11.12d, which has 0.2-mm-diameter surface pores and 0.35-mm-square subsurface tunnels. Assume that menisci of radius 0.15 mm exist in the corners of the tunnel. Calculate the limiting liquid superheat required for vapor existence if boiling is to occur at 4.4°C with 30 W/m² heat flux. For simplicity, assume that the liquid in the tunnel is uniformly superheated. Does the limiting superheat occur in the tunnel or at the cavity mouth? How does this value compare to the limiting cavity mouth superheat for a plain surface, whose cavity radius is assumed to be 0.006 mm? Explain why the ΔT of the enhanced surface is so much less than that of the plain surface. [Ans. Plain, 3.4 K; Enhanced, 0.17 K]

11.6 Porous Boiling Surface. Consider the Code C surface in Figure 11.10b, whose physical characteristics are given in Table 11.2. Predict $(T_w - T_s)$ at $q = 20$ kW/m² for R-11 at 1.0 atm using the Nishikawa correlation (Equation 11.33). Compare your results with the experimental value (0.79 K). The physical properties of R-11 at 1.0 atm are $\sigma = 0.0188$ N/m, $\lambda = 0.18\text{E-}6$ J/kg-K, $\mu_v = 10.5\text{E-}6$ kg/m-s, $k_l = 0.091$ W/m-K, $\rho_l = 1479.3$ kg/m³, $\rho_v = 5.87$ kg/m³, $m = dp/dT = 1.015$ kPa/K. The thermal conductivity of the copper matrix is $k_m = 197$ W/m-K. [Ans. 0.26 K]

11.7 Effect of Pore Pitch on Boiling Performance. Chien and Webb studied boiling on surfaces having subsurface tunnels and surface pores. Figure 11.37 shows their boiling map, which defines the effect of pore pitch and pore diameter for fixed tunnel size. Consider the effect of pore pitch for 0.23-mm pore diameter. Why does the 0.75-mm pore pitch yield the highest dryout heat flux? Why does the 0.75-mm pore pitch yield significantly lower boiling coefficient than 4.0-mm pore pitch for operation at 3.0 kW/m²?

11.8 Pore and Tunnel Boiling Surface. Water is boiled at 1.0 atm on a structured surface that has circular subsurface tunnels 0.45 mm wide. The pores are 0.15 mm in diameter and 0.45 mm in pitch. The contact angle of water for the given boiling conditions is 40°. Using the Chien and Webb model, predict the heat flux and the boiling coefficient for 4.0 K wall superheat (ΔT_{ws}). *Author note:* This is a quite complex problem and one must refer to the cited Chien and Webb publications.

12.1 Surface Tension Drained Condensation. Adamek [1981] defined a family of profile shapes, which are discussed in Section 12.4.2. The condensate interface geometry for these profiles is given by $r_o/r + (s/S_m)^\xi = 1$. Integrate this equation over $0 < \theta < \theta_m$ using $d\theta = ds/r$ to obtain Equation 12.27.

One may derive the equation for the condensation coefficient following a procedure similar to that used by Nusselt for a gravity drained vertical plate. Defining $\kappa' = d(1/r)/ds$, show that

$$\frac{3 v_l \kappa \Delta T}{\sigma \lambda} = \delta \frac{d}{ds}(\delta^3 \kappa')$$

Solve the differential equation to obtain Equation 12.26.

12.2 Design of Surface Tension–Drained Surfaces. Consider condensation of ammonia, whose properties at 8.9°C are ρ_l = 626 kg/m³, K_l = 0.531 W/m-K, v_l = 0.368 E-6 m²/s, σ = 0.0236 N/m, λ = 1226 kJ/kg. You plan to make a vertical fluted tube for which the convex profile will rotate through 90°, and the arc length (S_m) will be 1.5 mm. You will use the Gregorig profile shape (Adamek's ζ = 2.0 shape). What will be the tip radius? What is the value of ($P_L - P_V$) at $s = 0$ and $s = S_m$, where P_L and P_V are the pressures in the condensate film and the vapor, respectively?

The fluted tube will have 1.0-mm-wide drainage channels. Rather than using the Gregorig convex profile, consider use of the Kedzierski–Webb (K-W) profile. Assume that the K-W profile will have the same fin height as the Gregorig profile and 0.025-mm tip radius. Table 12.8 in the Webb book gives and compares the performance of these profiles. Assume that the profiles are applied to a tube having 25 mm diameter over the fin tips. Neglecting the condensation in the concave drainage channel, compare the hA/L ratio of the two geometries. Note that this requires you to account not only for the effect of fin shape, but also the number of fins on the tube circumference.

12.3 The Bond Number. The Adamek analysis used to obtain Equation 12.26 assumes that surface tension forces are much greater than the gravity force. Consider steam condensing at 60°C on a profile having $\zeta = 0$ and $S_m = 1.0$ mm. Calculate the Bond number at $s = S_m$ using Equation 12.28. At 60°C, σ = 66E-3 N/m, and ρ_l = 984 kg/m³. [Ans. Bo = 0.094]

12.4 Condensation on Vertical Fluted Surface. (a) The Figure 12.20 fluted surface concept is to be applied to condensation of ammonia at 10°C with ΔT_{vs} = 2.22°C on a 1.52-m-high vertical plate having 396 fins/m and the Gregorig profile (ζ = 2) with S_m = 1.52 mm. The ammonia properties are given in Problem 12.2. Neglecting the condensation in the concave drainage channels, calculate the condensation coefficient based on the projected surface area and compare with the value for a smooth plate (h_s). [Ans. $h = 46.7$ W/m²-K, $h/h_s = 9.08$]

(b) Now consider the benefit of using the Adamek $\zeta = -0.5$ with the same S_m for this application. Keep the same spacing between fin at the fin base (s_b) as exists for the Gregorig profile (1.0 mm). Study of the Gregorig profile shows that $t_b \cong S_m$. By scaling Figure 12.22b, find that, for the same S_m, the fin base thickness (t_b) thickness of the $\zeta = -0.5$ profile is 0.49 that of the $\zeta = 2$ profile. So, for $S_m = 1.52$ mm, the fin base thicknesses of the $\zeta = 2$ and $\zeta = -0.5$ profiles will be 1.52 and 0.74 mm, respectively. Thus, if a $\zeta = -0.5$ profile having $S_m = 1.52$ mm, the fin pitch for the $\zeta = -0.5$ profile will be $0.74 + 1.0 = 1.74$ mm, or 574 fins/m. How will the condensation coefficient based on the projected surface area for the Adamek profile compare to that of the Gregorig profile? [Ans. $h = 90.9$ W/m²-K, $h/h_s = 17.7$]

12.5 Condensation on Integral-Fin Tube. Ammonia condenses at 50°F on a horizontal integral-fin tube having 748 fin/m (19 fins/in.), $d_e = 19.1$ mm (0.75 in.), and $d_o = 15.88$ mm (0.625 in.) with $\Delta T_{vs} = 4°F$. The fins are trapezoidal shape having $t_b = 0.38$ mm (0.015 in.) and $t_t = 0.23$ mm (0.009 in.). Compute the condensing coefficient for $\eta_f = 1.0$, using (1) the Beatty and Katz gravity drained model and (2) the approximate model given in Section 12.5.3 assuming $\zeta = -0.8$ model. Assume rectangular fins ($t_{av} = 0.30$ mm) for calculation of c_b. What fraction of the surface is condensate flooded (c_b)? The ammonia properties at 8.9°C are $\rho_l = 626$ kg/m³, $k_l = 0.531$ W/m-K, $\nu_l = 0.368E{-}6$ m²/s, $\sigma = 0.0236$ N/m, $\lambda = 1226$ kJ/kg. [Ans. $A_o/L = 0.166$ m²/m, $h = 18.2$ kW/m²-K (Beatty and Katz model), $h = 21.1$ kW/m²-K (surface tension model), $c_b = 0.43$]

12.6 Beatty and Katz Model. The Beatty and Katz model assumes zero condensate retention. Would the accuracy of the Beatty and Katz model be improved (for all applications) if one multiplied the composite h- value by $(1 - C_b)$, where C_b is the fraction of the circumference bridged by condensate? Explain your reasoning.

12.7 Preferred Tube Geometry. Your supervisor has proposed that the performance (A based on $\pi D_o L$) of the 748 fins/m tube would be increased if you used 1378 fins/m). Assume that the fin profile shape, height, and thickness remain constant. He argues that the condensation coefficient on the fin would be the same, but the 1378 fins/m design has more fin area (A_f), where $A = A_f + A_r$. For the 748 and 1378 fins/m tubes, A_f/A is 0.77 (748 fins/m) and 0.88 (1378 fins/m), and $A/L = 0.156$ m²/m (748 fins/m) and 0.287 m²/m (1378 fins/m). Is your supervisor correct? Provide a *quantitative* basis for your answer. Use the ammonia properties in Problem 12.5 for the evaluation.

13.1 Vaporization in Tube Bundle. An evaporator is designed to boil R-11 at 40 kW/m² and 101.3 kPa on the outside of tubes in a tube bundle. The tubes are 19 mm diameter and arranged on 23.75-mm equilateral triangular pitch tubes. Consider operation at 40 kg/m²-s mass velocity (based on frontal area) and 50% vapor quality. Calculate the ΔT for plain tubes and for the High-Flux porous boiling surface. The pool boiling performance of the plain and enhanced tubes are shown on Figure 11.10a. A curve fit of the Figure 11.10a data gives $q = 0.0406\Delta T^{2.5}$ and $q = 48.8\Delta T^{3.0}$ (kW/m²) for the plain and enhanced tubes, respectively. Calculate the ΔT reduction possible from use of the High-Flux porous boiling surface. Use the superposition model and calculate the single-phase convection coefficient (h) using the Žukauskas tube bank correlation. The F factor is given by $(\phi_f^2)^{0.44}$ with ϕ^2 given by Equation 13.9. The pool boiling coefficient must be calculated at the ΔT that exists in the tube bundle, which is unknown. Hence, you must iteratively solve for this ΔT. First, calculate h_{cv}, assume ΔT, and calculate q_{cv} using an assumed ΔT. Then calculate $q_{nbp} = q - q_{cv}$ and use the pool boiling curve fits to calculate ΔT. Iterate on ΔT until it converges. After obtaining the ΔT values for the plain and enhanced tube bundles, calculate q_{nbp}/q and comment on the results. Use $\mu_l = 0.425E\text{-}3$ kg/m-s, $Pr_l = 4.21$, and the additional fluid properties given in Problem 11.3. For simplicity, neglect boiling suppression and assume that the porous coated tube acts as a plain tube for calculation of h_{cv}. [Ans. Plain, $\Delta T = 7.45$ K, $q_{nbp} = 0.16$; Enhanced, $\Delta T = 0.90$K, $q_{nbp} = 0.90$].

13.2 Single-Phase Flow in Microfin Tubes. Single-phase tests on tube Wolverine DX 75S have shown that the Nusselt number and friction for DX 75S can be expressed as a function of the Reynolds number based on the tube meltdown diameter (D_m) as:

$$f = 0.1524Re^{-0.248}$$

$$j = 0.028Re^{-0.1523}$$

where the heat transfer coefficient is based on the nominal diameter (D_i). The geometry dimensions of the tube are fin height ($e = 0.0128$ in.), tube inside nominal diameter ($D_i = 0.585$ in.), tube meltdown diameter ($D_m = 0.581$ in.), fin helix angle ($e = 21.5°$), fin include angle ($\beta = 45°$), number of fins ($n_f = 75$), fin pitch normal to fins (0.0233 in.).

1. Compare the single-phase heat transfer coefficient and friction factor with that of a plain tube of the same D_i for Re = 20,000 and Pr = 3.
2. Calculate the efficiency index, $\eta = (h/h_p)/(f/f_p)$, where subscript p is for a plain tube of the same diameter.
3. Calculate the surface area ratio, relative to a plain tube of the same D_i. The needed equation is given as $A_{tot}/A_p = 1 + (2e/p)[\sec(\beta/2) - \tan(\beta/2)]$.
4. How does h/h_p compare with the area ratio, A_{tot}/A_p? If it is smaller, comment on why this may be.

13.3 Convective Vaporization in a Smooth Tube. Consider R-22 flowing inside a 10-mm-diameter smooth tube at 3.0 kg/s and $T_{sat} = 5°C$. If $x = 0.30$ and $T_w - T_{sat} = 10$ K, calculate convective evaporation coefficient. The tube has a smooth inner surface, calculate h at 60% vapor quality.

13.4 Convective Vaporization in Tube Bundles. Figure 11.19 shows single tube test data for nucleate boiling on the Wolverine Turbo-B enhanced boiling tube, and on a standard 1024 fins/mintegral fin tube. Assume that these tubes were used to make separate refrigerant evaporator tube bundles, each having 150 tubes in ten rows. The tubes are 19.1 mm OD, and are spaced at 22.2 mm (0.875 in.) on an equilateral pitch. Duplicate Figure 11.19 and sketch on the figure your *qualitative* estimate of the h vs. heat flux curves based on the average heat flux. State your reasoning.

14.1 Convective Condensation in Plain and Microfin Tubes. The relationship between the condensing coefficient (h_{cond}) and the single-phase heat transfer coefficient (h_{sp}) is defined by Equation 14.4. The h_{sp} is the heat transfer coefficient for single-phase flow at the "equivalent all-liquid Reynolds number." Note that $T_\delta = T_{sat}$ for condensation of a pure vapor. Explain the physical meaning of the above equation. Use a sketch of the annular condensing flow and the all-liquid flow to assist in your explanation.

14.2 Equivalent Reynolds Number. To obtain the "equivalent all-liquid Reynolds number," one defines the all-liquid flow that gives the same wall shear stress as that of the condensing flow. This Re_{eq} is defined by the condition, $(dp/dz)_F = (dp/dz)_{Geq}$, where $(dp/dz)_F$ is the frictional pressure gradient for the condensing flow and $(dp/dz)_{Geq}$ = frictional pressure gradient for the equivalent all-liquid flow. Defining $Re_{eq} = G_{eq}D\mu$ show how one uses the above equation, to get Equation 14.6.

14.3 Contribution of Surface Tension Drainage. What conditions are required if surface tension drainage from the tips of the microfins is to contribute to the condensation coefficient in convective vaporization in a microfin tube? For a given tube diameter, fin height, mass velocity, and vapor quality, how would you calculate if this condition has been met? Give qualitative explanation (with figure) to explain.

16.1 Offset Wires in Tube A water film is used to enhance heat transfer to an air stream flowing inside a 73-mm-ID vertical tube. A water film 0.25 mm thick and $Re_\delta = 328$ drains down the wall of the tube and contacts air flowing upward at $Re_d = 41,180$. The experimentally measured mass transfer coefficient (K_w) is 0.0405 kg/s-m^2 for $y = 0.005$ kg steam/kg air.

Additional enhancement of the air boundary layer is provided by spaced wires, which are offset from the tube wall as shown in Figure 16.2. and 0.51-mm wire spaced at 5.1 mm and set at the air–water interface. Use the heat transfer similarity law of Section 9.3.3 to predict the air heat transfer coefficient. Then, use Equation 16.4 to predict the mass transfer coefficient. [Ans. $e^+ = 20$, $K_m = 0.0608$ kg/s-m^2).